INTERNATIONAL UNION OF CRYSTALLOGRAPHY
BOOK SERIES

IUCr Texts on Crystallography

Intermetallics

Structures, Properties, and Statistics

Walter Steurer
Julia Dshemuchadse

*Laboratory of Crystallography, Department of Materials, ETH Zürich,
Vladimir-Prelog-Weg 5, 8093 Zürich, Switzerland*

OXFORD
UNIVERSITY PRESS

UNIVERSITY PRESS

Great Clarendon Street, Oxford, OX2 6DP,
United Kingdom

Oxford University Press is a department of the University of Oxford.
It furthers the University's objective of excellence in research, scholarship,
and education by publishing worldwide. Oxford is a registered trade mark of
Oxford University Press in the UK and in certain other countries

First Edition published in 2016

Impression: 2

Published in the United States of America by Oxford University Press
198 Madison Avenue, New York, NY 10016, United States of America

British Library Cataloguing in Publication Data
Data available

Library of Congress Control Number: 2016932189

ISBN 978–0–19–871455–2

Printed and bound by
CPI Group (UK) Ltd, Croydon, CR0 4YY

Preface

It is a fascinating world, the world of intermetallics, and it is largely unexplored. The big picture of intermetallics still has many white spots, more than regions already painted. Imagine the sheer number of the roughly three thousand binary and eighty thousand ternary intermetallic systems, which can be formed by the about eighty metallic chemical elements. While in some of them ten or more intermetallic phases have been observed, in others not even a single compound has been found. So far, only about twenty thousand different intermetallic phases are known from the hundreds of thousands that may exist. It is not clear whether this rather small number of intermetallics represents just the tip of the iceberg or already the iceberg itself; whether the just six thousand ternary intermetallic phase diagrams studied so far, and which were probably selected by chemical intuition, were already the most relevant and interesting ones or whether the remaining seventy-four thousand will be as rich in intermetallic phases. Are there more surprises to be unveiled, such as quasicrystals, the existence of which nobody anticipated before 1982, or high-entropy alloys, which were introduced twenty years later?

Furthermore, the parameter space has been only partially explored so far. In particular, almost nothing is known about the structures and properties of binary and ternary intermetallics as functions of pressure, or of pressure and temperature. In contrast, a great many of allotropes of the chemical elements have been discovered under non-ambient conditions, some of them with very interesting and unexpected properties. We also know that properties of materials can drastically change when their dimensions are reduced down to the nanoscale. Only a few intermetallic phases have been studied in this way so far.

Think of the exciting physical properties and important technological applications of intermetallics, from magnetism to superconductivity; perhaps many more exciting materials might be waiting to be discovered. Keep in mind that not all metals are intermetallic phases and not all intermetallic phases are metals. Chemical bonding in intermetallics spans the full range between metallic, covalent, and ionic, and in some cluster-based complex intermetallics all of them may be present at the same time, leading to potentially interesting properties.

This said, we want to emphasize that the most challenging problem in writing this book was selecting the intermetallic crystal structure types to be discussed out of the more than two thousand ones known so far, and identifying the examples that are suitable for illustrating the most common structural building principles, and interesting structure/property relationships. While we think that for the metallic elements an encyclopedic discussion is feasible and appropriate, this would not be possible and meaningful for all intermetallic compounds. Our goal was to demonstrate in characteristic examples the fundamental structural building principles underlying the large variety of intermetallics. In particular, the most common crystal structure types will be discussed as well as those that underlie and govern the properties of intermetallic functional and structural materials, respectively. Of course, this attempt, and a particular focus on complex intermetallics, reflect the personal interests and views of the authors.

The focus of this book is clearly on the statistics, topology, and geometry of crystal structures and crystal structure types. The extensive use of data mining gives a fresh view on the relatively old topic of intermetallics. For the first time, the structure discussions follow largely the frequency of structure types, which themselves are analyzed from different points of view. This allows us to uncover important structural relationships and to illustrate the relative simplicity of most of the general structural building principles. It also allows us to show that a large variety of actual structures can be related to a rather small number of aristotypes.

Our main concern was writing a book that is readable and beneficial in one way or another for everyone interested in intermetallic phases—from graduate students to experts in solid state chemistry/physics/materials science. For that purpose we avoided using an enigmatic abstract terminology for the classification of structures. Our focus on the statistical analysis of structures and structure types should be seen as an attempt to draw the background of the big picture of intermetallics, and to point to the white spots in it, which could be worthwhile to be explored. We also want to emphasize that this book was not planned as a textbook; it should rather be a reference and guide through the incredibly rich world of intermetallic phases.

Finally, we want to add our usual *caveat*. Structural subunits of intermetallic phases (atomic environment types, coordination polyhedra, clusters, structure modules, etc., however they may be named) must not be seen as entities that could be stable outside their respective structures. The interactions between atoms within such subunits can, but in most cases do not, differ from the interactions between the atoms of the subunits and of their atomic environments. Thus, in most cases, these structural subunits should be just considered as purely geometrical objects, being quite useful for the description of complex structures and/or for the illustration of structural relationships. Furthermore, ball-and-stick structure models of intermetallics show the distribution of atomic sites and shortest interatomic distances, but usually they do not indicate chemical bonds as is the case for organic molecules. The choice of a particular crystallographic unit cell, which is always a parallelepiped by definition, is based on conventions and just allows a simple, lattice-based description of crystal structures.

Zurich
November 2015

Walter Steurer
Julia Dshemuchadse

Contents

Part I

Concepts and statistics

In this first part of the book, the basic concepts and tools are presented for the description of symmetry and structures of metallic elements and intermetallic phases (short "intermetallics"), periodic, and quasiperiodic ones, while in the second part the focus is on the discussion of their actual structures and properties. We preferentially use the term "phase" rather than "compound", in order to take into account the sometimes very wide compositional stability ranges of intermetallics. In Part I, we will also introduce the basic concepts explaining the stability of intermetallics such as electronic stabilization by the Hume-Rothery mechanism as well as the role of entropy in the case of high-entropy alloys (HEAs), for instance. Furthermore, we will discuss the distribution of intermetallics as a function of composition, symmetry and unit-cell size, in order to get an overview of what is possible, as well as what is more and what is less probable.

In Chapter 1, we introduce the general notation and terminology for the description of crystal structures as well as their graphical representation. We will shortly sketch the crystallographic concept of symmetry, and how it can be used for the comparative discussion of structures. In Chapter 2, we give an overview of the most common methods for the calculation of the stability and chemical bonding of crystal structures, and we discuss the most important factors that control their formation. In Chapter 3, we introduce the concept of lattices, tilings, coverings and packings, which are particularly important for the description of complex intermetallics, quasiperiodic structures as well as their approximants. Chapter 4 is dedicated to the higher-dimensional approach for the description of quasicrystals. The last chapter of this first part of the book, Chapter 5, deals with data mining, critically discussing the databases we use. It also presents a general statistics of intermetallics as function of symmetry, stoichiometry, chemical composition, and number of atoms per unit cell.

1

Introduction

In this chapter we introduce the general crystallographic concepts, notation and terminology we are using in this book. We also discuss the role symmetry is playing in the self-assembly of crystal structures. Furthermore, we introduce the different ways of describing crystal structures, and of illustrating their main features.

1.1 General notation and terminology

The information needed for uniquely describing a periodic 3D crystal structure comprises its symmetry and metrics, i.e., space group type and lattice parameters $a, b, c, \alpha, \beta, \gamma$; additionally, for each atom in the asymmetric unit the following parameters have to be given: its Wyckoff position with the set of coordinates x, y, z, the occupancy factor p, and the atomic displacement parameters (ADPs) U_{ij}, which describe the static and dynamic mean-square displacements of the atoms from their average positions. The ADPs are not needed if one is interested in just a geometrical structure model. However, in the case of thermoelectric materials with "rattling" atoms, for instance, the ADPs would be indicative of the amount of space that atoms have for their movements.

There are conventions for the choice of the unit cell and its origin, which can be found in the *International Tables for Crystallography* (IUCr, 2002), for instance. By applying the symmetry operations of the respective space group on the atomic positions within an asymmetric unit, the entire infinite crystal structure can be generated. In the case of intrinsically structurally (partially) disordered phases, disorder parameters are needed, describing the deviations from the respective ordered structure.

In the case of aperiodic crystal structures, i.e., those of incommensurately modulated phases, composite (host-guest) crystals, or quasicrystals, the parameters to be given depend on their respective ways of higher-dimensional (nD) embedding, which allows us to describe such structures in terms of nD unit cells and space group types (Janssen *et al.*, 2007; van Smaalen, 2007; Steurer and Deloudi, 2009). Then the full periodic nD crystal structure can be generated analogously by applying the symmetry operations of the respective nD space group on the

Intermetallics: Structures, Properties, and Statistics. First Edition. Walter Steurer and Julia Dshemuchadse.
© Walter Steurer and Julia Dshemuchadse 2016. Published in 2016 by Oxford University Press.

"hyperatoms" ("atomic surfaces", "occupation domains") within an nD asymmetric unit. The actual 3D structure follows as a special section of the nD crystal structure (see Chapter 4). Aperiodic crystal structures differ from general non-periodic structures, deterministic or non-deterministic ones, by their pure-point Fourier spectrum (Bragg peaks only). If they are disordered they show, additionally, continuous contributions to the Fourier spectrum (diffuse diffraction intensities) in the same way as disordered 3D periodic structures do.

For an illustrative and vivid description of a crystal structure, a list of atomic coordinates is insufficient, a geometrical structure model is needed for its visualization. The local atomic arrangements are reflected in the atomic environment types (AETs), which are also called coordination polyhedra. The general structural building principles of more complex structures can be elucidated by subdividing them into larger subunits (clusters, modules, etc.), which may or may not have a crystal-chemical meaning.

For the shorthand notation of crystal structures, we will use the Pearson symbol (Table 1.1) in combination with the chemical formula defining the structure type. For instance, Na at ambient conditions has the structure type $cI2$-W. Above 65 GPa, it transforms into a structure of the type $cF4$-Cu. In the case of the element structures, we will denote these two allotropes (modifications) in condensed form as $cI2$-Na and $cF4$-Na and in the formula index as Na ($cI2$-W) and Na ($cF4$-Cu), respectively.

Unfortunately, for the sequence of elements in the chemical formula of intermetallic compounds, there is no clear guidance by the IUPAC recommendations, which do not even define the term "intermetallic compound" or "metallic

Table 1.1 *Meaning of the letters in the Pearson symbol. It consists of one lower- and one upper-case italic letter denoting the crystal family and the Bravais type of the lattice, respectively, followed by the number of atoms per unit cell. In the case of a rhombohedral unit cell with n atoms, we write hR3n, indicating the number of atoms in the structurally equivalent rhombohedrally-centered hexagonal unit cell. It should be pointed out that some other authors such as Villars and Calvert (1991), for instance, write hRn instead. However, they use our notation in their database Pearson's Crystal Data (PCD) (Villars and Cenzual, 2011a), the basis of all our statistical analyses.*

Crystal family		Bravais lattice type	
a	triclinic (anorthic)	P	primitive
m	monoclinic	I	body centered
o	orthorhombic	F	all-face centered
t	tetragonal	S, C	side- or base-face centered
h	hexagonal, trigonal (rhombohedral)	R	rhombohedral
c	cubic		

element". The recommendations as of 2005 (Conelly *et al.*, 2005) just refer to those of 1990 (IUPAC, 1990), where either alphabetical ordering of the constituting elements or a sequence according to their electronegativity was suggested. Consequently, every author may find good reasons for writing the chemical formula of a new intermetallic compound in her/his preferred way. Alphabetical ordering is used in the database Pearson's Crystal Data (PCD) by Villars and Cenzual (2011*b*), for instance, as well as for the index in our book. Arranging the elements according to increasing electronegativity or to the closely related Mendeleev numbers, ($M_A < M_B < M_C \ldots$), makes more sense from a crystal-chemical point of view, in particular, if one wants to compare different intermetallics with the same structure type (see Section 5.1). This ordering scheme was employed by Ferro and Saccone (2008) in their book on intermetallic phases, for instance. We will adopt the usage found in the PCD, in order to be consistent with our fundamental database. The PCD is essentially taking over the sequence of elements as given in the papers reporting the new structures and structure types. For instance, in the text we write for the cubic and hexagonal Laves phases $cF24$-MgCu$_2$ and $hP12$-MgZn$_2$, respectively, and Cu$_2$Mg ($cF24$) and MgZn$_2$ ($hP12$) in the formula index. Based on Mendeleev numbers, the formulae would read: $cF24$-Cu$_2$Mg and $hP12$-MgZn$_2$.

1.1.1 The role of symmetry

The symmetry of all *idealized crystal structures* of intermetallic compounds in thermodynamic equilibrium can be described by nD space groups, where $n = 3$ in the case of periodic and $n > 3$ in the case of aperiodic crystal structures. By *idealized crystal structure* we mean an averaged, defect-free structure that is thought to be infinite, its atoms to be spherical, their thermal vibrations to be time-averaged and, like structural disorder, spatially averaged modulo one unit cell. It is obvious, but we want to emphasize that the space group symmetry of a structure is an output parameter resulting from the self-assembly of atoms, and not an input parameter as a kind of construction plan for the atoms. In the case of intermetallics that are in thermodynamic equilibrium, self-assembly leads to that arrangement of atoms, which gives the lowest Gibbs free energy, $G = H - TS$, with the enthalpy H, the temperature T, and the entropy of the system S. In the case of isotropic atomic interactions, proper atomic size ratios, and not too complex stoichiometries, the self-assembly of atoms can lead to the maximization of the packing density. Otherwise, less dense structures can be formed such as the simple cubic structure of $cP1$-Po, for instance, with a packing density of only $q = \pi/6 = 0.524$ compared to cubic-close packed (*ccp*) structures such as $cF4$-Cu, with $q = \pi/\sqrt{18} = 0.740$. In the case of covalent bonding the deviation from closest sphere packings can be even more drastic. For instance, $q = \pi\sqrt{3}/16 = 0.340$ for diamond, $cF8$-C. It has to be kept in mind that at temperatures $T \gg 0$ K, less dense structures may be dynamically more stable than denser ones allowing for higher contributions to the vibrational entropy.

Crystal structures can be geometrically decomposed into small subunits, each one centered by an atom, i. e., the atomic environment types (AETs) or coordination polyhedra. The number n of atoms in such a coordination polyhedron is called coordination number CNn. By definition, AETs always overlap one another, because the central atom of an AET is at the same time part of the AETs of each of its coordinating atoms. In the case of more complex structures, larger structural subunits (clusters, modules, etc.) can frequently be identified, which may give some insight into the way the structures are designed geometrically. In this way it can be shown, for instance, that a particular structure type can be geometrically constructed by fitting together subunits of two or more other structure types (modular structures).

Only if a structure shows nD space group symmetry does it have a discrete distribution of interatomic distances, a finite set of different AETs, and more or less densely populated lattice planes (atomic layers), a low-energy subset of which would be parallel to a crystal's facets.

In the following, the correspondence between the local symmetry (point group symmetry) of AETs or clusters and the space group symmetry of a structure will be discussed using the example of the space group 216 $F\bar{4}3m$ (Fig. 1.1 and Table 1.2), which corresponds to the symmetry of many Frank-Kasper (FK) phases. The local symmetry of a structure is reflected in the site symmetry of the respective Wyckoff position. A Wyckoff position of a space group $\mathcal{G}$ consists of all points X for which the site-symmetry groups are conjugate subgroups of $\mathcal{G}$. A Wyckoff position with variable parameters x, y, z forms a Wyckoff set. Depending on the values of the coordinates x, y, z, a variety of polyhedra, all with the same minimum point group symmetry corresponding to the site symmetry, may be generated within a given Wyckoff set.

If we choose special parameters for the coordinates, then we can obtain highly regular polyhedra, which have higher symmetry than the site symmetry of the Wyckoff position they are centered on. An example is shown in Fig. 1.2, with $x = 1/8$, $y = 0$, $z = 1/4$. The generated truncated octahedra (Kelvin polyhedra) are centered on all four Wyckoff positions with special parameters, $4a$-c, and fill space completely. The resulting unit cell has symmetry $Im\bar{3}m$, and half the lattice parameter. By varying the coordinates, the semi-regular truncated octahedra lose their high symmetry by an increasing tetrahedral distortion, thereby maintaining only the actual space group symmetry.

Each atom in the unit cell of a structure can be assigned to a Wyckoff position with a particular site symmetry. In the case of our example, space group no. 216, $F\bar{4}3m$, it constitutes, together with its symmetrically equivalent atoms, one of the polyhedra listed in Table 1.2. Some of them, generated by atoms occupying Wyckoff positions from $16e$ to $96i$, are of variable size, and can have alternative centers depending on the values of the coordinates. Frequently, the polyhedra around high-symmetry sites are considered as cluster shells in the purely geometrical meaning of the word. However, cluster shells can also be centered on low-symmetry sites, if they include symmetrically non-equivalent atoms from different Wyckoff positions.

F $\bar{4}3m$ **T_d^2** **$\bar{4}3m$** **Cubic**

No. 216 **F $\bar{4}3m$** Patterson symmetry *F m$\bar{3}$m*

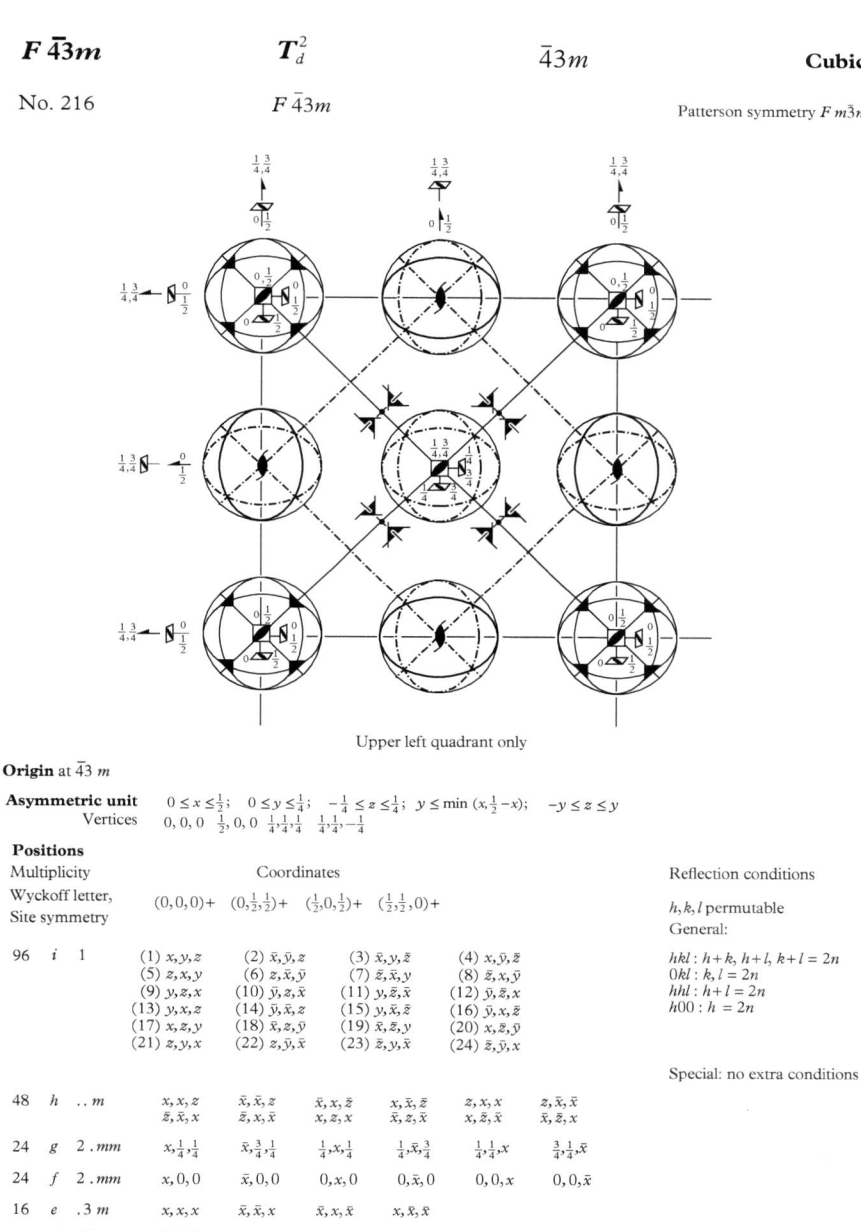

Upper left quadrant only

Origin at $\bar{4}3\,m$

Asymmetric unit $0 \le x \le \frac{1}{2}$; $\quad 0 \le y \le \frac{1}{4}$; $\quad -\frac{1}{4} \le z \le \frac{1}{4}$; $\quad y \le \min(x, \frac{1}{2}-x)$; $\quad -y \le z \le y$

Vertices $0, 0, 0 \quad \frac{1}{2}, 0, 0 \quad \frac{1}{4}, \frac{1}{4}, \frac{1}{4} \quad \frac{1}{4}, \frac{1}{4}, -\frac{1}{4}$

Positions

Multiplicity | Coordinates | Reflection conditions
Wyckoff letter, | |
Site symmetry | |

h, k, l permutable

$(0,0,0)+ \quad (0,\frac{1}{2},\frac{1}{2})+ \quad (\frac{1}{2},0,\frac{1}{2})+ \quad (\frac{1}{2},\frac{1}{2},0)+$

General:

96	i	1	(1) x,y,z	(2) $\bar{x},\bar{y},z$	(3) $\bar{x},y,\bar{z}$	(4) $x,\bar{y},\bar{z}$
			(5) z,x,y	(6) $z,\bar{x},\bar{y}$	(7) $\bar{z},\bar{x},y$	(8) $\bar{z},x,\bar{y}$
			(9) y,z,x	(10) $\bar{y},z,\bar{x}$	(11) $y,\bar{z},\bar{x}$	(12) $\bar{y},\bar{z},x$
			(13) y,x,z	(14) $\bar{y},\bar{x},z$	(15) $y,\bar{x},\bar{z}$	(16) $\bar{y},x,\bar{z}$
			(17) x,z,y	(18) $\bar{x},z,\bar{y}$	(19) $\bar{x},\bar{z},y$	(20) $x,\bar{z},\bar{y}$
			(21) z,y,x	(22) $z,\bar{y},\bar{x}$	(23) $\bar{z},y,\bar{x}$	(24) $\bar{z},\bar{y},x$

$hkl: h+k, h+l, k+l = 2n$
$0kl: k, l = 2n$
$hhl: h+l = 2n$
$h00: h = 2n$

Special: no extra conditions

48	h	$..m$	x,x,z	$\bar{x},\bar{x},z$	$\bar{x},x,\bar{z}$	$x,\bar{x},\bar{z}$	z,x,x	$z,\bar{x},\bar{x}$
			$\bar{z},\bar{x},x$	$\bar{z},x,\bar{x}$	x,z,x	$\bar{x},z,\bar{x}$	$x,\bar{z},\bar{x}$	$\bar{x},\bar{z},x$

24	g	$2.mm$	$x,\frac{1}{4},\frac{1}{4}$	$\bar{x},\frac{3}{4},\frac{1}{4}$	$\frac{1}{4},x,\frac{1}{4}$	$\frac{1}{4},\bar{x},\frac{3}{4}$	$\frac{1}{4},\frac{1}{4},x$	$\frac{3}{4},\frac{1}{4},\bar{x}$

24	f	$2.mm$	$x,0,0$	$\bar{x},0,0$	$0,x,0$	$0,\bar{x},0$	$0,0,x$	$0,0,\bar{x}$

16	e	$.3\,m$	x,x,x	$\bar{x},\bar{x},x$	$\bar{x},x,\bar{x}$	$x,\bar{x},\bar{x}$

4	d	$\bar{4}3\,m$	$\frac{3}{4},\frac{3}{4},\frac{3}{4}$
4	c	$\bar{4}3\,m$	$\frac{1}{4},\frac{1}{4},\frac{1}{4}$
4	b	$\bar{4}3\,m$	$\frac{1}{2},\frac{1}{2},\frac{1}{2}$
4	a	$\bar{4}3\,m$	$0,0,0$

Fig. 1.1 *Representation of the space group no. 216, F$\bar{4}$3m, in the International Tables for Crystallography (2006), Vol. A, Space group 216, pp. 658–659. Reproduced with permission of the International Union of Crystallography.*

Table 1.2 *Polyhedra resulting from point configurations generated by occupying different Wyckoff positions (first column) in space group no. 216, F$\bar{4}$3m. The polyhedra have the symmetry of the site they are centered on, which is given in the last column. In the case of the octahedra and just spherical atoms, the symmetry is even higher, m$\bar{3}$m or ∞/mm. Which one of the two Wyckoff positions given in the last column is centering a given polyhedron depends on the values of the coordinates. Wyckoff positions are given by their multiplicities followed by the respective Wyckoff letter. In the column "Edge length", a is the lattice parameter of the cubic unit cell.*

Wyckoff position	Site symmetry	Coordinates x	y	z	Type of polyhedron	Edge length	Center of polyhedron
4 *a*	$\bar{4}$3m	0	0	0	tetrahedron	$a\sqrt{2}/2$	4*c* or 4*d*
4 *a*	$\bar{4}$3m	0	0	0	octahedron	$a\sqrt{2}/2$	4*b*
4 *b*	$\bar{4}$3m	1/2	1/2	1/2	tetrahedron	$a\sqrt{2}/2$	4*c* or 4*d*
4 *b*	$\bar{4}$3m	1/2	1/2	1/2	octahedron	$a\sqrt{2}/2$	4*a*
4 *c*	$\bar{4}$3m	1/4	1/4	1/4	tetrahedron	$a\sqrt{2}/2$	4*a* or 4*b*
4 *c*	$\bar{4}$3m	1/4	1/4	1/4	octahedron	$a\sqrt{2}/2$	4*d*
4 *d*	$\bar{4}$3m	3/4	3/4	3/4	tetrahedron	$a\sqrt{2}/2$	4*a* or 4*b*
4 *d*	$\bar{4}$3m	3/4	3/4	3/4	octahedron	$a\sqrt{2}/2$	4*c*
16 *e*	.3m	x	x	x	tetrahedron	$2xa\sqrt{2}$	4*c* or 4*d*
24 *f*	2.mm	x	0	0	octahedron	$xa\sqrt{2}$	4*a* or 4*b*
24 *g*	2.mm	x	1/4	1/4	octahedron	$(1/8-x)a\sqrt{2}$	4*c* or 4*d*
48 *h*	..m	x	x	z	truncated tetrahedron		4*c* or 4*d*
96 *i*	1	x	y	z	edge/vertex-truncated tetrahedron		4*c* or 4*d*

Let us shortly discuss, for instance, the endohedral clusters in the structure of $cF444$-Ta$_{36.4}$Al$_{63.6}$ (Conrad *et al.*, 2009) (see Subsection 7.4.4). The outermost Al$_{76}$ fullerene-like cluster shell consists of the atoms Al(1–3), each one of them in Wyckoff position 48*h*, Al(4) in 16*e*, and Al(5) in 24*f*. All hexagon faces are part of either truncated tetrahedra (part of CN16 Friauf polyhedra) or CN15 Frank-Kasper polyhedra (see Subsection 7.4.1). From a crystal-chemical point of view, the fullerene-like shell can be considered as just the inside of a framework of CN16 Friauf polyhedra and CN15 Frank-Kasper polyhedra. Since Friauf polyhedra play a dominant role in this and the larger related structures,

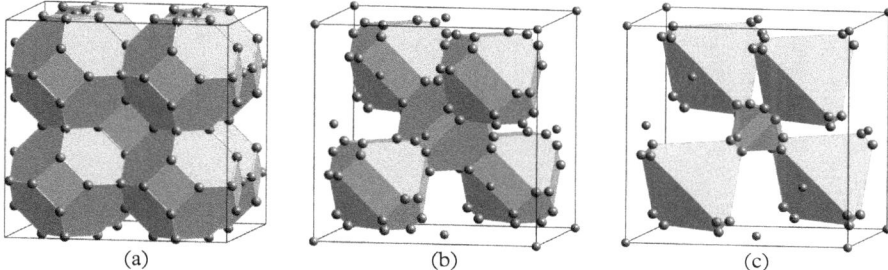

Fig. 1.2 *Polyhedra generated in space group no. 216, $F\bar{4}3m$, general Wyckoff position 96i, x, y, z, for the special parameters (a) x = 1/8, y = 0, z = 1/4, (b) x = 0.1, y = 0.05, z = 0.3, and (c) x = 0.05, y = 0.1, z = 0.4. In (a), a packing of truncated octahedra (Kelvin polyhedra) is created, with the actual symmetry increased to $Im\bar{3}m$ (space group no. 229), and at the same time the unit cell parameters are halved. In (b) and (c), the polyhedra get increasingly tetrahedrally distorted, thereby maintaining only the actual site symmetry corresponding to the space group.*

the symmetry of the Friauf polyhedron ($\bar{4}3m$) will determine the symmetry of the packing, i.e., $F\bar{4}3m$.

1.1.2   Wyckoff sets and lattice complexes

According to Fischer and Koch (2002), a "lattice complex" is defined as the set of all point configurations that may be generated within one type of Wyckoff set. A point configuration or crystallographic orbit is the infinite set of all points X that are symmetrically equivalent to a given point with respect to a certain space group $\mathcal{G}$. A Wyckoff set with respect to a space group $\mathcal{G}$ is the set of all points X for which the site-symmetry groups are conjugate subgroups of the normalizer $\mathcal{N}$ of $\mathcal{G}$ in the group of all affine mappings. For instance, the Wyckoff positions $4a$-d in space group no. 216, $F\bar{4}3m$ (see Fig. 1.1), form a Wyckoff set.

Example

The lattice complex F may be generated, among others, in the Wyckoff set $Fm\bar{3}m$ a and $F\bar{4}3m$ a, respectively; this means either in Wyckoff position $4a$ $0, 0, 0$ of space group No. 225 $Fm\bar{3}m$, or in Wyckoff position $4a$ $0, 0, 0$ of No. 216 $F\bar{4}3m$.

"Invariant lattice complexes" in their characteristic Wyckoff position, i.e., the one with the highest site symmetry, are represented by a capital letter, in some cases with a superscript in front of it, e.g., "+" or "−" for representatives of enantiomorphous pairs, or "$*$" indicating that the lattice complex is derived by combining two lattice complexes with the same symbol, but displaced from each other. If necessary, the crystal family is indicated by a lower case letter preceding the lattice-complex symbol.

Example

The lattice complex D may describe either the cubic diamond lattice complex 227 $Fd\bar{3}m$ a or the orthorhombic lattice complex 70 $Fddd$ a. Therefore, we have to specify additionally the crystal system cD or oD, respectively. oD may be generated by an orthorhombic distortion of cD.

The carbon atoms in the diamond structure, $oC8$-C, occupy the sites of the cD lattice complex, i.e., Wyckoff position $8a$ 1/8, 1/8, 1/8; 7/8, 3/8, 3/8, in space group no. 227, $Fd\bar{3}m$; so do the Mg atoms in the cubic Laves phase $cF24$-MgCu$_2$, while the Cu atoms constitute a cT lattice complex, resulting from Wyckoff position $16c$ 1/8, 1/8, 1/8; 7/8, 3/8, 5/8; 3/8, 5/8, 7/8; 5/8, 7/8, 3/8 in space group No. 227 $Fd\bar{3}m$.

The lattice complex N, 191 $P6/mmm$ f, describes a Kagomé net, and $E2z$, 194 $P6_3/mmc$ f, the hexagonal diamond sites, Wyckoff position $4f$ 1/3, 2/3, z; 2/3, 1/3, $z + 1/2$; 2/3, 1/3, $\bar{z}$; 1/3, 2/3, $\bar{z} + 1/2$ in space group No. 194 $P6_3/mmc$, occupied by the Mg atoms in the hexagonal Laves phase $hP12$-MgZn$_2$.

1.1.3 Structures and structure types

The atomic arrangement of any intermetallic compound can be described by its nD crystal structure, which is usually determined by diffraction methods. If standard techniques are employed, a time- and space-averaged structure is obtained. Models for the structural disorder, if any, can be derived by the evaluation of the diffuse diffraction intensities. Consequently, by taking the actual structure modulo one unit cell and integrating over the measurement time, dynamic and static atomic displacements, as well as all kinds of disorders and defects, are projected into the atomic (electron) density distribution functions. This can lead to partially occupied atomic sites, mixed positions, or split "atoms" in the averaged structure model. In the case of not fully occupied Wyckoff positions, the Pearson symbol is usually written as $fB(n - x)$, with the symbols fB for the crystal family and the Bravais-lattice type, respectively, the number of atoms per unit cell n in the case of fully occupied Wyckoff positions, and the number of atoms missing there due to partial occupancy, x.

If intermetallic phases have an extended compositional stability range, then this automatically implies the existence of substitutional disorder ("mixed atomic positions"), which sometimes can be accompanied by displacive disorder. Although the occupancy of particular atomic sites changes with the chemical composition, the "structure type" remains the same by definition.

The term "structure type" was originally defined by Pearson (1972), and refined later on. In order to understand this definition, we need to define a few additional terms:

Structure type Prototypic crystal structure representing a whole class of isoconfigurational structures. If there is no more than a single intermetallic phase with a structure assigned to a particular structure type then this structure type will be called unique.

Prototype Crystal structure of an element or compound used for the definition of a structure type, which represents the class of all materials with isoconfigurational structures.

Isotypic Two structures are crystal-chemically isotypic if they are isoconfigurational and the corresponding atoms and bonds (interactions) have similar physical/chemical characteristics. For instance, $cP2$-CoAl and $cP2$-NiAl are isotypic.

Isoconfigurational Two structures are configurationally isotypic if they are isopointal and both the crystallographic point configurations (crystallographic orbits) and their geometrical interrelationships are similar; all geometrical properties, such as axial ratios, angles between crystallographic axes, values of corresponding adjustable positional parameters (x, y, z), and coordinations of corresponding atoms (AETs) are similar. Isoconfigurational structures belong to the same structure type. For instance, the isoconfigurational metallic phase AlNi and the ionic compound CsCl are both representatives of the $cP2$-CsCl structure type, although they strongly differ in their atomic interactions, and are therefore not isotypic.

Isopointal Two structures are isopointal if they have the same space-group type, and the atomic (Wyckoff) positions, occupied either fully or partially at random, are the same in both structures; as there are no limitations on the values of the adjustable parameters of the Wyckoff positions or on the cell parameters, isopointal structures may have locally different geometric arrangements and atomic coordinations (AETs) and may belong to different structure types. An example for isopointal structure types are $tI2$-In and $tI2$-Pa (Fig. 3.2).

Homeotypic Two structures are homeotypic if one or more of the following conditions required for isotypism are relaxed:
   (i) Identical or enatiomorphic space-group types, allowing for group/subgroup or group/ supergroup relationships;
  (ii) Limitations imposed on the similarity of geometric properties, i.e., axial ratios, interaxial angles, values of adjustable positional parameters, and the coordination of corresponding atoms (AETs);
 (iii) Site occupancy limits, allowing given sites to be occupied by different atomic species. For instance, $cI2$-W and $cP2$-NiAl are homeotypic according to (i) and (ii).

We find many examples where binary intermetallic phases are assigned to unary structure types or ternary intermetallics to binary ones. How is this possible? Nothing is said in the above definitions about the number of different constituents, whether this number has to be the same in a structure and the structure type it is assigned to as long as they are isoconfigurational. However, we should distinguish the following two cases:

(i) Ternary compounds are partially inherently disordered, if the binary structure types are described with only two occupied Wyckoff positions in the respective space group, e.g., $cF24$-$MgCu_2$, $cP4$-Cu_3Au, $cP2$-$CsCl$, *etc.*

(ii) In the case of many binary structure types, however, three or more independent Wyckoff positions are occupied by the two different atomic species. This allows an ordered substitution of the two by three different atomic species, giving a homeotypic structure. Strictly speaking, this ordered structure variant could be seen as a new structure type, or, at least, named an ordered variant of a structure type. However, for that purpose, the above generally used definition would have to be modified. In the literature, such structures are sometimes assigned to a new structure type, sometimes not. For the sake of consistency, we follow the assignment of structures to structure types as it is handled in the PCD.

2

Factors governing structure and stability of intermetallics

Which factors are the most important ones governing formation and stability of the crystal structures of metallic elements and intermetallic phases? Why does an element or intermetallic compound adopt exactly this and not another crystal structure for a given chemical composition, temperature, and pressure? What kind of local interactions are responsible for the stability of the respective structure, and what is the influence of the global electronic band structure? How can it be that a slight change in chemical composition can turn a simple structure with just a few atoms per unit cell into a complex one with more than twenty thousand per unit cell or even into a quasiperiodic structure without any 3D unit cell, for instance?

Which approaches can be used to predict and identify binary, ternary, or multinary intermetallic systems that feature stable crystal structures, and not only to predict the systems but also the crystal structures themselves together with their stability range? In which cases are quantum-mechanical calculations necessary, and when can empirical approaches be useful such as quantum structure diagrams or M/M-plots (see Section 5.1)? How can statistical methods based on data mining contribute to our understanding of intermetallics? How can the structure determining factors, once identified, be described best, qualitatively and quantitatively, and used for structure prediction?

The determination of a (quasi)crystal structure, i.e., its geometrical description, is just the first step. It allows understanding of the packing principles, but not why the structure forms at all. This first step, however, is crucial for the second step: the study of the chemical bonding. While the first step is more or less routine for periodic crystals, nowadays, the second step needs more effort. In the case of quasicrystals, all approaches needing periodic boundary conditions are not directly applicable. For that purpose, periodic approximants are used. It has to be kept in mind, however, that the essence of what makes a structure quasiperiodic cannot then be captured in this way.

To get a full description of a crystal structure with regard to chemical bonding and stability, the Schrödinger equation has to be solved. This eigenvalue equation contains one wave function for both nuclear and electronic components. If one neglects the time dependence, it can be separated into these two parts based on

Intermetallics: Structures, Properties, and Statistics. First Edition. Walter Steurer and Julia Dshemuchadse.
© Walter Steurer and Julia Dshemuchadse 2016. Published in 2016 by Oxford University Press.

the Born-Oppenheimer approximation: due to the high mass of the nuclei, the electronic part of the equation can be solved for a fixed configuration of nuclei. The assumption makes use of the fact that the state of the nuclei cannot follow a change in electronic configuration on the time scale relevant to the motion of the much lighter electrons.

This approximation is also responsible for one of the main restrictions of such electronic structure calculations: they almost exclusively refer to the state of matter at a temperature of zero Kelvin. This is due to the fact that, usually, the state of the nuclei that is presumed for these calculations is the equilibrium configuration.

Chemical bonding on the local scale is commonly described employing either the valence-bond (VB) theory or the molecular orbital (MO) theory. In the former, the atomic orbitals of separate atoms are assumed to combine in a way that can be described as chemical bonds – valence bonds. The latter, however, regards electrons as being under the influence of the entire molecule and not as assigned to a specific atom. Consequently, all electrons are redistributed into common molecular orbitals. This redistribution refers mainly to the valence electrons, which play a role in the bonding state of a molecule.

The MO theory can be easily extended from single molecules to the whole crystal leading to dense (continuous) bands of electronic eigenstates, i.e., the electronic band structure. The size of the band gaps plays an important role for the physical properties of the crystals. For a more recent review on chemical bonding see, e.g., the review article by Gatti (2005) or the book by Frenking and Shaik (2014).

2.1 Quantum chemistry

In this section, we briefly review some basic approaches used for the calculation of the stability of crystal structures, and the understanding of the chemical bonding. In the case of intermetallic phases, the knowledge of the electronic band structure is crucial for understanding the role of its chemical composition and electron concentration.

2.1.1 Density functional theory

The density functional theory (DFT) is the standard approach for quantum-mechanical calculations of the electronic states of crystal structures. It is an *ab-initio* method, meaning that it does not require empirical information in addition to the structural parameters (type of atom and its coordinates). The DFT is based on the Hohenberg-Kohn theorem, whose corollary is that the electron density of the ground state uniquely determines a many-electron system. This reduces the N-electron problem with $3N$ spatial coordinates to a problem with just three coordinates. The defined energy functional has to be minimized in order to determine the correct ground-state electron density.

The Schrödinger equation is replaced by the Kohn-Sham equation for non-interacting electrons in an effective potential. The energy functional is now regarded as the sum of the kinetic energy of a system of non-interacting particles, the Coulomb interactions of the system, as well as the so-called exchange-correlation energy. The latter contains "the residual part of the true kinetic energy" and "non-classical electrostatic contributions". This means it contains all unknown terms and its approximation constitutes the challenge of employing DFT.

The exchange-correlation energy is separated into the sum of exchange and correlation energies. The most common approximation of exchange and correlation is the local-density approximation (LDA): it depends only on the electronic density at a given set of spatial coordinates and not on derivatives of the density or on orbitals. The generalized gradient approximations (GGAs) are also local, but additionally depend on the gradient of the electron density at a given point.

The non-interacting system that the Kohn-Sham equation models yields orbitals that, strictly speaking, carry no physical meaning. However, by choosing a suitable effective potential, the ground-state density of the system of interacting electrons can be reproduced. In order to perform DFT calculations, proper basis sets have to be chosen, upon which the Kohn-Sham orbitals can be expanded.

The DFT is not only used in order to calculate the (relative) stability of a specific geometric configuration—although this is the main purpose within structure science. Electric or magnetic properties and chemical reactivity are only a few of the applications that DFT has successfully been applied to. For further reading see Koch and Holthausen (2001), for instance.

2.1.2 Extended Hückel method

The extended Hückel method is a semi-empirical approach used for the description of molecular orbitals and their occupancy. It is also well-suited for the determination of relative energies of different structures (see, e.g., Berger *et al.* (2011)). The original Hückel molecular orbital (HMO) method (Hückel, 1931), based on a simple linear combination of atomic orbitals (LCAO-MO) only takes π-bonding in conjugated hydrocarbon systems such as benzene into account, while the extended Hückel method additionally includes σ-bonding as well (Hoffmann, 1963).

(Extended) Hückel calculations remain a powerful tool for analyzing the electronic state of solid compounds, which can be tuned to DFT data to utilize *ab-initio* results (Stacey and Fredrickson, 2012*b*) in order to better describe chemical bonding phenomena. They take advantage of the concept of valence atomic orbitals in order to provide a "shortcut" compared with highly complex DFT calculations.

If the Hückel method is combined with parameters refined against results from corresponding DFT calculations, it can be similarly accurate. In addition, it enables a direct connection between geometrical and electronic structure through the "method of moments"—hence, the DOS can be reconstructed from its moments using different moments inversion schemes (Gaspard and Cyrot-Lackmann, 1973; Burdett and Lee, 1985). If a short-range interatomic repulsion potential is added to the Hückel method, the expected dependence of bonding interactions on the interatomic distance can be reproduced. The μ_2-Hückel model uses the second moment of the DOS, μ_2, to approximate the repulsion energy of the system (Lee, 1991).

2.2   Electronic structure

Single-crystal X-ray diffraction is the method of choice for quantitative crystal structure analysis. Conveniently, it can be done in-house on automated diffractometers, not only at ambient conditions but also in a wide temperature range. Only in the case of very complex structures and/or structural disorder synchrotron radiation may be necessary, as well as in the case of high-pressure studies employing diamond anvil cells. The observable thereby is the electron density distribution, which is a probability distribution function that describes how the electrons are arranged on average. This is an indicator for the kind of chemical bonding present in a crystal. For a full analysis of the chemical bonding and the electronic density of states (eDOS), quantum mechanical calculations are necessary.

2.2.1   Electron density

Within the Quantum Theory of Atoms In Molecules (QTAIM), the electron density is used for analyzing the localized density at/around atoms and the chemical bonds within a structure. The stationary points of the electron density distribution, as well as their gradient paths, are interpreted for this purpose. As a first step, the space is divided into volumes that contain one atomic nucleus each – the attractors. Neighboring atomic basins are connected by ridges in electron density; the emerging interatomic surface can be quantified. Within each interatomic surface, a maximum is found, demarcating the saddle point in electron density that emerges between two atoms. This is the bond critical point, which lies at the minimum of the ridge of electron density between two atoms. The ridges themselves—the connecting lines between both attractors and the common bond critical point—constitute the bond path.

The complete, non-overlapping partitioning of space into atomic basins, which are linked by zero-flux surfaces in the gradient vector field of the electron density, allows for the rigorous calculation of several atomic properties. The volume and the electronic charge assigned to each atom, for example, are easily accessible. The connectivity can be analyzed, as well, by determining which atomic basins are connected by common interatomic surfaces and how large the solid angles covered by different connections are. For more on the synaptic order of bonds and multicenter bonding, see Silvi (2002). Consequently, physical properties like ionic charges and relative bond strengths become accessible through the topological analysis of the electron density—calculated or measured.

For a detailed account by a pioneer of QTAIM, see Bader (1994), for an elaborate treatise from a chemist's point of view, see Gillespie and Popelier (2001), and for an alternative way of space partitioning, see Pendás *et al.* (2012).

2.2.2 Electron localization

The Electron Localization Function (ELF) measures the probability of a second electron with the same spin being present in the neighborhood of a reference electron. Already upon its introduction, it revealed the atomic shell structure, as well as core, binding, and lone electron pairs for molecules (Becke and Edgecombe, 1990), with the first application to solids following shortly after (Savin *et al.*, 1992). In this context, the ELF was also generalized to electron densities obtained from DFT calculations, while the original derivation was based on a Hartree-Fock pair density. Subsequently, the equivalency of the ELF determined from semi-empirical Extended-Hückel calculations was also demonstrated (Burkhardt *et al.*, 1993).

The ELF is dimensionless and normalized with values between 0 and 1. Large values (ELF> 0.5) correspond to a high localization, meaning that for an electron located there, no second, same-spin electron can be found in its vicinity; in contrast, electron pairs with opposite spins in the same region indicate localization. If ELF = 0.5, the respective region resembles the homogeneous electron gas, which serves as a reference to the ELF.

In the past, the ELF was extensively used to characterize the chemical bonding situation, in particular the occurrence of covalent bonding in intermetallics (Kohout *et al.*, 2002). It represents paired-electron densities as they are observed in electron shells, bonding pairs, and non-bonding lone pairs (Silvi and Savin, 1994). For more on ELF topology see Savin *et al.* (1996); for more on ELF applied to delocalized bonds and basins see Savin (2005).

The calculation of the ELF was implemented as an option in many DFT software packages, most relevant for intermetallics is the Tight-Binding Linear Muffin-Tin Orbital Atomic Sphere Approximation (TB-LMTO-ASA) (Jepsen *et al.*, 2000). If the crystal structure of a compound is fully known and the electron density can be calculated using DFT, a subsequent calculation of the ELF can be easily performed.

The ELF topology holds information about the connectivity of covalent interactions in a compound, and can serve as a basis to generate bifurcation diagrams (Marx and Savin, 1997) (e.g., see Armbrüster *et al.* (2007)). Details and the exact shape of the ELF basins can be visualized in the same manner as the electron density, displaying the ELF values on planes as slices through a 3D structure or in a 3D manner in the form of isosurfaces.

ELF attractors can be found wherever covalent bonding plays a role. Their electron count, calculated from the integration of the electron density over the volume of the ELF basin, illustrates the interaction between the atoms. Their connectivity, calculated from the solid angles enclosed by this and the neighboring atomic basins, characterizes the interaction as two- or multi-center. If all interactions are taken into account, the dimensionality of the covalently bonded network can be determined, also. For 3D representations of the ELF see, for instance, the studies on $TiSb_2$ and VSb_2 by Armbrüster *et al.* (2007), on the series $Al \rightarrow CaAl_2 \rightarrow SrAl_2 \rightarrow BaAl_4 \rightarrow CaAl_2Si_2 \rightarrow Si$ by Häussermann *et al.* (1994), and on $CuAl_2$ (Grin *et al.*, 2006). For reviews on the ELF see Savin *et al.* (1997) and Grin *et al.* (2014).

2.2.3 Electron localizability

As an improved measure for the electron localizability, the Electron Localizability Indicator (ELI) was introduced about a decade ago (Kohout, 2004). It can be related to the ELF in the Hartree-Fock approximation, but differs from it for correlated wave functions (Kohout *et al.*, 2004; Kohout *et al.*, 2005; Kohout *et al.*, 2007). However, even where it is similar to the ELF, the ELI does not rely on the use of the uniform electron gas as a reference.

In the years following its initial formulation, the ELI was investigated with respect to its representation of atomic shells (Kohout *et al.*, 2006) and bonding (Kohout, 2007) (in both, direct and momentum space). On cursory inspection, ELI and ELF topologies can be regarded as largely equivalent. However, contrary to the ELF, the ELI can be decomposed into partial orbital contributions in an exact manner (Wagner *et al.*, 2007). The topological analysis of the ELI-D, in direct space, can therefore be interpreted on the level of electron pair formation and can help analyze chemical bonding on a more fundamental level (Wagner *et al.*, 2008).

The ELI was applied to both molecular and crystal structures during the last few years. In intermetallics, the ELI of specific compounds was analyzed, for instance, for antiferromagnetic $EuTM_2Ga_8$ (TM = Co, Rh, Ir) (Sichevych *et al.*, 2009), the stannides $TMSn_2$ (TM = Mn, Fe, Co) (Armbrüster *et al.*, 2010), the quasicrystal approximant Al_5Co_2 (Ormeci and Grin, 2011), and the clathrate $Rb_{8-x-t}K_x\square_tAu_yGe_{46-y}$ (Zhang *et al.*, 2013). More extensive studies cover entire structure classes, such as *hcp* element structures (Baranov and Kohout, 2008), intermetallic Laves phases (Ormeci *et al.*, 2010), or diborides crystallizing in the AlB_2 structure type (Wagner *et al.*, 2013).

Today, localization and delocalization indices for solids are the subject of ongoing studies (Baranov and Kohout, 2011) in order to help elucidate the chemical bonding from quantities that can be derived from the electron density of crystal structures.

2.3 Crystal structure interpretation

The methods described in the Section 2.2 all heavily rely on computational resources. This means that they not only became more and more generally applicable in recent years, but also that they require quite time-consuming calculations for each compound separately. Furthermore, quantum-mechanical calculations require some experience and background that not everybody has who is successfully performing standard single-crystal X-ray diffraction structure analyses. There are many pitfalls, because it needs some experience to correctly interpret details in the calculated electron density on one hand; on the other hand, artifacts can result from not properly chosen basis sets, **k**-mesh sizes, etc., which may lead to misinterpretations.

The classical concepts of chemical bonding, with the idealized limiting cases of ionic, covalent, metallic, van der Waals, and hydrogen bonding, can still be useful for a qualitative interpretation and understanding of crystal structures. The chemical bonding in intermetallics cannot be understood through the simple assumption of metallic bonding, only. Covalent and/or ionic interactions are frequently present to a differing extent as well. A prominent example for such heterodesmic compounds are the Zintl phases, which are dealt with in the next subsection. For general reviews on the structural chemistry of intermetallics see, for instance, Corbett (1996), Corbett (1997), Corbett (2000*a*), Corbett (2000*b*).

2.3.1 Concept of electronegativity

Frequently, the electronegativities, χ, of the constituents of an intermetallic compound are used for estimating the polarity of the chemical bonds between them. Electronegativity can be seen as a measure to attract valence electrons from more electropositive atoms. In contrast, electropositivity is related to the ability of an atom to donate its valence electron(s) to a more electronegative atom. The electronegativity of an atom depends on its position in the periodic table and on the distance of the shell of valence electrons from the atomic nucleus. This gives a general trend for increasing electronegativities along the rows of the periodic table, and decreasing values down the columns. One has to keep in mind that electronegativity is a classical concept with its limits, although it proves quite useful in many cases.

To some extent the electronegativity parameters are transferable to atoms in different atomic environments. Unfortunately, there are several electronegativity scales in use, which not only differ in their absolute values but also in some of their relative values within the respective scale. The most common scales are that of Pauling (1932) and Allred and Rochow (1958), but also those of Mulliken (1934), Pearson (1985), and Pearson (1988) are used sometimes.

> **Pauling electronegativity** Pauling (1932) based his semiempiric parameters on the concept of the additivity of the energies of covalent chemical bonds, $E_{A-B} = 1/2(E_{A-A} + E_{B-B})$, derived from the known experimental formation enthalpies of binary molecules and compounds. He assigned any additional stabilization energy to ionic contributions due to different electronegativities, $\Delta_{A-B} = (\chi_A - \chi_B)^2$. Due to the lack of an absolute scale, the electronegativity of H was first fixed to 2.1 and later to 2.20. This gives a range from 0.7 for Fr to 3.98 for F (see Section 6.2).

Mulliken electronegativity After the usefulness of Pauling's electronegativity parameter became obvious, Mulliken (1934) tried to put these values on an absolute scale. His electronegativity values are just proportional to the arithmetic mean of the first ionization energy, I, and the electron affinity, A, of an atom, $\chi = 1/2(I + A)/2$. All these parameters can be determined quantitatively on an absolute scale for free atoms. However, this approach does not take into account the influence of the respective atomic environments. There is a good but not perfect linear correlation between Mulliken and Pauling electronegativities. Not so many electronegativities have been determined in this manner thus far.

Allred and Rochow electronegativity Allred and Rochow (1958) tried to improve the concept of electronegativity by basing it on the force of electrostatic attraction between the nucleus and an electron from a bonded atom, both approximated by point charges. By choosing proper constants, an even closer linear relationship can be found between this scale and Pauling's scale. The values range here from 0.9 for Fr to 4.1 for F.

Pearson absolute electronegativity and absolute hardness Based on the MO-theory, Pearson (1985) defined the absolute electronegativity in the same way as Mulliken (1934), but now for general chemical systems (atoms, ions, molecules, ...), $\chi = 1/2(I + A) = -\mu$, with the chemical potential μ. He also uses the concept of chemical hardness, $\eta = 1/2(I - A)$. While the direction of the net flow of valence electrons between two atoms is determined by the electronegativity difference, the magnitude of the total electron transfer is given by the hardness. The smaller it is, the larger the covalency of a bond. In a chemical system with two constituents B and C, there will be a flow of valence electrons from the system with larger χ to that with smaller χ until the chemical potential becomes equal everywhere. The fractional number ΔN of electrons being transferred thereby is given by

$$\Delta N = \frac{\chi_C - \chi_B}{2(\eta_C - \eta_B)}. \tag{2.1}$$

In contrast to the local electronegativity, the local hardness can differ from the average global one, $\eta = 1/2(I - A)$. The electronegativity values are only given for some elements, and range from 2.0 eV for Sr to 10.41 eV for F (see Section 6.2).

2.3.2 Zintl-Klemm concept

The Zintl-Klemm concept describes a category of binary and ternary intermetallic line compounds with an appearance spanning the full range between metallic and ionic. Originally, this class comprised compounds between electropositive s-block metals on one side, and electronegative p-block metals and semimetals around the Zintl-line on the other side. Later on, the rare earth elements also were included as electropositive bonding partners, and, to some extent, also late TM elements as electronegative ones originally. In this concept, a complete charge transfer was assumed from the electropositive to the electronegative elements leading to full valence shells for both of them. Depending on the number of transferred electrons, either isolated anions or polyanions are formed. The atoms in the polyanions are bonded covalently. In this concept, the electron acceptors are thought to behave similarly to the isoelectronic elements—isoelectronic regarding the valence electrons, only. Of course, in spite of its usefulness, this is a rather simplistic assumption neglecting the complex electronic structure of these compounds.

A classical example of a Zintl phase is $cF16$-NaTl (Zintl and Dullenkopf, 1932), where, according to the Zintl-Klemm concept, Na donates one electron to Tl, a group 13 element, which consequently increases its valence electron number from 3 to 4. This makes Tl isoelectronic to a group 14 element, i.e., from a triel to a pseudo-tetrel element, which then adopts an anionic diamond network with covalently bonded Tl-atoms. The Na^+ cations fill the space in-between, forming a double-diamond structure (two mutually interpenetrating diamond networks). In spite of the character of the polar/covalent chemical bonding, NaTl still has metallic character, however. It has been shown by quantum mechanical calculations that besides considering just the effects of the covalent interactions in the polyanions, also the competition among metallic, ionic, and covalent interactions has to be taken into account (Wang and Miller, 2011).

An early review on intermetallics was published by Zintl (1939) where he presented his view of (polar) metallic bonding on several examples and, twenty years later, a summary on compounds made up of metalloids and alkali metals by Klemm (1958), following a number of manuscripts on this class of materials (e.g., (Klemm, 1950*a*; Klemm, 1950*b*; Klemm, 1950*c*)). A considerable number of reviews and new interpretations of the Zintl-Klemm concept have been published within the last few years (Sevov, 2002; Miller *et al.*, 2011; Nesper, 2014).

2.3.3 Electron counting

The Wade-Mingos rules were successively developed by Wade (1971), Wade (1976), Mingos (1972) and Mingos (1984), and allow the prediction of structures made up of clusters of atoms. The chemistry behind the rules is also called "polyhedral skeletal electron pair theory" and is based on the MO-description of bonding. This is an advanced way of counting electrons compared with the octet rule and the 18-electron rule for transition metals that are employed in the Zintl-Klemm approach described in Section 2.3.2.

The $4n$-rules apply to deltahedra, i.e., polyhedra with only triangular faces, that have 4–12 vertices. Examples are the tetrahedron, octahedron, and icosahedron, but also the trigonal bipyramid, pentagonal bipyramid, tricapped trigonal prism, bicapped square prism, etc. As "$4n$" indicates, 4 electrons are assigned to each vertex, as is the case for B_n- or $B_n C_m$-clusters. The clusters can be complete (*closo*-), or might be missing 1, 2, or 3 vertices (*nido*-, *arachno*-, and *hypho*-clusters, respectively). A closo-cluster has a total of $4n + 2$ electrons, while each missing vertex adds $+2$ to this balance and each additional capping vertex deducts -2.

The $5n$-rules apply to polyhedra with only 3-connected vertices such as the tetrahedron, trigonal prism, cube, dodecahedron, etc. These polyhedra are also necessarily duals of the deltahedra. They become relevant if around five electrons can be assigned to each structure. Then, a cluster is assigned $5n$ electrons in total, with each missing vertex deducting -1 from this number.

The $6n$-rules finally apply to rings, which can accommodate around six electrons per vertex. The number of electrons is $6n$ in total; $+2$ have to be added per broken bond and -2 have to be subtracted for each trans-annular bond that "shortcuts" the ring.

The work of Jemmis and Balakrishnarajan (2001) and Jemmis and Jayasree (2003) further improved the understanding of (car)boranes. A generalization of the Wade-Mingos rules are the Jemmis *mno* rules (*m* is the number of sub-clusters, *n* is the number of vertices, and *o* is the number of single-vertex shared condensations) that also contain the concept behind the Hückel rules for two-dimensional assemblies; see Balakrishnarajan and Jemmis (2000) and Jemmis *et al.* (2001).

2.3.4   Hume-Rothery electron concentration rule

The concept generally termed the "Hume-Rothery rule" describes which requirements have to be met for the formation of solid solutions instead of intermetallic compounds or just phase separation. That is, which requirement must another element meet in order to be incorporated into the structure of a chemical element without changing the structure type. In order for this to happen, the rule says that both elements in their pure form have to adopt the same structure type. Additionally, their radii should not differ by more than 15%. Similar values in electronegativity are also necessary to prevent compound formation, and complete mutual solubility can only be reached for elements that display the same valency as well.

If solid solutions cannot be formed by a mixture of elements, they may form a superstructure of a simple structure type (Hume-Rothery and Powell, 1935). Hume-Rothery himself found that the number of valence electrons per atom also affects the structure of the formed compound (Hume-Rothery *et al.*, 1940; Zintl and Brauer, 1933). The parameter he used to assess the electron concentration, e/a, was the number, e, of itinerant electrons per unit cell divided by the number of atoms, a.

This behavior can be illustrated by the two γ-brass-type compounds $cI52$-Cu_5Zn_8 and $cP52$-Cu_9Al_4, which crystallize in virtually the same structure type despite having significantly different compositions (compare with Subsection 7.3.1). Consequently, one structure can not simply be created through a replacement of one element by another, but a redecoration of the atomic sites has to take place, which also results in a change of the lattice symmetry from cI to cP. Despite this difference, however, the approximate geometry of the structures is very similar, and this was ascribed to their identical values of $e/a = 1.60$ (Mizutani *et al.*, 2010), also cocorroborated by *ab-initio* calculations (Asahi *et al.*, 2005).

The stabilization of complex structures of intermetallics obeying the Hume-Rothery electron concentration rule originates from the optimal nesting of the Fermi sphere in a near-spherical Brillouin zone. In such a case, the electronic density of states (eDOS) at the Fermi level is forming a pseudo-gap. Later, compounds that were thought to be stabilized in this way were named Hume-Rothery alloys or phases, and their structure and stabilization mechanism were investigated in greater detail (Massalski and Mizutani, 1978). In later studies, Mizutani *et al.* (2010) found the valence electron concentration (VEC) to be a more meaningful parameter if determined by the number of all valence electrons—including those assigned to d-orbitals—divided by the number of atoms. The difference would be in the case of $cF4$-Cu, for instance, $e/a = 1$ and VEC = 11. More on the Hume-Rothery electron concentration rule applied to complex periodic and quasiperiodic intermetallics can be found in Mizutani *et al.* (2010) and Mizutani *et al.* (2014).

2.3.5 μ_3-acids and -bases

In recent years, another theory to rationalize the formation of specific structures of intermetallic compounds was developed by Fredrickson and co-workers: an extension of the Lewis theory of acids and bases to intermetallics (Stacey and Fredrickson, 2012c). Therein, metallic elements are termed as acidic or basic, respectively, if they are prone to accept or donate electrons. The basis of this approximation is the third moment of the eDOS, μ_3. Electron-poor and -rich systems relative to the ideal μ_3-value are then identified as μ_3-acidic and μ_3-basic, respectively, and can neutralize each other upon compound formation. The parallel to Lewis acids and bases is the quantification of the reactivity of transition metals, i.e., their tendency to form intermetallic phases.

The first three moments of the eDOS correspond to the area under the eDOS curve (μ_0), the average energy value of the eDOS (μ_1), and the variance around $E = 0$ (μ_2), respectively. The third moment of the eDOS (μ_3) is responsible for the (a)symmetry of the eDOS around $E = 0$: the eDOS is symmetrically distributed around its average value for $\mu_3 = 0$, and tends to shift towards below-average values for $\mu_3 > 0$ and to values above average for $\mu_3 < 0$. The kurtosis, defined by the third and fourth moments via $\kappa = \mu_4 - \mu_3^2 - 1$ for standardized values of μ_0, μ_1, and μ_2, is related to the broadness of the eDOS peaks: the peaks correspond to δ-functions for $\kappa = 0$ and are increasingly broadened for higher values. This also means that smaller κ-values enable the formation of a deeper (pseudo-)gap.

A clear separation of filled and empty states is favorable for the energy balance of a phase. Therefore, the position of the energy minimum shifts in sync with the μ_3-values, depending on the number of available electrons, i.e., the amount of band-filling: $\mu_3 = 0$ is an ideal value for half-filled bands, while lower/higher fillings have ideal μ_3-values below/above zero. The preference for the ideal electron count is sharpened with lower κ-values, leading to the largest possible gap between filled and empty states.

The first such view on intermetallic phases was illustrated on examples of $cP2$-CsCl-type binary intermetallic compounds such as $cP2$-ScCu and $cP2$-TiFe, as well as on the system Ti–Ni (Stacey and Fredrickson, 2012c). Also the stability of $tP4$-TiCu and $hR276$-Ti$_{21}$Mn$_{25}$ can be explained by this method: in these systems, the formation of $cP2$-CsCl-type structures would lead to an imbalance of the acid/base pairing and therefore to an unstable configuration (Stacey and Fredrickson, 2013). Instead, the neutralization is accompanied by more complex structural features, partly forming regions of different compositions within a structure. Even a system containing very complex structures can be viewed in this light: the Mackay-type icosahedral clusters in Sc–Ir intermetallics emerge due to the maximization of Sc–Ir contacts in the Ir-poor region of the phase diagram (Guo *et al.*, 2014).

2.3.6 μ_2-chemical pressure

Factors that inhibit or promote the stability of a phase can be identified by investigating the "chemical pressure" that occurs at different locations within a structure. This method was devised in detail by Fredrickson *et al.* in the last few years. It can help to analyze the electronic packing frustration that occurs in intermetallic compounds, and arises from competing bonding mechanisms, and demands on the packing in a structure.

The basis for this approach are Hückel tight-binding calculations supplemented with short-range interatomic repulsion forces that are proportional to the second moment of the eDOS (i.e., μ_2-Hückel calculations). These are calibrated against values from DFT calculation—in the Fredrickson group by using their program *eHtuner* (Stacey and Fredrickson, 2012a)—to result in "an effective orbital-based rendition of the DFT electronic structure" (Harris *et al.*, 2011). The resulting model is not only accurate—due to it being based on *ab-initio* calculations—but its total energy can also be decomposed into a sum of onsite and pairwise interaction energies. This leads to the possibility of extracting pressure contributions from individual orbital interactions, which can be interpreted as chemical pressures. The overall pressure, however, is zero after energy minimization, balancing the competing terms.

Chemical pressures are usually illustrated as spheres around the respective atomic positions. Their radii represent the magnitude of the respective values and their color—white or black—the sign of the local pressure, which can be positive or negative, in analogy to the astronomic phenomena of white-hot stars radiating outward and black holes pulling in their surroundings (Fredrickson, 2011). The anisotropy of the chemical pressure can be illustrated by projecting the different chemical pressure contributions acting on an atom onto spherical harmonics, representing magnitude and sign in the same way as described above.

A first example was the occurrence of the $oS36$-Ca$_2$Ag$_7$-phase in the $oS36$-Yb$_2$Ag$_7$ structure type, which is a derivative of the $hP6$-CaCu$_5$ structure type (Fredrickson, 2012). By comparing the $hP6$-CaCu$_5$-type phase in the related Sr–Ag-system with a hypothetical $hP6$-CaAg$_5$-structure, the influence of the size of the atom occupying the Ca-site becomes apparent and the resulting alteration of the structure has to be the logical consequence (Fredrickson, 2011). The structural response to the increased chemical pressure that triggers the insertion of defect planes to form the $oS36$-Ca$_2$Ag$_7$-structure can also be suppressed or exacerbated by tuning the valence electron count: fewer valence electrons in the Ca–Pd-system lead to the $hP6$-CaCu$_5$-type structure $hP6$-CaPd$_5$, whereas a higher number in the Ca–Cd-system leads to the formation of a $hP68$-Gd$_{14}$Ag$_{51}$-type phase (Fredrickson, 2011).

The Ca–Ag-system also served as an example for demonstrating the applicability of the density-functional-theory chemical-pressure analysis (DFT-CP) as a means to gain direct insight into the local (in)stability of a structure. The CP-distribution is obtained from the comparison of electronic structure calculations at equilibrium volume, as well as at slightly expanded and slightly contracted unit cell volumes (Fredrickson, 2012). The space of numerically approximated voxel pressures then has to be divided between the atoms of the unit cell; the partitioning into Voronoi cells (assigning all points in space to the atom that they lie closest to) or volumes according to Bader's quantum theory of atoms in molecules, QTAIM, (Bader, 1994) have been deemed as the most useful methods (Fredrickson, 2012). Fredrickson (2012) also demonstrated that the CP anisotropy surfaces behave very similarly in the case of μ_2-Hückel- and DFT-CP analyses, as well as for different partitionings of space in the latter case (according to Voronoi cells or Bader volumes). Differences do occur and the values of net atomic chemical pressures depend highly on the way that space is partitioned.

In the Ca–Cu–Cd system (Harris *et al.*, 2011), the analysis of the chemical pressure μ_2 helped rationalize the stabilization of binary structure types in this ternary system by examining the corresponding—hypothetical—binary compounds. Therein, high values of positive and negative chemical pressure indicated locations within the structure, which stand in the way of these binaries to be stabilized. By replacing atoms in the respective positions by a different chemical element, selected according to the sign of the chemical pressures, these structures could be stabilized in their ternary versions.

The compounds $tI32$-Ca$_5$Cu$_2$Cd and $cP39$-Ca$_2$Cu$_2$Cd$_9$ adopt the binary structure types $tI32$-Cr$_5$B$_3$ and $cP39$-Mg$_2$Zn$_{11}$, respectively, without occurring in any of the respective binary systems (although the former can be found in a similar system as $tI32$-Ca$_5$Zn$_3$). This indicates that the ternary combination of elements is crucial to the phase stability. Structural fragments of unary and binary structures of the subsystems of Ca–Cu–Cd can be recognized in these ternary phases: $tI32$-Ca$_5$Cu$_2$Cd contains structural motifs from $cI2$-Ca (*bcc*), $tP20$-Ca$_3$Cd$_2$, and $oP12$-Ca$_2$Cu of $mP20$-CaCu, whereas $cP39$-Ca$_2$Cu$_2$Cd$_9$ contains motifs that also occur in $cF4$-Cd (*ccp*), $hP68$-Ca$_{14}$Cd$_{51}$, and $cI52$-Cu$_5$Cd$_8$ (Harris *et al.*, 2011).

The hypothetical structures Ca_5Cu_3 and Ca_5Cd_3 illustrate how Cu- and Cd-atoms occupying the same positions can have opposite effects on the structure (Harris *et al.*, 2011). It also becomes clear that the ternary compound is stable, as opposed to the rather unbalanced hypothetical binary variants. The distribution of the different elements over the structure type can be explained by evaluating the specific features of the chemical pressure. Similarly, the binary version Ca_2Cd_{11} of $cP39$-$Ca_2Cu_2Cd_9$ exhibits strong packing frustrations that can be relieved by substituting the smaller Cu-atoms on select Cd-positions. The intermediate composition $Ca_2Cu_xCd_{11-x}$ ($x = 1/3$) illustrates the tendency towards the Cu-substituted variant and how the change of one atom out of 39, from Cd to Cu, can have a decisively stabilizing effect on the structure.

An important aspect of the structural response to chemical pressure is the interplay between coordination numbers and bond lengths (Fredrickson, 2011): the usually high coordination of atoms in high-pressure phases and the higher number of bonds per atom results in weaker individual bonds and therefore larger bond lengths. This is true for pressure of physical, as well as chemical, nature. Therefore, if the chemical pressure in an intermetallic system with typically high coordination numbers is released through a structural rearrangement, the resulting phase—which is under lower chemical pressure now—will have lower coordination numbers and at the same time shorter, individually stronger, interatomic bonds.

The comparison of the compound $hP6$-$CaZn_5$ ($hP6$-$CaCu_5$ structure type) with the 1/1-Tsai-type approximant $cI184$-$CaCd_6$ reveals another mechanism to relieve chemical pressure (Berns and Fredrickson, 2013)—different from the introduction of a defect plane into the $hP6$-$CaCu_5$ structure type to form the $oS36$-Ca_2Ag_7-phase. The structure responds to the increasingly negative CP on the too-small Ca-atoms by replacing the hexagonal Zn-rings that surround them in $hP6$-$CaZn_5$ by pentagonal Cd-rings in $cI184$-$CaCd_6$. The tiling of space that is achieved by honeycomb and kagome nets in hexagonal $hP6$-$CaZn_5$ consequently changes from a stacking of these periodic planes to a curved arrangement in the quasicrystal approximant. The Zn-honeycomb network is replaced by a Cd-pentagonal dodecahedron and the hexagonal Ca-net is curled into a Ca-icosahedron around it. Instead of stacked layers, the resulting structure exhibits concentrically nested polyhedra (Berns and Fredrickson, 2013).

Another structure found in the Ca–Cd-phase diagram—$hP65$-$Ca_{14}Cd_{51}$—responds to the negative CPs that occur in this system close to the 1:5-composition by forming a more complex structure that exhibits three different Ca-sites with coordination numbers as low as 13 (Berns and Fredrickson, 2014). The structural motif from the $hP6$-$CaCu_5$-type is yet again adapted and distorted so that CPs are lowered.

Two recent methodological developments helped improve the interpretability of the DFT-CP analysis (Berns *et al.*, 2014). 1. Artifacts in the form of large, isotropic contributions to the CP values could be removed by adapting the voxel grid of DFT calculations for the equilibrated, as well as slightly contracted and expanded, versions of the structure under consideration, so that the relative positions of atoms and voxels remain connected. 2. The assignment of voxels to different interatomic contacts was improved by changing the scheme from a purely geometrical one (i. e., based on Voronoi contact volumes) to a Hirshfeld-inspired weighting of contact volumes that relates the local geometries in a compound to the respective free atom electron densities of the different elements, taking into account varying radii and—in combination with the first improvement—rendering CP results that corroborate experimental findings of stable intermetallic phases. The usefulness and applicability of the DFT-CP analysis following these amendments were demonstrated on $cF24$-$MgCu_2$-type Laves phases, as well as a comparison of the stabilities of the $cF24$-$AuBe_5$- and $hP6$-$CaCu_5$ structure types, which compete for intermetallic phases with 1:5-compositions but different relative radii (Berns *et al.*, 2014).

Additional studies were performed on $tP118$-$Ca_{36}Sn_{23}$—a superstructure of the $tI32$-W_5Si_3 structure type—in combination with a methodological survey of the developments of DFT-CP (Engelkemier *et al.*, 2013).

2.3.7   Topological analysis

The analysis of crystal structures through their decomposition into larger structural building blocks (clusters, modules, etc.) has a long tradition in structure research. Molecules form units that can be easily defined as being separate from one another: the intramolecular bonding is much stronger and the interatomic distances therefore much shorter than in the intermolecular case. Intermetallic structures, however, do not offer such an obvious and intuitive division into subunits.

One started early with the description of structures based on their local atomic environments (AETs). This kind of description does justice not only to the translational order that arises from the atomic arrangement, but also to the local motifs that occur in the structure. These could possibly hold clues as to how the neighboring atoms interact. Sometimes the shapes of AETs are clearly defined, and it is obvious when the nearest neighbor shell is complete and where the next shell starts. However, there are also more ambiguous cases.

The *ccp* packing of spheres is an example of a clear case of 12-fold coordinated atoms with cuboctahedral geometry. In contrast, a *bcc* packing is far more ambiguous. The closest neighbors—if strictly defined—are located on the vertices of a cube with respect to the centering atom. This would render a coordination number of eight, which is very small for a metal, and would produce even more issues when trying to track the next surrounding shells. The next-nearest neighbors are located in the centers of the neighboring cubic unit cells, forming an octahedron. There are six of them and their distance to the central atom is not much larger than the one of the nearest neighbors. Usually, the coordination polyhedron of *bcc*-structures is now regarded as being a rhombic dodecahedron, which is the compound shape of a cube and an octahedron, having a coordination number of $CN = 14$.

This concept can be transferred to outer shells of metallic and intermetallic structures, of course. The shapes emerging from this analysis of a compound's geometry are generally termed "clusters" if they are close to spherical. It is important to distinguish these geometrical constructs from molecular clusters, which can normally be unambiguously justified through the analysis of chemical bonding and interatomic distances. In intermetallics, clusters are usually just structural subunits, and only in some cases does this also coincide with a specific kind of chemical bonding (see above).

It becomes increasingly difficult and ambiguous to separate higher and higher cluster shells from one another and to define them in a rigorous manner. The "maximum-gap rule" (Brunner and Schwarzenbach, 1971), developed for the identification of AETs, can be used for the derivation of higher-order cluster shells to a limited extent only, as the shells become larger and usually less spherical. A deformation of the shells away from a spherical envelope leads to a broadening of their footprints in a distances histogram, leading to an overlap in the distances histogram.

Whereas it is often possible to identify suitable clusters and separate their shells just by visual inspection (see, e.g., Dshemuchadse *et al.*, 2011; Dshemuchadse and Steurer, 2014), a rigorous method seems to be necessary to investigate structures in a objective, comparable way. This is now possible owing to the software ToposPro (Blatov *et al.*, 2014; Blatov and Shevchenko, 2015) (formerly TOPOS (Blatov *et al.*, 2000; Blatov, 2014)). The program package was designed to analyze structures topologically and can also be used, for example, for identifying nets in metal-organic frameworks. If applied to intermetallics, it can describe any structure as being composed of rather large "nanoclusters" that are basically extensions of the cluster-shell concept to additional shells. The software calculates the "adjacency matrix" that determines the coordination shells for each atomic position in the unit cell. Consequently, the compound of all coordination shells of the atoms in the first shell are thought of as a two-shell nanocluster. The next shell then consists of the coordination shells of the atoms of the previous shell put together. For an in-depth discussion of the application of the "nanocluster analysis" to intermetallic structures, see Blatov (2012).

This algorithm has been applied to various intermetallic structures of different levels of complexity: in addition to a general piece on nanoclusters with Frank-Kasper polyhedral cores (Blatov *et al.*, 2011), the group around Blatov also covered $cF184$-$ZrZn_{22}$-type structures and their superstructures (Ilyushin and Blatov, 2009), $cF1192$-$NaCd_2$-type structures (Shevchenko *et al.*, 2009), the Mg–Al system (Blatov and Ilyushin, 2010; Blatov *et al.*, 2010; Blatov and Ilyushin, 2011; Blatov and Ilyushin, 2012), as well as studies, where specific cluster types are searched within a database of intermetallic compounds (Pankova *et al.*, 2012; Pankova *et al.*, 2013; Shevchenko *et al.*, 2013).

2.4 Crystal structure prediction (CSP)

The prediction of crystal structures just from the basic "ingredients" that constitute a compound has been keeping materials scientists busy for decades at this point (Maddox, 1988; Cohen, 1989; Hawthorne, 1990; Lommerse *et al.*, 2000; Motherwell *et al.*, 2002; Day *et al.*, 2005; Day *et al.*, 2009; Bardwell *et al.*, 2011; Kazantsev *et al.*, 2011). The process of crystal structure prediction involves the solution of two entangled problems. One is the exploration of the energy landscape of a system to determine possible structures. The other is weighing up these different arrangements and geometries against one another by ranking them with respect to their energies. Depending on the type of system one investigates, the former or latter part of the problem proves to be more challenging (Oganov *et al.*, 2011). In the case of molecular crystals, the determination of possible arrangements within a structure may be relatively simple, as there often are strong interactions competing with very weak ones, resulting in a rather straightforward outcome. There, the calculation of accurate energies of each possible state is still difficult. A project organized by the Cambridge Crystallographic Data Centre (CCDC), called the "Blind Test of Organic Crystal Structure Prediction Methods" (CCDC, 2015) has made increasingly striking improvements on predicting organic crystal structures, especially with the most recent, sixth blind test (Gibney, 2015). Inorganic systems—and among them, perhaps even more so, intermetallics—can be ranked more easily due to quite accurate relative energies. However, due to the multitude of interactions with comparable interaction strengths, the search for possible or probable structural arrangements is the major difficulty in this case.

Most methods for crystal structure prediction rely on the following approaches: simulated annealing (Kirkpatrick *et al.*, 1983; Pannetier *et al.*, 1990; Schön and Jansen, 1996; Salamon *et al.*, 2002), metadynamics (Laio and Parrinello, 2002; Martoňák *et al.*, 2003; Martoňák *et al.*, 2005), random sampling (Pickard and Needs, 2011), basin hopping (Wales and Doye, 1997; Doye and Wales, 1998; Wales and Scheraga, 1999; Wales, 2004), minima hopping (Goedecker, 2004; Amsler and Goedecker, 2010), data mining (Fischer *et al.*, 2006), evolutionary algorithms (Goldberg, 1989; Bush *et al.*, 1995; Woodley *et al.*, 1999; Woodley, 2004; Abraham and Probert, 2006; Trimarchi and Zunger, 2007), and particle-swarm optimization (Wang *et al.*, 2010).

Quite promising results in the case of inorganic materials (elements, minerals) have been obtained by the application program *USPEX* ("Universal Structure Predictor: Evolutionary Xtallography"), which is mainly based on evolutionary algorithms (Glass *et al.*, 2006; Oganov and Glass, 2006; Oganov *et al.*, 2007; Lyakhov *et al.*, 2010; Lyakhov *et al.*, 2013). Several reviews provide an overview of the applied methods and give examples for systems that have been studied successfully (Oganov *et al.*, 2010*a*; Oganov *et al.*, 2010*b*; Oganov *et al.*, 2011; Zhu *et al.*, 2014; Zhu *et al.*, 2015).

"Evolution" takes place by applying different variations to "parent structures", i.e., previous generations, during the execution of the evolutionary algorithm in order to produce "children", i. e., the new generation of structures. These are (Oganov *et al.*, 2011): heredity (combination of planar slabs of different parent structures), mutation (random deformation of the unit cell), permutation (swapping of chemically different atoms), special coordinate mutation (displacement of atoms along eigenvectors of the lowest-frequency phonon modes (Lyakhov *et al.*, 2010). The population that is being produced during the execution of the evolutionary algorithm is kept diverse, which is even more challenging for structures with larger unit cells (Oganov and Valle, 2009). Larger systems, indeed, lead to a narrower, more Gaussian-like maximum in the density of states (DOS) that is shifted towards higher energies.

USPEX was used, for example, to investigate high-pressure structures of metallic elements, such as Li, K, Rb (Ma *et al.*, 2008), Ca (Errea *et al.*, 2008; Oganov *et al.*, 2010c), Na (Ma *et al.*, 2009), and Mg (Li *et al.*, 2014), as well as intermetallic compounds such as $CaLi_2$ (Xie *et al.*, 2010) and others in the Zr–Ni system (Mukherjee *et al.*, 2015).

Other codes and packages were applied to the investigation of the high-pressure structures of Li (Lv *et al.*, 2011), Ca (Ishikawa *et al.*, 2008; Yao *et al.*, 2009; Ishikawa *et al.*, 2010), Fe (Cottenier *et al.*, 2011), Ta (Liu *et al.*, 2013), or the structure of the compounds NaAl (Feng *et al.*, 2010) and $BaGe_3$ (Zurek and Yao, 2015).

Other promising approaches to crystal structure prediction have been described in recent years, among them the Parrinello-Rahman Method (Laio and Parrinello, 2002; Martoňák *et al.*, 2003), *CALYPSO* (Wang *et al.*, 2010; Wang *et al.*, 2012b), *EVO* (Bahmann and Kortus, 2013), *MUSE* (Liu, 2014), *etc.* (Abraham and Probert, 2008; Fadda and Fadda, 2010).

Incorporating databases into the study and comparison of predicted structures can make use of data-mining and big-data approaches into the quest for reliably predicted structures (Le Bail, 2010). One approach to handle symmetry in structure prediction is, for example, Monte-Carlo-based symmetry building (Michel and Wolverton, 2014). Exploring the composition space will open the door to another world of possible findings (d'Avezac and Zunger, 2008; Meredig *et al.*, 2014). More reviews on structure prediction are available (Woodley and Catlow, 2008; Wang and Ma, 2014; Zurek and Grochala, 2015). Extensive overviews on recent developments in structure prediction can also be found in the books by Oganov (2011) and Atahan-Evrenk and Aspuru-Guzik (2014).

3

Crystallographic description of crystal structures

Crystal structures, be they periodic or aperiodic, can be described in many different ways: as lattices, tilings, or coverings decorated by atoms or clusters on the one hand, or as packings of structural subunits (AETs, clusters, modules, etc.) on the other hand. In the following, we start with the decomposition of crystal structures into AETs and clusters, then we move on to a discussion of tilings, from the 1D quasiperiodic Fibonacci sequence to examples of 2D and 3D periodic, as well as quasiperiodic, tilings. The standard unit cell description of periodic crystal structures corresponds to a specific tiling approach with just a single prototile with the shape of a parallelepiped, i.e., the crystallographic unit cell.

We will not discuss the well-known 2D and 3D crystallographic standard concepts and the familiar crystallographic nomenclature and conventions. For detailed information on these topics see either any crystallographic textbook, the *International Tables For Crystallography* (IUCr, 2002), or the educational website of the *International Union of Crystallography (IUCr)* http://www.iucr.org/education. In contrast, we will discuss in greater detail periodic and quasiperiodic tilings and packings, which can be quite helpful in understanding structural building principles of all kinds of intermetallics, periodic as well as quasiperiodic ones.

3.1 Coordination polyhedra, atomic environment types, and clusters

A crystal structure can be regarded as resulting as a compromise between the energetically most favorable local atomic arrangements and their most efficient packing on the global scale, thereby minimizing the Gibbs free energy. At non-zero temperatures, not only the energy but also the configurational and vibrational entropy can play a decisive role for the formation and stabilization of a particular structure (low- *vs.* high-temperature phases, for instance).

One has to keep in mind that every atom B belonging to AET(A), i.e., the co-ordination polyhedron (atomic environment type, AET) around atom A, also has its own AET, AET(B), with the atom A being part of it (Fig. 3.1). AETs overlap

Intermetallics: Structures, Properties, and Statistics. First Edition. Walter Steurer and Julia Dshemuchadse.
© Walter Steurer and Julia Dshemuchadse 2016. Published in 2016 by Oxford University Press.

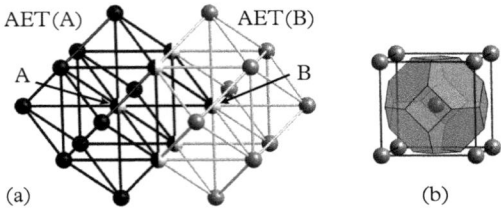

AET(A)   AET(B)

A   B

(a)   (b)

Fig. 3.1 *(a) Schematic representation of overlapping rhombic-dodecahedral AETs around atoms A and B, respectively, in the cI2-W structure. AET(A) is marked by black spheres, AET(B) is marked by gray ones, and atoms shared by both AETs are depicted half black/half white. The overlapping region is a distorted octahedron in this case, which can be decomposed into four Sommerville tetrahedra (space-filling distorted tetrahedra). (b) Unit cell with its central Voronoi polyhedron, a truncated octahedron.*

each other; therefore, each atom of an AET itself centers another AET. If the AET around an atom A contains n atoms, then the coordination number (CN) of atom A equals n, and the AET may be called a CNn coordination polyhedron. In most cases, atoms in intermetallic phases do not only interact with their nearest neighbors in the first coordination shell, but also with atoms of the second coordination shell, or even higher ones. This is reflected in atomic pair potentials in so-called Friedel oscillations, i.e., minima at specific interatomic distances, which favor particular local atomic arrangements.

In more complex structures it is not always clear which atom belongs to which AET. There are some conventions for the definition of atomic environment types (AETs), for instance, the maximum-gap rule (Brunner and Schwarzenbach, 1971), which says that all those atoms belong to a particular AET whose distances from the central atom fall into the range before the first large gap in the distance histogram. Its application is shown in Fig. 3.2 on the example of the structures of $tI2$-In and $tI2$-Pa. Both structures are isopointal, but show different AETs with coordination numbers 14 (CN14) and 12 (CN12), respectively. They belong to different structure types, consequently. Another approach would be to assign those atoms to a particular AET, which defines the Voronoi cell of its central atom (Fig. 3.1). In other words, these are those atoms whose Voronoi cells touch the Voronoi cell of the central atom of the desired AET. The Voronoi cell is dual to the related AET and vice versa.

The shape of an AET around a central atom depends on the directionality of the atomic interactions, if any, and on the size ratios of the central atom to the coordinating atoms. In the case of hard spheres touching each other, the ideal size ratios are $2\sqrt{3}/3 - 1 = 0.155$ in the case of triangular coordination (CN3), $\sqrt{3/2} - 1 = 0.225$ for a tetrahedral or square planar AET (CN4), $\sqrt{2} - 1 = 0.414$

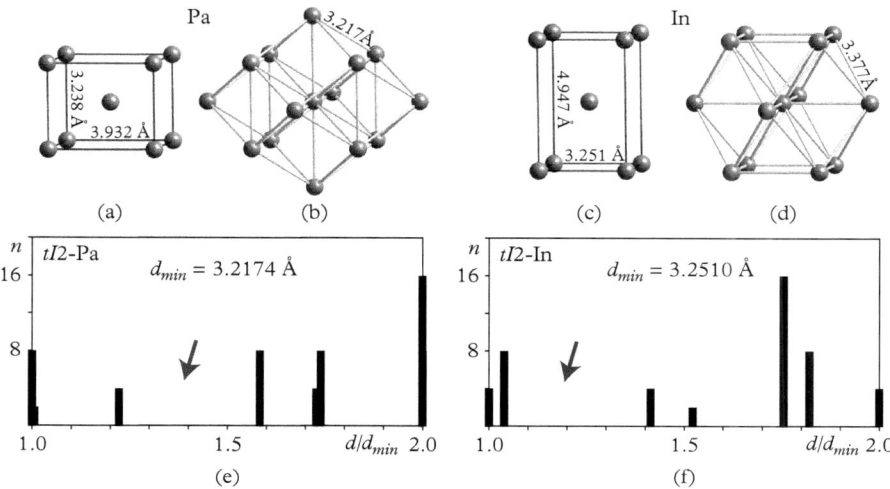

Fig. 3.2 *Crystal structures and AETs of tI2-Pa (a,b) and tI2-In (c,d) together with the respective distance histograms (e,f) with the maximum gaps marked by arrows. The distorted rhombic-dodecahedral AET (CN14) of tI2-Pa and the distorted cuboctahedral AET (CN12) of tI2-In are closely related to the respective undistorted AETs of cI2-W and cF4-Cu.*

for an octahedral AET (CN6), $\sqrt{3} - 1 = 0.732$ for a hexahedral AET (CN8), $\sin 2\pi/5 = 0.902$ for icosahedral coordination (CN12), 1 for cuboctahedral or disheptahedral (anticuboctahedral) AETs (CN12), and $\sqrt{3}/2(1 + \sqrt{5}) - 1 = 1.803$ for dodecahedral coordination (CN20).

There are no general rules for the visualization of a crystal structure in terms of structural building blocks (structure motifs, fundamental building units, clusters, etc.). To some extent, the maximum-gap rule can be applied also to larger structural units such as clusters. If the cluster shells were close to spherical then distinct gaps would separate them in the distance histogram. An attempt to identify "nanoclusters" in an unbiased way, just based on a set of geometrical input parameters, was put forward by the computer program ToposPro (Blatov *et al.*, 2014; Blatov and Shevchenko, 2015) (formerly TOPOS (Blatov *et al.*, 2000; Blatov, 2014)). However, even if one finds a neat-looking cluster-based description, this does not mean that it makes sense from a crystal-chemical point of view, i.e., that the intra-cluster atomic interactions differ from the inter-cluster ones.

Let us discuss the problem of cluster identification on the simple example of a cubic close packing (*ccp*) of hard spheres, as realized in the structure of face-centered cubic (*fcc*) aluminum (*cF4*-Al). It is usually described as a packing of hexagonal close-packed (*hcp*) layers with layer sequence ABC (Fig. 3.3(a)). Of course, an *fcc* packing of hard spheres is by no means a layer structure in the crystal-chemical meaning of the word, just in one of its geometrical descriptions.

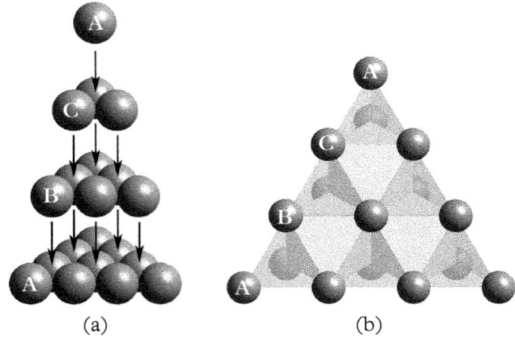

(a)   (b)

Fig. 3.3 *Structure of fcc aluminium (cF4-Al): (a) as composed from hexagonal close-packed (hcp) layers, and (b) built from edge-sharing octahedra and gap-filling tetrahedra (atoms are shown as non-space filling). Horizontal hexagonal close packed layers are marked A, B, and C, depending on their positions relative to one another.*

According to the cubic symmetry, the layer sequence ABC is realized in each of the four symmetrically equivalent [111]-directions. This means that the layers perpendicular to the four three-fold axes completely interpenetrate each other.

Another way to describe this simple *fcc* structure in terms of subunits would be a packing of non-overlapping uniform polyhedra, i.e., tetrahedra and octahedra (Fig. 3.3(b)). The gaps left in an edge-connected framework of octahedra are filled by tetrahedra or, vice versa, the gaps left in a corner-connected framework of tetrahedra are filled by octahedra. This description may be useful and crystal-chemically reasonable if the tetrahedral and/or the octahedral voids are filled by other, smaller, atoms. However, the then resulting octahedral and tetrahedral AETs are just a part of the story, because each atom of these AETs is also the center of other, less regular AETs, which also have to be taken into account.

The AET around each atom of *fcc* aluminum is a cuboctahedron (Fig. 3.4). The next larger AET (second coordination polyhedron) is an octahedron, and so on. These "cluster shells" look quite convincing. However, the crucial point is that every single Al atom in the structure is surrounded by exactly the same coordination polyhedron and that the distances between neighboring atoms are always exactly the same. Consequently, there is no crystal-chemical meaning at all in these kinds of "cluster shells". Perhaps, they could be seen as geometrical growth models of Al nanocrystals, which does not reflect reality as we know, however.

Another way of breaking down a crystal structure into its structural subunits is based on Voronoi cells, which are convex polyhedra containing just a single

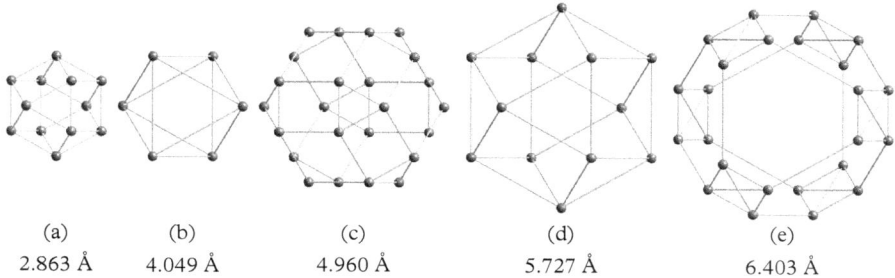

(a)	(b)	(c)	(d)	(e)
2.863 Å	4.049 Å	4.960 Å	5.727 Å	6.403 Å

Fig. 3.4 *The first five coordination polyhedra ("cluster shells") around any Al atom in fcc aluminum. Their radii, listed below each polyhedron, have ratios of $1 : \sqrt{2} : \sqrt{3} : \sqrt{4} : \sqrt{5}$ (Steurer, 2006b). A part or all of the atomic distances in each higher coordination shell are larger than in the first one, a cuboctahedron (CN12), which is the AET common to all atoms in the structure.*

atom each. Every point inside a Voronoi cell is closer to its centering atom than to any other atom outside of it. The Voronoi polyhedron around an atom is dual to its coordination polyhedron. This means that every face of the Voronoi cell is capped by an atom of the corresponding AET, and vice versa. The Voronoi-cell decomposition is a unique way of subdividing a structure.

A generalized cluster-Voronoi cell can be obtained if instead of single atoms only cluster centers are considered. In contrast to the determination of an AET around an atom by the maximum-gap method, the derivation of the Voronoi polyhedra is unique. The packing of all Voronoi polyhedra of a structure is called Voronoi tessellation or diagram, while the dual triangulated structure itself is called Delauney tessellation or triangulation (Fig. 3.5).

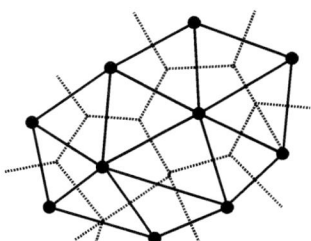

Fig. 3.5 *Triangulated arbitrary point set (Delauney tiling, black lines) and the dual Voronoi tiling (dotted lines). Each point inside a Voronoi cell around a vertex is closer to this vertex than to any other vertex.*

The geometrical analysis of a complex crystal structure may be performed in the following way:

(i) Identify the content of subunit cells that are defined around the high-symmetry points of the unit cell. Useful subunit cells can be (semi)regular polyhedra, which can be packed without gaps and overlaps. For instance, for cubic structures the truncated octahedron (Kelvin polyhedron) has been frequently suggested as subunit cell (Chieh, 1979; Chieh, 1980; Chieh, 1982; Chieh *et al.*, 1982) (see Fig. 3.15). In the case where a structure can be described as an *fcc* packing of one type of cluster, and the tetrahedral and octahedral voids are filled by other clusters or "glue atoms", a packing of truncated octahedra, cuboctahedra, and truncated tetrahedra may be a useful subunit-cell description as well.

(ii) Identify recurrent structural subunits (AETs, clusters, modules, etc.) and the way they are linked. Choose those clusters that are closest to units (e.g., polyanions), which can be distinguished crystal-chemically from their environment (matrix atoms or other clusters).

(iii) Analyze the packing of the subunit cells and clusters, respectively. A good starting point can be the well-known packing principles of uniform polyhedra or of tilings.

(iv) The structure of clusters and/or their packing principles as well as structural relationships can sometimes be well-described based on the higher-dimensional approach for aperiodic structures, even if the structure itself is periodic (Berger *et al.*, 2008).

3.2 Tilings (packings) and coverings

Tilings (tessellations) are infinite arrangements (packings) of copies of unit tiles (prototiles) without gaps and overlaps (for an exhaustive presentation of 2D tilings, see Grünbaum and Shephard (1986)). 2D tilings share the edges of the unit tiles; 3D tilings share the edges and faces. In contrast, coverings fill the space without gaps but with partial, well-defined overlaps. There is always a one-to-one correspondence between coverings and tilings, and each covering can be represented by a tiling decorated at its vertices with the center of a covering cluster. However, not every tiling can be represented by a covering based on a finite number of covering clusters, which usually consist of a patch (subset) of tiles on a smaller scale.

Euler's equation reformulated for tilings can be written in the form $v + f - e = 0$, with v, f, and e the normalized number of vertices, faces, and edges. In the case of a hexagon tiling, $v = 2$ (6 vertices/hexagon, each shared by 3 hexagons), $f = 1$, and $e = 3$ (6 edges/hexagon, each shared by two hexagons), yielding $v + f - e = 2 + 1 - 3 = 0$ (O'Keeffe and Hyde, 1996). It can also be shown that this equation holds: $1/\langle n \rangle + 1/\langle i \rangle = 1/2$, with $\langle n \rangle = \sum n\phi_n$ the average ring size of the polygons, and $\langle i \rangle = \sum i f_i$ the average connectivity i of the vertices in the net. If we have, for instance, a 3-connected ($\langle i \rangle = 3$) pentagon/hexagon/heptagon tiling with ϕ_5, ϕ_6, and ϕ_7 now the fraction of the respective n-gons ($\phi_5 + \phi_6 + \phi_7 = 1$), we obtain $5\phi_5 + 6(1 - \phi_5 - \phi_7) + 7\phi_7 = 6$. This leads to $\phi_5 = \phi_7$ and $\phi_6 = 1 - 2\phi_5$.

Euler's equation for 3D tilings reads $p - f + e - v = 0$, with p the number of polyhedra, f of faces, e of edges, and v of vertices (O'Keeffe and Hyde, 1996). For instance, for dense sphere packings, there are two tetrahedra and one octahedron per vertex. Each vertex is connected to 12 others (the AET is a cuboctahedron), giving $e = 12/2 = 6$, and we have three face-sharing polyhedra ($p = 3$), $f = (8 + 2 \times 4)/2 = 8$: $p - f + e - v = 3 - 8 + 6 - 1 = 0$.

In the case of a topologically close-packed (*tcp*) structure, which is constituted from distorted face-sharing tetrahedra, only, we get $f = 2p$ and so $p = e - v$. If there are N_n vertices that are n-coordinated, then e is half the sum of N_n

$$p = \sum_n nN_n/2 - v \tag{3.1}$$

and dividing both sides by p and rearranging leads to

$$p/v = \sum_n nN_n/2v - 1. \tag{3.2}$$

This means that the number of tetrahedra per vertex is half the average coordination number minus one.

Crystallographically relevant tilings can be periodic or aperiodic. nD periodic tilings can always be reduced to a packing of copies of a decorated single unit cell, an nD parallelotope (parallelepiped in 3D, parallelogram in 2D). In the case of quasiperiodic tilings at least two different prototiles are needed (with some exceptions not relevant here).

While periodic tilings can be generated by simple translation operations, the generation of quasiperiodic tilings is more complex. There are several methods in use: (i) the substitution method, (ii) tile assembly guided by matching rules, (iii) the generalized dual-grid method, and (iv) the higher-dimensional (nD) approach.

We start with a simple example of a 1D quasiperiodic tiling, the Fibonacci sequence (FS), which can be found in some quasiperiodic, incommensurately modulated and host/guest structures. For examples of 2D tilings, related to the structure of layers, we will first discuss the Archimedean tilings and their duals, the Laves tilings, followed by a short introduction into quasiperiodic tilings, in particular the Penrose tiling.

3.2.1 1D tilings: the quasiperiodic Fibonacci sequence and its periodic approximants

The Fibonacci sequence (FS), a 1D quasiperiodic substitutional sequence (see, e.g., Luck *et al.* (1997)), can be obtained by iterative application of the substitution rule $\sigma : L \mapsto LS, S \mapsto L$ on the two-letter alphabet $\{L, S\}$. The substitution rule can be alternatively written employing the substitution matrix S

$$\sigma : \begin{pmatrix} S \\ L \end{pmatrix} \mapsto \underbrace{\begin{pmatrix} 0 & 1 \\ 1 & 1 \end{pmatrix}}_{=\mathsf{S}} \begin{pmatrix} S \\ L \end{pmatrix} = \begin{pmatrix} L \\ LS \end{pmatrix}. \tag{3.3}$$

The substitution matrix also gives the relative frequencies of the letters L and S in the resulting words w_n, which are finite strings of these letters. Longer words can be created by multiple action of the substitution rule. Thus, $w_n = \sigma^n(L)$ means the word resulting from the n-th iteration of σ (L): $L \mapsto LS$. The action of the substitution rule is also called inflation operation as the number of letters is inflated by each step. The FS can as well be created by recursive concatenation of shorter words according to the concatenation rule, $w_{n+2} = w_{n+1}w_n$. The generation of the first few words is shown in Table 3.1.

The substitution rule applied to a word w_n leaves this word invariant and adds a word w_{n-1} to it. This means that the FS is self-similar and shows scaling symmetry by factors of τ^n, with the irrational algebraic number $\tau = (1 + \sqrt{5})/2 = 2\cos(\pi/5) = 1.618. . . $, the golden mean or golden ratio. The frequencies $v_n^L = F_{n+1}, v_n^S = F_n$ of letters L, S in the word $w_n = \sigma^n(L)$, with $n \geq 1$, result from the $(n-1)$th power of the transposed substitution matrix to

$$\begin{pmatrix} v_n^L \\ v_n^S \end{pmatrix} = (\mathsf{S}^T)^{n-1} \begin{pmatrix} 1 \\ 1 \end{pmatrix}. \tag{3.4}$$

Table 3.1 *Generation of words $w_n = \sigma^n(L)$ of the quasiperiodic Fibonacci sequence by repeated action of the substitution rule $\sigma(L) = LS$, $\sigma(S) = L$. v_n^L and v_n^S denote the frequencies of L and S in the words w_n; F_n are the Fibonacci numbers.*

n	$w_{n+2} = w_{n+1}w_n$	v_n^L	v_n^S
0	L	1	0
1	LS	1	1
2	LSL	2	1
3	LSLLS	3	2
4	LSLLSLSL	5	3
5	LSLLSLSLLSLLS	8	5
6	$\underbrace{\text{LSLLSLSLLSLLS}}_{w_5} \underbrace{\text{LSLLSLSL}}_{w_4}$	13	8
$\vdots$	$\vdots$	$\vdots$	$\vdots$
n		F_{n+1}	F_n

The Fibonacci numbers $F_{n+2} = F_{n+1} + F_n$, with $n \geq 0$ and $F_0 = 0, F_1 = 1$, form a series with $\lim_{n \to \infty} F_n/F_{n-1} = \tau = 1.618. \ldots$ Arbitrary Fibonacci numbers can be calculated directly by Binet's formula

$$F_n = \frac{(1 + \sqrt{5})^n - (1 - \sqrt{5})^n}{2^n \sqrt{5}}. \tag{3.5}$$

If we assign a long and a short line segment to L and S, respectively, with $L = \tau S$, then we get a 1D quasiperiodic tiling (Fig. 3.6). Decorating the vertices of the tiling, and/or the line segments with atoms, yields a model of a 1D quasiperiodic structure. It is self-similar and shows scaling symmetry by factors of τ or $1/\tau$.

The FS has a periodic average structure (PAS) with a mean vertex distance, d_{av}, which results in

$$d_{av} = \lim_{n \to \infty} \frac{F_{n+1}L + F_nS}{F_{n+1} + F_n} = \left\{ \frac{F_{n+1}}{F_{n+2}}\tau + \frac{F_n}{F_{n+2}} \right\} S = (3 - \tau)S, \tag{3.6}$$

yielding a vertex point density $D_p = 1/d_{av}$. There is a one-to-one relationship between the vertices of the FS and its PAS. The total length of a finite subset of the FS for n line segments reads (in units of S)

$$x_n = (n + 1)(3 - \tau) - 1 - \frac{1}{\tau}\left\{ \left[\frac{n+1}{\tau} \right] \pmod 1 \right\}. \tag{3.7}$$

It should be mentioned here that the FS can be equally well described as a modulated structure, with an incommensurate saw-tooth modulation wave due to the one-to-one relationship between the vertices of the FS and its PAS.

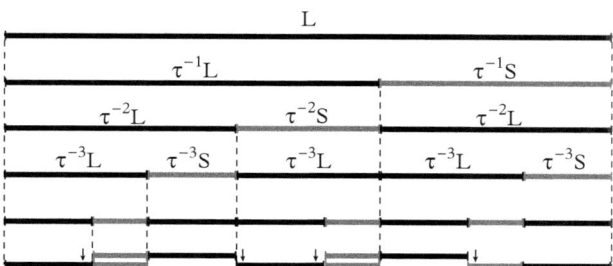

Fig. 3.6 *Graphical representation of the substitution rule* σ *of the Fibonacci sequence. Rescaling by a factor of $1/\tau$ at each step keeps the total length constant. Shown is a deflation of the line segment lengths corresponding to an inflation of letters. At the bottom of the figure, the decomposition of the FS into covering clusters of the type (LS) is indicated.*

Approximants are periodic structures, which have basic building elements (clusters) in common with quasiperiodic structures. An example would be, for instance, the sequence $(LSSLSSL)_n$. If approximants can be derived employing the nD approach, which is not the case in the previous example, then they are called rational approximants. As is illustrated in Fig. 4.8, a rational approximant of the FS can be obtained by a rational cut of the 2D structure, while the FS itself results along an irrational cut with a slope related to τ. Examples are, for instance: $(L)_n$, $(LS)_n$, $(LSL)_n$, $(LSLLS)_n$, etc. with $n \rightarrow \infty$.

3.2.2   2D Archimedean (Kepler) tilings

Structural subunits of particular classes of intermetallic phases can be described as decorated Archimedean tilings. Examples are the σ-phase, $tP30$-$Cr_{46}Fe_{54}$, the Laves phase $cF24$-$MgCu_2$, and other Frank-Kasper phases. The 11 Archimedean tilings (also called Kepler tilings) were derived a long time ago by Kepler (1619) in analogy to the Archimedean solids. Three of them are regular, i.e., they consist of congruent regular (equilateral and equiangular) polygons of one kind (mono-hedral) with just one type of vertex configuration: the triangle tiling 3^6, the square tiling 4^4, and the hexagon tiling 6^3. A vertex configuration can be described either by the *Cundy-and-Rollet symbol* n^m (Cundy and Rollet, 1952) or by the *Schläfli symbol* $\{n, m\}$, meaning that m n-gons meet at a vertex. The Archimedean tilings can be colored (decorated) uniformly in exactly 32 different ways. A coloring is uniform if it maintains the vertex transitivity, i.e., symmetry equivalence between the vertices.

The eight semi-regular Archimedean tilings (see Fig. 3.7 and Table 3.2) are uniform (symmetrically equivalent) like the regular ones, i.e., they have only one type of vertex configuration (vertex transitive), but consist of two or more different regular polygons as prototiles (di- or trihedral).

In the case of structures that can be geometrically described as stackings of layers, subsequent layers are frequently dual to each other. A tiling dual to another one can be obtained by putting vertices in the centers of the unit tiles and connecting them properly by lines. If a tiling is regular, so will be its dual tiling. The dual to a square tiling is a square tiling again, so it is called self-dual, while the dual to the hexagon tiling is the triangle tiling and vice versa. A tiling and its dual can be seen analogous to a lattice and its reciprocal lattice; both have the same point group symmetry. The unit tiles of the dual tiling correspond to the Voronoi cells of the tiling.

The duals to the Archimedean tilings are the Catalan or Laves tilings, which are isohedral (monohedral, face transitive with just one type of unit tile) but which have more than one vertex configuration. These tilings are, therefore, described by their face configuration. It can be given by the Cundy-and-Rollet symbol $Vm_1.m_2\ldots$, which lists the number of tiles meeting at each vertex along a circuit around a unit tile. For example, the Cairo pentagon tiling $V3^2.4.3.4$, consisting of smashed pentagons, is the dual of the Archimedean snub square tiling $3^2.4.3.4$.

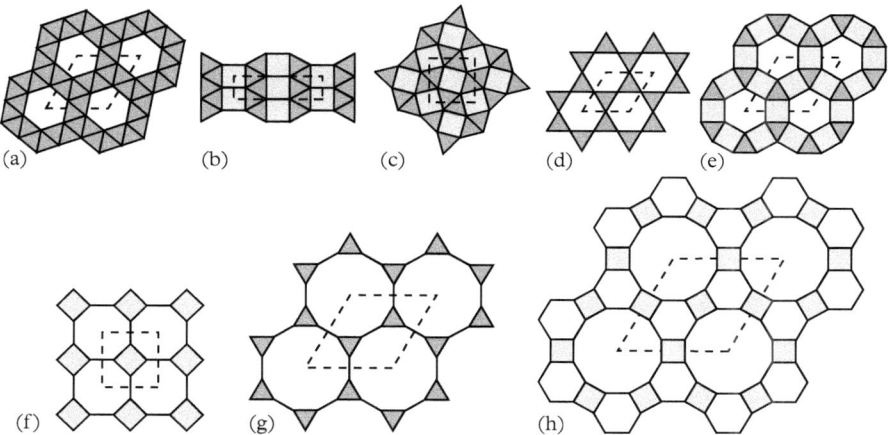

Fig. 3.7 *The eight semi-regular Archimedean tilings: (a) snub hexagonal tiling $3^4.6$, (b) elongated triangular tiling $3^3.4^2$, (c) snub square tiling $3^2.4.3.4$, (d) trihexagonal tiling 3.6.3.6 (Kagomé tiling), (e) small rhombitrihexagonal tiling 3.4.6.4, (f) truncated square tiling 4.8^2, (g) truncated hexagonal tiling 3.12^2, and (h) great rhombitrihexagonal tiling 4.6.12. The unit cells are outlined by dashed lines.*

Its face configuration means that each unit tile has five vertices where 3, 3, 4, 3, 4 tiles of the same kind meet. The symbols for face and vertex configuration of dual tilings are the same except that the former is preceded by the letter V.

3.2.3   2D quasiperiodic Penrose tilings and the Gummelt decagon covering

The Penrose tiling (PT) is named after its discoverer, Roger Penrose, a British mathematician, theoretical physicist, and philosopher (Penrose, 1974). It became well-known through Martin Gardner's article in the popular scientific journal *Scientific American* (Gardner, 1977). In the book *Tilings and Patterns* by Grünbaum and Shephard (1986), three different versions of the PT are presented: one based on pentagons (P1 tiling, PPT), one on kites and darts (P2 tiling), and one on thick and thin rhombs (P3 tiling, RPT) (Fig. 3.8).

All three of them belong to the Penrose local isomorphism (PLI) class, i.e., they are mutually locally derivable. All PLI-class tilings have matching rules forcing quasiperiodicity. It should be kept in mind that matching rules are no growth rules in contrast to the substitution (inflation) rules. This means it is not possible to construct a PT just by obeying the matching rules, because one would always run into situations where the PT could not be continued. In contrast, if one encounters a tiling obeying the matching rules, it belongs to the PLI class for sure. Relaxing the matching rules leads to random tilings or even periodic tilings.

Table 3.2 *Characteristic data for the eight semi-regular Archimedean tilings. The number of vertices n_V per unit cell is given; the density is calculated for a close packing of equal circles at the vertices. In the second lines, the lattice parameter a is given for a tile edge length of 1 and the Wyckoff positions of the plane group occupied for generating the tiling are listed (O'Keeffe and Hyde, 1980).*

Names of the Archimedean tiling and its dual, the respective Catalan tiling			
Vertex configuration	n_V	Plane group a	Density Wyckoff position
Snub hexagonal tiling[a] – Floret pentagonal tiling			
$3^4.6$	6	$p6$	$\pi\sqrt{3}/7 = 0.7773$
		$a = \sqrt{7}$	$6d\ x = 3/7, y = 1/7$
Elongated triangular tiling – Prismatic pentagonal tiling			
$3^3.4^2$	4	$c2mm$	$\pi/(2 + \sqrt{3}) = 0.8418$
		$a = 1, b = 2 + \sqrt{3}$	$4e\ y = (1 + \sqrt{3})/(4 + 2\sqrt{3})$
Snub square tiling – Cairo pentagonal tiling			
$3^2.4.3.4$	4	$p4gm$	$\pi/(2 + \sqrt{3}) = 0.8418$
		$a = (2 + \sqrt{3})^{1/2}$	$4c\ x = 1 - 1/4[(2 - \sqrt{3})(2 + \sqrt{3})]^{1/2}$
Trihexagonal tiling[b] – Rhombille tiling			
$3.6.3.6$	3	$p6mm$	$\pi\sqrt{3}/8 = 0.6802$
		$a = 2$	$3c$
Small rhombitrihexagonal tiling – Deltoid trihexagonal tiling			
$3.4.6.4$	6	$p6mm$	$\pi\sqrt{3}/(4 + 2\sqrt{3}) = 0.7290$
		$a = 1 + \sqrt{3}$	$6e\ 4x = 1/(3 + \sqrt{3})$
Truncated square tiling – Tetrakis square tiling			
4.8^2	4	$p4mm$	$\pi/(3 + 2\sqrt{2}) = 0.5390$
		$a = 1 + \sqrt{2}$	$4e\ x = 1/(2 + 2\sqrt{2})$
Truncated hexagonal tiling – Triakis triangular tiling			
3.12^2	6	$p6mm$	$\pi\sqrt{3}/(7 + 4\sqrt{3}) = 0.3907$
		$a = 2 + \sqrt{2}$	$6e\ x = (1 - 1/\sqrt{3})$
Great rhombitrihexagonal tiling – Kisrhombille tiling			
$4.6.12$	12	$p6mm$	$\pi/(3 + 2\sqrt{3}) = 0.4860$
		$a = 3 + \sqrt{3}$	$12f\ x = 1/(3\sqrt{3} + 3), y = x + 1/3$

[a] Two enantiomorphs. [b] Kagomé net; quasiregular tiling because all edges are shared by equal polygons.

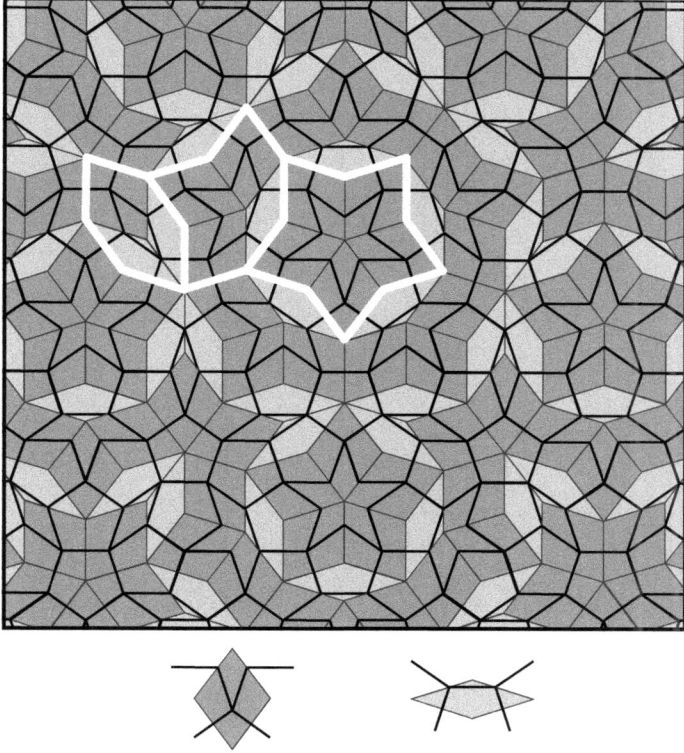

Fig. 3.8 *Penrose pentagon tiling (PPT) (thick black lines) with underlying rhomb Penrose tiling (RPT) (thin black lines). At the bottom, the decoration of the rhomb prototiles is shown that produces the PPT. Hexagon, boat, and star (HBS) supertiles are outlined by thick white lines (from Steurer and Deloudi (2009), Fig. 1.10. With kind permission from Springer Science+Business Media.)*

The rhomb PT (RPT) is based on a set of two unit tiles: a skinny rhomb (acute angle $\alpha_s = \pi/5$) and a fat rhomb (acute angle $\alpha_f = 2\pi/5$) with equal edge lengths a_r, and with areas $A_s = a_r^2 \sin \pi/5$ and $A_f = a_r^2 \sin 2\pi/5 = \tau A_s$. The total area they are covering and their frequencies in the RPT both have the ratio $1 : \tau$. An inflation rule exists, which replaces each rhomb tile with a set of four τ times smaller tiles (Fig. 3.9). By iterative application of this rule, an infinite RPT can be generated. The number of unit tiles is inflated by this substitution, but the size of the unit tiles is deflated by τ. The RPT has a scaling symmetry, which can be represented by the matrix **S**. Applying the scaling matrix to the set of vertices M_{RPT} of the RPT yields an RPT dual to the original RPT, but blown up by a factor τ.

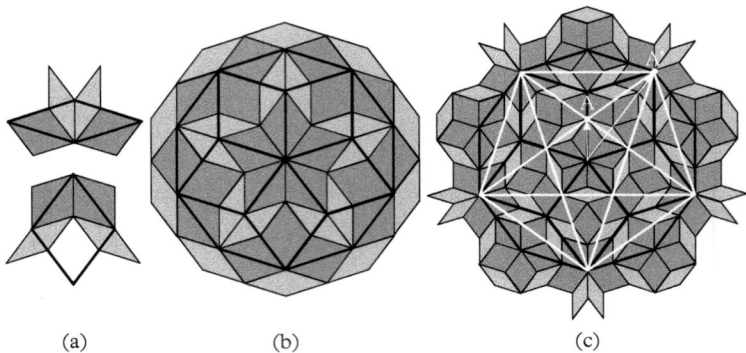

(a)   (b)   (c)

Fig. 3.9 *Scaling properties of the rhomb Penrose tiling (RPT). (a) The substitution (inflation) rule for the rhomb prototiles. In (b) a RPT (thin lines) is superimposed with another RPT (thick lines) scaled by* **S**, *and in (c) scaling by* **S**^{2} *is shown. A subset of the vertices of the scaled tilings coincide with the vertices of the original tiling. The rotoscaling operation* **S**^{2} *is also a symmetry operation of a pentagram (white lines), mapping each vertex of a pentagram onto another one. This is demonstrated in (c) on the example of the vertex A which is mapped onto A' by* **S**^{2} *(from Steurer and Deloudi (2009), Fig. 1.6. With kind permission from Springer Science+Business Media.)*

The set of vertices of the RPT, M_{RPT}, is a subset of the vector module

$$M = \left\{ \mathbf{r} \sum_{i=0}^{4} n_i a_r \mathbf{e}_i \middle| \mathbf{e}_i = (\cos 2\pi i/5, \sin 2\pi i/5, 0) \right\}. \tag{3.8}$$

M_{RPT} consists of five subsets

$$M_{RPT} = \cup_{k=0}^{4} M_k \text{ with } M_k = \left\{ \pi^{\parallel}(\mathbf{r}_k) \middle| \pi^{\perp}(\mathbf{r}_k) \in T_{ik}, i = 0, \dots, 4 \right\} \tag{3.9}$$

and $\mathbf{r}_k = \sum_{j=0}^{4} \mathbf{d}_j \left(n_j + \frac{k}{5} \right), n_j \in Z$. The i-th triangular subdomain T_{ik} of the k-th pentagonal occupation domain corresponds to

$$T_{ik} = \left\{ \mathbf{t} = x_i \mathbf{e}_i + x_{i+1} \mathbf{e}_{i+1} \middle| x_i \in [0, \lambda_k], x_{i+1} \in [0, \lambda_k - x_i] \right\} \tag{3.10}$$

with λ_k the radius of a pentagonally shaped occupation domain: $\lambda_0 = 0$, for $\lambda_{1,4} = a_r/\tau^4$ and $\lambda_{2,3} = a_r/\tau^3$, with a_r the rhomb unit tile edge length.

$$S = \begin{pmatrix} 0 & 1 & 0 & \bar{1} \\ 0 & 1 & 1 & \bar{1} \\ \bar{1} & 1 & 1 & 0 \\ \bar{1} & 0 & 1 & 0 \end{pmatrix}_D = \begin{pmatrix} \tau & 0 & 0 & 0 \\ 0 & \tau & 0 & 0 \\ 0 & 0 & -\frac{1}{\tau} & 0 \\ 0 & 0 & 0 & -\frac{1}{\tau} \end{pmatrix}_V = \begin{pmatrix} S^{\parallel} & 0 \\ 0 & S^{\perp} \end{pmatrix}_V \tag{3.11}$$

The matrix-subscript D refers to the 4D crystallographic basis (D-basis), while subscript V indicates that the vector components refer to a Cartesian coordinate system (V-basis) (see Section 4.5). Only scaling by S^{4n} results in an identical RPT (with unit-tile edge-lengths increased by a factor τ^{4n}) of original orientation. Then the relationship $S^{4n}M_{RPT} = \tau^{4n}M_{RPT}$ holds. S^2 maps the vertices of an inverted, and by a factor τ^2 enlarged, PT upon the vertices of the original RPT. The rotoscaling operation $\Gamma(10)S^2$ leaves the subset of vertices forming a pentagram invariant (Fig. 3.9).

By a particular decoration of the unit tiles with line segments, infinite lines (Ammann lines) are created forming a Fibonacci penta-grid (5-grid, "Ammann quasilattice" (Levine and Steinhardt, 1986)) (Fig. 3.10). The dual of the Ammann quasilattice is the deflation of the original RPT. The line segments can act as matching rules forcing strict quasiperiodicity. In the case of simpleton flips (phason flips and hexagon flips), the Ammann lines are broken. A simpleton flip is a jump between two positions of the inner vertex in a hexagon-arrangement

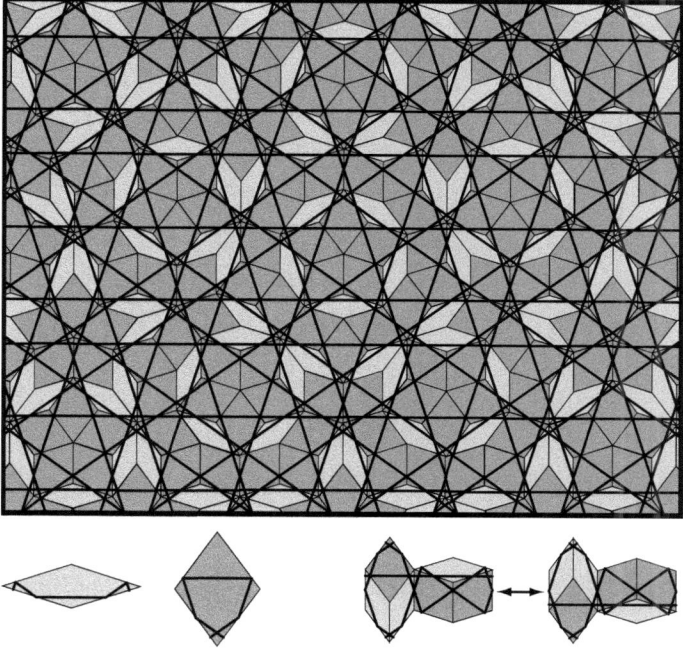

Fig. 3.10 *The rhomb Penrose tiling with Ammann lines drawn in. The decoration of the unit tiles by Ammann line segments and the action of simpleton (phason) flips are shown at the bottom (from Steurer and Deloudi (2009), Fig. 1.8. With kind permission from Springer Science+Business Media.)*

of two skinny and one fat rhomb or two fat rhombs and one skinny rhomb. In a quasicrystal, such flips are low energy excitations.

Particular quasiperiodic tilings, including some with 8-, 10-, and 12-fold symmetry that are relevant for real QCs, can be fully covered by one or more covering clusters. By covering cluster we mean a patch of tiles of the respective tiling, which is called Gummelt decagon in the case of the PT. In Fig. 3.11(h) and (i), the decoration of the Gummelt decagon with patches of the RPT and the PPT, respectively, are shown. The Gummelt decagon is a single, mirror-symmetrical, decagonal cluster with overlap rules that force perfectly ordered structures of the PLI class (Gummelt, 1996) (Fig. 3.11(a)). There are different ways of marking the overlap rules. In 3.11(a)–(e), the rocket decoration is used, where the colors of the overlap areas of two Gummelt decagons must match. There are nine different allowed coordinations of a central Gummelt decagon by other decagons so that all decagon edges are fully covered. The coordination numbers are 4, 4, 4, 4, 5, 5, 5, 5, 6.

The centers of the Gummelt decagons form a PPT when the overlap rules are obeyed. Its dual is the so-called τ^2-HBS supertiling. The H(hexagon) tiles contain 4 Gummelt decagon centers, the B(oat) tiles 7, and the S(tar) tiles 10. The HBS tile edge length is τ^2 times that of the decagon, which itself is equal to τ times the edge length of the underlying RPT (Fig. 3.10).

For periodic tilings the number of different vertex surroundings within a coordination sphere of any size is limited. This is not the case for quasiperiodic tilings. In the RPT, for instance, one finds eight different vertex configurations in the

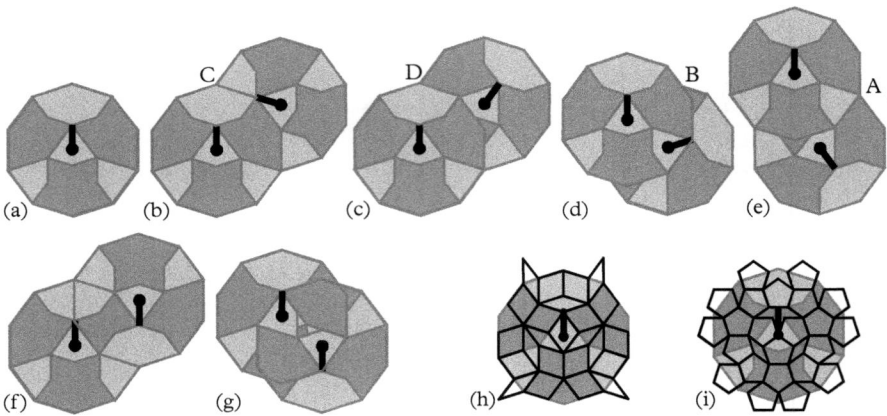

Fig. 3.11 *Gummelt-decagon (a) and its overlap rules for the construction of perfect tilings of the PLI class (b)–(e). With relaxed (unoriented) overlap rules, random decagonal coverings can be obtained (f)–(g). The relationship of the Gummelt decagon to the rhomb and the pentagon Penrose tiling is shown in (h) and (i), respectively (from Steurer and Deloudi (2009), Fig. 1.13. With kind permission from Springer Science+Business Media.)*

first coordination sphere, which includes the nearest neighbors (NN). This number increases to 23, if one increases the coordination sphere to include also the next-nearest neighbors (NNN), and to approximately 280 if the NNNN vertices are included as well (Peng and Fu, 2015). The range of the frequencies of the different vertex environments widens with the coordination sphere. The eight NN vertex frequencies range from $1/5(\tau^{-6} + \tau^{-8}) = 0.01540$ to $\tau^{-2} = 0.38197$, and the NNN frequencies from $1/5(\tau^{-8} + \tau^{-10}) = 0.00588$ to $2\tau^{-6} = 0.11146$.

This kind of complexity can be put into perspective taking into account the self-similarity of the RPT as well as its repetitivity. The self-similarity symmetry operation maps the vertices of a RPT onto the vertices of a copy of the RPT scaled by powers of τ^{-1}. Repetitivity means that any bounded patch of the RPT (a Gummelt decagon, for instance) can be found again in the RPT within a distance of less than two diameters of that patch. Decorating such a patch of the RPT with atoms, we can get a structural building block (cluster) of a quasiperiodic structure.

3.2.4 Sphere packings and polytypism

Atoms with isotropic interaction potentials will preferentially form structures related to closest packings of hard spheres. A good example would be the noble gases, which all crystallize either in *ccp* or *hcp* structures. However, really isotropic interaction potentials are rare in the case of metallic elements although a significant number of them crystallizes in close-packed structures, at least at ambient conditions. And these are not the simple s-metals such as the alkali metals, which show *bcc* structures at ambient pressure, and a large variety of rather complex structures at higher pressures.

In a close sphere packing, any pair of spheres is connected via a chain of spheres with mutual contact. If all spheres are symmetrically equivalent, it is called a homogenous sphere packing, otherwise a heterogeneous one (Koch and Fischer, 1992). In the latter case, the non-symmetrically-equivalent spheres can have different radii and can occupy different crystallographic orbits. In Table 3.3, examples of different sphere packings are listed with the highest and lowest densities and contact numbers. The number k of contacts per sphere is in the range $3 \leq k \leq 12$.

The packing densities, i.e., the fraction of space occupied by the spheres, with $q = \pi/\sqrt{18} = 0.74048\ldots$, are the highest for the well-known cubic (*ccp*) and hexagonal close(st) packings (*hcp*) (Fig. 3.12). In both cases, the coordination numbers CN are 12 and the distances to the nearest neighbors equal. They start to differ not before the third coordination shell. The two closest packings, *ccp* and *hcp*, can be seen as stackings of *hcp* layers (3^6 tilings decorated by spheres), with the sequence ABC along [111] in the cubic case, and AB along [001] in the hexagonal case (Fig. 3.12). The *hcp* layers are named A, B, and C depending on their relative positioning to each other: A (0, 0), B (1/3, 2/3), and C (2/3, 1/3).

Theoretically, an infinite number of structures exists with other stacking sequences of the *hcp* layers, called polytypes, which all have exactly the same packing

Table 3.3 *Characteristics of some homogenous sphere packings with high (low) contact numbers k and high (low) fractional packing densities q; a, b, and c are lattice parameters, d is the distance between the centers of neighboring spheres (Wilson and Prince, 1999).*

k	Space group	Wyckoff position	Parameters	d	q
12	194 $P6_3/mmc$	$2c\ \frac{1}{3}, \frac{2}{3}, \frac{1}{4}$	$\dfrac{c}{a} = \dfrac{2}{3}\sqrt{6} = 1.6330$	a	0.7405
12	225 $Fm\bar{3}m$	$4a\ 0, 0, 0$		$\dfrac{a}{2}\sqrt{2}$	0.7405
11	12 $C2/m$	$4i\ x, 0, z$	$x = \dfrac{1}{2}\left(\sqrt{2}-1\right),$ $z = 3\sqrt{2}-4,$ $\dfrac{b}{a} = \dfrac{1}{3}\sqrt{3},$ $\dfrac{c}{a} = \dfrac{1}{6}\left(\sqrt{6}+2\sqrt{3}\right)$ $\cos\beta = \dfrac{1}{6}\left(\sqrt{6}-2\sqrt{3}\right)$	b	0.7187
10	139 $I4/mmm$	$2a\ 0, 0, 0$	$\dfrac{c}{a} = \sqrt{6}/3 = 0.81650$	c	0.6981
3	214 $I4_132$	$24h\ \dfrac{1}{8}, y, \dfrac{1}{4}-y$	$y = \dfrac{1}{8}\left(2\sqrt{3}-3\right)$	$\dfrac{a}{4}\left(2\sqrt{6}-3\sqrt{2}\right)$	0.0555

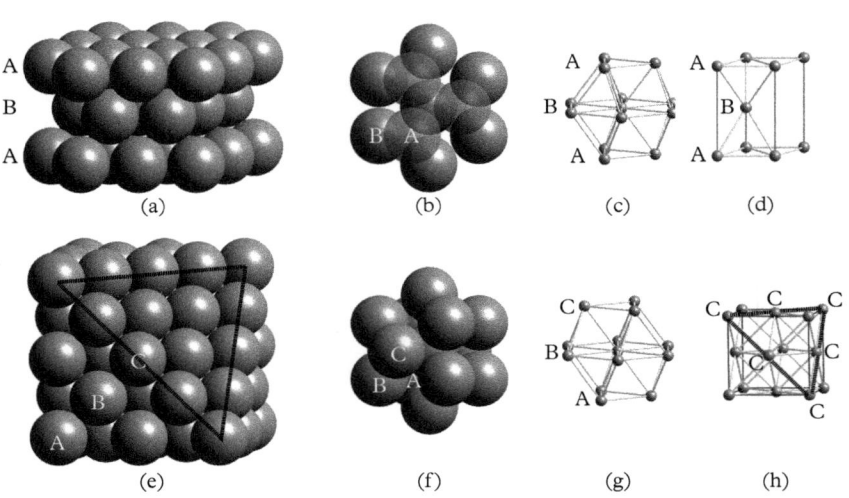

Fig. 3.12 *Characteristics of the (a)–(d) hexagonal close (hcp) and (e)–(h) the cubic close (ccp) sphere packing. In the case of the hcp structure, the hcp layers are packed with the sequence AB along the [001]-direction. For the ccp structure, the sequence is ABC along [111]. The AET corresponds to a non-centrosymmetric disheptahedron (anticuboctahedron) in the case of the hexagonal structure (c), and to a centrosymmetric cuboctahedron in the case of the cubic structure (g). The respective unit cells are shown in (d) and (h).*

density, and either rhombohedral or hexagonal symmetry. Actually, experimentally observed examples are the structures of *hP*4-La (ACAB) and *hR*9-Sm (ABABCBCAC). For the description of polytypic structures, different notations have been used. Frequently employed is the one introduced by Jagodzinski (1954), which characterizes the layer sequence by the either *c*(ubic) or *h*(exagonal) surrounding of each layer. For instance, in the stacking sequence ABC, B would be characterized as *c* as well as A and B, because A is surrounded by C and B, and C by B and A. *hR*9-Sm, ABABCBCAC would then be written as *chhchhchh*. Its degree of hexagonality, i.e., the relative number of the letter *h*, results to 2/3. In the less common Ramsdell notation, just the number of layers per period and the Bravais type lattice type are given. For instance, for *hP*4-La, ACAB, and *hR*9-Sm, ABABCBCAC, we would get 4H and 9R.

It is remarkable that the monoclinic sphere packing with space group 12 *C*2/*m* and $k = 11$ exhibits a packing density only $\approx 3\%$ lower than the highly symmetric cubic and hexagonal close packings. Among the low-symmetry structures, this space group shows, with 300, by far the most representatives, indeed (see Subsection 5.3.1). Table 3.4 shows space filling values for the structures of a number of elements, ranging from the values known from close sphere packings, 0.74, to ones as low as 0.285. Very low packing densities like that for *cF*8-C (diamond), for instance, indicate that a hard-sphere packing model is not an adequate description of such a structure. The hard constraints of directional covalent bonding dominate the structure formation overriding the maximization of packing density. As we will see in Chapter 6, most of the metallic elements show structures corresponding to quite simple sphere packings. At elevated pressures, in many cases, the

Table 3.4 *Fractional packing densities q of elemental structures under the assumption of hard spherical atoms (Pearson, 1972).*

Element	Pearson symbol c/a	Space-filling value q	Element	Pearson symbol c/a	Space-filling value q
Cu	*cF*4	0.740	Po	*cP*1	0.523
Mg	*hP*2, 1.63	0.740	Bi	*hR*2, 2.60	0.446
Zn	*hP*2, 1.86	0.650	Sb	*hR*2, 2.62	0.410
Pa	*tI*2	0.696	As	*hR*2, 2.80	0.385
In	*tI*2	0.686	Ga	*oC*8	0.391
W	*cI*2	0.680	Te	*hP*3	0.364
Hg	*hR*1	0.609	C	*cF*8	0.340
Sn	*tI*4	0.535	P	*oC*8	0.285
U	*oC*4	0.534			

isotropic atomic interactions governing structure formation at ambient pressures are replaced by more anisotropic ones leading to low-symmetry structures.

3.3 Polyhedra and packings

The representation of crystal structures as packings of polyhedra is quite common (wherever it is possible), because in this way the building principles of a complex crystal structure can often be better illustrated. There are infinitely many different polyhedra possible. Since we want to show the principles of this kind of representation only, we restrict our discussion to convex regular and semi-regular polyhedra. They are called regular if their faces are all equal and regular (equilateral and equiangular), and surround each vertex (corner) in the same way, with the same solid angles. Consequently, regular polyhedra are vertex-transitive and face-transitive. Furthermore, we discuss also their space-filling packings (3D tilings or tessellations) with cubic symmetry, and we demonstrate on one example how quasiperiodic packings can be arranged.

We have to add again the *caveat* that a polyhedral description of a structure generally does not mean that the polyhedra have any crystal-chemical meaning. Even if nested polyhedra are sometimes termed clusters, they are, at least in most cases, purely geometrical entities, structural subunits. Only for certain classes of intermetallics such as the Zintl phases, for instance, they may bear some physical relevance as polyanions, for instance, or for quasicrystals with covalent bonding contributions.

3.3.1 Platonic, Archimedean, and Catalan solids

There are five regular polyhedra in 3D space, the Platonic solids: the tetrahedron, with point symmetry $\bar{4}3m$, and Schläfli symbol[1] {3,3}, the octahedron, $m\bar{3}m$: {3,4}, the hexahedron (cube), $m\bar{3}m$: {4,3}, the icosahedron, $m\bar{3}\bar{5}$: {3,5}, and the dodecahedron, $m\bar{3}\bar{5}$: {5,3}. The orientational relationship to the cubic symmetry in each case is indicated by a circumscribed cubic unit cell (Fig. 3.13). The dual to a Platonic solid $\{p, q\}$ is again a Platonic solid $\{q, p\}$. The tetrahedron is its own dual, that of the cube is the octahedron, and the icosahedron is the dual of the dodecahedron (and *vice versa*).

If polyhedra are only vertex-transitive such as the 13 semi-regular Archimedean solids (Fig. 3.5(a)–(m) and Table 3.5), they are called uniform. Semi-regular polyhedra are characterized by faces that are all regular polygons but of at least two different kinds. There is also an infinite number of prisms and antiprisms with n-fold symmetry that belong in this class. The prisms consist of two congruent n-gons plus n squares, $4^2.n$, and have point symmetry N/mmm (N denotes an n-fold rotation axis). The antiprisms consist of two twisted congruent n-gons plus

---

[1] For a polyhedron, the *Schläfli symbol* $\{n, m\}$ means that m n-gons meet at each vertex.

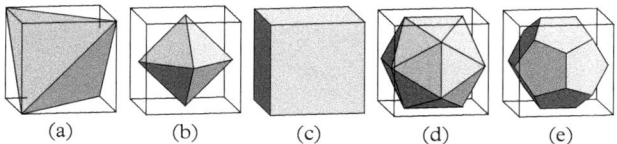

Fig. 3.13 *The five Platonic solids with their vertex configuration inscribed in cubic unit cells to show their orientational relationships to the 2-, 3-, and 4-fold axes of the cube: tetrahedron, {3, 3} 3^3, octahedron, {3, 4} 3^4, hexahedron (cube), {4, 3} 4^3, icosahedron, {3, 5} 3^5, and dodecahedron, {5, 3} 5^3.*

n equilateral triangles, $3^3.n$, with point symmetry $\overline{(2N)}m2$. Consequently, the only antiprism with crystallographic symmetry is the octahedron, 3^4. The square antiprism, $3^3.4$, has point symmetry $\overline{8}m2$ and the hexagonal antiprism, $3^3.6$, $\overline{12}m2$, both "non-crystallographic" symmetries.

Most of the Archimedean polyhedra can be related by duality (see above, Platonic solids) or by truncation. For instance, by successive truncation of the cube, first the truncated cube results (Fig. 3.14(c)), then the cuboctahedron (Fig. 3.14(b)), followed by the truncated octahedron (Fig. 3.14(d)), and finally, the octahedron.

The Archimedean solids can all be inscribed in a sphere and one of the Platonic solids. Their duals are called Catalan solids, which have in-spheres. In contrast to the Archimedean solids, they are face-transitive and non-uniform. Examples are the rhombic dodecahedron, $V(3.4)^2$, the dual of the cuboctahedron, $(3.4)^2$, and the rhombic triacontahedron, $V(3.5)^2$, which is dual to the icosidodecahedron, $(3.5)^2$ (Fig. 3.14(n) and (o), respectively). The number of faces, edges, and vertices of the Archimedean solids corresponds to the number of vertices, edges, and faces of the dual Catalan solids. The rhombic dodecahedron is the AET of the atoms in a $cI2$-W- or $cP2$-CsCl-type structure, and the triacontahedron is part of the fundamental structural subunits of icosahedral quasicrystals. It should also be mentioned here that endohedral clusters frequently consist of cluster shells that are mutually dual to one another.

Space filling packings of regular and semi-regular polyhedra always require at least two kinds of polyhedra except in the case of the cube and the truncated octahedron (Kelvin polyhedron, Voronoi cell of the bcc lattice) (Fig. 3.15(a)). Euler's formula can be reformulated for polyhedra packings: $p - f + e - v = 0$, with p, f, e, and v the number of polyhedra, faces, edges, and vertices, respectively (O'Keeffe and Hyde, 1996). For instance, a ccp packing of atoms can be described as a packing of octahedra and tetrahedra, with eight and four faces, respectively. There are one octahedron and two tetrahedra per atom (vertex, $v = 1$, $p = 3$). Taking face-sharing into account, we get eight faces per vertex ($f = 8$). Since the coordination

Table 3.5 *Characteristic data for the 13 Archimedean solids and of two of their duals (below the horizontal dashed line). Faces are abbreviated tri(angle), squ(are), pen(tagon), hex(agon), oct(agon), dec(agon), and rho(mb). In the last column, the ratio of the edge length a_s of the faces to the edge length of the circumscribed polyhedron (Platonic solid) a_p is given, where $p = c(ube)$, t(etrahedron), o(ctahedron), i(cosahedron), d(odecahedron), and m(idsphere radius).*

Name Vertex configuration	Faces	Edges	Vertices	Point group	Typical ratios $p : a_s/a_p$
Truncated tetrahedron 3.6^2	8: 4 tri, 4 hex	18	12	$\bar{4}3m$	$t : 1/3$
Cuboctahedron $(3.4)^2$	14: 8 tri, 6 squ	24	12	$m\bar{3}m$	$c : 1/\sqrt{2}$
Truncated cube 3.8^2	14: 8 tri, 6 oct	36	24	$m\bar{3}m$	$c : \sqrt{2}-1$
Rhombicuboctahedron 3.4^3	26: 8 tri, 18 squ	48	24	$m\bar{3}m$	$c : \sqrt{2}-1$
Truncated cuboctahedron $4.6.8$	26: 12 squ, 8 hex, 6 oct	72	48	$m\bar{3}m$	$c : 2/7(\sqrt{2}-1)$
Truncated octahedron 4.6^2	14: 8 hex, 6 oct	36	24	$m\bar{3}m$	$c : 1/2\sqrt{2}$
Snub cube[a] $3^4.4$	38: 32 tri, 6 squ	60	24	432	$c : 0.438$
Icosidodecahedron $(3.5)^2$	32: 20 tri, 12 pen	60	30	$m\bar{3}\bar{5}$	$i : 1/2$
Truncated dodecahedron 3.10^2	32: 20 tri, 12 dec	90	60	$m\bar{3}\bar{5}$	$d : 1/\sqrt{5}$
Truncated icosahedron 5.6^2	32: 12 pen, 20 hex	90	60	$m\bar{3}\bar{5}$	$i : 1/3$
Rhombicosidodecahedron $3.4.5.4$	62: 20 tri, 30 squ, 12 pen	120	60	$m\bar{3}\bar{5}$	$d : \sqrt{5}+1/6$
Truncated icosidodecahedron $4.6.10$	62: 30 squ, 20 hex, 12 dec	180	120	$m\bar{3}\bar{5}$	$d : \sqrt{5}+1/10$
Snub dodecahedron[a] $3^4.5$	92: 80 tri,12 pen	150	60	235	$i : 0.562$
Rhombic dodecahedron $V(3.4)^2$	12 rho	24	14	$m\bar{3}m$	$m : 3\sqrt{2}/4$
Rhombic triacontahedron $V(3.5)^2$	30 rho	60	32	$m\bar{3}\bar{5}$	$m : (5-\sqrt{5})/4$

[a]Two enantiomorphs each

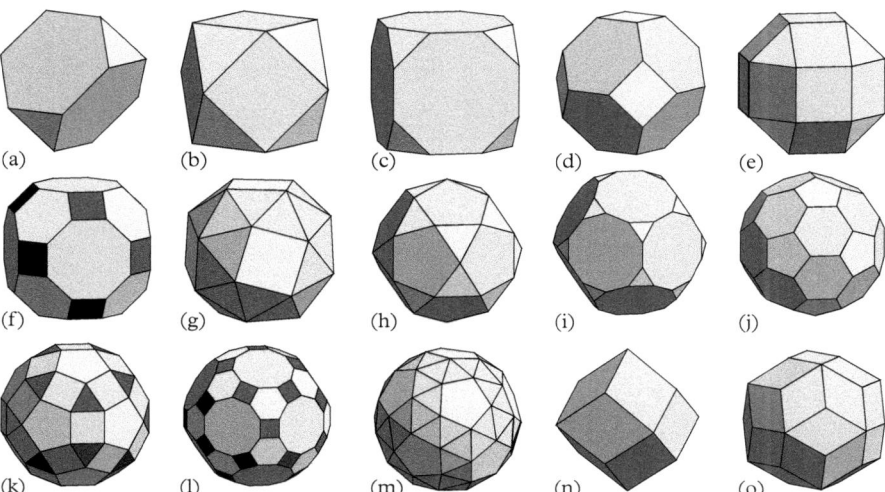

Fig. 3.14 *The 13 vertex-transitive Archimedean solids (a)–(m) and two of their duals (n)–(o) with their vertex configuration: (a) truncated tetrahedron (Friauf polyhedron), 3.6^2, (b) cuboctahedron, (3.4)^2, (c) truncated cube, 3.8^2, (d) truncated octahedron, 4.6^2, (e) (small) rhombicuboctahedron, 3.4^3, (f) truncated cuboctahedron (great rhombicuboctahedron), 4.6.8, (g) snub cube, 3^4.4, only one enantiomorph shown, (h) icosidodecahedron, (3.5)^2, (i) truncated dodecahedron, (3.10)^2, (j) truncated icosahedron, 5.6^2, (k) (small) rhombicosidodecahedron, 3.4.5.4, (l) truncated icosidodecahedron (great rhombicosidodecahedron), 4.6.10, and (m) snub dodecahedron, 3^4.5, only one enantiomorph shown. The rhombic dodecahedron, V(3.4)^2 (n), and the rhombic triacontahedron, V(3.5)^2 (o), are duals of the cuboctahedron (b) and the icosidodecahedron (h), and belong to the face-transitive Catalan solids.*

number of each atom equals twelve, the number of edges per vertex results to $e = 6$. In summary, we get $p–f+e–v = 3–8+6–1 = 0$. This allows us to derive all space-filling packings of the Archimedean solids (Table 3.6 and Fig. 3.15).

Packings of truncated cubes sharing their octagon faces leave voids, which have to be filled by octahedra (Fig. 3.15(b)). Octahedra are also needed to fill the empty spaces in packings of square-sharing cuboctahedra (Fig. 3.15(c)); however, the gaps can also be filled with truncated octahedra and truncated tetrahedra 3.15(j)); those left in packings of edge-connected octahedra have to be closed by tetrahedra (Fig. 3.15(d)). The same is true for truncated tetrahedra connected via their hexagon faces (Fig. 3.15(e)).

The following packings need three or four different Archimedean solids, respectively: rhombicuboctahedra plus cubes combined with cuboctahedra lead to a primitive cubic packing (Fig. 3.15(g)), while with tetrahedra an *fcc* packing results (Fig. 3.15(h)); together with truncated cubes and octagonal prisms a primitive lattice is obtained again (Fig. 3.15(l)).

Table 3.6 *Space-filling packings of regular and semi-regular polyhedra with resulting cubic symmetry. The trivial packing of cubes is not listed.*

Polyhedra	Fig. 3.15	Space group: Vertex symbols
Truncated octahedra	(a)	229 $Im\bar{3}m$: 4.6^2
Truncated cubes + octahedra	(b)	221 $Pm\bar{3}m$: $3.8^2 + 3^4$
Cuboctahedra + octahedra	(c)	221 $Pm\bar{3}m$: $3.4.3.4 + 3^4$
Octahedra + tetrahedra	(d)	225 $Fm\bar{3}m$: $3^3 + 3^4$
Truncated tetrahedra + tetrahedra	(e)	227 $Fd\bar{3}m$: $3.6^2 + 3^3$
Truncated cuboctahedra + octagonal prisms	(f)	229 $Im\bar{3}m$: $4.6.8 + 4^2.8$
Rhombicuboctahedra + cuboctahedra + cubes	(g)	229 $Pm\bar{3}m$: $3.4^3 + 3.4.3.4 + 4^3$
Rhombicuboctahedra + cubes + tetrahedra	(h)	225 $Fm\bar{3}m$: $3.4^3 + 4^3 + 3^3$
Truncated cuboctahedra + truncated octahedra + cubes	(i)	221 $Pm\bar{3}m$: $4.6.8 + 4.6^2 + 4^3$
Truncated octahedra + cuboctahedra + truncated (Friauf) tetrahedra	(j)	225 $Fm\bar{3}m$: $4.6^2 + 3.4.3.4 + 3.6^2$
Truncated cuboctahedra + truncated cubes + truncated (Friauf) tetrahedra	(k)	225 $Fm\bar{3}m$: $4.6.8 + 3.8^2 + 3.6^2$
Rhombicuboctahedra + truncated cubes + octagonal prisms + cubes	(l)	221 $Pm\bar{3}m$: $3.4^3 + 3.8^2 + 4^2.8 + 4^3$

The gaps in tilings based on truncated cuboctahedra can be filled by octagonal prisms leading to a *bcc* lattice (Fig. 3.15(f)), by truncated octahedra and cubes given a primitive cubic packing (Fig. 3.15(i)), or by truncated cubes and truncated tetrahedra yielding an *fcc* structure (Fig. 3.15(k)).

The structures of icosahedral quasiperiodic structures and their approximants are essentially based on the packing of triacontahedra. In Fig. 3.16, a few examples are shown. Characteristic for quasiperiodic structures is that the unit clusters have to partially overlap. Depending on the overlap direction, different overlaps of the triacontahedra are allowed (Fig. 3.16(a)–(c)). In the simple rational approximant shown in Fig. 3.16(d), the triacontahedra just share faces along the two-fold directions.

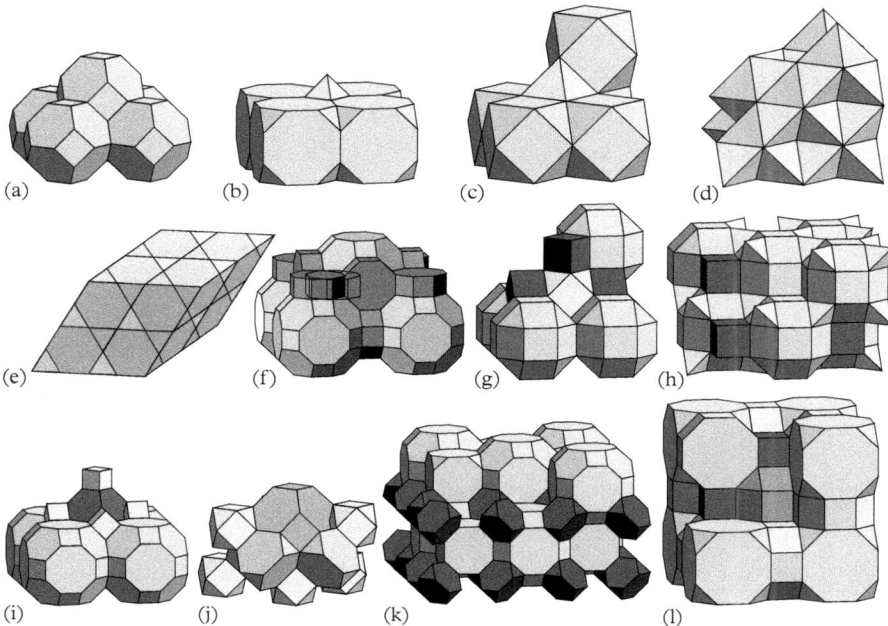

Fig. 3.15 *Packings of regular and semi-regular polyhedra with resulting cubic symmetry (see also Table 3.6). (a) Truncated octahedra, (b) truncated cubes + octahedra, (c) cuboctahedra + octahedra, (d) octahedra + tetrahedra, (e) truncated tetrahedra + tetrahedra, (f) truncated cuboctahedra + octagonal prisms, (g) rhombicuboctahedra + cuboctahedra + cubes, (h) rhombicuboctahedra + cubes + tetrahedra, (i) truncated cuboctahedra + truncated octahedra + cubes, (j) truncated octahedra + cuboctahedra + truncated tetrahedra, (k) truncated cuboctahedra + truncated cubes + truncated tetrahedra, and (l) rhombicuboctahedra + truncated cubes + octagonal prisms + cubes. The trivial packing of cubes is not shown.*

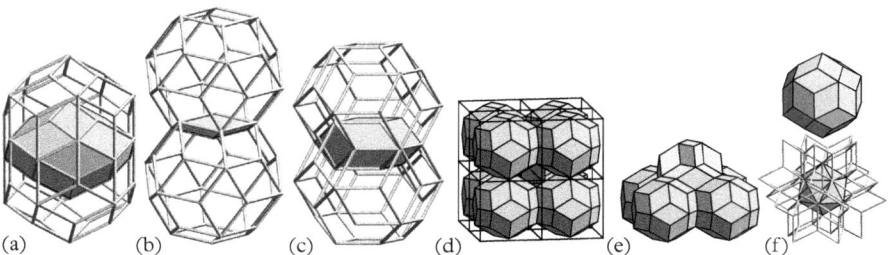

Fig. 3.16 *Triacontahedra overlapping along the (a) 5-, (b) 3-, and (c) 2-fold directions. The shared volumes, a rhombic icosahedron (a), an oblate golden rhombohedron and a rhombic dodecahedron (c), respectively, are marked. (d) Packing of triacontahedra by sharing a face along each of the eight 2-fold directions. (e) The remaining empty space has the shape of a dimpled triacontahdron, i.e., a triacontahedron with eight oblate rhombohedra removed. (f) Packing of a triacontahedron into one of the twelve pentagonal dimples of a rhombic hexecontahedron.*

3.3.2 Fullerenes and Frank-Kasper polyhedra

There are many intermetallic phases known, which can be described (geometrically) by endohedral clusters, which are constituted from alternating fullerene and Frank-Kasper (FK) polyhedra. Fullerenes[2] are polyhedra with twelve regular pentagon faces and a variable number $h > 1$ of regular hexagon faces. They are usually described by their number of vertices $v = 2h + 20$. We will use a description, which additionally gives the total number $f = 12 + h$ of faces, F_v^f. Frank-Kasper polyhedra are bounded by (not necessarily regular) triangular faces, only. Since they are dual to the fullerenes, their number of vertices and faces corresponds to the number of faces and vertices of the fullerenes: $FK_f^v \leftrightarrow F_v^f$. With the application of the Euler characteristic, $f + v = e + 2$, i.e., the number of faces and vertices is equal to the number of edges plus two, one finds that a fullerene and its dual FK-polyhedron have the same number of edges $e = 30 + 3h$.

The well-known C_{60} molecule (F_{60}^{32}), the smallest fullerene with isolated pentagons, has icosahedral symmetry. However, this is not the only possible point group symmetry for fullenes in general. Fullerenes can have 28 different point group symmetries, 6 non-crystallographic ones and 22 crystallographic ones (Fowler *et al.*, 1993): 1, $\bar{1}$, 2, m, $2/m$, 222, $mm2$, mmm, $\bar{4}$, $\bar{4}2m$, 3, $\bar{3}$, 32, $3m$, $\bar{3}m$, $\bar{6}$, 622, $\bar{6}2m$, $6/mmm$, 23, $m\bar{3}$, $\bar{4}3m$; 52, $\bar{5}m$, 235, $m\bar{3}5$, $\overline{10}2m$, $\overline{12}2m$. One has to keep in mind that isomers exist for all fullerenes with $h > 3$. There are, for instance, 1812 hypothetical isomers known for F_{60}^{32}, spanning 15 of the 28 possible point groups, but most of them (1508) are chiral with point symmetry 1. Generally, isolated pentagons can be found in isomers of fullerenes with $v = 60$, 70, and all even values $v > 70$.

The three-connected vertices of the fullerenes are located opposite the triangular faces of the dual FK-polyhedra. On the other hand, the five- and six-connected vertices of the FK-polyhedra sit across the pentagonal and hexagonal faces of the dual fullerene. The nesting of FK/fullerene shells constituting an endohedral cluster can be described in a rather straightforward way. Let us start with the innermost polyhedron, one of the common FK-polyhedra (see Table 3.7), FK_f^v with v vertices, $e = [12 \times 5 + (v - 12) \times 6]/2$ edges, and $f = e + 2v$ faces. Then, the next polyhedral shell corresponds to its dual fullerene, with $v(F) = f(FK)$, $e(F) = e(FK)$, and $f(F) = v(FK)$. In the case of intermetallic phases, this fullerene cluster shell is only a half-shell (not all atoms on the vertices have bonding distances) and will often be complemented by additional atoms capping the pentagonal and hexagonal faces, resulting in yet another—now larger—FK-polyhedron with $v(FK) = v(F) + f(F)$ (see also Alvarez (2006)). The sequences of cluster shells arising according to these rules are given for the four basic FK-polyhedra in Table 3.7. The fullerenes dual to them are the only ones with only pentagons, F_{20}^{12} ($m\bar{3}5$), and with isolated hexagons, F_{24}^{14} ($\overline{12}2m$), F_{26}^{15} ($\bar{6}2m$), and F_{28}^{16} ($\bar{4}3m$), respectively (Fig. 3.17).

---

[2] We use the term "fullerene" as a generic term for clusters that have a similar shape as the carbon-based fullerenes.

Table 3.7 *Sequences of typical nested FK- and fullerene-like polyhedra, starting with the innermost cluster being one of the common small FK-polyhedra. Given are the number of the cluster shell in the left superscript, the type of polyhedron (FK or F), as well as the numbers of vertices (v) and faces (f).*

^1FK		^2F		^2FK		^3F		^3FK	
v	f	v	f	v	f	v	f	v	f
12	20	20	12	32	60	60	32	92	180
14	24	24	14	38	72	72	38	110	216
15	26	26	15	41	78	78	41	119	234
16	28	28	16	44	84	84	44	128	252

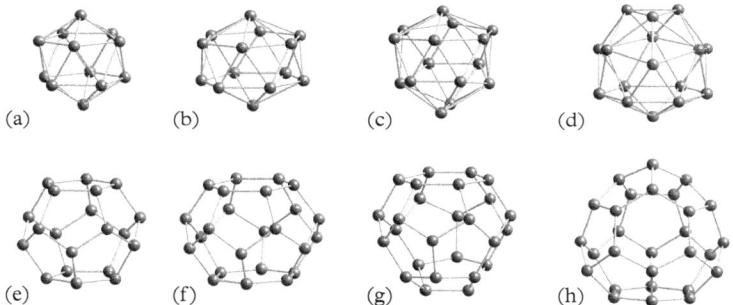

Fig. 3.17 *(a) – (d) Frank-Kasper polyhedra and (e) – (f) their dual fullerenes: (a) FK_{20}^{12}, (b) FK_{24}^{14}, (c) FK_{26}^{15}, (d) FK_{28}^{16}, (e) F_{12}^{20}, (f) F_{14}^{24}, (g) F_{15}^{26}, (h) F_{16}^{28}.*

3.4 Complexity in intermetallics

As we will see in Chapter 5, the number of atoms per unit cell of the known intermetallic compounds can range between one and more than twenty thousand. And there may even not be a unit cell in the structure as in quasicrystals or in the incommensurate host/guest structures of some chemical elements or intermetallic compounds. So, the question is: what governs the complexity of a structure?

The driving force for the formation of a crystal structure is always the minimization of the Gibbs free enthalpy. This requires:

- Maximum space filling for a given stoichiometry under the constraint of the optimization of attractive and repulsive interactions. The contribution of the electrons is reflected in the electronic band structure; a (pseudo)gap at the Fermi energy, for instance, can result from energy minimization either by structural optimization (Hume-Rothery) or by cluster hybridization.

- Entropy maximization by creation of the structural basis for phonon-based atomic vibrations and structural disorder. In most cases, entropy maximization is counteracting energy minimization. However, at sufficiently high temperature, the entropic contribution can be crucial for the stability of a structure.

In more crystallographic terms, a crystal structure is always a compromise between the energetically most favorable short-range order (AETs and higher coordination shells) and the long-range order resulting from the energetically most favorable packing of the overlapping AETs. The packing optimization usually requires a modification of the AETs, which may lower their symmetry. If the chemical composition (stoichiometry, atomic size ratios, electronegativity differences, directionality of bonding, etc.) requires a larger number of different AETs, their optimum packing may become quite complex. It is not necessarily the number of different chemical elements that is decisive for the formation of complex structures, as is demonstrated by the existence of high-entropy alloys with up to ten different constituting elements. This is discussed below for the case of unary and binary structures. Decisive are mainly the kinds of local atomic interactions and globally the electronic band structure. One of the driving forces for long-range ordering is the formation of narrow distance distribution functions (histograms), which are usually related to a more efficient packing of the atoms and, thereby, to low-energy configurations. Periodicity and quasiperiodicity are just a consequence of keeping the distance distribution function discrete.

What is characteristic for the structures of complex intermetallics is the existence of a kind of substructure or of subunits. These may be clusters, or subunits separated by atomic layers, which may be flat or slightly puckered (thick atomic layers, TALs). The latter can be seen as forming interfaces between structural subunits and the crystal facets.

3.4.1   Unary phases A

As we will see in Chapter 6, at ambient conditions the structures of the elements are mostly simple, but quite a few are also rather complex. In the case of non-directional bonding, each atom can be equally densely coordinated by the other atoms in the first coordination shell—close sphere packings are the consequence. Small energy differences decide between *ccp* (ABC) and *hcp* (AB) or *dhcp* (ACAB) packings. The resulting AETs can be cuboctahedra (*ccp*) or disheptahedra (anticuboctahedra) (*hcp*), with CN = 12 in both cases. In the case where the AETs are all equal but the constituting atoms are not all in close contact to each other due to some directional bonding contributions, structures can result, which can be described by either vertex-decorated Bravais type lattices (*cI2*-W, *cP1*-Po, etc.) or by structures such as, for instance, the diamond structure, *cF8*-C.

More complex element structures can result in the case of interactions resulting from different kinds of chemical bonding and/or magnetic interactions present in a phase (cI58-Mn, mP16-Pu, etc.). Under pressure, even disproportionation can take place leading to ionic interactions (hP4-Na, e.g.). The largest number of structural changes (six) as a function of temperature have been found for Pu with its almost half-filled 5f–orbitals. The most complex structures under compression have been observed for some of the alkali and alkaline earth metals, respectively. There, incommensurate host/guest structures form, e.g., tI19.3-Na$_{inc}$, where the periodicity of the Na (guest) atoms inside the channels is different from that of the host structure. Consequently, no common unit cell exists for the two substructures. The most complex structures of unary phases, i.e., of the elements, are high-pressure phases such as oC84-Cs, in the periodic case, or the incommensurate host/guest structures of the alkali metals, for instance.

3.4.2   Binary phases A–B

In a binary intermetallic phase A$_x$B$_y$, each atom or cluster of atoms (e.g., tetrahedra or even icosahedra) of one kind has to be surrounded by as many atoms as possible of the other kind in order to maximize the number of attractive interactions, otherwise the pure element phases would separate. The most extreme stoichiometry of an intermetallic compound known so far is that of cF184-ZrZn$_{22}$, i.e., with 95.7% Zn content (see Section 7.11 and Fig. 7.30). This structure can be described as packing of Zn-centered Zn-icosahedra and Zr-centered CN16-FK-polyhedra.

Atomic size ratios and stoichiometry determine the coordination number and shape of AETs. If each of the two kinds of atoms has several different AETs, then more complex structures are likely to be formed. If the first coordination shells of each atom A and each atom B, respectively, are always the same, then periodicity and simple structures result. This can be generalized to the condition that the environment (coordination) of clusters or unit cells always have to be the same.

The structures of binary compounds can be simple superstructures of unary ones, such as cP2-CsCl and tP2-AuCu, to name just two of the most frequent simple binary structure types. Many representatives also have the structure type cF96-Ti$_2$Ni, which is quite complex in spite of its simple stoichiometry. Close to simple stoichiometries, but only close, are the structures of the complex intermetallics cF1832-Mg$_{28}$Al$_{45}$ and cF1456-Eu$_4$Cd$_{25}$, for instance. Aperiodic complex binary intermetallics are known for a couple of binary compounds such as incommensurately modulated LiZn$_{3.175}$ and NiBi, respectively, or for binary icosahedral quasicrystals such as i-Cd$_{84}$Yb$_{16}$, for instance.

3.4.3   Ternary phases A–B–C

In the case of ordered ternary intermetallics, the unit cells get larger in order to account for the packing of at least three different AETs around A, B, and C atoms, respectively. The most frequent simple structure types are those of $hP9$-ZrNiAl and $oP12$-TiNiSi. There are many more complex ternary than binary intermetallics, with $cF23\,256$-$\text{Ta}_{39.1}\text{Cu}_{5.4}\text{Al}_{55.4}$ featuring by far the largest number of atoms per unit cell.

To our knowledge, there are no ternary incommensurately modulated or host/guest structures of intermetallics known. Incommensurabilities between substructures can be compensated more easily if more degrees of freedom exist in their packing, as it is the case for ternary compounds. In contrast, there are much more ternary than binary decagonal and icosahedral quasiperiodic structures known, such as those in the systems Al–Mn–Pd, Mg–Zn–RE (RE...rare earth elements), and Cd–Mg–RE, for instance. However, in these cases the third element might be just necessary for the electronic stabilization of metastable binary QCs.

One way to avoid frustration in packing is the formation of an endohedral cluster, which then can be packed very efficiently either periodically or quasiperiodically. In the latter case, the clusters have to overlap in a systematic way. Another way for efficient packing is to organize the structure in flat or puckered layers.

3.4.4   Multinary phases

The larger the number of constituting elements, the more difficult the formation of a low-energy structure. Due to the increasing number of different AETs, their packing gets more and more difficult. On the other hand, the configurational entropy increases if structural disorder is present, leading to a more or less statistical distribution of the different elements on the sites of relatively simple structures leading, finally, to high-entropy alloys (HEAs). Ordered structures of intermetallic phases with more than three elements are rare, with more than five we are not aware of any examples. Sometimes, clusters existing in the binary phases are also used as fundamental structural units in multinary phases (Dong *et al.*, 2007).

Examples of ordered quaternary phases are $cF16$-LiPdMgSn, an *fcc* $(2 \times 2 \times 2)$-fold superstructure of the $cI2$-W structure type, and $tP14$-$\text{Ce}_2\text{CoGa}_9\text{Ge}_2$, with a somewhat more complex, modular structure. HEAs are, by definition, solid solutions of five or more metallic elements, most of them with simple *bcc* or *fcc* average structures, respectively. It seems that the easiest way of obtaining multinary structures is to substitute atoms in flexible simple structures such as the Heusler phases.

3.4.5   Definition and description of complex structures

How can we define or characterize structural complexity? Complexity is reflected in:

- broad distance distribution functions (histograms)
- a large number of different AETs for each of the constituting elements
- a large number of independent parameters for the description of a structure.

Complexity can result from:

- unfavorable size ratios of atoms geometrically hindering optimum interactions
- a preference of specific first shell coordinations (AETs) hindering optimum packings (e.g. 5-fold symmetry)
- parameters that are close to, but not quite, optimum (pseudosymmetry).

Complexity can frequently be described as a modulation or superstructure of a rather simple basic structure. There are two classes of modulations:

- simple modulation, i.e., just a correlated displacement or substitution leading to a comparatively small deviation from a basic structure
- complex modulation, i.e., a displacement or substitution leading to the local formation of clusters.

Complexity can sometimes result in a hierarchical structure (atom $\to$ cluster $\to$ cluster of clusters, ...), which may, additionally, show self-similarity of structure motifs.

Description of complex structures

There are two steps in the description of a structure. First, the derivation of an idealized (topologically equivalent) model structure, second the description of the difference structure (model structure minus actual structure). The second step is rarely done in standard structure analysis.

Periodic average structures (PAS), with bases defined by the sets of strongest Bragg reflections, allow for the identification of substructures (basic structures). If such basic structures are well defined, then complex structures can be described as superstructures (modulated structures). In other cases, complex structures can be described as composite structures or approximants of quasicrystals.

For studying the way of packing, the derivation of Voronoi cells of individual atoms and/or of clusters may be useful. The symmetry of a structure results from the way of packing of atoms or AETs. The more isotropic the structural building units are the higher the resulting symmetry can be.

Information necessary to describe a structure in a unique way includes:

- In the case of a structure with a 3D space group, the set of generating symmetry operators as well as the coordinates x, y, and z of the N unique atoms per unit cell. It is not simpler to say, for instance, the structure corresponds to a *ccp* packing with half of the tetrahedral interstices filled in an ordered way, because one would have to explain what a *ccp* packing is and what tetrahedral voids are. However, packings may better characterize the crystal-chemical origin of a structure and structural relationships than the purely geometrical description does.

- In the case of a QC structure with an nD space group, the set of generating symmetry operators as well as the coordinates and shapes of the N unique occupation domains per nD unit cell. In the 3D tiling decoration description, tiling generation rules and atomic decoration have to be given, if the 3D approach is feasible.

Periodicity and quasiperiodicity:

- If each atom of an A-type structure has exactly the same coordination then the structure is periodic; this is also true if we replace the term atom by cluster or unit cell (set of points).
- If the structure has a finite number (> 1) of different local arrangements of atoms within a radius R then it can be quasiperiodic as well.

For quickly comparing the structures of complex intermetallics with similar unit cell dimensions, projections along different directions can be quite useful (Fig. 3.18). If the 2D-projected structures are closely related to each other, this does not mean that the 3D structures are always similar also. If they are different, the actual 3D structures can be seen as differently "expanded" versions of the 2D projected structures (average structures).

Measures of complexity

A measure of complexity of a structure can be the amount of information needed for its description. It comprises the:

- lattice parameters
- space group symmetry operators in their action on the atoms of just one unit cell
- coordinates of the N atoms in the asymmetric unit.

Additional degrees of freedom to be considered are the vibrational degrees of freedom (Debye-Waller factors) and the chemical flexibility (substitutional disorder related to the phase width). Other indicators for complexity are the:

- distribution of atomic distances (distance histograms; number of maxima and their FWHM)
- number of topologically different AETs or Voronoi domains and degree of their deviation from idealized polyhedra
- configurational entropy of the system.

How to compare the complexity of different structures, for instance, that of a complex intermetallic with a giant unit cell with that of a quasicrystal? One way could be to make the comparison in reciprocal space. A measure of complexity

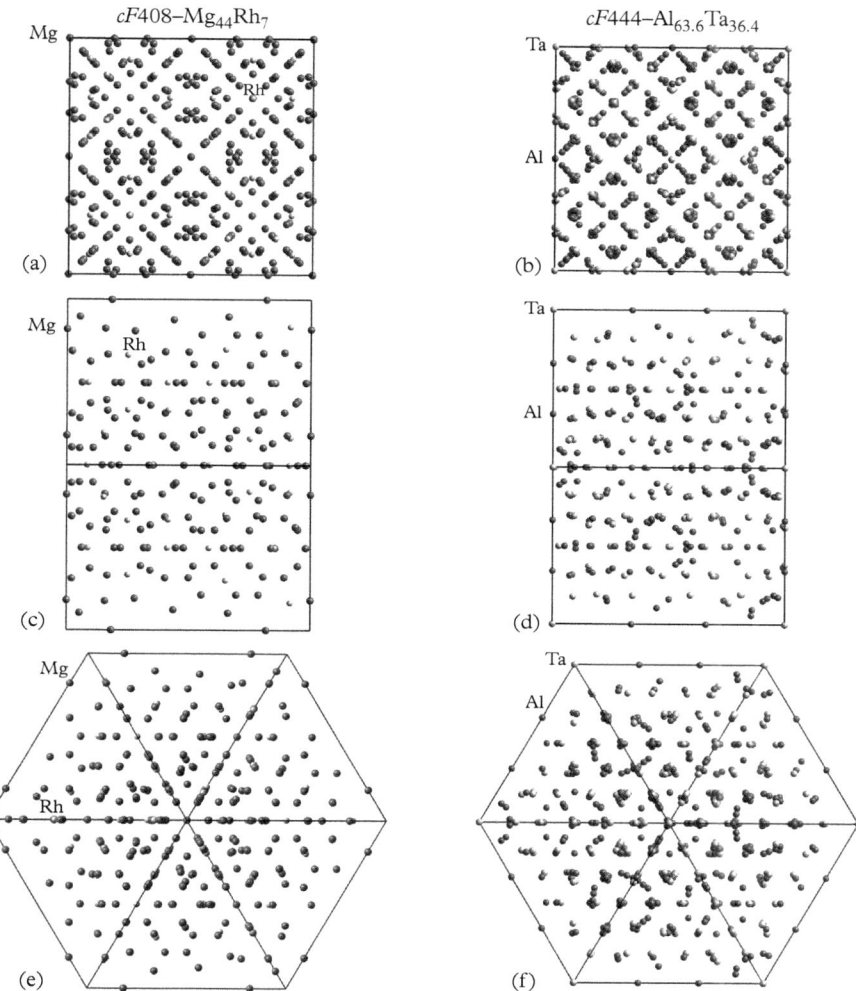

Fig. 3.18 *Projections of the structures of cF408-Mg$_{44}$Rh$_7$ (a, c, e) and cF444-Al$_{63.6}$Ta$_{36.4}$ (b, d, f) along [001] (a, b), [110] (c, d), and [111] (e, f). The seemingly small differences in the projected structures lead to quite different coordination polyhedra. cF408-Mg$_{44}$Rh$_7$: the* Rh *atoms, forming a network of octahedra and stellated tetrahedra, center interpenetrating icosahedral clusters of* Mg *atoms; cF444-Al$_{63.6}$Ta$_{36.4}$: cubic close packing of large fullerene clusters with the voids filled by truncated tetrahedra. One of the reasons for the different cluster formation may be the different ratios of atomic radii:* $r_{Mg}/r_{Rh} = 1.599/1.345 = 1.19$, $r_{Al}/r_{Ta} = 1.432/1.430 = 1.00$. *The structures are shown on 2/3 of the usual scale.*

could be the Bragg reflection density above given threshold values, as well as the Bragg intensity variation as a function of the reciprocal space vector (diffraction vector) $|\mathbf{H}|$. For instance, a Penrose tiling (PT) decorated by one type of atom at each vertex has the same infinite reflection density as one that is decorated by a complex large cluster, at least theoretically. However, the intensity distribution will be different as well as the reciprocal basis. The complexity of a structure increases if atoms are substituted by larger units such as clusters or if a uniform atom type is replaced by several kinds of atoms.

A similar approach, but now in direct space, would be to compare the Patterson functions (autocorrelation functions, vector maps) of structures. One could count, for instance, the number of Patterson maxima within a sphere of radius R. Additionally, the variation of the heights of the maxima could be included in the comparison. For the above example, the Patterson functions would considerably differ.

Another measure of complexity would be the number of parameters and the algorithms needed for a full description of a structure. This would allow even to compare the complexities of periodic and quasiperiodic structures.

Factors governing complexity

There are many factors governing complexity, the most important being:

- packing problems of 1D (chains), 2D (layers), or 3D (AETs and/or clusters) structural units; in order to efficiently pack, AETs or other structural building units may have to be modified (e. g., adopt higher or lower symmetries) compared to their isolated form.
- chemical order due to different attractive and repulsive interactions between atoms of different types
- electronic band structure (pseudo-gap formation due to Fermi-surface/Brillouin-zone interactions or due to spd-hybridization in clusters, for instance).

Consequently, the formation of a complex intermetallic with a particular structure is a complex interplay between stoichiometry, optimization of atomic interactions, and efficient packing under the constraint of minimum free enthalpy.

3.4.6   Layers, clusters, and interfaces

From a purely geometrical point of view, a complex cluster-based intermetallic structure can be seen as having interfaces between the clusters and their surrounding. To some extent it can be seen as a packing of nanoparticles (i.e., the clusters) embedded in a matrix (i.e., "glue" atoms between clusters). One has to keep in mind that the properties of embedded nanoparticles can differ considerably from that of free-standing ones (see, e.g., Mei and Lu (2007)). How can we understand this cluster/nanoparticle comparison? The following hypothetical scenario should illustrate it.

Scenario: Nucleation of cF444-Al$_{63.6}$Ta$_{36.4}$ from a melt of same composition.

In the melt, close to the solidification temperature, Ta atoms are always surrounded by more Al than Ta atoms, simply due to the given stoichiometry. Although the cohesion energy of Ta–Ta is much higher than that of Ta–Al and Al–Al, phase separation does not take place and elementary Ta does not solidify first (Ta: $T_m = 3020°C$, Al: $T_m = 660.5°C$, cF444-Al$_{63.6}$Ta$_{36.4}$: $T_m = 1548°C$), as it is the case for Ta compositions larger than 63.2%. This is due to the fact that the total free energy of the two-phase system Al–Ta with the given composition is higher than that of the compound cF444-Al$_{63.6}$Ta$_{36.4}$. The formation of small Ta-clusters (hexagonal bipyramids) and Ta-cluster shells (bifrusta and Ta$_{28}$ polyhedra), which are linked to a Ta framework, is the compromise. The porous Ta substructure allows for strong Ta–Ta interactions and at the same time it includes Al atoms and cluster shells maximizing the number of Al–Ta bonds and minimizing that of Al–Al bonds for the given composition.

An example how this can be understood is shown in Fig. 3.19. Each flat atomic layer in a structure can be seen as an (intra-unit-cell) interface between two parts of a structure (Fig. 3.19(a)). Similarly, cluster shells can be seen as (intra-cluster) interfaces between the interior and exterior of clusters (see also Subsection 2.3.6). In light of this, the surface energy of a smaller inner cluster shell must be higher than that of an outer shell with a larger radius of curvature. Even if one considers the larger surface area of cluster shells with larger radii, clusters would be forced to

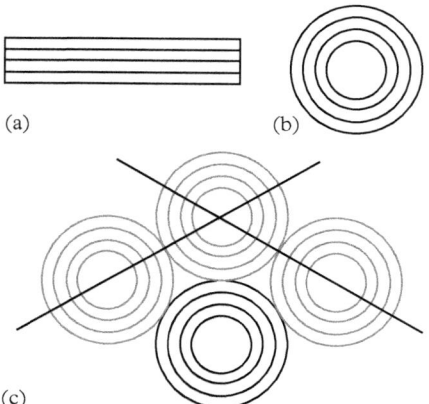

(a)   (b)

(c)

Fig. 3.19 *Schematic representation of atomic layers inside a structure as interfaces: (a) flat layers, (b) curved layers (cluster shells), and (c) aggregation of clusters with net planes (flat atomic layers) indicated.*

grow to reduce their surface energy (mechanism I). Once the first cluster reaches its full size, it grows further in the conventional way via flat atomic layers (mechanism II) (Fig. 3.19(c)) and faceting already takes place. Once the next cluster core is formed, the growth of the cluster follows mechanism I again.

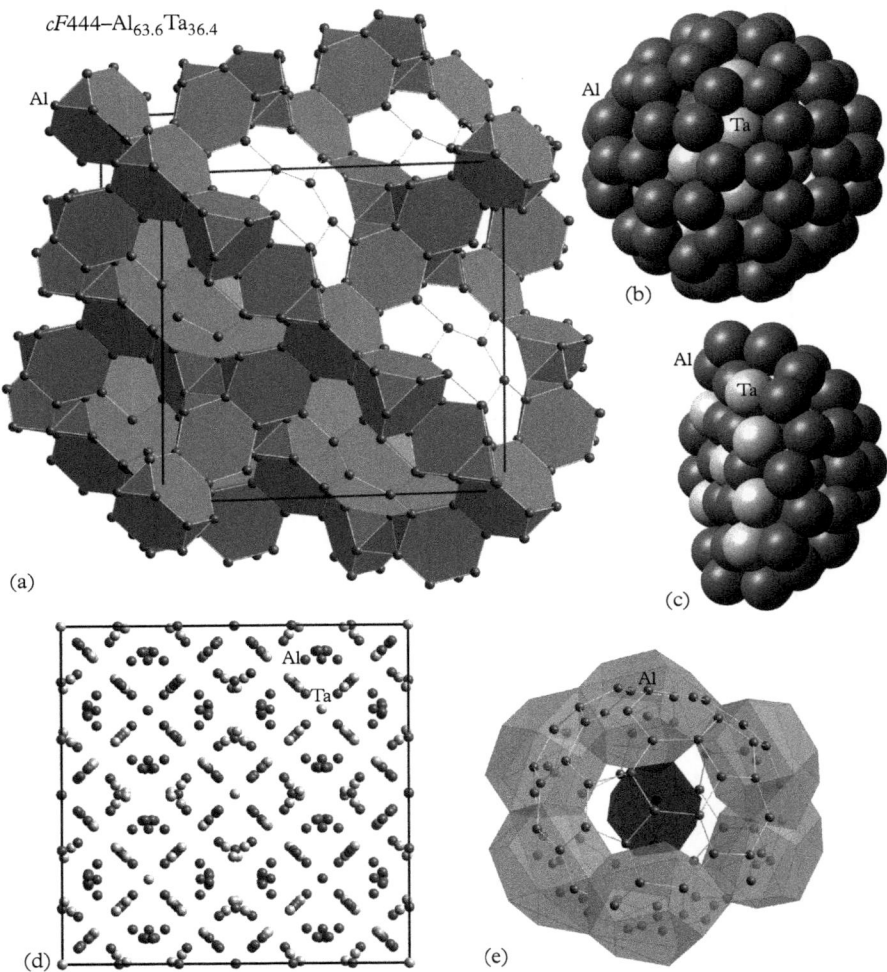

Fig. 3.20 *The structure of cF444-Al$_{63.6}$Ta$_{36.4}$ (Conrad et al., 2009) shown in projection as (a) cluster packing (fullerenes plus truncated tetrahedra) and (d) projected along [001]. A single fullerene cluster with space-filling atoms is shown in (b). In (c) one half-cluster is shown with its surface parallel to (110). (e) Architecture of the endohedral fullerene cluster, which is embedded in the shell of bifrusta, and centered by a rhombicdodecahedron. For more details, see Subsection 7.4.4.*

A realistic example of a cluster and the flat atomic (110) layer cutting it in half is shown on the example of $cF444$-$Al_{63.6}Ta_{36.4}$ in Fig. 3.20. It is also an example for how clusters can be seen as resulting from a packing of polyhedra, which is not equally possible with flat layers.

3.4.7 Cluster decomposition of complex structures

For intermetallics, there is no general rule as to how to decompose a crystal structure into structural subunits. Of course, an obvious but trivial structural subunit is the unit cell itself, which usually is not very illustrative for understanding the architecture of a crystal structure. Depending on the complexity of a structure, one or several of the following actions can be helpful:

- Look at different projections of the unit cell to identify any kind of layer or columnar structure, or other characteristic features.
- In a (geometrically) layered structure, characterize the layers and their mutual relationships.
- In the case of other characteristic features in the projected structures, identify the underlying atomic arrangements.
- Check all AETs and how they are linked.
- Check the coordination polyhedra around high-symmetry Wyckoff positions from the first coordination shell to higher ones.
- The atoms on each coordination shell should be in bonding distance ($d_{A-B} = r_A + r_B + \Delta$) to one another and to the atoms in the neighboring lower/higher shell.
- In the case of a polyhedra packing, ensure that there are no gaps left.
- Choose structural subunits that differ from their environment by chemical bonding, if there are differences anyway.

4

Higher-dimensional approach

For the understanding of the structural ordering of aperiodic crystals, i.e., incommensurately modulated phases (IMS), composite crystals (CS), quasicrystals (QCs), and their structurally closely related rational approximants, the higher-dimensional (nD) approach is essential. The rationale behind this approach is that in a properly selected nD space, aperiodic crystal structures can be described as 3D sections/projections of nD periodic "hypercrystal" structures. Furthermore, "non-crystallographic" symmetries of QCs can become compatible with nD lattice symmetry. For instance, 5-, 8-, 10-, and 12-fold rotational symmetries are proper symmetry operations in 5D and icosahedral point group symmetries in 6D hypercubic lattices. This permits a description of structures that are quasiperiodic in dD physical or par(allel) space (parspace) as sections, or window-bounded projections of lattice-periodic structures in nD space ($n > d$). We will first use the so-called "strip-projection" (or "cut-and-project") method for introducing the nD approach, because it allows a more intuitive understanding of when and how clusters order quasiperiodically. Thereby, we follow closely the description by Steurer and Deloudi (2012). Then we introduce the nD section method, which allows us to nicely illustrate the different kinds of aperiodic crystal structures in comparison. The latter is also the method of choice for nD structure analysis, because it can make use of the reciprocal space information experimentally accessible by diffraction methods. This discussion essentially follows the respective chapter in the book by Steurer and Deloudi (2009).

The projection of nD polytopes and hyperlattices, respectively, can be valuable in describing 3D endohedral clusters and how they can be packed best for given stoichiometries, not only for quasiperiodic but also for periodic structures (Coxeter, 1973; Berger *et al.*, 2008). For instance, every 3D zonohedron with octahedral or icosahedral symmetry can be described as an orthogonal projection of an nD hypercube (Coxeter, 1973), and its packing by the projection of a subset of an nD hypercubic lattice. The dimension n is defined by the number of edge directions of the respective zonohedron. For the triacontahedron, the most important cluster for icosahedral QCs, $n = 6$. The assembly of such zonohedra with optimum density (i.e., frequency in a given volume) for a given stoichiometry is a non-trivial problem that can be tackled much more easily in higher dimensions,

Intermetallics: Structures, Properties, and Statistics. First Edition. Walter Steurer and Julia Dshemuchadse.
© Walter Steurer and Julia Dshemuchadse 2016. Published in 2016 by Oxford University Press.

where the packing of partially overlapping clusters is reduced to selecting a strip out of a hypercubic lattice. Columnar clusters with, in projection, octagonal, decagonal, or dodecagonal shape can be related to 2D projections of nD ($n = 4, 5, 6$) hypercubes.

In the following, we will demonstrate the basic concepts of the nD approach, first on the simple example of the 1D Fibonacci sequence, followed by applications based on this approach for the 2D PT and the 3D PT. For a more detailed discussion see Steurer and Deloudi (2009) and Steurer and Deloudi (2012).

4.1 2D description of the 1D quasiperiodic Fibonacci sequence by the strip-projection method

For the description of the Fibonacci sequence (FS) by the strip-projection method, we need a properly oriented 2D hypercubic (square) lattice. It can be defined by the basis vectors $\mathbf{d}_1 = a(\tau, -1)_V$ and $\mathbf{d}_2 = a(1, \tau)_V$. The "golden mean" is defined as $\tau = 2 \cos \pi/5 = (1 + \sqrt{5})/2 = 1.618\ldots$. The vectors $\mathbf{d}_1$ and $\mathbf{d}_2$ form the D-basis, while the index V refers to the Cartesian V-basis. The parameter x_1^V is the V-basis coordinate in the 1D physical or par(allel)-space, $V^{\parallel}$ and x_2^V refers to the 1D perp(endicular)-space, $V^{\perp}$ (Fig. 4.1).

Since the 2D square lattice has an irrational slope relative to the 1D par-space, projecting all the vertices of the lattice onto it would give a dense distribution of points. If only those vertices are projected onto par-space that are inside a strip of width W (the acceptance window), then the FS is obtained. The projection of the vector $\mathbf{d}_1 = a(\tau, -1)_V$ onto par-space gives an interval L of length $a\tau$; that of $\mathbf{d}_2 = a(1, \tau)_V$ gives an interval S of length a. The ratio L/S then amounts to $\tau/1$, which is typical for the FS.

The projection onto par-space of just the part of a single unit cell, which is inside the strip, gives the covering cluster (LS). Going along the strip and projecting all the vertices that are within it leads to a covering without any gaps, but with overlapping S intervals allowed. Depending on the slope of the strip, we end up with different overall compositions, L$_x$S (Fig. 4.2). If x is a rational number, then a rational approximant results.

The slope of the strip not only defines the sequence of the intervals (letters) L and S as well as the overall stoichiometry, it also constrains the respective local compositions to stay as close as possible to the overall one. Consequently, the strip-projection method mimics the experimental method of preparing quasicrystals and their approximants by choosing the respective chemical compositions. If we want to create a periodic (rational) approximant, we just have to adjust the slope of the strip to the respective rational number. Since the sequences are created by selective projection of lattice vertices, and lattices are the densest possible arrangements of their unit cells, the projected, edge- or vertex-connected, unit cells give the densest possible arrangement of the clusters for a given stoichiometry.

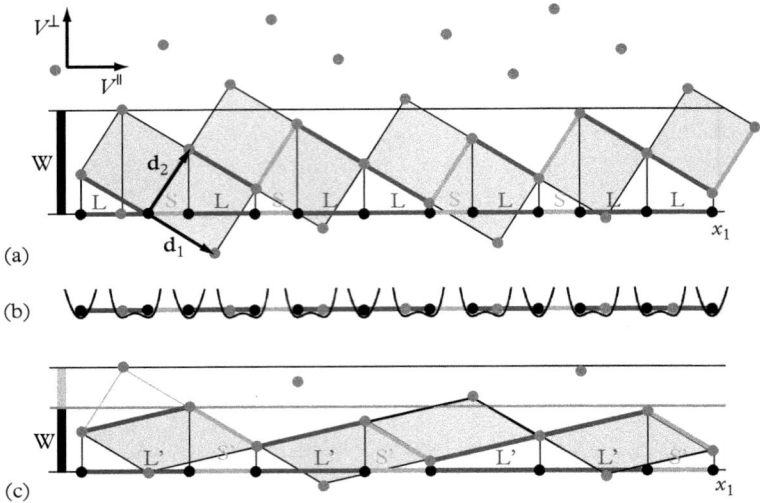

Fig. 4.1 *Fibonacci sequence in the 2D description. (a) Projection of a sequence of vertex- and edge-connected gray-shaded squares gives the FS (black dots) plus some additional vertices (gray dots) at flip positions in a double-well potential (b). The size W of the strip (acceptance window) defines the minimum distance between projected lattice points as well as the unit tile sequence; in our case, W is chosen so that the topmost vertex of the first gray-shaded square is outside the strip. Decreasing the acceptance window by a factor τ^{-1} leads to a scaling of the intervals L and S by a factor τ yielding the sequence L'S'L'S'L'S', with L' = L + S and S' = L (c) (from Steurer and Deloudi (2012), Fig. 2. With kind permission from Springer Science+Business Media).*

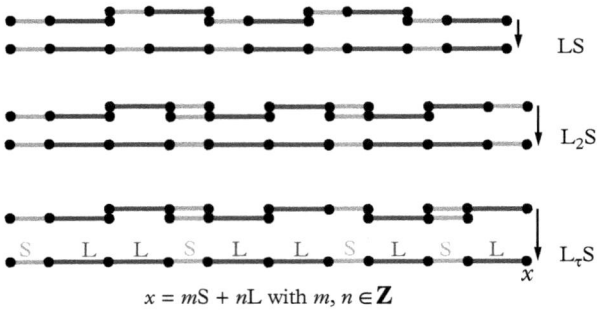

$$x = mS + nL \text{ with } m, n \in \mathbf{Z}$$

Fig. 4.2 *Sequences with overall compositions LS (top), L_2S (middle), and $L_\tau S$, i.e., the Fibonacci sequence (bottom), with the decomposition into covering clusters of the type (LS) in both orientations. The upper two sequences are rational approximants of the Fibonacci sequence (after Steurer and Deloudi (2012), Fig. 1. With kind permission from Springer Science+Business Media).*

Generally, the vertices of quasiperiodic tilings correspond to a subset of a ℤ-module (vector module with integer coefficients) of rank n, depending on the number n of basis vectors needed to index the vertices with integers. Any ℤ-module can be seen as a proper projection of an nD lattice onto the dD par-space. The minimum distance between the tiling vertices defines the width W of the strip that selects the lattice nodes to be projected. The acceptance window W is defined in the $(n-d)$D perp-space. A change in the strip width W always entails a change in the unit cell shape without changing the unit cell volume. This results in a scaling of the quasiperiodic structure as shown in Fig. 4.1(c).

4.2   5D description of the quasiperiodic 2D Penrose tiling by the strip-projection method

The 2D Penrose tiling (PT) can be covered by partially overlapping decagonal clusters—copies of the Gummelt cluster (Gummelt, 1996). It can be seen as the projection along its periodic axis of the decaprismatic clusters building decagonal QCs. Such a 2D decagonal cluster can be obtained as a projection of a 5D hypercube along one of its fivefold axes. Thereby, 22 out of its altogether $2^5 = 32$ vertices are projected into the interior of the decagon formed by the 10 remaining vertices (Fig. 4.3).

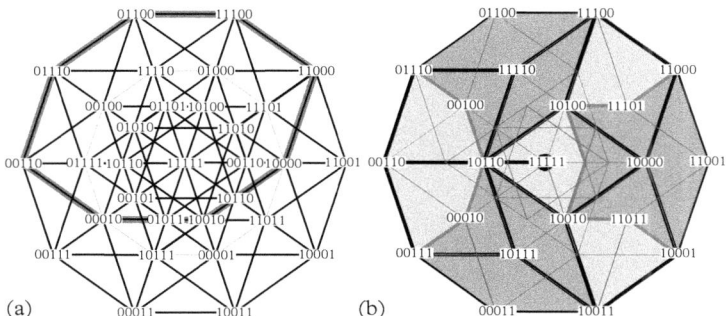

Fig. 4.3 *5D hypercubic unit cell projected along its body diagonal [11111] onto 2D par-space. The vertices are indexed based on the five-star of basis vectors. The thick outlined polygon in (a) corresponds to the projection of the 4D hyperface (subcube) of the 5D hypercube with the last index equal to zero. When applying the minimum distance criterion by adjusting the acceptance window, we obtain a vertex distribution related to a decorated RPT, for instance (b). The light- and dark-gray-shaded regions in (b) mark the overlap regions of the Gummelt decagon, a covering cluster of the PT (after Steurer and Deloudi (2012), Fig. 3. With kind permission from Springer Science+Business Media).*

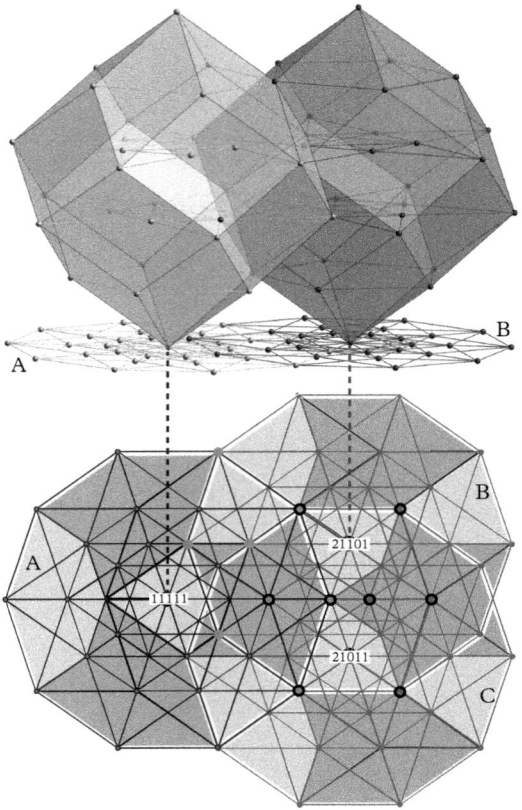

Fig. 4.4 *(top) 5D hypercubic unit cells, separated by the vector* $(10\bar{1}00)_D$, *in combined par-/perp-space projections giving elongated rhombic icosahedra and overlapping decagons A and B in the subspaces spanned by vectors* $(10000)_V$, $(01000)_V$, *and* $(00001)_V$ *and by vectors* $(10000)_V$, *and* $(01000)_V$, *respectively. (bottom) Three Gummelt decagons A, B, and C, with two different types of permitted overlaps. The centers of the decagons are separated by the vectors between A and B,* $(100\bar{1}0)_D$, *A and C,* $(10\bar{1}00)_D$, *and the* τ^{-1} *times smaller one for the vector between B and C,* $(00\bar{1}10)_D$. *The corresponding 5D hypercubes share 3D faces containing eight lattice points each. Gray dots mark the shared vertices of the hypercubes related to the decagons A and B, while black circles mark those of B and C (from Steurer and Deloudi (2012), Fig. 5. With kind permission from Springer Science+Business Media).*

The *D*-basis of the 5D hyperlattice can be written in the form

$$\mathbf{d}_i = a\left(\cos 2\pi i/5, \sin 2\pi i/5, \cos 4\pi i/5, \sin 4\pi i/5, 1/\sqrt{2}\right)_V, 1 \leq i \leq 5. \qquad (4.1)$$

The parameters x_i^V with $i = 1, 2$ are Cartesian coordinates in par-space, and those with $i = 3, 4, 5$ are perp-space coordinates. Again, the projection of the nD unit cell onto perp-space gives the acceptance window W for the generation of the 2D PT. It corresponds to a rhombic icosahedron, which is elongated along its fivefold axis compared to the zonohedron with the same name. Furthermore, its faces with a vertex on the fivefold axis are not congruent to the other ones. All hyperlattice points that fall into this 3D acceptance window if projected onto perp-space, are located inside the strip and generate the 2D PT by par-space projection.

The 5D hypercubes can be connected within the strip in the following five ways: they can share 4D, 3D, or 2D faces (subcubes) with 16, 8, or 4 joint vertices, respectively, as well as 1D edges or 0D vertices. The overlap rules of the decagons resulting in projection depend on the connectivity of the neighboring hypercubes. The rocket decoration shown in dark gray indicates the allowed overlap regions for the PT. They all result from the projection of hypercubes sharing 3D faces. A higher decagon-cluster density would result from the projection of hypercubic unit cells sharing 4D faces (16 lattice nodes), which are shifted against each other by one lattice translation of the type $(10000)_D$. Such a shared area is shown in Fig. 4.4. It is not allowed in the ideal PT, but it can be observed as a defect in decagonal quasicrystals.

4.3 6D description of the quasiperiodic 3D Ammann tiling in the strip-projection method

The 3D Ammann tiling (AT), also called 3D Penrose tiling, can be generated from the projection of a 6D hypercubic lattice with the 3D acceptance window W corresponding to a triacontahedron. The projection of a 6D hypercube along a five-fold axis onto 3D par-space gives three nested polyhedra: 32 of its $2^6 = 64$ vertices project onto the vertices of a triacontahedron, next to it a smaller dodecahedron is formed by 20 vertices, and the innermost icosahedron by the remaining 12 vertices of the projected 6D hypercube (Fig. 4.5(a)).

The projection of two unit cells – neighboring cells in the 6D hyperlattice – indicates the way the triacontahedra can overlap. There are different overlapping volumes, depending on the axis parallel to which the overlap takes place. The projection along a five-fold axis of two 6D hypercubes sharing a 5D face with 32 vertices, the origins of which are separated by a translation vector of the type $(100000)_D$, leads to a rhombic icosahedron (22 vertices, plus vertices inside) as shared volume (Fig. 4.5(b)). The vectors $(00110\bar{1})_D$ and $(00101\bar{0})_D$ connect 6D

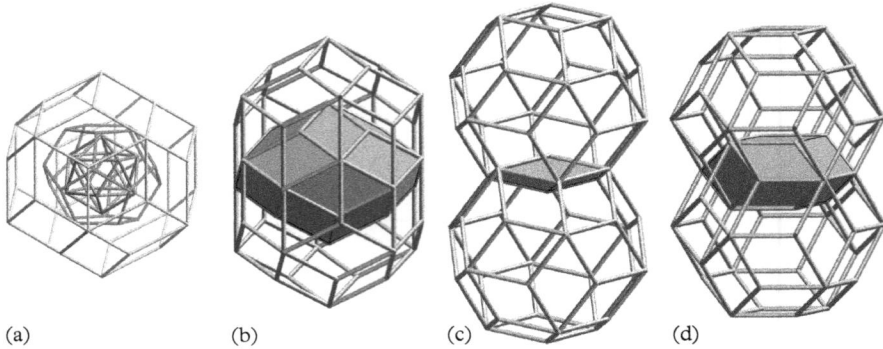

(a)   (b)   (c)   (d)

Fig. 4.5 *(a) 6D hypercube projected onto 3D par-space. Its 64 vertices project onto the 32 vertices of the third shell, a triacontahedron, to the 20 vertices of the second shell, a dodecahedron, and to the 12 vertices of the innermost shell, an icosahedron. (b)–(d) Shared regions of triacontahedra overlapping along a fivefold, threefold, and twofold axis, respectively: (b) rhombic icosahedron, (c) oblate rhombohedron, and (d) rhombic dodecahedron (after Steurer and Deloudi (2012), Fig. 7. With kind permission from Springer Science+Business Media).*

hypercubes along three- and two-fold axes, sharing 3D faces with 8 vertices, and 4D faces with 16 vertices, respectively. This leads to an oblate rhombohedron (8 vertices) (Fig. 4.5(c)) and a rhombic dodecahedron (14 vertices, plus vertices inside) (Fig. 4.5(d)), respectively, as shared volumes.

4.4   Example: Periodic complex intermetallics resulting in projection from the 4D 600-cell polytope

Some complex periodic structures can be described employing the higher-dimensional approach in the one form or another. On the one hand, they can be rational QC approximants, on the other hand these can be the structures of particular complex intermetallics, which are not directly related to QCs. One example of the latter category, are some complex, γ-brass related structures (see also Section 7.3) crystallizing in the space group 216 $F\bar{4}3m$ (Berger *et al.*, 2008), all of which exhibit around 400 atoms per unit cell, among them $cF444$-$Ta_{36.4}Al_{63.6}$. The diffraction patterns of these compounds show pseudo-decagonal symmetry, and their structures feature three pairs of perpendicular pseudo-fivefold rotation axes. Since no 3D point group contains such pairs of fivefold rotation axes, this suggests their description as projections from 4D space, where orthogonal five-fold symmetry axes are allowed in some point groups. Indeed, one of the six 4D Platonic solids, the 600-cell shows this kind of point symmetry, and, if properly projected, reproduces many features of the above-listed structures. The 600-cell 3,3,5 is a polytope with 120 vertices, 720 edges, 1200 faces, and 600 name-giving tetrahedra.

4.5 Aperiodic crystal structures in the nD section method

The nD section method considers the dD diffraction pattern (with the dimension of par-space d) of an aperiodic crystal structure as a proper projection of an nD diffraction pattern ($n > d$). The dimensionality n of the embedding space is given by the number of reciprocal basis vectors $\mathbf{a}_i^*$ necessary to index the diffraction pattern with integer indices. In other words, the Fourier spectrum M^* of an aperiodic crystal structure consists of δ-peaks supported by a $\mathbb{Z}$-module:

$$M^* = \left\{ \mathbf{H} = \sum_{i=1}^{n} h_i \mathbf{a}_i^* \middle| h_i \in \mathbb{Z} \right\} \tag{4.2}$$

of rank n ($n > d$) with basis vectors $\mathbf{a}_i^*$, $i = 1, \ldots, n$, and reciprocal space (diffraction) vectors $\mathbf{H}$. The nD embedding space V consists of two orthogonal subspaces both preserving the point group symmetry according to the nD space group

$$V = V^{\parallel} \oplus V^{\perp}, \tag{4.3}$$

with the dD par-space $V^{\parallel} = \mathrm{span}(\mathbf{v}_1, \ldots, \mathbf{v}_d)$ and the $(n-d)$D perp-space $V^{\perp} = \mathrm{span}(\mathbf{v}_{d+1}, \ldots, \mathbf{v}_n)$. If not indicated explicitly, the basis defined by the vectors $\mathbf{v}_i$ (V-basis) will refer to a Cartesian coordinate system. The n-star of rationally independent vectors defining the $\mathbb{Z}$-module M^* can be considered as an appropriate projection $\mathbf{a}_i^* = \pi^{\parallel}\left(\mathbf{d}_i^*\right)$ ($i = 1, \ldots, n$) of the basis vectors $\mathbf{d}_i^*$ (D-basis) of an nD reciprocal lattice Σ^* with

$$M^* = \pi^{\parallel}\left(\Sigma^*\right). \tag{4.4}$$

As a simple example, the relationship between the 1D reciprocal space of the Fibonacci sequence and its 2D embedding space is shown in Fig. 4.6(c). For comparison, embedding of a 1D incommensurately modulated structure (IMS) (Fig. 4.6(a)) and a 1D composite (host/guest) structure (CS) (Fig. 4.6(b)) are shown as well. Additionally, beside the standard way of embedding a quasiperiodic structure (QC-setting), an alternative way, the IMS-setting, is demonstrated (Fig. 4.6(d)). The latter one can be particularly useful for the study of structural phase transitions of QCs. In direct space, the aperiodic crystal structure results from a cut of a periodic nD hypercrystal structure with dD par-space $V^{\parallel}$ (Fig. 4.7). An nD hypercrystal structure corresponds to an nD lattice Σ decorated with nD hyperatoms (atomic surfaces, occupation domains), which correspond to the windows $\mathbf{W}$ in the strip-projection approach. The basis vectors of Σ are obtained via the orthogonality condition of direct and reciprocal space

$$\mathbf{d}_i \cdot \mathbf{d}_j^* = \delta_{ij}. \tag{4.5}$$

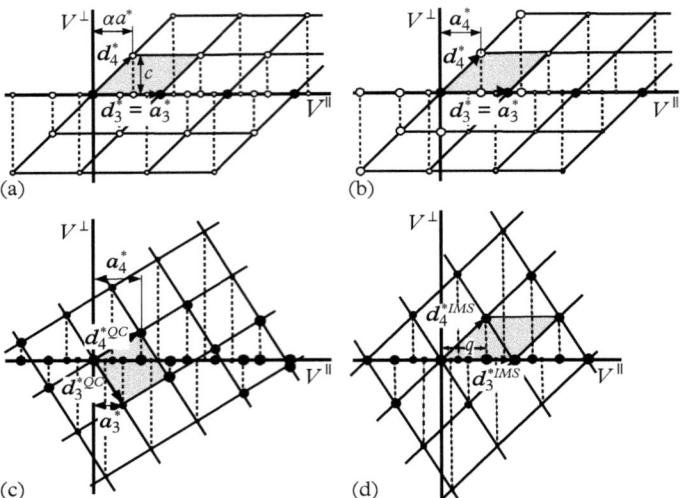

Fig. 4.6 *2D reciprocal space embedding of the aperiodic structures shown in Fig. 4.7. (a) Incommensurately modulated structure (IMS), (b) mutually modulated composite structure (CS), and (c) Fibonacci sequence in the standard QC-setting and in the (d) IMS-setting. The par-space diffraction patterns result from the projection of the 2D reciprocal space along the perp-space. Dashed lines indicate the projection directions, vectors $\mathbf{d}_i^*$ refer to the nD reciprocal basis (D-basis), a^* and a_i^* are the lattice parameters in reciprocal par-space, $q = \alpha a^*$ is the modulus of the wave vector of an incommensurate modulation (after Steurer and Deloudi (2009), Fig. 3.1. With kind permission from Springer Science+Business Media).*

The various types of aperiodic crystal structures differ from one another by the shape of their atomic surfaces AS (occupation domains). QCs show discrete atomic surfaces while those of IMS and CS are essentially continuous. With the amplitudes of the modulation function going to zero, a continuous transition to a periodic structure (basis structure) can be performed in the case of IMSs. CSs consist of two or more substructures, which themselves may be mutually modulated as in our example (Figs. 4.6(b) and 4.7(b)). If the substructures were not modulated, then only reflections of the types $h_1 h_2 h_3 0$ and $000 h_4$ would be present, and none of the general type $h_1 h_2 h_3 h_4$.

In reciprocal space, the characteristics of IMS and CS are the crystallographic point symmetry of their Fourier modules M^* and the existence of large Fourier coefficients on a distinct subset $\Lambda^* \subset M^*$ related to the reciprocal lattice of their periodic average structures (PASs). There is a one-to-one relationship between each atom of an IMS or the FS and its PAS in contrast to what is the case for 2D or 3D QCs with non-crystallographic symmetry.

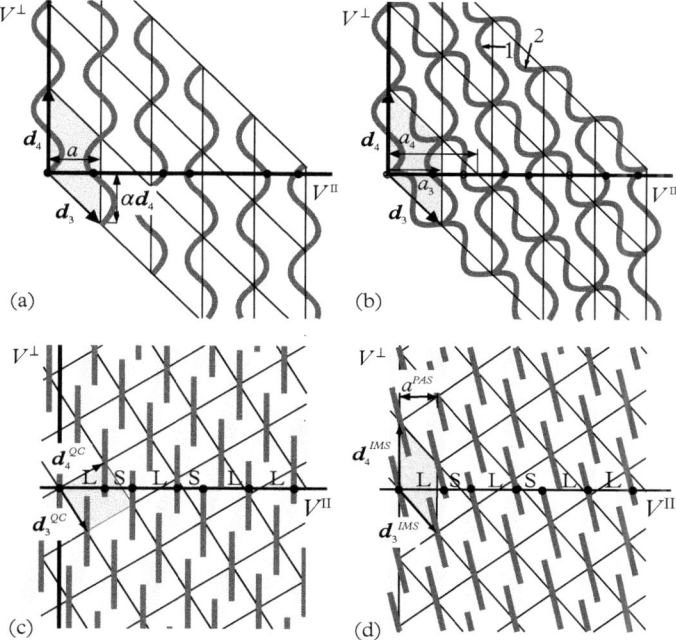

Fig. 4.7 *2D direct-space embedding of the three fundamental types of aperiodic structures: (a) IMS, (b) CS with modulated subsystems (marked 1 and 2), the FS in the (c) QC-setting, and (d) IMS-setting. The atomic surfaces (hyperatoms) are indicated by thick gray lines. The 1D aperiodic crystal structures result at the intersection of atomic surfaces with the par-space. Vectors $\mathbf{d}_i$ mark the nD basis vectors while a and a^{PAS} refer to the lattice parameters of the periodic average structures. L and S denote the long and short unit tiles of the Fibonacci sequence (after Steurer and Deloudi (2009), Fig. 3.2. With kind permission from Springer Science+Business Media).*

In the section method, the relationship of rational approximants of quasicrystals also can be nicely visualized. While a general periodic approximant of a QC just contains structural building units that can also be found in the structure of the QC, a rational (periodic) approximant shows a much closer relationship. Their structure directly results from shearing the nD QC structure properly along perp-space (Fig. 4.8). The term "rational" means that an irrational number such as τ, which is related to five-fold symmetry, is replaced by one of its approximations such as 2/1, 3/2, 5/3, 8/5, . . . , etc.

In the nD description, the hyperatoms decorate the Wyckoff positions in the nD unit cell. By hyperatom, we denote the nD entity consisting of the $(n-d)D$ atomic surface AS (occupation domain) defined in perp-space and the dD atom

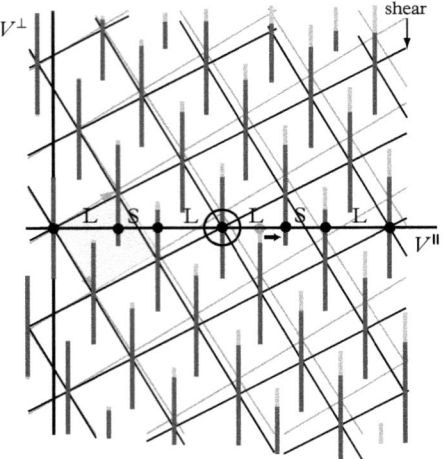

Fig. 4.8 *Embedded Fibonacci chain ...*
LSLSLL... *(semi-opaque in the background)
and its rational (LSL) approximant (black in
the foreground). The encircled lattice node is
shifted to par-space by shearing the 2D lattice
along perp-space. Thereby, one par-space cut
disappears in the drawing and a new one
appears, locally changing* SL *into* LS *(a
so-called phason flip marked by a horizontal
arrow). After Steurer and Deloudi (2009),
Fig. 3.4. With kind permission from Springer
Science+Business Media.*

existing in dD par-space. Generally, Wyckoff positions correspond to equivalence classes of points within a unit cell (modulo lattice translations). The orbit of such symmetrically equivalent points is the set of points generated by the action of the nD space group. All sites belonging to an orbit are occupied by the same type of hyperatom. In the case of the FS, the hyperatom is centered at the origin $(0, 0)$ of the unit cell. All atomic positions generated by its AS belong to the same orbit and are symmetrically equivalent.

In the case of a more complex structure consisting of different types of atoms, either the nD unit cell will be occupied by different hyperatoms or the AS is partitioned in some way, with the different partitions decorated by different atoms. This is shown schematically in Fig. 4.9. The partitions marked a, b, and c mark different vertex coordinations: a: L$|$S, b: L$|$L, c: S$|$L. If we decorate the light gray areas of the AS with A atoms and the dark gray ones with B atoms, we obtain a structure where the short intervals S always correspond to A–A neighbors, while the long ones can have A–B and A–A ones.

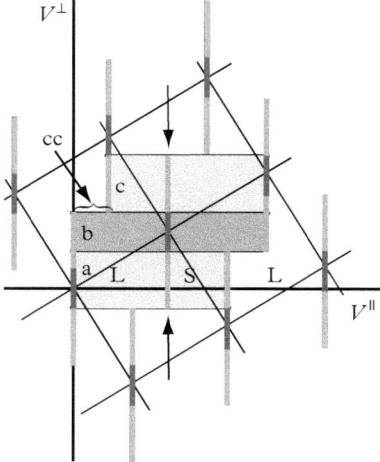

Fig. 4.9 *By projecting all nearest neighbors onto one AS (marked by arrows), it gets partitioned in a way that the different line segments (light and dark gray) correspond to different vertex coordinations: a: L|S, b: L|L, c: S|L. Their frequencies exhibit a ratio of 1:1/τ:1. The nearest neighbors show the closeness condition (cc), i.e., that by shifting the par-space along perp-space, the type of atoms does not change during a phason flip (after Steurer and Deloudi (2009), Fig. 3.6. With kind permission from Springer Science+Business Media).*

In the case of the RPT in the irreducible 4D description, the two unique ASs correspond to one small and one τ-times larger pentagon parallel to perp-space (Fig. 4.10). Fortunately for the visualization of the content of the 4D hyperrhombohedral unit cell, four of its vertices as well as its long body diagonal, with the centers of the AS, lie on the same hyperplane. The small pentagons occupy the Wyckoff position $(1/5, 1/5, 1/5, 1/5)_D$ and the inversion related one $(4/5, 4/5, 4/5, 4/5)_D$; the τ-times larger pentagons are centered on $(2/5, 2/5, 2/5, 2/5)_D$ and on inversion related $(3/5, 3/5, 3/5, 3/5)_D$. The site symmetry is in both cases $5m$, while the symmetry at the origin $(0, 0, 0, 0)_D$ corresponds to $10mm$. If only this special site is occupied by a decagonal AS of proper size, the PPT results. The projections of the 4D unit cell onto par- and perp-space are equal. However, since the AS only have an extension in perp-space, they are only visible in the perp-space projection, yielding points when projected onto par-space.

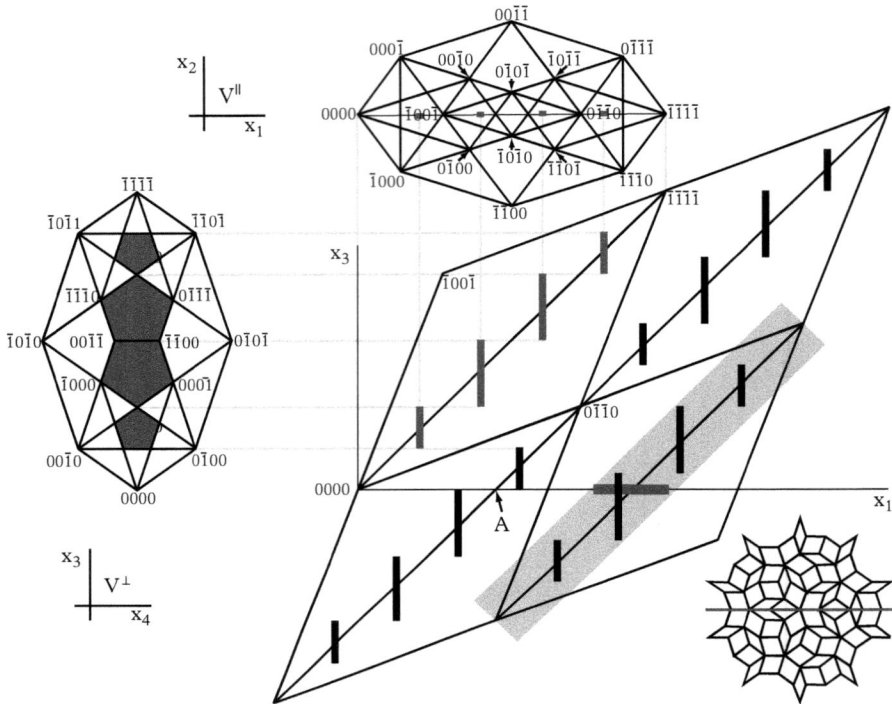

Fig. 4.10 *Characteristic* $(1001)_V$ *section of four 4D unit cells of the PT together with its par-(top) and perp-space (left) projections, respectively. The projected ASs are shown as gray-shaded pentagons. In the section they appear as gray bars. In the lower right corner the direction of the cut is marked on a PT. The shaded area in the lower right unit cell indicates the projection direction leading to the PAS of the PT (modified from Steurer and Deloudi (2009), Fig. 3.30. With kind permission from Springer Science+Business Media).*

The RPT shows eight different vertex coordinations, each one corresponding to a different partition of the AS (Fig. 4.11). Consequently, only those atoms in a quasiperiodic structure are symmetrically equivalent, which not only belong to symmetrically equivalent hyperatoms but also to the same partition on them. The area of the partitions corresponds to the frequency of the respective vertex coordinations. If occupied by different atomic species, the chemical composition (stoichiometry) is also related to these areas. There exists, also, a kind of generalization of the RPT, which consists of the same kind of rhomb unit tiles but with different matching rules and more different vertex coordinations (Pavlovich and Klèman, 1987). Depending on the kind of tiling, not only the shape and size of the then five ASs changes but also their partitioning. Consequently, the task of a QC structure analysis is the determination of the Wyckoff positions of the hyperatoms, their shape, and their partition and chemical decoration.

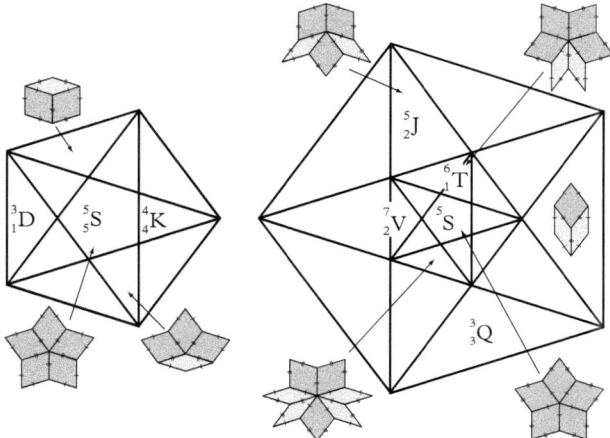

Fig. 4.11 *Partitioning of the AS of the RPT corresponding to the eight different vertex coordination of the RPT. The ASs in $(1/5, 1/5, 1/5, 1/5)_D$ and $(2/5, 2/5, 2/5, 2/5)_D$ are depicted, the other two are related by an inversion center. The arrows at the edges of the unit tiles illustrate the matching rules of the PT (modified from Steurer and Deloudi (2009), Fig. 3.33. With kind permission from Springer Science+Business Media).*

4.6 What is the physics underlying the *n*D approach?

In quasiperiodic structures, the position vectors of the atoms and/or the cluster centers form a $\mathbb{Z}$-module. A $\mathbb{Z}$-module of rank n can be seen as projection of an nD hyperlattice. QCs are 3D objects consisting of atoms interacting in the same way as they do in periodic crystals. Any structure can be described as a vector map of atomic distances. In the special case that only one A–B and one A|A–B|B distance exist (e.g., resulting in a rhomb or pentagon tiling), the vector map corresponds to a restricted $\mathbb{Z}$-module. A hard constraint is the existence of a minimum distance (A–B). Other constraints are chemical composition and packing density, as well as the formation of clusters, netplanes, and PASs.

The nD approach allows an elegant, self-contained description of a quasiperiodic structure based on an nD lattice-periodic hypercrystal structure. One has to keep in mind that the nD "hyperatoms" are by no means physical objects interacting in a way that 3D atoms do. Consequently, their perp-space components are called "atomic surfaces" or "occupation domains", with the meaning of a geometrical selection rule determining which points of a $\mathbb{Z}$-module of rank n are occupied by atoms.

Conversely, reciprocal space images (Fourier transforms) of quasiperiodic structures are purely point diffractive, i.e., they show Bragg reflections only. Reciprocal space vectors pointing to Bragg peaks are perpendicular to the related netplanes in direct space. These sets of netplanes are lattice planes of the nD hyperlattice and their intersections with par-space are the netplanes of one of the periodic average structures (PASs). For each set of Bragg reflections (harmonics) a different PAS and corresponding set of netplanes exists. These netplanes are relevant for the propagation of electrons and phonons analogous to those of periodic crystals. The waves can form interference patterns (i.e., standing waves) with maxima/minima at the netplanes (contrary to amorphous structures).

Thus, it is more likely that a structure is energetically favorable if Bragg reflections exist. Periodicity is a prerequisite of Bragg reflections, be it in 3D or nD. Consequently, quasiperiodic structures are in the intersection of an infinite number of PAS with their sets of netplanes. The "thickness" of the netplanes of the PASs leads to a rapid decrease of the intensities of higher harmonics. The smaller the AS, the simpler the nD unit cell, the better the PAS and the more and stronger higher harmonics (disfavoring random tiling-based structures), the more distinct the band gaps.

Physics is in the short-range order of the atomic arrangements (clusters), their packing and the electron wave interference (band structure) that result from the long-range order. The origin of quasiperiodicity lies in:

- the existence of two incompatibilities: non-crystallographic cluster symmetry vs. periodic packing
- the existence of two different length scales, i.e., that of interatomic distances and of the cluster diameters
- the difference between the local chemical composition of the clusters and the average composition of the intermetallic phase, which can only be achieved by particular kinds and frequencies of cluster overlaps
- a particular, favorable valence-electron concentration resulting in an incommensurate stoichiometry.

The strip/projection approach reflects the way clusters have to be packed in order to achieve an arrangement with minimum deviation of the local chemical composition from the global one. It also shows that this constraint produces the densest packing of clusters (see Steurer and Deloudi (2012) and Steurer (2012)).

Finally, what is the difference between a quasicrystal and its high-order approximants from a physical point of view? Why have only low-order approximants been experimentally observed so far, although a "devil's staircase" of rational approximants of arbitrary order would be theoretically possible? The main reasons may be the following ones: (i) the non-crystallographic symmetry of the QCs is broken in approximants, however high its order may be; (ii) the cohesion energies of a QC and its high approximant are almost the same; (iii) since the chemical compositions of a QC and its approximant differ, also their valence-electron concentrations differs, and the electronic stabilization via pseudo-gaps, for instance, can be maximum just for one composition; (iv) the configurational entropies of both complex intermetallics are different.

5

Statistical description and structural correlations

The following chapter presents a statistical analysis of all structures and structure types of the intermetallic compounds contained in the reduced database *Pearson's Crystal Data* (PCD) (Villars and Cenzual, 2011a). The goal is to identify, analyze, and discuss structural regularities and relationships for improving our understanding of the formation and stability of intermetallics, in order to get a flavor of the big picture. The most important ordering parameters used in our analysis are the Mendeleev numbers, introduced here. Their assignment to the different chemical elements is based on an empirical chemical scale χ, introduced by Pettifor (1984) (Fig. 5.1). Based on the Mendeleev numbers, we will analyze the distributions of chemical compositions, stoichiometries, and symmetries of the binary and ternary intermetallic compounds separately, and discuss the occurrence and characteristics of the most common structure types. Also of interest is the distribution of the number of representatives of the structure types, which varies in a wide range.

5.1 The Mendeleev numbers

For revealing structural relationships and predicting the stability of crystal structures, the atomic numbers are not very helpful as an ordering parameter. Therefore, many other indicators such as atomic radii, valence electron concentrations, electronegativities, Zunger's pseudopotential radii, etc. have been used frequently in quantum structure diagrams and structure stability maps (see, e.g., Daams and Villars (2000), and references therein). Unfortunately, atomic parameters such as radii or electronegativities are not structure-invariant quantities but, to some extent, depend on atomic environment and chemical bonding. An alternative simple phenomenological approach was introduced by Pettifor (1984) with the Mendeleev numbers, M, which proved quite successful in identifying stability fields of binary intermetallic compounds, A_aB_b, in $M(A)/M(B)$ plots (for short: M/M-plots) (Pettifor, 1986; Pettifor and Podloucky, 1986; Pettifor, 1988; Pettifor, 1995).

Intermetallics: Structures, Properties, and Statistics. First Edition. Walter Steurer and Julia Dshemuchadse.
© Walter Steurer and Julia Dshemuchadse 2016. Published in 2016 by Oxford University Press.

																H	He
																103	1
Li											Be	B	C	N	O	F	Ne
12											77	86	95	100	101	102	2
Na											Mg	Al	Si	P	S	Cl	Ar
11											73	80	85	90	94	99	3
K	Ca	Sc	Ti	V	Cr	Mn	Fe	Co	Ni	Cu	Zn	Ga	Ge	As	Se	Br	Kr
10	16	19	51	54	57	60	61	64	67	72	76	81	84	89	93	98	4
Rb	Sr	Y	Zr	Nb	Mo	Tc	Ru	Rh	Pd	Ag	Cd	In	Sn	Sb	Te	I	Xe
9	15	25	49	53	56	59	62	65	69	71	75	79	83	88	92	97	5
Cs	Ba		Hf	Ta	W	Re	Os	Ir	Pt	Au	Hg	Tl	Pb	Bi	Po	At	Rn
8	14		50	52	55	58	63	66	68	70	74	78	82	87	91	96	6
Fr	Ra																
7	13																

La	Ce	Pr	Nd	Pm	Sm	Eu	Gd	Tb	Dy	Ho	Er	Tm	Yb	Lu
33	32	31	30	29	28	18	27	26	24	23	22	21	17	20
Ac	Th	Pa	U	Np	Pu	Am	Cm	Bk	Cf	Es	Fm	Md	No	Lr
48	47	46	45	44	43	42	41	40	39	38	37	36	35	34

Fig. 5.1 *Assignment of the Mendeleev numbers to the chemical elements in the periodic table (Pettifor, 1984). Note the position of Be and Mg in the column together with Zn, Cd, and Hg, as well as the numbering of Yb and Eu, which are placed between Ca and Sc, because they are frequently divalent in contrast to the predominantly trivalent other lanthanoids. Of the 103 elements shown here, 81 are considered metallic and 22 non-metallic (shaded gray) according to our definition.*

The Mendeleev numbers start with the least electronegative element, He, $M(\text{He}) = 1$, $\chi = 0.00$, and end with one of the most electronegative ones, H, $M(\text{H}) = 103$, $\chi = 5.00$, in this way establishing an empirical chemical scale χ, with $0.00 \leq \chi \leq 5.00$. On this *a posteriori* scale, elements from the same homologous group (column) in the periodic table of the chemical elements, which are known to have similar chemical properties, are assigned values close to one another, in contrast to the sequence of atomic numbers, which jumps from row to row. By employing this approach, a good structural separation could be achieved in $M(A)/M(B)$ plots for stoichiometries AB, AB_2, AB_3, AB_4, AB_5, AB_6, AB_{11}, AB_{12}, AB_{13}, A_2B_3, A_2B_5, A_2B_{17}, A_3B_4, A_3B_5, A_3B_7, A_4B_5, and A_6B_{23} (Pettifor and Podloucky, 1985; Pettifor, 1986). The Mendeleev numbers have also proven useful for the identification of stability fields of quasicrystals and their approximants (Ranganathan and Inoue, 2006).

In order to explain this phenomenological observation, a microscopic theory was developed (Pettifor and Podloucky, 1984; Pettifor and Podloucky, 1986) based on the reasoning "that the angular momentum character of the valence orbitals is of prime importance as the relative stability of different building blocks is determined by whether we have sp-sp, p-d, d-d, [. . .] bonding" (Pettifor and Podloucky, 1985).

Originally, the scale was closely related to Pauling's electronegativities (Pauling, 1960), but it was subsequently refined and developed to its final form of integer values M (Pettifor, 1988, 1995), ordering the elements in a crystal-chemical useful way instead of proposing a questionable quantitative scale (Fig. 5.2).

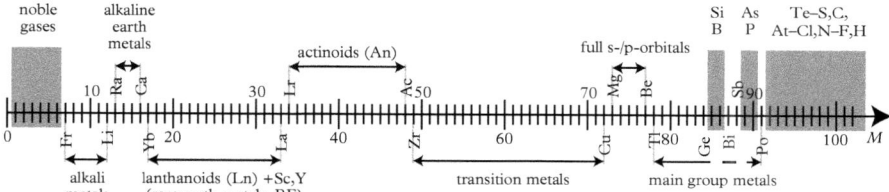

Fig. 5.2 *The arrangement of the chemical elements according to the Mendeleev scale, M. Non-metallic elements (according to our working definition) are shaded gray. Keep in mind that Be and Mg are located in the column together with Zn, Cd, and Hg, as well as the numbering of Yb and Eu, which are placed between Ca and Sc.*

As an example for an M/M-plot, a general map, $M(A)$ vs. $M(B)$, for all binary intermetallics A_aB_b ($a \geq b$) in the database is depicted in Fig. 5.3, with the stability fields of complex intermetallics (CIMs) marked by gray dots. It clearly shows which element combinations do not form binary compounds (in the case they were studied experimentally at all). At first glance, the diagram looks mirror-symmetric with respect to the diagonal; however, usually $A_aB_b \neq A_bB_a$, as we will see later. One has to keep in mind that each dot in the plot just refers to a binary intermetallic system representing an arbitrary number of different binary phases therein (≥ 1).

There are relatively few compounds in the range where both constituents have Mendeleev numbers smaller than $M = 58$. If the majority constituent has $M(A) < 58$ then the minority component mostly has $M(B) > 58$. In contrast, if the majority component has $M(A) > 58$, the minority component can have more or less any Mendeleev number. The small gaps at $M(A)$ or $M(B) = 13, 29$, or 48 just indicate that the database does not contain binary compounds of radioactive Ra, Pm, and Ac. Those at $M(A)$ or $M(B) = 85, 86, 89$, and 90 correspond to compounds with the non-metallic elements Si, B, and As, respectively, and which have been excluded from our analysis. The diagonal $M(A) = M(B)$ corresponds to unary phases (chemical elements), which are not included in this plot.

There is a more extended gap in the range $34 \leq M \leq 38$, indicating that there are no binary compounds of the late actinoids in the database. This does not mean that they could not form, it rather reflects the difficulty to experimentally study binary systems with actinoids as components. The early and intermediate actinoids, with $39 \leq M \leq 48$, however, form quite a few compounds with most element classes except the alkali and alkaline earth metals. Also significant are the white regions with $M(A) = 52–66$ and $M(B) = 71–75, 78$, and 79, i. e., compounds of A = transition metal as majority components with B = Ag, Cu, Mg, Hg, Cd, Tl, and In as minority constituents, as well as the other way around.

The largest, almost contiguous blocks in the diagram refer to systems where one constituent is either an alkali, alkaline earth metal, or a lanthanoid (Ln) element, and the other a late TM or main group (M) element. This can be understood

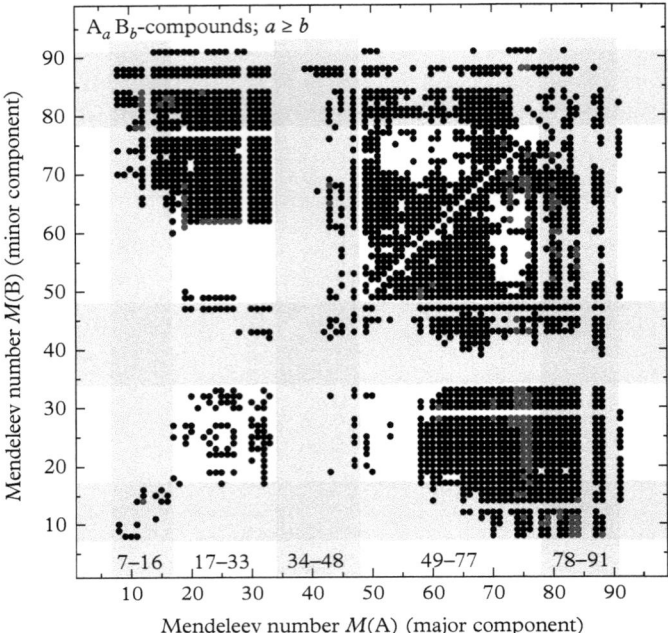

Fig. 5.3 $M(A)/M(B)$ *plot of all 6441 binary intermetallic compounds in the database with compositions* A_aB_b $(a \geq b)$. *Complex intermetallics are highlighted as gray dots.* $M(A)$ *refers to the majority component. In the case of* $a = b$, *the compounds are shown twice, as* A_aB_b *and* B_aA_b. *Areas containing alkali metals (7–12), alkaline earth metals (13–16), actinoids (34–48), and (semi)metallic main group elements (78–91), are shaded light-gray.*

by taking into account the chemical similarity of the rare earth elements. These elements can largely substitute each other in the structures of the binary phases without changing the structure type. This may be the main reason for the large extension along the $M(B)$-axis of the complex intermetallics with $M(A) = 75$ and 76, i. e., the elements Cd and Zn.

Due to their chemical similarity, the binary Ln–Ln phases are all solid solutions crystallizing in unary structure types. In most cases, these are just the structure types of the allotropes of the constituting elements, in some cases they correspond to that of $hR9$-Sm. The few binary An–An compounds marked in the M/M-plot also refer to either solid solutions or to still unidentified compounds with larger unit cells.

Complex intermetallics (CIMs) have been observed in the regions with Mendeleev numbers for the majority component $M(A) = 12$, 15–19, 22–28, 70, 72–76, and 78–84, i.e., the elements *Li*, (Sr, Ca, *Yb, Eu, Sc*), (*Er, Ho, Dy, Y, Tb, Gd, Sm*), *Au*, (Cu, *Mg*, Hg, *Cd*, Zn), (*Tl*, In, Al, Ga, Pb, and *Sn, Ge*). Elements that can occur in CIMs as both majority and minority components are written here in italics. The distribution of minority elements differs significantly with $M(B) = 8–12$, 14, 15, 17–28, 30–33, 43, 45, 49–52, 54, 56, 57, 60–70, 73, 75, 78, 82–84, and 88, i.e., the elements (Cs, Rb, K, Na, *Li*), Ba, *Sr*, (*Yb, Eu, Sc*), (Lu, Tm, *Er, Ho, Dy, Y, Tb, Gd, Sm*), (Nd, Pr, Ce, La), Pu, U, (Zr, Hf, Ti, Ta), V, Mo, Cr, (Mn, Fe, Ru, Os, Co, Rh, Ir, Ni, Pt, Pd, *Au*), Mg, Cd, Tl, (Pb, *Sn, Ge*), and Sb.

One has to keep in mind, however, that not all binary intermetallic phase diagrams have been reliably explored so far, so that quite a few more data points may have to be added in the future. This refers, in particular, to phase diagrams including not only temperature but also pressure as a parameter. In the case of the chemical elements, for instance, there are much more high-pressure allotropes known than modifications that are stable at ambient pressure. It is not clear, however, how Pettifor's chemical scale would have to be modified for high-pressure phases since not only atomic radii but also electronegativities can significantly vary with pressure.

5.2   Mining the database Pearson's Crystal Data (PCD)

We extracted a complete dataset of all fully described structures of intermetallics from the database *Pearson's Crystal Data* (PCD) (Villars and Cenzual, 2011*a*). Of the 227 145 entries for structures of inorganic compounds, 47 192 ($\approx 20.7\%$) can be regarded as being structures of intermetallics (for our definition of metallic elements see Fig. 5.1) and in turn 46 071 of these are completely described, i.e., with all atomic coordinates (Dshemuchadse and Steurer, 2015). We filtered this dataset once more by excluding duplicates and most of the entries just representing solid solutions. For this purpose, we correlated the items for the intermetallic system and the structure type: if these two features were identical for a number of entries, only one of them was kept in the dataset. This reduced the number of structures of intermetallics to 20 829, belonging to 2166 structure types, in total. This means that there are on average ≈ 9.6 intermetallic phases per structure type (Table 5.1). Among them, solid solutions will still be present to some extent. This is due to the fact that, for example, in the case of the solution of Al in $cF4$-Cu, two entries will be found: both with structure type $cF4$-Cu, but one in the binary system Al–Cu, and the other one in the "unary system" Cu.

It is important to understand how in the PCD crystal structure types are assigned to crystal structures. Thereby, the so-called "entry prototype" is assigned to a crystal structure by comparison of the space groups, cell parameter ratios, and occupied Wyckoff positions, including their atomic coordinates (Villars and Cenzual, 2011*a*). This is usually a sufficient definition of a structure prototype, but gets more complicated if different numbers of components are involved. For instance, the $cF4$-Cu structure type has been assigned to some binary compounds A_aB_b. Consequently, the structure must be inherently disordered, since the structure type features only one atomic position ($4a\ 0, 0, 0$ in space group 225 $Fm\bar{3}m$). Such a binary phase is a solid solution, therefore, and should be termed phase

Table 5.1 *Classification of intermetallic compounds in the PCD with respect to the number of their constituting elements (N_c). The numbers for quaternary and higher systems are not representative at all since these systems have not been studied to a similar extent as the binary and ternary ones. Duplicates and – for the most part – solid solutions were excluded from this dataset. Given are the numbers of theoretically possible systems (N_{sys}^{theo}), of the experimentally studied ones listed in the ASM Alloy Phase Diagram Database (N_{sys}^{ASM}) (of the 6499 ternary systems in the database, 4438 also contain non-metallic elements and were omitted (Fleming, 2014)) and of those listed in the PCD (N_{sys}^{PCD}), as well as the number of different structures (N_s) and structure types (N_{st}) in the PCD. The number of structure types (N_{st}), 2611, obtained by summarizing the entries in the table, contains double counts, the true number in the PCD being 2166. The number N_c of constituents of an intermetallic compound is not necessarily equal to the number of constituents of the respective structure type; it can be higher.*

N_c	N_{sys}^{theo}	N_{sys}^{ASM}	N_{sys}^{PCD}	N_s	N_{st}	N_s/N_{st}	N_s/N_{sys}
1	$\binom{81}{1} = 81$		75	277	86	3.2	3.42
2	$\binom{81}{2} = 3240$		1401	6441	943	6.8	1.99
3	$\binom{81}{3} = 85\,320$	2061	5109	13\,026	1391	9.4	0.15
4	$\binom{81}{4} = 1\,663\,740$			973	212	4.6	
5	$\binom{81}{5}$			65	22	3.0	
6	$\binom{81}{6}$			24	10	2.4	
7	$\binom{81}{7}$			13	4	3.3	
8	$\binom{81}{8}$			8	2	4.0	
9	$\binom{81}{9}$			2	1	2.0	
Total				20\,829	2166 (2611)	9.6	

rather than compound. We consequently marked the respective structure types in the following tables (with flag "s") and excluded them altogether in the treatment of binary intermetallics in Section 5.4.

Matters are more complicated for the analogous discussion of those compounds among the ternaries, which are described in binary structure types. Here, two cases are possible. In analogy to the unary-binary relationship, a ternary structure could be assigned to the $cF24$-MgCu$_2$ structure type, for instance. In this structure type, only two symmetrically independent atomic sites are occupied (227 $Fd\bar{3}m$, Mg in $8a\,0,0,0$ and Cu in $16d\,5/8,5/8,5/8$), which would lead to intrinsic disorder if three elements were to be distributed on these two sites. In this case, the marker refers to a supposed solid solution again (flag "s"). Another case can be illustrated based on the $hP24$-$MgNi_2$ structure type, for instance. It features five symmetrically independent atomic positions (194 $P6_3/mmc$, Mg1 in $4e\,0, 0, z$, Mg2 in $4f\,1/3, 2/3, z$, Ni1 in $4f\,1/3, 2/3, z$, Ni2 in $6g\,1/2, 0, 0$, and Ni3 in $6h\,x, 2x, 1/4$), which theoretically allows three (or up to five) elements to occupy separate sites in an ordered way. However, strictly speaking, the resulting structure would represent a substitutional derivative structure type, due to the different atomic decoration of the sites. As a result, these cases are marked as derivatives of the structure type (flag "d"). It should be noted, however, that all these structures – similar to most structures in the database – can show substitutional disorder, which cannot be extracted from the database in an automated and consistent way.

One main problem for the reliability of the assignment of structure types are older sources that are based just on the visual inspection and comparison of Debye-Scherrer X-ray diffraction (XRD) patterns. Superstructures can easily be overlooked in this case, in particular if elements are substituted, which differ by only a few atomic numbers, which leads to a weak scattering contrast for X-rays. This can even pose a problem for state-of-the-art single-crystal XRD structure analyses. Another problem is that the given stoichiometry and that of the assigned structure type often significantly differ. This is so because, particularly in older studies, frequently just the nominal composition of a sample is given and not that of the single crystal taken from it. Another reason may be a very broad compositional stability range of the structure type.

Caveat: Although we did our best to identify a unique subset of intermetallic phases in the PCD, it will still contain some unreliable entries due to the above mentioned problems. Consequently, the number of intermetallic compounds and their assignment to structure types will show some uncertainties. We think, however, that the number of structure types listed in our reduced subset of the PCD is less questionable, and that the trend in the data would not change once unreliable data sets are replaced by reliable ones.

The 20 829 entries of crystal structures in our reduced PCD include those of unary (pure chemical elements) and multinary compounds as listed in Table 5.1 (Dshemuchadse and Steurer, 2015). The number of structure types, 2611, resulting in this table by just summing up the entries is larger than the actual number, 2166, given above, because one and the same structure type can refer to compounds with different numbers of constituents (solid solutions and derivative structure types included). This is also the reason for the fact that the average number of representatives per structure type for all intermetallics is, with 9.6, larger than any such number for the individual n-constituents systems. Of the 2166 (1087 unique) structure types there are 80 (46) unary, 902 (436) binary, 1095 (547) ternary, 87 (56) quaternary, and 2 (2) quinary compounds.

While the number of structures for unary systems (allotropes of the chemical elements) is, with 277, 3.4 times larger than the number 81 of metallic elements, and almost two times larger than the number of binary systems in the case of binary intermetallics, it is, with a factor of 0.15, much smaller than the number of ternary systems in the case of ternary intermetallic compounds. The gap between the number of intermetallic systems and the number of known compounds drastically widens with the increasing number N_c of constituents. However, ternary and higher systems have only been studied to a very small extent, probably not reflecting the true trend.

We have to take into account that, according to the PCD, only for 1401 out of the theoretically possible 3240 binary intermetallic systems has at least one binary compound been observed so far. In the case of ternary systems, this amounts to 5109 out of 85 320. Consequently, for each binary and ternary intermetallic system with at least one intermetallic compound, there are on average 4.7 and 2.6 representatives, respectively. These numbers do not differ much from the factor 3 for the element structures.

For 4041 out of the 5109 ternary phases included in the PCD, at least one binary phase has been reported for each of the three binary subsystems. For 1053 systems, one binary subsystem did not exhibit any binary phases, while in the 15 remaining ternary systems two of their three binary subsystems are not exhibiting any intermetallic phases. These are the systems Al–Cs–Tl, Bi–Fe–Zn, Bi–Li–V, Ca–Co–Pb, Ca–Cr–Pb, Ca–Pb–Ru, Cr–La–Pb, Cu–Ta–V, Ge–Np–Tc, Ge–Tc–U, Hf–V–Y, K–Tc–Tl, La–Mn–Pb, Mn–Rh–Tl, and Mn–Sn–W. None of

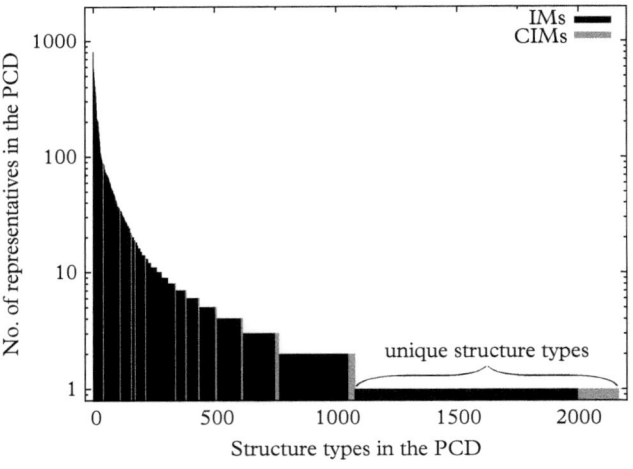

Fig. 5.4 *Distribution of the 20 829 intermetallics on the 2166 different structure types. Shown is the number of representatives (on a logarithmic scale) vs. the rank of the structure type, which is determined by the number of representatives. The fraction of structure types of complex intermetallics (CIMs) is highlighted in gray. The total number of unique structure types is 1087.*

the 5109 ternary systems exhibiting at least one ternary compound only featured elements that did not form any binary phases in their binary subsystems.

The number of representatives per structure type is illustrated in Fig. 5.4. The average numbers increase from 3.2 for unary structure types to 6.8 for binary and 9.4 for ternary ones (Table 5.1). The distribution of the actual numbers of representatives per structure type, however, spans a much wider range between ≈ 800 and 1. Consequently, quite a few structure types are substitutionally very flexible, being able to accommodate many different elements in their structures, while approximately half of all structure types are not flexible at all (unique structure types). Remarkably, 1087 structure types out of the 2166 ones known so far have just one representative, i.e., they are unique. The percentage of complex intermetallics in each bin of the histogram (equal to the number of representatives per structure type) is approximately constant, except in the case of unique structure types. There, the fraction of complex intermetallics is—with approximately 16%—significantly higher.

5.3   Stability maps and composition diagrams of intermetallics

As we will show in the following, the frequency distributions of the structure types of intermetallics and their representatives are very heterogeneous, and can vary significantly. The distribution of CIMs is similar to the general one. Somehow surprising is the prevalence of the rather low-symmetric monoclinic and orthorhombic structure types. Amazingly, the number of intermetallics with structures and structure types with an even number of atoms per primitive unit cell seems to be significantly larger than those where the number is odd.

In the following, we will discuss first the symmetry and size distributions of structure types, respectively, then some special cases of structure types such as with very low or very high symmetry or with just one atom per primitive unit cell. Finally, we will look into the most frequent stoichiometries of intermetallics and their chemical compositions (M/M-plots).

5.3.1   Symmetry and size distributions, fractions of unique structure types

As shown in Fig. 5.4, approximately half of all 2166 structure types (1087) are represented by just a single intermetallic compound each, which consequently makes them unique. Partly, this can be explained by the lack of sufficient reliable data, i.e., the fact that not all phase diagrams, especially the ternary ones, have been studied thoroughly or at all so far; and only a few of them have been explored as a function of both temperature and pressure. However, this does not necessarily bias the general trend too much. Therefore, the questions arise-how can the existence of such a large fraction of unique structure types be understood? What is so special about them that their structures cannot be realized in more than a single intermetallic phase, and only for one particular chemical composition? Why are they so compositionally inflexible in contrast to the Laves or Heusler phases, for instance, with each of them showing hundreds of representatives?

For a given stoichiometry, the possibility of the formation of an intermetallic compound and its stability range primarily depend on the complex interplay of atomic size ratios, electronegativity differences, valence electron concentrations and the kind of chemical bonding emerging therefrom. The symmetry of the resulting structure is easily accessible experimentally, and it is indicative of directional bonding and structural distortions of other origin such as the Peierl's distortion. Consequently, we will first explore the occurrence and distribution of unique structure types based on symmetry (Table 5.2, Figs. 5.5 and 5.6) and composition statistics (Figs. 5.9 and 5.24), respectively. This may give some hints as to why some structure types are less compositionally flexible than others.

Table 5.2 *Distribution over the 14 Bravais type lattices of all structure types and their representatives contained in the PCD. In addition to the number of representatives and structure types, also their ratio is given.*

Bravais lattice	No. of representatives	No. of structure types	Av. no. of representatives per structure type
aP	62	31	2.0
mP	231	108	2.1
mS	448	171	2.6
oP	1849	295	6.3
oS	1440	247	5.8
oI	899	99	9.1
oF	79	33	2.4
tP	1434	174	8.2
tI	2165	183	11.8
hP	5154	401	12.9
hR	940	141	6.7
cP	1713	91	18.8
cI	972	81	12.0
cF	3443	111	31.0
All	20 829	2166	9.6

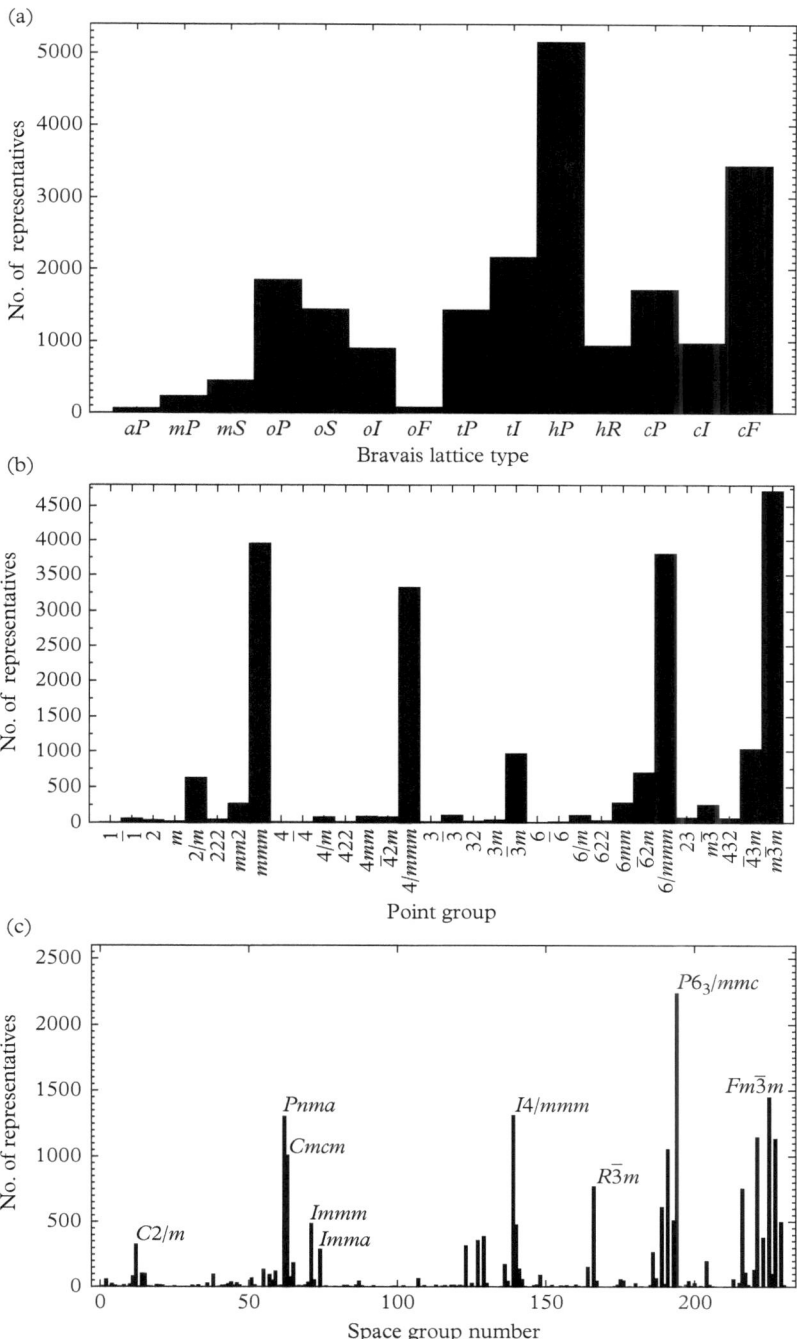

Fig. 5.5 *Symmetry of the structures of all the 20 829 intermetallic compounds listed in the PCD: distribution over (a) the 14 Bravais type lattices, (b) the 32 crystallographic point groups, and (c) the 230 space groups. The space groups are ordered by their numbers, where the most frequent ones are additionally labeled with their Hermann-Mauguin symbols.*

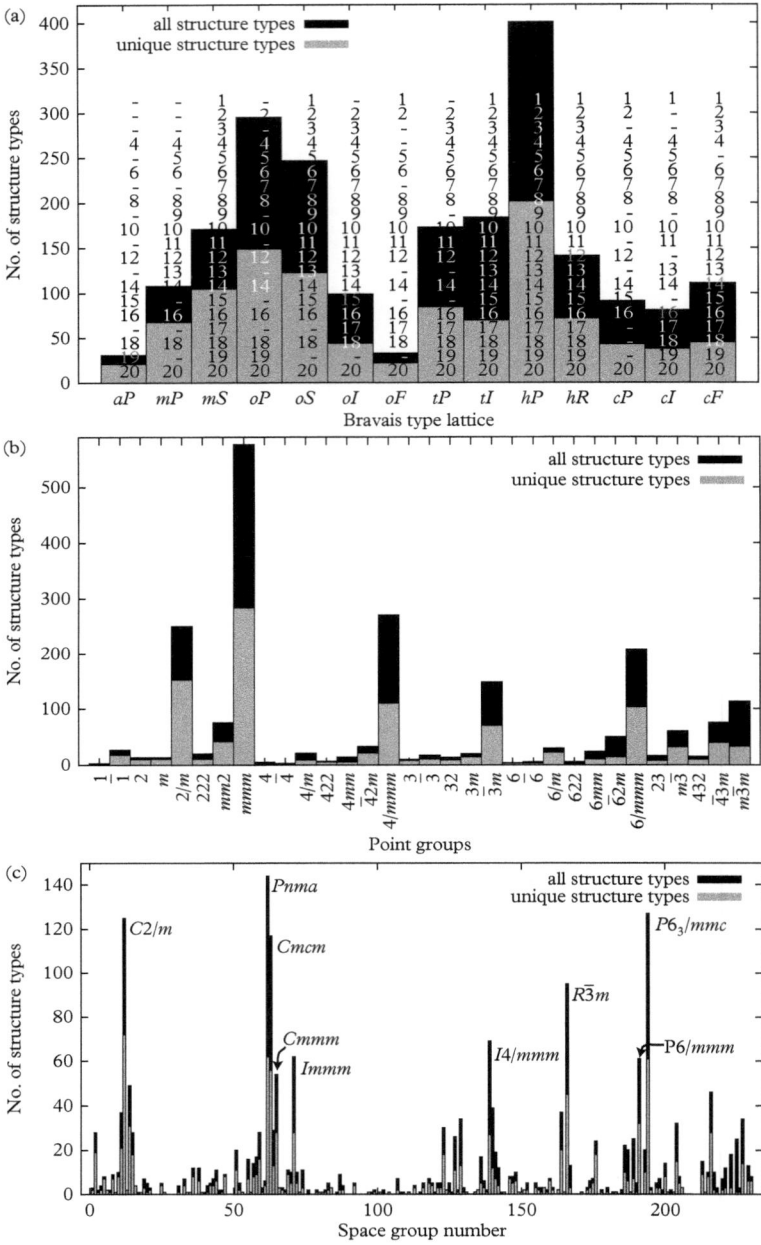

Fig. 5.6 *Symmetry of the intermetallic structure types listed in the PCD: distribution over (a) the 14 Bravais type lattices, (b) the 32 crystallographic point groups, and (c) the 230 space group types. The histograms for all the 2166 structure types are shown in black, while those for the subset of 1087 unique structure types, with only one representative each, are overlaid in gray. The space groups are ordered by their numbers, where the most frequent ones – with more than 50 representatives – are labeled with their Hermann-Mauguin symbols, additionally. In (a), the actually observed number of atoms per primitive unit cell ranging from 1 to 20 is listed for each Bravais type lattice type.*

The distribution of the structure types and, to a lesser amount, of their representatives over the 14 Bravais type lattices is close to bimodal, peaking around primitive and one-face-centered orthorhombic Bravais type lattices, *oP* and *oS*, on one hand, and at hexagonal primitive lattices, *hP*, on the other hand (Table 5.2 and Fig. 5.6(a)). Remarkable is the very small number of intermetallics with structures with Bravais type lattice type *oF*, while this is expected for the lowest Bravais type lattice symmetry *aP*. The structures of only 5008 out of the 20 829 compounds (24.0%) have the low symmetries *a, m, o*. This matches our intuitive assumption of intermetallics being generally "highly-symmetric" due to the isotropy of the (idealized) metallic bonding. However, not even all high-symmetry structures, let alone the low-symmetric ones, can be regarded as being "simple", as will be detailed below. In contrast, the number of low-symmetry (*a, m, o*) structure types is with 984 (45.4%) similar to that of high-symmetry (*t, h, c*) structure types, 1184 (54.6%). This means that the low-symmetry structure types have much less representatives than the high-symmetry ones, i.e., they are less flexible with respect to substitution on particular atomic sites. Indeed, the fraction of unique structure types is significantly higher for the Bravais type lattice types *aP, mP, mS*, and *oF* than for the other ones. The numbers overlaid on the Bravais type lattice distribution histogram in Fig. 5.6(a) refer to the actually observed number of atoms per primitive unit cell in the range 1 and 20. Only in the cases of *tI, hP*, and *hR*, have all numbers between 1 and 20 been observed so far.

For the distribution of the structure types and their representatives over the 32 crystallographic point groups, one finds that the unique structure types accumulate in those point groups that have only a few structure types anyway (Fig. 5.6(b)). A similar trend can be found for the distribution over the 230 space groups (Fig. 5.6(c)). The most frequent point groups are the holohedral ones, in particular 2/*m*, *mmm* (remarkably frequent), 4/*mmm*, and 6/*mmm*. In comparison with the large fraction of intermetallics in cubic point groups, the fraction of their structure types in cubic point groups is remarkably small, if we take into account that we are discussing intermetallic phases, with – in the ideal case – isotropic metallic bonds. The distribution clearly shows that in most cases the chemical bonding will not be isotropic but will contain significant covalent and/or ionic contributions or distortions of electronic origin.

There is also a clear preference for a few lower-symmetry centrosymmetric space groups such as 12 *C2/m*, 62 *Pnma*, 63 *Cmcm*, 65 *Cmmm*, and 71 *Immm*. The prevalence of mirror and glide planes in the structures can be seen as a kind of structure-inherent local twinning, which can help minimizing local strains. The maximum site symmetries are, with the order $k = 2$, lowest for 62 *Pnma* ($\bar{1}$, *m*), then follow with $k = 4$, 12 *C2/m* (2/*m*), and 63 *Cmcm* (2/*m*, *mm*2), with $k = 8$, 65 *Cmmm* (*mmm*) and 71 *Immm* (*mmm*), with $k = 12$, 166 *R$\bar{3}$m* ($\bar{3}$*m*), and 194 *P6₃/mmc* ($\bar{3}$*m*, $\bar{6}$2*m*), with $k = 16$, 139 *I4/mmm* (4/*mmm*), and, with $k = 24$, 191, *P6/mmm* (6/*mmm*). Although low-symmetric AETs (e.g., with site symmetry 1) can be accommodated in any of the 230 space groups, the arrangement of these AETs will be least constrained if the maximum site symmetry is low.

The most common space groups (representing more than 100 structure types or more than 500 compounds) are 12 $C2/m$, 62 $Pnma$, 63 $Cmcm$, 139 $I4/mmm$, 166 $R\bar{3}m$, 189 $P\bar{6}2m$, 191 $P6/mmm$, 193 $P6_3/mcm$, 194 $P6_3/mmc$, 216 $F\bar{4}3m$, 221 $Pm\bar{3}m$, 225 $Fm\bar{3}m$, and 227 $Fd\bar{3}m$. Among them, only the two space groups 189 $P\bar{6}2m$ and 216 $F\bar{4}3m$ are non-centrosymmetric, which may be of interest in the search for materials with specific physical properties requiring non-centrosymmetry. Furthermore, 8 of the 13 space groups are symmorphic (contain only point group symmetry operators besides the translational ones creating the lattice) and only five non-symmorphic. In general, symmorphic space groups can have the lowest multiplicities of special Wyckoff positions of one (normalized to a primitive unit cell, i.e., the multiplicity of the equipoint position divided by the multiplicity of the type of lattice centering), i.e., an odd number. In contrast, the five non-symmorphic ones, 62 $Pnma$, 63 $Cmcm$, 193 $P6_3/mcm$, 194 $P6_3/mmc$, and 227 $Fd\bar{3}m$ have minimum multiplicities of 4, 2, 2, 2, and 2, respectively-all even numbers.

In Fig. 5.7, the frequency of intermetallics as a function of the number of atoms per primitive unit cell is given. The black histogram bars indicate the number of structure types (2166 in total) in each bin (bin size 4, i.e., from 1 to 4 atoms per unit cell, from 5-8, etc.), while the gray ones mark the fraction of unique structure types (1087 in total) among them. The fraction of unique structure types in each histogram bar is roughly the same as the overall average ($\approx$ 50%). Histogram bars related to unique or non-unique structure types only are rare. The number of structure types as a function of unit cell size rapidly decays from the peak value of more than 200 in the second bin (5 to 8 atoms per primitive unit cell), and gets one

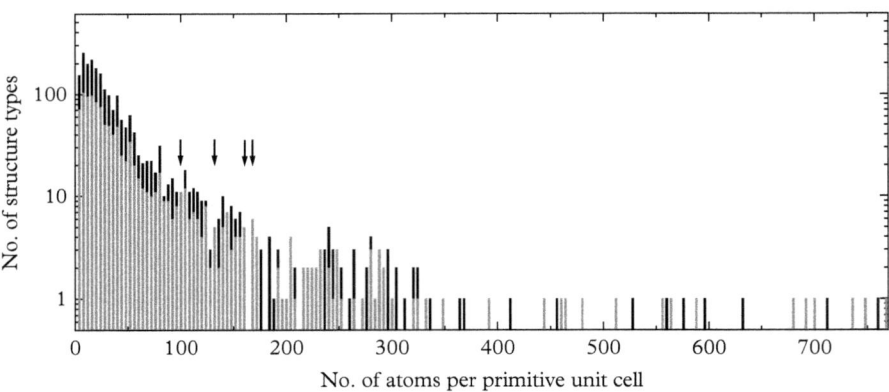

Fig. 5.7 *Frequency of intermetallics as a function of the number of atoms per primitive unit cell. The black histogram bars indicate the number of structure types (2166 in total) in each bin (of size 4), while the gray ones mark the fraction of unique structure types (1087 in total) thereof. Histogram bars related to unique structure types only are rare. The four of them representing more than five structure types are marked by arrows. Note the logarithmic y-axis.*

order of magnitude smaller for unit cell sizes above 100 atoms per primitive unit cell, and one more magnitude smaller beyond 300 atoms per primitive unit cell.

It is worthwhile to have a closer look at this distribution. In Fig. 5.8, only the range up to 100 atoms per primitive unit cell (bin size 1) is shown. With bin size 1, we do not average over odd and even numbers of atoms per primitive unit cell, which uncovers new features. One clearly sees that structure types and their representatives show both significantly higher frequencies in the case of unit cells containing an even number of atoms compared to those with an odd number of atoms per unit cell. Keep in mind that we always refer here to primitive unit cells. This means that $cF4$-Cu would be represented by its rhombohedral subcell containing just one atom. Consequently, there is no bias from lattice centering.

An odd number of atoms in a unit cell can only result from Wyckoff positions with odd multiplicities, if we assume that all atomic sites are fully occupied. Such odd equipoint positions only exist for special sites in the 73 symmorphic space groups. All non-symmorphic space groups, except the six containing 3_1 or 3_2 screw axes, and the four with 6_2 or 6_4 screw axes, have only equipoint positions with even multiplicities, i.e., 147 out of 157. For an equal distribution of structure types on the 230 space groups (which is not at all the case), the ratio of their frequencies for odd to even primitive unit cell contents would be $83/147 \approx 0.56$, if we assume that all structures with symmorphic space groups have an odd number of atoms per unit cell, which is not the case, of course. If we compare this number with that resulting for the 20 most frequent structure types (Table 5.3), then we find a very good agreement with a ratio of $2693/4718 \approx 0.57$. Of the space groups of the 20 most frequent structure types, 12 are symmorphic and 8 non-symmorphic. Of the 12 structures with symmorphic space groups only 8 have an odd number of atoms per primitive unit cell.

If we compare the total number of intermetallics with structures with an odd number of atoms per primitive unit cell with that with an even number of atoms per primitive unit cell, then we get an even more distinct ratio, $5532/15297 \approx 0.36$. For the respective ratio of the structure types we obtain $401/1765 \approx 0.23$. This means that, in the case of structures with symmorphic space groups if any, there must be an even number of odd-numbered Wyckoff positions occupied, in order to show an even number of atoms per unit cell. Examples are $cP4$-Cu$_3$Au and $cP2$-CsCl crystallizing in the space group $Pm\bar{3}m$ with occupied Wyckoff positions $1a$ & $3c$ and $1a$ & $1b$, respectively (Table 5.3).

So why does it make such a difference for the frequency of structure types and their representatives whether we have an even or an odd number of atoms? Why does an intermetallic compound prefer forming a structure with an even number of atoms per unit cell? Why would it be so difficult to accommodate one atom more or less in a structure? Do non-symmorphic space groups, centered Bravais type lattices, or centrosymmetric structure types allow better local strain compensation, a higher degree of chemical homogeneity? This has to remain an open question for now.

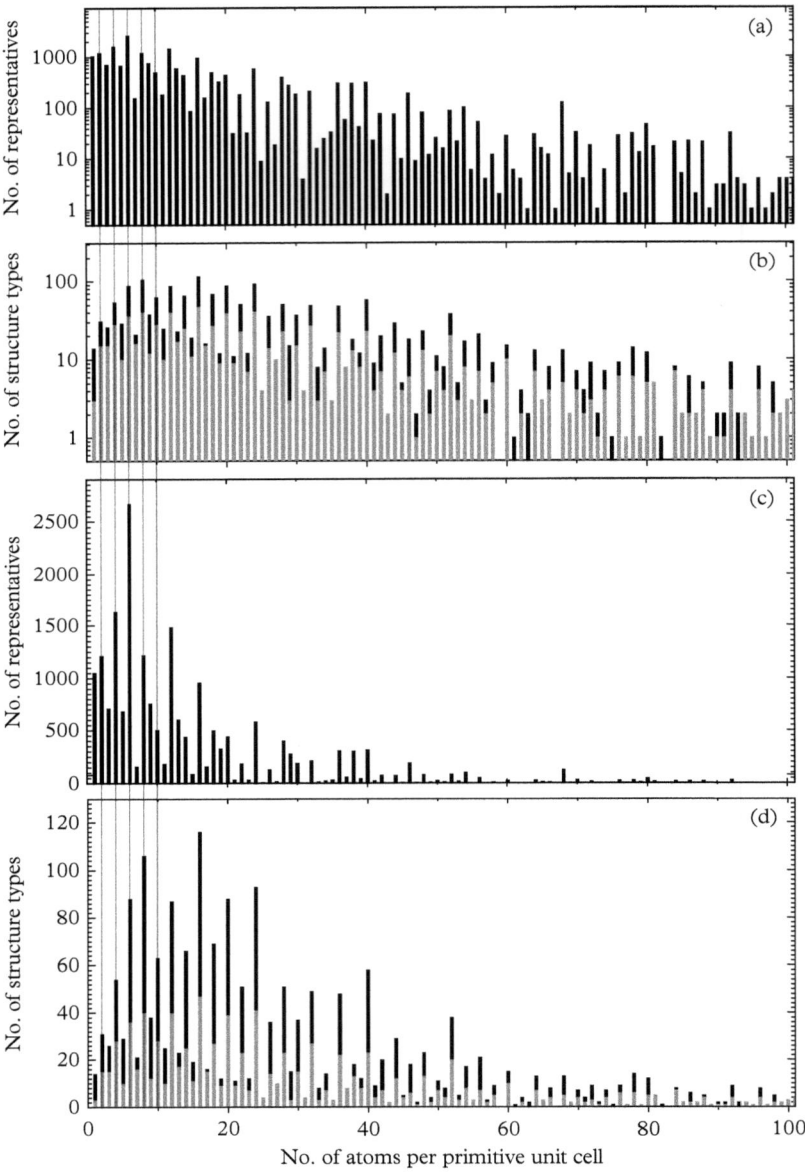

Fig. 5.8 *Frequency of intermetallics as a function of the number of atoms per primitive unit cell (bin size 1). The black histogram bars indicate the number of representatives (a,c) and structure types (b,d), respectively, in each bin, while the gray ones in (b,d) mark the fraction of unique structure types among them. As a guide for the eyes, the histogram bars related to 2, 4, 6, 8, and 10 atoms per primitive unit cell are indicated by thin gray lines connecting the subfigures. Note the logarithmic scale on the vertical axes in (a) and (b).*

Table 5.3 *The 20 most common structure types among the 2166 structure types of intermetallics in the PCD (Dshemuchadse and Steurer, 2015). Each one represents at least 200 compounds and, therewith, 1.0% of all intermetallics, altogether 7411 compounds (35.6%).*

Rank	Structure type	Space group	Wyckoff positions	No. of representatives	% of all representatives
1.	$cF24$-MgCu$_2$	227 $Fd\bar{3}m$	$8a\,16d$	806	3.9%
2.	$cF4$-Cu	225$Fm\bar{3}m$	$4a$	581	2.8%
3.	$cP4$-Cu$_3$Au	221 $Pm\bar{3}m$	$1a\,3c$	544	2.6%
4.	$cP2$-CsCl	221 $Pm\bar{3}m$	$1ab$	512	2.5%
5.	$hP9$-ZrNiAl	189 $P\bar{6}2m$	$1a\,2d\,3fg$	490	2.4%
6.	$hP12$-MgZn$_2$	194 $P6_3/mmc$	$2a\,4f\,6h$	456	2.2%
7.	$cF16$-Cu$_2$MnAl	225 $Fm\bar{3}m$	$4abcd$	414	2.0%
8.	$oP12$-TiNiSi	62 $Pnma$	$4c^3$	403	1.9%
9.	$cI2$-W	229 $Im\bar{3}m$	$2a$	375	1.8%
10.	$hP6$-CaCu$_5$	191 $P6/mmm$	$1a\,2c\,3g$	366	1.8%
11.	$tI10$-CeAl$_2$Ga$_2$	139 $I4/mmm$	$2a\,4de$	338	1.6%
12.	$hP2$-Mg	194 $P6_3/mmc$	$2c$	338	1.6%
13.	$tI26$-ThMn$_{12}$	139 $I4/mmm$	$2a\,8fij$	294	1.4%
14.	$oI12$-KHg$_2$	74 $Imma$	$4e\,8i$	252	1.2%
15.	$hP6$-CaIn$_2$	194 $P6_3/mmc$	$2b\,4f$	215	1.0%
16.	$hP16$-Mn$_5$Si$_3$	193 $P6_3/mcm$	$4d\,6g^2$	209	1.0%
17.	$hR57$-Zn$_{17}$Th$_2$	166 $R\bar{3}m$	$6c^2\,9d\,18fh$	209	1.0%
18.	$cF12$-MgAgAs	216 $F\bar{4}3m$	$4abc$	206	1.0%
19.	$tP10$-Mo$_2$FeB$_2$	127 $P4/mbm$	$2a\,4gh$	203	1.0%
20.	$hP3$-AlB$_2$	191 $P6/mmm$	$1a\,2d$	200	1.0%
			Total	7411	35.6%

Then, with regard to the number of atoms per primitive unit cell, most flexible Bravais type lattice types appear to be mS, tI, hP, and hR (see Fig. 5.6(a)). For the triclinic Bravais type lattice, aP, the trend is strongest to an even number of atoms per unit cell (at least for the range from one to twenty atoms per unit cell). This may just be due to the fact that only centrosymmetric structures (space group 2 $P\bar{1}$) have been observed so far for these low-symmetry compounds. Of course, centrosymmetric structures can also have an odd number of atoms per primitive unit cell if an odd number of atoms occupy a special Wyckoff position on an inversion center. For instance, in space group 2 $P\bar{1}$, these are the eight Wyckoff positions $1a$-$1h$ with site symmetry $\bar{1}$.

Is there any characteristic trend in the distribution of the unique structure types with regard to the chemical composition? Let's have a look at that on the example of binary intermetallics. Although the number of binary intermetallics with unique structure types, 523, is even slightly larger than that with non-unique ones, 420, their distribution in Fig. 5.9 appears to be much less dense. This is clear, however, because the 420 non-unique structure types have 5918 representatives distributed over much more binary systems than the 523 representatives of the 523 unique structure types. This leads to an approximately 5918/523 = 11.3 times larger number of dots, which, however, overlap with one another to a large extent (representing several different intermetallics in one and the same binary system).

The M/M-plot of the 6441 binary intermetallics just indicates that for Ln–Ln, An–An, and Ln–An compounds no unique structure types are known so

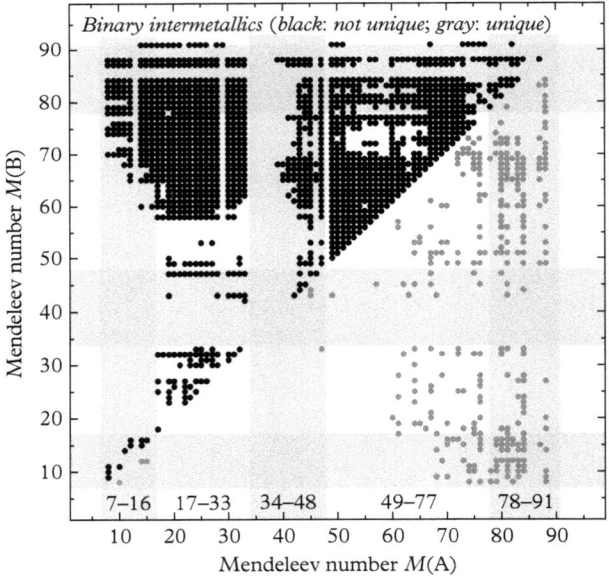

Fig. 5.9 *Chemical compositions of the 6441 binary intermetallics, which crystallize in 943 structure types, 523 (55.5%) of them unique ones. These are shown in gray with ($M(A) > M(B)$), while the distribution of the remaining 5918 intermetallics with non-unique structure types is shown in black (with $M(A) < M(B)$). Areas containing alkali metals, alkaline earth metals, actinoids, and (semi)metallic main group elements are shaded gray.*

far, except for those in the systems U–Np and U–Pu (Fig. 5.9). This can be explained largely by the similar properties of lanthanoids and actinoids, respectively, which just form solid solutions instead of compounds. The densest distribution of unique structure types can be found for compounds of alkali and alkaline earth metals $(8 \leq M(\mathrm{B}) \leq 18)$ with main group elements $(74 \leq M(\mathrm{A}) \leq 84$ and 87), i.e., mostly Zintl phases, as well as for compounds of late transition elements $(60 \leq M(\mathrm{B}) \leq 72)$ and main group elements $(74 \leq M(\mathrm{A}) \leq 84$ and 87).

5.3.2   Structure types with very low or very high symmetry

As we have seen in Table 5.2 and Fig. 5.5, a significant number of structure types (31) and their representatives (69) shows triclinic symmetry (space group $P\bar{1}$). Why do intermetallics adopt such low-symmetry structures, and what do these structures look like? Is there a general principle that allows us to understand the origin of this low symmetry? We will discuss this aspect on several examples of triclinic intermetallic phases:

- aP8-KHg ($P\bar{1}$, $2i^4$; isotypic to CsHg and EuGa) (Biehl and Deiseroth, 1996; Deiseroth and Strunck, 1987): the structure consists of square-planar Hg_4 units forming with ten K atoms a distorted *fcc*-unit-cell-like structural subunit (Fig. 5.10 (a)–(c)). These subunits are linked by edges and vertices, respectively. The formation of the Hg_4 clusters together with the large difference in atomic radii ($r_{\mathrm{K}} = 2.27$ Å, $r_{\mathrm{Hg}} = 1.50$ Å) seems to be the main reason for the distortion of the structure leading to the low symmetry.

- aP12-RhBi$_2$ ($P\bar{1}$, $2i^6$) (Ruck, 1996): the structure consists of almost flat Bi-decorated triangle/square ($3^3.4^2$) nets, which are shifted parallel to each other in a way so that the triangles and squares are forming the opposite faces of polyhedra, which are each centered by one Rh atom each (Fig. 5.10 (d)–(f)). The Rh atoms ($r_{\mathrm{Rh}} = 1.35$ Å) form pairs along the [010] direction, with their distances alternating between 3.114 Å and 3.916 Å, compared to 2.70 Å in the element. The Bi layer is modulated in order to accommodate these varying distances. The chains of Bi atoms ($r_{\mathrm{Bi}} = 1.55$ Å) in this direction show alternating distances of 3.565 Å and 3.503 Å, which are larger than Bi–Bi single bonds in the element (3.10 Å), and shifts (± 0.3 Å) out of the layer. Rh and Bi chains with ideal atomic distances would be incommensurate to each other.

- aP15-Re$_4$Al$_{11}$ (2 $P\bar{1}$, $1a2i^7$; isotypic for Cr, Mn instead of Re) (Kontio *et al.*, 1980): the structure can be described as a layer structure, consisting of distorted triangle/square ($3^2.4.3.4$) nets (Fig. 5.11 (d)–(f)). It consists of two layers per asymmetric unit, one almost flat and one strongly puckered, but both of the $3^2.4.3.4$ type. These layers are shifted relative to each other forming more or less distorted tetrahedra, octahedra, and trigonal prisms.

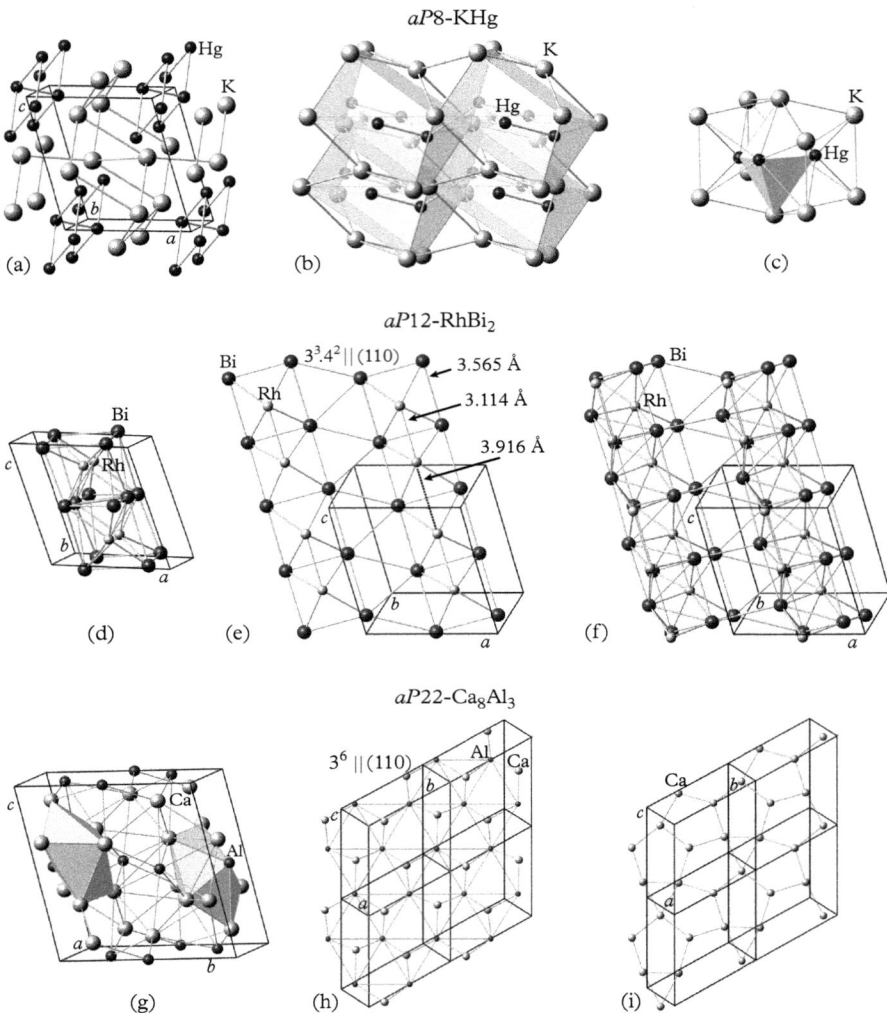

Fig. 5.10 *Structures of (a)–(c) aP8-KHg, (d)–(f) aP12-RhBi₂, and (g)–(i) aP22-Ca₈Al₃. In addition to one (expanded) unit cell (a, d, g), some characteristic structural features are shown. (b) Packing of pseudo fcc unit cells, centered by a K₂Hg₄ octahedron (c). (e) Bi-decorated triangle/square (3³.4²) net, with the squares capped by Rh atoms. (f) Another triangle/square net added so that now the triangles are capped by the underlying Rh atoms. In the unit cell of Ca₈Al₃ (g), pentagonal bipyramids are shown, which are oriented differently than the pentagons in the pentagon/triangle tiling shown in (i). (h) The Al atoms form a triangle tiling with every other triangle capped by Ca atoms leading to a vertex-sharing 2D 3-connected network of tetrahedra. The structures shown in (h) and (i) are on half the usual scale.*

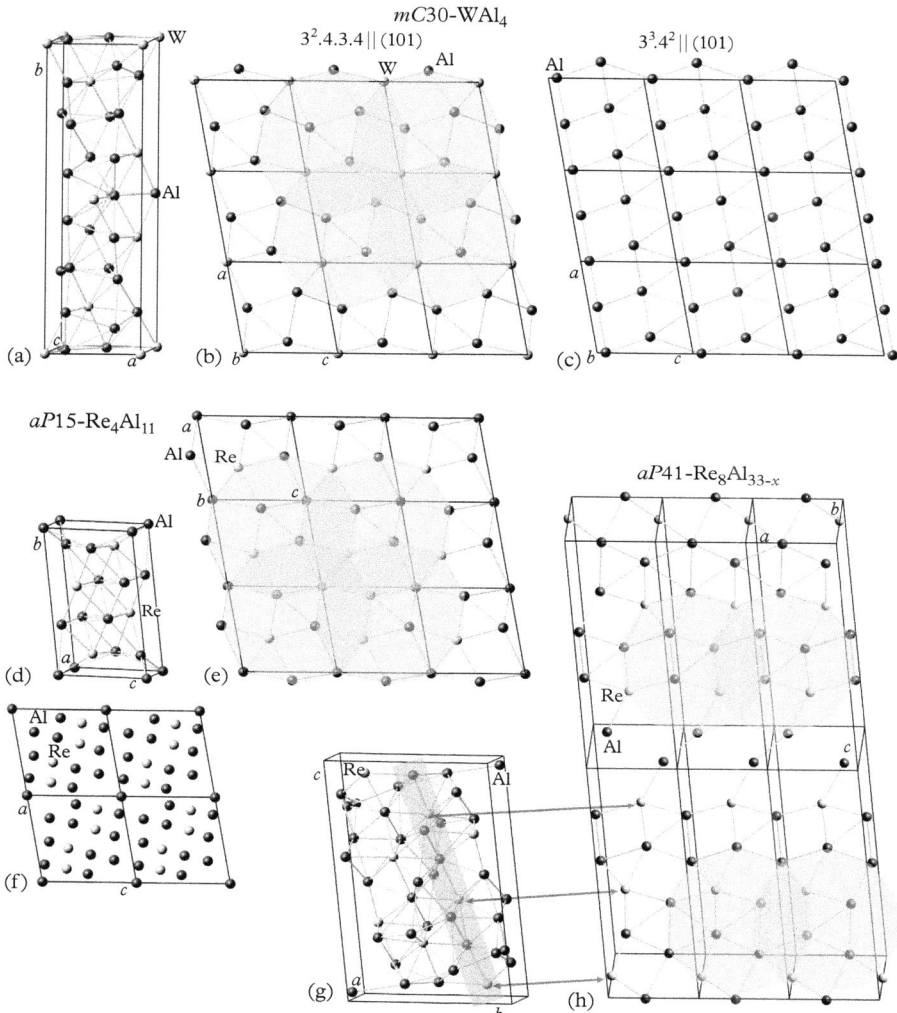

Fig. 5.11 *Structures of (a)–(c) mC30-WAl$_4$, (d)–(f) aP15-Re$_4$Al$_{11}$, and (g)–(h) aP40-Re$_8$Al$_{33-x}$. Besides the unit cells, some characteristic features are also shown. (b) Slightly puckered Al/W-decorated triangle/square (3^2.4.3.4) net in y ≈ 1/4, with the overlapping Gummelt decagons shaded gray. The overlaid Gummelt decagons are regular and undistorted to show the high regularity of the local structure. (c) Almost flat Al-decorated 3^3.4^2 net in y ≈ 0. The squares in these nets are all compressed to rhombs, while the triangles remain approximately equilateral. In (e), the strongly puckered layer at 0 < y < 0.14 is depicted with Gummelt decagons marked showing the close resemblance to the layer in (b). The oblique projection of the structure along [010] shows a quite regular 2D average structure. In the unit cell of aP41-Re$_8$Al$_{33-x}$ (g) those atoms are marked that form the puckered atomic layer shown in (h). Again, Gummelt decagons are marked, which are, however, incomplete in this case.*

It should be mentioned here that the $3^2.4.3.4$ tiling is particularly well-suited for accommodating strains by a kind of origami-like folding (Silverberg *et al.*, 2015). It should also be mentioned that the distortions of the unit tiles lead to almost regular decagonal structural units resembling the Gummelt decagon (see Subsection 3.2.3). Pairs of decagons share either the smashed hexagon (H) units (i.e., the allowed overlap in the PT) or a rhomb. Similar layers are found in the eight-layer structure of $mC30$-WAl$_4$. However, of the four layers in the asymmetric unit of $mC30$-WAl$_4$, there is one, containing only Al atoms, which shows a different kind of triangle/square tiling, namely one of the $3^3.4^2$ type like those in the structure of $aP12$-RhBi$_2$. Furthermore, the squares are much more distorted into rhombs, in this way increasing the in-plane coordination number from five to six, and thereby approaching an *hcp* arrangement in the layer. This additional Al-layer accounts for the larger Al content of $mC30$-WAl$_4$ compared to $aP15$-Re$_4$Al$_{11}$ (ReAl$_{2.75}$). It acts as a kind of buffer between the two strongly puckered $3^3.4^2$-like layers which it is sandwiched in-between.

- $aP19$-Fe$_{6.41}$Al$_{12.59}$ (FeAl$_2$) (2 $P\bar{1}$, $1a2i^9$) (Chumak *et al.*, 2010): the structure can be described with Fe sitting in the origin plus an arrangement of distorted trigonal prisms. It has several icosahedral AETs. The structure can be derived from the $oP16$-YPd$_2$Si structure type by adding three more atoms and rotating the trigonal prisms. The stability of the structure was confirmed by first-principles calculations and explained by the contribution of vibrational entropy at sufficiently high temperatures (> 360 K). This was attributed to the enhanced density of low-frequency phonons due to the comparatively large mean atomic volume (Mihalkovic and Widom, 2012). At lower temperature, this compound was shown to adopt a structure of the $tI6$-MoSi$_2$ type, which can be described as a $cP2$-CsCl type derivative.

- $aP20$-RESn$_3$ (RE = La–Nd, Sm) (2 $P\bar{1}$, $2i^{10}$) (Fornasini *et al.*, 2003): these structures are all of the $aP20$-Nd$_2$Sn$_3$ type, which has been described to have similarities with the $oS12$-ZrSi$_2$ structure type. As is illustrated in Fig. 7.77 in Subsection 7.15.6, the structure can be described as a stacking of equal layers that are shifted parallel to each other. The layers correspond to triangle/square/pentagon tilings with decagonal structural subunits.

- $aP20$-Ce$_3$Pt$_5$Al$_2$ (2 $P\bar{1}$, $2i^{10}$) (Tursina *et al.*, 2014): the structure has been described as packing of Ce-centered irregular polyhedra. The shortest Ce–Ce distances with 3.68 Å are comparable to the sum of their radii (3.65 Å).

- $aP22$-Ca$_8$In$_3$ (2 $P\bar{1}$, $2i^{11}$; isotypic to Ca$_8$Al$_3$, Yb$_8$Tl$_3$) (Fornasini, 1987; Marsh and Slagle, 1988): the structure can be derived by distorting the $cF16$-BiF$_3$ structure type, with In in the Bi positions and Ca being more or less displaced from the F sites, in this way allowing for a different stoichiometry (Fig. 5.10 (g)–(i)).

- $aP37$-RE$_4$TM$_9$Al$_{24}$ (TM:Pd, RE: Gd–Tm; TM: Pt, RE: Y, Gd–Lu) ($P\bar{1}$, $1a2i^{18}$) (Thiede *et al.*, 1999): the structure has pseudo-trigonal symmetry and can be described by a stacking of layers A, A', and B. While the layers A and A' can be described as puckered triangle tilings, layer B alludes to a pentagon triangle tiling.

- $aP41$-Re_8Al_{33-x} (2 $P\bar{1}$, $1a2i^{21}$) (Grin and Schuster, 2007): the structure of this low-temperature phase (Fig. 5.11(g)–(h)) is closely related to $aP71$-$Re_{14}Al_{57-x}$.

- $aP71$-$Re_{14}Al_{57-x}$ (2 $P\bar{1}$, $1a2i^{35}$) (Schuster and Parthé, 1987): the rather complex structure can be described as a stacking of two $mC30$-WAl_4 units along [001], one cut to a slab parallel to (110) and another one parallel to ($\bar{1}$10).

It is remarkable that a significant fraction of these triclinic structure types can be found among Mn and Re aluminides. In the case of the latter one, as many as three phases have triclinic structures. The pseudo-decagonal arrangements in the layers of these structures indicate that these intermetallics are quasicrystal approximants. In the case of the system Al–Mn, metastable decagonal and icosahedral quasicrystals have been observed, indeed, which can be stabilized by adding Pd. In the case of Al–Pd–Re only stable icosahedral quasicrystals have been identified so far.

After the discussion of structures with very low symmetry, we move now to that of structures with very high symmetry. By very high-symmetry structures, we mean cubic structures in which at least one Wyckoff position with the highest multiplicity is occupied, i.e., its orbit centered on a site with symmetry $m\bar{3}m$ and order $k = 48$. Such Wyckoff positions exist in ten space groups, only, i.e., those from no. 221 to no. 230. However, a kind of cluster shell with the highest point symmetry (a truncated cuboctahedron, see Fig. 3.14 (f)) is only generated in the symmorphic space groups 221 $Pm\bar{3}m$, 225 $Fm\bar{3}m$, and 229 $Im\bar{3}m$.

Amazingly, intermetallics with such very high point-group-symmetry structures seem to be even rarer than those with very low symmetry. We could not identify any intermetallic compound with the symmetry of one of the three space groups mentioned above, where the Wyckoff position with the highest symmetry is occupied. The reason may be that large cluster shells need some degrees of freedom, which are not compatible with a single high-symmetry orbit. This can be due to chemical composition (more than one kind of atom forming such a cluster shell) or small distortions due to the symmetry of underlying shells. See, for instance, the Al_{76} fullerene cluster shell identified in the structure of $cF444$-$Al_{63.6}Ta_{36.4}$ (Fig. 7.17(d)). This cluster shell is centered at $4d$ $3/4, 3/4, 34$, and constituted by five symmetrically independent Al atoms: Al(1)–Al(3), each in $48h$ x, x, z, Al(4) in $16e$ x, x, x, and Al(5) in $24f$ $x, 0, 0$. The cluster center has only site symmetry $\bar{4}3m$, with $k = 24$.

Furthermore, the truncated cuboctahedron is bounded by squares, hexagons, and octagons of equal edge lengths. If the edge length is d_{edge}, then the squares have to be decorated with atoms A of equal size at the vertices, the hexagons at the vertices and the center, in order for all atoms to touch each other. The void in the center of the octagons amounts to $1.307 d_{edge}$, and had to be occupied by 30% larger atoms B. If A corresponds to TM atoms, for instance, then B could be a RE, an alkali, or alkaline earth element. A cluster shell consisting, for instance, of 48 + 8 A atoms and 6 B atoms, or 48 A atoms, 8 B atoms, and 6 C atoms is not very likely.

The low symmetry appears to mainly originate from the preferred formation of specific AETs and structural subunits together with size incompatibilities and/or odd stoichiometries requiring degrees of freedom only available in triclinic lattices.

Caveat: one has to check carefully the quality of the low-symmetry structures listed in the PCD, in particular that of the non-centrosymmetric ones determined by powder diffraction. In quite a few cases it was found later that the actual symmetry is higher (see, e.g., Fornasini, 1987; Marsh and Slagle, 1988). This particularly refers to all non-centrosymmetric low-symmetry structures in the PCD.

5.3.3 Structures with one atom per primitive unit cell

As we can see from Fig. 5.8, there are 14 structure types listed in the PCD, three of them unique ones, with just one atom per primitive unit cell: nine of them are element structures $oF4$-In, $tI2$-In, $tI2$-Pa, $hR3$-Hg, $hR3$-Te (unique), $hR3$-Po, $cP1$-Po, $cI2$-W, and $cF4$-Cu, and one is a disordered binary intermetallic, $hP1$-HgSn$_9$. There are four more structure types in the PCD, which are either incommensurately modulated ($R\bar{3}m$ (00γ), $hR3$-SnSb), metastable ($hR3$-Zn$_{0.29}$Al$_{0.71}$; $mS2$-Ce, unique), or not confirmed ($oS2$-Sn, unique).

It is worthwhile to have a closer look at the ten confirmed structure types, which are the simplest observed ones, in order to see what kinds of structures can form with just one atom per unit cell, and how they differ from one another. In the following, the space group number and space group symbol, occupied Wyckoff position, invariant lattice complex, first and second coordination shells, AET-1 and AET-2, and the histograms of interatomic distances of all these structure types are given (Fig. 5.12):

> $oF4$-**In** (69 *Fmmm*, $4a$, oF): AET-1: distorted cuboctahedron, AET-2: distorted octahedron.
>
> $tI2$-**In** (139 *I4/mmm*, $2a$, tI): AET-1: tetragonally distorted cuboctahedron, AET-2: tetragonally distorted octahedron; $c/a = 1.521$.
>
> $tI2$-**Pa** (139 *I4/mmm*, $2a$, tI): AET-1: tetragonally distorted rhombicdodecahedron, AET-2: tetragonally distorted cube; $c/a = 0.825$.
>
> $hR3$-**Hg** (166 $R\bar{3}m$, $1a$, hR): AET-1: along [001] distorted cuboctahedron, AET-2: along [001] distorted octahedron; $c/a = 1.928$.
>
> $hR3$-**Po** (166 $R\bar{3}m$, $1a$, hR): AET-1: along [001] distorted octahedron, AET-2: along [001] distorted bicapped disheptahedron; $c/a = 0.967$.
>
> $hR3$-**Te** (166 $R\bar{3}m$, $1a$, hR): AET-1: along [001] distorted rhombicdodecahedron, AET-2: along [001] distorted cuboctahedron; $c/a = 0.757$.

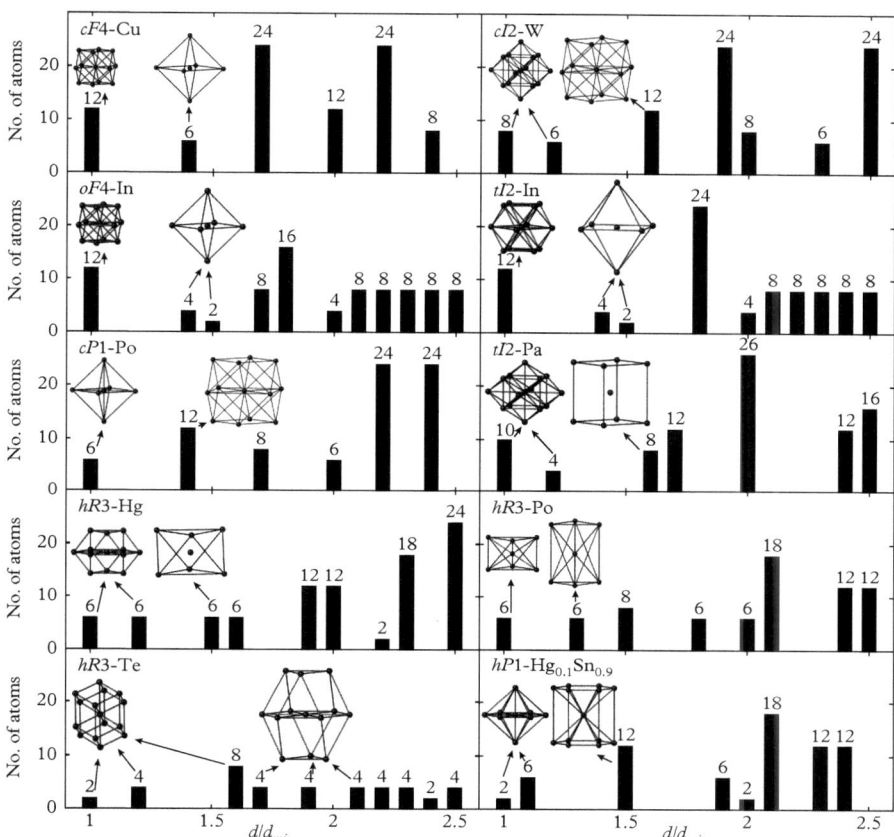

Fig. 5.12 *Distance histograms and first (at left) and second (at right) coordination spheres of the structure types with one atom per unit cell. cF4-Cu: cuboctahedron, octahedron; cI2-W: rhombicdodecahedron, cuboctahedron; oF4-In: distorted cuboctahedron, distorted octahedron; tI2-In: distorted cuboctahedron, distorted octahedron; cP1-Po: octahedron, cuboctahedron; tI2-Pa: distorted rhombicdodecahedron, cube; hR3-Hg: distorted cuboctahedron, distorted octahedron; hR3-Po: distorted octahedron (trigonal antiprism), distorted octahedron (trigonal antiprism); hR3-Te: distorted rhombicdodecahedron, distorted cuboctahedron; hP1-Hg$_{0.1}$Sn$_{0.9}$: hexagonal bipyramid, hexagonal prism. The histogram bars related to the AETs are marked by arrows.*

$hP1$-**HgSn$_9$** (191 $P6/mmm$, 1a, hP): AET-1: hexagonal bipyramid (hbp), AET-2: hexagonal prism; $c/a = 0.931$.

$cP1$-**Po** (221 $Pm\bar{3}m$, 1a, cP): AET-1: octahedron, AET-2: cuboctahedron.

$cI2$-**W** (229 $Im\bar{3}m$, 2a, cI): AET-1: rhombicdodecahedron, AET-2: cuboctahedron.

$cF4$-**Cu** (225 $Fm\bar{3}m$, 4a, cF): AET-1: cuboctahedron, AET-2: octahedron.

AET-1 in $cF4$-Cu corresponds to a cuboctahedron, characteristic for all atoms in a cubic close packing. The same is the case for AET-1 of $tI2$-In, $oF4$-In, and for $hR3$-Hg in an increasingly distorted manner. AET-2 is of octahedral shape in all these cases, more or less distorted except in the case of $cF4$-Cu. The atoms on the vertices of these octahedra cap the quadrangular faces of the cuboctahedra, forming edge-centered octahedra and thereby resulting in close packings. In the case of $cP1$-Po, the reverse scenario is the case: AET-1 is an octahedron and AET-2 a cuboctahedron. The octahedron centers the square faces of the cuboctahedron in this case, which leads to a cubic primitive structure in the end.

In the case of $cI2$-W and $hR3$-Te, AET-1 corresponds to a rhombicdodecahedron, AET-2 is of cuboctahedral shape, and both AETs are distorted in the case of $hR3$-Te. The rhombicdodecahedron and the cuboctahedron are dual polyhedra, i.e., each face of the one polyhedron is centered by a vertex of the other. Although AET-1 of $tI2$-Pa corresponds to a distorted rhombicdodecahedron, AET-2 is not a cuboctahedron but just a tetragonally distorted cube. The vertices of the cube center the upper and lower four rhomb faces of the rhombicdodecahedron, which finally results in a body-centered structure.

The AET-1 of $hR3$-Po is a distorted octahedron (or trigonal antiprism), which is surrounded by a bicapped disheptahedron (anticuboctahedron) for AET-2. From another point of view, the 14 atoms closest to the central atom form, together with the central atom, two rhombohedral unit cells sharing one vertex.

All these changes resulting in different AETs for these smallest possible structure types are reflected in the distance histograms shown in Fig. 5.12. The question remains, why no other structure type with just one atom per unit cell has been experimentally observed so far. These could have lattice complexes P (2 $P\bar{1}$, 1a), mP (10 $P2/m$, 1a), mC (12 $C2/m$, 2a), oP (47 $Pmmm$, 1a), oC (65 $Cmmm$, 2a), oI (71 $Immm$, 2a), and tP (123 $P4/mmm$, 1a).

5.3.4   Most common crystal structure types

The 20 most common structure types, altogether representing more than 35% of all intermetallics in the PCD, are given in Table 5.3. Each of them represents at least 1% of all the structures of intermetallics examined thus far. Among the top 20 structure types, there are 2 unary ones, 11 binary ones, and 6 ternary ones. Their symmetries are mostly cubic ($4 \times cF$, $1 \times cI$, $2 \times cP$) or hexagonal ($7 \times hP$), a few are tetragonal ($2 \times tI$, $1 \times tP$), orthorhombic ($1 \times oI$, $1 \times oP$), or trigonal ($1 \times hR$). None of the 310 triclinic or monoclinic structure types listed in the PCD are among the top 20. The number of atoms per unit cell for the 20 most common structure types ranges from 2 to 57. Five of the 11 binary structure types have more than two occupied Wyckoff positions, and can be used as ternary derivative structure types (ordered structures), which has been observed, indeed, 12 of the 20 structure type have an even number of atoms per primitive unit cell (2, 4, 6, 10, 12, 16 atoms), and 8 an odd number (1, 3, 5, 9, 13, 19 atoms).

The most common structure types are those of the Laves phases (see Subsection 7.4.3), with the cubic one, $cF24$-$MgCu_2$ (806 representatives, 3.9%), in the first place, and the hexagonal ones in the sixth place, $hP12$-$MgZn_2$ (456 representatives, 2.2%), and in the 66th place, $hP24$-$MgNi_2$ (63 representatives, 0.3%), respectively. The structures of the Laves phases can be described as topologically close-packings (*tcp*) of spheres of two different sizes with a well-defined size ratio or, alternatively, as packings of Friauf polyhedra.

Many of the most common structure types, however, correspond to simpler close sphere packings, for instance $cF4$-Cu on rank 2 (581 representatives, 2.8%) and some of its derivatives such as $cP4$-Cu_3Au on rank 3 (544 representatives, 2.6%), a substitutional superstructure. The next ranked, less-dense, sphere packing is the $cI2$-W structure type on rank 9 (375 representatives, 1.8%), with its derivative structures, $cP2$-$CsCl$ on rank 4 (512 representatives, 2.5%), the Heusler phase $cF16$-Cu_2MnAl on rank 7 (414 representatives, 2.0%), and the half-Heusler phase $cF12$-$MgAgAs$ on rank 18 (206 representatives, 1.0%). The hexagonal close sphere packing, $hP2$-Mg, only ranks as the 12th most common structure type (338 representatives, 1.6%) among intermetallics.

5.4 Stability maps and composition diagrams of binary intermetallics

Taking into account the 6441 binary intermetallics only – out of the in total 20 829 – the ranking of the most common structure types changes a little bit compared to that for all intermetallics shown before (Table 5.3). All in all, 943 unary and binary structure types can be found among binary intermetallics, and the 20 most common ones are listed below in Table 5.4 representing, in total, 43.1% of all binary compounds.

Obviously, no ternary structure types are contained in this list. In addition, the importance of sphere-packings compared with that of the Laves phases is enhanced: $cF4$-Cu on rank 1 (385 representatives, 6.0%) is now the most common structure type. Of course, a unary structure type for a binary compound implies a statistical distribution of the constituting elements. $cP2$-$CsCl$ has now rank 2 (290 representatives, 4.5%), and $cP4$-Cu_3Au keeps rank 3 (263 representatives, 4.1%), while $cI2$-W, again a structure type leading to a solid solution for a binary intermetallic phase, climbed to rank 6 (210 representatives, 3.3%). One more unary structure type, $hP2$-Mg ranks now fourth (248 representatives, 3.9%). The smaller number of representatives for the structure types compared to Table 5.3 indicates that a number of ternary compounds have been assigned to binary structure types, again indicating solid solutions or not properly defined derivative structures.

It is striking that the number of Laves phases in binary systems amounts to just 223 for $cF24$-$MgCu_2$ and 154 for $hP12$-$MgZn_2$ compared to 806 and 456, respectively, when including ternary and higher systems as well. This means that

Table 5.4 *Most common structure types among binary intermetallics (Dshemuchadse and Steurer, 2015). The top 20 structure types – representing more than 50 intermetallics each – out of 943 are given. Note, that the structure types cF4-Cu, hP2-Mg, and cI2-W are included in this list, although they represent mainly solid solution phases or – more generally – inherently disordered and pseudo-binary phases. They are marked by an "s"-entry (solid solution) in the flag column.*

Rank	Structure type	Space group	Wyckoff positions	No. of structures	% of all structures	Flag
1.	$cF4$-Cu	225 $Fm\bar{3}m$	$4a$	385	6.0%	s
2.	$cP2$-CsCl	221 $Pm\bar{3}m$	$1ab$	290	4.5%	
3.	$cP4$-Cu$_3$Au	221 $Pm\bar{3}m$	$1a\,3c$	263	4.1%	
4.	$hP2$-Mg	194 $P6_3/mmc$	$2c$	248	3.9%	s
5.	$cF24$-MgCu$_2$	227 $Fd\bar{3}m$	$8a\,16d$	223	3.5%	
6.	$cI2$-W	229 $Im\bar{3}m$	$2a$	210	3.3%	s
7.	$hP12$-MgZn$_2$	194 $P6_3/mmc$	$2a\,4f\,6h$	154	2.4%	
8.	$hP16$-Mn$_5$Si$_3$	193 $P6_3/mcm$	$4d\,6g^2$	153	2.4%	
9.	$hP6$-CaCu$_5$	191 $P6/mmm$	$1a\,2c\,3g$	109	1.7%	
10.	$oS8$-TlI	221 $Pm\bar{3}m$	$1ab$	99	1.5%	
11.	$oP16$-Fe$_3$C	62 $Pnma$	$4c^2\,8d$	90	1.4%	
12.	$oP12$-Co$_2$Si	62 $Pnma$	$4c^3$	67	1.0%	
13.	$hP38$-Th$_2$Ni$_{17}$	194 $P6_3/mmc$	$2bc\,4f\,6g\,12jk$	67	1.0%	
14.	$cP8$-Cr$_3$Si	223 $Pm\bar{3}n$	$2a\,6c$	66	1.0%	
15.	$oI12$-KHg$_2$	74 $Imma$	$4e\,8i$	64	1.0%	
16.	$hP8$-Mg$_3$Cd	194 $P6_3/mmc$	$2d\,6h$	64	1.0%	
17.	$cF8$-NaCl	225 $Fm\bar{3}m$	$4ab$	61	0.9%	
18.	$oP36$-Sm$_5$Ge$_4$	62 $Pnma$	$4c^3\,8d^3$	58	0.9%	
19.	$tP2$-CuAu	123 $P4/mmm$	$1ad$	53	0.8%	
20.	$hP6$-Co$_{1.75}$Ge	194 $P6_3/mmc$	$2acd$	52	0.8%	
			Total	2776	43.1%	

the structure of these ternary phases are disordered, at least in the case of $cF24$-MgCu$_2$. In the case of $hP12$-MgZn$_2$ an ordered distribution of three different atomic species is possible. Also, large discrepancies are found for $cP4$-Cu$_3$Au with 263 representatives in binary systems vs. 544 overall, and some other structure types. In this case, three different constituents can occupy the atomic sites of this structure type in a disordered way, only.

The number of binary phases assigned to the three unary structure types, which, of course, are solid solutions, amounts to a remarkable 843 (30.4%) in total. The symmetries of the 17 true binary ones out of the top 20 structure types are either hexagonal ($6 \times hP$), cubic ($2 \times cF$, $3 \times cP$), orthorhombic ($1 \times oS$, $1 \times oI$, $3 \times oP$), or tetragonal ($1 \times tP$). No triclinic, monoclinic, or trigonal structure types are among these top 17 binary structure types. The number of atoms per prmitive unit cell ranges from 1 to 38.

The composition histograms of the binary intermetallics are given in Fig. 5.13 for (a) all 6441 binary phases, for (b) only those 385 assigned to the unary structure type $cF4$-Cu, and for (c) the subset of the 5505 compounds without those, which were assigned to unary structure types; these are mainly solid solutions with structure types such as $cF4$-Cu, $cI2$-W, $hP2$-Mg, or, e.g., HT-phases of binary element mixtures, where the constituting elements have a different structure such as $cI2$-(Ag, Al) (β-phase) and many others. The frequencies of the 43 most frequently occurring compositions are listed in Table 5.5. They each feature a minimum of ten representatives and, in total, represent already 5041 binary intermetallics and therefore more than 90% of the here-examined 5598 compounds.

It is obvious from Fig. 5.13(c) that the frequency distribution is not symmetric around the composition AB. Compounds with the compositions A_2B and AB_2, A_3B and AB_3 show similar frequencies in contrast to those with stoichiometries A_4B and AB_4, A_6B and AB_6, A_9B and AB_9, for instance (note the logarithmic scale). This means that the elements A and B, with $M(A) < M(B)$, cannot simply swap their sites in a given structure type or that they form different structure types with lesser probability. Typical examples are the Laves phases with more than 200 representatives of the type AB_2 and less than ten of the type A_2B. However, not only the frequencies of structures for intermetallics with stoichiometries A_mB_n and A_nB_m differ but so does the overall distribution density of different compositions. It is also obvious from Fig. 5.13(c) that the density of the distribution on the left-hand side of the figure is smaller than that on the right-hand side. Compare, for instance, the densities in the range A_6B–A_4B with that in the range AB_4–AB_6. The distribution of binary phases with the unary $cF4$-Cu structure type (Fig. 5.13(b)) could be symmetric, basically. These asymmetries most probably originate from the arbitrary distribution of nominal compositions of the samples studied experimentally. The solid solubility of an element B in a phase A with $cF4$-Cu structure type can strongly differ from the solid solubility of an element A in a phase B with $cF4$-Cu structure type.

In Fig. 5.13(d), those intermetallic phases are marked in gray that are representatives of unique structure types. The histogram shows that there are only a few stoichiometries where no unique structure types are present. These are to a large extent concentrated in the regions with very high contents of one or the other of the constituents. Furthermore, most unique structure types are found to have rather simple stoichiometries. The histogram in Fig. 5.13(e) illustrates the distribution of unique structure types in comparison with the general structure types as a function of stoichiometry. On one hand, there are quite a few compositions that feature just one structure type, which is not necessarily unique, however.

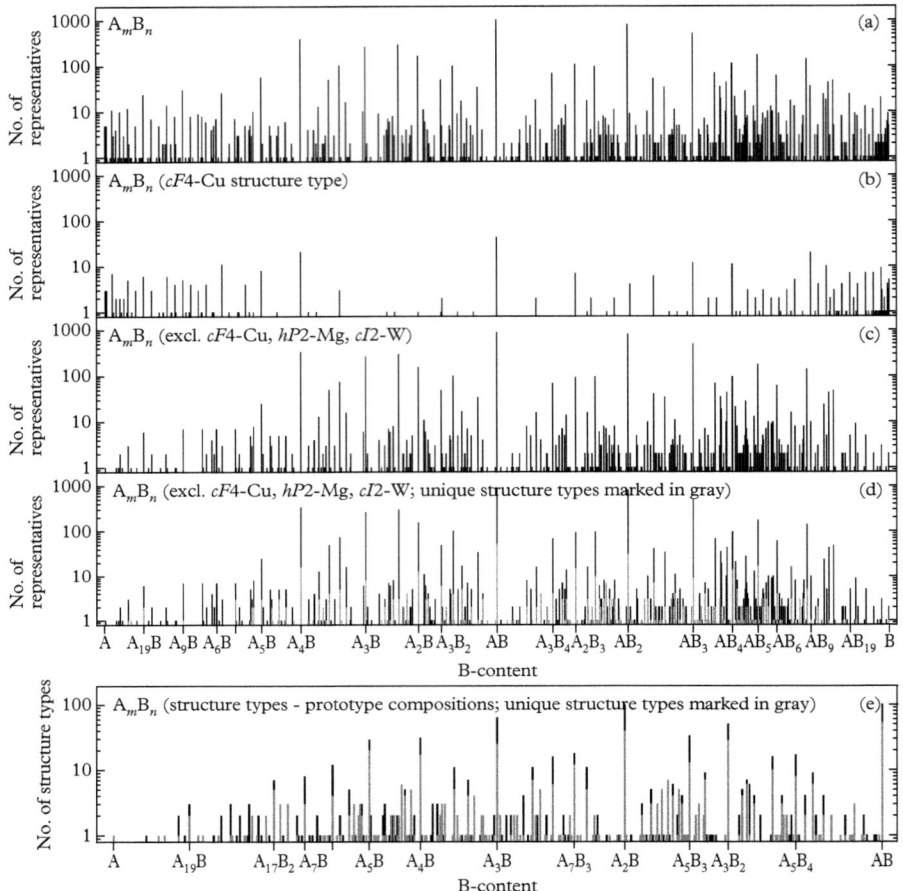

Fig. 5.13 *Frequency of binary intermetallic phases with compositions* A_mB_n *and* $M(A) < M(B)$: *(a) all 6441 binary intermetallic phases, (b) only those 385 binary phases assigned to the unary structure type cF4-Cu. In (c), the same data are shown as in (a), with just those binary phases (solid solutions) excluded that were assigned to unary structure types, with 5505 binary intermetallics remaining. (d) Intermetallic representatives of the unique structure types are marked in gray on the same type of histogram as shown in (c). (e) Frequency of unique structure types marked on the distribution histogram of structure types in general as a function of stoichiometry. Since the prototype structures of these structure types are not all intermetallics, the histogram was merged into the range* A–AB, *not differentiating between* $M(A)$ *and* $M(B)$. *Note the logarithmic vertical scale.*

On the other hand, there are quite a few stoichiometries as well, which only feature unique structure types.

The frequency distribution of the number of unique structure types over the number of the non-unique structure types of the 43 most frequently occurring binary chemical compositions listed in Table 5.5 is shown in Fig. 5.14. The ratio

Table 5.5 *The top 43 compositions of binary intermetallics, each representing 10 or more IMs for each pair* A_mB_n | A_nB_m. *Excluded are structure types cF4-Cu, cI2-W, and hP2-Mg, as well as all other intrinsically unary structure types, resulting in 5505 binary IMs. The number of different structure types (STs) and their representatives is given for each composition m, n* $(M(A) < M(B))$. *In addition, the total number of representatives, different STs, and unique STs (uSTs) is given for all pairs* A_mB_n | A_nB_m.

Rank	A_mB_n	No. of repr.	No. of STs	A_nB_m	No. of repr.	No. of STs	No. of repr.	No. of STs	No. of uSTs
1.	A_2B	268	38	AB_2	794	81	1062	106	49
2.	AB	872	120				872	120	69
3.	A_3B	332	46	AB_3	493	54	825	80	34
4.	A_5B_3	294	17	A_3B_5	96	26	390	40	18
5.	A_3B_2	153	43	A_2B_3	87	39	240	73	42
6.	A_5B	8	6	AB_5	174	28	182	33	23
7.	$A_{17}B_2$	0	0	A_2B_{17}	140	15	140	15	6
8.	A_4B	19	13	AB_4	91	36	110	47	33
9.	A_4B_3	49	10	A_3B_4	68	13	117	22	14
10.	A_7B_3	74	16	A_3B_7	39	19	113	31	21
11.	A_5B_4	98	13	A_4B_5	6	5	104	17	9
12.	A_5B_2	50	10	A_2B_5	34	9	84	19	9
13.	A_7B_2	6	5	A_2B_7	67	9	73	13	7
14.	A_6B	7	1	AB_6	61	13	68	14	6
15.	$A_{13}B$	4	1	AB_{13}	45	2	49	2	0
16.	$A_{23}B_6$	1	1	A_6B_{23}	43	1	44	1	0
17.	$A_{12}B$	0	0	AB_{12}	43	3	43	3	1
18.	$A_{51}B_{14}$	0	0	$A_{14}B_{51}$	35	1	35	1	0
19.	$A_{11}B_{10}$	34	4	$A_{10}B_{11}$	0	0	34	4	3
20.	$A_{58}B_{13}$	0	0	$A_{13}B_{58}$	28	5	28	5	3
21.	A_9B_4	19	9	A_4B_9	7	7	26	12	5
22.	$A_{11}B$	1	1	AB_{11}	23	2	24	2	0
23.	A_8B_3	13	7	A_3B_8	11	8	24	15	10
24.	A_7B	9	6	AB_7	14	7	23	11	7
25.	$A_{45}B_{11}$	0	0	$A_{11}B_{45}$	21	4	21	4	3
26.	$A_{11}B_9$	5	5	A_9B_{11}	16	10	21	14	10
27.	$A_{11}B_3$	1	1	A_3B_{11}	19	4	20	5	4

continued

Table 5.5 *continued*

Rank	A_mB_n	No. of repr.	No. of STs	A_nB_m	No. of repr.	No. of STs	No. of repr.	No. of STs	No. of uSTs
28.	A_8B_5	4	4	A_5B_8	16	7	20	11	8
29.	$A_{10}B_7$	4	4	A_7B_{10}	14	5	18	9	7
30.	A_6B_5	17	10	A_5B_6	1	1	18	11	7
31.	$A_{24}B_5$	2	2	A_5B_{24}	13	2	15	3	2
32.	A_7B_6	8	4	A_6B_7	7	3	15	6	4
33.	A_7B_4	6	2	A_4B_7	8	5	14	7	5
34.	A_9B	6	6	AB_9	6	5	12	9	6
35.	$A_{16}B_{11}$	11	2	$A_{11}B_{16}$	1	1	12	3	2
36.	$A_{12}B_7$	8	4	A_7B_{12}	3	3	11	7	6
37.	$A_{23}B_4$	1	1	A_4B_{23}	9	4	10	4	2
38.	$A_{17}B_3$	2	2	A_3B_{17}	8	4	10	6	4
39.	$A_{11}B_2$	3	3	A_2B_{11}	7	4	10	6	3
40.	$A_{17}B_4$	8	5	A_4B_{17}	2	2	10	7	5
41.	$A_{19}B_5$	2	1	A_5B_{19}	8	2	10	3	0
42.	$A_{10}B_3$	5	4	A_3B_{10}	5	5	10	9	8
43.	$A_{13}B_7$	4	3	A_7B_{13}	6	6	10	9	8

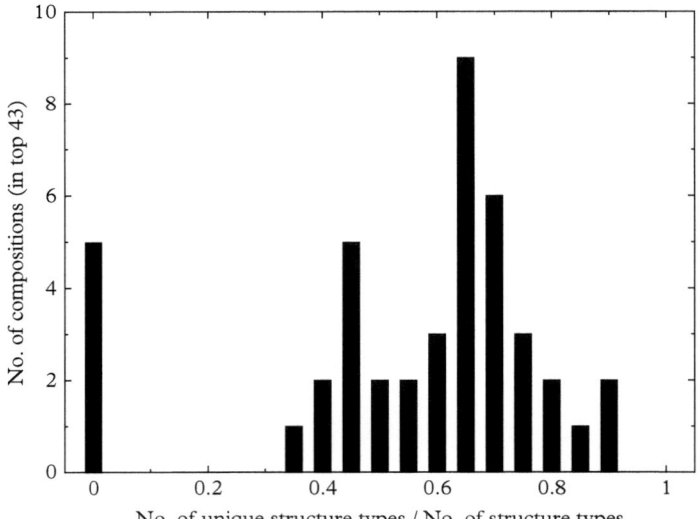

Fig. 5.14 *Frequency distribution of the numbers of unique structure types/numbers of non-unique structure types over the 43 compositions listed in Table 5.5.*

of unique structure types to all 2166 unique and non-unique structure types amounts to ≈ 0.5. There are five stoichiometries that do not feature unique structure types at all (No. 15, 16, 18, 22, and 41), and two compositions that feature $\approx 90\%$ unique structure types (No. 42 and 43). Since 28 stoichiometries out of the 43 most frequently occurring binary ones ($\approx 65\%$) show ratios > 0.5, they contain a disproportionately large number of unique structure types.

It is remarkable in how many different structure types binary intermetallic compounds with the same stoichiometry can crystallize, thereby adopting unit cell sizes from very small to very large. For the simple stoichiometry AB, 120 different structure types are listed in the PCD – an impressive number. 57.5% of them are unique structure types, i.e., seemingly specifically designed for a particular element combination. Regarding the number of atoms per unit cell, in the case of A_2B and AB_2, for instance, unit cells have been observed containing from 3 up to 96 atoms ($hP3$-AlB_2 and $cF96$-Ti_2Ni, respectively). This means that the energetically best AETs cannot be simply realized and packed, but the structure needs more different AETs than atomic types in order to compensate for sterical mismatches. But even when the number of atoms per unit cell is the same as well as the stoichiometry, completely different structure types can result.

In the following, certain interesting regions of the frequency diagrams in Fig. 5.13, as well as the most frequent stoichiometries, will be discussed in greater detail. Keep in mind that the nominal stoichiometry of the intermetallic phases studied and of the structure type do not necessarily agree for reasons mentioned in Section 5.2.

5.4.1 A_2B/AB_2

The 1062 intermetallics with compositions A_2B/AB_2 crystallize in 106 different structure types, i.e., with on average 10.0 representatives per structure type. The ten most common structure types together comprise 711 (66%) out of these, which corresponds to 71.1 representatives per structure type on average. The number of atoms per unit cell ranges from 3 up to 96 ($hP3$-AlB_2 and $cF96$-Ti_2Ni, respectively).

As shown in the following list, the two most frequent structure types are the cubic and the two hexagonal Laves phases, respectively. Five other structure types can be considered derivatives of the $hP3$-AlB_2 structure type, which itself can be found on rank ten.

> $cF24$-$MgCu_2$ (227 $Fd\bar{3}m$, $8a\,16d$; 223 representatives): ccp packing of Mg-centered Friauf polyhedra (FK_{16}^{28}) with composition $Mg@Cu_{12}Mg_4$. Cu is coordinated by icosahedra FK_{12}^{20} with composition $Cu@Cu_6Mg_6$. The Mg atoms form an expanded ccp structure with half of the tetrahedral voids occupied by the large Mg atoms and half by Cu tetrahedra. Mg atoms and Cu atoms form mutually interpenetrating networks with the shortest distances between like atoms. See Fig. 7.15.

$hP12$-**MgZn₂** (194 $P6_3/mmc$, $2a\,4f\,6h$; 154 representatives): *hcp* packing of Mg-centered Friauf polyhedra (FK_{16}^{28}) with composition Mg@Zn₁₂Mg₄. Zn is coordinated by icosahedra (FK_{12}^{20}) with composition Zn@Zn₆Mg₆. The Mg atoms form an expanded *hcp* structure with half of the tetrahedral voids occupied by the large Mg atoms and half by Zn tetrahedra. Mg atoms and Zn atoms form mutually interpenetrating networks with the shortest distances between like atoms. See Fig. 7.15.

$oP12$-**Co₂Si** (62 *Pnma*, $4c^3$; 67 representatives): packing of self-dual nets consisting of pentagons, quadrangles, and triangles in equal numbers. It can also be seen as a distorted $hP6$-Ni₂In type structure, which itself is considered an $hP3$-AlB₂-derivative structure. See Fig. 7.48.

$oI12$-**KHg₂** (74 *Imma*, $4e\,8i$; 64 representatives): 4-fold superstructure of $hP3$-AlB₂, with layers of edge-connected hexagonal bipyramids (*hbps*), as basic building units. See Fig. 7.23.

$tI12$-**CuAl₂** (140 $I4/mcm$, $4a\,8h$; 46 representatives): stacking of $\pi/4$ rotated $3^2.4.3.4$ triangle/square nets, with the vertices decorated by Al and the squares capped by Cu atoms. It contains Al-centered Frank-Kasper polyhedra FK_{15}^{26}, and FK_{10}^{16} around Cu, with compositions Cu@Cu₄Al₁₁ and Cu@Cu₂Al₈, respectively. See Fig. 7.26.

$tI6$-**MoSi₂** (139 $I4/mmm$, $2a\,4e$; 38 representatives): 3-fold superstructure of the $cI2$-W type, compressed along [001], resulting in CN10 coordination (top- and bottom-capped square prism) of Mo by Si with composition Mo₈Si₂.

$tI6$-**Zr₂Cu** (139 $I4/mmm$, $2a\,4e$; 33 representatives): 3-fold superstructure of the $cI2$-W structure type with two differently decorated rhombicdodecahedra for AETs, i.e., Cu@Cu₅Zr₉ and Zr@Zr₉Cu₅. See Fig. 7.7.

$hP6$-**Co₁.₇₅Ge** (194 $P6_3/mmc$, $2acd$; 31 representatives): 2-fold superstructure of the $hP3$-AlB₂ type with Co sitting on all Al- and half of the B-sites, and Ge occupying the remaining half of the B-sites. Ge centers pentacapped trigonal prisms of eleven Co atoms, Ge@Co₁₁, half of Co shows a topologically identical coordination with composition Co@Co₆Ge₅, while the other half is coordinated by bicapped hexagonal prisms Co@Co₈Ge₆. See Fig. 7.60.

$cF96$-**Ti₂Ni** (227 $Fd\bar{3}m$, $16c\,32e\,48f$; 28 representatives): icosahedral coordination of all 32 Ni, Ni@Ti₉Ni₃ and of 16 Ti atoms, Ti@Ti₆Ni₆. The remaining 48 Ti atoms show 14-fold coordination in a hybrid form between a bicapped pentagonal prism and an icosahedron, Ti@Ti₁₀Ni₄. See Fig. 7.38.

$hP3$-**AlB₂** (191 $P6/mmm$, $1a\,2d$; 27 representatives): layers of edge-connected hexagonal bipyramids (*hbps*) sharing the apical vertices with the *hbps* of the adjacent layers as basic building units. Primitive hexagonal structure, decorated by Al at the vertices and by B in the trigonal Al-prisms. It can also be described as a $hP2$-Mg-type derivative structure with Al replacing Mg at the unit cell vertices, and B replacing Mg in the resulting Al-trigonal prisms and also occupying the empty ones. See Fig. 7.23.

$M(A)/M(B)$ plots of all intermetallics with composition A_2B/AB_2, and those with the six most common structure types highlighted in gray, are depicted in Fig. 5.15. The stability fields of these six structure types are clearly separated from one another, with $cF24$-$MgCu_2$, $hP12$-$MgZn_2$, $oI12$-KHg_2, and $tI6$-$MoSi_2$ being adjacent to one another. Only the stability fields of $cF24$-$MgCu_2$ and $hP12$-$MgZn_2$ coincide partially for $60 \leq M(A) \leq 62$.

In comparison with the $M(A)/M(B)$ plot of all binary intermetallics (Fig. 5.3), the distribution of the compounds with stoichiometry A_2B/AB_2 is significantly different, with the exception of the lower right part defined by $58 \leq M(A) \leq 88$ and $8 \leq M(B) \leq 50$. This means, that most compounds of the type A_2B/AB_2 have for the majority component a late transition metal or main group element, and for the minority component an alkali or alkaline earth element or one of the lanthanoids.

The stability field of intermetallics with the $cF24$-$MgCu_2$-type (cubic Laves phase) structures ranges from $M(A) = 54$–73 to $M(B) = 11$–33, 40–53, as well as $M(A) = 80$ and $M(B) = 14$–33. Compounds with $hP12$-$MgZn_2$-type (hexagonal Laves phase) structures are concentrated in the area defined by $M(A) = 57$–63 and $M(B) = 17$–33, 41–56, as well as, e.g., $M(A) = 73$ and $M(B) = 14$–26. Most intermetallics with $oP12$-Co_2Si-type structures can be found with $M(B) = 68$–88 and $M(A) = 14$–33, 64–69. With one exception, compounds with $oI12$-KHg_2-type structures only occur in systems with $M(A) = 70$–81 and $M(B) = 8$–33. Those with $tI12$-$CuAl_2$-type structures are rather spread out, but mostly occur within $M(A) = 47$–52 and $M(B) = 61$–84 or $M(A) = 78$–88 and $M(B) = 51$–72. Phases with $tI6$-$MoSi_2$-type structures are concentrated within $M(A) = 67$–74 and $M(B) = 17$–27, 49–52. Also, intermetallics with most other structure types occur in connected M/M areas, but usually also have a few representatives with other element combinations.

There are 106 different structure types with composition A_2B/AB_2 ($M(A) < M(B)$). The existence of stability fields for representatives of the more common ones illustrates the strong correlation between chemical composition, atomic properties, and structure type. The question is to what extent structure types with a specific stoichiometry are structurally related, e.g., have structural building units in common as the Laves phases do, which can be described as different packings of Friauf polyhedra. No other obvious structural similarities can be identified, except that the three structure types mainly appearing in the lower right of the $M(A)/M(B)$ plot – $cF24$-$MgCu_2$, $hP12$-$MgZn_2$, and $oI12$-KHg_2 – exhibit 12-coordinated A- and 16-coordinated B-atoms, albeit with different polyhedral shapes.

5.4.2 AB

The 888 intermetallics with compositions AB crystallize in 127 different structure types, i.e., with on average 7.0 representatives per structure type. The ten most common ones comprise 646 (73%) out of these, this corresponds to 64.6 representatives per structure type on average. The number of atoms per unit cell ranges from 3 up to 304 ($cP2$-CsCl and $oS304$-IrMg, respectively). $M(A)/M(B)$ plots

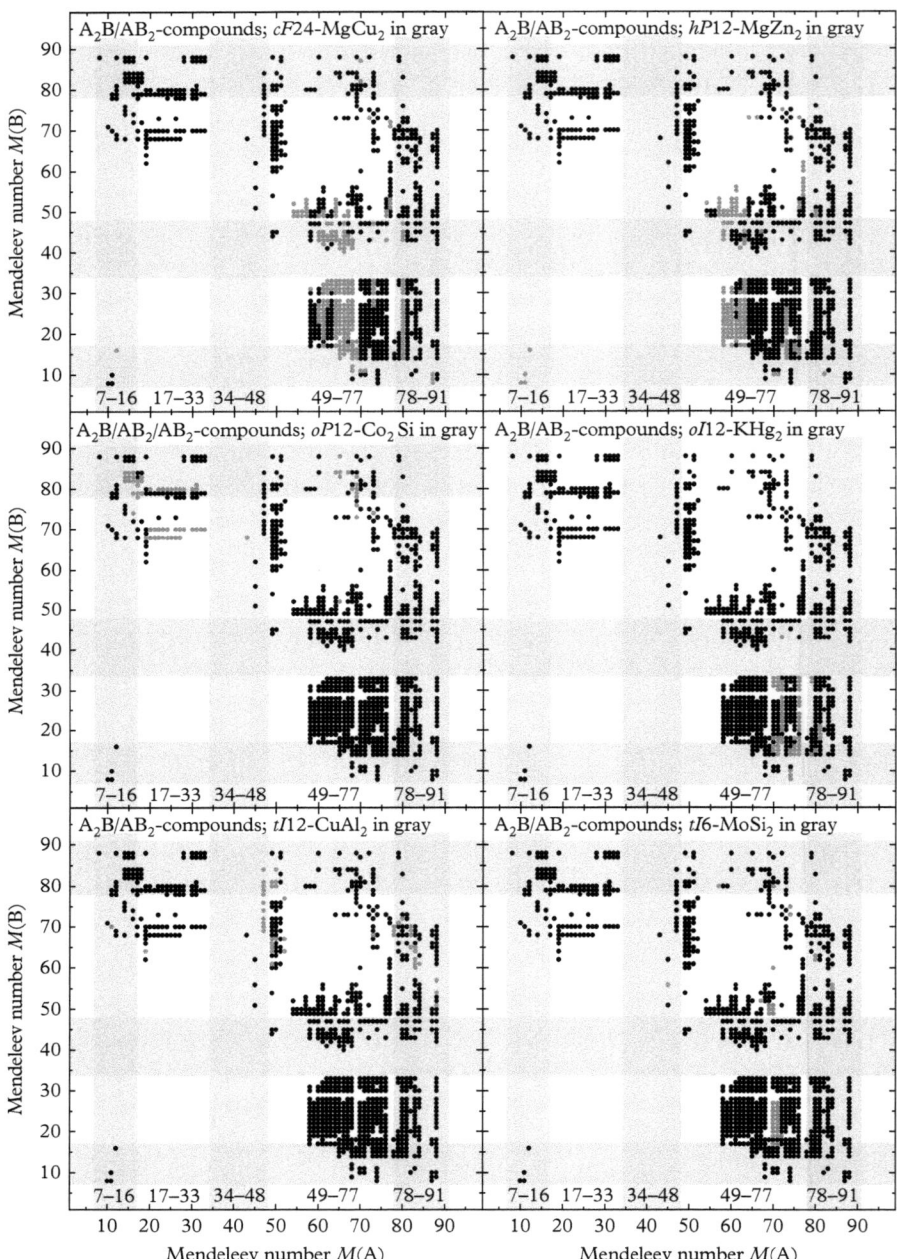

Fig. 5.15 $M(A)/M(B)$ *plots of the 1062 binary intermetallic compounds with composition* A_2B/AB_2 *($M(A) < M(B)$). In the plots, however, $M(A)$ always refers to the majority component. The six most common structure types (in total 711 representatives) are highlighted in gray. Areas containing alkali metals, alkaline earth metals, actinoids, and (semi)metallic main group elements are shaded gray.*

$(M(A) < M(B))$, with the six most common structure types highlighted in gray, are given in Fig. 5.16. The stability fields are well-separated and adjacent to one another with only a few overlapping parts.

$cP2$-**CsCl** (221 $Pm\bar{3}m$, 1ab; 278 representatives): primitive cubic plus hexahedral voids occupied; packing density 0.72901 for $r_s/r_l = (\sqrt{3} - 1)/2$; superstructure of the $cP2$-W type. See Fig. 7.7.

$oS8$-**TlI** (63 $Cmcm$, 4c^2; 99 representatives): orthorhombically distorted $cF8$-NaCl structure type. See Fig. 7.54.

$cF8$-**NaCl** (225 $Fm\bar{3}m$, 4ab; 61 representatives): ccp plus octahedral voids occupied; packing density 0.79308 for $r_s/r_l = (\sqrt{2} - 1)$. See Fig. 7.87.

$oP8$-**FeB** (62 $Pnma$, 4c^2; 50 representatives): distorted $cF8$-NaCl structure type.

$tP2$-**CuTi** (123 $P4/mmm$, 1ad; 48 representatives): more commonly called $tP2$-MnHg structure type, since $tP2$-CuTi is just a metastable phase; tetragonally distorted $cP2$-CsCl structure type, for $c/a \approx \sqrt{2}$ it is also related to the $cF4$-Cu structure type (Bain transformation). See Fig. 7.65.

$tP2$-**CuAu** (123 $P4/mmm$, 1ad; 39 representatives): tetragonally distorted $cP2$-CsCl structure type, for $c/a \approx \sqrt{2}$ it is also related to the $cF4$-Cu structure type (Bain transformation). See Fig. 7.60.

$hP4$-**NiAs** (194 $P6_3/mmc$, 2ac; 27 representatives): hcp plus octahedral voids occupied; packing density 0.79308 for $r_s/r_l = (\sqrt{2} - 1)$. See Fig. 7.87.

$oP4$-**AuCd** (51 $Pmma$, 2ef; 17 representatives): distorted $hP2$-Mg derivative structure.

$cP8$-**FeSi** (198 $P2_13$, 4a^2; 16 representatives): distorted $cF8$-NaCl structure.

$tI64$-**NaPb** (142 $I4_1/acd$, 16ef 32g; 11 representatives): polyanionic Zintl phase featuring Pb-tetrahedra.

Nine of the top ten structure types with 1:1 stoichiometry can be considered rather simple close packed structures or distorted derivative structures. The origin of the distortions is in most cases electronic (Peierls distortion). Number ten, the $tI64$-NaPb structure type, is a Zintl phase, and all of its 32 representatives are Zintl phases as well.

Representatives of the $cP2$-CsCl structure type are rather broadly distributed over the occurrence region of all AB-compounds, but are especially dominant in the region $M(B) = 70$–80. $oS8$-TlI, on the other hand, dominates for values $M(A) = 67$–70 and $M(B) = 81$–84. $cF8$-NaCl is – apart from one outlier – restricted to values $M(B) = 87$, 88, and 91. Most phases with $oP8$-FeB-type among AB-structures are found in a similar $M(B)$-range as those with $oS8$-TlI-type structures, but for a big part with smaller $M(A)$-values. Intermetallics with structures belonging to types $tP2$-CuTi and $tP2$-CuAu, on the other hand, occur in a more scattered manner than the above-mentioned ones.

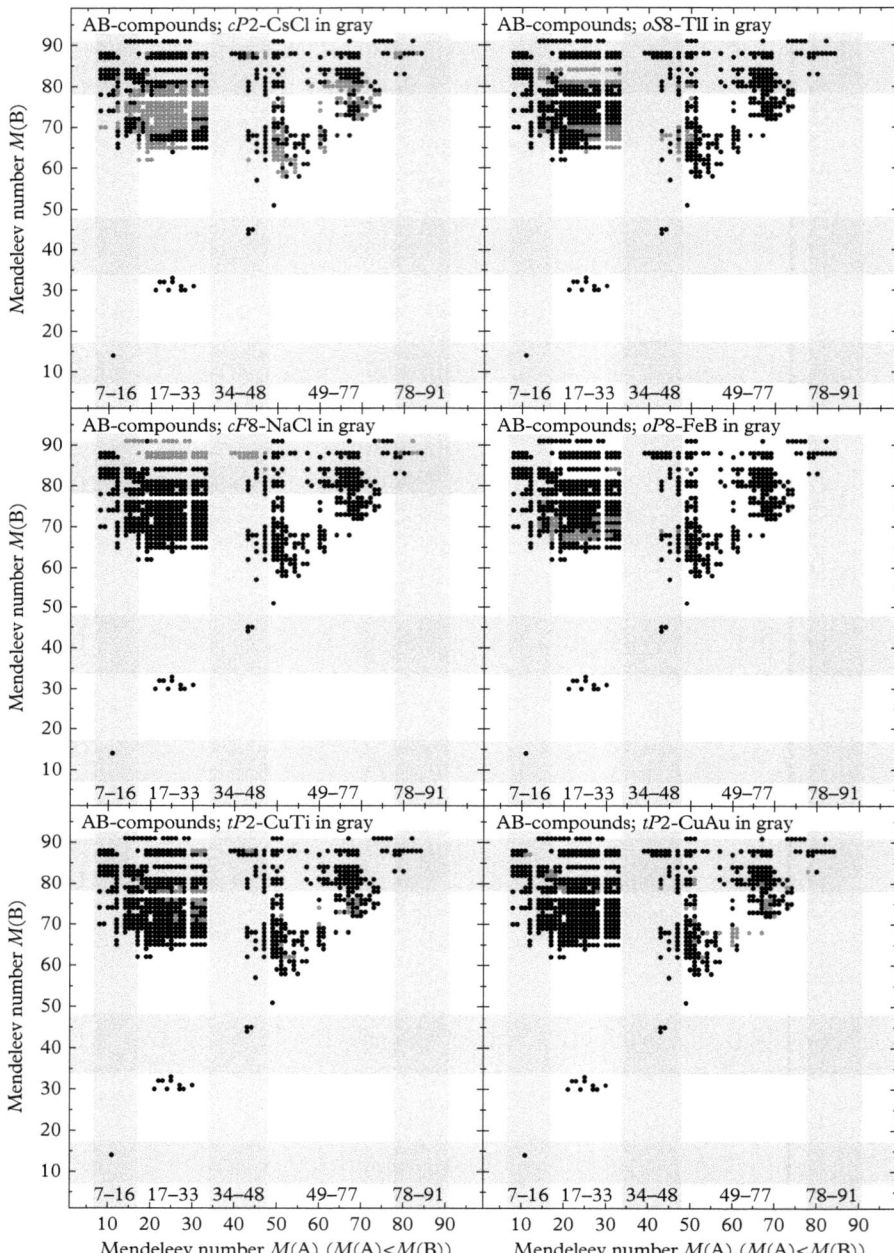

Fig. 5.16 $M(A)/M(B)$ *plots ($M(A) < M(B)$) of the binary intermetallic compounds with composition AB. The six most common structure types are highlighted. Areas containing alkali metals, alkaline earth metals, actinoids, and (semi)metallic main group elements are shaded gray.*

An interesting series is formed by the Zintl-phases $cP2$-LiTl, $cF16$-NaTl, $oS48$-KTl, and $oF96$-CsTl (Fig. 5.17) (Dong and Corbett, 1996). Depending on the cation size, structures with different Tl-networks/clusters are formed. In $cF16$-NaTl, we have a diamond-type Tl-network, in $oS48$-KTl the distorted naked Tl^{6-} clusters form an approximate ccp arrangement, while in $oF96$-CsTl the

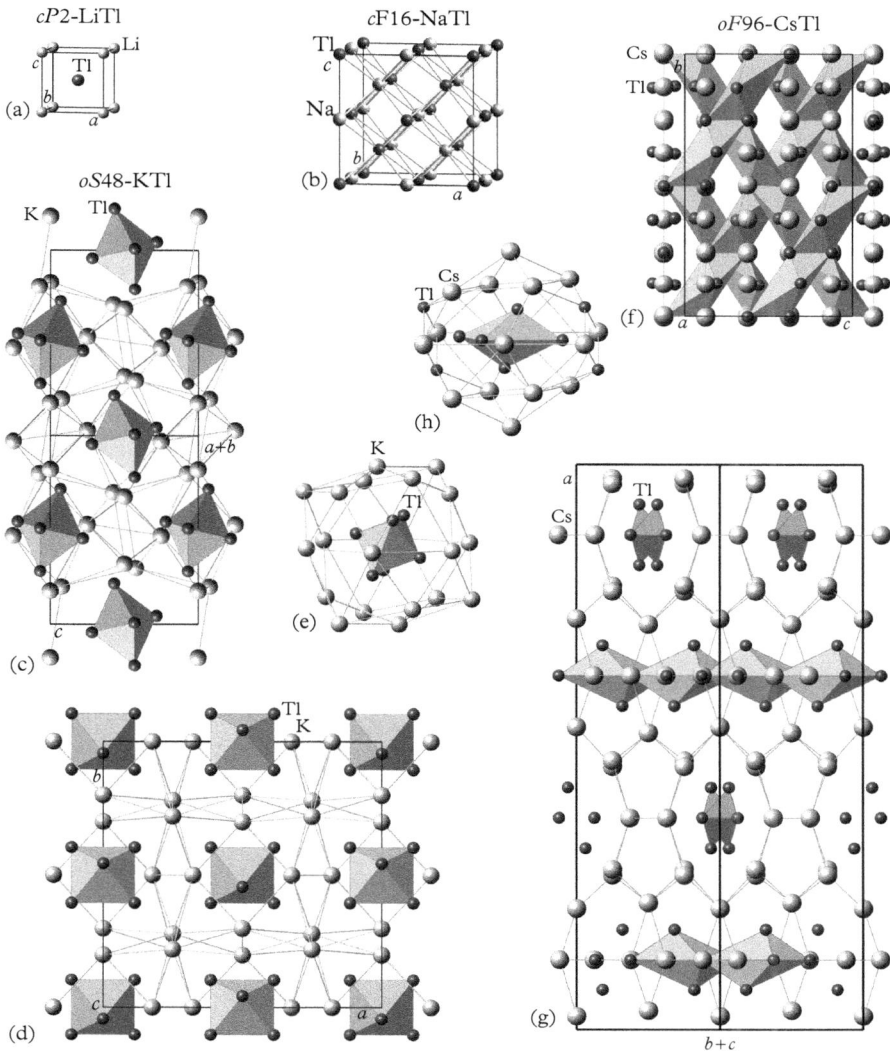

Fig. 5.17 *The structures of the Zintl-phases (a) $cP2$-LiTl, (b) $cF16$-NaTl, (c)–(e) $oS48$-KTl and (f)–(h) $oF96$-CsTl in different projections. The Tl^{6-} clusters in $oS48$-KTl and $oF96$-CsTl are shown as distorted octahedra. The Tl^{6-} cluster in $oS48$-KTl is coordinated by 20 Cs (e), that in $oF96$-CsTl by 16 Cs and 4 Tl from the edge-sharing adjacent octahedra.*

Tl^{6-} clusters are even more deformed building a network by sharing two opposite edges in each case.

To summarize, AB-type structures have been observed with 2, 4, 6, 8, 12, 16, 24, 32, 40, 48, 64, 78, 96, and 304 atoms per unit cell. Those with two atoms per unit cell, for instance, can be described as distorted variants of either the $cP2$-CsCl or the $hP2$-LiPt structure type. In contrast, those with 96 or 304 atoms per unit cell are defect structures, probably of electronic origin (pseudogap at the Fermi edge). For instance, the unit cell of $cF96$-CdNi (227 $Fd\bar{3}m$) contains 48 Cd, 44 Ni, and 4 vacancies (Critchley and Jeffery, 1965). In the case of $oS304$-MgIr (*Cmca*), one Mg site (16g) is not fully occupied, and two Ir sites (16g) contain a fraction of Mg (Cerny *et al.*, 2004).

5.4.3 A$_3$B/AB$_3$

The 830 intermetallics with compositions A$_3$B/AB$_3$ crystallize in 82 different structure types, i.e., with on average 10.1 representatives per structure type. The number of atoms per unit cell ranges from 4 up to 320 ($cP4$-Cu$_3$Au and $oS320$-Cu$_3$Pd, respectively). The ten most common structure types, with 611 representatives (61.1%), are the following:

$cP4$-**Cu$_3$Au** (221 $Pm\bar{3}m$, 1a 3c; 249 representatives): $cF4$-Cu derivative structure. See Fig. 7.3.

$oP16$-**Fe$_3$C** (62 *Pnma*, 4c^2 8d; 88 representatives): 9-fold coordination based on a tricapped prism; $hP2$-Mg derivative structure.

$hP8$-**Mg$_3$Cd** (194 $P6_3/mmc$, 2d 6h; 61 representatives): the system Mg–Cd forms a continuous solid solution with strongly varying c/a ratio. At temperatures below 200°C, $hP8$-Mg$_3$Cd, a $hP2$-Mg superstructure, orders from the solid solution. See Fig. 7.41.

$cP8$-**Cr$_3$Si** (223 $Pm\bar{3}n$, 2a 6c; 54 representatives): the A atoms (Si) form a cI lattice through which chains of B atoms (Cr) are running parallel to the edges of the cubic cell; A atoms are icosahedrally coordinated by B atoms. See Fig. 7.28.

$hR36$-**PuNi$_3$** (166 $R\bar{3}m$, 3ab 6c^2 18h; 50 representatives): a combination of structure motifs of the $hP6$-CaCu$_5$ type and the $cF24$-MgCu$_2$ type.

$cF16$-**BiF$_3$** (225 $Fm\bar{3}m$, 4c^2 8d; 32 representatives): ($2 \times 2 \times 2$)-fold superstructure of $cI2$-W. See Fig. 7.7.

$oP8$-**Cu$_3$Ti** (59 *Pmmn*, 2ab 4e26 representatives): $hP2$-Mg superstructure with close packed layers parallel to (010), and no Ti–Ti nearest neighbors.

$tI8$-**TiAl$_3$** (139 $I4/mmm$, 2ab 4e; 20 representatives): 2-fold superstructure of $cF4$-Cu with no Ti–Ti nearest neighbors. See Fig. 7.3.

$tP4$-**SrPb$_3$** (123 $P4/mmm$, 1ac 2e; 16 representatives): $cF4$-Cu derivative structure.

*h*P16-**TiNi**₃ (194 *P*6₃/*mmc*, 2*ad* 6*gh*; 15 representatives): *h*P2-Mg super-structure, double-hexagonal packing with sequence ABAC along [001].

While the representatives of the *c*P4-Cu₃Au structure type are distributed over nearly the entire range of A₃B-compounds, the region occupied by *o*P16-Fe₃C-type structures is restricted to the area of $M(A) = 15$–33 and $M(B) = 62$–74, with only three exceptions and only one compound falling in the same area and be-longing to a different structure type. Phases with *h*P8-Mg₃Cd-type structures are again rather widely spread over the range of elements. The *c*P8-Cr₃Si-type representatives are all located within $M(A) = 49$–57 and $M(B) = 49$–88. Most *h*R36-PuNi₃-type structures have values of $M(A) = 61$–67 and $M(B) = 16$–33 or 43–47. Intermetallics with structures of the *c*F16-BiF₃ type, again, are spread over disparate regions of M/M space. $M(A)/M(B)$ plots of all structures at this com-positions and highlights of the respective most common structure types, are given in Fig. 5.18. The stability fields are well-separated and partially adjacent to one another.

5.4.4   A₅B₃/A₃B₅

The 391 intermetallics with compositions A₅B₃/A₃B₅ exhibit 41 different struc-ture types, i.e., with on average 9.5 representatives per structure type. The number of atoms per unit cell ranges from 16 up to 128 (*h*P16-Mn₅Si₃ and *o*I128-Au₅Zn₃, respectively). The ten most common structure types, with 330 representatives (84%), are the following:

*h*P16-**Mn₅Si₃** (193 *P*6₃/*mcm*, 4*d* 6*g*²; 148 representatives). One of the No-wotny phases; it can be described as a packing of face-sharing CN16 FK-polyhedra, leaving empty spaces corresponding to columns of face-sharing octahedra running along [001]. See Fig. 7.54.

*o*S32-**Pu₃Pd₅** (63 *Cmcm*, 4*c*² 8*efg*; 39 representatives). A puckered layer, with non-bonding distances between the atoms, is sandwiched between two sym-metrically equivalent flat layers stacked along [001]. The puckering is caused by fitting this layer in the best way between the flat triangle/square/hexagon layers. See Fig. 7.65.

*t*I32-**W₅Si₃** (140 *I*4/*mcm*, 4*ab* 8*h* 16*k*; 35 representatives). Two symmetrically equivalent layers of the type $3^2.6.3.4+3^2$ are stacked in anti-orientation along [001]. In-between the layers, W and Si atoms center the resulting hexagonal and square antiprisms, respectively.

*t*I32-**Cr₅B₃** (140 *I*4/*mcm*, 4*ac* 8*h* 16*l*; 33 representatives). Cr-centered Cr cubes are edge-connected so that their base and top squares form a $3^2.4.3.4$ triangle/square tiling. The B atoms cap the sides of the cubes, and center the square antiprisms formed by the top and bottom squares of along [001] neighboring Cr cubes.

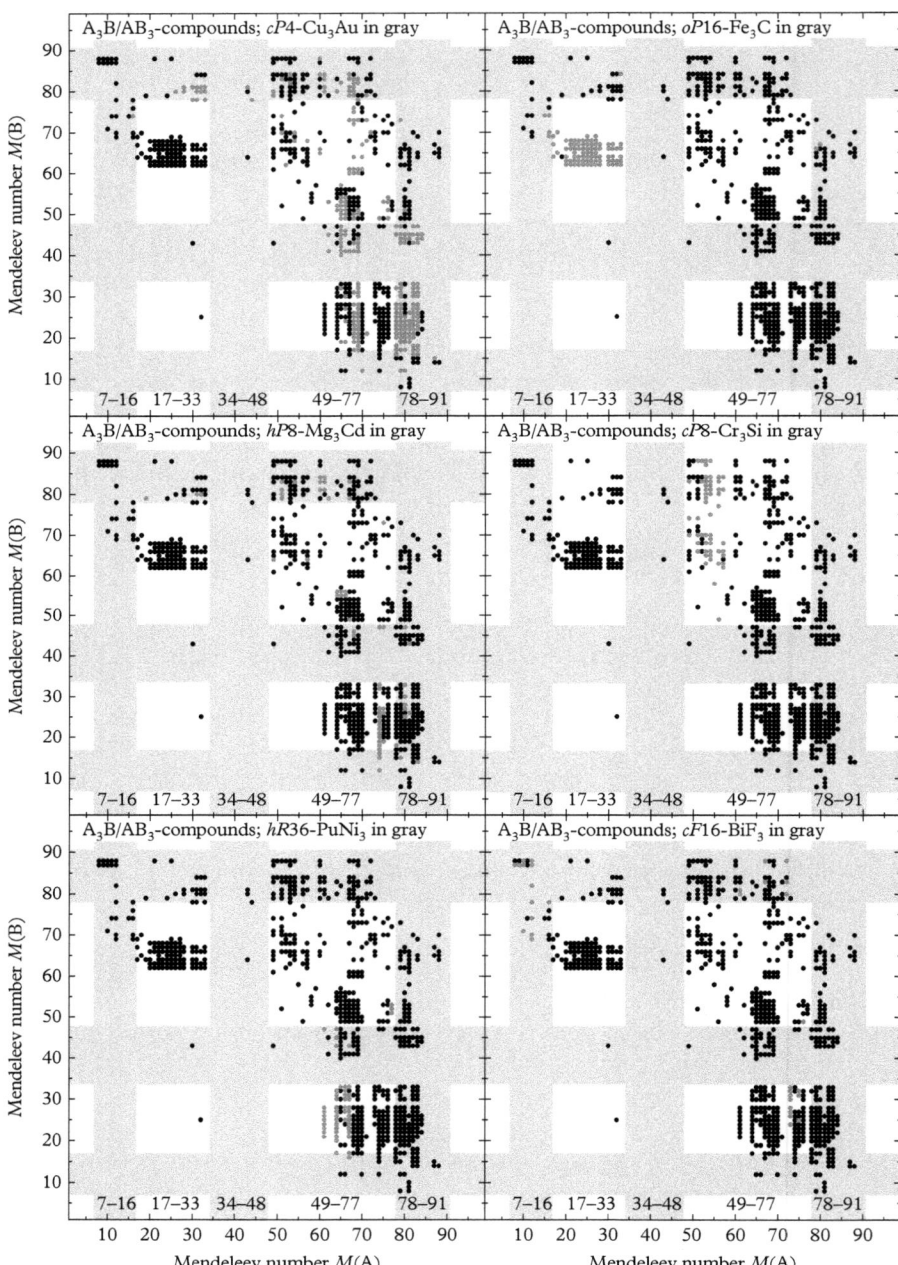

Fig. 5.18 $M(A)/M(B)$ *plot of the binary intermetallic compounds with composition* A_3B/AB_3 *(*$M(A) < M(B)$*). In the plots, however,* $M(A)$ *always refers to the majority component. The six most common structure types are highlighted. Areas containing alkali metals, alkaline earth metals, actinoids, and (semi)metallic main group elements are shaded gray.*

*t*P32-**Pu₅Rh₃** (130 *P*4/*ncc*, 4*bc* 8*f* 16*g*; 16 representatives). This structure type is related to that of *t*I32-W₅Si₃, and can be obtained from the other by cutting the structures into slices and shifting these along [001].

*o*P32-**Yb₅Sb₃** (62 *Pnma*, 4*c*⁴ 8*d*²; 16 representatives). This structure type is closely related to that of *o*P32-Y₅Bi₃, listed below.

*o*P32-**Y₅Bi₃** (62 *Pnma*, 4*c*⁴ 8*d*²; 14 representatives). This structure type is closely related to that of *o*P32-Yb₅Sb₃, listed above. The structure shows similar pseudo-cubic units, centered by octahedra. Along [010], in the corners of the unit cell, there are columns of trigonal Y-prisms centered by Bi atoms, which share edges forming deformed hexagonal channels. The octahedra share two of their vertices forming chains along [010]. At the same time the pseudo-cubes, which enclose them, share faces.

*o*P32-**Tm₃Ga₅** (62 *Pnma*, 4*c*⁴ 8*d*²; 11 representatives). The structure type is related to that of *o*S32-Pu₃Pd₅, listed above. The structure motifs are similar, just the packing principle differs. There is also a close structural relationship to *o*P32-Yb₅Sb₃ and *o*P32-Y₅Bi₃, also listed above (Yatsenko *et al.*, 1983).

*o*F64-**Y₃Ge₅** (43 *Fdd*2, 8*a*² 16*b*³; 10 representatives). The structure can be described as a defect *t*I12-ThSi₂ derivative structure (Venturini *et al.*, 1999).

*t*P32-**Ba₅Si₃** (130 *P*4/*ncc*, 4*c*² 8*f* 16*g*; 8 representatives). The structure is related to the above-listed *t*I32-Cr₅B₃ structure type.

The 148 intermetallics with the *h*P16-Mn₅Si₃ structure type dominate the main region with values $M(A) = 14$–33 or 43–56 and $M(B) = 60$–88. Phases with *o*S32-Pu₃Pd₅-type structures are concentrated within $M(A) = 78$–84 and $M(B) = 14$–33 or 45–49 with only two exceptions. Intermetallics belonging to the types *t*I32-W₅Si₃ and *t*I32-Cr₅B₃ are spread over the general *h*P16-Mn₅Si₃-region specified above. Compounds with *t*P32-Pu₅Rh₃-type structures can only be found within $M(A) = 17$–33 or 43 and $M(B) = 65$–66. Those with the type *o*P32-Yb₅Sb₃ are concentrated within $M(A) = 15$–27 (with only one outlier at $M(A) = 51$) and $M(B) = 87$–88. $M(A)/M(B)$ plots of all structures at this composition and highlights of the respective most common structure types, are given in Fig. 5.19.

5.4.5   A₃B₂/A₂B₃

The 247 intermetallics with composition A₃B₂/A₂B₃ exhibit 78 different structure types, i.e., with on average 3.2 representatives per structure type. The number of atoms per unit cell ranges from 5 up to 140 (*h*P5-Ni₂Al₃ and *t*I140-Y₃Rh₂, respectively). The eight most common structure types, with 110 representatives (44.5%), two of them with different stoichiometry of the structure type, are the following:

*h*R45-**Er₃Ni₂** (148 *R*3̄, 3*a* 6*c* 18*f*²; 24 representatives). The structure can be described as a stacking of polyhedra along [001] in the hexagonal setting: an

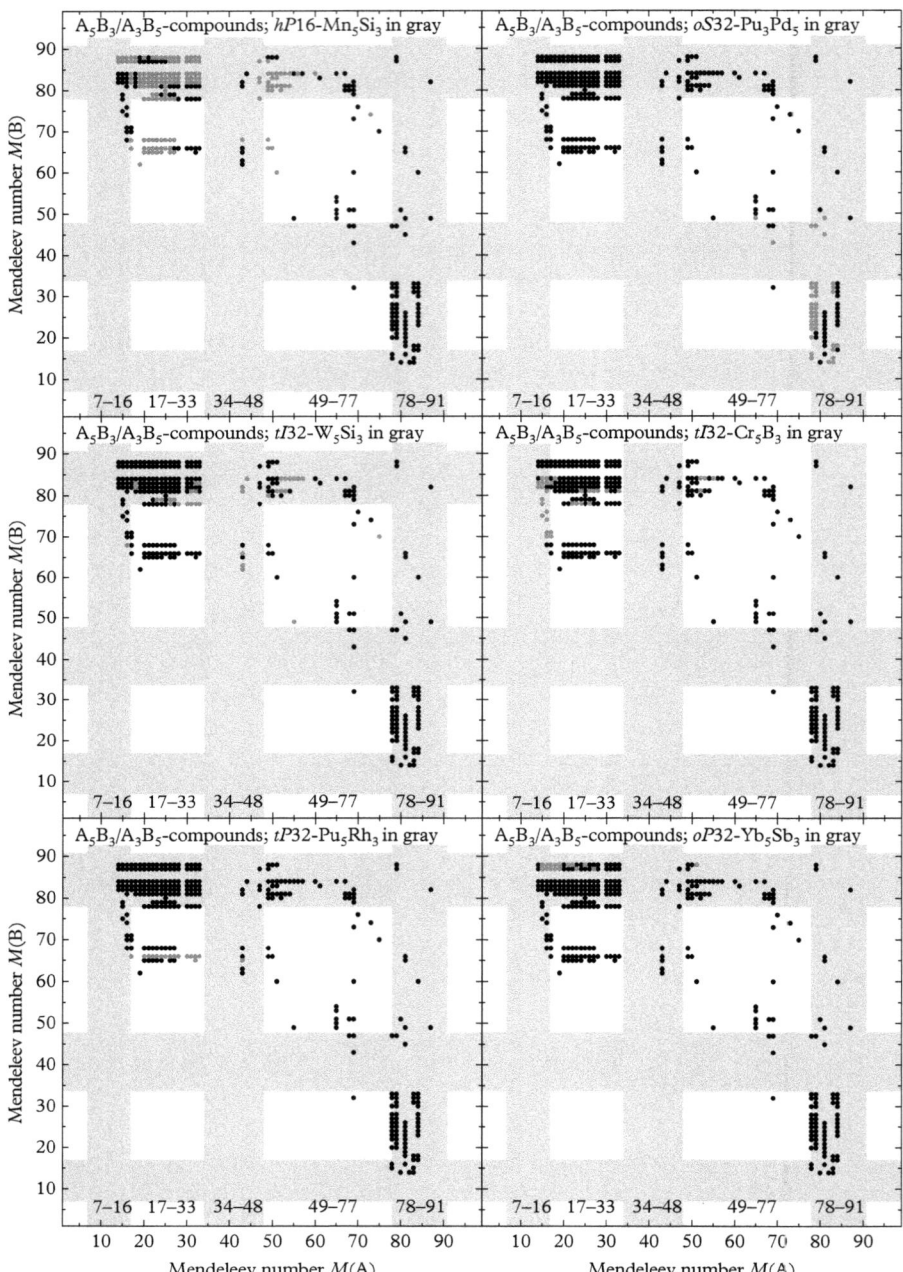

Fig. 5.19 $M(A)/M(B)$ *plot of the binary intermetallic compounds with composition* A_5B_3/A_3B_5 *(M(A) < M(B)). In the plots, however, M(A) always refers to the majority component. The six most common structure types are highlighted. Areas containing alkali metals, alkaline earth metals, actinoids, and (semi)metallic main group elements are shaded gray.*

octahedron shares one face with a CN16 polyhedron, which is interpenetrated by a CN14 polyhedron, then, symmetrically equivalent by an inversion center, the CN16 polyhedron and the octahedron (Moreau *et al.*, 1974).

*t*P10-**U₃Si₂** (127 *P*4/*mbm*, 2*a*4*gh*; 17 representatives). Stacking variant of $3^2.4.3.4$ triangle/square nets. The net in $z = 0$ is decorated on its vertices by Si and in the centers of the squares by the much larger U atoms, while the one in $z = 1/2$ is decorated by U atoms only, which leads to almost regular triangles. The structure can also be seen as a framework of vertex-sharing U-octahedra sandwiched between the dual triangle square nets. See Fig. 7.26.

*t*P30-**Cr₀.₄₉Fe₀.₅₁** (136 *P*4₂/*mnm*, 2*a*4*f*8*i²j*; 13 representatives). Frank-Kasper σ-phase. See Fig. 7.13. The stoichiometry of the prototype structure type differs from the representative structures discussed here, A_3B_2/A_2B_3.

*h*P5-**Ni₂Al₃** (162 *P*3̄*m*1, 1*a*2*d²*; 12 representatives). Defect derivative structure of the *c*P2-CsCl type.

*t*I140-**Y₃Rh₂** (140 *I*4/*mcm*, 4*abc*8*fgh²*16*l²*32*m²*; 12 representatives). This complex structure is related to the *t*I32-W₅Si₃, which was discussed above (Moreau *et al.*, 1976).

*h*P3-**AlB₂** (191 *P*6/*mmm*, 1*a*2*d*; 11 representatives). Stacking of honeycomb nets with Al at the vertices, and B in the centers of the hexagonal prisms formed by the Al atoms of adjacent layers. See Fig. 7.23. The stoichiometry of the prototype structure differs from the representative structures discussed here, A_3B_2/A_2B_3.

*t*I80-**Gd₃Ga₂** (140 *I*4/*mcm*, 4*ac*8*gh²*16*l*32*m*; 11 representatives). The structure can be described by a stacking of parts of the *t*I32-W₅Si₃ structure and *t*I140-Y₃Rh₂ (Yatsenko *et al.*, 1986).

*t*P20-**Zr₃Al₂** (136 *P*4₂/*mnm*, 4*dfg*8*j*; 10 representatives). The structure is related to that of *t*I12-CuAl₂, which can be described as stackings of $3^2.4.3.4$ triangle/square nets located in $z = 0$ and 1/2, with the squares capped by Cu in $z = 1/4$ and 3/4 (see Fig. 7.26(a)–(d)).

The A_3B_2-structures are generally spread over a large space of element-combinations. The ones belonging to different structure types often occur in rather small areas, on the other hand. A good example are those in structure type *h*R45-Er₃Ni₂ bearing values $M(A) = 14$–33 and $M(B) = 65$–71. *t*P10-U₃Si₂-type structures occur in two small regions: either within $M(A) = 15$–27 and $M(B) = 69$–74 or in $M(A) = 47$–53 and $M(B) = 77$–84. *t*P30-Cr₀.₄₉Fe₀.₅₁ dominates the rather central area with $M(A) = 52$–61 and $M(B) = 56$–69. The region of *h*P5-Ni₂Al₃-type structures is also quite compact: $M(A) = 70$–81 and $M(B) = 59$–79. *t*I140-Y₃Rh₂-type structures cover an even smaller and exclusive area with $M(A) = 22$–27 and $M(B) = 65$–66. All *h*P3-AlB₂-type representatives of structures with composition A_3B_2 have the same major element $M(A) = 84$ (Ge) and $M(B) = 17$–28 and 44. An $M(A)/M(B)$ plot of all structures at this compositions is shown in Fig. 5.20, where the mostly non-overlapping regions occupied by the first six structure types are highlighted.

5.4.6 A_5B/AB_5

The 182 phases with composition A_5B/AB_5 crystallize in 33 different structure types, i.e., with on average 5.5 representatives per structure type. The number of atoms per unit cell ranges from 5 up to 140 ($hP6$-CaCu$_5$ and $cF448$-Mg$_5$Gd, respectively). The five most common ones – $hP6$-CaCu$_5$ (107 structures) (Fig. 7.31), $cF24$-Be$_5$Au (22 structures), $hP8$-Cu$_{5.44}$Tb$_{0.78}$ (7 structures), $oI12$-LaGe$_5$ (6 structures), and $hP36$-Zn$_5$Er (6 structures) – comprise 148 out of these, i.e., 81%. $hP6$-CaCu$_5$ itself makes up 59% of all A_5B-structures and is spread over almost all of its compositional space, while especially $oI12$-LaGe$_5$- and $hP36$-Zn$_5$Er-type structures are found in extremely narrow regions only. An $M(A)/M(B)$ plot of all structures at these compositions is shown in Fig. 5.21.

5.4.7 $A_{17}B_2/A_2B_{17}$

The 140 compounds with composition $A_{17}B_2/A_2B_{17}$ crystallize in 15 different structure types, i.e., with on average 9.3 representatives per structure type. The number of atoms per unit cell ranges from 38 up to 72 ($mS38$-Th$_2$Fe$_{17}$ and $hR72$-Sm$_2$Fe$_{17}$, respectively). The three most common ones – $hP38$-Th$_2$Ni$_{17}$ (66 structures), $hR57$-Zn$_{17}$Th$_2$ (43 structures), and $hP80$-Lu$_{1.82}$Fe$_{17.35}$ (6 structures) – comprise 109 out of these, i.e., 78%. The first two are both spread over nearly all element combinations in the stability field, and also the distribution of the few $hP80$-Lu$_{1.82}$Fe$_{17.35}$-type structures is broad. An $M(A)/M(B)$ plot of all structures at this compositions is shown in Fig. 5.21.

How different can structures be with such a specific complex composition? This will be explored using the example of $A_{17}B_2/A_2B_{17}$ type structures. As we can see from Table 5.5, we have 0 structure types/representatives for the composition $A_{17}B_2$, with $M(A) < M(B)$, and 17 structure types with 140 representatives for the composition A_2B_{17}. This means that for these stoichiometries only structure types are possible, which do not allow an exchange of atom types A and B. In the following, the ten structure types are discussed that appear to be closest to structure types with the required stoichiometry for full occupancy of the respective Wyckoff positions. However, not all of them have been confirmed. We list these structure types anyway, because this is also characteristic for some of the data in the PCD. There are several more entries for this composition in the PCD, however, which are assigned to the structure types $hP6$-CaCu$_5$ or $cI58$-Mn, etc., which do not have the required stoichiometry.

$hP38$-Th$_2$Ni$_{17}$ 194 $P6_3/mmc$ $2bc\,4f\,6g\,12jk$, 66 representatives: Al–(Dy, Er, Gd), Be–(Hf, Sc, Ti), Cd–(Ce–La), Ce–(Co, Fe, Mg, Zn), Co–(Dy, Er, Gd,

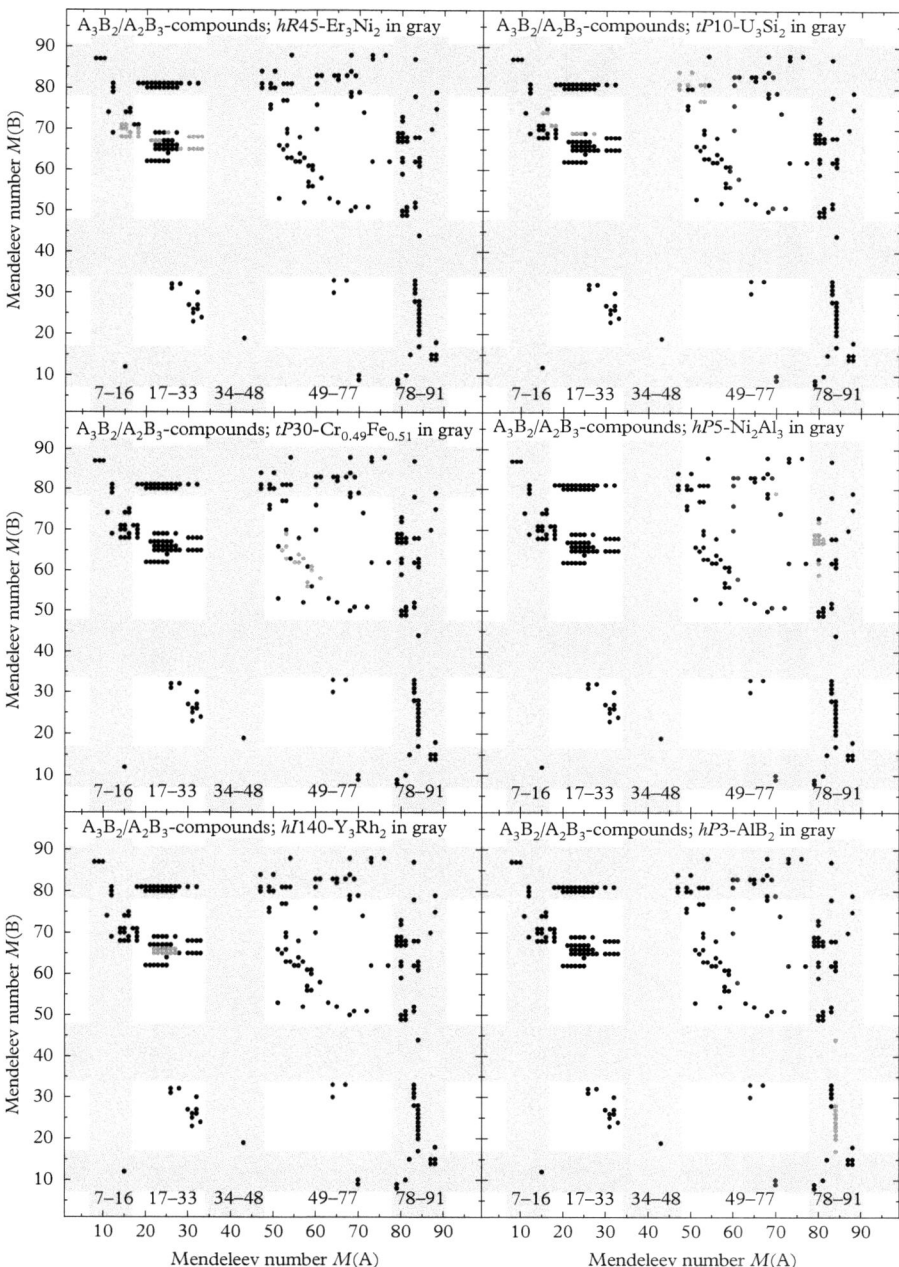

Fig. 5.20 $M(A)/M(B)$ *plot of the binary intermetallic compounds with composition* A_3B_2/A_2B_3 *(*$M(A) < M(B)$*). In the plots, however,* $M(A)$ *always refers to the majority component. The six most common structure types are highlighted.*

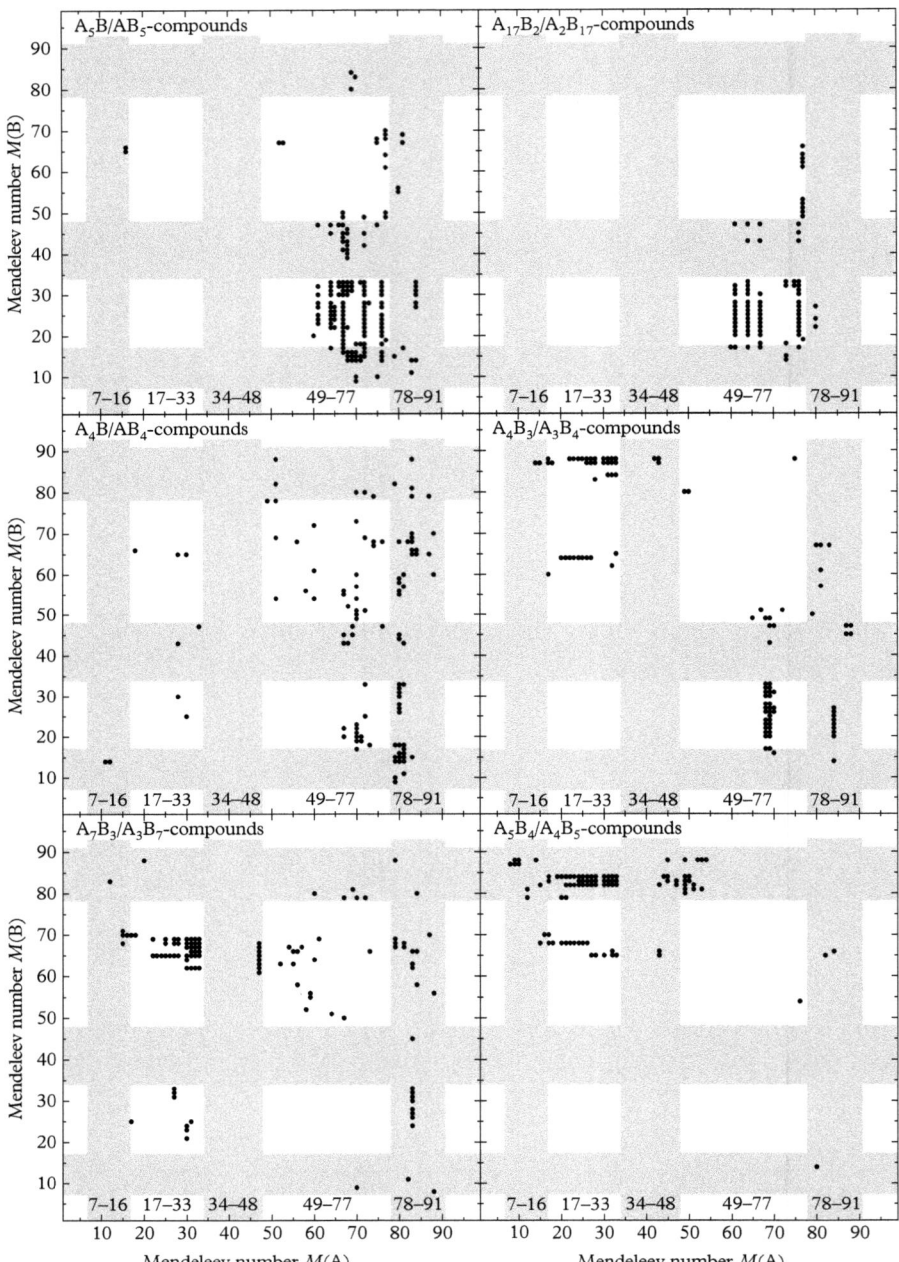

Fig. 5.21 $M(A)/M(B)$ *plot of the binary intermetallic compounds with compositions* A_5B, $A_{17}B_2$, A_4B, A_4B_3, A_7B_3, *and* A_5B_4. *In the plots,* $M(A)$ *always refers to the majority component. Areas containing alkali metals, alkaline earth metals, actinoids, and (semi)metallic main group elements are shaded gray.*

Ho, Lu, Pu, Sm, Tb, Tm, Y, Yb), (Dy, Er)–(Fe, Ni, Zn), Eu–(Mg, Ni), Fe–(Gd, Ho, Lu, Tb, Tm, Y, Yb), (Gd, Ho, Lu. Nd)–(Ni, Zn), La–(Mg, Zn), Mg–Sr, Mn–Yb, Ni–(Pu, Sm, Tb, Th, Tm, Y, Yb), (Pr, Pu, Sm, Tb, Th, Tm, U, Y, Yb)–Zn. The structure (Fig. 5.22(a,b)) can be described as two-fold superstructure of $hP6$-CaCu$_5$ (Fig. 5.22(e,f)). $hP6$-CaCu$_5$ is a stacking

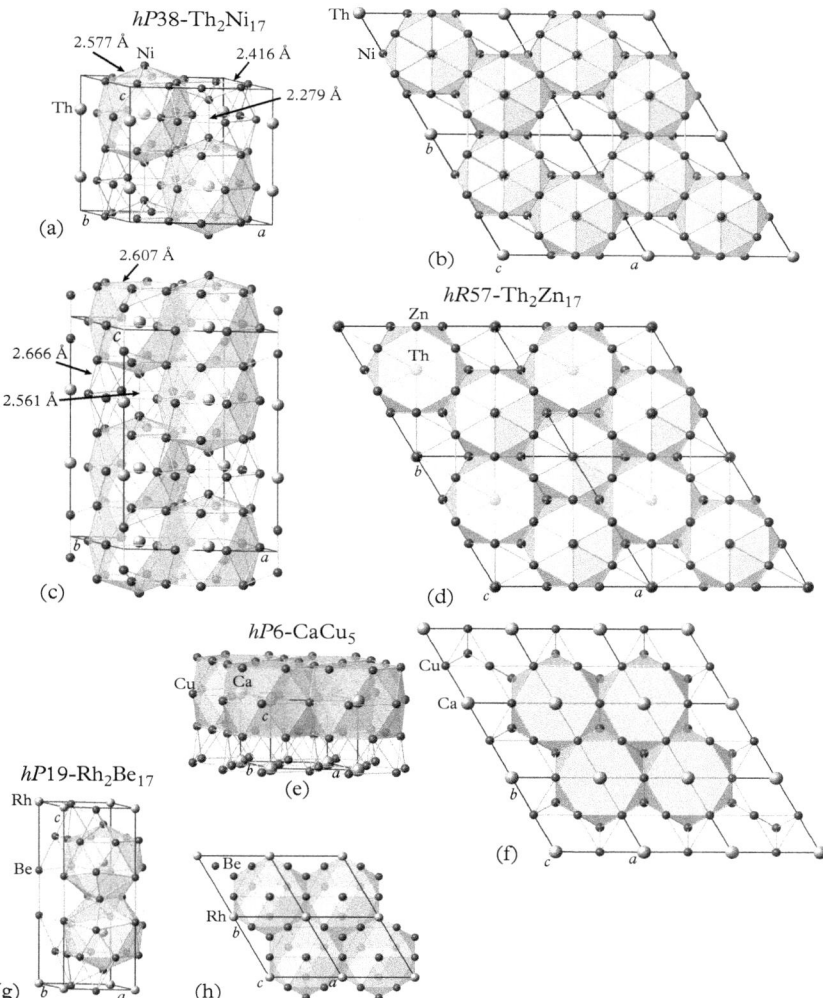

Fig. 5.22 *Most common structures of compounds with stoichiometry* A$_{17}$B$_2$/A$_2$B$_{17}$: *(a, b) $hP38$-Th$_2$Ni$_{17}$, (c, d) $hR57$-Zn$_{17}$Th$_2$, (e, f) $hP6$-CaCu$_5$, and (g, h) $hP19$-Rh$_2$Be$_{17}$. All compounds can be described as variants of stackings of honeycomb and Kagomé layers, and are related to the structures of $hP6$-CaCu$_5$ and $hP7$-Al$_3$Zr$_4$.*

of Cu-honeycomb layers, with the hexagon centers occupied by Ca, and of Cu-Kagomé layers. $hP38$-Th_2Ni_{17}, a Ni-Kagomé layer in $z = 0$ is capped by Ni atoms in $x = 1/3$, $y = 2/3$, i.e., in one third of the hexagons. Then a Ni-honeycomb layer follows in $z = 1/4$, with one third of the hexagons centered by Th atoms in $x = 0$, $y = 0$. Now, again, one third of the hexagons is capped by Ni atoms, which form dumbbells with the other capping Ni atoms below this layer. The Ni–Ni distances in these dumbbells are rather short, 2.279 Å, compared with the Ni–Ni distances in the hexagons of 2.790 Å and 2.417 Å in the honeycomb and the Kagomé layers, respectively ($r_{Ni} = 1.246$ Å). The next layer is again a Kagomé layer, etc. Each of the two sandwiches of Kagomé and honeycomb layers contribute 19 atoms to the structure within one unit cell, in total 38.

$hR57$-$Zn_{17}Th_2$ 166 $R\bar{3}m$ $6c^2$ $9d$ $18fh$, 43 representatives: Ba–Mg, Be–(Hf, Nb, Ta, Ti, Zr), (Ce, Dy)–(Co, Fe), Co–(Gd, Ho, La, Nd, Pr, Sm, Tb, Th,Y), Fe–(Gd, Nd, Pr, Sm, Tb, Th, Y), (Ce, Dy, Er, Gd, Ho, La, Lu, Nd, Pr, Pu, Sm, Tb, Th, Tm, U, Y, Yb)–Zn. The structure (Fig. 5.22(c, d)) can be described as threefold superstructure of $hP6$-$CaCu_5$. The Zn–Zn distances in the dumbbells are short, 2.561 Å, but comparable to other Zn–Zn distances in the hexagon layers of 2.607 Å ($r_{Zn} = 1.335$ Å). Each of the three sandwiches of Kagomé and honeycomb layers contribute 19 atoms to the structure within one unit cell, in total 57.

$hP19$-Rh_2Be_{17} ($hP19$-$Rh_{2.36}Be_{15.34}$) 187 $P\bar{6}m2$ $1ac$ $2ghi^2$ $3k$ $6n$, 5 representatives: Be–(Co, Fe, Ir, Os, Ru). The structure (Fig. 5.22(g, h)) can be described as a stacking of three Kagomé layers per unit cell, shifted so that each hexagon of one layer is capped by a triangle of the other one, with a Rh atom in between. The Rh atoms capping from both sides the triangles of the Kagomé layer in $z = 1/2$ are centering the Be-hexagons in the puckered honeycomb layer, and capping the hexagons of the Kagomé layer in $z = 0.183$. These are capped by Be atoms from the other side, which also cap the Be/Rh honeycomb tiling in $z = 0$. The shortest Be–Be distances, in the triangles bicapped by Rh, amount to 2.038 Å ($r_{Be} = 1.113$ Å), indicating covalent bonding contributions.

$hP38$-Lu_2Fe_{17} ($hP38$-$Lu_{1.82}Fe_{17.35}$) 194 $P6_3/mmc$ $2bcd$ $4ef$ $6g$ $12j^3k^2$, 6 representatives: Co–Er, Fe–(Lu, Y, Yb), Ni–(Th, Y). This is a rather disordered derivative of the $hP38$-Th_2Ni_{17} structure type.

$hP44$-$Mg_{17}Ce_2$ ($hP38$-$Mg_{17.58}Ce_{1.71}$) 194 $P6_3/mmc$ $2bcd$ $4ef$ $6g$ $12jk$ disordered $hP38$-Th_2Ni_{17} structure type with additional, partially occupied sites.

$hP114$-$Zn_{17}U_2$ has not been confirmed as $hP114$. The RT structure type of $Zn_{17}U_2$ is a representative of the $hR57$-$Zn_{17}Th_2$ structure type, and the HT phase of the $hP38$-Th_2Ni_{17} type.

$hR60$-Pr_2Fe_{17} ($Pr_{2.33}Fe_{16.33}$) has not been confirmed as $hR60$. The structure is a partially disordered variant of the $hR57$-$Zn_{17}Th_2$ type.

$hP56$-$\mathbf{Y_2Fe_{17}}$ has not been confirmed as $hP60$. There have been reported an RT structure of the $hR57$-$\mathrm{Zn_{17}Th_2}$ type and a HT structure of the $hP38$-$\mathrm{Th_2Ni_{17}}$ type.

$hR72$-$\mathbf{Sm_2Fe_{17}}$ ($hR57$-$\mathrm{Sm_{1.97}Fe_{17.06}}$) $hR72$ just reflects the sum of the site multiplicities not taking into account their partial occupancies. The structure is a partially disordered variant of the $hR57$-$\mathrm{Zn_{17}Th_2}$ type.

$mS38$-$\mathbf{Th_2Fe_{17}}$ has not been confirmed as $mS38$. Actually, this structure is of the $hR57$-$\mathrm{Zn_{17}Th_2}$ structure type.

5.4.8 $\mathbf{A_4B/AB_4}$

The 121 intermetallics with composition $\mathrm{A_4B/AB_4}$ exhibit 55 different structure types, on average 2.2 representatives per structure type. The number of atoms per unit cell ranges from 10 up to 102 ($tI10$-$\mathrm{BaAl_4}$ and $hP102$-$\mathrm{PtAl_4}$, respectively). The three most common structure types – $tI10$-$\mathrm{BaAl_4}$ (19 structures), $tI10$-$\mathrm{MoNi_4}$ (14 structures) (Fig. 7.3), and $oI20$-$\mathrm{UAl_4}$ (6 structures) – comprise only 39 out of these, i.e., 32%, while 33 structures with composition $\mathrm{A_4B}$ are assigned to structure types unique for this composition, and 22, 15, and 12 structures belong to structure types with 2, 3, or 4 representatives, respectively. All three most common structure types occur over very narrow ranges with respect to the majority element, while the minority elements come from a larger range. An $M(\mathrm{A})/M(\mathrm{B})$ plot of all structures at these compositions are shown in Fig. 5.21.

5.4.9 $\mathbf{A_4B_3/A_3B_4}$

The 117 phases with composition $\mathrm{A_4B_3/A_3B_4}$ exhibit 22 different structure types i.e., on average 5.3 representatives per structure type. The number of atoms per unit cell ranges from 7 up to 276 ($hP7$-$\mathrm{Zr_4Al_3}$ and $mP276$-$\mathrm{Cd_4Sb_3}$, respectively). The five most common ones – $hR42$-$\mathrm{Pu_3Pd_4}$ (38 structures), a $cI2$-W derivative structure, $cI28$-$\mathrm{Th_3P_4}$ (35 structures) (Fig. 7.87), $hP22$-$\mathrm{Ho_6Co_{4.5}}$ (9 structures), $oS28$-$\mathrm{Er_3Ge_4}$ (8 structures), and $oS32$-$\mathrm{Gd_3Ge_4}$ (7 structures) – comprise 97 out of these, i.e., 83%. The first four are very well-separated in an M/M-plot: both $hR42$-$\mathrm{Pu_3Pd_4}$- and $oS28$-$\mathrm{Er_3Ge_4}$-type structures are formed by majority elements with large M-values (65–70 and 84, respectively) and rather low minority element M-values (16–51 and 20–27, respectively). $cI28$-$\mathrm{Th_3P_4}$- and $hP22$-$\mathrm{Ho_6Co_{4.5}}$-type structures, on the other hand, mostly exhibit high M-values for their minority elements (83–88 and 60–64, respectively) and lower ones for the majority elements (14–43 and 17–27, respectively). In both cases, the range of low M-values is large and the one of high values is narrow, independent of the proportion of the respective element in the compound. (Four structures with structure type $cI28$-$\mathrm{Th_3P_4}$ have a reverse combination of elements compared with the above-noted trend – $M(\mathrm{A}) = 87$–88 and $M(\mathrm{B}) = 45$–47 – and another one falls outside those ranges altogether with $M(\mathrm{A}) = 33$ and $M(\mathrm{B}) = 65$.) An $M(\mathrm{A})/M(\mathrm{B})$ plot of all structures at this compositions is shown in Fig. 5.21.

5.4.10 A_7B_3/A_3B_7

The 115 intermetallics with composition A_7B_3 exhibit 33 different structure types i.e., on average 3.5 representatives per structure type. The number of atoms per unit cell ranges from 20 up to 80 ($hP20$-Th_7Fe_3 and $oP80$-Ca_7Au_3, respectively). The five most common ones – $hP20$-Th_7Fe_3 (45 structures), $cI40$-Ru_3Sn_7 (11 structures), $tP30$-$Cr_{0.49}Fe_{0.51}$ (8 structures) (Fig. 7.13), $oS28$-Tb_3Sn_4 (6 structures), and $hR9$-Sm (5 structures) – comprise 75 out of these, i.e., 65%. They are well-separated and occur in quite compact regions: $hP20$-Th_7Fe_3 – $M(A) = 15$–47 and $M(B) = 61$–70, $cI40$-Ru_3Sn_7 – $M(A) = 73$–88 and $M(B) = 56$–69, $tP30$-$Cr_{0.49}Fe_{0.51}$ – $M(A) = 52$–59 and $M(B) = 55$–67, $oS28$-Tb_3Sn_4 – $M(A) = 83$ and $M(B) = 24$–33, and $hR9$-Sm – $M(A) = 27$–33 and $M(B) = 21$–33. $M(A)/M(B)$ plots of all structures at these compositions are shown in Fig. 5.21.

5.4.11 A_5B_4/A_4B_5

The 104 phases with composition A_5B_4/A_4B_5 exhibit 17 different structure types i.e., on average 6.1 representatives per structure type. Most of them are Zintl phases. The number of atoms per unit cell ranges from 9 up to 72 ($hP9$-Li_5Ga_4 and $oF72$-Rh_4Pb_5, respectively). The three most common ones—$oP36$-Sm_5Ge_4 (58 structures) (Fig. 7.71), $hP18$-Ti_5Ga_4 (17 structures), and $hP16$-Mn_5Si_3 (5 structures) (Fig. 7.54)—comprise 80 out of these, i.e., 77%. While the first two are mostly separated, the few $hP16$-Mn_5Si_3-type compounds (a structure type, which does not have the required stoichiometry, anyway) at this composition are spread out over the A_5B_4 compositional range. $oP36$-Sm_5Ge_4 is mostly concentrated at $M(A) = 15$–33 and $M(B) = 65$–68 and $M(A) = 15$–33 and $M(B) = 82$–84. $hP18$-Ti_5Ga_4 occurs nearly only within $M(A) = 43$–53 and $M(B) = 80$–88. An $M(A)/M(B)$ plot of all structures at these compositions are shown in Fig. 5.21.

How diverse can structures with such a specific complex composition be? This will also be explored here on the example of A_5B_4/A_4B_5 type structures. As we can see from Table 5.5, we have 98/13 structure types/representatives for the composition A_5B_4, with $M(A) < M(B)$, and 6/5 structure types/representatives for the composition A_4B_5. This means that for these stoichiometries mainly structure types are possible that do not allow an exchange of atom types A and B. In the following, the 14 structure types are discussed that appear to be closest to structure types with the required stoichiometry for full occupancy of the respective Wyckoff positions.

> $oP36$-Sm_5Ge_4 62 *Pnma* $4c^3\, 8d^3$, 58 representatives: Au–Yb, Ba–Sb, Ir–(Ce, Pu), Rh–(Ce, Gd, La, Nd, Pu, Sm), Ge–(Ce, Dy, Er, Gd, Hf, Ho, La, Lu, Nd, Pr, Sc, Sm, Tb, Tm, Y, Yb), Pb–(Ce, Dy, Er, Gd, Ho, La, Nd, Pr, Sm, Sr, Tb, Tm), Pt–(Dy, Er, Eu, Ho, Lu, Sr, Tb, Tm, Y, Yb), Sn–(Ce, Dy, Gd, La, Nd, Pr, Sm, Tb, Y, Yb). The structure can be described as stacking of slabs that are shifted against each other (Fig. 5.23(a, b)). The slabs consist of

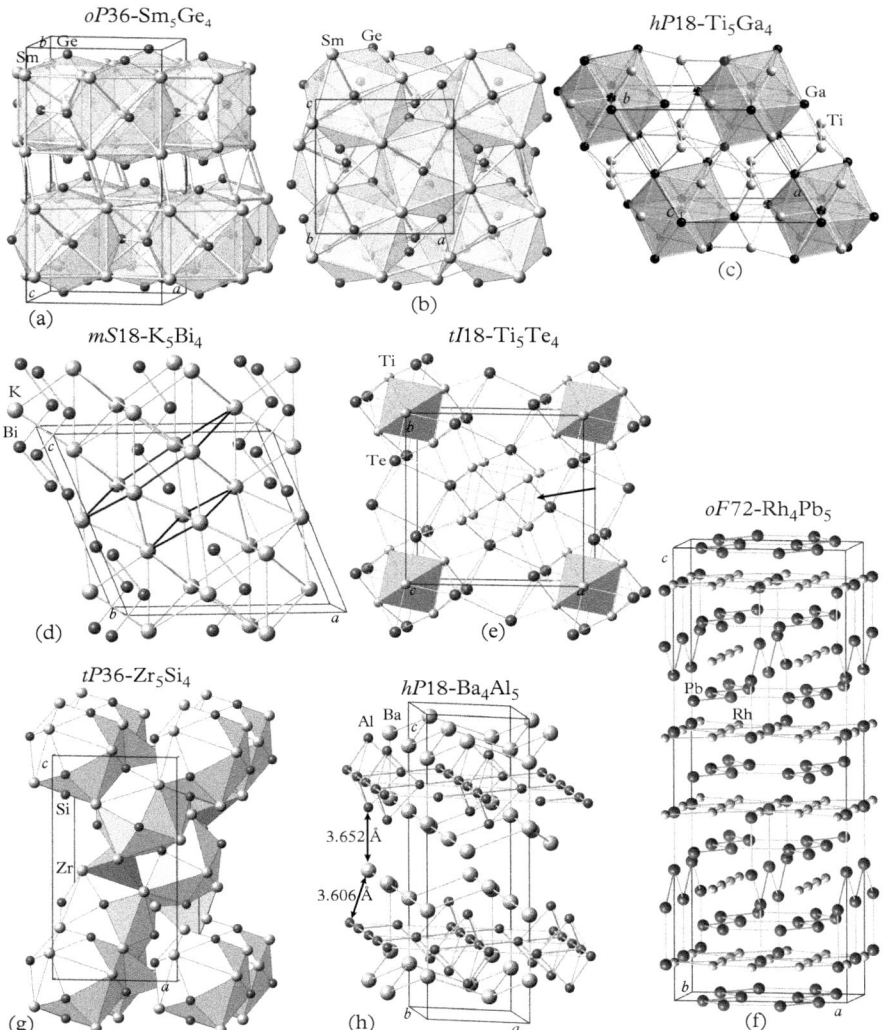

Fig. 5.23 *The structure types of intermetallics with composition* A_5B_4/A_4B_5: *(a, b)*
oP36-Sm_5Ge_4, with the all-side-Ge-capped bcc Sm_8-cubes shaded gray; (c, d)
hP18-Ti_5Ga_4, with the all-side Ti-capped Ga_8-cubes shaded gray; (e) mS18-K_5Bi_4: note the
hexagon and square outlined in black as well as the Bi_4^{4-} polyanions; (f) tI18-Ti_5Te_4: the
4-side Te-capped bcc tetragonally distorted Ti_8 cube is marked by an arrow;
(g) oF72-Rh_4Pb_5; (h) tP36-Zr_5Si_4: the Si-capped bcc Zr-cubes are shaded gray;
(i) hP18-Ba_4Al_5.

face-sharing all-side-Ge-capped *bcc* Sm_8-cubes. The top and bottom faces of the cubes form a $4.3.4^2$ triangle-square tiling.

*hP*18-**Ti₅Ga₄** 193 *P*6₃/*mcm* $2b\,4d\,6g^2$, 17 representatives: Ga–(Nb, Ti), Ge–Np, Hf–Sn, In–(Lu, Tm), Pb–(Pu, Th), Sn–Th, U–(Ge, Sb, Sn), Zr–(Al, Ga, Pb, Sb, Sn). *hP*6-AlB₂ derivative structure that can also be considered a packing of all-side Ti-capped Ga_8-cubes (shaded gray) and Ti-chains running along [001], with a period of 2.726 Å (Fig. 5.23(c, d)). Alternatively, the structure could be seen as a packing of *stellae octangulae*, with all-side Ga-capped Ti-octahedra face-sharing along [001].

*mS*18-**K₅Bi₄** 12 *C*2/*m* $2a\,4i^4$, 6 representatives: (K, Rb, Cs)–(Bi, Sb). The structure can be described as two interpenetrating stackings of K-hexagon-square tilings with the Bi_4^{4-} polyanions crossing through the centers of the elongated hexagons (Fig. 5.23(e)).

*oP*36-**Gd₅Si₄** 62 *Pnma* $4c^3\,8d^3$, 4 representatives: Ge–(Gd, Ho, Yb), Pb–Y. This structure is closely related to the *oP*36-Sm₅Ge₄ structure type.

*tI*18-**Ti₅Te₄** 87 *I*4/*m* $2a\,8h^2$, 3 representatives: Sb–(Nb, Ta, V). The structure can be described as vertex-sharing Ti-octahedra running along [001] (Fig. 5.23(f)). All octahedra faces are capped by Te atoms forming a *stella octangula*, which link these units to the neighboring ones. The Ti–Te distances are with 2.772 Å significantly shorter than the sum of atomic radii 2.880 Å ($r_{Te} = 1.432$ Å, $r_{Ti} = 1.448$ Å) indicating covalent bonding contributions. Te–Te distances are, with 3.782 Å, significantly larger. The structure could also be interpreted as a packing of all-side capped cubes.

*hP*9-**Li₅Ga₄** 162 *P*3̄*m* $1a\,2cd^3$, 2 representatives: Li–(Ga, In). The structure can be derived from the *cI*2-W structure type. The AETs of all atoms are cubes.

*mS*20-**Rb₅As₄** 12 *C*2/*m* $4i^5$, 2 representatives: Cs–Bi, K–Sb. Identical to the *mS*18-K₅Bi₄ structure type, just showing a K split-position.

*oF*72-**Rh₄Pb₅** 69 *Fmmm* $8fhi^3\,16jm$, 1 representative, Rh–Pb. Rh is arranged in six equidistant Kagomé layers, four complete ones and two with every other column of atoms missing (Fig. 5.23(g)). There are six equidistant Pb-honeycomb layers in-between the Rh-layers. Each Pb-hexagon is bicapped by Pb atoms. On one side, the capping atoms center the hexagons of a Rh-Kagomé layer, on the other hand, they form square-antiprismatic channels for a column of Rh atoms. The next layer is a Rh-Kagomé layer again, with the Rh-hexagons centered by Pb. Between this sandwich and an equivalent one, one Pb-honeycomb layer is inserted. Another way to look at the structure identifies patches of a $4.3.4.3^2$ triangle-square-tiling of Pb atoms parallel to the (101) plane.

*tP*36-**Zr₅Si₄** 92 *P*4₁2₁2 $4a\,8b^4$, 2 representatives: (Ge, Si)–Zr. In the structure, there are Si-capped *bcc* Zr-cubes, which form face-sharing infinite units along the 4₁ screw axis, and are connected otherwise via tetrahedra (Fig. 5.23(h)).

*m*P18-**Ca₅Au₄** 14 $P2_1/c$ $2a\,4e^4$, 1 representative: Au–Ca. The structure is closely related to that of *o*P36-Sm₅Ge₄.

*o*S36-**Eu₅As₄** 64 *Cmce* $4a\,8df$ $16g$, 2 representatives: As–Eu, Ba–Sb. The structure is a more symmetrical, but closely related, derivative of *o*P36-Sm₅Ge₄.

*h*P18-**Ba₄Al₅** 194 $P6_3/mmc$ $4ef^2$ $6h$, 1 representative: Al–Ba. Chair structure of Ba atoms in puckered layers connected by Al. Al forms Kagomé nets, where half of the triangles are bicapped with Al and half with Ba (Fig. 5.23(i)). The Al hexagons are bicapped with Ba atoms.

*t*P36-**Ir₄Ge₅** 194 $P6_3/mmc$ $4ef^2$ $6h$, 1 representative: Ir–Ge. This is a Nowotny chimney ladder structure (see Section 7.9).

*t*I18-**V₄Zn₅** 139 $I4/mmm$ $2a\,8hj$, 3 representatives: V–Zn. $(3 \times 3 \times 1)$ *c*I2-W superstructure with a kind of distorted sublattice (see Fig 7.7).

The 14 structure types discussed here can be classified into five groups. The top group already contains 50% of the structure types and more than 80% of all intermetallics with stoichiometry A_5B_4/A_4B_5. The structures belonging to a group have common structural subunits: (i) the B-capped *bcc* A-cubes can be found in the structures of *o*P36-Sm₅Ge₄, *o*P36-Gd₅Si₄, *m*P18-Ca₅Au₄, *o*S36-Eu₅As₄, *t*I18-Ti₅Te₄, *t*P36-Zr₅Si₄, and A-capped *bcc* B-cubes in *h*P18-Ti₅Ga₄; (ii) stacking of B-Kagomé nets in *h*P18-Ba₄Al₅ and *o*F72-Rh₄Pb₅; (iii) *c*I2-W derivative structures in *h*P9-Li₅Ga₄ and *t*I18-V₄Zn₅; (iv) hexagon/square tilings in *m*S18-K₅Bi₄ and *m*S20-Rb₅As₄; and (v) quite singular is *t*P36-Ir₄Ge₅ with a Nowotny chimney ladder structure.

5.5 Stability maps and composition diagrams of ternary intermetallics

Only a few ternary intermetallic systems, compared to the large number of possible ones, have been studied thoroughly so far. Indeed, although the 13 026 ternary compounds in the PCD originate from 5109 different ternary systems (see Table 5.6 and Fig. 5.24), phase diagrams of only 2061 of the 85 320 possible different ternary systems have been studied well enough to be included in the ASM Alloy Phase Diagram Database[1] (Fleming, 2014). Furthermore, until now only the chemical elements have been studied systematically as a function of temperature as well as a function of pressure, also, only a few of them as a function of pressure and temperature at the same time. High-pressure studies have rarely been done for binary or ternary intermetallic compounds.

Looking at Fig. 5.24, one has to keep in mind that most of the volume spanned by the three coordinates $M(A)$, $M(B)$, and $M(C)$ is empty, because it contains

---

[1] Altogether, the ASM database contains 6499 different systems, 4438 of which also contain nonmetallic elements.

Table 5.6 *Number N of elements out of specific Mendeleev-number ranges that constitute the* 13 026 *ternary intermetallics, which have been observed in* 5109 *intermetallic systems so far. The numbers are given also for compounds with unique structure types for both truly ternary ones and including binary and unary ones. Mendeleev numbers M = 7–16 correspond to alkali and alkaline earth metals, 17–33 to rare-earth elements, 34–48 to actinoids, 49–77 to transition metals as well as Mg and Be with M = 73 and 77, respectively, and 78–91 to metallic main group elements.*

N	M = 7 – 16	17 – 33	34 – 48	49 – 77	78 – 91
Non-unique structure types (ternary & binary & unary)					
0	4361	2083	4819	574	1367
1	673	2755	283	2693	3326
2	73	269	7	1480	407
3	2	2	0	362	9
Unique structure types (ternary & binary & unary)					
0	432	423	646	150	67
1	192	238	21	334	471
2	42	6	0	162	128
3	1	0	0	21	1
Unique truly ternary structure types, only					
0	352	343	543	132	46
1	175	217	19	294	408
2	34	2	0	126	108
3	1	0	0	10	0

only 5109 data points out of the 85 320 possible ones. The projected M/M-plots of the 13 026 ternary compounds (Fig. 5.24) differ only marginally from that of the 6441 binary intermetallics (Fig. 5.9). The main difference is that the gap from $M(A) = 17$ (Yb) to $M(A) = 33$ (La) and $M(B) = 51$ (Ti) to $M(B) = 58$ (Re) is closed now in the projection along $M(C)$. There are also significantly more unique structure types for compounds with $8 \leq M(A,B) \leq 33$ in this projection. The big gap in the distribution for $34 \leq M \leq 48$ results from the small number of ternary intermetallics with actinoids as constituents (except $M = 43$ Pu, $M = 45$ U, and $M = 47$ Th), which is larger in the case of binary intermetallics. The small gaps at $M = 13$, 29, and 59 mark the locations of the not-yet-studied ternary intermetallics containing radioactive Ra, Pm, and Tc, respectively. In contrast to

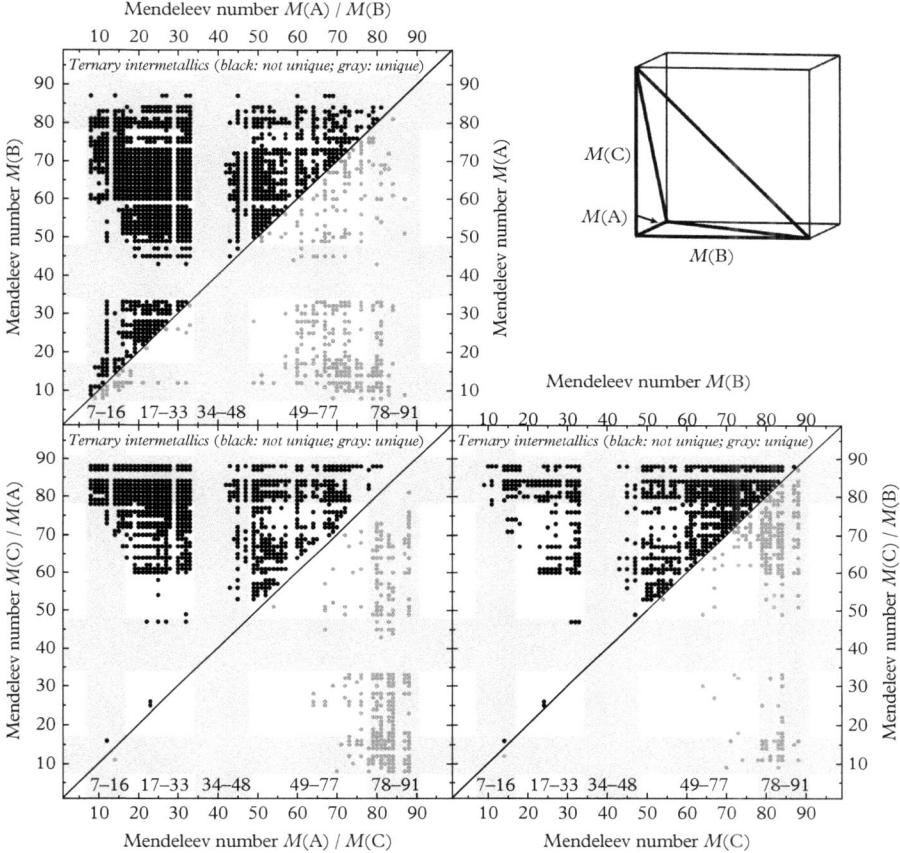

Fig. 5.24 *Chemical compositions of ternary intermetallics. Among the* 13 026 *ternary intermetallics crystallizing in 1391 structure types, 667 (48.0%) structure types occur only once. In the projected M/M-plots, the compounds with unique structures are shown in gray with M*(A) > *M*(B) > *M*(C), *while the remaining* 12 359—*not unique—ones are shown in black with M*(A) < *M*(B) < *M*(C). *Each 2D M/M-plot is projected along the third coordinate. Areas containing alkali metals, alkaline earth metals, actinoids, and (semi)metallic main group elements are shaded gray. The total asymmetric volume (one sixth of the volume of the cube) is marked by bold lines in the cube of the M-parameter space (upper right).*

binary intermetallics, no ternary intermetallics are known with Po ($M = 91$) as one of the constituents. It is remarkable that there are compounds for mere four systems with $7 \leq M(A), M(B), M(C) \leq 33$. If $7 \leq M(A), M(B) \leq 33$, i.e., the alkali and alkaline earth metals and the RE elements. It seems to be necessary to have $60 \leq M(C) \leq 88$, i.e., the late TM elements as well as the main group elements, to find stable ternary compounds in this case.

The distribution of intermetallics with unique structure types (marked gray in Fig. 5.24) is significantly less dense than expected from their overall fraction. There are 13 026 ternary intermetallic phases known in 5109 different ternary systems crystallizing in 1391 structure types (1095 ternary prototypes) out of which 667 types are unique (562 unique ternary prototypes) and their representatives occur in 489 different systems. Although more than half of all ternary structure types are unique ($562/1095 \approx 0.51$), this is not true for the number of ternary systems featuring ternary intermetallic compounds with unique structure types compared to the number of ternary systems featuring ternary compounds in general ($489/5109 \approx 0.10$). Furthermore, there are on average about two ternary intermetallic compounds with non-unique structure types per system studied ($12\,359/5014 \approx 2.46$) compared to, on average, approximately one intermetallic with unique structure type per system ($667/489 \approx 1.36$).

As is obvious from Fig. 5.24, in most cases ternary compounds with unique structure types are observed in the same systems as those with non-unique structure types. So, it seems that it is mainly the stoichiometry that makes a structure unique. From the chemical point of view, most compounds with unique structure types contain any element except actinoids for A, a few alkali and alkaline earth metals plus a few rare earth (RE) elements, but mainly late TM elements for B and late TM and main group elements for C.

It is remarkable that RE elements ($17/81 = 0.210$) provide one of the three constituents in as much as 2755 of the 5109 ternary intermetallic systems ($2755/5109 = 0.539$) (see Table 5.6), slightly more often than the 29 transition metals (TMs) (including Be and Mg) ($29/81 = 0.358$) with 2693 cases ($2693/5109 = 0.527$), but both significantly less frequently than the 10 main group elements ($10/81 = 0.123$) with 3326 cases ($3326/5109 = 0.651$). That all three constituents come out of the same class of elements is rare except for TMs with 362 cases. Only 574 ternary intermetallic systems (11.2%) do not contain TM elements. The distribution for intermetallic compounds with unique structure types is similar.

Increasing the number of constituents can make the formation of intermetallic compounds more and more difficult. Depending on atomic size ratios and stoichiometry, complex structures may be necessary to maximize attractive and minimize repulsive atomic interactions. Furthermore, it is less likely that a ternary compound is formed by adding a third element to a binary system if it does not form any binary compounds with either of the two other elements. The situation is even more unfavorable if none of the three binary subsystems form a binary compound. Indeed, no ternary intermetallics are known for this case, as we will see in Subsection 5.5.1.

Out of the 3240 theoretically possible binary intermetallic systems, there are 1401 (43.2%) known forming at least one intermetallic compound. Ternary intermetallics have been found so far in only 5109 (6%) out of the 85 320 possible ones. For 4041 of these ternary systems, binary phases have been reported in all three binary subsystems, for 1053 only in two subsystems, and for the remaining

15 only in one subsystem. An example for the second case with three ternary intermetallics known is the system Al–Cu–Ta, where Cu and Ta are immiscible (Conrad *et al.*, 2009; Weber *et al.*, 2009; Dshemuchadse *et al.*, 2013). In the following 15 systems, ternary compounds have been observed although two of the three binary subsystems do not form any intermetallic phase: Al–Cs–Tl, Bi–Fe–Zn, Bi–Li–V, Ca–Co–Pb, Ca–Cr–Pb, Ca–Pb–Ru, Cr–La–Pb, Cu–Ta–V, Ge–Np–Tc, Ge–Tc–U, Hf–V–Y, K–Tc–Tl, La–Mn–Pb, Mn–Rh–Tl, and Mn–Sn–W.

5.5.1 Stoichiometries of ternary intermetallics

In Fig. 5.25, a concentration diagram is shown reflecting the stoichiometries of the 13 026 ternary intermetallic phases listed in the PCD. Each dot corresponds to a particular intermetallic compound with composition $A_a B_b C_c$, where A, B, and C denote elements with Mendeleev numbers $M(A) < M(B) < M(C)$. Of course most dots coincide exactly, representing all compounds of the same structure type and those of other structure types but with the same stoichiometry. For instance, the dot at ABC represents 462 intermetallics with structures of the $hP9$-ZrNiAl type, 388 of the $oP12$-TiNiSi type, 166 of the $cF12$-MgAgAs type, etc. In the lower figure, the concentration triangle is merged into one asymmetric unit applying point group symmetry $3m$. Now each dot again corresponds to one particular compound, however, only reflecting its stoichiometry and no longer the chemical composition.

It should be kept in mind that symmetry $3m$ does apply approximately only. For a given stoichiometry $A_a B_b C_c$ and with $M(A) < M(B) < M(C)$, the sequence of elements cannot be arbitrarily exchanged against one another in the formula. While an intermetallic compound $A_a B_b C_c$ may exist, one with composition $B_a A_b C_c$ may not, for instance (also, see Table 5.7). This is quite obvious if one compares tie lines parallel to the sides of the triangles. The dots are rather densely distributed on tie lines parallel to A–B and B–C in contrast to those parallel to A–C.

It is remarkable that most ternary intermetallic phases have stoichiometries placing them onto tie lines between particular binaries with end points such as AB (888 different structures), $A_3 B_2$ (247), $A_2 B$ (1064), and $A_3 B$ (830). The ternary stoichiometries on these tie lines peak at ABC (< 1000), $A_2 BC$ (< 500), $A_2 B_2 C$ (< 300), and $A_4 BC$ (< 100). The accumulation of stoichiometries along the tie lines give the impression that these ternary compounds are actually a kind of pseudobinary. The tie lines are mostly parallel to the edges of the concentration triangle, indicating that along the tie line the concentration of one component remains constant while the other two vary at mutual cost, i.e., replacing each other in the structure.

There are a few tie lines connecting a binary composition with the opposite corner, such as AB–C or AC_2–B, marked by gray lines in Fig. 5.25. The latter, however, already ends at the ternary composition ABC_2. But what are the implications of this finding? For instance, the tie line AC_2–B can be written as $(AC_2)_x B_{1-x}$.

Table 5.7 *Most common structure types among ternary intermetallic phases. The top 20 structure types out of 1391 are given. Note, that the inherently disordered and pseudo-ternary structure types are again marked by entries in the flag column. Those with less atomic sites than components (i.e. max. 2) are marked "s" (solid solution) and those with more components than the structure type are marked "d" (derivative).*

Rank	Structure type	Space group	Wyckoff positions	No. of repr.	% of all repr.	Flag
1.	$cF24$-MgCu$_2$	227 $Fd\bar{3}m$	$8a\,16d$	523	4.0%	s
2.	$hP9$-ZrNiAl	189 $P\bar{6}2m$	$1a\,2d\,3fg$	462	3.5%	
3.	$oP12$-TiNiSi	62 $Pnma$	$4c^3$	388	3.0%	
4.	$cF16$-Cu$_2$MnAl	225 $Fm\bar{3}m$	$4abcd$	333	2.6%	
5.	$tI10$-CeAl$_2$Ga$_2$	139 $I4/mmm$	$2a\,4de$	289	2.2%	
6.	$cP4$-Cu$_3$Au	221 $Pm\bar{3}m$	$1a\,3c$	274	2.1%	s
7.	$hP12$-MgZn$_2$	194 $P6_3/mmc$	$2a\,4f\,6h$	265	2.0%	d
8.	$cP2$-CsCl	221 $Pm\bar{3}m$	$1ab$	212	1.6%	s
9.	$tI26$-ThMn$_{12}$	139 $I4/mmm$	$2a\,8fij$	210	1.6%	d
10.	$hP6$-CaIn$_2$	194 $P6_3/mmc$	$2b\,4f$	203	1.6%	s
11.	$tP10$-Mo$_2$FeB$_2$	127 $P4/mbm$	$2a\,4gh$	193	1.5%	
12.	$hP6$-CaCu$_5$	191 $P6/mmm$	$1a\,2c\,3g$	192	1.5%	d
13.	$oS16$-CeNiSi$_2$	63 $Cmcm$	$4c^4$	187	1.4%	
14.	$oI12$-KHg$_2$	74 $Imma$	$4e\,8i$	187	1.4%	s
15.	$cF12$-MgAgAs	216 $F\bar{4}3m$	$4abc$	166	1.3%	
16.	$hP3$-AlB$_2$	191 $P6/mmm$	$1a\,2d$	153	1.2%	s
17.	$hR57$-Zn$_{17}$Th$_2$	166 $R\bar{3}m$	$6c^2\,9d\,18fh$	151	1.2%	d
18.	$cF184$-CeCr$_2$Al$_{20}$	227 $Fd\bar{3}m$	$8a\,16cd\,48f\,96g$	143	1.1%	
19.	$cF96$-Gd$_4$RhIn	216 $F\bar{4}3m$	$16e^3\,24f^2$	138	1.1%	
20.	$hP18$-CuHf$_5$Sn$_3$	193 $P6_3/mcm$	$2b\,4d\,6g^2$	135	1.0%	
	Total			4804	36.9%	

For the point ABC$_2$, we get $x = 0.5$, for AC$_2$B$_2$ $x = 1/3$. We may have any ratio between x and $1 - x$, but the ratio A/C must always equal 2. It is remarkable that in the range $1 \leq x \leq 0.5$ the stoichiometries are quite densely populated while they are rather sparse for $x > 0.5$. This is quite exceptional since for the tie lines parallel to the edges of the concentration triangle the distribution is rather homogenous over

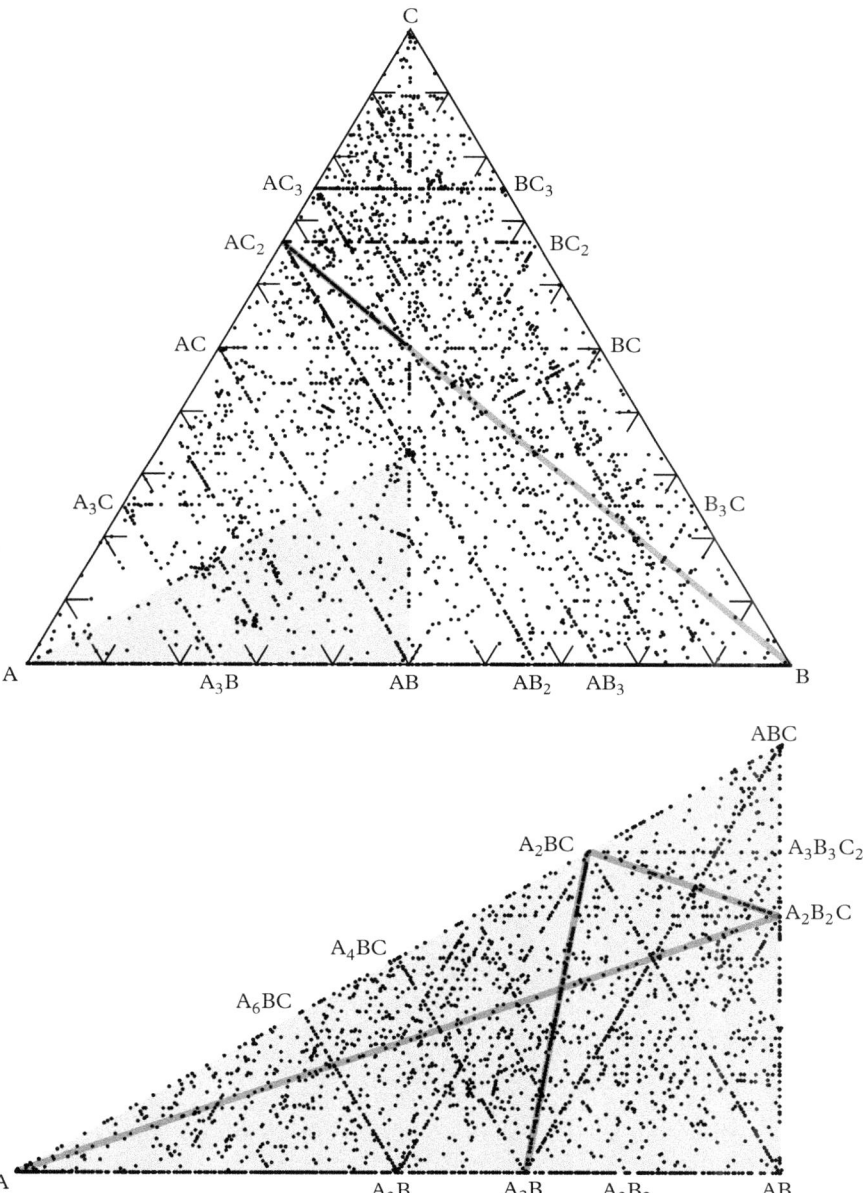

Fig. 5.25 *(top) Concentration diagram reflecting the stoichiometries $A_a B_b C_c$ of the 13 026 ternary intermetallic compounds ($M(A) < M(B) < M(C)$). (bottom) Enlarged triangular section (generic asymmetric unit) out of the concentration diagram above (shaded region) after merging under point group 3m. The gray line marks the stoichiometry band $(AC_2)_x B_{1-x}$ in the upper drawing and its symmetrically equivalent one in the lower.*

the full range. What is the origin of this limitation? It seems to be mainly a kind of solid solubility of a third element in one of the binary structure types (Laves phases, AlB_2, etc., see, also, Fig. 5.29), since 310 ternary intermetallic phases listed in the PCD have been assigned to unary and 4543 to binary structure types. This means, for instance, that although we have an almost continuous distribution of stoichiometries for binary intermetallics, only very few of the most frequent ones can be almost continuously extended into another parameter dimension: AB, A_3B_2, A_2B, and A_3B.

In order to find out whether there are preferred chemical compositions along different tie lines as marked in Fig. 5.25, the respective M/M maps have been plotted and are depicted on the following pages: the 3382 intermetallic compounds along the line A–BC in Fig. 5.26, the 486 intermetallic compounds along the line AB–AC in Fig. 5.27, the 2814 intermetallic compounds along the line AB_2–AC_2 in Fig. 5.28, and the 645 intermetallic compounds along the line B–AC_2 in Fig. 5.29.

The plots in Figs. 5.26 and 5.28 representing a comparable number of intermetallics are almost indistinguishable. In comparison, the distribution of dots in Figs. 5.27 and 5.29 is much sparser because their numbers are much smaller, but they significantly differ from one another. While the $M(B)/M(C)$ plots are very similar, there are clear differences in the region $19 < M(A) < 33$ (Sc and RE elements), $M(A) = 69$ (Pd), 70 (Au), 72 (Cu), and $M(A) = 73$ (Mg), 75 (Cd), 77 (Be), 79 (In), 82 (Pb), and 83 (Sn).

The distribution of the compounds with the 667 unique structure types among the 13 026 ternary intermetallic phases is reflected in Fig. 5.30. There is no obvious difference between these two distributions; also, their density on the tie lines shows a similar trend. The only clearly visible difference is on the tie line AC_2–ABC_2, where the unique structure types are quite sparsely distributed. This indicates that the dense distribution of intermetallics is just an artifact resulting from the broad stability range of some structure types as will be confirmed below.

If we exclude ternary compounds, which have been assigned to unary (310) or binary (4543) structure types then we get the distribution shown in Fig 5.31. It decreases the density of structure types on some specific concentration lines. In particular, the dots on the line A–B, representing binary compounds, disappear completely, naturally.

However, there is still a significant accumulation along directions such as AC_2–B, AB–ABC, AB–A_2BC, A_2B–A_2BC, and AB–A_3BC. Superimposing the compounds with unique structure types on top of the non-unique ones leads to Fig. 5.32. One sees that most stoichiometries featuring unique structure types also feature non-unique ones. So, the stoichiometry is certainly not the most important factor controlling the compositional flexibility of a structure type.

However, if only the stoichiometries of the structure types are considered, the density of points along the tie lines AC_2–B and AC–B (marked gray in Figs. 5.31 and 5.32) is drastically reduced. This means that the stoichiometry of many intermetallic phases with given structure type scatters around its ideal composition.

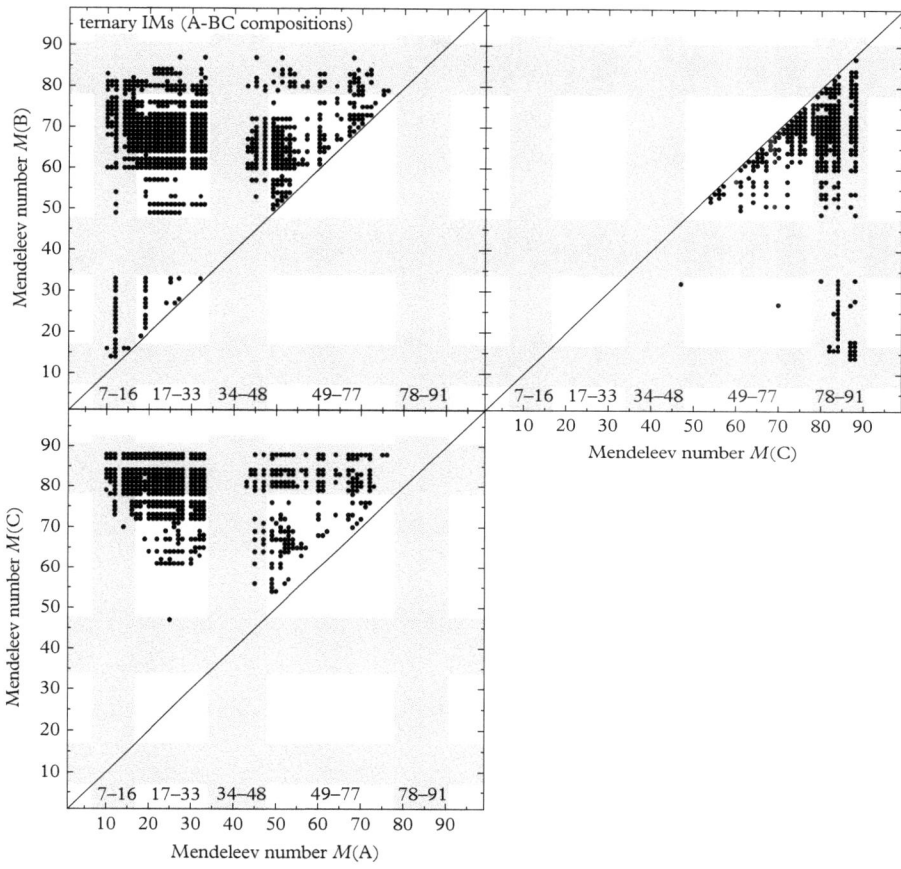

Fig. 5.26 *(top) Characteristic projections of the ternary M(A)/M(B)/M(C) plot of the 3382 intermetallic compounds along the line A–BC of the ternary concentration diagram depicted in Fig. 5.25 (M(A) < M(B) < M(C)). Areas containing alkali metals, alkaline earth metals, actinoids, and (semi)metallic main group elements are shaded gray.*

This may result from a broad compositional stability range of the intermetallic phase or just from an experimental nominal composition, which differs from the equilibrium composition of the structure type. In contrast, the densities on the tie lines AC–AB, AC–BC, and AB–C are still significantly higher than elsewhere, for both unique- and non-unique structure types. The very short distances between the data points may be due to stoichiometries, which differ from simple rational numbers due to partial occupancy of some atomic sites.

To summarize, the concentration diagram taking into account all intermetallic phases (Fig. 5.25) just reflects their experimental stability ranges. It is more informative, therefore, to look at the concentration diagram of the structure types

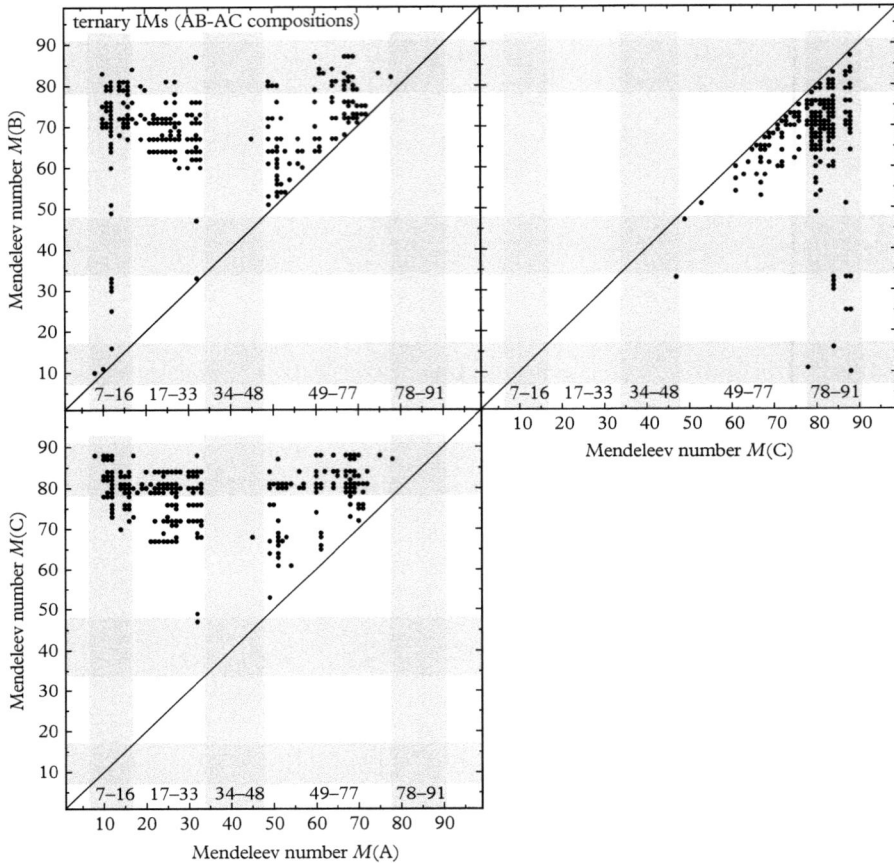

Fig. 5.27 *(top) Characteristic projections of the ternary M(A)/M(B)/M(C) plot of the 486 intermetallic compounds along the line AB–AC of the ternary concentration diagram depicted in Fig. 5.25 (M(A) < M(B) < M(C)). Areas containing alkali metals, alkaline earth metals, actinoids, and (semi)metallic main group elements are shaded gray.*

instead (Fig. 5.32), if one wants to get an idea about which stoichiometries intermetallic compounds can have.

5.5.2 Quasicrystals

Quasicrystals (QCs) constitute a special class of intermetallics (for more information see Chapter 9). They are not covered by common databases such as the PCD, as their structural information cannot be represented by a unit cell in three dimensions. Most quasicrystals are ternary compounds—but quite a few binary icosahedral phases are known in the systems Cd–(Ca, RE) and Zn–Sc. One may

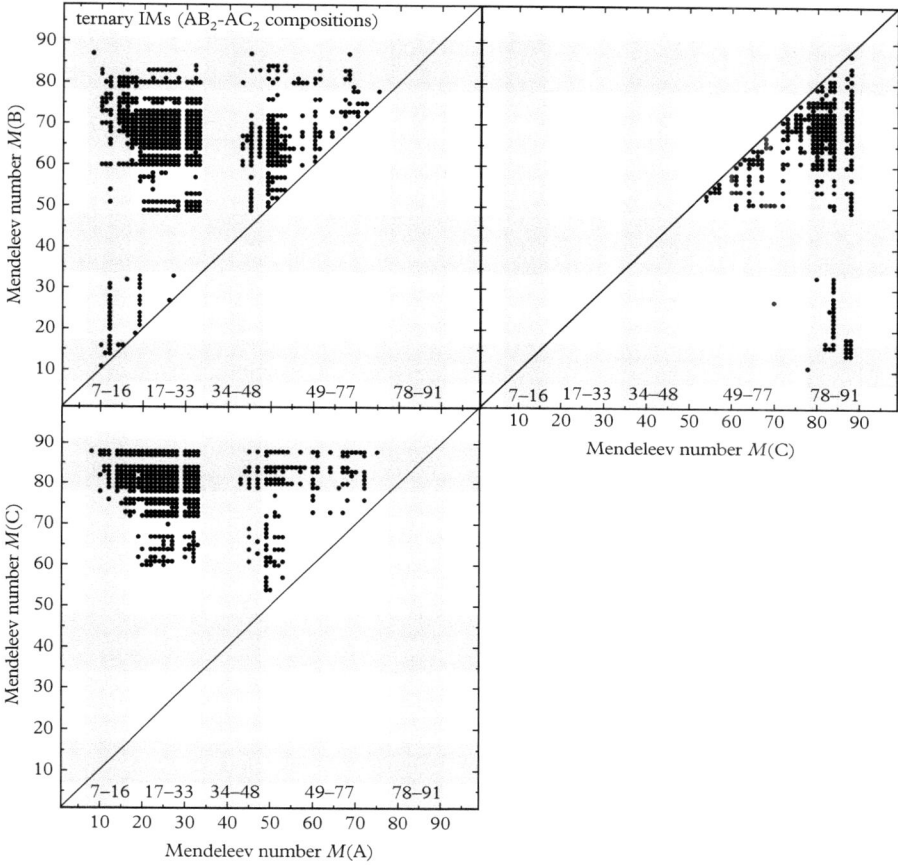

Fig. 5.28 *(top) Characteristic projections of the ternary M(A)/M(B)/M(C) plot of the 2814 intermetallic compounds along the line* AB_2-AC_2 *of the ternary concentration diagram depicted in Fig. 5.25 (M(A) < M(B) < M(C)). Areas containing alkali metals, alkaline earth metals, actinoids, and (semi)metallic main group elements are shaded gray.*

ask now, where the compositional stability fields of QCs are located relative to those of periodic intermetallics. In some more recent reviews, such composition diagrams have been presented (Steurer and Deloudi, 2008; Steurer and Deloudi, 2009). Fig. 5.33 is based on the same data, updated with some quasicrystalline phases discovered over the past few years. Therein, the compositions of ternary quasicrystals are shown analogous to Fig. 5.24. The rather few different structure types are marked by symbols as described in the figure. Decagonal quasicrystals (DQCs) are classified according to the number of quasiperiodic atomic layers per translation period along the tenfold axis, which can amount to 2, 4, 6, or 8. In the case of icosahedral quasicrystals (IQCs), the structure type is defined by the

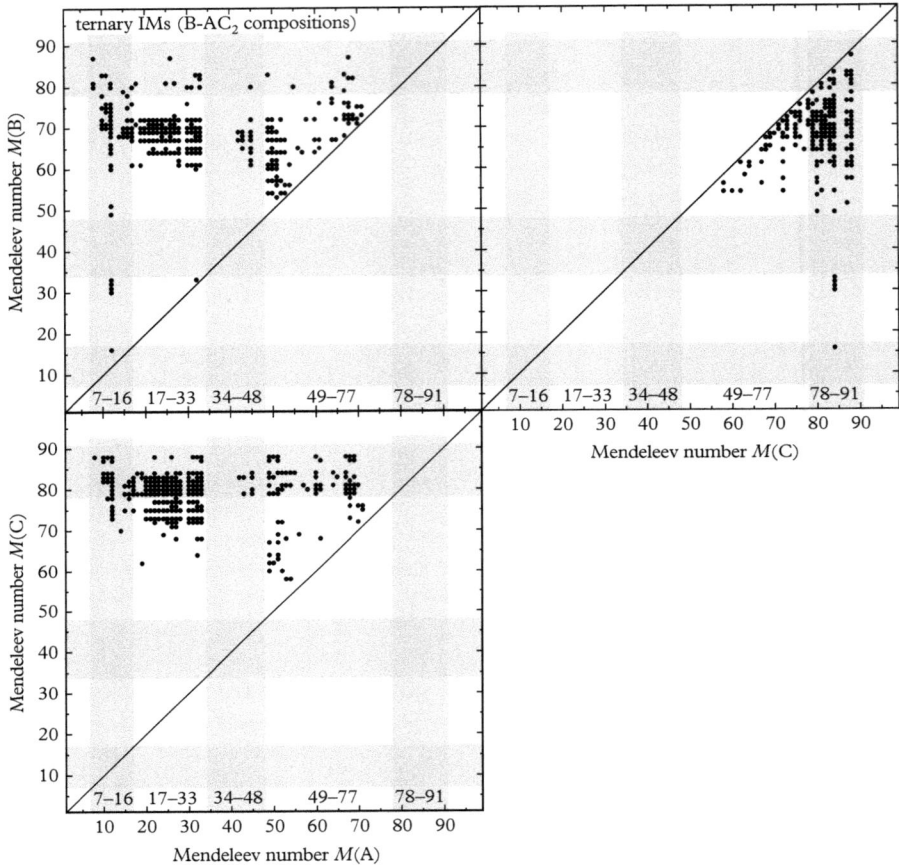

Fig. 5.29 *(top) Characteristic projections of the ternary $M(A)/M(B)/M(C)$ plot of the 645 intermetallic compounds along the line B–AC$_2$ of the ternary concentration diagram depicted in Fig. 5.25 ($M(A) < M(B) < M(C)$). Areas containing alkali metals, alkaline earth metals, actinoids, and (semi)metallic main group elements are shaded gray.*

kind of fundamental building cluster, which can be of the Mackay-, Bergman-, or Tsai-type. There are more systems exhibiting IQCs than DQCs, and a few systems contain both of them. These are mainly the Al- and Zn-Mg-based ones, respectively.

The distribution of QCs on the M/M-plots is very sparse. However, what is significant is that there are "lines" of elements, indicating that in a structure type elements can be substituted by other elements with slightly larger or smaller Mendeleev numbers. For instance, for DQCs, element A can be an RE element ($20 \leq M \leq 25$), then B corresponds to Mg ($M = 73$), and C to Zn ($M = 76$).

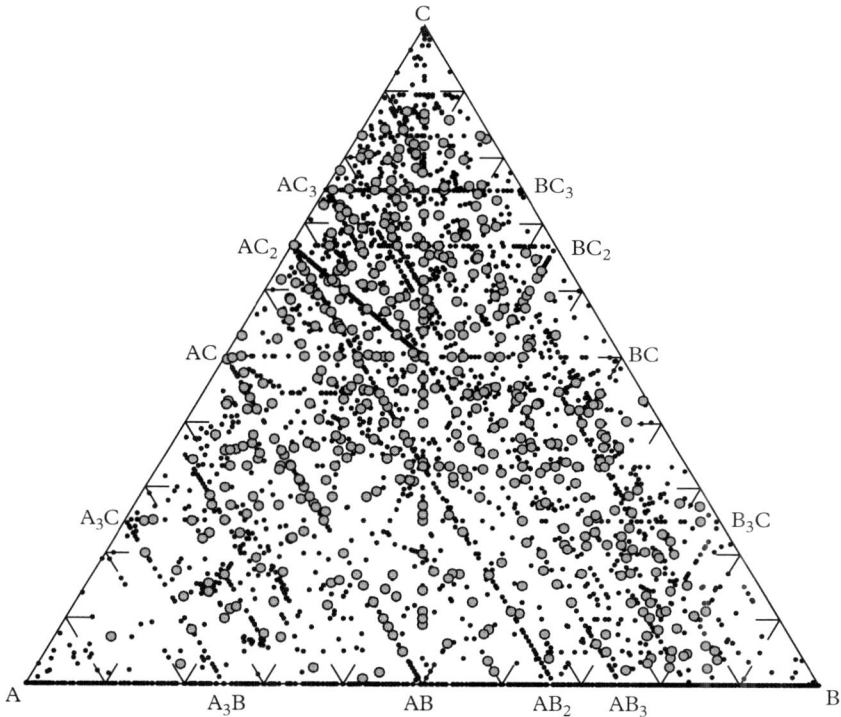

Fig. 5.30 *Concentration diagram reflecting the stoichiometries $A_a B_b C_c$ of the 667 intermetallics with unique structure types (large gray dots) among the 13 026 ternary intermetallic phases $(M(A) < M(B) < M(C))$. The compounds with unique structure types show a similar distribution as those with non-unique ones. However, their density on some tie lines such as AC_2–B and AC–B, for instance, is much lower.*

The diagrams for DQCs and IQCs are quite mirror-symmetric with respect to the diagonal, which means that both classes of QCs have a similar distribution in the projected M/M-plots. The main differences are the Sc- and Cd-based IQCs, which do not have counterparts in the case of DQCs.

There are four classes of DQCs named after their period along the tenfold axis (for detailed information see Chapter 9). Their preferred compositions are:

2-layer: Al–Ni–Co, Zn–Mg–Dy, etc.

4-layer: Al–(Ni, Cu)–(Co, Fe, Ir, Rh), etc.

6-layer: Al–(Mn, Re)–Pd

8-layer: Al–Os–(Pd, Ir), Al–Ni–Ru

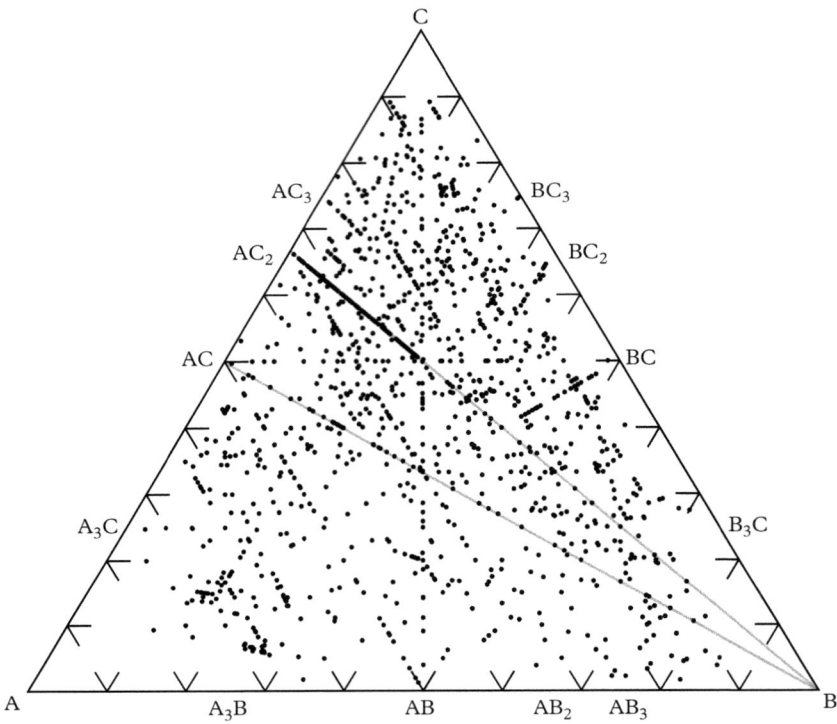

Fig. 5.31 *Concentration diagram reflecting the stoichiometries $A_aB_bC_c$ of the 8173 ternary intermetallic compounds remaining after the elimination of all ternary intermetallics assigned to unary (310) or binary (4543) structure types ($M(A) < M(B) < M(C)$). The array of dots on the tie line AC_2–ABC_2 (marked by gray lines) is still very dense, indicating intermetallic phases with broad stability ranges.*

There are three classes of IQCs named after the constituting clusters (for detailed information see Chapter 9). Their preferred compositions are:

Mackay-cluster: Al–TM(1)–TM(2), etc.

Bergman cluster: Zn–Mg–(Y, Zr, Hf, RE, Al, Ga), Ti–Zr–Ni, Al–Cu–Li, Mg–Al–Pd, etc.

Tsai cluster: Zn–(Sc, Mg)–TM, (Ag, Au)–(Al, In)–(Yb, TM), Cd–Mg–(Ca, RE), etc.

It is not surprising that quasicrystals occur in the same systems as periodic crystals, because frequently periodic rational and non-rational approximants, respectively, exist with similar chemical compositions as their quasiperiodic

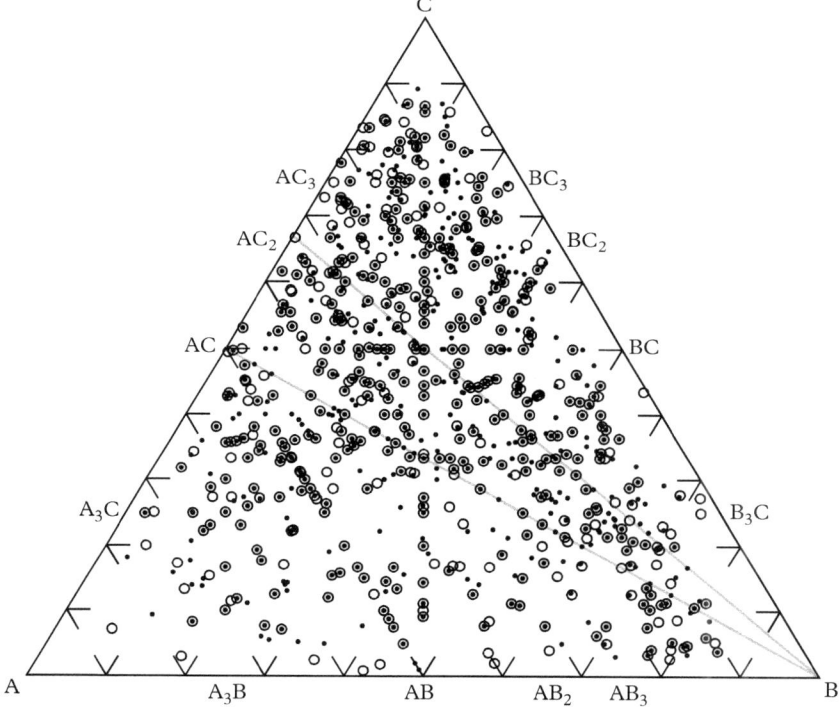

Fig. 5.32 *Concentration diagram reflecting the stoichiometries $A_aB_bC_c$ of the 1095 ternary structure types ($M(A) < M(B) < M(C)$) (keep in mind that the prototype structures are not all intermetallics). The unique structure types, which are all intermetallics in our case, are marked by open circles. A dot in a circle means that at this specific stoichiometry both unique and non-unique structure types exist. The gray lines mark the tie lines AC_2–B and AC–B.*

counterparts. However, in most cases several other periodic phases are observed in the respective systems, which have no structural relationship at all to quasicrystals.

5.5.3 Four lines of ternary stoichiometries

Fig. 5.34 shows cuts through the composition diagram of the 1095 ternary intermetallic structure types (see, also, Fig. 5.32). In Fig. 5.34(a)–(d), their frequencies are shown along the four main lines: AB–A_2BC, A_2B–ABC, A_2B–A_2BC–A_2B$_2$C–A, as well as A–ABC–AB, which runs along two edges of the triangle. Their location in the full ternary composition diagram, as well as the resulting trajectories in the reduced plot, are depicted in Fig. 5.34(e). This graph

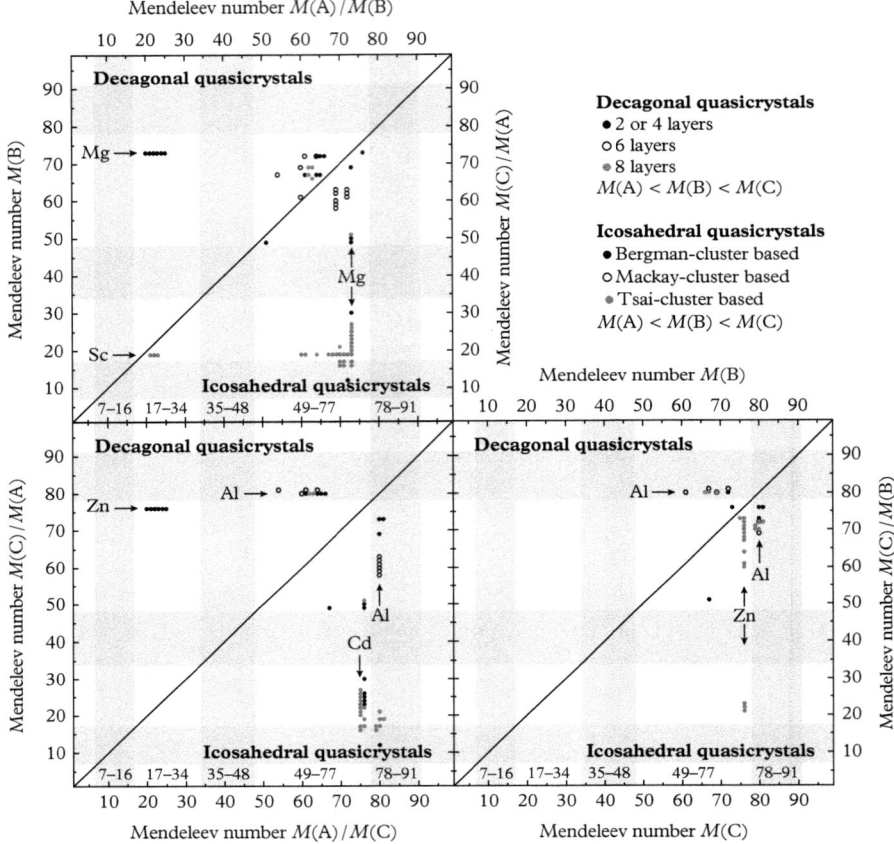

Fig. 5.33 *Projected M/M-plots illustrating the chemical compositions of ternary intermetallic quasicrystals: 21 decagonal and 53 icosahedral ones. The components of all compounds have been assigned to elements A, B, and C according to $M(A) < M(B) < M(C)$, but for better illustration, the plots are shown with reversed axes for the icosahedral phases. Decagonal phases with 2 or 4 layers are shown as full black circles, those with 6 or 8 layers as open circles or gray dots, respectively. Icosahedral phases that are based on Bergman-, Mackay-, and Tsai-type clusters are shown as full black circles, as open circles, or gray dots, respectively. Each 2D M/M-plot results from a projection along the third coordinate. Mendeleev numbers 7–16 mark alkali and alkaline earth metals, 17–33 rare earth elements, 34–48 actinoides, 49–77 transition metals plus Be (77) and Mg (73), and 78–91 metallic main group elements.*

illustrates, for instance, the path along the line A_2B–A_2BC–A_2B_2C–A. For all tie lines in the reduced concentration diagram the law of reflection applies when they touch the two boundaries A–ABC and ABC–AB.

The horizontal axes in Fig. 5.34(a)–(d) display the content of the respective varying element(s). The compositions AB–A_2BC all contain 50% of element A,

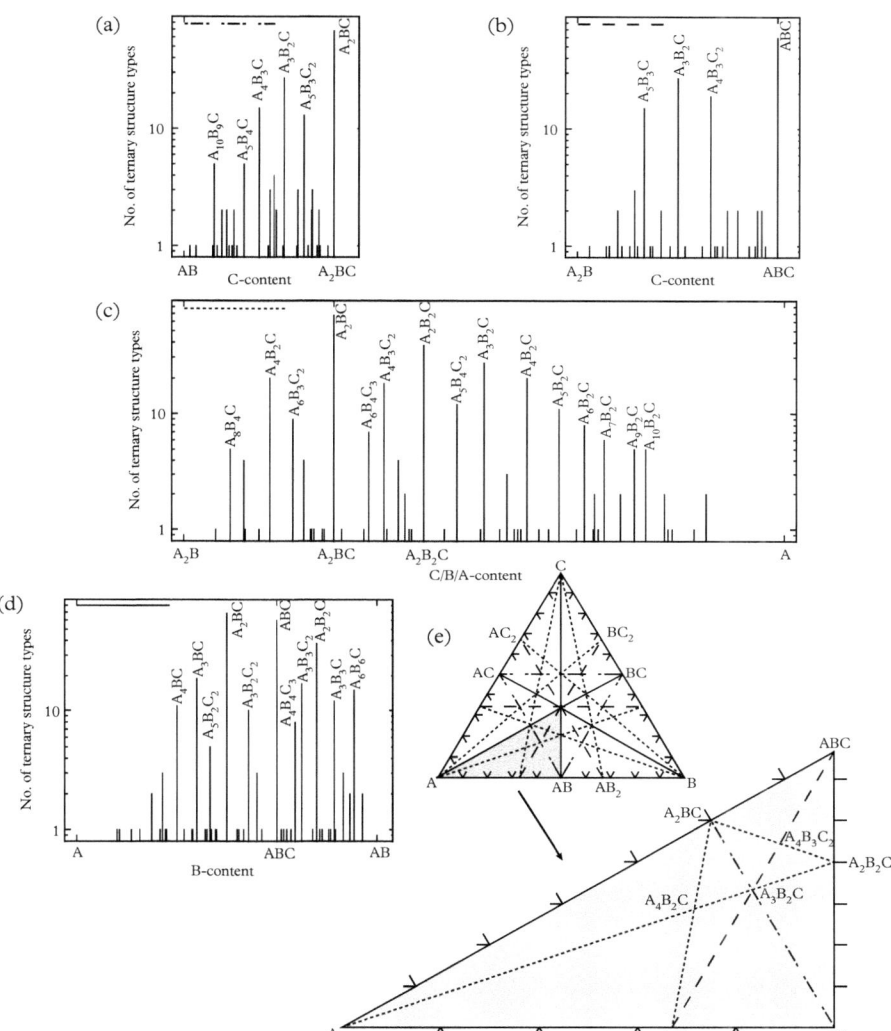

Fig. 5.34 *Cuts through the concentration diagram of the 1095 ternary structure types of the 13 026 ternary intermetallic compounds. Shown are the lines (a) AB–A$_2$BC, (b) A$_2$B–ABC, (c) A$_2$B–A$_2$BC–A$_2$B$_2$C–A, and (d) A–ABC–AB, as well as (e) a schematic view of where the respective lines lie in the full and reduced (shaded gray) ternary composition diagrams. Please note the logarithmic scale of the y-axes in (a–d) and the differently dashed lines.*

while the C-content increases from 0% to 25% and the B-content decreases from 50% to 25%. The compositions A$_2$B–ABC all contain 33.3% of element B, while the C-content increases from 0% to 33.3% and the A-content decreases from 66.7% to 33.3%. The compositions A$_2$B–A$_2$BC all have A- and B-contents that correspond to the ratio $a/b = 2/1$, while the C-content increases from 0% to 25%. The compositions on the consecutive line segment $A_2BC–A_2B_2C$ all have A- and C-contents that correspond to the ratio $a/c = 2/1$, while the B-content increases from 25% to 40%. The third and last line segment $A_2B_2C–A$ contains compositions with B- and C-contents in a ratio of $b/c = 2/1$ and A-contents increasing from 40% to 100%. For the fourth line, the ratio of elements B and C is $b/c = 1$ in the range A–ABC, where the B- and C-contents increase from 0% to 33.3%. The second part of the plot contains the same amount of the A- and B-components ($a/b = 1$) with values ranging from 33.3% to 50%.

Along the line $AB–A_2BC$, 172 ternary structure types can be found at 30 different stoichiometries, representing a total of 1570 ternary intermetallics. The maximum value in Fig. 5.34(a) corresponds to 68 structure types (STs) at composition A_2BC. Additional high values are found at A_3B_2C (27 STs), A_4B_3C (15 STs), and $A_5B_3C_2$ (13 STs). Smaller numbers of structure types are reported at the following compositions: $A_{10}B_9C$ and A_5B_4C (both 5 STs), $A_{10}B_7C_3$ (4 STs), $A_7B_5C_2$, $A_{21}B_{13}C_8$, and $A_7B_4C_3$ (all 3 STs), with six compositions being featured in 2 STs each, and 14 in one ST each, only. This means 14 out of the total 30 different stoichiometries along this line refer to unique structure types.

Along the line $A_2B–ABC$, 153 ternary structure types are found at 28 different stoichiometries, which represent 2162 ternary intermetallics. The maximum value in Fig. 5.34(b) corresponds to 60 structure types at composition ABC. Additional high values are found at A_3B_2C (27 STs–where this line, $A_2B–ABC$, intersects with line $AB–A_2BC$), $A_4B_3C_2$ (19 STs), and A_5B_3C (15 STs). Composition $A_{12}B_7C_2$ is still featured in three structure types, while six more compositions are adopted by 2 STs each, and 17 compositions occur only in one structure type each. This means 17 out of the altogether 28 different stoichiometries along this line refer to unique structure types.

The line $A_2B–A_2BC–A_2B_2C–A$ consists of three segments coinciding with a total of 291 ternary structure types at 51 different stoichiometries, representing altogether 3539 ternary intermetallic compounds (referring to Fig. 4(c)). The turning points at A_2BC and A_2B_2C, which are also the two highest maxima in Fig. 5.34(c), correspond to groups of 68 and 38 structure types, respectively. The intermediate segments contain the following numbers of structure types: 50 in $A_2B–A_2BC$ (excluding A_2BC), 36 in $A_2BC–A_2B_2C$ (excluding both, A_2BC and A_2B_2C), and 99 in $A_2B_2C–A$ (excluding A_2B_2C). Additional significant values are found at A_3B_2C (27 STs—where the respective line segment, $A_2B_2C–A$, intersects with both lines, $AB–A_2BC$ and $A_2B–ABC$), A_4B_2C (20 STs—where the two line segments $A_2B–A_2BC$ and $A_2B_2C–A$ intersect), $A_4B_3C_2$ (18 STs), $A_5B_4C_2$ (12 STs), $A_5B_2C_2$ (11 STs), $A_6B_3C_2$ (9 STs), A_6B_2C (8 STs), $A_6B_4C_3$ (7 STs), A_7B_2C (6 STs), A_8B_4C (5 STs), A_9B_2C (5 STs), $A_{10}B_2C$ (5 STs), A_6B_3C (4 STs), $A_8B_4C_3$ (4 STs), $A_6B_5C_3$ (4 STs), and $A_7B_4C_2$ (3 STs), with five more compositions being adopted by 2 STs and another 27 by one structure type each. This means 27 out of the 51 different stoichiometries along this line correspond to unique structure types.

The fourth and last line A–ABC–AB contains a total of 307 ternary structure types, which occur at 45 different stoichiometries and represent 4882 ternary intermetallic compounds. The first segment, A–ABC, contains 140 structure types, excluding ABC itself, which occurs in 60 structure types, while the second segment, ABC–AB, contains 107 structure types. The maximum values in Fig. 5.34(d) correspond to the following compositions: A_2BC (68 STs), ABC (60 STs), A_2B_2C (38 STs), A_3BC (19 STs), A_3B_3C (17 STs), A_6B_6C (15 STs), A_3B_3C (12 STs), A_4BC (11 STs), $A_3B_2C_2$ (10 STs), $A_4B_4C_3$ (8 STs), and $A_5B_2C_2$ (5 STs). Of the remaining compositions, three occur in 3 STs, three in 2 STs, and 29 in only one structure type each.

It is remarkable how many unique structure types ($\approx 50\%$) are located on these relatively densely occupied lines in the concentration triangle. It is also amazing how many different structure types can be found for a given stoichiometry. A_2BC and ABC top the list with 68 and 60 structure types, respectively, which is less than the values for the binary structure types, which are almost twice that much.

5.6 Statistics of crystal structure types

Ternary intermetallics are usually described based on their ternary prototype structures. However, according to the usual definition of a structure type, it is also possible that a well-ordered ternary intermetallic compound is assigned to a binary structure type, if both structures, that of the ternary structure and the binary structure type, are isoconfigurational (see Table 5.7). This means that they are isopointal and both the crystallographic point configurations and their geometrical interrelationships are similar. The definition does not mean that isoconfigurational compounds have to have the same number of different constituting elements. Indeed, a considerable number of structure types that ternary intermetallics have been assigned to are binary. However, these ternary compounds are, at least partially, inherently disordered, if the binary structure types can be described with only two independent Wyckoff positions in the respective space group, e.g., $cF24$-$MgCu_2$, $cP4$-Cu_3Au, $cP2$-$CsCl$, etc. Many binary structure types, however, have their atoms occupying three or more independent atomic sites, making them not inherently binary. In a ternary representative structure crystallizing in such a binary structure type exactly the same Wyckoff positions are occupied in an ordered way, but now by three instead of two elements. We think, however, that this is a shortcoming of the current definition of a "structure type". Strictly speaking, a chemically ordered variant of a structure is a derivative structure (ordered structure variant), even if the symmetry does not change.

To conclude, the majority of the representative ternary compounds of the most common binary structure types with three Wyckoff positions feature a high degree of disorder with respect to the distribution of the three chemical elements on the different positions. Some can be considered pseudo-binary compounds, with two of the three constituents occupying the site(s) of one of the binary components in a purely statistical manner or by slightly different mixing ratios of the same elements. However, a few structures that are ordered derivatives of binary prototype structures are also known. One example is that of two ternary representatives of the $hP12$-$MgZn_2$ structure type exhibiting fully ordered occupancies of the Wyckoff positions ($2a$, $4f$, $6h$—occupied by Zn, Mg, and Zn, respectively, in the prototype structure): $hP12$-Lu_2CoAl_3 and $hP12$-Er_2CoAl_3 (Oesterreicher, 1973). Also, four ordered variants of the $cF24$-Be_5Au structure type ($4a$, $4c$, $16e$—Au, Be, Be, in the prototype structure) were found among ternary intermetallics: $REMgNi_4$ with RE = Y, Ce, Pr, and Nd (Kadir *et al.*, 2002) (RE on $4a$, Mg on $4c$, Ni on $16e$).

Of the 20 829 intermetallics, 13 026 are ternary compounds crystallizing in 1391 different structure types. The twenty most common structure types are given in Table 5.7. They represent more than 130 intermetallics each, in total 36.9% of all ternary intermetallics covered in this study. Contrary to the binary compounds, no pseudo-unary sphere-packings are found among the most common structure types. However, the ten binary structure types among the top twenty, which already comprise 49.4% of the structures, have to be considered as solid solutions or derivative structures. There are also quite a few ternary compounds that crystallize in binary structure types although none of the three subsystems feature this structure type. This means that a meta- or unstable binary compound crystallizing in a binary structure type can be stabilized by the addition of a third component, changing the electron concentration decisively, for instance. There are several known examples of ternary compounds crystallizing in the binary Laves phase prototype $hP12$-$MgZn_2$, with three occupied Wyckoff positions (Stein, Palm, and Sauthoff, 2005), for instance. The binary structure types $tI26$-$ThMn_{12}$, $hP6$-$CaCu_5$, and $hR57$-$Zn_{17}Th_2$ also feature more than two occupied Wyckoff positions, which can be substituted by the three different atomic species of ternary intermetallic phases.

The symmetries of the ten ternary ones out of the top 20 structure types are mostly cubic (cF: 4), a few are hexagonal (hP: 2), tetragonal (tI: 1, tP: 1), or orthorhombic (oS: 1, oP: 1). No triclinic, monoclinic, or trigonal structure types are among these ten ternary structure types. The number of atoms per unit cell ranges from 9 to 184 ($hP9$-$ZrNiAl$ and $cF184$-$CeCr_2Al_{20}$, respectively).

The distribution of unit cell sizes of all 13 026 ternary compounds, as well as of all 8145 ternary compounds crystallizing in truly ternary structure types, and the 1095 ternary structure types themselves, are shown in Fig. 5.35. The histograms for all 13 026 ternary compounds and that for the only 8145 ternary compounds crystallizing in ternary structure types differ only marginally. With a few exceptions, there are hardly more than ten compounds for given unit cell sizes larger than 100 atoms per primitive unit cell. The distribution of structure types with given unit cell sizes peaks at around 14 atoms per primitive unit cell with a value of ca. 50 and falls off to less than 10 beyond $\approx$ 50 atoms per primitive unit cell. Remarkably, almost all structure types become unique ones beyond $\approx$ 170 atoms per primitive unit cell.

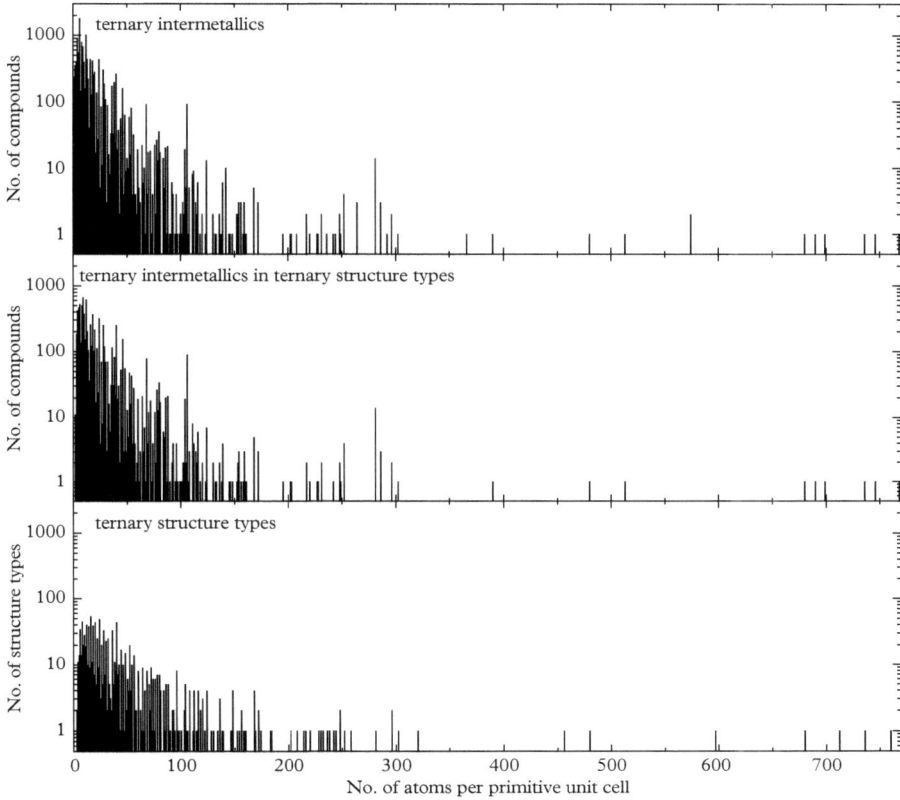

Fig. 5.35 *Unit cell size distributions of the* 13 026 *ternary compounds (top),* 8145 *ternary compounds crystallizing in ternary structure types (*middle*), and 1095 ternary structure types (*bottom*). All plots have been truncated at a maximum of 770 atoms per unit cell, excluding four intermetallics:* $hP1164$-$Cr_{10.7}Fe_{8.6}Al_{80.8}$, $hP1192$-$Cr_{10.7}Fe_{8.7}Al_{80.6}$, $cF5908$-$Ta_{39.5}Cu_{3.9}Al_{56.6}$, $cF23\,134$-$Ta_{39.1}Cu_{5.4}Al_{55.4}$.

5.6.1   Common stoichiometries of ternary intermetallics

If the compositions of all 8145 ternary intermetallics with ternary structure types are normalized and rounded off to three digits after the decimal point, 998 different stoichiometries result. 671 occur exactly once, 327 more than once: 253 occur 2–9 times, 29 occur 10–19 times, and 25 occur 20–70 times. The 671 unique stoichiometries are not equivalent to the 562 unique ternary structure types. In other words, a unique structure type can have a stoichiometry that is also adopted by a non-unique one, and a single non-unique structure type can have representatives with different stoichiometries.

The top 45 of all compositions are given in Table 5.8. The asymmetry is remarkable, i.e., the assignment of *a*, *b*, and *c* in the formula $A_aB_bC_c$ ($M(A) < M(B) < M(C)$). For instance, there are 159 compounds crystallizing in 22 structure types with the composition 3:1:1; for 153 of them AB_3C applies, and for the other 6 compounds ABC_3 does. This also means that no compound A_3BC with $M(A) < M(B,C)$ is present in the database.

Some of the compounds with compositions A_4BC (rank 4), $A_{20}B_2C$ (rank 9), A_6B_6C (rank 10), $A_9B_3C_2$ (rank 13), $A_{43}B_6C_4$ (rank 15), and A_8B_4C (rank 20) have structure types and structures with interesting magnetic properties (Thiede, Jeitschko, Niemann, and Ebel, 1998; Wolff, Niemann, Ebel, and Jeitschko, 2001), all of them aluminides containing RE and early TM elements. For instance, $A_{43}B_6C_4$ exists in just one structure type with compositions $A_6B_4C_{43}$ ($M(A) < M(B) < M(C)$), the $hP106$-Ho$_6$Mo$_4$Al$_{43}$ type. In the 86 different representatives, Ho can be replaced by other RE elements, and Mo by other early TM elements.

There are 1495 compounds with composition ABC that are known. Since the PCD contains ternary intermetallics out of 5109 ternary systems, this means that for only slightly less than every third intermetallic system a phase is known with the equiatomic stoichiometry 1:1:1. The most frequent structure types for this composition are $hP9$-ZrNiAl with 462 representatives, $oP12$-TiNiSi with 388, and $cF12$-MgAgAs, the structure type of the half-Heusler phases, with 166 compounds. Of these, 161 are reported as stoichiometric, with $M(A) = 12$–72, $M(B) = 54$–80, and $M(C) = 80$–88. A large subset of the ABC intermetallics has been named REME phases, with RE a rare earth metal (in most cases), an actinoid, or a group 1–4 element. M is a late transition metal from groups 8–12, and E is an element of groups 13–15 (Bojin and Hoffmann, 2003*a*; Bojin and Hoffmann, 2003*b*). Some of the REME phases are insulators, semiconductors, or semimetals, others have unusual magnetic and electronic properties. For instance, some representatives of the most common $oP12$-TiNiSi, as well as of the $hP9$-ZrNiAl structure types are heavy-fermion compounds. A large subclass of the REME phases belongs to structural derivatives of the $hP3$-AlB$_2$ structure type, such as $oP12$-TiNiSi, for instance.

The largest group of compounds with composition A_2BC is formed by the Heusler phases, $cF16$-Cu$_2$MnAl, with 333 representatives, some of them with interesting magnetic, thermoelectric, or superconducting properties. Of these, 309 are reported to be stoichiometric with 50% of element A with $M(A) = 10$–81 and 25% for each of elements B and C with $M(B, C) = 8$–88 (or $M(B) = 8$–77 and $M(C) = 62$–88 with $M(B) < M(C)$). Since most of their structures show great compositional flexibility, partial substitution of elements allows for fine-tuning of the valence electron concentration and therewith of specific physical properties. For reviews see, for instance, Trudel *et al.* (2010) or Graf *et al.* (2009). Heusler and half-Heusler phases are superstructures of the $cI2$-W structure type.

Table 5.8 *The top* 45 *compositions of ternary intermetallics,* $A_aB_bC_c$ *(with* $a > b > c$*), representing* 20 *or more IMs each. The number of IMs and of different structure types (STs) is given for the general stoichiometry and the number of IMs for each permutation of* a, b, c *with* $M(A) < M(B) < M(C)$ *for* $A_aB_bC_c$*:* $a \geq b \geq c$ *(I),* $a \geq c \geq b$ *(II),* $b \geq a \geq c$ *(III),* $c \geq a \geq b$ *(IV),* $b \geq c \geq a$ *(V),* $c \geq b \geq a$ *(VI). Values in italics represent the cases where the compounds are counted twice or six times due to equalities* $a = b$*,* $a = c$*,* $b = c$*, or* $a = b = c$*, respectively. For instance, the values for composition ABC are equal in columns I to VI.*

Rank	$A_aB_bC_c$ ($a > b > c$)	No. of IMs	No. of STs	I	II	III	IV	V	VI	Comment
1.	ABC	1495	62	*1495*	*1495*	*1495*	*1495*	*1495*	*1495*	*a = b = c*
2.	A_2BC	841	80	*119*	*119*	*334*	*388*	*334*	*388*	*a > b = c*
3.	A_2B_2C	677	41	*257*	*70*	*257*	*70*	*350*	*350*	*a = b > c*
4.	A_4BC	324	14	*130*	*130*	*95*	*99*	*95*	*99*	*a > b = c*
5.	A_6B_2C	186	9	40	56	0	88	0	2	–
6.	A_5B_3C	161	18	2	96	1	37	18	7	–
7.	A_3B_2C	160	32	3	23	14	17	62	41	–
8.	A_3BC	159	22	*0*	*0*	6	*153*	6	*153*	*a > b = c*
9.	$A_{20}B_2C$	141	3	0	0	0	1	0	140	–
10.	A_6B_6C	131	17	*0*	*4*	*0*	*4*	*127*	*127*	*a = b > c*
11.	A_5B_2C	108	13	64	29	2	1	5	7	–
12.	$A_{13}B_4C_3$	106	6	0	0	0	3	0	103	–
13.	$A_9B_3C_2$	104	7	1	0	0	2	10	91	–
14.	A_5BC	94	6	*0*	*0*	29	65	29	65	*a > b = c*
15.	$A_{43}B_6C_4$	86	1	0	0	0	86	0	0	–
16.	A_4B_2C	86	22	0	6	0	19	29	32	–
17.	$A_5B_3C_2$	82	17	6	1	1	1	0	73	–
18.	$A_4B_3C_2$	82	22	0	7	4	23	0	48	–
19.	$A_4B_4C_3$	76	9	*1*	6	*1*	6	*69*	*69*	*a = b > c*
20.	A_8B_4C	75	7	0	2	1	8	0	64	–
21.	$A_{10}B_2C$	67	7	0	0	0	4	0	63	–
22.	$A_{12}B_4C$	66	4	0	0	0	32	0	34	–
23.	$A_4B_3C_3$	64	3	*4*	*4*	*0*	60	*0*	60	*a > b = c*
24.	$A_{16}B_7C_6$	60	1	0	1	0	0	18	41	–

continued

Table 5.8 *continued*

Rank	$A_aB_bC_c$ $(a > b > c)$	No. of IMs	No. of STs	I	II	III	IV	V	VI	Comment
25.	A_9B_2C	58	7	0	0	1	3	40	14	–
26.	$A_{14}B_3C_2$	58	4	0	8	0	0	0	50	–
27.	A_8B_2C	56	2	0	0	0	32	0	24	–
28.	$A_3B_2C_2$	56	12	*31*	*31*	*2*	*23*	*2*	*23*	$a > b = c$
29.	$A_{23}B_7C_4$	49	2	49	0	0	0	0	0	–
30.	$A_5B_4C_2$	49	15	0	35	0	5	1	8	–
31.	$A_{13}B_6C$	48	2	0	0	48	0	0	0	–
32.	$A_3B_3C_2$	47	20	*8*	*21*	*8*	*21*	*18*	*18*	$a = b > c$
33.	$A_6B_3C_2$	42	10	0	1	18	7	6	10	–
34.	$A_{10}B_5C_4$	37	3	0	0	0	37	0	0	–
35.	$A_7B_6C_4$	36	3	0	3	0	0	33	0	–
36.	$A_8B_4C_3$	31	6	0	0	0	6	24	1	–
37.	$A_{12}B_6C$	29	1	29	0	0	0	0	0	–
38.	$A_{12}B_4C_3$	27	2	0	0	0	0	0	27	–
39.	A_4B_3C	27	17	0	5	5	10	3	4	–
40.	A_9B_4C	22	4	12	0	0	0	10	0	–
41.	A_6B_4C	22	10	0	3	0	3	5	11	–
42.	$A_7B_3C_2$	21	8	0	0	0	13	7	1	–
43.	A_3B_3C	21	12	*7*	*8*	*7*	*8*	*6*	*6*	$a = b > c$
44.	A_8B_3C	20	2	0	9	11	0	0	0	–
45.	$A_{21}B_{10}C_4$	20	2	0	0	0	0	0	20	–

5.6.2   Symmetry vs. composition

Is there any correlation between the symmetry group of a structure type and its stoichiometry? Before we focus on this discussion let us first have a look on the general distribution of binary and ternary intermetallics over the 14 Bravais type lattices, which is quite heterogeneous (Table 5.9 and Figs. 5.36 and 5.37). There are only minor differences between the distributions of binary and ternary intermetallic compounds: the frequencies of ternary intermetallics with symmetries aP, oF, and cI are only half of that of binary compounds, small numbers anyway.

Table 5.9 *Binary and ternary intermetallics (IMs) and their distribution over the 14 Bravais type lattices.*

Bravais type lattice	Binary IMs		Ternary IMs		Total
aP	33	0.5%	27	0.2%	0.3%
mP	93	1.4%	128	1.0%	1.1%
mS	155	2.4%	249	1.9%	2.1%
oP	567	8.8%	1225	9.4%	9.2%
oS	397	6.2%	974	7.5%	7.0%
oI	152	2.4%	724	5.6%	4.5%
oF	44	0.7%	30	0.2%	0.4%
hP	1538	23.9%	3280	25.2%	24.8%
hR	346	5.3%	532	4.1%	4.5%
tP	383	6.0%	981	7.5%	7.0%
tI	593	9.2%	1397	10.7%	10.2%
cP	738	11.5%	891	6.8%	8.4%
cI	391	6.1%	471	3.6%	4.4%
cF	1011	15.7%	2117	16.7%	16.1%
All	6441	100%	13026	100%	100%

It is amazing that just $\approx 0.3\%$ of all structures have triclinic and $\approx 3.2\%$ monoclinic crystal symmetry compared to the $\approx 21.1\%$, which show orthorhombic symmetry, for instance. This means, getting rid of two of the three oblique angles of the unit cell, allowing not just inversion but also for rotation and screw axes as well as mirror and glide planes, increases the frequency of intermetallics with such a symmetry by one order of magnitude. Losing the third oblique angle, and allowing for symmetry elements in all three space dimensions, increases the frequency of intermetallics with such a symmetry by another order of magnitude. It is also remarkable that F-centering is with 16.1% quite beneficial in the case of cubic structures; however, with only 0.4% detrimental in the case of orthorhomic ones.

The most frequent symmetries are cubic with $\approx 28.9\%$ and hexagonal with $\approx 24.8\%$. Why are such high-symmetry crystal structures so much more frequent than low-symmetry ones? The symmetry adopted by a crystal structure depends on the symmetry of its AETs, the number of different AETs, and the way they pack. One has to keep in mind that all AETs overlap with all neighboring ones, since each atom of the coordination polyhedron, is at the same time the center of

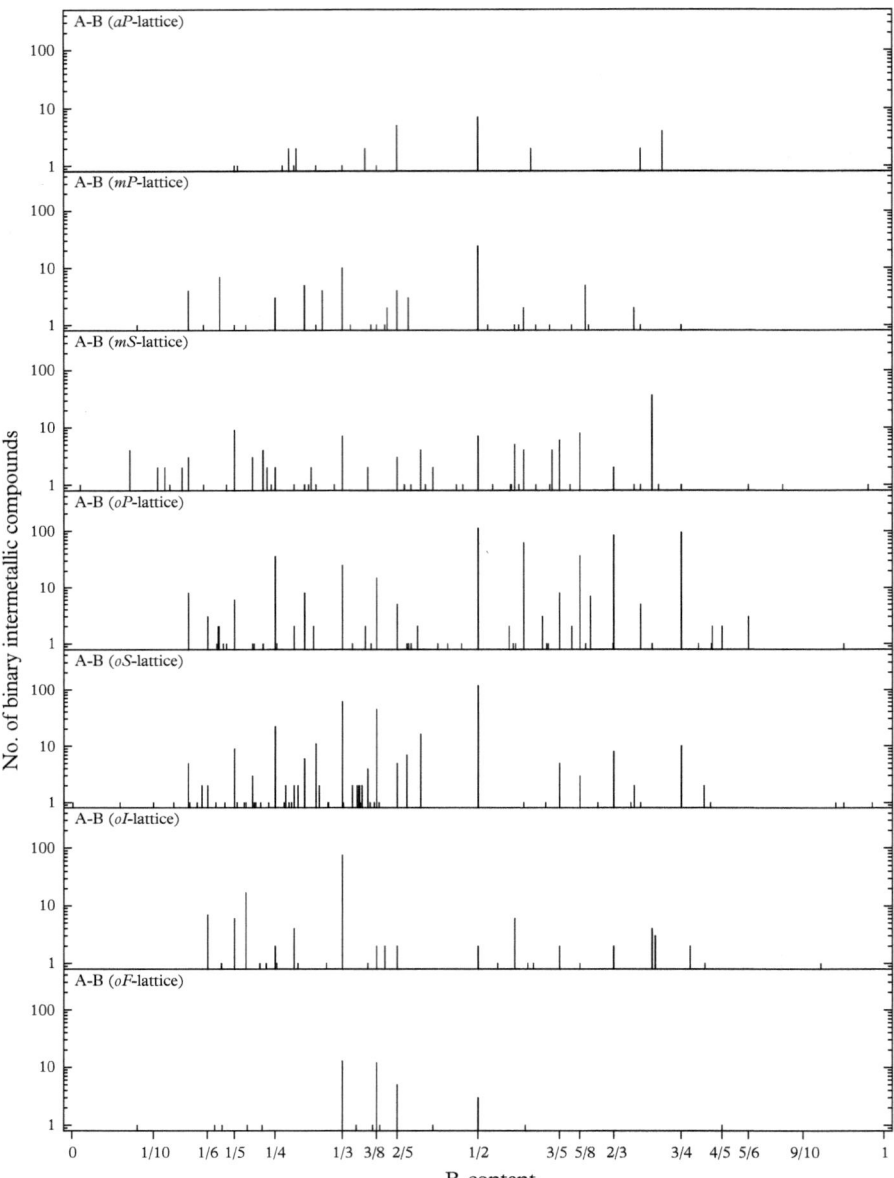

Fig. 5.36 *Distribution of binary compositions A_mB_n, with $M(A) < M(B)$, within the Bravais type lattices aP, mP, mS, oP, oS, oI, and oF. Note the logarithmic scale for the number of representatives.*

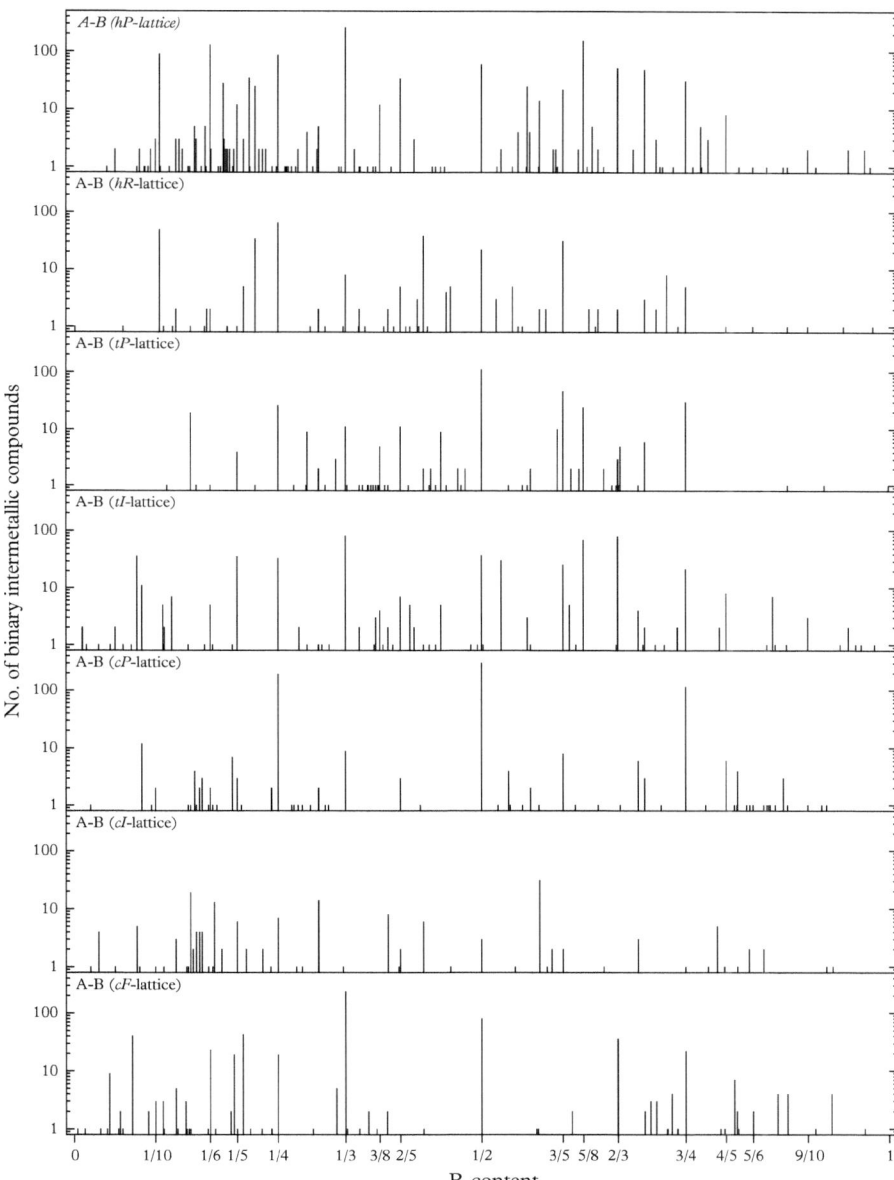

Fig. 5.37 *Distribution of binary compositions A_mB_n, with $M(A) < M(B)$, within the Bravais type lattices hP, hR, tP, tI, cP, cI, and cF. Structure types cF4-Cu, cI2-W, and hP2-Mg are excluded in these histograms in order to dispose of probable sphere packing entries. Note the logarithmic scale for the number of representatives.*

another AET. A small number of different high-symmetry AETs leads to high-symmetry structures. For instance, if the preferred coordination polyhedron of each atom of a structure is mirror-symmetric then there will be a high probability that the structure itself is mirror-symmetric as well. However, the symmetry of the AETs may be lower than that of their packing. The simplest examples are the AETs of the structure type $hP2$-Mg, which are disheptahedra (anticuboctahedra) and therefore non-centrosymmetric in contrast to the crystal structure itself.

The compositions of the binary intermetallics, as they are distributed over the Bravais type lattices, are illustrated in Figs. 5.36 and 5.37. From the underlying data, the entries belonging to the sphere packings $cF4$-Cu, $cI2$-W, and $hP2$-Mg were excluded in order to get rid of their bias on the distribution. The histograms for the lattices hP, cI, and cF—including the structure types $cF4$-Cu, $cI2$-W, and $hP2$-Mg—are shown in Fig. 5.37. The distributions clearly differ from one another; however, a more detailed interpretation is difficult.

The histograms for the Bravais type lattices oP, hP, tI, cP, and cF are quite symmetric around 1/2 B-content, while the others are rather asymmetric, in particular the centered ones oS, oI, oF, and cI. In the latter case, this means that if intermetallic compounds A_mB_n, with $M(A) < M(B)$, crystallize with these symmetries, compounds with composition A_nB_m cannot. It is also remarkable that the composition 1/2 is very unfavorable in the case of the centered Bravais type lattices oI, oF, and cI.

The compositions of the ternary intermetallics, as they are distributed over the Bravais type lattices, are illustrated in Figs. 5.38, 5.39, and 5.40. In contrast to the distribution of binary intermetallics, that of ternary ones shows significant differences as functions of lattice symmetry. In the case of the rarely occurring Bravais symmetry aP, most compositions appear to be pseudobinary since the datapoints are close to the edges of the concentration triangle, probably indicating that a few percent of a third component stabilize the compound.

In the case of the concentration triangles for symmetries mP, oP, oS, cP, and cF, the tie lines AC–AB and AC–BC stand out clearly. This means that intermetallics with compositions AB_xC_{1-x} ($0 \leq x \leq 1$) and $A_xB_{1-x}C$ ($0 \leq x \leq 1$), respectively, have some prevalence for these types of Bravais type lattices. Also, in the case of oS, the line AC_2–ABC_2 is very densely populated, which is the signature of intermetallic phases with compositions differing in a broad range from that of their structure types. One prominent example is the binary structure type $oS8$-Tll, which has been assigned to many (disordered) ternary phases with compositions varying in this range. Another example is the structure type $oS16$-CeNiSi$_2$ with a similarly broad stability range in many ternary systems.

In the case of oI, the lines AB_2–AC_2 stand out clearly as well as two additional lines with A contents around $\approx 24\%$ and $\approx 22\%$, respectively. Due to the lack of data, not much can be said about oF except that there is also a similar trend as in oS. For Bravais symmetry hP, the densest populated lines are AB_2–AC_2 and AC_2–BC_2 as well as AB_3–AC_3 and AC_3–BC_3, while for hR mainly the lines AB_3–AC_3 and AC_3–BC_3 as well as those with $\approx 10\%$ A or C play a role.

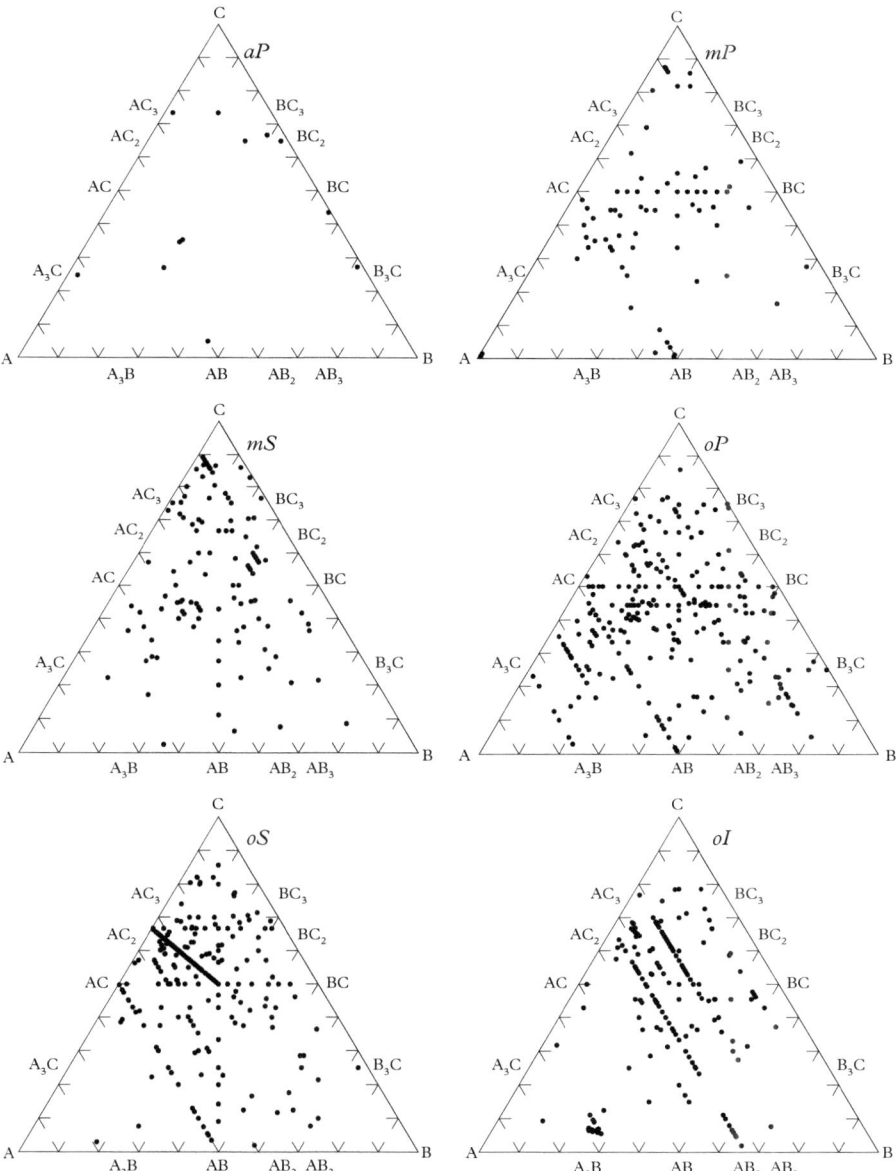

Fig. 5.38 *Occurrence of ternary compounds with compositions ABC within the Bravais type lattices aP, mP, mS, oP, oS, and oI. The elements are assigned to A, B, and C by* $M(A) < M(B) < M(C)$.

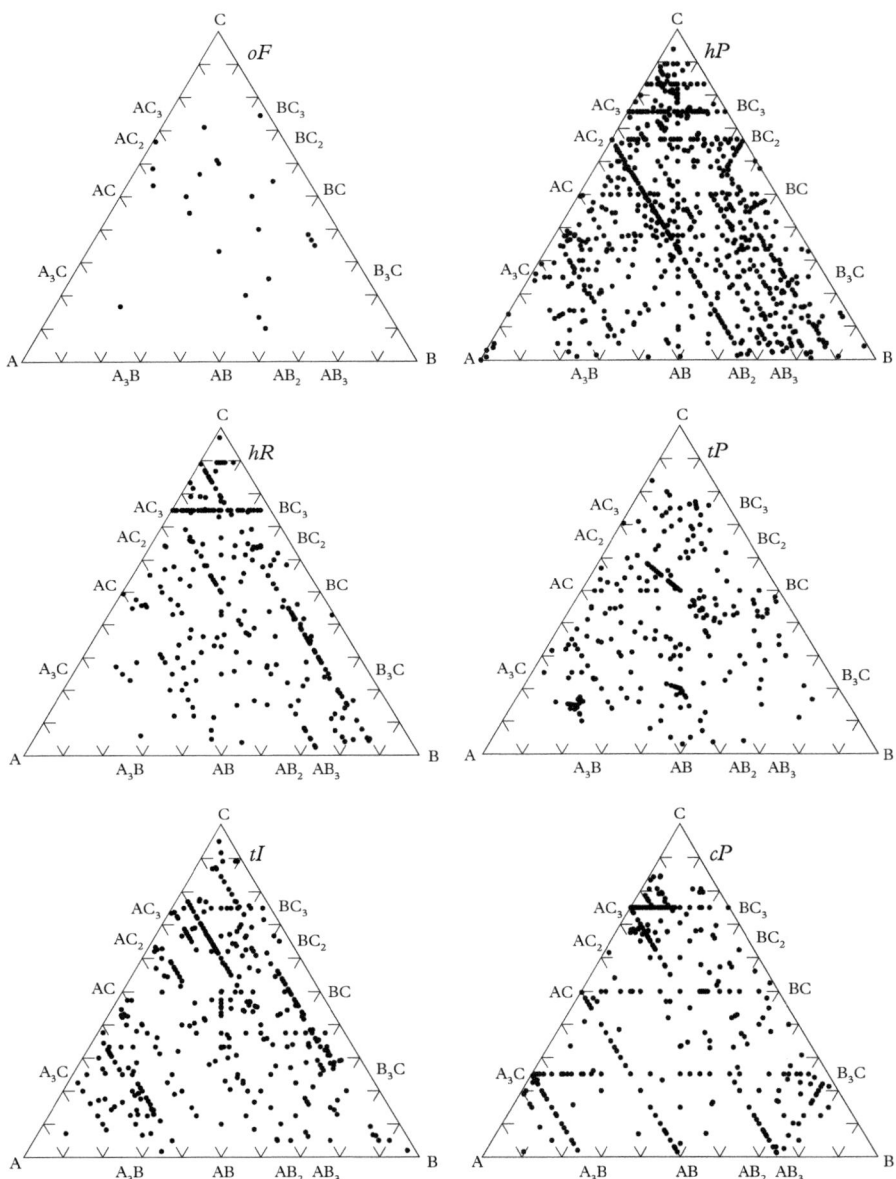

Fig. 5.39 *Occurrence of ternary compounds with compositions ABC within the Bravais type lattices oF, hP, hR, tP, tI, and cP. The elements are assigned to A, B, and C by* $M(A) < M(B) < M(C)$.

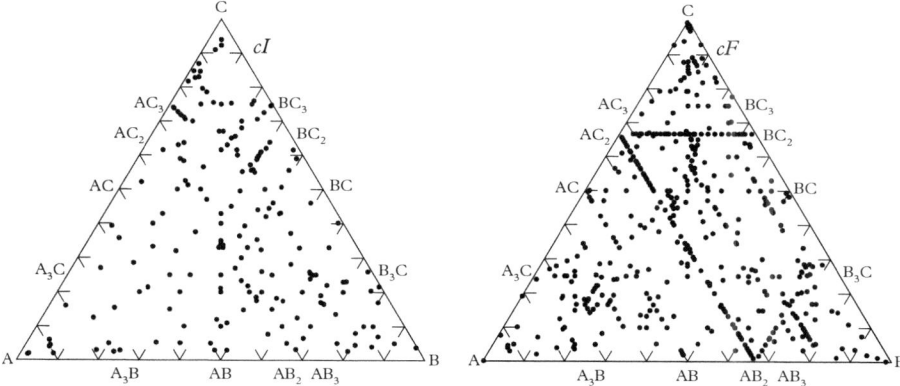

Fig. 5.40 *Occurrence of ternary compounds with compositions ABC within the Bravais type lattices cI and cF. The elements are assigned to A, B, and C by M(A) < M(B) < M(C).*

In the case of tP, a part of the line AC_2–ABC_2 is densely occupied as was the case for oS. The most significant features for tI are the lines with A $\approx$ 20% and $\approx$ 9%, respectively. The population of the lines is clearly distinct for cP and cF, while for cI the distribution is completely different and does not follow lines. The diagram of cP shows quite a regular line structure every 25 at.%, while that for cF features mainly the lines AB_2–AC_2 and AC_2–BC_2. What is also remarkable is that the distributions in the ternary concentration diagrams do not show threefold symmetry. Closest to it are the diagrams for cP and cI.

Overall, one can conclude that for most Bravais symmetries, the concentration of the element with the lowest Mendeleev number, A, remains constant while B and C can vary via mutual cost. In many less cases, the element B can take over the role of A. Only for symmetry cP, all three elements, A, B, and C, can play this role equivalently. In the diagram of cP, some tie lines are very densely occupied, mainly due to the assignment of $cP4$-Cu_3Au and $cP8$-Cr_3Si to ternary phases. Furthermore, it seems that particular compositions are not equally likely in the different Bravais type lattices:

aP Most phases are close to the edges of the concentration triangle, indicating that the ternary phases are just binary phases stabilized by a small amount of a third element.

mP Most phases are located on the tie lines AC–AB and AC–BC; there are no phases with more than 70–80% of A or B. The few dots close to the corner of A seem to be solid solutions of two elements in A.

mS No phases contain more than 75% A. Most phases have a high B and medium to high C content.

oP Most phases are located on the tie lines AC–AB and AC–BC; there are no phases with more than 75% A.

oS Most phases are located on the tie lines AC–AB and AC–BC; very densely populated tie line AC_2–ABC_2 due to disordered phases. Almost no phases beyond the densely occupied tie line AC–AB towards a higher A content.

oI Most phases are located on tie lines with constant A ratio. Most phases have a low concentration of A.

oF Very few ternary phases, just one system with more than 40% A.

hP Most phases on tie lines with constant A or C, less significant with constant B. The tie line AC_2–BC_2 seems to include many (disordered) phases with broad stability ranges. Phases with more than 75% A seem to be solid solutions of one or two elements in binary or unary phases, respectively.

hR Similar to the *hP* diagram with a relatively higher importance of the tie line AC_3–BC_3.

tP Rather scattered distribution with some point agglomerations and a part of the tie line AC_2–ABC_2 densely occupied, indicating disordered phases.

tI Some similarities to *oI*, but much less data points.

cP Almost all data points on the tie lines A_3C–B_3C, AC–BC, AC_3–BC_3 and symmetrically equivalent ones.

cI Almost homogeneous distribution over all compositions.

cF Similar to *hP*, but less data points.

The distributions of binary and ternary intermetallics over the point groups are given in Table 5.10. One clearly sees an extremely heterogeneous distribution of the frequencies of intermetallic compounds with these symmetries. There are only a few significant differences between the distributions of the binary and ternary intermetallics; one is found for point group $\bar{6}m2$ with 40 (0.6%) and 622 (4.8%) representatives, respectively. Amazingly, the most and second most frequent point groups for binary (ternary) intermetallics are $m\bar{3}m$ (*mmm*) with 28.0% (21.0%), and $6/mmm$ ($m\bar{3}m$) with 19.7% (19.4%), respectively. The by far most frequent non-centrosymmetric point group is $\bar{4}3m$ with 3.5% (5.6%).

There are only ten point groups, each containing more than 1% of intermetallic compounds. Within the crystal systems the by far highest frequencies are found for the respective holohedral groups $\bar{1}$, $2/m$, *mmm*, $4/mmm$, $\bar{3}m$, $6/mmm$, and $m\bar{3}m$. Furthermore, the 11 centrosymmetric point groups (Laue groups) have higher frequencies than their respective non-centrosymmetric subgroups, and already apply to 16 586 (85.1%) of all intermetallics.

The distributions of binary and ternary intermetallics over the space groups are given in Tables 5.11, 5.12, and 5.13. They are extremely heterogeneous without too many disparities between them. The largest ones are for the tetragonal space groups 127 $P4/mbm$, with 34 binary and 306 ternary representatives, and 129 $P4/nmm$, with 29 binary and 329 ternary representatives, respectively, and the cubic space group 216 $F\bar{4}3m$, with 97 binary and 595 ternary representatives. The 22 chiral space groups, which are part of the 65 Sohncke groups, are marked.

Table 5.10 *Binary and ternary intermetallics (IMs) and their distribution over the 32 point groups (PGs).*

PG	Binary IMs		Ternary IMs		PG	Binary IMs		Ternary IMs	
1	3	0.05%	2	0.02%	$\bar{3}$	77	1.2%	28	0.2%
$\bar{1}$	30	0.5%	25	0.2%	32	12	0.2%	9	0.1%
2	11	0.2%	19	0.2%	$3m$	19	0.3%	19	0.2%
m	11	0.2%	7	0.1%	$\bar{3}m$	319	5.0%	584	4.5%
$2/m$	226	3.5%	351	2.7%	6	3	0.1%	2	0.0%
222	21	0.3%	22	0.2%	$\bar{6}$	2	0.0%	12	0.1%
$mm2$	64	1.0%	202	1.6%	$6/m$	62	1.0%	39	0.3%
mmm	1075	16.7%	2729	21.0%	622	12	0.2%	18	0.1%
4	3	0.1%	5	0.1%	$6mm$	65	1.0%	207	1.6%
$\bar{4}$	8	0.1%	1	0.0%	$\bar{6}m2$	40	0.6%	622	4.8%
$4/m$	54	0.8%	18	0.1%	$6/mmm$	1267	19.7%	2265	17.4%
422	2	0.0%	7	0.1%	23	31	0.5%	39	0.3%
$4mm$	10	0.2%	77	0.6%	$m\bar{3}$	53	0.8%	159	1.2%
$\bar{4}2m$	43	0.7%	29	0.2%	432	28	0.4%	27	0.2%
$4/mmm$	856	13.3%	2241	17.2%	$\bar{4}3m$	227	3.5%	733	5.6%
3	6	0.1%	7	0.1%	$m\bar{3}m$	1801	28.0%	2521	19.4%
					All	6441	100.0%	13026	100.0%

The Sohncke groups contain only symmetry operations of the first kind (translations, rotation, and screw axes, respectively). A chiral space group is a space group whose group structure is chiral. Every chiral space group type occurs in pairs of the two enantiomorphic variants. Chiral (enantiomorphic) crystal structures can occur not only in the chiral space groups but in any of the Sohncke groups.

For comparison, the number of structures in the Cambridge Structural Database (CSD) (Cambridge Crystallographic Data Centre, 2015) in the respective space groups is listed as well. The CSD contains 754 897 entries (16 February 2015) of small-molecule organic and metal-organic crystal structures. There, we can see large differences in the frequency distribution compared to that of the intermetallics listed in the PCD. The largest differences are in the space groups 2 $P\bar{1}$, 14 $P2_1/c$, and some other low-symmetry space groups. As was demonstrated by Kitaigorodsky (1973), these two and a few other low-symmetry groups allow the best close packings of molecules of rather arbitrary shapes and point group symmetry 1, while for intermetallics the relevant symmetries are those of close

Table 5.11 *Distribution of the binary and ternary intermetallics over the 230 space groups (SGs). For comparison, in columns "CSD", the number of compounds in the Cambridge Structural Database (Cambridge Crystallographic Data Centre, 2015) in the respective SGs is listed, showing large differences in the frequency distribution. The CSD contains in total 754 897 entries (16 February 2015) of small-molecule organic and metal-organic crystal structures. The asterisk, ∗, marks structures that are listed in the PCD, but not confirmed by later structure analyses. The 65 Sohncke groups are marked by No.[S], the 22 chiral space groups among them by No.[c].*

No.	SG	Binary	Ternary	CSD	No.	SG	Binary	Ternary	CSD
1[S]	$P1$	3*	2*	7136	41	$Aea2$	6	3	815
2	$P\bar{1}$	30	25	184 087	42	$Fmm2$	2	9	68
3[S]	$P2$	0	1	134	43	$Fdd2$	20	1	2570
4[S]	$P2_1$	3	18	39 092	44	$Imm2$	6	29	69
5[S]	$C2$	8	0	6352	45	$Iba2$	0	9	445
6	Pm	1	1	21	46	$Ima2$	0	28	110
7	Pc	3	0	3212	47	$Pmmm$	5	6	30
8	Cm	7	6	281	48	$Pnnn$	0	0	56
9	Cc	0	0	7910	49	$Pccm$	0	0	17
10	$P2/m$	6	8	102	50	$Pban$	0	0	73
11	$P2_1/m$	30	48	3817	51	$Pmma$	29	33	51
12	$C2/m$	90	197	3883	52	$Pnna$	9	3	795
13	$P2/c$	2	0	4897	53	$Pmna$	1	0	101
14	$P2_1/c$	48	52	261 358	54	$Pcca$	0	0	362
15	$C2/c$	50	46	63 007	55	$Pbam$	15	115	213
16[S]	$P222$	0	4	35	56	$Pccn$	1	0	2680
17[S]	$P222_1$	0	3	76	57	$Pbcm$	18	69	758
18[S]	$P2_12_12$	0	0	3096	58	$Pnnm$	15	30	537
19[S]	$P2_12_12_1$	6	9	55 146	59	$Pmmn$	40	76	237
20[S]	$C222_1$	8	4	1332	60	$Pbcn$	2	3	6461
21[S]	$C222$	7	2	57	61	$Pbca$	7	2	25 376
22[S]	$F222$	0	0	25	62	$Pnma$	409	853	8276
23[S]	$I222$	0	0	173	63	$Cmcm$	266	691	741
24[S]	$I2_12_12_1$	0	0	57	64	$Cmce$	35	36	936
25	$Pmm2$	3	3	11	65	$Cmmm$	49	125	100

Table 5.11 *continued*

No.	SG	Binary	Ternary	CSD	No.	SG	Binary	Ternary	CSD
26	$Pmc2_1$	1	0	126	66	$Cccm$	2	1	85
27	$Pcc2$	0	0	15	67	$Cmme$	0	6	56
28	$Pma2$	0	0	13	68	$Ccce$	4	2	354
29	$Pca2_1$	0	0	5575	69	$Fmmm$	5	5	56
30	$Pnc2$	0	0	104	70	$Fddd$	17	15	815
31	$Pmn2_1$	1	8	481	71	$Immm$	53	415	85
32	$Pba2$	0	0	144	72	$Ibam$	11	40	305
33	$Pna2_1$	4	8	10 438	73	$Ibca$	0	0	215
34	$Pnn2$	1	0	244	74	$Imma$	82	203	148
35	$Cmm2$	0	0	6	75^S	$P4$	0	0	42
36	$Cmc2_1$	8	18	1071	76^c	$P4_1$	2	2	700
37	$Ccc2$	0	0	97	77^S	$P4_2$	0	0	80
38	$Amm2$	11	22	84	78^c	$P4_3$	0	0	578
39	$Aem2$	1	0	48	79^S	$I4$	0	3	212
40	$Ama2$	0	2	133	80^S	$I4_1$	1	0	192

Table 5.12 *Binary and ternary intermetallics and their distribution over the 230 space groups (continued).*

No.	SG	Binary	Ternary	CSD	No.	SG	Binary	Ternary	CSD
81	$P\bar{4}$	0	0	171	121	$I\bar{4}2m$	3	9	143
82	$I\bar{4}$	8	1	1018	122	$I\bar{4}2d$	3	5	494
83	$P4/m$	4	3	38	123	$P4/mmm$	136	110	118
84	$P4_2/m$	1	2	90	124	$P4/mcc$	0	2	70
85	$P4/n$	0	0	661	125	$P4/nbm$	16	9	20
86	$P4_2/n$	9	2	1005	126	$P4/nnc$	0	2	161
87	$I4/m$	34	9	499	127	$P4/mbm$	34	306	62
88	$I4_1/a$	6	2	2718	128	$P4/mnc$	0	5	79
89^S	$P422$	0	0	7	129	$P4/nmm$	29	329	167
90^S	$P42_12$	0	0	56	130	$P4/ncc$	25	1	342
91^c	$P4_122$	0	0	59	131	$P4_2/mmc$	0	0	31
92^c	$P4_12_12$	2	4	1502	132	$P4_2/mcm$	0	0	13

continued

Table 5.12 *continued*

No.	SG	Binary	Ternary	CSD	No.	SG	Binary	Ternary	CSD
93[S]	$P4_222$	0	0	8	133	$P4_2/nbc$	0	0	31
94[S]	$P4_22_12$	0	0	136	134	$P4_2/nnm$	0	1	43
95[c]	$P4_322$	0	0	54	135	$P4_2/mbc$	1	0	76
96[c]	$P4_32_12$	0	0	1290	136	$P4_2/mnm$	79	89	131
97[S]	$I422$	0	1	46	137	$P4_2/nmc$	2	40	95
98[S]	$I4_122$	0	2	81	138	$P4_2/ncm$	5	1	89
99	$P4mm$	1	3	3	139	$I4/mmm$	280	891	140
100	$P4bm$	0	0	3	140	$I4/mcm$	161	298	64
101	$P4_2cm$	0	1	6	141	$I4_1/amd$	66	65	141
102	$P4_2nm$	6	0	23	142	$I4_1/acd$	22	29	383
103	$P4cc$	0	0	21	143[S]	$P3$	2	1	177
104	$P4nc$	0	1	90	144[c]	$P3_1$	1	0	558
105	$P4_2mc$	0	0	2	145[c]	$P3_2$	0	0	532
106	$P4_2bc$	0	0	79	146[S]	$R3$	3	6	952
107	$I4mm$	3	59	13	147	$P\bar{3}$	7	7	891
108	$I4cm$	0	3	28	148	$R\bar{3}$	70	21	4850
109	$I4_1md$	0	10	35	149[S]	$P312$	1	0	9
110	$I4_1cd$	0	0	278	150[S]	$P321$	1	0	73
111	$P\bar{4}2m$	1	0	6	151[c]	$P3_112$	1	1	23
112	$P\bar{4}2c$	0	0	27	152[c]	$P3_121$	6	4	680
113	$P\bar{4}2_1m$	11	0	207	153[c]	$P3_212$	0	0	15
114	$P\bar{4}2_1c$	0	0	951	154[c]	$P3_221$	0	0	499
115	$P\bar{4}m2$	0	4	4	155[S]	$R32$	3	4	344
116	$P\bar{4}c2$	8	1	27	156	$P3m1$	0	6	8
117	$P\bar{4}b2$	0	0	53	157	$P31m$	3	6	13
118	$P\bar{4}n2$	10	0	147	158	$P3c1$	3	0	70
119	$I\bar{4}m2$	7	4	27	159	$P31c$	1	1	254
120	$I\bar{4}c2$	0	6	73	160	$R3m$	9	5	229

Table 5.13 *Binary and ternary intermetallics and their distribution over the 230 space groups (continued).*

No.	SG	Binary	Ternary	CSD	No.	SG	Binary	Ternary	CSD
161	$R3c$	3	1	739	196^S	$F23$	0	0	70
162	$P\bar{3}1m$	0	0	17	197^S	$I23$	3	12	133
163	$P\bar{3}1c$	0	2	290	198^S	$P2_13$	23	21	428
164	$P\bar{3}m1$	61	86	82	199^S	$I2_13$	3	2	51
165	$P\bar{3}c1$	0	1	516	200	$Pm\bar{3}$	8	11	14
166	$R\bar{3}m$	238	469	314	201	$Pn\bar{3}$	1	0	27
167	$R\bar{3}c$	20	26	1172	202	$Fm\bar{3}$	1	5	37
168^S	$P6$	0	0	22	203	$Fd\bar{3}$	1	0	81
169^c	$P6_1$	0	0	471	204	$Im\bar{3}$	36	132	91
170^c	$P6_5$	0	0	414	205	$Pa\bar{3}$	5	10	682
171^c	$P6_2$	1	0	58	206	$Ia\bar{3}$	1	1	88
172^c	$P6_4$	0	0	44	207^S	$P432$	0	0	5
173^S	$P6_3$	2	2	498	208^S	$P4_232$	0	5	
174	$P\bar{6}$	2	12	22	209^S	$F432$	0	0	29
175	$P6/m$	40	16	30	210^S	$F4_132$	0	0	36
176	$P6_3/m$	22	23	890	211^S	$I432$	0	0	22
177^S	$P622$	0	0	7	212^c	$P4_332$	0	0	29
178^c	$P6_122$	2	0	213	213^c	$P4_132$	28	26	39
179^c	$P6_522$	0	0	165	214^S	$I4_132$	0	1	23
180^c	$P6_222$	7	18	53	215	$P\bar{4}3m$	15	12	49
181^c	$P6_422$	0	0	33	216	$F\bar{4}3m$	97	595	49
182^S	$P6_322$	3	0	96	217	$I\bar{4}3m$	71	36	258
183	$P6mm$	0	0	4	218	$P\bar{4}3n$	4	8	115
184	$P6cc$	0	0	9	219	$F\bar{4}3c$	0	0	80
185	$P6_3cm$	9	2	25	220	$I\bar{4}3d$	40	82	234
186	$P6_3mc$	56	205	105	221	$Pm\bar{3}m$	574	524	108
187	$P\bar{6}m2$	16	47	15	222	$Pn\bar{3}n$	0	0	100
188	$P\bar{6}c2$	1	0	11	223	$Pm\bar{3}n$	78	275	45

continued

Table 5.13 *continued.*

No.	SG	Binary	Ternary	CSD	No.	SG	Binary	Ternary	CSD
189	$P\bar{6}2m$	14	563	23	224	$Pn\bar{3}m$	0	0	23
190	$P\bar{6}2c$	9	12	121	225	$Fm\bar{3}m$	554	740	437
191	$P6/mmm$	248	666	43	226	$Fm\bar{3}c$	45	44	44
192	$P6/mcc$	0	0	91	227	$Fd\bar{3}m$	313	733	116
193	$P6_3/mcm$	176	335	50	228	$Fd\bar{3}c$	0	0	107
194	$P6_3/mmc$	843	1264	166	229	$Im\bar{3}m$	229	201	120
195[S]	$P23$	2	4	14	230	$Ia\bar{3}d$	8	4	73
					All		6441	13026	682 999

sphere packings, and their derivative structures. For some higher-symmetric space groups, the number of representatives of intermetallics is significantly higher, for instance, for 191 $P6/mmm$, 194 $P6_3/mmc$, 216 $F\bar{4}3m$, 221 $Pm\bar{3}m$, 225 $Fm\bar{3}m$, and 227 $Fd\bar{3}m$.

On average, there are 84.6 intermetallics assigned to each space group; however, there are remarkably 63 space groups (27.4%), to which not a single intermetallic compound is assigned: triclinic 0 out of 2 (0%), monoclinic 1 out of 13 (7.6%), orthorhombic 15 out of 59 (25.4%), tetragonal 23 out of 67 (34.3%), trigonal 4 out of 25 (16%), hexagonal 9 out of 27 (33.3%), and cubic 11 out of 36 (30.5%).

There are only six space groups each representing more than 5% (973) of all binary and ternary intermetallics (62 $Pnma$, 139 $I4/mmm$, 194 $P6_3/mmc$, 221 $Pm\bar{3}m$, 225 $Fm\bar{3}m$, and 227 $Fd\bar{3}m$), and out of them just a single one, 194, $P6_3/mmc$, with more than 10% (1947) assigned intermetallics (843 binary and 1264 ternary representatives, in total 10.8%).

The above-mentioned most frequent space groups also describe the symmetry of some of the stable homogenous packings of symmetrically equivalent spheres (O'Keeffe and Hyde, 1996; Wilson and Prince, 1999) (see, also, Table 3.3). They are called *stable* if each sphere is in contact with four others. In the following examples of homogenous sphere packings for these space groups are described according to Wilson and Prince (1999). Additionally, for each of these space groups the number of representatives in the PCD are given and up to five of the most frequent structure types with more than 50 representatives. This shows that a few very common structure types are responsible for the peaks in the space group symmetries:

62 $Pnma$: Wyckoff position $4c$ $x, 1/4, z$; $x = 7/20$, $z = 7/8$, $b/a = 4/5$, $c/a = 2\sqrt{15}/15$; inter-sphere distance c, contact number $k = 10$, net (010) 4^4, stacking 3 (contacts to the one neighboring layer), 3 (contacts to the other neighboring layer), 2 (layers per translation period), density 0.6981. For the

Pnma (oP4): close
sphere packing   oP12-Co₂Si   oP12-TiNiSi

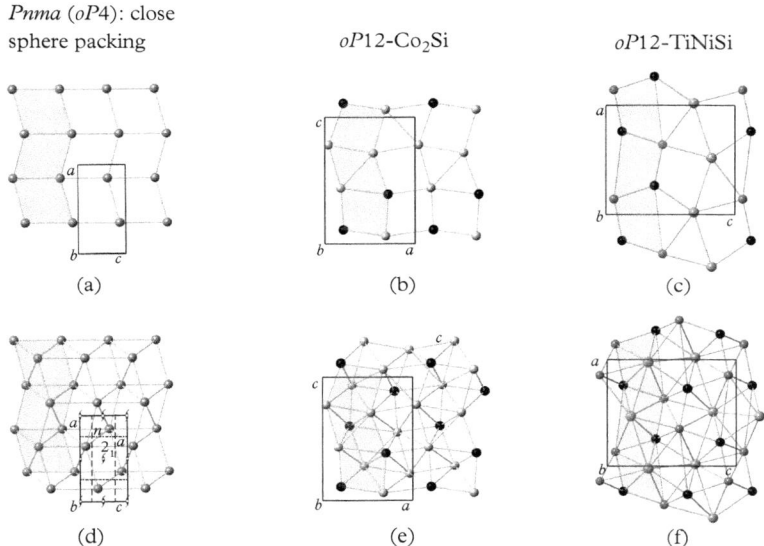

(a)   (b)   (c)

(d)   (e)   (f)

Fig. 5.41 *A single flat layer of the structures of (a) the close sphere packing (oP4) in the space group 62 Pnma, (b) oP12-Co₂Si, and (c) oP12-TiNiSi. In (d)–(f) the respective second layer is shown as well. In (d) the framework is shown of the symmetry elements of space group 62 Pnma. The a glide plane acts within each layer, the n glide plane and the 2₁ screw axis connect the self-dual layers, which are flat and lying on mirror planes. The quadrangles (shaded gray) are quite regular in (a) and (b), and strongly distorted in (c), in order to account for the larger ratio of atomic radii.*

structure, see Fig. 5.41 (a, d). There are 1262 binary and ternary intermetal-lic phases, listed in the PCD.

oP12-TiNiSi: 403 representatives: 289 contain a RE element, the others alkali metals or alkaline earth metals or early TM elements as first com-ponent; most of them a TM element as second constituent, and all of them either Au, Mg, Cd, Zn, Tl, In, Ga, Al, Pb, Sn, Ge, Bi, or Sb as the third constituent), hP3-AlB₂ derivative structure; stacking along [010] of distorted pentagon nets (Fig. 5.41 (c, f); tetrahedral coordina-tion of Ni by Si; the structure is very flexible and is closely related to oP12-Co₂Si (Landrum *et al.*, 1998).

oP16-Fe₃C: 97 representatives: most contain a RE element and a late TM element; see, also, Fig. 5.18.

oP8-FeB: 79 representatives: most contain a RE element and a late TM element; see, also, Fig. 5.16.

oP36-Sm₅Ge₄: 75 representatives: most contain a RE element and a late TM element, some also a main group element.

*o*P12-Co$_2$Si: 73 representatives: most contain a RE element and a late TM element, some also Ge, Sn, or Pb; see Fig. 5.41(b, e) and also Fig. 5.15.

139 *I*4/*mmm*: Wyckoff position 2*a* 0, 0, 0; *c*/*a* = $\sqrt{6}$/3, inter-sphere distance *c*, contact number *k* = 10, net {110} 3^6, stacking 2, 2 2, density 0.6981. There are 1171 binary and ternary intermetallic phases, listed in the PCD.

  *tI*10-CeAl$_2$Ga$_2$: 338 representatives (most contain a RE or TM element and two main group elements).

  *tI*26-ThMn$_{12}$: 294 representatives.

  *tI*26-CeMn$_4$Al$_8$: 87 representatives (most contain Al or In, a TM and a RE element).

  *tI*10-BaAl$_4$: 70 representatives (most contain Al or Ga, a TM and a RE element).

  *tI*28-TiAl$_3$: 56 representatives (most contain Al, Ga or In, one or two TM, and one RE element).

194 *P*6$_3$/*mmc*: Wyckoff position 4*f* 1/3, 2/3, *z*; *z* = 3/4 − $\sqrt{6}$/4, *c*/*a* = 2$\sqrt{6}$/3 + 2, inter-sphere distance *a*, contact number *k* = 10, net (001) 3^6, stacking 3, 1 4, density 0.6657. There are 2107 binary and ternary intermetallic phases, listed in the PCD.

  *h*P12-MgZn$_2$: 456 representatives (see distribution of Laves phases, Fig. 7.16).

  *h*P2-Mg: 338 representatives.

  *h*P6-CaIn$_2$: 215 representatives (most contain a main group element a TM or a RE element).

  *h*P38-Th$_2$Ni$_{17}$: 185 representatives (most contain a TM and a RE element).

  *h*P8-Mg$_3$Cd: 123 representatives (most contain a main group and a TM and a RE element; see, also, Fig. 5.18).

221 *Pm*$\bar{3}$*m*: Wyckoff position 3*c* 0, 1/2, 1/2, inter-sphere distance *a*$\sqrt{2}$/2, contact number *k* = 8, . . . , density 0.5554. There are 1098 binary and ternary intermetallic phases, listed in the PCD.

  *c*P2-CsCl: 512 representatives (see distribution of *c*P2-CsCl type phases, Figs. 7.1) and 5.16.

  *c*P4-Cu$_3$Au: 544 representatives (see distribution of *c*P4-Cu$_3$Au type phases, Fig. 5.18).

No 225 *Fm*$\bar{3}$*m*: Wyckoff position 4*a* 0, 0, 0, inter-sphere distance *a*$\sqrt{2}$/2, contact number *k* = 12, net {111} 3^6 and {001} 4^4, stacking 4, 4 2, density 0.7405. There are 1294 binary and ternary intermetallic phases, listed in the PCD.

  *c*F4-Cu: 581 representatives.

  *c*F16-Cu$_2$MnAl: 414 representatives (see distribution of Heusler phases, Fig. 7.9).

$cF116$-Th_6Mn_{23}: 92 representatives (most contain a TM and RE element, the ternary phases, and also Al or Ga).

$cF16$-BiF_3: 86 representatives (most are Zintl phases, the ternary representatives also contain TM or RE elements; see also Fig. 5.18).

$cF16$-$Mg_6Cu_{16}Si_7$: 71 representatives.

227 $Fd\bar{3}m$: Wyckoff position $32e$ x,x,x; $x = (3 - \sqrt{6})/8$, inter-sphere distance $(3\sqrt{2}-2\sqrt{3})a/4$, contact number $k = 4$, density 0.1235. There are 1046 binary and ternary intermetallic phases, listed in the PCD.

$cF24$-$MgCu_2$: 806 representatives (see distribution of Laves phases, Fig. 7.16).

$cF184$-$CeCr_2Al_{20}$: 143 representatives (most contain Al or Zn, a TM, and a RE element).

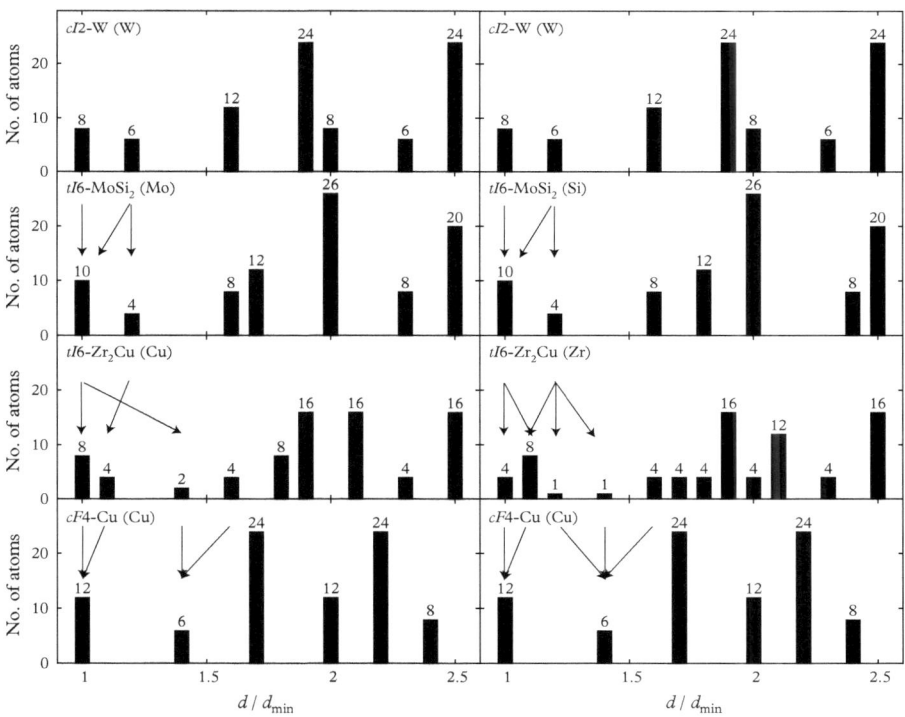

Fig. 5.42 *Distance histograms of tI6-MoSi$_2$ and tI6-Zr$_2$Cu illustrating their relationships to the cI2-W and cF4-Cu structure types. The two columns show the atomic environments of the different atomic positions in tI6-MoSi$_2$ and tI6-Zr$_2$Cu, while equivalent plots are displayed for cI2-W and cF4-Cu. The number of coordinating atoms is plotted over the relative distance from the respective central atom, i.e., the distance between two atoms divided by the minimum distance between the respective central atom and the closest atom of the AET. The binning of the histograms is 0.1.*

5.6.3   Statistics based on atomic distances and environment types

It can be quite difficult to derive AETs or larger structural subunits that make sense from a crystal-chemical point of view without taking into account chemical bonding. Furthermore, it is not clear in all cases what we can learn from the coordination number and shape of AETs, which both can vary considerably for

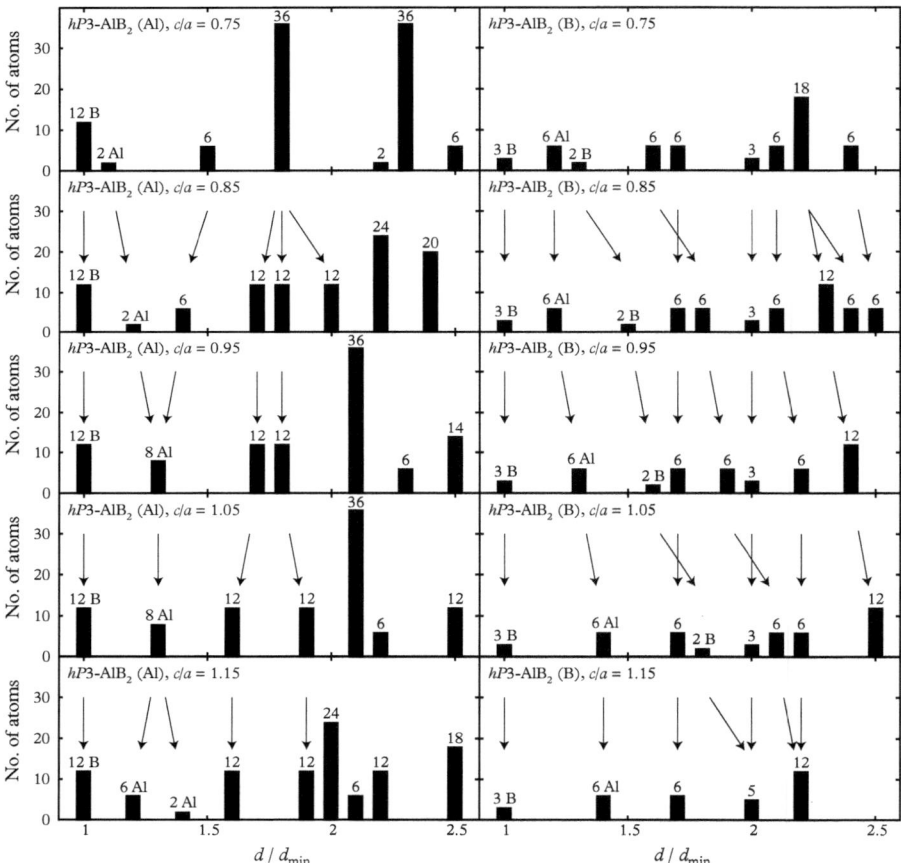

Fig. 5.43 *Variation of hP3-AlB$_2$ distance histograms as a function of the c/a-ratio (c/a = 0.75-1.15; 5 values). The Al-environment is shown in the left column, that of the B-environment in the right one. The shifts of specific atomic positions relative to one another are indicated by arrows (pointing in the direction of increasing values of c/a). The number of coordinating atoms is given over the relative distance from the respective central atom, i.e., the distance between two atoms divided by the minimum distance between the respective central atom and the closest atom of the AET. The binning of the histograms is 0.1.*

the representatives of a given structure type (see Fig. 5.43). In cases such as the Frank-Kasper phases, however, the number of different AETs is small despite the complexity of many of these structures, and easier to interpret. In the following we restrict ourselves to the analysis of a few distance histograms in order to illustrate some examples of structural relationships.

In the simplest case, the symmetry of an AET corresponds to the site symmetry of the Wyckoff position the central atom is sitting in. The resulting coordination polyhedra for one and the same coordination number may look quite different. For instance, even if generated from a single Wyckoff position, CN6 can mean a trigonal prism in point group $\bar{6}2m$, a twisted trigonal antiprism in 32, or a regular trigonal antiprism (octahedron) in $\bar{3}m$.

In periodic crystal structures, regular icosahedral (CN12) coordination can only be achieved based on lower site symmetries such as $mm2..$, e.g., on Wyckoff position $6f$ in space group $Pm\bar{3}$ choosing a special value for the freely variable x coordinate. Regular dodecahedral coordination (CN20) needs the occupancy of two Wyckoff positions, e.g., $6f\ mm2..$ and $8i\ .3.$ in space group $Pm\bar{3}$ choosing special values for the freely variable x coordinates.

The coordination number mainly depends on the atomic size ratios if no directed bonds come into play. In the case of hard spheres, the optimum size ratios of the central sphere to the surrounding spheres amount to $(\sqrt{6} - 2)/2 = 0.22474$ for tetrahedral AETs, $\sqrt{2} - 1 = 0.41421$ for octahedral AETs, $\sqrt{3} - 1 = 0.73205$ for hexahedral AETs, $(\sqrt{2(5 + \sqrt{5})} - 2)/2 = 0.90211$ for icosahedral AETs, and $(\sqrt{3}(1 + \sqrt{5}) - 2)/2 = 1.43649$ for dodecahedral AETs.

While AETs only provide information about the first coordination shell, distance histograms also illustrate sizes, thicknesses, and occupancies of higher coordination polyhedra although in a 1D projection only. As an example, Fig. 5.42 shows the transition from a *bcc*- to an *fcc*-type structure. The two intermediate

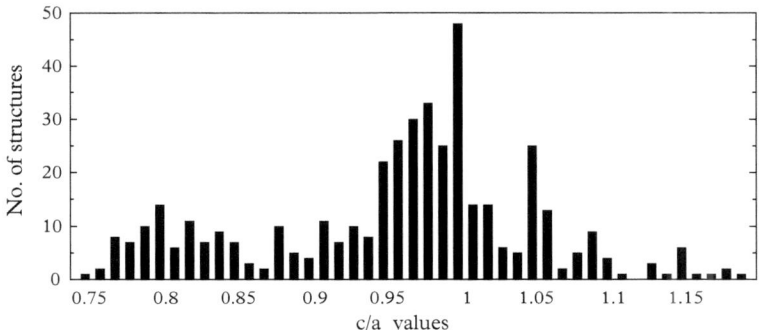

Fig. 5.44 *c/a-value distribution of the hP3-AlB$_2$-type structures listed in the PCD (439 out of 46 071 entries of intermetallics including solid solutions and off-stoichiometric compounds).*

structure types—between $cI2$-W and $cF4$-Cu—are both 3-fold superstructures of the $cI2$-W structure type. The increasing distortion via $tI6$-MoSi$_2$ and $tI6$-Zr$_2$Cu can be regarded as a deformation of the AETs from a rhombic dodecahedral (CN14) coordination towards the cuboctahedral (CN12) one.

Fig. 5.43 shows the development of the distance histograms of $hP3$-AlB$_2$-type structures with varying c/a-values with the unit cell volume kept constant. The displayed c/a-range reflects the values found for the structures assigned to this structural prototype within the *Pearson's Crystal Data* (PCD) database. It varies in the range from 0.75 to 1.19 (see Fig. 5.44).

If we consider the $hP3$-AlB$_2$-type structure of a compound AB$_2$ as close ...AA... packing of hcp A-atom layers with B atoms in the trigonal prismatic voids, then the ideal lattice parameters would correspond to $a = c = 2r_A$, with r_A the radius of an A atom. The radius ratio of the B atom ideally fitting in the void would have to be $r_B/r_A = (\sqrt{7/3} - 1) = 0.52753$. The c/a ratio would be just one. This is almost exactly the case for $hP3$-SrPtSn ($P\bar{6}m2$, $a = 4.504$ Å, $c = 4.507$ Å, Sr in $1a$ 0, 0, 0, Pt in $1d$ 1/3, 2/3, 1/2, Sn in $1f$ 2/3, 1/3, 1/2).

A c/a ratio greater than one means that the A atoms no longer touch each other along [001], while a c/a ratio less than one means that they are no more in contact with each other in the (001) plane. The B atoms, forming a honeycomb structure, touch each other in the (001) plane if their radius has the value $r_B = a\sqrt{3}/6$. An example for this case is $hP3$-BaAlGe ($P\bar{6}m2$, $a = 4.349$ Å, $c = 5.136$ Å, Ba in $1a$ 0, 0, 0, Al in $1d$ 1/3, 2/3, 1/2, Ge in $1f$ 2/3, 1/3, 1/2), an ordered $hP3$-AlB$_2$ derivative structure of the $hP3$-SrPtSn type. The $c/a = 1.18$ indicates hcp layers of Ba atoms ($r_{Ba} = 2.174 \approx a/2$) expanded along [001]. This is due to the rather large Al ($r_{Al} = 1.432$ Å) and Ge ($r_{Ge} = 1.225$ Å) atoms compared to the size of the trigonal prism void ($r_{void} = 1.147$ Å). Furthermore, the Ge and Al atoms, alternatingly occupying a honeycomb structure, are in close contact in the (001) plane as well.

In contrast, $hP3$-LuMn$_{0.67}$In$_{1.33}$ (191 $P6/mmm$, $a = 4.841$ Å, $c = 3.375$ Å) with a disordered $hP3$-AlB$_2$-type structure, is with a c/a ratio of 0.697 an example for the opposite extreme. Based on the radii $r_{Lu} = 1.718$ Å, $r_{Mn} = 1.367$ Å, and $r_{In} = 1.626$ Å, we can conclude that the Lu atoms touch each other along [001] forming isolated chains, and have much larger distances in the (001) layer, no longer forming an hcp arrangement. The In and Mn atoms, statistically distributed over the honeycomb structure, on the other hand are in close contact with one another.

The distribution of c/a-ratios of $hP3$-AlB$_2$-type structures is depicted in Fig. 5.44. It shows a large maximum around $c/a = 1$, and a smaller peak centered at $c/a = 0.8$. Only a few structures show c/a ratios significantly larger than 1. This means that structures where the A layers do not touch each other are energetically unfavorable.

Part II

Structures and properties

In this second part of the book, Part II, we discuss crystal structures and their variation as a function of pressure/temperature or of chemical composition/stoichiometry. While in the first case the interatomic potentials or the continuously changing band structures are probed, in the latter case it is the influence of geometrical parameters (atomic size ratios) and chemical bonding on the structure. The simplest model systems to study the influence of temperature and pressure on the structures are metallic elements, which are discussed in Chapter 6. Their compositional simplicity does not necessarily imply structural simplicity. In particular as a function of pressure, quite a few elements develop very complex structures. However, since we have to deal with only one kind of atoms in each case, the origin of their structure and complexity is easier to understand.

In the case of binary and ternary intermetallic compounds (Chapter 7), we focus on some typical classes of structures, which are either of interest due to their abundance or because they allow us to illustrate important principles of structure formation. The influence of the chemical composition on the evolution of particular crystal structures is discussed systematically in this chapter as well as in Part I, Chapter 5, on the statistics of crystal structures. High-entropy alloys (HEAs), i.e., solid solutions of five or more metallic elements and their interesting properties, are discussed at the end of this chapter. Cluster-based complex structures are the main topic of Chapter 8, and quasiperiodic structures with decagonal and icosahedral symmetry, respectively, as well as their approximants are in the focus of Chapter 9.

Finally, in Chapter 10, we discuss structure/property relationships of some selected functional intermetallics. We start with the discussion of magnetic properties, continue with electrical ones, and conclude with superconducting intermetallics. Since intermetallic compounds are, with a few exceptions such as high-entropy alloys, poor structural materials, we do not include a discussion of structural materials in this part of the book. The few of them that may have some technologically important mechanical properties have been discussed already under their structure class.

All crystal structures are shown on the same scale if not indicated otherwise.

6

Crystal structures of the metallic elements

In this chapter, the crystal structures of the metallic elements (for our definition see Figs. 6.1 and 6.2) will be discussed in an almost encyclopedic manner. Conceptually, it is based on one of the authors' contribution to the *Fifth Edition of Physical Metallurgy*, Vol. I (Steurer, 2014*b*). There is no clear definition as to which element is metallic and which is not. We know that at very high pressures, typical metals such as sodium can become ionic crystals (electrides), or non-metallic elements such as hydrogen or the noble gases can become metals. Also as a function of temperature, the metallicity may change. An example is metallic gray tin which transforms at below 286.2 K to non-metallic white tin. Also, the electrical conductivity cannot be a clear measure of metallicity, otherwise graphite could be classified as metallic. Since no clear dividing line is possible between metallic and non-metallic elements, we decided to choose the Zintl line for that purpose.

All modifications (allotropes) will be treated exemplarily, either as a function of temperature at ambient pressure (RP) or as a function of pressure at ambient temperature (RT). If not otherwise indicated, the data have been extracted from Massalski (1990), Tonkov (1996), Tonkov and Ponyatovsky (2005), Villars and Calvert (1991), Young (1991), or *Pearson's Crystal Data* (Villars and Cenzual, 2011*b*). In the case of strongly scattering data, we tried to identify the most reliable ones. If we could not find trustworthy data, we marked this with a question mark. Some of the high-pressure data are based on low-quality X-ray diffraction data, and may be revised in the future. The focus of our discussion is on the relationships between the structures of the different allotropes of an element, trying to understand what they have in common, what they do not have in common, and why.

In many cases, the high-pressure modifications of lighter elements in a column of the periodic table show structures of low-pressure polymorphs of the heavier elements (corresponding states rule) (Young, 1991). Under pressure, the atomic distances get shorter, the energy of electronic states and the widths of energy bands are altered and so are the interatomic interactions; the character of the chemical bonding can change as well as its degree of anisotropy. Metals may

Intermetallics: Structures, Properties, and Statistics. First Edition. Walter Steurer and Julia Dshemuchadse.
© Walter Steurer and Julia Dshemuchadse 2016. Published in 2016 by Oxford University Press.

1	2	3	4	5	6	7	8	9	10	11	12	13	14	15	16
3 Li cI2-W	4 Be hP2-Mg														
11 Na cI2-W	12 Mg hP2-Mg											13 Al cF4-Cu			
19 K cI2-W	20 Ca cF4-Cu	21 Sc hP2-Mg	22 Ti hP2-Mg	23 V cI2-W	24 Cr cI2-W	25 Mn cI58-Mn	26 Fe cI2-W	27 Co hP2-Mg	28 Ni cF4-Cu	29 Cu cF4-Cu	30 Zn hP2-Mg	31 Ga oC8-Ga	32 Ge cF8-C		
37 Rb cI2-W	38 Sr cF4-Cu	39 Y hP2-Mg	40 Zr hP2-Mg	41 Nb cI2-W	42 Mo cI2-W	43 Tc hP2-Mg	44 Ru hP2-Mg	45 Rh cF4-Cu	46 Pd cF4-Cu	47 Ag cF4-Cu	48 Cd hP2-Mg	49 In tI2-In	50 Sn tI4-Sn	51 Sb hR2-As	
55 Cs cI2-W	56 Ba cI2-W	57* La hP4-La	72 Hf hP2-Mg	73 Ta cI2-W	74 W cI2-W	75 Re hP2-Mg	76 Os hP2-Mg	77 Ir cF4-Cu	78 Pt cF4-Cu	79 Au cF4-Cu	80 Hg	81 Tl hP2-Mg	82 Pb cF4-Cu	83 Bi hR2-As	84 Po cP1-Po
87 Fr	88 Ra cI2-W	89+ Ac cF4-Cu	104 Rf	105 Db	106 Sg	107 Bh	108 Hs	109 Mt	110 Ds	111 Rg	112 Cn				

* Lanthanoids	58 Ce cF4-Cu	59 Pr hP4-La	60 Nd hP4-La	61 Pm hP4-La	62 Sm hR3-Sm	63 Eu cI2-W	64 Gd hP2-Mg	65 Tb hP2-Mg	66 Dy hP2-Mg	67 Ho hP2-Mg	68 Er hP2-Mg	69 Tm hP2-Mg	70 Yb hP2-Mg	71 Lu hP2-Mg
+ Actinoids	90 Th cF4-Cu	91 Pa tI2-Pa	92 U oC4-U	93 Np oP8-Np	94 Pu mP16-Pu	95 Am hP4-La	96 Cm hP4-La	97 Bk hP4-La	98 Cf hP4-La	99 Es	100 Fm	101 Md	102 No	103 Lr

Fig. 6.1 *Metallic elements (according to our definition) with their structure types at ambient conditions (atomic number on top in each box). Most of the elements have close-packed (ccp, hcp, or dhcp) or body-centered (bcc) structures. Even the huge-unit-cell structure cI58-Mn, for instance, can be seen as just a (3 × 3 × 3) superstructure of the cI2-W type with four atoms added.*

1	2	3	4	5	6	7	8	9	10	11	12	13	14	15	16
12 Li 3.01 / 1.0 1.52	77 Be 4.90 / 1.5 1.11														
11 Na 2.85 / 0.9 1.86	73 Mg 3.75 / 1.2 1.60											80 Al 3.23 / 1.5 1.43			
10 K 2.42 / 0.8 2.27	16 Ca 2.20 / 1.0 1.97	19 Sc 3.34 / 1.3 1.61	51 Ti 3.45 / 1.4 1.45	54 V 3.60 / 1.5 1.31	57 Cr 3.72 / 1.6 1.25	60 Mn 3.72 / 1.5 1.37	61 Fe 4.06 / 1.8 1.24	64 Co 4.30 / 1.8 1.25	67 Ni 4.40 / 1.8 1.25	72 Cu 4.48 / 1.9 1.28	76 Zn 4.45 / 1.6 1.34	81 Ga 3.20 / 1.6 1.22	84 Ge 4.60 / 1.8 1.23		
9 Rb 2.34 / 0.8 2.48	15 Sr 2.00 / 1.0 2.15	25 Y 3.19 / 1.2 1.78	49 Zr 3.64 / 1.4 1.59	53 Nb 4.00 / 1.6 1.43	56 Mo 3.90 / 1.8 1.36	59 Tc / 1.9 1.35	62 Ru 4.50 / 2.2 1.33	65 Rh 4.30 / 2.2 1.35	69 Pd 4.45 / 2.2 1.38	71 Ag 4.44 / 1.9 1.45	75 Cd 4.33 / 1.7 1.49	79 In 3.10 / 1.7 1.63	83 Sn 4.30 / 1.8 1.41	88 Sb 4.85 / 1.9 1.45	
8 Cs 2.18 / 0.7 2.66	14 Ba 2.40 / 0.9 2.17	33* La 3.10 / 1.1 1.88	50 Hf 3.80 / 1.3 1.56	52 Ta 4.11 / 1.5 1.43	55 W 4.40 / 1.7 1.37	58 Re 4.02 / 1.9 1.37	63 Os 4.90 / 2.2 1.34	66 Ir 5.40 / 2.2 1.36	68 Pt 5.60 / 2.2 1.37	70 Au 5.77 / 2.4 1.44	74 Hg 4.91 / 1.9 1.50	78 Tl 3.20 / 1.8 1.70	82 Pb 3.90 / 1.8 1.75	87 Bi 4.69 / 1.9 1.55	91 Po / 2.0 1.67
7 Fr / 0.7	13 Ra / 0.9 2.15	48+ Ac / 1.1 1.88													

* Lanthanoids	32 Ce / 1.1 1.87	31 Pr / 1.1 1.82	30 Nd / 1.1 1.81	29 Pm / 1.1 1.63	28 Sm / 1.2 1.62	18 Eu / 1.2 2.00	27 Gd / 1.2 1.79	26 Tb / 1.1 1.76	25 Dy / 1.2 1.75	24 Ho / 1.2 1.74	23 Er / 1.2 1.73	22 Tm / 1.3 1.72	17 Yb / 1.1 1.94	20 Lu / 1.3 1.72
+ Actinoids	47 Th / 1.3 1.88	46 Pa / 1.5 1.80	45 U / 1.4 1.39	44 Np / 1.4 1.30	43 Pu / 1.3 1.51	42 Am / 1.1	41 Cm / 1.3	40 Bk / 1.3	39 Cf / 1.3	38 Es / 1.3	37 Fm / 1.3	36 Md / 1.3	35 No / 1.3	34 Lr / 1.3

Fig. 6.2 *Mendeleev numbers (on top in each box), element symbol, Pearson absolute electronegativity [eV] (Pearson, 1985) next to the element symbol, Pauling electronegativities χ (relative to χ_F = 4.0) (bottom left in each box) (Pauling, 1932), and atomic radii (half of the shortest distance between atoms in the crystal structure at ambient conditions) (bottom right in each box) of the metallic elements.*

Table 6.1 *Schematic representation of the structure types that metallic elements adopt at ambient pressure as a function of temperature, which increases from left to right. In the first column, the Mendeleev number and element symbol are given and in the black-outlined column the structure at ambient temperature (RT) is given; LT and HT mean low and high temperatures, respectively; inc denotes an incommensurate phase (Table continued on the next page.)*

Element	LT allotropes		RT	HT allotropes				
8 Cs			*cI*2-W					
9 Rb			*cI*2-W					
10 K			*cI*2-W					
11 Na		*hR*9-Sm	*cI*2-W					
12 Li		*hR*9-Sm	*cI*2-W					
13 Ra			*cI*2-W					
14 Ba			*cI*2-W					
15 Sr			*cF*4-Cu	*hP*2-Mg	*cI*2-W			
16 Ca			*cF*4-Cu	*cI*2-W				
17 Yb			*hP*2-Mg	*cF*4-Cu	*cI*2-W			
18 Eu			*cI*2-W					
19 Sc			*hP*2-Mg	*cI*2-W				
20 Lu			*hP*2-Mg					
21 Tm			*hP*2-Mg	*cI*2-W				
22 Er			*hP*2-Mg					
23 Ho			*hP*2-Mg	*cI*2-W				
24 Dy			*hP*2-Mg					
25 Y			*hP*2-Mg	*cI*2-W				
26 Tb		*oC*4-Dy	*hP*2-Mg	*cI*2-W				
27 Gd			*hP*2-Mg					
28 Sm			*hR*9-Sm	*cI*2-W				
29 Pm			*hP*4-La	*cI*2-W				
30 Nd			*hP*4-La	*cI*2-W				
31 Pr			*hP*4-La	*cI*2-W				
32 Ce	*cF*4-Cu	*hP*4-La	*cF*4-Cu	*cI*2-W				
33 La			*hP*4-La	*cF*4-Cu	*cI*2-W			
35 No								
36 Md								
37 Fm								
38 Es								
39 Cf			*hP*4-La					
40 Bk			*hP*4-La	*cF*4-Cu				
41 Cm			*hP*4-La	*cF*4-Cu				
42 Am			*hP*4-La	*cF*4-Cu	*cI*2-W			
43 Pu			*mP*16-Pu	*mC*34-Pu	*oF*8-Pu	*cF*4-Cu	*tI*2-In	*cI*2-W
44 Np			*oP*8-Np	*tP*4-Np	*cI*2-W			
45 U		*mP*4-inc	*oC*4-U	*tP*30-CrFe	*cI*2-W			
46 Pa			*tI*2-Pa	*cI*2-W				
47 Th			*cF*4-Cu	*cI*2-W				
48 Ac			*cF*4-Cu					

Table 6.2 *(Table continued from the previous page.) Schematic representation of the structure types that metallic elements adopt at ambient pressure as a function of temperature, which increases from left to right. In the first column, the Mendeleev number and element symbol are given and in the black-outlined column the structure at ambient temperature (RT) is given; LT and HT mean low and high temperatures, respectively.*

Element	LT allotropes	RT	HT allotropes			
49 Zr		hP2-Mg	cI2-W			
50 Hf		hP2-Mg	cI2-W			
51 Ti		hP2-Mg	cI2-W			
52 Ta		cI2-W				
53 Nb		cI2-W				
54 V		cI2-W				
55 W		cI2-W				
56 Mo		cI2-W				
57 Cr		cI2-W				
58 Re		hP2-Mg				
59 Tc		hP2-Mg				
60 Mn		cI58-Mn	cP20 – Mn	cF4-Cu	cI2-W	
61 Fe		cI2-W	cF4-Cu	cI2-W		
62 Ru		hP2-Mg				
63 Os		hP2-Mg				
64 Co		hP2-Mg	cF4-Cu			
66 Ir		cF4-Cu				
67 Ni		cF4-Cu				
68 Pt		cF4-Cu				
69 Pd		cF4-Cu				
70 Au		cF4-Cu				
71 Ag		cF4-Cu				
72 Cu		cF4-Cu				
73 Mg		hP2-Mg				
74 Hg	hR3-Hg	liquid				
75 Cd		hP2-Mg				
76 Zn		hP2-Mg				
77 Be		hP2-Mg	cI2-W			
78 Tl		hP2-Mg	cI2-W			
79 Mg		tI2-In				
80 Al		cF4-Cu				
81 Ga		oC8-Ga				
82 Pb		cF4-Cu				
83 Sn	cF8-C	cF4-Cu				
84 Ge		cF8-C				
87 Bi		hR6-As				
88 Sb		hR6-As				
91 Po		cP1-Po				

Table 6.3 *Schematic representation of the structure types that metallic elements adopt at ambient temperature as a function of pressure, which increases from left to right. In the first column, the Mendeleev number and element symbol are given and in the black-outlined column the structure at ambient pressure (RP) is given; HP means high pressure and inc denotes an incommensurate phase. (Table continued on the next page.)*

Element		RP	HP allotropes				
8	Cs	cI2-W	cF4-Cu	oC84-Cs	tI4-Sn	oC16-Cs	hP4-La
9	Rb	cI2-W	cF4-Cu	oC52-Rb	tI19.3-inc	oC16-Cs	hP4-La
10	K	cI2-W	cF4-Cu	tI19.2-inc	oP8-MnP	tI4-Sn	oC16-Cs
11	Na	cI2-W	cF4-Cu	cI16-Li	oP8-MnP	tI19.3-inc	hP4-La
12	Li	cI2-W	cF4-Cu	liquid	oC40-Li	oC24-Li	cP4-Li
13	Ra	cI2-W					
14	Ba	cI2-W	hP2-Mg	tI10.8-inc	hP2-Mg		
15	Sr	cF4-Cu	cI2-W	tI4-Sn	mC12-?	tI10.8-inc	
16	Ca	cF4-Cu	cI2-W	cP1-Po	tP8-?	oC8-?	oP4-?
17	Yb	hP2-Mg	cI2-W	hP2-Mg	cF4-Cu	hP3-Nd	
18	Eu	cI2-W	hP2-Mg	hR24-Pr			
19	Sc	hP2-Mg	tI10.6-inc				
20	Lu	hP2-Mg	hR9-Sm	hP4-La	hP8-Pr		
21	Tm	hP2-Mg	hR9-Sm	hP4-La			
22	Er	hP2-Mg	hR9-Sm	hP4-La	cF4-Cu		
23	Ho	hP2-Mg	hR9-Sm	hP4-La	cF4-Cu	hR24-Pr	
24	Dy	hP2-Mg	hR9-Sm	hP4-La	cF4-Cu	oS8-Dy	
25	Y	hP2-Mg	hR9-Sm	hP4-La	hP6-Sc		
26	Tb	hP2-Mg	hR9-Sm	hP4-La	hR24-Pr	mC4-Ce	
27	Gd	hP2-Mg	hR9-Sm	hP4-La	cF4-Cu	hR24-Pr	mC4-Ce
28	Sm	hR9-Sm	hP4-La	hR24-Pr	hP3-Nd	mC4-Ce	
29	Pm	hP4-La	cF4-Cu	hR24-Pr			
30	Nd	hP4-La	cF4-Cu	hR24-Pr	hP3-Nd	mC4-Ce	oC4-U
31	Pr	hP4-La	cF4-Cu	hR24-Pr	oI16-?	oC4-U	oP4-?
32	Ce	cF4-Cu	mC4-Ce	tI2-In			
33	La	hP4-La	cF4-Cu	hR24-Pr	cF4-Cu		
35	No						
36	Md						
37	Fm						
38	Es						
39	Cf	hP4-La	cF4-Cu	oC4-U			
40	Bk	hP4-La	cF4-Cu	oC4-U			
41	Cm	hP4-La	cF4-Cu	mC4-?	oF8-Am	oP4-Am	
42	Am	hP4-La	cF4-Cu	oF8-Pu	oP4-Am		
43	Pu	mP16-Pu	oP4-Pu				
44	Np	oP8-Np					
45	U	oC4-U					
46	Pa	tI2-Pa	oC4-U				
47	Th	cF4-Cu	tI2-In				
48	Ac	cF4-Cu					

Table 6.4 *(Table continued from the previous page.) Schematic representation of the structure types that metallic elements adopt at ambient temperature as a function of pressure, which increases from left to right. In the first column, the Mendeleev number and element symbol are given and in the black-outlined column the structure at ambient pressure (RP) is given; HP means high pressure and inc denotes an incommensurate phase, respectively.*

Element		RP	HP allotropes				
49	Zr	$hP2$-Mg	$hP3$-AlB$_2$	$cI2$-W			
50	Zr	$hP2$-Mg	$hP3$-AlB$_2$	$cI2$-W			
51	Ti	$hP2$-Mg	$hP3$-AlB$_2$	$oC4$-Ti	$oP4$-Ti		
52	Ta	$cI2$-W					
53	Nb	$cI2$-W					
54	V	$cI2$-W	$hR3$-Hg				
55	W	$cI2$-W					
56	Mo	$cI2$-W					
58	Re	$hP2$-Mg					
59	Tc	$hP2$-Mg					
60	Mn	$cI58$-Mn	$cI2$-W				
61	Fe	$cI2$-W	$hP2$-Mg				
62	Ru	$hP2$-Mg					
63	Os	$hP2$-Mg					
64	Co	$hP2$-Mg					
65	Rh	$cF4$-Cu					
66	Ir	$cF4$-Cu	$hP14$-Ir				
67	Ni	$cF4$-Cu					
68	Pt	$cF4$-Cu					
69	Pd	$cF4$-Cu					
70	Au	$cF4$-Cu					
71	Ag	$cF4$-Cu					
72	Cu	$cF4$-Cu					
73	Mg	$hP2$-Mg	$cI2$-W				
74	Hg	liquid	$tI2$-Pa	$mC6$-Hg			
75	Cd	$hP2$-Mg					
76	Zn	$hP2$-Mg					
77	Be	$hP2$-Mg	$oP4$-?				
78	Tl	$hP2$-Mg	$cF4$-Cu				
79	In	$tI2$-In	$oF4$-In				
80	Al	$cF4$-Cu	$hP2$-Mg				
81	Ga	$oC8$-Ga	$oC104$-Ga	$hR18$-Ga	$tI2$-In	$cF4$-Cu	
82	Pb	$cF4$-Cu	$hP2$-Mg	$cI2$-W			
83	Sn	$tI4$-Sn	$tI2$-Pa	$cI2$-W			
84	Ge	$cF8$-C	$tI4$-Sn	$oI4$-Sn	$hP1$-BiIn	$oC16$-Si	$hP2$-Mg
87	Bi	$hR6$-As	$mC4$-Bi	$tI?$-inc	$cI2$-W		
88	Sb	$hR6$-As	$mI?$-inc	$tI?$-inc	$cI2$-W		
91	Po	$cP1$-Po	$hR3$-Hg				

become semiconducting or insulators and non-metals semi-metallic or metallic. The change in free energy ($\int dVp$) resulting from a compression to 100 GPa can reach hundreds of meV per atom, while that obtained by cooling to low temperature ($\int dTS$) may amount to ≈ 25 meV, only (Hamlin, 2015).

While atomic potentials are changed under pressure, they are just probed by increasing the temperature. The hardening or softening of phonons can play a role for the evolution of particular crystal structures as a function of temperature as well as entropic contributions in general. While under pressure, structures in most cases get lower symmetric, the opposite is true going to higher temperatures, because orbital symmetries become less important with increasing atomic distances. While with increasing temperature phase transformations involving diffusion becomes possible, this is hardly the mechanism for structural transitions at high pressures.

The different structures (modifications and allotropes) that the elements adopt as a function of temperature are shown in Tables 6.1 and 6.2, and as a function of pressure in Tables 6.3 and 6.4. There appear to be much less allotropes between zero K and the melting point than as a function of pressure. This is due to the fact that changes as a function of temperature are less dramatic. While as a function of pressure the atomic volumes of some elements can be easily decreased by 50% and more, their increase as a function of temperature will hardly exceed 6–7% up to the melting temperature. The nonmagnetic transition metals are most resistant to changes of their modifications.

6.1 Groups 1 and 2: Alkali and alkaline earth metals

6.1.1 Group 1: Alkali metals Li, Na, K, Rb, and Cs

At ambient conditions, the *bcc* alkali metals (AM) can be described as typical metals, with their single valence electrons and almost spherical Fermi surface conforming closely to the free-electron-gas model of metals. This makes the large number of allotropes they exhibit as a function of pressure even more surprising, under which even some very complex structures form (see Table 6.5). Furthermore, at high pressures the melting temperatures run through deep minima and even superconductivity can appear. Also quite remarkable is the drastic decrease in atomic volume with pressure due to s-p orbital mixing for Li and Na or ns to $(n-1)$d electron transfer for the heavier alkali metals. In the case of lithium, the atomic volume $V_{at} = 24.31$ Å^3 at ambient conditions decreases to 6.5 Å^3 at 88 GPa, and, according to theoretical calculations, to 3.42 Å^3 at 300 GPa, i.e., by 85% (Ma *et al.*, 2008). In terms of atomic radii and interatomic distances, this corresponds to a reduction by approximately 50% at 300 GPa. It has been shown by quantum-mechanical calculations that at high pressures electron density can be increasingly pushed away by Coulomb repulsion from the relatively incompressible, each other closely approaching, atomic cores and accumulated at interstitial

Table 6.5 *Structural data for the Group-1 elements, the alkali metals. First line: group, atomic number Z, element abbreviation and full name, and electronic ground-state configuration. Following lines: prototype structure (inc means incommensurate), trivial name of the allotrope (if any), temperature* T *and pressure* P *limiting the stability range of this allotrope, and atomic volumes* $V_{at} = V_{uc}/n_{at}$ *for the given* T|P, *with* V_{uc} *the unit cell volume and* n_{at} *the number of atoms therein. With some exceptions, only those phases are listed that are stable at either RT or RP.*

Prototype	Allotrope	T [K]	P [GPa]	V_{at} [Å³]	T|P	References
1 3 Li			Lithium		[He]2s¹	
$hR9$-Sm	α-Li	<74	RP	20.99	20|RP	[1], [2]
$cI2$-W	Li-I, β-Li	RT	RP	24.32	RT|RP	
$cF4$-Cu	Li-II	RT	>7.5	14.83	RT|8	[3]
liquid		RT	>40		RT|40	[4]
$oC40$-Li	Li-VII	RT	>67	6.7	240|75	[4], [5]
$oC24$-Li	Li-VIII	RT	>86	6.5	RT|88	[4], [5], [6]
$cP4$-Li	IX	RT	>300	3.42	RT|400	[7]
1 11 Na			Sodium		[Ne]3s¹	
$hR9$-Sm	α-Na	<36	RP	37.74	20|RP	[8]
$cI2$-W	Na-I, β-Na	RT	< 65	39.50		
$cF4$-Cu	Na-II	RT	<105			
$cI16$-Li	Na-III	RT	<118	9.77	RT|115	[9]
$oP8$-MnP	Na-IV	RT	<125	9.45	RT|119	[9]
$tI19.3$-Na$_{inc}$	Na-V	RT	<200	8.62	RT|147	[10]
$hP4$-La	Na-VI	RT	>200	7.88	RT|200	[11]
1 19 K			Potassium		[Ar]4s¹	
$cI2$-W	K-I	RT	<11.4	75.72	301|RP	
$cF4$-Cu	K-II	RT	<23	34.11	RT|12.4	
$tI9.2$-K$_{inc}$	K-III	RT	>20	23.48	RT|22.2	[12]
$oP8$-MnP	K-IV	RT	>54	14.79	RT|58	[13]
$tI4$-Sn	K-V	RT	>90			[13]
$oC16$-Cs	K-VI	RT	>96			[13]
1 37 Rb			Rubidium		[Kr]5s¹	
$cI2$-W	Rb-I	RT	<7	87.10	RT|RP	
$cF4$-Cu	Rb-II	RT	<13	37.63	RT|9.0	[14]
$oC52$-Rb	Rb-III	RT	<16.6	31.42	RT|14.3	[15]
$tI19.3$-Na$_{inc}$	Rb-IV	RT	<20	28.85	RT|16.8	[16], [17]
$tI4$-Sn	Rb-V	RT	<48	25.96	RT|20.2	[16]
$oC16$-Cs	Rb-VI	RT	>48	17.97	RT|48.1	[18]
1 55 Cs			Cesium		[Xe]6s¹	
$cI2$-W	Cs-I	RT	<2.3	110.45	RT|RP	
$cF4$-Cu	Cs-II	RT	>2.3	53.57	RT|4.1	
$oC84$-Cs	Cs-III	RT	>4.2	50.21	RT|4.3	[19]
$tI4$-Sn	Cs-IV	RT	>4.3	35.01	RT|8	
$oC16$-Cs	Cs-V	RT	>10	26.09	RT|25.8	[20]
$hP4$-La	Cs-VI	RT	>72	11.48	RT|92	[14]

[1] Ernst *et al.* (1986), [2] Berliner and Werner (1986), [3] Hanfland *et al.* (2000), [4] Guillaume *et al.* (2011), [5] Marqués *et al.* (2011), [6] Rousseau *et al.* (2005), [7] Ma *et al.* (2008), [8] Berliner and Werner (1986), [9] Gregoryanz *et al.* (2008), [10] Lundegaard *et al.* (2009a), [11] Ma *et al.* (2009), [12] McMahon *et al.* (2006b), [13] Lundegaard *et al.* (2009b), [14] Takemura *et al.* (2000), [15] Nelmes *et al.* (2002), [16] Schwarz *et al.* (1999a), [17] McMahon *et al.* (2006c), [18] Schwarz *et al.* (1999b), [19] McMahon *et al.* (2001), and [20] Schwarz *et al.* (1998c).

sites (electride formation) leading to a rather large band gap. This has been experimentally observed for Li and Na, as will be discussed below, and predicted for Mg, Al, Si, Tl, In, and Pb for pressures in the range between approximately one and five TPa (Miao and Hoffmann, 2014).

At RT and under compression, $cI2$-Li (Li-I) first transforms to a *ccp* phase (Li-II) (tetragonal Bain path, Fig. 6.3) due to a pressure-induced instability of the tetragonal shear elastic constant C' and a softening of the transverse acoustic phonons along the $[0\xi\xi]$-direction near the Brillouin zone center (Xie *et al.*, 2008). At above 39 GPa it becomes liquid until it solidifies again at 67 GPa to semimetallic $oC40$-Li (Li-VII) (Guillaume *et al.*, 2011; Marqués *et al.*, 2011), and transforms at 86 GPa into poorly metallic $oC24$-Li (Li-VIII). At pressures around 50 GPa, lithium has with $T_m \approx 190$ K the lowest melting temperature of any known material, much lower even than helium or hydrogen. A metallic, cubic modification, $cP4$-Li (Li-IX), was predicted to be formed beyond 300 GPa, with a structure characterized by six-fold coordinated Li (Ma *et al.*, 2008).

At low temperatures, Li adopts the structure of the $hR9$-Sm type, i.e., a nine-layer stacking of *hcp* atomic layers with the sequence ABCBCACAB and, at 20 K, a c/a ratio of 1.631, close to the ideal one of 1.633 for *hcp* structures (Berliner and Werner, 1986). Non-superconducting Li becomes superconducting at above 20.3 GPa with $T_c = 5.47$ K, probably correlated with the transition of $hR9$-Li to $cF4$-Li (Li-I) (Deemyad and Schilling, 2003). The critical temperature changes discontinuously with the structural phase transitions taking place at low temperatures. It reaches its maximum with $T_c \approx 14$ K at 30 GPa, where another transition takes place to $cI16$-Li (Li-V), a low-temperature, high-pressure phase.

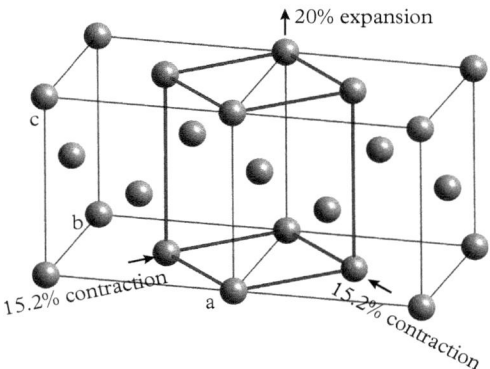

Fig. 6.3 *The ccp to bcc phase transition via a continuous change in the c/a ratio from 1 to $\sqrt{2}$ (tetragonal Bain path). The orientational relationship between the two structures is schematically shown, with the changes in the lattice parameters necessary. On twice the usual scale.*

No superconductivity was found beyond 63 GPa, after the phase transformation to $oC88$-Li (Li-VI) (Guillaume *et al.*, 2011). It is remarkable that the temperature difference $\Delta T = T_m - T_c \approx 175$ K, with T_m the melting temperature around 40 GPa, is the smallest known in any system (Guillaume *et al.*, 2011). From this point of view, lithium could be seen as an HT-superconductor.

Up to 125 GPa, sodium shows a similar sequence of allotropes as Li (if the LT/HP allotropes are taken into account as well). Beyond this pressure, $cI16$-Na (Na-III), which can be seen as a $(2 \times 2 \times 2)$-fold superstructure of $cI2$-Na (Na-I), transforms, via semimetallic $oP8$-Na (Na-IV), which is closely related to the $hP4$-NiAs type (Degtyareva and Degtyareva, 2009), to $tI19.3$-Na (Na-V), an incommensurate host-guest structure (Lundegaard *et al.*, 2009a). In contrast to the *bct* host lattice, the sublattice of guest chains is monoclinic, and the interchain correlation length is, with approximately six interchain spacings, rather short (≈ 28 Å at 147 GPa). Around ≈ 200 GPa, Na becomes an optically transparent wide-band-gap dielectric (electride) (Ma *et al.*, 2009) due to p-d hybridization of the valence electrons and their repulsion by core electrons into interstitial sites of the strongly compressed $hP4$-La type structure (Na-VI). The value $c/a = 1.391$ at 320 GPa corresponds to less than one half of the ideal value $2\sqrt{8/3} = 3.266$, leading to only 6-fold instead of 12-fold coordinated atoms. The $hP4$-La type structure (Na-VI) can also be seen as related to the $hP6$-Ni$_2$In structure type: the ionic cores correspond to the Ni sublattices and the interstital electron density maxima to that of In (Rousseau *et al.*, 2011).

The phase sequence of potassium looks largely different from that of sodium. Here, s-d electron transfer takes place under high pressure, which is typical for the heavier alkali metals. At ≈ 20 GPa, $cF4$-K (K-II) transforms into a *bct* incommensurately modulated host-guest structure, $tI19.2$-K (K-III). The square anti-prismatic host framework contains linear chains of K atoms related by C-centering (K-IIIa). The guest atoms undergo intraphase transitions at 30 GPa to an A-centered orthorhombic structure (K-IIIb) and back to the C-centered one (K-IIIa) at 39.7 GPa (reentrant phase transition) (Lundegaard *et al.*, 2013). Although the c_{host}/c_{guest} ratio of both host/guest-structures passes through commensurate values, such as 8/5 at 22.0 GPa, or is close to 5/3 at 54 GPa where the transformation to $oP8$-K (K-IV) takes place (McMahon and Nelmes, 2006), no lock-in transition has been observed. The ratios 8/5 and 5/3 correspond to approximants of the golden mean, $\tau = (1 + \sqrt{5})/2$. This means that at ≈ 19 GPa and at ≈ 44 GPa, the ratio c_{host}/c_{guest} equals τ. τ is also called the "most irrational" number, since it has the slowest converging chain fraction expansion, which consists of ones only. This could be interpreted in the way that this host/guest structure tends to avoid a lock-in transition to a simple rational ratio.

Apart from $oP8$-K (K-IV) ($oP8$-MnP type), two other high-pressure modifications have been observed (K-V and K-VI). The $oP8$-MnP type structure (K-IV), which is also present in the phase sequence of sodium, can be seen as a distorted version of an $hP4$-La-type structure.

oC52-Rb (Rb-III)

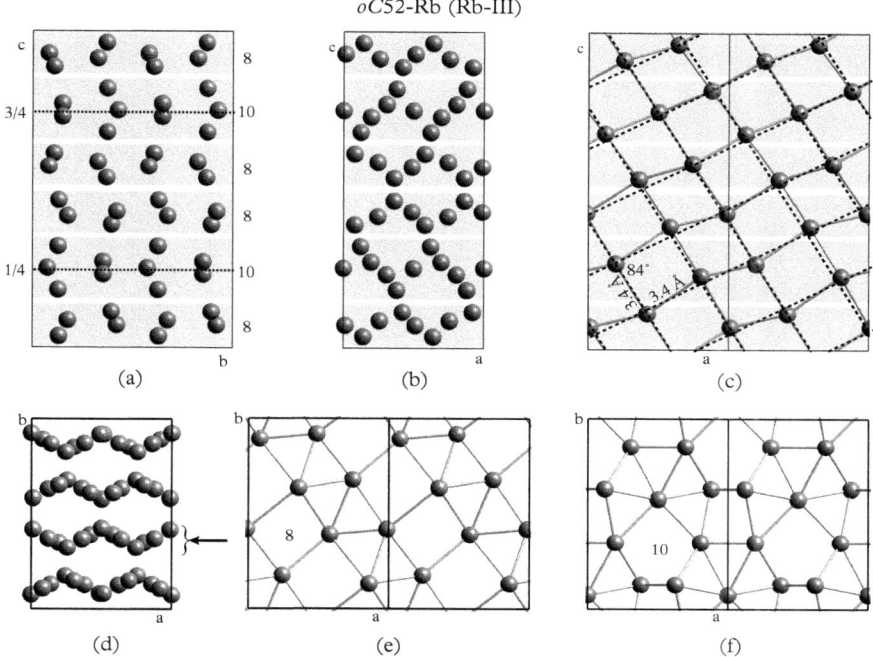

Fig. 6.4 *The structure of oC52-Rb (Rb-III) in different views: projections along a (a), b (b), and c (d), respectively. The puckered 8- and 10-layer sections shown in (e) and (f), respectively, correspond to the respective gray-shaded areas depicted in (a)–(c). One of the corrugated layers perpendicular to the b direction, with $0.3 < y < 0.5$, is shown in (c) (marked by an arrow in (d)). The atoms in this layer are only slightly shifted away from the vertices of the dashed oblique lattice overlaid with the structure.*

Rubidium shows a similar series of phase transformations as potassium. The transition from $cF4$-Rb (Rb-II) to $tI19.3$-Na$_{inc}$ (Rb-IV), however, runs via $oC52$-Rb (Rb-III), with a complex structure that can be described as a six-layer stacking of 8- and 10-atom layers with a sequence 8-10-8-8-10-8 (Nelmes *et al.*, 2002) (Fig. 6.4). Since $oC84$-Cs has a ten-layer sequence 8-8-10-8-8-8-8-10-8-8 (Nelmes *et al.*, 2002), these different stacking sequences can be seen as polytypic structural variants.

This phase transforms at ≈ 17 GPa into the *bct* incommensurate host/guest structure $tI19.3$-Rb (Rb-IV) (McMahon *et al.*, 2006c) (Fig. 6.5). While Na, K, and Rb all have the same type of host structures, they differ in their guest structures, which are monoclinic, *C*-centered, *A*-centered, and body-centered, respectively. Close to the pressure for the transition to $tI4$-Rb (Rb-V), the ratio of host and guest periods along the *c*-direction approaches the commensurate value $c_H/c_G = 5/3$, which is also a τ-approximant. The highest-pressure modification known so far is $oC16$-Rb (Rb-VI), which can be obtained by small atomic

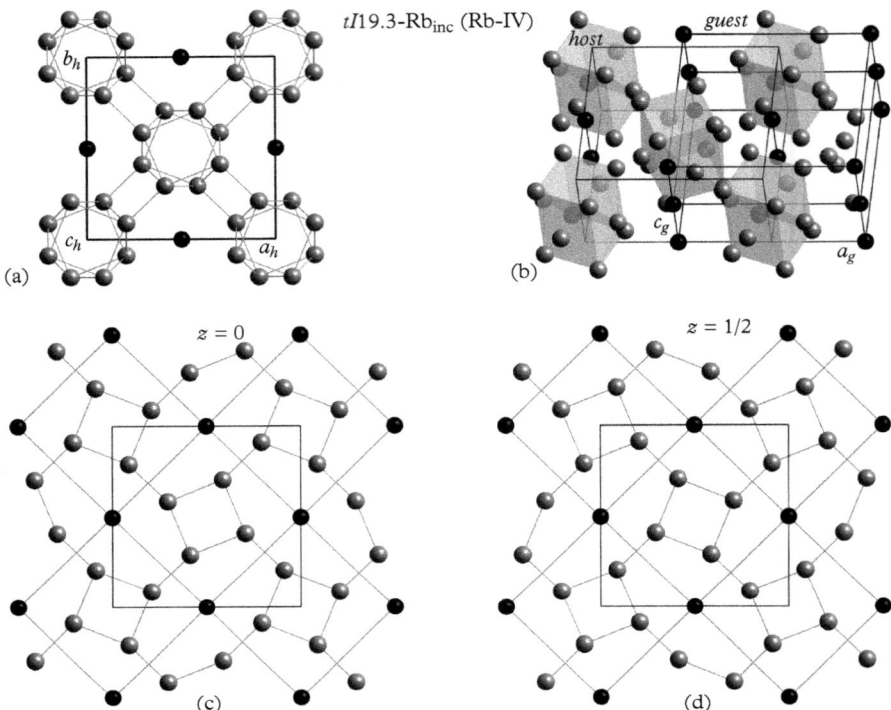

Fig. 6.5 *The structure of the incommensurate composite structure tI19.3-Rb$_{inc}$ (Rb-IV). (a) Projection along the [001]-direction of the bct host-unit-cell (host substructure ... gray spheres, guest substructure ... black spheres). (b) Perspective view with the bct host substructure in polyhedral representation with both subunit cells (both are bct) drawn in (c, d). The atomic layers of the bct host substructure with $z = 0$ and $z = 1/2$; the positions of the guest atoms at arbitrary z values are marked by black spheres.*

shifts from tI4-Rb (Rb-V) or from cF4-Rb (Rb-II) (Fig. 6.6). Its structure is characterized by layers of corner-sharing and rather regular empty octahedra.

The structural relationships and mechanisms of the displacive phase transitions between all six Rb modifications were discussed by Katzke and Toledano (2005) by associating particular critical instabilities of the ambient-pressure phase cI2-Rb (Rb-I). Accordingly, oC52-Rb (Rb-III) can be described as a modulated structure with a modulation period of $\approx 13a$ along the initial [100] direction of cI2-Rb (Rb-I), and tI4-Rb (Rb-V) can be considered a lock-in phase of tI19.3-Rb (Rb-IV). It is also shown that the assumed deformation of the atomic shells due to hybridization of s- and d-electron wave functions allows a closer packing within the tI19.3-Rb (Rb-IV) and the tI4-Rb (Rb-V) structures.

Cesium runs through quite a similar phase sequence as K and Rb. The main difference is that cF4-Cs (Cs-II) transforms into an even more complex structure

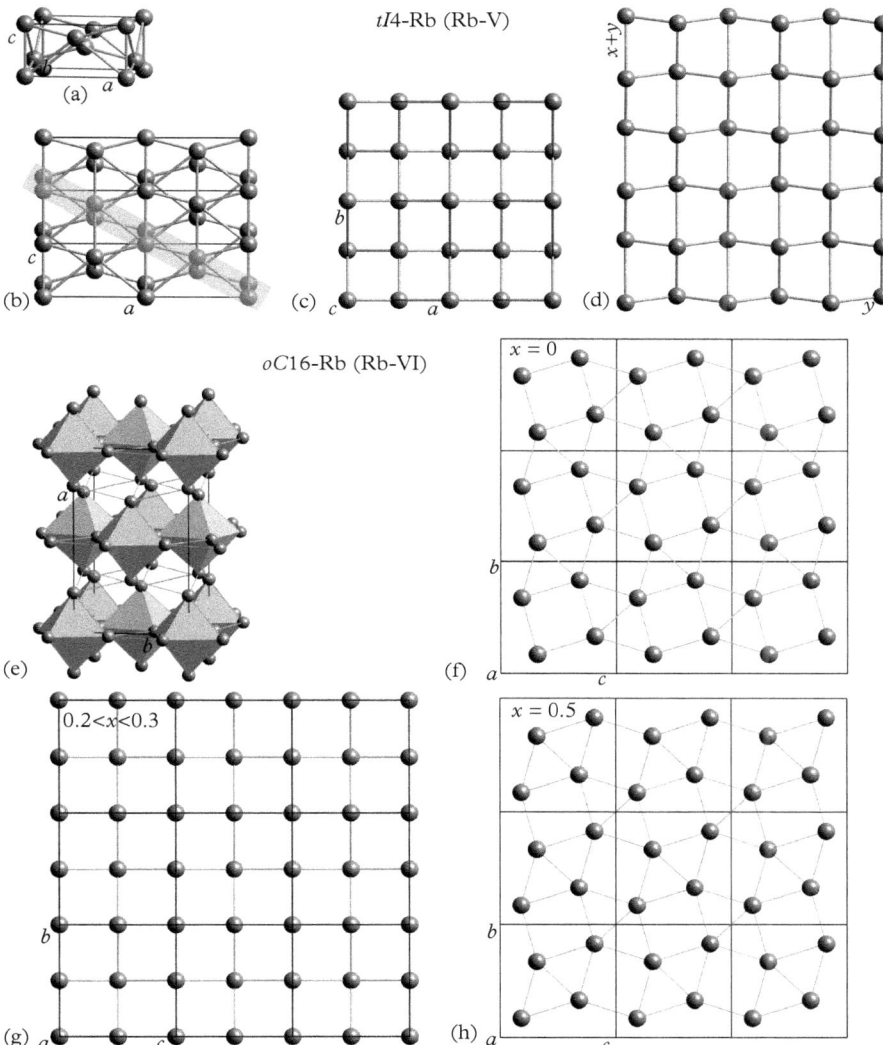

Fig. 6.6 *The structure of tI4-Rb (Rb-V) (a)–(d) in comparison with that of oC16-Rb (Rb-VI) (e)–(h). (a) One unit cell of tI4-Rb in perspective view; projection of unit cells along (b) [010] and (c) [001]; (d) one puckered section around the (101)-plane (marked by a rectangle in (b)). The structure of oC16-Rb (e) consists of octahedra, the apices of which form a puckered square tiling (g). In (f) and (g) cuts are depicted through the arrangement of octahedra.*

than $oC52$-Rb (Rb-III), namely $oC84$-Cs (Cs-III). The 8- and 10-atom layers are ordered now in the ten-layer sequence 8-8-10-8-8-8-8-10-8-8 (Nelmes *et al.*, 2002). Cs shows by far the strongest dependence of the atomic volume on pressure of all elements. With increasing compression, the valence electrons move from the 6s to the unfilled 5d band, i.e., Cs becomes a "transition metal", until the s-d electron transfer is complete at about 10 GPa. This corresponds to a reduced volume $V/V_0 \approx 0.235$, with V_0 the volume of Cs at ambient conditions, when the bottom of the 6s band rises above the Fermi energy E_F (McMahan, 1985). The d-character of the valence band changes when it merges with the broadening 5p core below $V/V_0 \approx 0.13$. Due to the unique 5d (or 5d–5p hybridized) monovalent electronic states, directional bonding leads to the rather complex structures observed (Takemura *et al.*, 1991). At even higher pressures, 5p core repulsion dominates over 5d bonding resulting in the *hcp* structure of $hP4$-Cs (Cs-VI).

6.1.2   Group 2: Alkaline earth metals Be, Mg, Ca, Sr, Ba, and Ra

The alkaline earth metals (AEM) partly show similar phase sequences as the alkali metals (compare Tables 6.5 and 6.6). With the exception of *bcc* Ba and Ra, the modifications at ambient conditions are either *ccp* or *hcp*. At higher pressures, Sr and Ba adopt the same incommensurate structure type, $tI10.8$-Sr$_{inc}$. While the alkali elements are good metals, the smaller than ideal c/a ratio of 1.56 indicates covalent bonding contributions for Be at ambient conditions. This value increases to 1.60 at 180 GPa (Evans *et al.*, 2005). Close to the melting point of 1560 K, $hP2$-Be transforms into $cI2$-Be. Such a transition was also predicted at 506 GPa at room temperature (Cheng *et al.*, 2013).

According to theoretical calculations, Mg is expected to transform from *bcc* to simple hexagonal at ≈ 750 GPa, then to simple cubic at ≈ 1 TPa, and finally to an orthorhombic structure at ≈ 30 TPa (Pickard and Needs, 2010). All these theoretically predicted modifications would be electrides, i.e., charge would be shifted from the Mg atoms into interstitial sites. Experimentally observed so far was merely the transformation from *hcp* to *bcc* at ≈ 50 GPa.

There has been a long-lasting discussion about the structure that $cI2$-Ca (Ca-II) transforms into above 32 GPa, which was experimentally clearly observed to be of the $cP1$-Po type (Ca-III) (Gu *et al.* (2009) and references therein). Its dynamical stability at 300 K was confirmed recently by several independent DFT calculations (DiGennaro *et al.*, 2013; Yao *et al.*, 2010), and a study by Tse *et al.* (2012). Accordingly, $cP1$-Ca is only on average cubic, featuring large dynamic Jahn-Teller lattice distortions. The transition to $tP8$-Ca (Ca-IV) is driven by s-d electron transfer and the resulting static Jahn-Teller distortion. After several more steps, Ca transforms into $tP128$-Ca (Ca-VII), a $(2 \times 2 \times 1)$-fold supercell of a tetragonal basic structure of the host/guest kind with a commensurate host/guest ratio of 4/3 along the c-axis (Fujihisa *et al.*, 2013). This Ca allotrope has with $T_c = 29$ K the highest superconducting transition temperature of all elements.

Table 6.6 *Structural data for the Group-2 elements, the alkaline earth metals. First line: group, atomic number Z, element abbreviation and full name, and electronic ground-state configuration. Following lines: prototype structure (inc means incommensurate), temperature* T *and pressure* P *limiting the stability range of this allotrope, and atomic volumes* $V_{at} = V_{uc}/n_{at}$ *for the given* T | P, *with* V_{uc} *the unit cell volume and* n_{at} *the number of atoms therein. With some exceptions, only those phases are listed that are stable at either RT or RP.*

Prototype	Allotrope	T [K]	P [GPa]	V_{at} [Å³]	T	P	References
2 4 Be		Beryllium		[He]2s²		
hP2-Mg	α-Be	<1523	RP	8.11	RT	RP	
cI2-W	β-Be	>1523	RP	8.30	1528	RP	
2 12 Mg		Magnesium		[Ne]3s²		
hP2-Mg	Mg-I	*RT*	<50	23.24	RT	RP	
cI2-W	Mg-II	*RT*	>50	12.88	RT	58	
2 20 Ca		Calcium		[Ar]4s²		
cF4-Cu	Ca-I	<721	RP	43.63	RT	RP	
cI2-W	Ca-II	>721	*or* > 20	44.95	740	58	
cP1-Po	Ca-III	RT	>32	18.99	RT	42	[1], [2]
tP8	Ca-IV	RT	>119	11.62	RT	130	[3]
oC8	Ca-V	RT	>143	10.50	RT	154	[3]
oP4	Ca-VI	RT	>158	9.94	RT	172	[4]
tP128-Ca	Ca-VII	RT	>210	8.96	RT	212	[5]
2 38 Sr		Strontium		[Kr]5s²		
cF4-Cu	Sr-I	<504	*or* <3.5	56.08	RT	RP	
cI2-W	Sr-II	>820	*or* >3.5	57.75	901	58	
tI4-Sn	Sr-III	RT	>24.4	22.42	RT	34.8	[6]
mC12	Sr-IV	RT	>37.7	20.53	RT	41.7	[7]
tI10.8-Sr$_{inc}$	Sr-V	RT	>46.3	17.74	RT	56	[8]
2 56 Ba		Barium		[Xe]6s²		
cI2-W	Ba-I	*RT*	<5.5	62.99	RT	RP	
hP2-Mg	Ba-II	*RT*	<12.6	38.44	RT	6.9	[9]
tI10.8-Sr$_{inc}$	Ba-IV	RT	<45	31.17	RT	12.0	[8]
hP2-Mg	Ba-V	RT	>45	20.34	RT	53	[10]
2 88 Ra		Radium		[Rn]7s²		
cI2-W		RT	RP	68.22	RT	RP	

[1] Mao *et al.* (2010), [2] Yao *et al.* (2010), [3] Fujihisa *et al.* (2008), [4] Nakamoto *et al.* (2010), [5] Fujihisa *et al.* (2013), [6] Allan *et al.* (1998), [7] Bovornratanaraks *et al.* (2006), [8] McMahon *et al.* (2000), [9] Takemura (1994), and [10] Nelmes *et al.* (1999).

At 3.5 GPa, Sr transforms from a *ccp* into a *bcc* structure via a reverse Bain path. At higher pressures, it goes through the phase sequence tI4-Sr → mC12-Sr → tI10.8-Sr$_{inc}$. The allotrope mC12-Sr (or in another setting mI12-Sr) can be described as a monoclinic superstructure (helical distortion) of tI4-Sr (Fig. 6.7), while tI10.8-Sr$_{inc}$ corresponds to an incommensurate *bct* host/guest structure, with

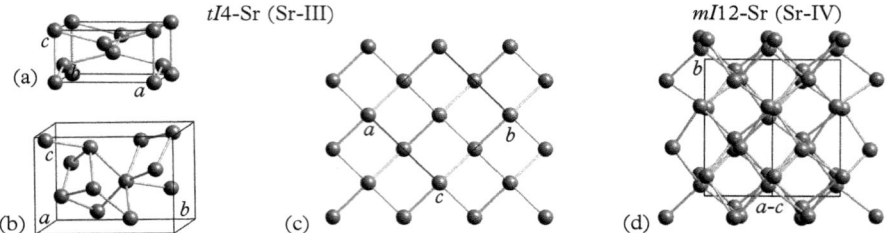

Fig. 6.7 *The structure of tI4-Sr (Sr-III) (a) and (c) in comparison with that of mI12-Sr (Sr-IV) (b) and (d). One unit cell of (a) tI4-Sr (Sr-III) and (b) mI12-Sr (Sr-IV) in perspective view, and their respective projections along [001] and [10$\bar{1}$]. In these projections the structure of mI12-Sr (Sr-IV) (d) results from a helical distortion of tI4-Sr (Sr-III) (c).*

the guest being C-centered. Ba shows the same type of self-hosting structure as an intermediate phase between two stability regions of $hP2$-Ba, which exists at below 12.6 GPa and at above 45 GPa (Reed and Ackland, 2000). At a closer look, three related complex structures exist in this pressure range. For instance, at 19 GPa a commensurate host/guest structure with 768 atoms/asymmetric unit has been identified (Loa *et al.*, 2012), which can be seen as a kind of modulated structure.

Low-pressure $hP2$-Ba shows a strong decrease in the c/a ratio from 1.58 to 1.50 due to the s-d electron transfer. In contrast, that of the isostructural high-pressure is with a value of 1.575 (similar to $hP2$-Be at ambient conditions) pressure-independent up to 90 GPa. This observation has been explained by the s-d transfer being finished already at about 40 GPa (Takemura, 1994).

6.2   Groups 3 to 12: Transition metals (TM)

The transition elements are all good metals, and their nonmagnetic representatives show a structural trend from $hcp \rightarrow bcc \rightarrow hcp \rightarrow fcc$, going from left to right in the periodic table (Skriver, 1989). They have in common that their d-orbitals are filled up more and more going from Group 3 to Group 12. Since they are only slightly screened by the outer filled s-orbitals, their chemical properties vary with increasing occupancy of the d-orbitals. The atomic volumes decrease when filling the bonding d-orbitals and increase again when occupying the antibonding ones ("parabolic behavior", see Fig. 6.15 in Subsection 6.4.1). This parabolic behavior results from the participation of the itinerant d-electrons in the metallic bonding. Under pressure, the sp-orbitals rise faster in energy than the d bands, resulting in an sp- to d-band electron transfer.

6.2.1   Group 3: Sc, Y, La, Ac and Group 4: Ti, Zr, Hf

Scandium, yttrium, lanthanum, and actinium (see Table 6.7) show similar phase sequences to some extent, where the high-pressure phases of the light elements

Table 6.7 *Structural data for the Group-3 and Group-4 elements. First line: group, atomic number Z, element abbreviation and full name, and electronic ground-state configuration. Following lines: prototype structure (inc means incommensurate), temperature T and pressure P limiting the stability range of this allotrope, and atomic volumes $V_{at} = V_{uc}/n_{at}$ for the given $T|P$, with V_{uc} the unit cell volume and n_{at} the number of atoms therein. Only those phases are listed that are stable at either RT or RP.*

Prototype	Allotrope	T [K]	P [GPa]	V_{at} [Å^3]	T|P	References
3 21 Sc			Scandium		[Ar]3d^{1}4s^2	
$hP2$-Mg	Sc-I, α-Sc	RT	RP	24.96	RT|RP	[1]
$cI2$-W	β-Sc	>1610	RP	26.41	1623|RP	[1]
$tI10.6$-Sc$_{inc}$	Sc-II	RT	>20.5	18.65	RT|23	[2]
?	Sc-III		>104		RT	[3]
?	Sc-IV		>140		RT	[3]
$hP6$-Sc	Sc-V	RT	>240	7.96	RT|297	[3]
3 39 Y			Yttrium		[Kr]4d^{1}5s^2	
$hP2$-Mg	Y-II, α-Y	RT	RP	33.01	RT|RP	
$cI2$-W	Y-I, β-Y	>1755	RP			
$hR9$-Sm	Y-III	RT	>12			
$hP4$-La	Y-IV	RT	>25			
$hR24$-Pr	Y-V	RT	>50			[4]
$mC4$-Ce	Y-VI	RT	>99	27.84	RT|120	[4]
3 57 La			Lanthanum		[Xe]5d^{1}6s^2	
$hP4$-La	La-III, α-La	RT	RP	37.17	RT|RP	
$cF4$-Cu	La-II, β-La	>533	or >2.3	34.55	RT|2.3	
$cI2$-W	La-I, γ-La	>1153	RP	38.65	1160|RP	
$hR24$-La	La-IV	RT	>7			[5], [6]
$cF4$-Cu	La-V	RT	>60			[5]
3 89 Ac			Actinium		[Rn]6d^{1}7s^2	
$cF4$-Cu		RT	RP	37.45	RT|RP	[7]
4 22 Ti			Titanium		[Ar]3d^{2}4s^2	
$hP2$-Mg	Ti-II, α-Ti	RT	RP	17.65	RT|RP	
$cI2$-W	Ti-I, β-Ti	>1155	RP	18.15	1193|RP	
$hP3$-AlB$_2$	Ti-III, ω-Ti	RT	>2	17.23	RT|4	
$oC4$-Ti?	Ti-IV, γ-Ti	RT	>116	10.30	RT|130	[8]
$oP4$-Ti?	Ti-V, δ-Ti	RT	>145	9.22	RT|178	[8]
4 40 Zr			Zirconium		[Kr]4d^{2}5s^2	
$hP2$-Mg	Zr-II, α-Zr	RT	RP	23.28	RT|RP	
$cI2$-W	Zr-I, β-Zr	>1136	RP			
$hP3$-AlB$_2$	Zr-III, ω-Zr	RT	>2.2			
$cI2$-W	Zr-IV	RT	>35			
4 72 Hf			Hafnium		[Xe]4f^{14}5d^{2}6s^2	
$hP2$-Mg	Hf-II, α-Hf	RT	RP	22.31	RT|RP	
$cI2$-W	Hf-I, β-Hf	>2013	RP			
$hP3$-AlB$_2$	Hf-III	RT	>38			
$cI2$-W	Hf-IV	RT	>71			

[1] Kammler *et al.* (2008), [2] McMahon *et al.* (2006*a*), [3] Akahama *et al.* (2005*a*), [4] Samudrala *et al.* (2012), [5] Porsch and Holzapfel (1993), [6] Seipel *et al.* (1997), [7] Farr *et al.* (1961), and [8] Akahama *et al.* (2005*b*).

occur as ambient pressure phases of the heavy homologues. Sc is the only element in Group 3 to show an incommensurate host/guest structure, $tI10.6$-Sc$_{inc}$. The *bct* host structure accommodates a C-centered guest structure (McMahon *et al.*, 2006a). It can be described by the superspace group $I4/mcm(00\gamma)$ and is isostructural to the high-pressure modification of strontium, $tI10.8$-Sr$_{inc}$ (see Table 6.6 in Subsection 6.1.2). Between 23 and 101 GPa, the modulation wavevector component γ varies between 1.28 and 1.36, passing through the commensurate value of 4/3 at 72 GPa.

Sc, which is non-superconducting at ambient pressure, becomes superconducting when transforming into $tI10.8$-Sr$_{inc}$. Upon further compression, it reaches its highest critical temperature, $T_c = 8.2$ K, at 74 GPa. In contrast, T_c of Y continues from the onset of superconductivity at ≈ 10 GPa to increase until the highest pressures investigated, $T_c \approx 20$ K at 120 GPa (Hamlin and Shilling, 2007). The emergence of superconductivity under pressure has been explained by the pressure-increased d-electron concentration due to (partial) 4s-3d electron transfer.

The *hcp* to *bcc* phase transitions for Sc, Y, Ti, Zr, and Hf at elevated temperature has been described as martensitic, with the orientation relationships $(110)_{bcc} \| (0001)_{hcp}$ and $[\bar{1}11]_{bcc} \| [\bar{2}110]_{hcp}$ (Sanati *et al.*, 2001).

At the highest pressures applied, at above 240 GPa, $hP6$-Sc forms with a structure showing 6_1-screw helical chains (Akahama *et al.*, 2005a) (Fig. 6.8). It can be

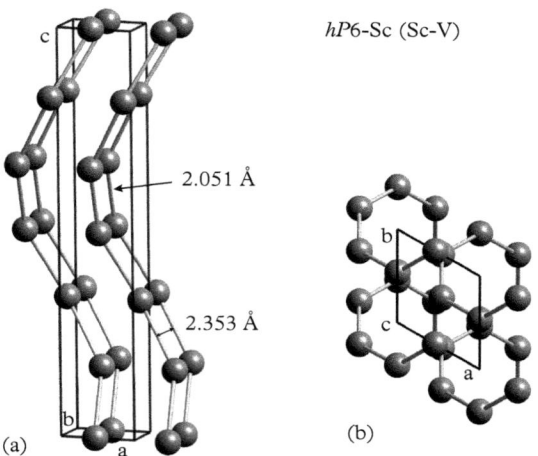

$hP6$-Sc (Sc-V)

2.051 Å

2.353 Å

(a)   (b)

Fig. 6.8 *The structure of the high-pressure phase hP6-Sc (Sc-V) at 242 GPa. (a) Considering the shortest Sc–Sc distances only, the structure can be described as a packing of helical chains around the 6_1-screw axes along [001]. The projection along [001] is shown in (b). The scale is twice the usual one.*

explained as resulting from a modulation of the stacking sequence of the (111)-planes in an *ccp* sequence due to 3d-orbital interactions along the helical chains; the 3d-orbital is considered to be hybridized with the 4s-orbital. Due to the s-d electron transfer, the compressibility of Sc is rather high; at 297 GPa, the atomic volume is squeezed down to only 32% of its value at ambient conditions.

In contrast to Sc, with increasing pressure Y shows largely the typical rare-earth sequence of close-packed structures $hP2$-Mg $\rightarrow$ $hR9$-Sm $\rightarrow$ $hP4$-La $\rightarrow$ $hR24$-Pr $\rightarrow$ $mC4$-Ce (Samudrala *et al.*, 2012). Thereby, $hR24$-Pr can be considered a distorted *ccp* structure as well as $mC4$-Ce. At ultrahigh pressures, yttrium becomes quite incompressible, similar to the heavier lanthanoids where this effect has been related to f-electron delocalization. This may indicate electron transfer from s- and d-bands into the f-band at ultrahigh pressures.

$hP4$-La, with the sequence of *hcp* layers ACAB, represents one of the simpler polytypic *hcp* structures common for the lanthanoids compared to more complex $hR9$-Sm with stacking sequence ABABCBCAC. The second order transition at 7 GPa from $cF4$-La to $hR24$-La, is driven by a pressure-induced softening of a transverse acoustic phonon mode at the L-point of the Brillouin zone, trigonally distorting the *ccp* phase (Gao *et al.*, 2007). At 60 GPa the structure transforms back to $cF4$-La (Porsch and Holzapfel, 1993).

At ambient conditions, titanium, zirconium, and hafnium (see Table 6.7) all crystallize in slightly compressed *hcp* structures, which transform martensitically to bcc at higher temperatures. Thereby, the (0001)-plane of the LT phase becomes equivalent to one of the (110)-planes of the HT phase, and $\langle 11\bar{2}0\rangle_{hP2}\|\langle 1\bar{1}1\rangle_{cI2}$. For the also martensitic $hP2 \rightarrow hP3$ transformation, an orientation relationship $(0001)_{hP2}\|(11\bar{2}0)_{hP3}$ was identified for Ti and Zr (Wenk *et al.*, 2013).

Under pressure, first a modification with a structure of the $hP3$-AlB$_2$ type, $hP3$-Ti (ω-Ti), is obtained by a martensitic transformation (Fig. 6.9). Its packing density is with ≈ 0.57 only slightly larger than that for simple-cubic $cP1$-Po (≈ 0.52), but significantly lower than for $cI2$-W (≈ 0.68) or *ccp* and

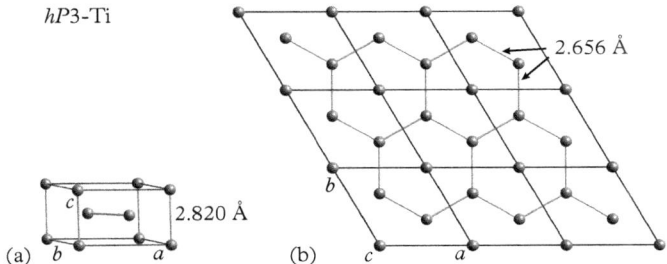

Fig. 6.9 *The structure of the high-pressure phase $hP3$-Ti (ω-Ti). Considering the shortest Ti–Ti distances with covalent bonding contributions only, a hexagon net can be seen at $z = 1/2$ with each hexagon centered by a rod of Ti atoms perpendicular to it.*

hcp (≈ 0.74) structures. The stability of *hP*3-Ti mainly results from covalent bonding contributions due to electron transfer from the broad sp-band to the narrow d-band. This is reflected in the *c*/*a* ratio, which amounts to ≈ 1.587 for *hP*2-Ti, and ≈ 0.613 for *hP*3-Ti, all over the pressure range from 0 to 8.1 GPa (Zhang *et al.*, 2008).

Zr and Hf show the same phase sequence at all pressures and temperatures studied so far. The modification at highest pressure shows the *cI*2-W type structures, while Ti first transforms to *oC*4-Ti and finally to *oP*4-Ti (Akahama *et al.*, 2005*b*). *oC*4-Ti can be seen as a distorted *hcp* structure, while *oP*4-Ti rather corresponds to a modified *bcc* structure; however, these transitions are still controversially discussed (Verma *et al.*, 2007), and a path-dependent direct transition *hP*3-Ti → *cI*2-W has been proposed (Ahuja *et al.*, 2004).

6.2.2 Group 5: V, Nb, Ta and Group 6: Cr, Mo, W

At ambient pressure and all temperatures, vanadium, niobium, tantalum, molybdenum, and tungsten have simple *cI*2-W type structures (see Table 6.8). Only for V has a phase transformation as a function of pressure been observed so far (Ding *et al.*, 2007). The structural transition to the high-pressure phase of the *hR*3-Hg

Table 6.8 *Structural data for the Group-5 and Group-6 elements. First line: group, atomic number Z, element abbreviation and full name, and electronic ground-state configuration. Following lines: prototype structure (inc means incommensurate), temperature T and pressure P limiting the stability range of this allotrope, and atomic volumes $V_{at} = V_{uc}/n_{at}$ for the given $T \mid P$, with V_{uc} the unit cell volume and n_{at} the number of atoms therein. Only those phases are listed that are stable at either RT or RP.*

Prototype	T [K]	P [GPa]	V_{at} [Å^3]	T$\mid$P	References
5 23 V		Vanadium	[Ar]3d^{3}4s^2		
*cI*2-W	RT	RP	13.82	RT$\mid$RP	
*hR*3-Hg	RT	>69	10.46	RT$\mid$90	
5 41 Nb		Niobium	[Kr]4d^{4}5s^1		
*cI*2-W	RT	RP	17.98	RT$\mid$RP	
5 73 Ta		Tantalum	[Xe]4f^{14}5d^{3}6s^2		
*cI*2-W	RT	RP	18.02	RT$\mid$RP	
6 24 Cr		Chromium	[Ar]3d^{5}4s^1		
*cI*2-W	RT	RP	11.99	RT$\mid$RP	
6 42 Mo		Molybdenum	[Kr]4d^{5}5s^1		
*cI*2-W	RT	RP	15.58	RT$\mid$RP	
6 74 W		Tungsten	[Xe]4f^{14}5d^{4}6s^2		
*cI*2-W	RT	RP	15.85	RT$\mid$RP	

type was explained by a band Jahn–Teller mechanism (Verma and Modak, 2008). Cr shows two antiferromagnetic phase transitions, which modify the structure only slightly (Young, 1991).

6.2.3   Group 7: Mn, Tc, Re and Group 8: Fe, Ru, Os

Only two out of the six elements of these groups can adopt different modifications, Mn and Fe (see Table 6.9). The four others crystallize in the $hP2$-Mg type. The at low temperatures antiferromagnetic modification, $cI58$-Mn, transforms isostructurally into a paramagnetic phase at the Néel temperature $T_N = 95$ K. Its structure can be described as a $(3 \times 3 \times 3)$-fold superstructure of $cI2$-W, with 20 atoms slightly shifted and four atoms added (Fig. 6.10(a),(b)). Further increasing the temperature leads to a phase with a $cI20$-Mn type structure, which is

Table 6.9 *Structural data for the Group-7 and Group-8 elements. First line: group, atomic number Z, element abbreviation and full name, and electronic ground-state configuration. Following lines: prototype structure (inc means incommensurate), temperature T and pressure P limiting the stability range of this allotrope, and atomic volumes $V_{at} = V_{uc}/n_{at}$ for the given T | P, with V_{uc} the unit cell volume and n_{at} the number of atoms therein. Only those phases are listed that are stable at either RT or RP.*

Prototype	Allotrope	T [K]	P [GPa]	V_{at} [Å³]	T|P	References
7 25 Mn			Manganese		[Ar]$3d^5 4s^2$	
$cI58$-Mn	Mn-IV, α-Mn	RT	RP	12.21	RT|RP	
$cP20$-Mn	Mn-III, β-Mn	>980	RP	13.61	1008|RP	
$cF4$-Cu	Mn-II, γ-Mn	>1360	RP			
$cI2$-W	Mn-I, δ-Mn	>1410	or >165	14.62	1422|RP	
7 43 Tc			Technetium		[Kr]$4d^5 5s^2$	
$hP2$-Mg		RT	RP	14.26	RT|RP	
7 75 Re			Rhenium		[Xe]$4f^{14} 5d^5 6s^2$	
$hP2$-Mg		RT	RP	14.71	RT|RP	
8 26 Fe			Iron		[Ar]$3d^6 4s^2$	
$cI2$-W	Fe-I, α-Fe	RT	or RP	11.78	RT|RP	
$cF4$-Cu	Fe-II, γ-Fe	>1185	RP	12.13	1189|RP	
$cI2$-W	Fe-III, δ-Fe	>1667		12.64	1712|RP	
$hP2$-Mg	Fe-IV, ε-Fe	RT	>13	10.66	RT|15	
8 44 Ru			Ruthenium		[Kr]$4d^7 5s^1$	
$hP2$-Mg		RT	RP	13.57	RT|RP	
8 76 Os			Osmium		[Xe]$4f^{14} 5d^6 6s^2$	
$hP2$-Mg		RT	RP	13.99	RT|RP	

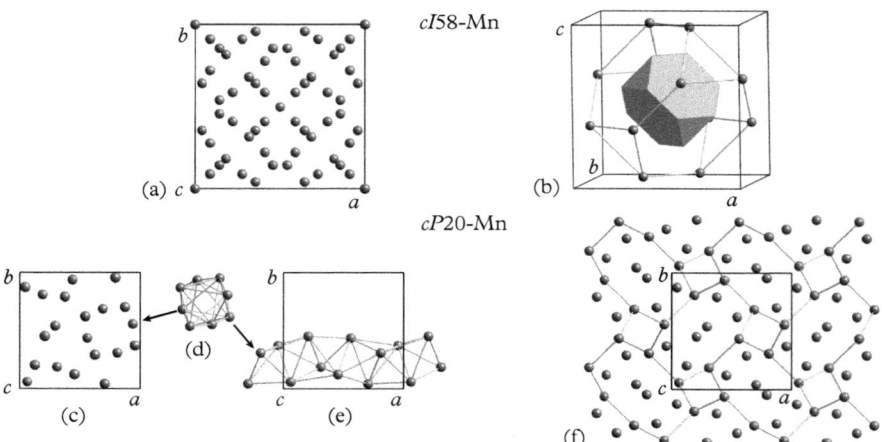

Fig. 6.10 *The complex structures of (a,b) cI58-Mn and (c)–(f) cP20-Mn. cI58-Mn can be seen as a (3 × 3 × 3)-fold superstructure of the cI2-W type plus four atoms added. Its projection is shown in (a), with a perspective view highlighting the innermost truncated tetrahedron, which is surrounded by another by an ≈ τ-times distorted larger one. In (c) the structure of cP20-Mn is shown in projection. In (d) and (e), a structural subunit in form of a tetrahelix is shown in different projections. In (f), the Mn atoms at Wyckoff position 8c generate 4- and 8-spirals, which form, in projection along the 4-fold screw axis, a tiling of squares and squashed octagons. The "bonds" shown all have the length 2.365 Å.*

electronically stabilized (Hume-Rothery mechanism). This structure can be seen as a rod packing of Mn tetrahelices with the interstices filled by Mn zigzag chains (Nyman and Hyde, 1991) (Fig. 6.10(c)–(f)). In contrast to a tetrahelix of regular tetrahedra, one of distorted tetrahedra can be periodic along the rod axis. In the case of cI20-Mn, the repeat unit is eight tetrahedra. Increasing the temperature further, cI20-Mn transforms to cI2-Mn. This phase can also be obtained directly from cI58-Mn at pressures above 165 GPa due to the loss of the magnetic moments (Fujihisa and Takemura, 1995).

The technically most important metallic element and the main constituent of the earth's core, iron, shows four allotropes (see Table 6.9): ferromagnetic cI2-Fe, which transforms to a paramagnetic isostructural (β-)phase at the Curie temperature $T_c = 1043$ K; with further increasing temperature first cF4-Fe forms, which is then transformed back into cI2-Fe. At ambient temperature, there is only one high-pressure phase known so far, hP2-Fe, which conforms to a slightly compressed *hcp* structure.

The most incompressible metal osmium has been studied up to static pressures of 770 GPa (Dubrovinsky *et al.*, 2015). No structural phase transition was observed; however, anomalies in the c/a ratios of hP2-Os indicate electronic transitions. One, at 150 GPa, originates from a topological change in the Fermi surface, and at 440 GPa a core-level crossing (CLC) transition takes place.

6.2.4 Group 9: Co, Rh, Ir and Group 10: Ni, Pd, Pt

All elements of Groups 9 and 10 are typical metals with *hcp* or *ccp* structures or their superstructures (see Table 6.10). In contrast to *ccp* Rh and Ir, at ambient conditions Co is *hcp* and ferromagnetic. It has been shown that the magnetism stabilizes the *hcp* structure (Söderlind *et al.*, 1994), which transforms martensitically, without any apparent volume change, to likely nonmagnetic *ccp* at higher temperatures or pressures (Yoo *et al.*, 2000). By a special thermal treatment, stacking disorder can be created in the *hcp* phase. Thereby, the *hcp* sequence AB is statistically substituted by a *ccp* sequence ABC, with a frequency of at most one to ten, giving rise to sequences of the kind ABABABABCBCBCBC (see Frey and Boysen (1981), and references therein).

All other elements of these groups, Rh, Ir, Ni, Pd, and Pt, crystallize in the *cF*4-Cu structure type. No phase transformations to other allotropes have been observed so far except for Ir. The high-pressure modification *hP*14-Ir can be

Table 6.10 *Structural data for the Group-9 and Group-10 elements. First line: group, atomic number Z, element abbreviation and full name, and electronic ground-state configuration. Following lines: prototype structure (inc means incommensurate), temperature* T *and pressure* P *limiting the stability range of this allotrope, and atomic volumes* $V_{at} = V_{uc}/n_{at}$ *for the given* $T | P$, *with* V_{uc} *the unit cell volume and* n_{at} *the number of atoms therein. Only those phases are listed that are stable at either RT or RP.*

Prototype	T [K]	P [GPa]	V_{at} [Å³]	T|P	References
9 27 Co		Cobalt		[Ar]3d⁷4s²	
*hP*2-Mg	RT	RP	11.08	RT|RP	
*cF*4-Cu	>673	RP	11.36	793|RP	
*cF*4-Cu	RT	>105	7.40	RT|202	[1]
9 45 Rh		Rhodium		[Kr]4d⁸5s¹	
*cF*4-Cu	RT	RP	13.75	RT|RP	
9 77 Ir		Iridium		[Xe]4f¹⁴5d⁷6s²	
*cF*4-Cu	RT	RP	14.15	RT|RP	
*hP*14-Ir	RT	>59	12.42	RT|65	[2]
10 28 Ni		Nickel		[Ar]3d⁸4s²	
*cF*4-Cu	RT	RP	10.94	RT|RP	
10 46 Pd		Palladium		[Kr]4d¹⁰5s⁰	
*cF*4-Cu	RT	RP	14.72	RT|RP	
10 78 Pt		Platinum		[Xe]4f¹⁴5d⁹6s¹	
*cF*4-Cu	RT	RP	15.10	RT|RP	

[1] Yoo *et al.* (2000) and [2] Cerenius and Dubrovinsky (2000).

understood as a 14-layer close packed structure (Cerenius and Dubrovinsky, 2000), with so far unknown layer sequence.

6.2.5 Group 11: Cu, Ag, Au and Group 12: Zn, Cd, Hg

The *ccp* coinage metals Cu, Ag, and Au are typical metals like the alkali metals but with a nobler character (see Table 6.11). Their single ns-electron is less screened by the filled $(n-1)$d-orbitals than the ns-electron by the filled noble gas shell below, and their d-electrons also contribute to the metallic bonding. This gives some similarities to the transition elements with which they are usually grouped. No phase transformations have been observed so far. The color variation from Cu to Ag results from the increasing width of the gap between the d band and the sp conduction band, and the color of Au can be explained by relativistic effects decreasing this band gap again.

Table 6.11 *Structural data for the Group-11 and Group-12 elements. First line: group, atomic number Z, element abbreviation and full name, and electronic ground-state configuration. Following lines: prototype structure (inc means incommensurate), temperature T and pressure P limiting the stability range of this allotrope, and atomic volumes $V_{at} = V_{uc}/n_{at}$ for the given T|P, with V_{uc} the unit cell volume and n_{at} the number of atoms therein. Only those phases are listed that are stable at either RT or RP.*

Prototype	T [K]	P [GPa]	V_{at} [Å³]	T|P	References
11 29 Cu		Copper	[Ar]3d¹⁰4s¹		
*cF*4-Cu	RT	RP	11.81	RT|RP	
11 47 Ag		Silver	[Kr]4d¹⁰5s¹		
*cF*4-Cu	RT	RP	17.05	RT|RP	
11 79 Au		Gold	[Xe]4f¹⁴5d¹⁰6s¹		
*cF*4-Cu	RT	RP	16.96	RT|RP	
12 30 Zn		Zinc	[Ar]3d¹⁰4s²		
*hP*2-Mg	RT	RP	15.20	RT|RP	
12 48 Cd		Cadmium	[Kr]4d¹⁰5s²		
*hP*2-Mg	RT	RP	21.60	RT|RP	
12 80 Hg		Mercury	[Xe]4f¹⁴5d¹⁰6s²		
*hR*3-Hg	<234.32	or >1.2	23.07	83|RP	
*tI*2-Pa	RT	>3.4	19.04	RT|15	
*mC*6-Hg	RT	>12	18.29	RT|20	[1]
*hP*2-Mg	RT	>37	16.91	RT|35.2	

[1] Takemura *et al.* (2007).

The deviations from the ideal value $c/a = \sqrt{8/3} \approx 1.633$ for highly anisotropic *hcp* zinc, $c/a = 1.856$, and cadmium, $c/a = 1.886$, have been explained by covalent bonding contributions, i.e., by the hybridization of the filled d-band with the conduction band (Takemura, 1997). With increasing pressure, this ratio continuously decreases approaching the ideal value with $c/a = 1.633$ at ≈ 23 GPa for Zn, and ≈ 50 GPa for Cd (Schulte and Holzapfel, 1996), and $c/a = 1.76$ at 46.8 GPa for Hg (Schulte and Holzapfel, 1993). A further increase of the pressure decreases c/a below the ideal value, to ≈ 1.59 at the highest pressures applied. It should be mentioned that at $c/a = \sqrt{3} \approx 1.732$, an anomaly of the volume dependence of c/a was observed. At this value, the unit cell in its base-face centered orthohexagonal representation ($\mathbf{a}_{oh} = 2\mathbf{a}_h + \mathbf{b}_h$, $\mathbf{b}_{oh} = \mathbf{b}_h$, $\mathbf{c}_{oh} = \mathbf{c}_h$) becomes pseudotetragonal ($\mathbf{b}_{oh} = \mathbf{c}_{oh}$), which has implications for the shape of the Brillouin zone. In contrast to Hg, no phase transformations have been observed for pressures up to 126 GPa for Zn and 174 GPa for Cd (see Table 6.11).

Mercury is the only metallic element that is liquid at ambient conditions ($T > 234.32$ K), caused by relativistic effects (Calvo *et al.*, 2013). The structures of all mercury modifications can be seen as distorted variants of *ccp* or *hcp* structures. *h*R3-Hg, for instance, can be obtained by compressing a *ccp* structure along a threefold axis. Its $c/a = 1.457$ is smaller than the ideal ratio, while that of the high-pressure phase (stable at least up to 193 GPa) is with $c/a = 1.75$ at 35.2 GPa larger and comparable to that of zinc and cadmium (Takemura *et al.*, 2007). The ideal value was estimated to be reached at ≈ 90 GPa (Schulte and Holzapfel, 1996).

6.3 Groups 13 to 16: (Semi)metallic main group elements

The heavier elements from the main Groups 13 to 16 differ from those of Groups 1 and 2 mainly by filled d-bands. Thus, s-d electron transfer, which is important for the high-pressure phases of the alkali and alkaline earth metals, is not possible for these elements. Instead s-p orbital mixing may take place as it has been shown for the light alkali metals. However, in the case of Ga the 3d-core electrons may become delocalized, forming valence electrons by mixing the 3d-band with the, under pressure widened, 4s- and 4p-bands. The critical pressure has been calculated to 79 GPa (Takemura and Masao, 1998).

6.3.1 Group 13: Al, Ga, In, Tl and Group 14: Ge, Sn, Pb

Aluminum behaves like a typical sp-bonded metal and only close-packed structures have been identified experimentally so far, with an *fcc* to *hcp* phase transformation at 217 GPa (see Table 6.12) (Akahama *et al.*, 2006). Up to 333 GPa, the c/a-ratio is constant, and with 1.618 close to the ideal value of $\sqrt{8/3} = 1.633$. The partial filling of the empty 3d-band was assumed to play a key role for the stability of the high-pressure structure.

Table 6.12 *Structural data for the Group-13 elements. First line: group, atomic number Z, element abbreviation and full name, and electronic ground-state configuration. Following lines: prototype structure (inc means incommensurate), temperature T and pressure P limiting the stability range of this allotrope, and atomic volumes $V_{at} = V_{uc}/n_{at}$ for the given T|P, with V_{uc} the unit cell volume and n_{at} the number of atoms therein. Only those phases are listed that are stable at either RT or RP.*

Prototype	Allotrope	T [K]	P [GPa]	V_{at} [Å³]	T|P	References
13 13 Al			Aluminium		$[Ne]3s^23p^1$	
$cF4$-Cu	Al-I	RT	RP	16.60	RT|RP	
$hP2$-Mg	Al-II	RT	>217	8.31	RT|222	[1]
13 31 Ga			Gallium		$[Ar]3d^{10}4s^24p^1$	
$oC8$-Ga	Ga-I, α-Ga	RT	RP	19.58	RT|RP	
$oC104$-Ga	Ga-II	RT	>2	17.62	RT|2.8	[2]
$hR18$-Ga	Ga-V	RT	>10.5	15.76	RT|12.2	[2]
$tI2$-In	Ga-III	RT	>14	15.27	RT|15.6	[2]
$cF4$-Cu	Ga-IV	RT	>120			[3]
13 49 In			Indium		$[Kr]4d^{10}5s^25p^1$	
$tI2$-In	In-I	RT	RP	26.16	RT|RP	
$oF4$-In	In-II	RT	>45	15.00	RT|93	
13 81 Tl			Thallium		$[Xe]4f^{14}5d^{10}6s^26p^1$	
$hP2$-Mg	Tl-II, α-Tl	RT	RP	28.59	RT|RP	
$cI2$-W	Tl-I, β-Tl	>503		29.00	523|RP	
$cF4$-Cu	Tl-III, γ-Tl	RT	>4	27.27	RT|6	[4]

[1] Akahama *et al.* (2006), [2] Degtyareva *et al.* (2004*a*), [3] Takemura and Masao (1998), and [4] Staun-Olsen and Gerward, (1994).

However, at TPa pressures, phases with more open complex structures have been predicted theoretically (Pickard and Needs, 2010). At 380 GPa a *bcc* phase should form; then at 3.2 TPa a modification is formed, which is related to the incommensurate host/guest structures of Ba-IV and Rb-IV, followed at 8.8 TPa by a simple hexagonal structure, which transforms to an orthorhombic structure at 10 TPa. Further calculations showed that even at highest pressures up to 30 TPa no close-packed structures are stable. This may be explained by the formation of electrides, i.e., positively charged Al ions and interstitial "electron blobs" (a kind of "anions").

Gallium shows a much more complex phase diagram already at moderate pressures. At ambient conditions, the structure of $oC8$-Ga (Ga-I) is characterized by a stacking of layers, each consisting of a 6^3 (honeycomb) network of slightly distorted hexagons parallel to (100) (Fig. 6.11). The interlayer bonds are significantly weaker than the partially covalent intralayer ones. Increasing the pressure leads to a liquid phase at 0.5 GPa and ambient temperature, before it solidifies

oC8-Ga

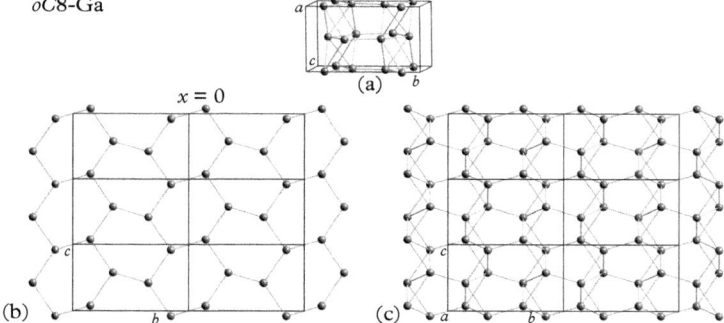

Fig. 6.11 *The structure of oC8-Ga (Ga-I) in different views: (a) unit cell of oC8-Ga in perspective view and projections along [100] of (b) the layer in x = 0, and (c) the layers in x = 0 and x = 1/2, respectively.*

oC104-Ga (Ga-II)

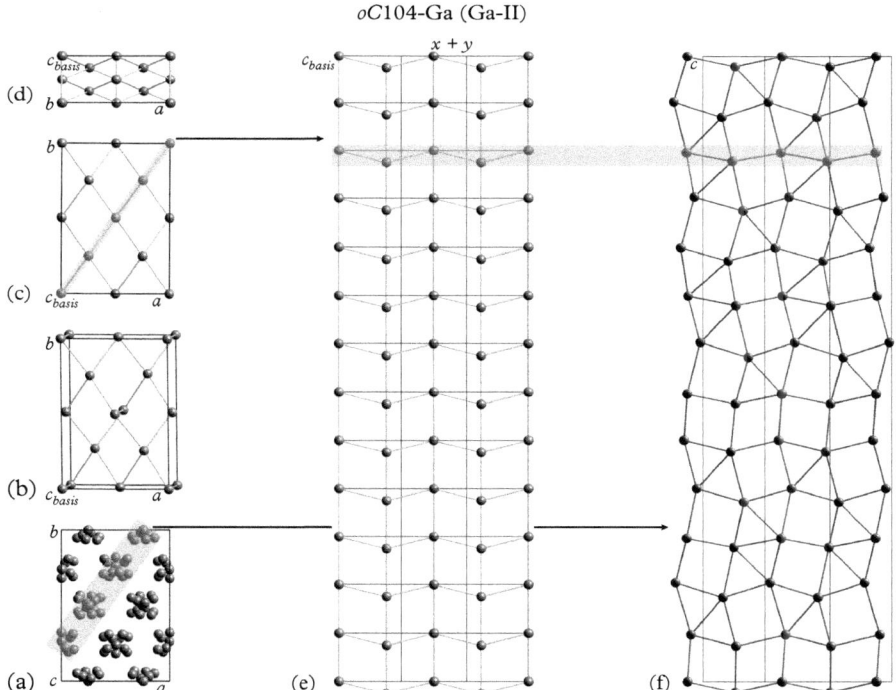

Fig. 6.12 *The structure of oC104-Ga (Ga-II) in different views: (a) unit cell of oC104-Ga projected down [001], (b) one unit cell of its basic structure (oF8) in perspective view, and (c) projected along [001] as well as (d) along [010]. Projections of structure slices (inside the rectangles) of (a) and (c) are shown in (f) and (e), respectively.*

again into a $tI2$-In-type structure, which is metastable at this pressure and stable above 14 GPa. The stable phase between 2 and 10.5 GPa is $oC104$-Ga (Ga-II) (Degtyareva *et al.*, 2004*a*), which can be described as a commensurately modulated 13-fold superstructure of a simple eight-atom basic structure ($oF8$), with $a_{basis} = a$, $b_{basis} = b$, and $c_{basis} = a/13$ (Fig. 6.12) (Perez-Mato *et al.*, 2006). This structure is closely related to that of $oC84$-Cs (Cs-III), $oC52$-Rb (Rb-III), $cI16$-Li, and $cI16$-Na (Na-III). The $tI2$-In-type structure (Ga-III), which can be seen as a tetragonally distorted *fcc* structure, transforms to a *ccp* structure (Ga-IV) at 120 GPa. In the process, the c/a-value of the tetragonal modification changes continuously from 1.558 at 15.6 GPa, for instance, to $\sqrt{2} = 1.414$ for the *ccp* structure at 120 GPa, indicating a second order phase transition (Takemura and Masao, 1998).

The open $tI2$-In structure can be considered a distorted *ccp* one. This can be seen more clearly if we compare $c/a = 1.076$ of the *fcc* setting, $tF4$, with that of the

Table 6.13 *Structural data for the Group-14 elements. First line: group, atomic number Z, element abbreviation and full name, and electronic ground-state configuration. Following lines: prototype structure (inc means incommensurate), temperature T and pressure P limiting the stability range of this allotrope, and atomic volumes $V_{at} = V_{uc}/n_{at}$ for the given $T | P$, with V_{uc} the unit cell volume and n_{at} the number of atoms therein. Only those phases are listed that are stable at either RT or RP.*

Prototype	Allotrope	T [K]	P [GPa]	V_{at} [Å³]	T|P	References
14 32 Ge		Germanium		$[\text{Ar}]3\text{d}^{10}4\text{s}^24\text{p}^2$		
$cF8$-C		RT	RP	22.63	RT|RP	
$tI4$-Sn		RT	>9	16.71	RT|12	
$oI4$-Sn		RT	>75	12.34	RT|81	[1]
$hP1$-BiIn		RT	>85	12.04	RT|83	
$oC16$-Si		RT	>102	10.67	RT|135	[2]
$hP2$-Mg		RT	>160	9.73	RT|180	[2]
14 50 Sn		Tin		$[\text{Kr}]4\text{d}^{10}5\text{s}^25\text{p}^2$		
$cF8$-C	Sn-I, α-Sn	<286.2	RP	34.16	285|RP	
$tI4$-Sn	Sn-II, β-Sn	>286.2	RP	27.05	298|RP	
$tI2$-Pa	Sn-III, γ-Sn	RT	>9.4	20.25	RT|24.5	
$oI2$-Sn?	Sn-IV	RT	>32	19.64	RT|32.5	[3]
$cI2$-W	Sn-V	RT	>40	17.76	RT|53	[3]
$hP2$-Mg	Sn-VI	RT	>157	13.71	RT|167	[4]
14 82 Pb		Lead		$[\text{Xe}]4\text{f}^{14}5\text{d}^{10}6\text{s}^26\text{p}^2$		
$cF4$-Cu		RT	RP	30.32	RT|RP	
$hP2$-Mg		RT	>13.9	28.02	RT|15.2	
$cI2$-W		RT	>110	16.11	RT|127	

[1] Nelmes *et al.* (1996), [2] Takemura *et al.* (2001), [3] Salamat *et al.*, (2013), and [4] Salamat *et al.*, (2011).

ccp one, $c/a = 1.000$. At 45 GPa, $tI2$-In transforms by a simple deformation into the densely packed high-pressure phase $oF4$-In. According to Simak *et al.* (2000), the *fcc* structure of Al already results from an optimum s-p hybridization, whereas for Ga and In this happens only under high pressure.

Thallium shows one HT and one HP phase. The HP transformation takes place with a volume change of just 0.75%.

Germanium shows several different modifications (see Table 6.13). There are some parallels with the high-pressure phases of Si; however, the influence of the 3d core states in Ge, which are absent in Si, leads to significant differences in the expected transition pressures (Takemura *et al.*, 2001). At ambient conditions, semiconducting Ge adopts the diamond structure, $cF8$-C, due to strong covalent bonding contributions. At elevated pressure, it first transforms into a phase with the structure of metallic white tin, $tI4$-Sn. Its coordination number changes from 4 to 6 if the c/a-ratio approaches the value 0.528. Then, if the pressure is further increased, $oI4$-Ge is formed by a slight distortion. However, this phase transformation is of first order as is the subsequent transformation to $hP1$-Ge. The atoms in the latter structure, with $c/a = 0.930$, are quasi-eightfold coordinated. The ideal ratio for CN8 would be $c/a = 1$. With increasing compression, $hP1$-Ge transforms into $oC16$-Ge, with CN10.5 on average (Takemura *et al.*, 2001). Finally, at the highest pressure so far, *hcp* Ge is formed, with CN12. Thus, with increasing pressure, germanium runs through structures with coordination numbers 4, 6, 8, 10.5, and 12.

Metallic white tin, $cF8$-Sn (α-Sn), transforms at below 286.2 K into semiconducting gray tin ("tin pest" or "tin disease"), $tI4$-Sn (β-Sn) (Fig. 6.13) (see Table 6.13). This transformation is accompanied by a volume increase of 26%. Due to the incomplete ionization of the single ns-electron, the effective radii of tin atoms in $tI4$-Sn and of lead atoms in $cF4$-Pb are large compared to those of other typical metals with large atomic numbers. This has implications for the chemical bonding. In the case of $cF8$-Sn, for instance, the electron configuration is [Kr]$4d^{10}5s^15p^3$ allowing sp³ hybridization and covalent, tetrahedrally coordinated bonding, whereas in the case of $tI4$-Sn, with [Kr]$4d^{10}5s^25p^2$, only two p-orbitals are available for covalent bonding and the third one is used for metallic bonding. With increasing pressure, at ambient conditions stable $tI4$-Sn transforms first into

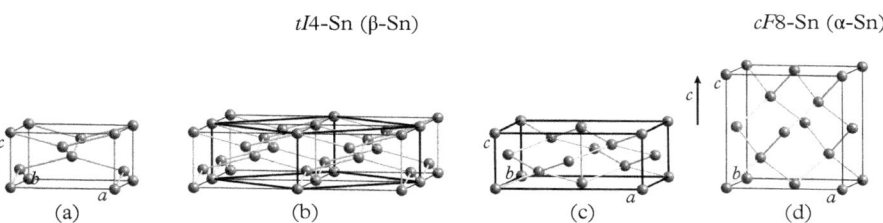

$tI4$-Sn (β-Sn) $cF8$-Sn (α-Sn)

(a)   (b)   (c)   (d)

Fig. 6.13 *The structure of (a,b) tI4-Sn (β-Sn) in relation to that of (c,d) cF8-Sn (α-Sn).*

*tI*2-Sn and then into a slightly orthorhombically distorted phase, *oI*2-Sn. From about 40 GPa *cI*2-Sn starts forming, and up to 70 GPa both phases coexist, perhaps due to some deviations from hydrostatic conditions (Salamat *et al.*, 2013). Finally, at 157 GPa, *cI*2-Sn transforms into *hcp* *hP*2-Sn by a first-order transition (Salamat *et al.*, 2011). According to theoretical predictions, beyond 1.3 TPa a reentrant phase transition to *cI*2-Sn should take place due to an electride-like behavior similar to that predicted for Al (Pickard and Needs, 2010).

Lead shows just two phase transformations, both as a function of pressure. It is remarkable that the *ccp* to *hcp* transition takes place with a larger volume reduction (1.5%) than that from *ccp* to *bcc* (0.3%) (Mao *et al.*, 1990).

6.3.2   Group 15 metallic pnictogens: Sb, Bi, and Group 16 metallic chalcogens: Po

Both, antimony and bismuth have structures of the *hR*6-As type at ambient conditions (see Table 6.14). This structure can be derived from a primitive cubic structure such as *cP*1-Po, in which the atomic distance d_1 within the covalently bonded layers perpendicular to [111] equals the distance d_2 between the layers. The metallic character of the elements increases if the ratio d_2/d_1 approaches 1.

Table 6.14 *Structural data for the Group-15 and -16 elements. First line: group, atomic number Z, element abbreviation and full name, and electronic ground-state configuration. Following lines: prototype structure (inc means incommensurate), temperature T and pressure P limiting the stability range of this allotrope, and atomic volumes $V_{at} = V_{uc}/n_{at}$ for the given T | P, with V_{uc} the unit cell volume and n_{at} the number of atoms therein. Only those phases are listed that are stable at either RT or RP.*

Prototype	Allotrope	T [K]	P [GPa]	V_{at} [Å^3]	T|P	References
15 51 Sb		Antimony		[Kr]4d^{10}5s^{2}5p^3		
*hR*6-As	Sb-I	RT	RP	30.21	RT|RP	
mI-Sb$_{inc}$	Sb-IV	RT	> 8.2	24.79	RT|6.9	[1]
tI-Sb$_{inc}$	Sb-II	RT	> 9.0	23.95	RT|10.3	[1]
*cI*2-W	Sb-III	RT	> 28	20.45	RT|28.8	[1]
15 83 Bi		Bismuth		[Xe]4f^{14}5d^{10}6s^{2}6p^3		
*hR*6-As	Bi-I	RT	RP	35.39	RT|RP	[1]
*mC*4-Bi	Bi-II	RT	> 2.60	31.37	RT|2.7	[1]
tI-Bi$_{inc}$	Bi-III	RT	> 2.75	28.46	RT|6.8	[1]
*cI*2-W	Bi-V	RT	> 8.5	27.43	RT|8.5	
16 84 Po		Polonium		[Xe]4f^{14}5d^{10}6s^{2}6p^4		
*cP*1-Po	α-Po	RT	RP	38.14	311|RP	
*hR*3-Hg	β-Po	> 348	RP	36.61	?|RP	

[1] Degtyareva *et al.* (2004*b*).

At 8.2 GPa $hR6$-Sb transforms into a body-centered monoclinic, incommensurate host-guest structure, mI-Sb$_{inc}$, with a volume change of 5.0%. This structure, the most strongly modulated composite structure yet observed in the elements (Degtyareva *et al.*, 2004*b*), can be seen as a distorted version of the tetragonal host-guest structure, tI-Sb$_{inc}$, which is stable at pressures above 8.6 GPa (Degtyareva *et al.*, 2004*b*). While the guest atoms are arranged in zigzag chains in the former structure, they form linear chains modulated along the [001]-direction in the latter (quasipairing of atoms). In contrast, the host framework structure is modulated perpendicular to this direction. The mI-Sb$_{inc}$ to tI-Sb$_{inc}$ phase transformation is of first order and shows a volume change of 0.5% while that to $cI2$-Sb is related to a volume change of 3.3%.

Bismuth shows a similar phase sequence with one exception (see Table 6.14). Instead of the monoclinic incommensurate phase, a monoclinic commensurate one is formed, which can be seen as a distorted simple cubic structure (Fig. 6.14). The tetragonal body-centered incommensurate structure is not only found in high-pressure modifications of Sb and B, but also of K, Rb, Ba, Sr, Sc, and As (McMahon *et al.*, 2007). It is noteworthy that in both Sb and Bi, the transition to the incommensurate phase is accompanied by either the onset or the enhancement of superconductivity.

Radioactive polonium has—as the only element at ambient conditions—the exceptionally simple cubic structure, $cP1$-Po. This structure type is adopted by Ca and As only at elevated pressures. With increased temperature, $cP1$-Po transforms by a slight compression along the [111]-direction into a modification with the $hR3$-Hg structure, which can be seen as intermediate between simple cubic and bcc. This unusual symmetry lowering with increasing temperature has been explained by the growing influence of vibrational entropy (Verstrate, 2010).

The stability of $cP1$-Po has been shown to result from a relativistic spin-orbit coupling leading to a hardening of most phonon frequencies and reducing the

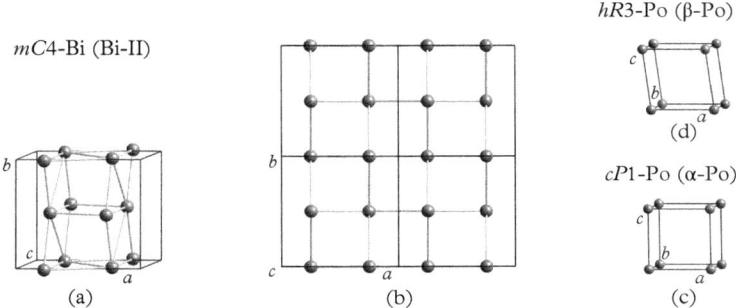

Fig. 6.14 *The structure of mC4-Bi (Bi-II) in (a) perspective view and (b) projected along [001] in relation to that of (c) cP1-Po (α-Po) and (d) hR3-Po (β-Po). The structures of mC4-Bi and hR3-Po, respectively, can be seen as differently distorted cubic-primitive structures.*

Fermi surface/Brillouin zone nesting (Verstrate, 2010). Qualitatively, this can be explained in the way that the p-orbitals do not hybridize with the s-orbitals because of the strong relativistic repulsion of the s-orbitals, allowing orthogonal bonding, forcing a primitive cubic structure. At elevated pressures (1–3 GPa), Po is predicted to become trigonal (Legut *et al.*, 2007).

6.4   Lanthanoids and actinoids

Lanthanoids (Ln) (Tables 6.15, 6.16, and 6.17), together with Sc and Y also called rare earth metals (RE), although they are not so rare (Ce is similarly abundant as Cu), and actinoids (An) (Tables 6.18 and 6.19), which all are radioactive, are characterized by the fact that their valence electrons are increasingly filling the f-orbitals in the fourth and fifth shell, respectively. The chemical properties of the lanthanoids are rather uniform since their 4f-orbitals are largely screened from atomic interactions by their 5s- and 5p-electrons. The lanthanoids are all trivalent except for Eu and Yb, which are divalent at ambient conditions because half- or completely-filled f-orbitals are energetically most favorable. The chemical behavior of the actinoids is somewhere in-between that of the 3d transition metals and that of the lanthanoids, since the 5f-orbitals are shielded from the surrounding atoms to a much lesser extent by the 6s- and 6p-electrons. However, the heavier actinoids following americium show quite a similar chemistry.

The atomic radii of the trivalent lanthanoids monotonically decrease slightly with increasing occupancy of the f-orbitals ("lanthanoide contraction") due to only a partial screening of the f-electrons from the nuclear charge by themselves, as well as by s- and p-orbitals (see Fig. 6.15). This leads to smaller than expected values for the subsequent Group 4 elements Hf and Rf, respectively. The atomic volumes of divalent Eu and Yb are so anomalously large because these two elements have only two (spd)-type electrons. There is no parabolic behavior as is the case for transition elements due to largely localized d-electrons not strongly participating in metallic bonding, which is mainly taking place via the $(spd)^3$-electrons.

The light actinoids from Th to Pu show a similar parabolic behavior as the 5d transition metals indicating that the delocalized f-electrons are forming bands and are participating in the metallic bonding. Relativistic effects additionally contribute to this contraction. Starting with Am, the f-electrons become localized, with the high-temperature δ-phase of Pu ($cF4$-Cu) midways between localization and delocalization. The localized f-electrons do not participate in metallic bonding anymore, leaving it mainly to the $(spd)^3$-electrons, thus reducing the cohesive energy leading to larger interatomic distances and, correspondingly, Wigner-Seitz atomic radii. Their slow decrease follows a similar monotonic decrease as in the case of the transition metals.

It should be kept in mind that the electronic configurations given in the tables refer to the ground state (isolated atoms), and that in the chemically bonded state (solid state) the electron configuration can vary.

Table 6.15 *Structural data for the lanthanoids* Ce, Pr, Nd, Pm, *and* Sm. *First line: atomic number* Z, *element abbreviation and full name, and electronic ground-state configuration. Following lines: prototype structure (inc means incommensurate), temperature* T *and pressure* P *limiting the stability range of this allotrope, and atomic volumes* $V_{at} = V_{uc}/n_{at}$ *for the given* T | P, *with* V_{uc} *the unit cell volume and* n_{at} *the number of atoms therein. Only those phases are listed that are stable at either RT or RP.*

Prototype	T [K]	P [GPa]	V_{at} [Å^3]	T|P	References
58 Ce		Cerium		[Xe]$4f^1 5d^1 6s^2$	
cF4-Cu	<96	RP	28.52	76|RP	
hP4-La	>96	RP	34.78	?|RP	
cF4-Cu	>220	RP	34.37	?|RP	
cI2-W	>999	RP	34.97	1030|RP	
cF4-Cu	RT	>0.76	28.00	RT|1.05	[1]
mC4-Ce	RT	>5.1	23.59	RT|8.3	[2]
tI2-In	RT	>12.2	12.76	RT|208	[3]
59 Pr		Praseodymium		[Xe]$4f^3 6s^2$	
hP4-La	RT	RP	34.56	RT|RP	
cI2-W	>1069	RP	35.22	1094|RP	
cF4-Cu	RT	>4	29.05	RT|4	
hR24-Pr	RT	>7.4	25.09	RT|12.5	[4]
oI16-Pr	RT	>13.7	21.99	RT|19	[4]
oC4-U	RT	>20.5	19.06	RT|21.8	[4]
oP4-Pr	RT	>147	11.82	RT|313	[5]
60 Nd		Neodymium		[Xe]$4f^4 6s^2$	
hP4-La	RT	RP	34.15	RT|RP	
cI2-W	>1128	RP	35.22	1156|RP	
cF4-Cu	RT	>5	27.65	RT|5	[6]
hR24-Pr	RT	>17	25.09		[4], [6]
hP3-Nd	RT	>35			[6]
mC4-Ce	RT	>75	14.61	RT|89	[6], [7]
oC4-U	RT	>113	12.58	RT|155	[6]
61 Pm		Promethium		[Xe]$4f^5 6s^2$	
hP4-La	RT	RP	33.60	RT|RP	
cI2-W	>1163	RP			
cF4-Cu	RT	>10			
hR24?-Pr	RT	>18			
62 Sm		Samarium		[Xe]$4f^6 6s^2$	
hR9-Sm	RT	RP	33.23	RT|RP	
hP2-Mg	>1007	RP	33.79	980|RP	
cI2-W	>1195	RP			
hP4-La	RT	>6			
hR24?-Pr	RT	>11			
hP3-Nd	RT	>37.4	15.11	RT|77	
mC4-Ce	RT	>105	13.74	RT|132	

[1] Franceschi and Olcese (1969), [2] McMahon and Nelmes (1997), [3] Vohra and Beaver (1999), [4] Evans *et al.* (2009), [5] Velisavljevic and Vohra (2004), [6] Chesnut and Vohra (2000), and [7] Akella *et al.* (1999).

Table 6.16 *Structural data for the lanthanoids* Eu, Gd, Tb, Dy *and* Ho. *First line: atomic number Z, element abbreviation and full name, and electronic ground-state configuration. Following lines: prototype structure (inc means incommensurate), temperature* T *and pressure* P *limiting the stability range of this allotrope, and atomic volumes* $V_{at} = V_{uc}/n_{at}$ *for the given* T | P, *with* V_{uc} *the unit cell volume and* n_{at} *the number of atoms therein. Only those phases are listed that are stable at either RT or RP.*

Prototype	T [K]	P [GPa]	V_{at} [Å³]	T|P	References
63 Eu		Europium		[Xe]4f⁷6s²	
cI2-W	RT	RP	48.07	RT|RP	
hP2-Mg	RT	>12.5	27.39	RT|12.5	[1]
mC4-Eu$_{inc}$	RT	>31.5	19.30	RT|33.9	[1]
?	RT	>37			[1]
64 Gd		Gadolinium		[Xe]4f⁷5d¹6s²	
hP2-Mg	RT	RP	33.00	RT|RP	
hR9-Sm	RT	>1.5	30.00	RT|3.5	[2]
hP4-La	RT	>6.5	27.43	RT|10	[2]
cF4-Cu	RT	>26			[3]
hR24-Pr	RT	>33	20.64	RT|39	[3]
mC4-Ce	RT	>60.5	17.28	RT|65	[3]
65 Tb		Terbium		[Xe]4f⁹6s²	
oC4-Dy	<223	RP	32.11	195|RP	
hP2-Mg	>223	RP	31.97	RT|RP	
cI2-W	>1560				
hR9-Sm	RT	>3	27.41	RT|6	[4]
hP4-La	RT	>10			[4]
hR24-Pr	RT	>30	18.61	RT|40.2	[4]
mC4-Ce	RT	>51	12.48	RT|155	[4]
66 Dy		Dysprosium		[Xe]4f¹⁰6s²	
hP2-Mg	<1654	RP	31.60	RT|RP	
cI2-W	>1654	RP			
hR9-Sm	RT	>7	27.34	RT|7	[5]
hP4-La	RT	>17	21.67	RT|26.3	[5]
hR24-Pr	RT	>42	16.09	RT|70	[6]
mC4-Ce	RT	>82	11.74	RT|167	[6]
67 Ho		Holmium		[Xe]4f¹¹6s²	
hP2-Mg	<1521	RP	31.09	RT|RP	
cI2-W	>1521				
hR9-Sm	RT	>7	25.87	RT|8.5	
hP4-La	RT	>19.5			
cF4-Cu	RT	>54			
hR24-Pr	RT	>58			
mC4-Ce	RT	>102			

[1] Husband *et al.*, (2012), [2] Akella *et al.* (1988), [3] Errandonea *et al.* (2007), [4] Cunningham *et al.* (2007), [5] Shen *et al.* (2007), and [6] Samudrala and Vohra (2012).

Table 6.17 *Structural data for the lanthanoids* Er, Tm, Yb, *and* Lu. *First line: atomic number* Z, *element abbreviation and full name, and electronic ground-state configuration. Following lines: prototype structure (inc means incommensurate), temperature* T *and pressure* P *limiting the stability range of this allotrope, and atomic volumes* $V_{at} = V_{uc}/n_{at}$ *for the given* T | P, *with* V_{uc} *the unit cell volume and* n_{at} *the number of atoms therein. Only those phases are listed that are stable at either RT or RP.*

Prototype	T [K]	P [GPa]	V_{at} [Å³]	T	P	References
68 Er		Erbium		[Xe]4f¹²6s²	
$hP2$-Mg	RT	RP	30.71	RT	RP	
$hR9$-Sm	RT	>12.4			
$hP4$-La	RT	>24			
$hR24$-Pr	RT	>58	15.05	RT	80	[1]
$mC4$-Ce	RT	>118	12.67	RT	151	[1]
69 Tm		Thulium		[Xe]4f¹³6s²	
$hP2$-Mg	RT	RP	30.27	RT	RP	
$cI2$-W	>1800	RP			
$hR9$-Sm	RT	>9	25.01	RT	11.6	
$hP4$-La	RT	>32	19.34	RT	35	[2]
$hR24$-Pr	RT	>61	13.53	RT	102	[2]
$mC4$-Ce	RT	>124	11.54	RT	195	[2]
70 Yb		Ytterbium		[Xe]4f¹⁴6s²	
$hP2$-Mg	RT	RP	41.62	296	RP	
$cF4$-Cu	>310	RP	41.28	298	RP	
$cI2$-W	>1065	or >3.5	43.76	1047	RP	
$hP2$-Mg	RT	>26	18.98	RT	34	
$cF4$-Cu	RT	>53	16.36	RT	53	[3]
$hP3$-Nd	RT	>98	10.85	RT	202	[4]
71 Lu		Lutetium		[Xe]4f¹⁴5d¹6s²	
$hP2$-Mg	RT	RP	29.90	RT	RP	
$hR9$-Sm	RT	>25	21.13	RT	23	[5]
$hP4$-La	RT	>45				[5]
$hP8$-Pr	RT	>88	12.52	RT	142	[5]

[1] Samudrala *et al.* (2011), [2] Montgomery *et al.* (2011), [3] Zhao *et al.* (1994), [4] Chesnut and Vohra (1999), and [5] Chesnut and Vohra (1998).

6.4.1   Lanthanoids La – Lu

All lanthanoids, with the exception of Eu, exhibit close packed structures at ambient conditions: either simple *hcp* (*ccp* in the case of Ce) with stacking sequence AB ($n = 1$) or double *hcp* (*dhcp*) with stacking sequence ACAB ($n = 2$) (Fig. 6.16). The stacking sequence of Sm is even more complex with ABABCB-CAC ($n = 4.5$). The ratio c/a is in all cases close to n times the ideal value of 1.633. The large volume decrease under slight compression can be explained by

Table 6.18 *Structural data for the actinoids* Th, Pa, U, Np, Pu, Am, *and* Cm. *First line: atomic number Z, element abbreviation and full name, and electronic ground-state configuration. Following lines: prototype structure (inc means incommensurate), temperature* T *and pressure* P *limiting the stability range of this allotrope, and atomic volumes* $V_{at} = V_{uc}/n_{at}$ *for the given* T|P, *with* V_{uc} *the unit cell volume and* n_{at} *the number of atoms therein. Only those phases are listed that are stable at either RT or RP.*

Prototype	T [K]	P [GPa]	V_{at} [Å³]	T|P	References
90 Th		Thorium		$[Rn]6d^2 7s^2$	
$cF4$-Cu	RT	RP	32.86	RT|RP	
$cI2$-W	>1673	RP	34.01	1720|RP	
$tI2$-In	RT	>63	17.56	RT|102	[1]
91 Pa		Protactinium		$[Rn]5f^2 6d^1 7s^2$	
$tI2$-Pa	RT	RP	24.94	RT|RP	[2]
$cI2$-W	1443	RP	27.65	1443|RP	[3]
$oC4$-U	RT	>77	15.40	RT|130	[2]
92 U		Uranium		$[Rn]5f^3 6d^1 7s^2$	
$mP4$-U$_{inc}$	<43	RP	20.58	4|RP	[4]
$oC4$-U	RT	RP	20.75		
$tP30$-CrFe	>941	RP	21.81	955|RP	
$cI2$-W	>1048	RP	22.06	1060|RP	
93 Np		Neptunium		$[Rn]5f^4 6d^1 7s^2$	
$oP8$-Np	<554	RP	19.22	RT|RP	
$tP4$-Np	>554	RP	20.31	586|RP	
$cI2$-W	>850	RP	21.81	873|RP	
94 Pu		Plutonium		$[Rn]5f^5 6d^1 7s^2$	
$mP16$-Pu	<394	RP	20.43	RT|RP	
$mC34$-Pu	>394	RP	22.44	463|RP	
$oF8$-Pu	>478	RP	23.14	508|RP	
$cF4$-Cu	>588	RP	24.89	653|RP	
$tI2$-In	>741	RP	24.78	750|RP	
$cI2$-W	>754	RP	24.07	773|RP	
$oP4$-Am	RT	>37	14.56	RT|37	[5]
95 Am		Americium		$[Rn]5f^6 6d^1 7s^2$	
$hP4$-La	<1043	RP	29.27	RT|RP	
$cF4$-Cu	>1043	or >6.1	25.23	RT|7.8	[6]
$cI2$-W	>1350	RP			
$oF8$-Am	RT	>10	22.34	RT|10.9	[6]
$oP4$-Am	RT	>17	18.04	RT|17.6	[6]
96 Cm		Curium		$[Rn]5f^7 6d^1 7s^2$	
$hP4$-La	<1550	RP	29.98	RT|RP	
$cF4$-Cu	>1550	or >17			[7]
$mC4$-Cm	RT	>37			[7]
$oF8$-Am	RT	>56	16.23	RT|81	[7]
$oP4$-Am	RT	>95	13.65	RT|100	[7]

[1] Ghandehari and Vohra (1992), [2] Haire *et al.* (2003), [3] Marples (1965), [4] van Smaalen and George (1987), [5] Sikka (2005), [6] Heathman *et al.* (2000), and [7] Heathman *et al.* (2005).

Table 6.19 *Structural data for the actinoids Bk, Cf, Es, Fm, Md, No, and Lr. First line: atomic number Z, element abbreviation and full name, and electronic ground-state configuration. Following lines: prototype structure (inc means incommensurate), temperature T and pressure P limiting the stability range of this allotrope, and atomic volumes $V_{at} = V_{uc}/n_{at}$ for the given T | P, with V_{uc} the unit cell volume and n_{at} the number of atoms therein. Only those phases are listed that are stable at either RT or RP.*

Prototype	T [K]	P [GPa]	V_{at} [Å^3]	T | P	References
97 Bk		Berkelium		$[\mathrm{Rn}]5f^9 7s^2$	
hP4-La	RT	RP	27.97	RT | RP	
cF4-Cu	>1203	or >7			
oC4-U	RT	>25	14.49	RT | 45.9	
98 Cf		Californium		$[\mathrm{Rn}]5f^{10} 7s^2$	
hP4-La	RT	RP	27.27	RT | RP	
cF4-Cu					
oC4-U	RT	>41	14.29	RT | 46.6	
99 Es		Einsteinium		$[\mathrm{Rn}]5f^{11} 7s^2$	
cF4-Cu	RT				[1]
100 Fm		Fermium		$[\mathrm{Rn}]5f^{12} 7s^2$	
101 Md		Mendelevium		$[\mathrm{Rn}]5f^{13} 7s^2$	
102 No		Nobelium		$[\mathrm{Rn}]5f^{14} 7s^2$	
103 Lr		Lawrencium		$[\mathrm{Rn}]5f^{14} 6d^1 7s^2$	

[1] Haire and Baybarz (1979).

sp-electron transfer into the d-band similar as to yttrium, which has no adjacent f-states (Cunningham *et al.*, 2007). At higher pressures, a sequence of low-symmetry structures appears indicative of a transition of localized non-bonding f-electrons to itinerant f-electrons contributing to the metallic bonding. At this transition, a volume collapse was also observed in Ce (16% at 0.7 GPa), Pr (9.1% at 21 GPa), Eu (3% at 12 GPa), Gd (5% at 59 GPa), Tb (5% at 53 GPa), Dy (6% at 73 GPa), Ho (3% at 103 GPa), Tm (1.5% at 120 GPa), and Lu (5% at 90 GPa) (Fabbris *et al.*, 2013).

The phase sequence of La has already been discussed above (see Table 6.7 in Subsection 6.2.1). cF4-Ce performs an isostructural transition at 0.76 GPa where a 4f–5d electron transfer gives rise to a drastic volume collapse of ≈ 15%. As can be seen in Table 6.7, this collapsed *ccp* structure is also adopted by the allotrope stable at ambient pressure and at below 96 K. Further compression leads at 5.1 GPa to mC4-Ce, a superstructure of the *ccp* phase, and finally, beyond 12.2 GPa to tI2-Ce, which is stable at least up to the highest pressure applied, 208 GPa (Vohra and Beaver, 1999). This *bct* modification shows a rather constant

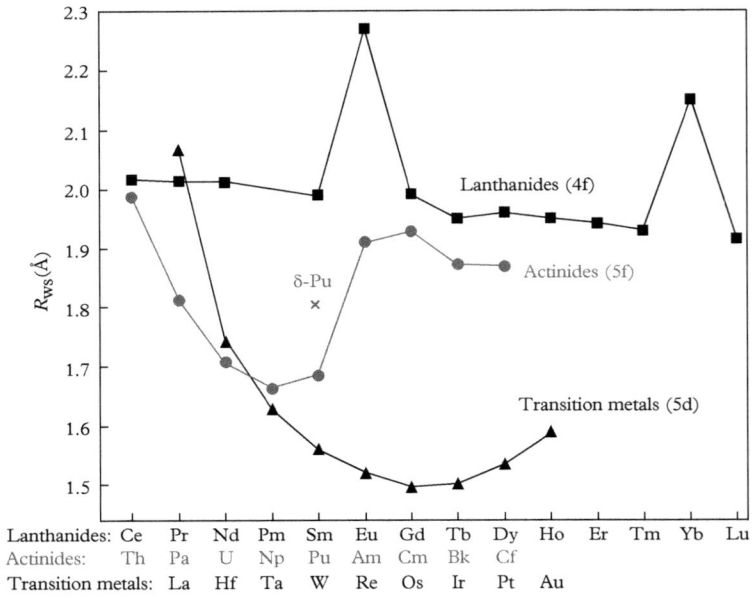

Fig. 6.15 *Wigner Seitz radii (calculated from the atomic volume as defined by the volume of the unit cell divided by its number of atoms) of 5d transition elements, the lanthanoids (Ln) and actinoids (An) (Hecker, 2000). Reprinted courtesy of "Los Alamos Science", Los Alamos National Laboratory.*

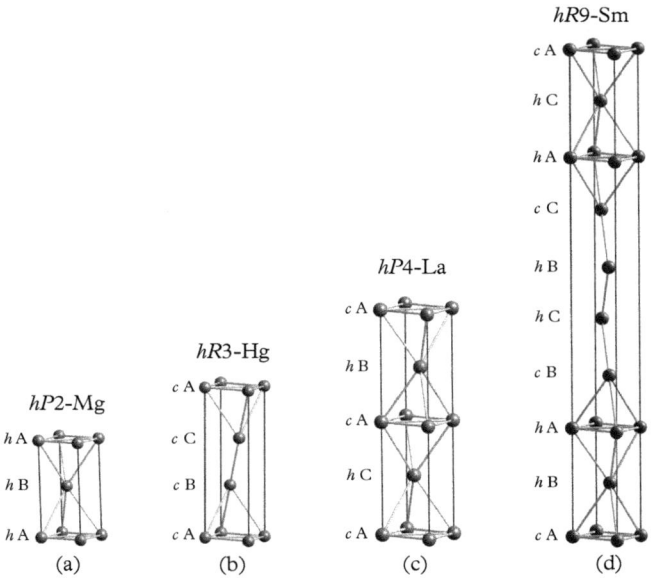

Fig. 6.16 *The polytypic packing sequences (a) hP2-Mg, (b) hR3-Hg, (c) hP4-La (double-hexagonal close packed, dhcp), and (d) hR9-Sm.*

$c/a = 1.68$ in the range between 90 GPa and, at least, 208 GPa, similar to Th, which is isostructural at high pressure and temperature (see Table 6.18).

As already mentioned, the low-temperature modification $cF4$-Ce shows a collapsed *ccp* structure. At 96 K, it first transforms into a *dhcp* one, and then martensitically back into a *ccp* structure (McHargue and Yakel, 1960), but now in uncollapsed form. Finally, at 999 K a *bcc* phase forms that has a lower density than the melt ($T_m = 1068$ K).

Ce has one electron in the f-band, which is localized and does not directly contribute to the metallic bonding in the way that the $5d^1 6s^2$-electrons do. Since the radius of the 4f-orbital is much smaller than that of the 5d- and 6s-electrons, it is effectively shielded from its surroundings. However, there is some coupling with the itinerant spd-electrons to decrease Coulomb repulsion. The volume collapse has been explained by the above transition of localized non-bonding f-electrons to itinerant, but still correlated, f-electrons. At low pressure, $cF4$-Ce behaves similar to the other lanthanoids; at high pressure, after its isostructural transition, it is chemically more similar to $cF4$-Th (Johansson *et al.*, 2014).

$cF4$-Ce gets slightly distorted with further increasing pressure, thereby forming $mC4$-Ce. The structural relationship is depicted in Fig. 6.17. Finally, it transforms into the $tI2$-In structure type.

In the case of the other lanthanoids, the isostructural transition observed in Ce is replaced by an almost isostructural one, namely from the *ccp* to the $hR24$-Pr structure type, which is a kind of distorted *ccp* structure. Pr is the first of this series. The subsequent transformations are connected with another volume collapse due to the delocalization of f-electrons. Nd and Sm run through a largely similar phase

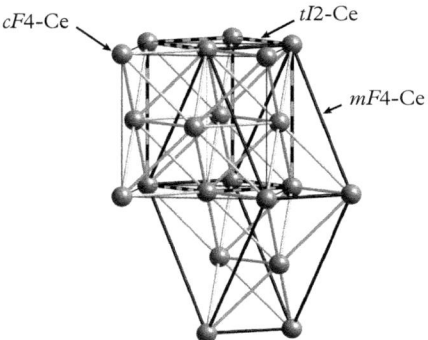

Fig. 6.17 *Structural relationship between the structures of the low-pressure modification cF4-Ce, the intermediate-pressure allotrope mC4-Ce (black thick bonds), and the high-pressure phase tI2-Ce (black/gray multiband bonds) (after McMahon and Nelmes (1997)). On twice the usual scale.*

sequence at slightly higher pressures; however, no such abrupt volume collapse has been observed.

Divalent Eu shows quite a different behavior at ambient conditions due to the stability of its half-filled 4f-orbitals. It has more similarities to the alkaline earth metals than to the other lanthanoids, its phase diagram being closer to that of Ba. This is also indicated by the existence of a monoclinic, incommensurately modulated phase in the pressure range between 31.5 GPa and 37 GPa. The structure is closely related to an *hcp* one. It is the only known incommensurate phase among the lanthanoids so far.

A similar situation is observed for Yb, which is divalent due to the stability of the completely filled 4f-orbitals, and whose phase diagram resembles that of Sr. The anomalously large compressibility of both elements, Eu and Yb, is attributed to a mixing between di- and trivalent 4f-configurations.

When Gd is cooled to below 298 K then its *c* lattice parameter shows an anomalous expansion due to a change in the magnetic properties. Several other lanthanoids behave similarly. Tb behaves in a largely similar way to Ce, and Dy, Ho, Er, and Tm show a similar phase sequence as Gd except that the *ccp* phase could not be observed in the most recent studies (Samudrala and Vohra, 2012; Samudrala *et al.*, 2011; Montgomery *et al.*, 2011). Basically, the structures of all these phases are either close packed or at least related to close packing. The phase transitions from the *hP*2-Mg structure type up to the *hR*24-Pr type are mainly driven by s-d electron transfer, while the transformation to the *mC*4-Ce structure type is attributed to f-electron contributions to chemical bonding. f-electron delocalization is thought to be the reason for the increased stiffness of this low-symmetric modification. It is also noteworthy that the *c*/*a*-ratio for Ho increases with pressure from 1.570 for *hP*2-Ho to almost the ideal value of $n = 1.633$ for *hR*24-Ho, at about 60 GPa.

Divalent Yb has two *ccp* phases in its phase sequence, one at slightly elevated temperatures (at above 310 K) and one in the pressure range between 53 and 98 GPa. At higher pressures, an allotrope with the *hP*3-Nd structure type was found, which was also observed in trivalent Nd and Sm. It can be described as a distorted *ccp* structure. The rather high compression by 74% at 202 GPa is attributed to $4f^{14}$-$4f^{13}$ valence fluctuations and s-d electron rearrangements (Chesnut and Vohra, 1999).

6.4.2   Actinoids Ac – Lr

The actinoids (Tables 6.18 and 6.19) can be divided into two subgroups: the light actinoid elements from Th to Pu have itinerant 5f-electrons, contributing to the metallic bonding, whereas in the elements from Am onwards the 5f-electrons are more localized, similar to electrons in inner shells. This allows superconductivity for Th, Pa, and U, for instance, and magnetic ordering for Cm, Bk and Cf (Dabos-Seignon *et al.*, 1993). Furthermore, the atomic volume, which continuously decreases from Th to Pu, shows a big jump upwards from Am onwards, because the now localized 5f-electrons do not contribute to the bonding anymore (see Fig. 6.15). Consequently, from Th to Pu the melting temperatures decrease continuously by more than 50% and increase again towards Am. The contribution of 5f-electrons to bonding not only leads to rather small atomic volumes and, related therewith, to a high mass density but also to low structural symmetry. In contrast, the heavier actinoids, as far as they have been studied until now, all show highly symmetric *hcp* structures at ambient conditions, and become similar to the light actinoids at high pressure with the increasing delocalization of the f-electrons forming narrow bands, which are prone to more directional bonding (Söderlind *et al.*, 1995*a*).

Overall, the many crystal structure types of the actinoids can be derived from a few simple prototype structures: *cF*4-Cu, *hP*2-Mg, and *cI*2-W. The deviations from the high-symmetry structures mainly arise from Peierls distortions originating from strong bonding interactions of the electrons in the narrow f-bands.

With its, in contrast to the other light actinoids, highly symmetric structure at ambient conditions, *cF*4-Th is behaving more like a transition metal element. It partly shows a phase sequence similar to Ce pointing to a comparable electronic structure at high pressures, which was confirmed by theoretical calculations (Hu *et al.*, 2010). Under pressure, accompanied by sd-f electron transfer, the population of the f-band approaches that of the following element Pa, which already shows features of the *tI*2-Pa structure at ambient conditions (Ce: > 12.2 GPa; Th: > 90 GPa). The transition from *cF*4-Th ($c/a = \sqrt{2} = 1.414$ at 61 GPa) to *tI*2-Th ($c/a = 1.572$ at 86 GPa) can be achieved just by a Bain-type distortion of the structure.

Pa is the first actinoid element with f bonding character already at ambient conditions. Compared to the *tI*2-In structure, which can be seen as a uniaxially expanded *bcc* structure ($c/a = 1.66$ for Ce and 1.57 for Th), *tI*2-Pa is a strongly compressed one ($c/a = 0.825$ for Pa) (see Fig. 3.2 in Section 3.1). The AET is a rhombic dodecahedron (CN14) in the case of *tI*2-Pa and a cuboctahedron (CN12) in the case of *tI*2-In. The transformation from *tI*2-Pa to *oC*4-Pa triggers the promotion of spd- to 5f-states. Therewith, Pa approaches the electron configuration of U with one more 5f-electron than Pa at ambient conditions.

No phase transformation as a function of pressure at ambient temperature has been found so far for U (< 100 GPa), which is indicative of the fully delocalized and strongly bonding f-electrons. However, several transitions have been observed as a function of temperature. In the case of U, an incommensurately modulated structure was observed at below 43 K, which was found to originate from a charge-density wave state due to a Peierls-type transition (van Smaalen and George, 1987). The modification stable at ambient conditions, *oC*4-U, can be described as an *hcp* structure with a reduced displacement vector between neighboring layers, and being contracted by covalent bonding in the orthorhombic [100]- and [001]-directions (see Fig. 6.18). In a packing model, the non-spherical shape due to covalent bonding leaves approximately 20% less interstitial space than in an *hcp* structure of perfectly spherical atoms, explaining the missing primary solubilities for C, N, and O in *oC*4-U (Blank, 1998).

oC4-U

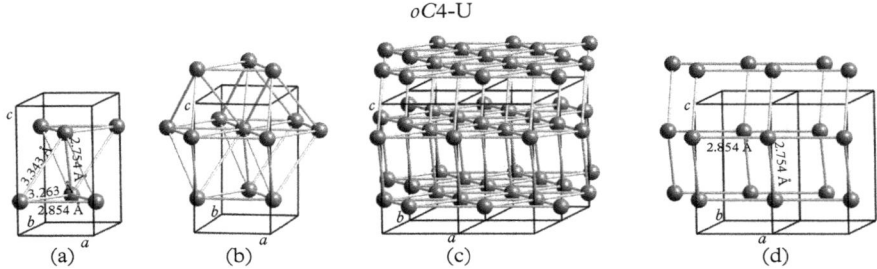

Fig. 6.18 *The crystal structure of oC4-U in different views: (a) unit cell, (b) disheptahedral AET; perspective views emphasizing (c) the distorted hcp layers perpendicular to the c-direction and (d) just showing bonds for interatomic distances ≤ 2.854 Å. On 3/2 of the usual scale.*

Np is the first light actinoid element showing f-electron localization at ambient conditions. No phase transformations under compression (< 50 GPa) have been observed so far, but they have been predicted for much higher pressures (Söderlind *et al.*, 1995b). oP8-Np transforms first to tP4-Np and finally to cI2-Np as a function of temperature. oP8-Np and tP4-Np can be seen as distorted cI2-Np structures.

The position of Pu at the border of itinerant and localized 5f-states (and itinerant 6d-band) causes its unusually complex phase diagram as a function of temperature, with structures typical for both cases. The energy differences between the allotropes are small due to the narrow width of the partially filled 5f-bands with a high density of states at the Fermi energy in combination with a broader 6d-band, which is close in energy and also incompletely filled (Moore and van der Laan, 2009).

Monoclinic mP16-Pu can be considered as a distorted *hcp* structure with a more than 20% higher packing density than cF4-Pu due to covalent bonding contributions from 5f-electrons (Ek *et al.*, 1993). This ratio is quite similar to the aforementioned one of the isostructural transition in *ccp* Ce. Pu runs through a sequence of phase transitions between allotropes, which are characterized by increasing f-electron localization and partly large volume changes (see Fig. 6.19). Their decreasing contribution to metallic bonding and cohesion leads to the low melting temperature of Pu, $T_m = 913$ K, where the deep minimum in the melting temperatures of the actinoids in general is centered. The only high-pressure allotrope known has the oP4-Am structure type and transforms diffusionless from mP16-Pu.

Finally, the phase diagram of Am is very similar to those of La, Pr, and Nd. Due to the localization of 5f-electrons, it is the first lanthanoid-like actinoid element. The f-electron localization also leads to an approximately 50% larger atomic volume for Am compared to the preceding element Pu. However, with increasing pressure, a fraction of the f-electrons becomes delocalized again. This is also

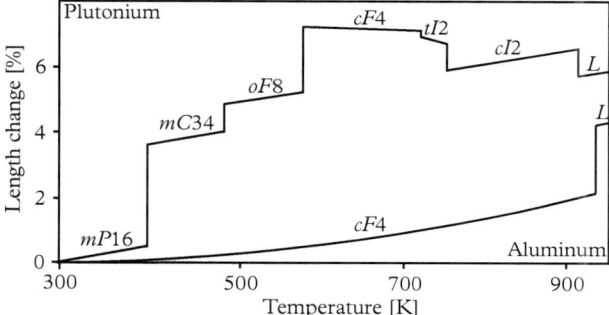

Fig. 6.19 *Length changes connected with the phase transformations between the six different allotropes of* Pu *between ambient temperature and the melt. For comparison, the thermal expansion of aluminum is shown as well (Hecker, 2000). The melt of* Pu (Al) *is denser (less dense) than the solid phase. Reprinted courtesy of "Los Alamos Science", Los Alamos National Laboratory).*

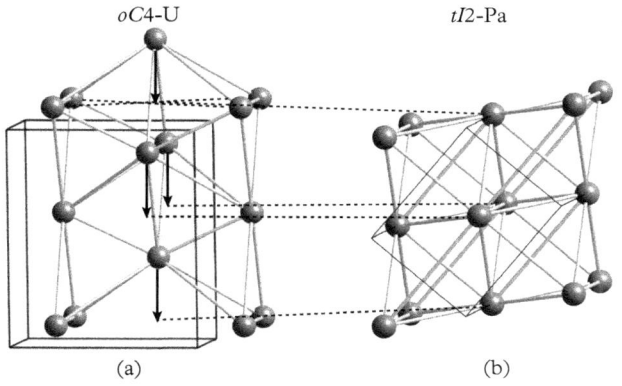

Fig. 6.20 *The closely related structures of (a) oC4-U and (b) tI2-Pa. The arrows indicate the shifts necessary to transform one structure type into the other. On twice the usual scale.*

indicated by volume jumps of 2% and 7% during the transitions at 10 GPa and 17 GPa, respectively.

The structures of *o*P4-Am, *o*C4-U (*o*C4-Pa), and *t*I2-Pa are topologically equivalent (see Fig. 6.20), while *o*F8-Am is closely related to the *h*P4-Am structure. According to Lindbaum *et al.* (2003), the different allotropes—except for the high-temperature *bcc* one—can be seen as different packings of *hcp* layers, with some distortions of the layers to *o*F8-Am and some buckling of the layers in

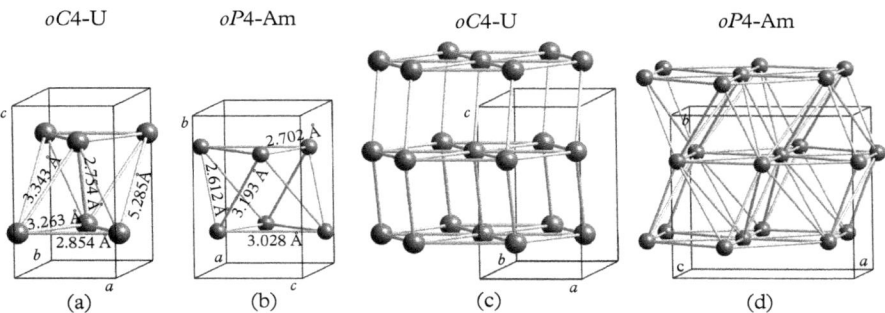

Fig. 6.21 *The topologically equivalent structures of (a) and (c) oC4-U and (b) and (d) oP4-Am in comparison. The octahedral arrangement of the U atoms in (a) is much more distorted than that of the Am atoms in (b). On twice the usual scale.*

the case of *o*P4-Am (see Fig. 6.21). *o*F8-Am is structurally equivalent to *o*F8-Pu, indicating that at this pressure f-electrons participate in chemical bonding. The same is true for Am at highest pressures where it adopts the *o*P4-Am-type structure, a lower-symmetric variant (62 *Pnma*) of the *o*C4-U structure (63 *Cmcm*) (Heathman *et al.*, 2000).

Cm, with its half-filled 5f-band, shows a similar phase sequence as Am. However, the phase transitions are shifted to higher pressures, which are necessary to force the 5f-electrons to take part in the metallic bonding, in addition to the already itinerant 6d- and 7s-electrons. The modification *m*C4-Cm, which does not occur in Am, has f-electrons that are still not fully delocalized. It has an ABA stacking of distorted and shifted *hcp* layers, and is stabilized by antiferromagnetic interactions (Heathman *et al.*, 2005).

7

Crystal structures of intermetallic compounds

Our immodest ambition is to give a representative picture of the overwhelmingly rich world of intermetallic compounds. Unfortunately, this will not and cannot be the "big picture" explaining everything related to intermetallics, although for a few classes of intermetallics "little pictures" already exist. With the "big picture" already in place, we would be able not only to predict the existence of intermetallics in any binary, ternary, or multinary system but also their structures and properties. Unfortunately, we are far away from this ultimate goal. Our more descriptive approach is suited only to increase our knowledge base.

We will show the distribution of intermetallics as a function of chemical composition, stoichiometry, and structure types. In some cases it may go beyond, providing an explanation of the principles underlying their formation and stability. And what do we mean by "representative"? Within the scope of this book, there is no such encyclopedic approach possible as we used for the metallic elements. We have to restrict ourselves to an exemplary discussion of typical representatives of structure types we consider important or illustrative from one point of view or another. It is an attempt to gain more insight into the structural ordering principles of intermetallics, a better understanding in what way intermetallic phases adopt their structures from simple periodic to complex quasiperiodic ones. It is also an attempt to present a broad overview of which structures are more or less common, which are the simplest possible and the most complex ones known so far, from one atom per unit cell to more than twenty thousand, or even without a 3D unit cell.

Which structure types do we consider important and/or illustrative? Certainly, these are the most frequently occurring ones and also those that are helping us to understand fundamental structural formation principles, or those that show structural similarities reducing the large number of structure types to a few basic (prototypic) and many derivative ones, and, of course, those structure types that have representatives with interesting physical properties and (potential) technological applications, and which allow us to illustrate structure/property relationships.

Intermetallics: Structures, Properties, and Statistics. First Edition. Walter Steurer and Julia Dshemuchadse.
© Walter Steurer and Julia Dshemuchadse 2016. Published in 2016 by Oxford University Press.

According to Pearson (1972), a powerful way for classifying and discussing the structures of intermetallics is by describing them as stackings of atomic layers, whenever this is possible. He found that this works well for 590 of the then known ≈ 650 structure types. We follow his suggestion wherever it proves useful, but employ a cluster-based approach where it is more illustrative. For particular symmetries, cluster-based structures also show distinct layer structures anyway. We will also show characteristic projections of crystal structures, which can be especially useful for comparing structures of complex intermetallics. In properly chosen projections, layer packings or substructures may also become visible. In the cases where complex intermetallics can be described as superstructures of underlying basic structures we will also depict their average structures. Of course, we will also employ the approach of derivative structures, i.e., either by formally filling specific interstitial sites or by selectively substituting atoms in prototype structures, for instance.

In the reduced database *Pearson's Crystal Data* (PCD) (Villars and Cenzual, 2011*a*), there are altogether 2166 different structure types listed. However, summing up just the 86 structure types given for the elements, the 943 for binary and the 1391 for ternary phases one gets 2420 structure types altogether. This means that a significant number of binary or ternary compounds are just considered to be pseudo-binaries or pseudo-ternaries, respectively, because their structures can be (or have been, at least) described by unary or binary structure types, respectively (see discussion in Chapter 5).

It has to be kept in mind that our discussion of crystal structures is purely geometrical in most cases. The description in terms of structural subunits (layers, clusters, . . .) can, but does not necessarily need to, reflect the relevant atomic short-range interactions (chemical bonding); also not taken into account are long-range interactions such as the electronic and dynamic structure as reflected in the electronic and vibrational density of states, respectively. There is simply not enough reliable information of this kind in the literature except in the more recent literature. For us, a feasible way to include a minimum amount of chemical information was the use of M/M-plots, showing which elements are prone to form particular structure types, thus indicating at least some of the factors controlling their formation (electronegativity differences, atomic radii ratios, etc.).

The validity of our approach to focus on a geometrical discussion is corroborated by the recent experimental and theoretical studies of mesoscopic model systems (colloids, nanoparticles, polyhedra, etc.) (see, e.g., Lifshitz (2014)). Those can adopt rather complex long-range-ordered structures, even quasiperiodic ones, based on interactions that can be described in computer simulations by rather simple isotropic pair potentials with or without Friedel oscillations. Consequently, we will discuss crystal structures mainly by packing principles for given structural subunits. The prevalence of particular atomic configurations is assumed to be controlled by chemical bonding within the boundary conditions of chemical composition, stoichiometry, and atomic size ratios.

This chapter will start with three sections on the statistics of structures and structure types as well as with a detailed discussion of structure types that can be derived from simple sphere packings by selective occupation of interstitial sites or by the substitution of specific atoms with another kind of atom. The subsequent section is devoted to topologically close-packed (*tcp*) (Frank-Kasper, FK) phases, which have only tetrahedral voids.

Three more sections are devoted to large subclasses of intermetallics, the Zintl phases, the equiatomic REME phases, and the $hP3$-AlB_3 derivative structures. The section on topological layer structures deals with partly more complex structures, which can be nicely decomposed (geometrically) into layer stackings. The following section on long-period structures illustrates their alternative descriptions as superstructures, modulated or host/guest structures. In the section on hierarchical and modular structures, it is demonstrated how complex structures can be derived by replacing atoms in simple structures by groups of atoms (clusters), i.e., applying the same packing principles on different length scales. The section on structures with one dominating element shows the complexity of structures resulting from extremely unbalanced stoichiometries.

Finally, a systematic overview will be given about different classes of intermetallic compounds. First, alkali/alkaline earth metal compounds (groups 1 and 2, only) are discussed, then alkali/alkaline earth metal compounds with transition metals (groups 1 and 2 with groups 3–12), followed by transition metal/transition metal compounds (groups 3–12, only). These sections are followed by a discussion of intermetallic compounds with at least one (semi)metallic element from groups 13–16, i.e., the trielides (Al, Ga, In, and Tl), tetrelides (Ge, Sn, and Pb), pnictides (Sb, and Bi) and polonides (Po). The last two sections deal with lanthanoid/lanthanoid and actinoid/actinoid compounds, and high-pressure phases of selected intermetallic compounds, respectively.

7.1 Statistics of structures derived from simple sphere packings

The database *Pearson's Crystal Data* (PCD) (Villars and Cenzual, 2011*a*) was searched for structures with cuboctahedral (co_{12}), anticuboctahedral (ac_{12}), also called disheptahedral, and rhombic-dodecahedral (rd_{14}) AETs, respectively, in order to identify *ccp*, *hcp*, and *bcc*-based derivative structures, which have a lower symmetry than their parent structures in most cases. Out of the 20 829 entries, there are 4369 compounds (21 %) crystallizing in 174 structure types, where such AETs were identified: 83 for rd_{14}, 82 for co_{12}, and 19 for ac_{12}. Of course, this does not mean that these AETs are the only ones in these structure types. The overall number of different structure types, 174, is smaller than the sum of the numbers of structure types combined, $83 + 82 + 19 = 184$. This is due to structure types constituted from more than one type of these three AETs. Either only rd_{14} or only co_{12} are found in compounds with structures assigned to the $tP2$-CuAu type (103 compounds), to the $tI6$-CuZr$_2$ type (37), to the $oF4$-In type (2), $tI2$-In (44), to the $tI8$-TiAl$_3$ type (55), and to the $tI8$-VRh$_2$Sn type (16); rd_{14} or ac_{12} in the $oP4$-AuCd type (22) and the $oP8$-Cu$_3$Ti type (51); and co_{12} or ac_{12} in the $hP2$-Mg type (338) and the $hP4$-Nd type (52).

The distribution of binary intermetallics with such structures is illustrated in the M/M-plots depicted in Fig. 7.1. The stability areas for compounds with *bcc* derivative structures are well separated from those of the other ones. *bcc* structures are mainly found for the element compositions with $M(A)/M(B)$ in the ranges

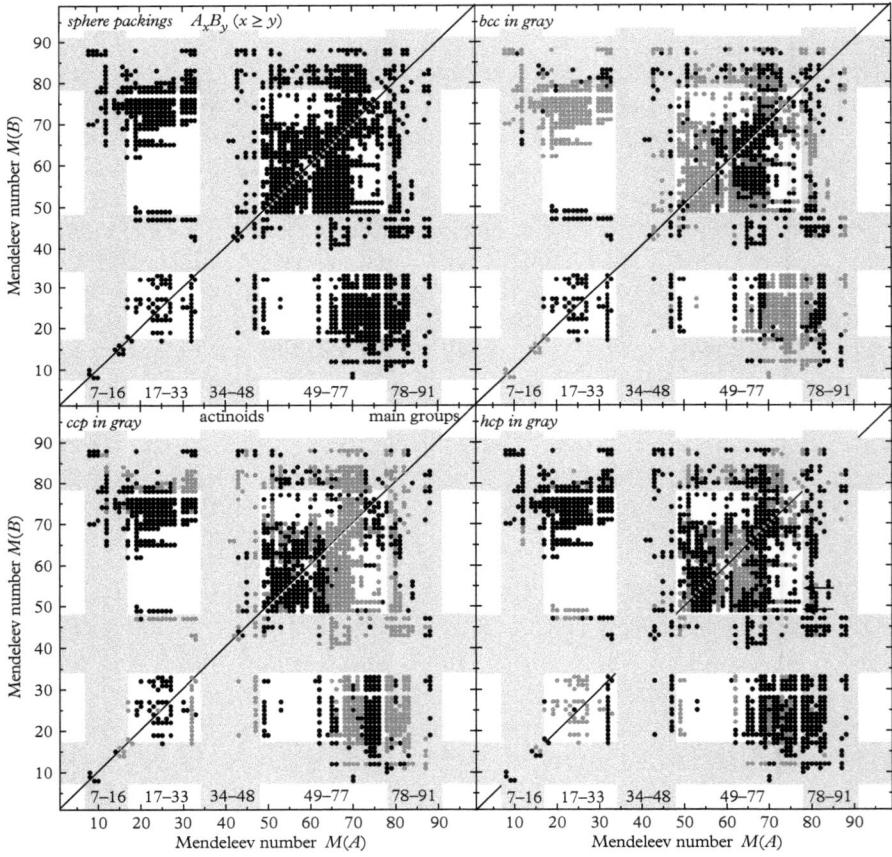

Fig. 7.1 *Binary element combinations, A_xB_y ($x \geq y$), forming compounds with derivative structures based on bcc, ccp, and hcp sphere-packings. $M(A)$ and $M(B)$ denote the Mendeleev numbers of the majority and minority elements, A and B, respectively. For AB-compounds, both element combinations are plotted. The upper left subfigure shows all element combinations of the 4369 structures (black dots). In the other subfigures, the element combinations referring to bcc, ccp, or hcp derivative structures are marked by gray dots.*

(7–33)/(60–91) and (49–70)/(49–70). The distribution of compounds with *ccp* and *hcp* structures is almost complementary to it, while both also differ quite significantly from each other. The agglomeration of compounds with *hcp* structures in the field of phases containing two lanthanoids may be due to the formation of solid solutions between two *hcp* lanthanoids. Another accumulation of gray dots in the *hcp* subfigure can be found for compositions (55–65)/(49–70). Compared to the $M(A)/M(B)$ plots for *bcc* and *hcp* structures, that for *ccp* structures is the least symmetric one with respect to the majority component, $A_xB_y \neq A_yB_x$ $(x \geq y)$.

Caveat: One has to keep in mind that these AET-assignments are not always without ambiguity. The atomic environments of the structures in the PCD are automatically classified with the maximum-gap method (Brunner and Schwarzenbach, 1971; Daams and Villars, 1997). Most structures exhibit a clear gap in interatomic distances after the first coordination sphere (Villars and Cenzual, 2011*a*), containing all positions considered for the determination of an AET. However, a pure distance-based approach is not always successful and also does not account for the completeness of a coordination shell. A more powerful approach relies on the Voronoi-Dirichlet polyhedra (Blatov, 2006), which basically takes into account all neighbors of a central atom that represent a certain minimum solid angle of its coordination sphere. This method, as well as alternative approaches to determine an atom's connectivity, are implemented in the program package "TOPOS" (Blatov, 2012), for instance.

7.2 Close-packed structures and their derivatives

By selective substitution (also by vacancies) of subsets of atoms forming close-packed structures and/or by the selective filling of tetrahedral or octahedral voids, families of substitutionally ordered structures, as well as of other derivative structures, can be obtained (Fig. 7.2). It should be kept in mind that some of the structures that can be described as derivatives of *ccp* structures may be equally well considered as *bcc* superstructures. Usually, a derivative structure has a lower symmetry and/or a larger unit cell than its reference (parent) structure and there is always a dedicated group/subgroup relationship. The AETs in the case of parent *ccp* and *hcp* structures are cuboctahedra and disheptahedra (anticuboctahedra), respectively. Both AETs have an ideal ratio of the radius of the central atom to the one of the surrounding atoms, $r_c/r_s = 1$. Ideal means in this case that all spheres touch each other ("kissing spheres"). The parent structure types *cF*4-Cu and *hP*2-Mg are centrosymmetric, and so is the cuboctahedron, while the disheptahedron itself is non-centrosymmetric.

The schematics for selectively filling the tetrahedral and octahedral voids, respectively, in the *cF*4-Cu structure type (225 $Fm\bar{3}m$, Cu in 4*a* $0,0,0$) in different ways is illustrated in Fig. 7.2. If half of the tetrahedral voids are filled with the same kind of atoms, the *cF*8-C (diamond) structure type is obtained (227 $Fd\bar{3}m$, C in 8*a* $0,0,0$); if another type of atom is used, the *cF*8-ZnS (sphalerite, zincblende)

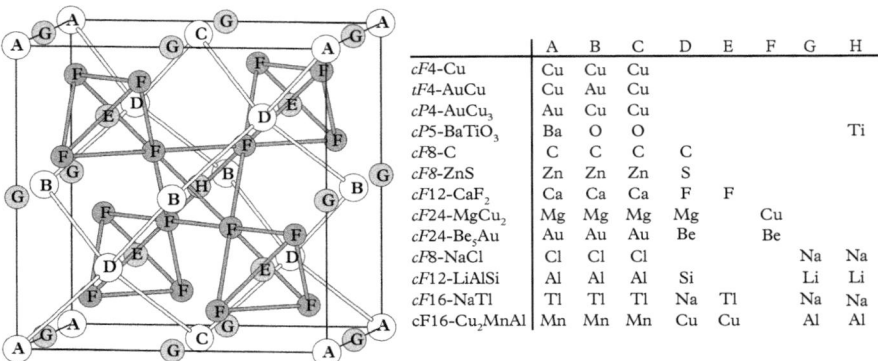

	A	B	C	D	E	F	G	H
cF4-Cu	Cu	Cu	Cu					
tF4-AuCu	Cu	Au	Cu					
cP4-AuCu₃	Au	Cu	Cu					
cP5-BaTiO₃	Ba	O	O					Ti
cF8-C	C	C	C	C				
cF8-ZnS	Zn	Zn	Zn	S				
cF12-CaF₂	Ca	Ca	Ca	F	F			
cF24-MgCu₂	Mg	Mg	Mg	Mg			Cu	
cF24-Be₅Au	Au	Au	Au	Be			Be	
cF8-NaCl	Cl	Cl	Cl				Na	Na
cF12-LiAlSi	Al	Al	Al	Si			Li	Li
cF16-NaTl	Tl	Tl	Tl	Na	Tl		Na	Na
cF16-Cu₂MnAl	Mn	Mn	Mn	Cu	Cu		Al	Al

Fig. 7.2 *Site occupancies of several structure types derived from the cubic close packing by selective occupancy of tetrahedral and octahedral voids, respectively:* A $(0,0,0)$, B $(1/2,0,1/2)$, C $(1/2,1/2,0)$, D $(1/4,1/4,1/4)$, E $(1/4,3/4,1/4)$, G $(1/2,0,0)$, H $(1/2,1/2,1/2)$, *and symmetrically equivalent ones.*

structure type results (216 $F\bar{4}3m$, Zn in $4a$ $0,0,0$; S in $4c$ $1/4,1/4,1/4$). With all tetrahedral interstices occupied by another kind of atoms, we get the $cF12$-CaF₂ (fluorite) structure type (225 $Fm\bar{3}m$, Ca in $4a$ $0,0,0$; F in $8c$ $1/4,1/4,1/4$). If we replace a subset of one kind of atoms in the $cF4$-Cu type by a subset of another kind, we obtain either the $tP4$-CuAu type (123 $P4/mmm$, Au in $1a$ $0,0,0$ and $1c$ $1/2,1/2,0$; Cu in $2e$ $0,1/2,1/2$) or the $cP4$-Cu₃Au type (221 $Pm\bar{3}m$, Au in $1a$ $0,0,0$; Cu in $3c$ $0,1/2,1/2$) (Fig. 7.3). $cP5$-BaTiO₃ (221 $Pm\bar{3}m$, Ba in $1a$ $0,0,0$; Ti in $1b$ $1/2,1/2,1/2$; O in $3c$ $0,1/2,1/2$), with the central position (octahedral void in the $cF4$-Cu type) occupied additionally, can be seen as a filled $cP4$-Cu₃Au type.

The binary system Au–Cu shows full miscibility of the two components above 683 K. Below this temperature, several ordered phases with broad stability ranges form around the compositions Au₃Cu, CuAu, and Cu₃Au. The liquidus curve shows a significant minimum (1183 K) at ≈ 44 at.% Cu, indicating that Au–Cu interactions are weaker than those between Au–Au and Cu–Cu. The melting temperature of Au, 1337.4 K, is slightly lower than that of Cu, 1357.9 K, respectively. It is not surprising, therefore, that $tP4$-CuAu forms a structure with alternating Au and Cu layers along [001], with the Au–Au distance of 2.811 Å even slightly shorter than in $cF4$-Au with 2.884 Å (Fig. 7.3(a–c)). The Cu–Cu distances are determined by the larger Au atoms, and lead to a shrinking of the unit cell along [001] resulting in a ratio $c/a = 0.927$. $tP4$-CuAu can either be described by a stacking of alternate 4^4 layers of Au and Cu atoms along [001] or as ABC-stacking of hcp layers along [111]. In the case where the ratio c/a is around 1, the description as a $cF4$-Cu superstructure is appropriate; if it is around $\sqrt{2}/2$, a description based on a $cI2$-W-derivative structure is more favorable. In a description of the structures as a stackings of 3^6 layers, these are more regular in the

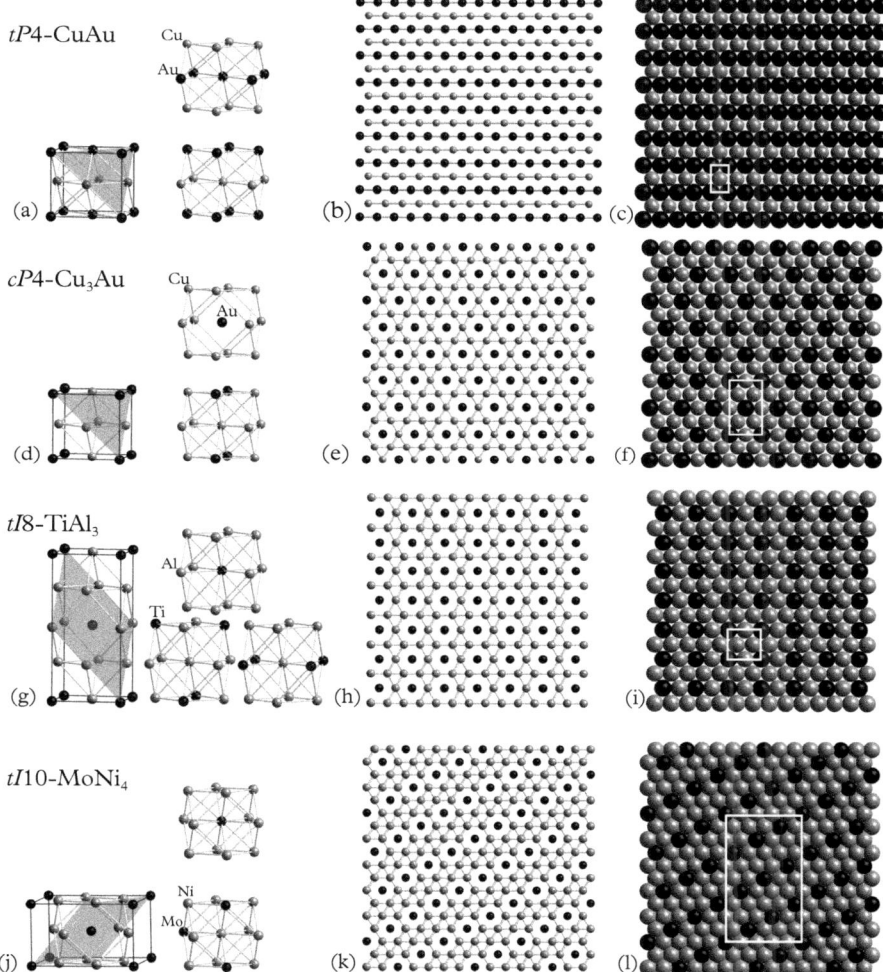

Fig. 7.3 *Substitutional derivative (super)structures of the cF4-Cu structure type. In each case, the following is shown: the unit cell, AETs around each symmetrically independent atom (all are cuboctahedra), one hexagonal close-packed layer in the ball-and-stick model (only bonds between like atoms are drawn for clarity), as well as with space-filling atoms (the rectangle marks the unit cell of the layer). The black spheres always correspond to the heavier atoms. In the cases where the orientation of the hcp layer is not perpendicular to a lattice parameter, its orientation is indicated by a gray plane. (a)–(c) tP4-CuAu, (d)–(f) cP4-Cu₃Au, (g)–(i) tI8-TiAl₃, (j)-(l) tI10-MoNi₄. For subfigures (b, c), (e, f), (h, i), and (k, l), the scale is one third the usual one.*

case of $cF4$-Cu derivatives and more distorted in the case of $cI2$-W derivatives, where the stacking runs along [110].

In the case of $cP4$-Cu_3Au, the only continuous monoatomic layers are 4^4 layers constituting Cu along [001] (Fig. 7.3(d–f)). The Cu–Cu distances are smaller than in $tP4$-CuAu, but with 2.647 Å in both the Cu-layers as well in the Kagomé net layer they are still larger than in $cF4$-Cu (2.556 Å). This is a consequence of the larger Au atoms expanding the lattice. The hcp layers in $cP4$-Cu_3Au, stacked along [111] with the sequence ABC, consist of Cu decorated 3.6.3.6 Kagomé nets, while the Au atoms form a 3^6 triangle net centering the hexagons. The distance histograms of these and other ccp derivative structures are shown in Fig. 7.5.

$tI8$-$TiAl_3$ is an incongruently melting line-compound in the complex Al–Ti system. The substitution of 25 at.% Al by Ti raises the melting temperature steeply from 933.5 K, i.e., that of Al, to 1160 K for $tI8$-$TiAl_3$. Its structure can be described as a two-fold superstructure along [001] of $cF4$-Cu, with the hcp layers in the same orientation as in the pseudo-cubic subcells (Fig. 7.3(g–i)). The sequence of 4^4 layers along [001] is equal to that in two unit cells of $cP4$-Cu_3Au, except that the layer in the middle is shifted by 1/2, 1/2, 0. This leads to a smaller rectangular unit cell of the hcp layer. The Al–Ti distances in the 4^4 layers with 2.712 Å are shorter than the distance calculated from their atomic radii, 2.880 Å, and of those in other directions (2.877 Å). This also shortens the Al–Al distances to 2.712 Å, compared to those in $cF4$-Al (2.864 Å). The c/a ratio is with 1.120 (referred to one subcell) significantly larger than that for $cP4$-Cu_3Au, which is obviously equal to 1. The doubling of the unit cell compared to $cP4$-Cu_3Au replaces one vertex of the pure Al octahedron by a Ti increasing the attractive forces along [001]. This results in shorter Al–Al distances in the pure Al layers below and above the cell center. Indeed, *ab initio* calculations revealed a strong hybridization of Al-p- and Ti-d-states and a strong directionality in bonding (Hong *et al.*, 1990) explaining the stability of this structure.

$tI10$-$MoNi_4$ ($\mathbf{a}_t = 3/2\mathbf{a}_c - 1/2\mathbf{b}_c$, $\mathbf{b}_t = 3/2\mathbf{b}_c + 1/2\mathbf{a}_c$, $\mathbf{c}_t = \mathbf{c}_c$) is a low-temperature ordering state in the system Mo–Ni, its phase field neighboring that of the solid solution $cF4$-(Ni,Mo). There are no more layers with only Mo or Ni in bonding distance as in the previous three cases, just bands of Ni atoms in the hcp layers. The rectangular unit cell of these layers is very large now. The larger unit cell is needed to accommodate the stoichiometry, which is incompatible with the $cF4$-Cu unit cell. Along [111], the repeat unit of the stacking sequence amounts to 15 hcp layers (Fig. 7.3(j–l)).

More complex superstructures with larger unit cells are, $tI16$-$ZrAl_3$ (4-fold superstructure along [001]) (Fig. 7.4(a)–(c)), $oP20$-$ZrAu_4$ (close packed layers along [010]) (Fig. 7.4(d)–(f)), $hP12$-WAl_5 (close packed layers along [001]) (Fig. 7.4(g)–(i)) as well as $oI6$-$MoPt_2$ ($\mathbf{a}_o \times 1/2(\mathbf{a}_c - \mathbf{b}_c)$, $\mathbf{b}_o \times 3/2(\mathbf{a}_c + \mathbf{b}_c)$, $\mathbf{c}_o \times \mathbf{c}_c$), and $hP8$-Ni_3Sn ($2 \times 2 \times 1$)-fold superstructure of $hP2$-Mg), for instance. $hP8$-Ni_3Sn is the hcp analogue to ccp $cP4$-Cu_3Au, with the same kind of layers stacked in the sequence AB along [001]. One principle underlying all these superstructures is that direct contacts of like atoms are only allowed for one kind of atoms.

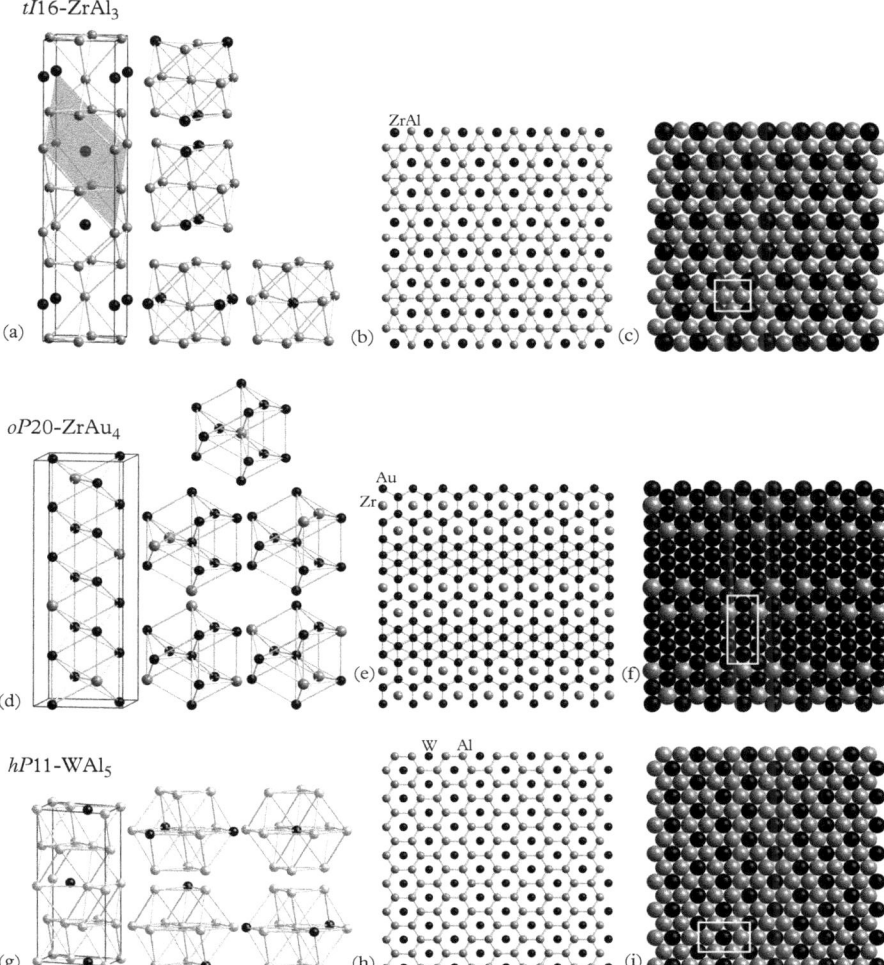

tI16-ZrAl₃

oP20-ZrAu₄

hP11-WAl₅

Fig. 7.4 *Substitutional derivative (super)structures of the cF4-Cu structure type. In each case, the following are shown: one unit cell, AETs around each symmetrically independent atom (cuboctahedra in (a) and disheptahedra in (d) and (g), one hexagonal close-packed layer in the ball-and-stick model (only bonds between like atoms are drawn for clarity), as well as with space-filling atoms (the rectangle marks the unit cell of the layer). The black spheres always correspond to the heavier atoms. In the cases where the orientation of the hcp layer is not perpendicular to a lattice parameter, its orientation is indicated by a gray plane. (a)–(c) tI16-ZrAl₃, (d)–(f) oP20-ZrAu₄, (g)–(i) hP12-WAl₅. For subfigures (b, c), (e, f), and (h, i)), the scale is one third the usual one.*

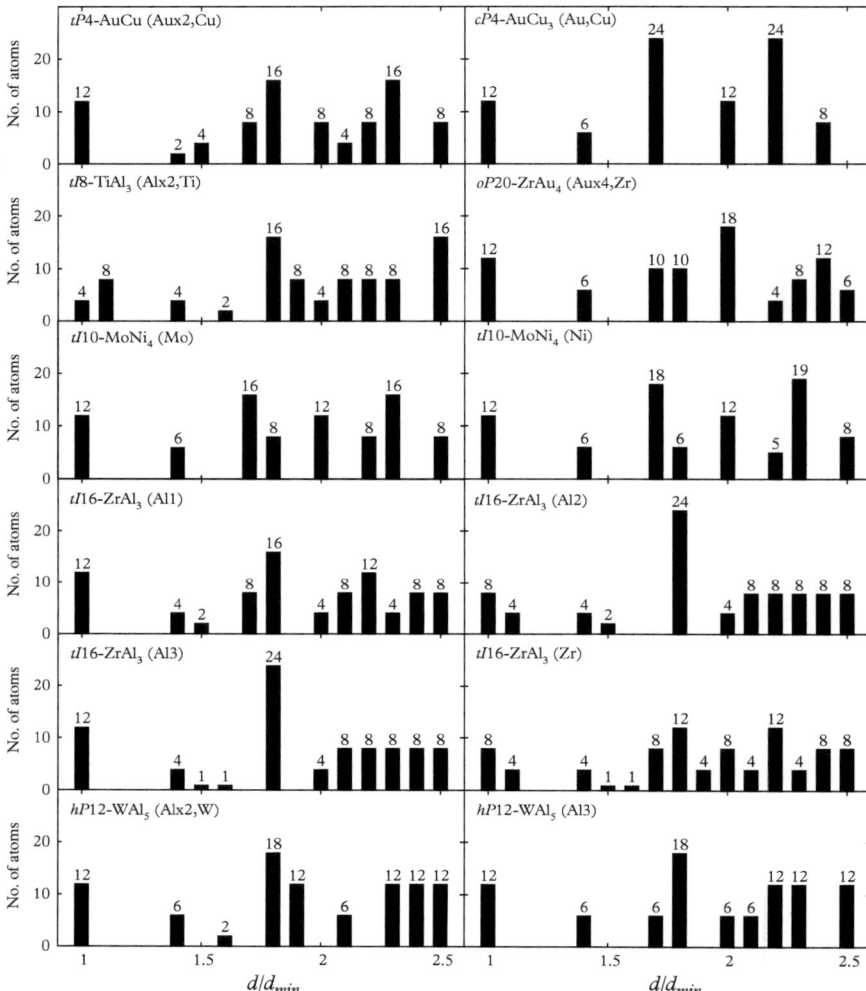

Fig. 7.5 *Distance histograms of the substitutional derivative (super)structures of the cF4-Cu structure type shown in Figs. 7.3 and 7.4: tP4-CuAu, cP4-Cu₃Au, tI8-TiAl₃, oP20-ZrAu₄, tI10-MoNi₄, tI16-ZrAl₃, and hP12-WAl₅. In parentheses, the kind and number of atoms involved in the histogram calculations are given. The last three structure types exhibit varying coordinations for the different atomic sites, which are shown in separate plots (tI10-MoNi₄: Mo and Ni; tI16-ZrAl₃: Al1, Al2, Al3, and Zr; hP12-WAl₅: Al1, Al2, W, and Al3). The applied binning does not allow us to see individual histogram bars for differences < 0.1dₘᵢₙ.*

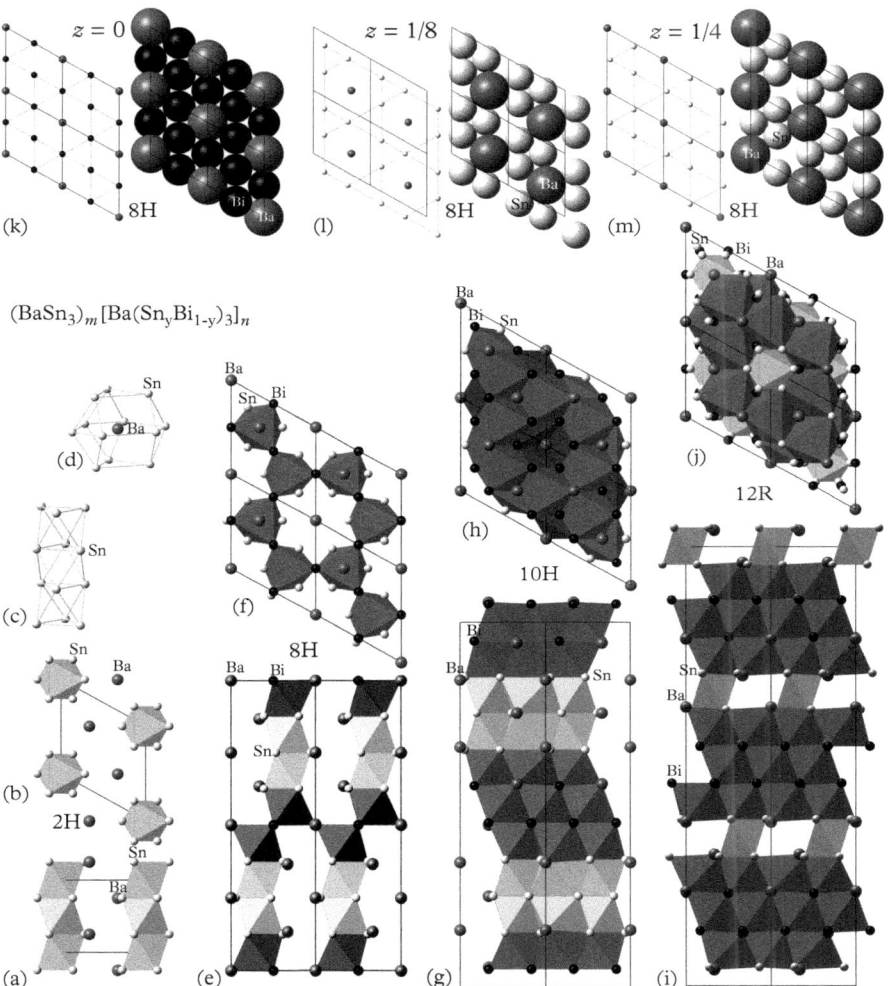

Fig. 7.6 *Structures of the compounds in the homologous series* $(BaSn_3)_m[Ba(Sn_yBi_{1-y})_3]_n$ *(Sn. . . light gray, Ba. . . dark gray, Bi. . . black; Sn-octahedra. . . light gray, Sn/Bi-octahedra. . . dark gray): (a)–(b)* $y = 0$, $m = 1$, $n = 0$, hh, 194 $P6_3/mmc$, 2H, and hP8, *projected along [010] and [001], respectively; the structure consists of columns of face-sharing Sn-octahedra along [001] (c), while the* Ba *atoms are coordinated disheptahedrally (d); (e)–(f)* $y = 0.43$, $m = 3$, $n = 1$, $(hhhc)_2$, 194 $P6_3/mmc$, 8H, hP32; (g)–(h) $y = 0.39$, $m = 3$, $n = 2$, $(hhcc)_2$, 194 $P6_3/mmc$, 10H, HP40; and (i)–(j) $y = 0.33$, $m = 2$, $n = 2$, $(hhcc)_3$, 166 $R\bar{3}m$, 12R, hR48 (Ponou et al., 2008). The layers of the structure shown in (e)–(f) are depicted in the ball-and–stick as well as the space-filling model in (k) for $z = 0$, in (l) for $z = 1/8$ and in (m) for $z = 1/4$.

The phase diagrams of the systems Al–Zr and Au–Zr are—with more than ten phases each—quite complex. $tI16$-ZrAl$_3$ and $oP20$-ZrAu$_4$ are the Al- and Au-richest (line) compounds with high melting temperatures indicating strong Al-Zr- and Au-Zr-interactions. The latter is also true for the Al-W-interactions in $hP12$-WAl$_5$. $hP8$-Ni$_3$Sn is the low-temperature ordering state for this stoichiometry while at high temperature $cF16$-Ni$_3$Sn ($cF16$-Li$_3$Bi type) is the stable phase. The latter can be described as a filled $cF4$-Cu structure type or as a $(2 \times 2 \times 2)$-fold $cI2$-W superstructure. $oI6$-MoPt$_2$ corresponds to a low-temperature ordering state below the quite extended (Mo, Pt) solid solution stability field.

$hP8$-BaSn$_3$ ($hP8$-Ni$_3$Sn structure type) is a *hcp* derivative structure that belongs to the class of Zintl phases, which will be discussed in greater detail in Section 7.5 (Fig. 7.6). By partial substitution of Sn by Bi in the quasibinary system BaSn$_{3-x}$Bi$_x$ ($0.4 \leq x \leq 1$), a series of polytypic superstructures is formed, $(BaSn_3)_m[Ba(Sn_yBi_{1-y})_3]_n$, with periods along [001] of up to 39 atomic layers (Ponou *et al.*, 2008) (Fig. 7.6). The driving force behind the superstructure formation is its partition into polar (BaSn$_3$) and non-polar (Ba(Sn$_y$Bi$_{1-y}$)$_3$) substructures.

The structures are characterized by columns of face-sharing octahedra with 12-coordinated Ba atoms in–between (Fig. 7.6(c)). If adjacent octahedra form an *h*-layer (Jagodzinski notation), i.e., the face-sharing octahedra are related by mirror symmetry, then the AET of Ba corresponds to a distorted disheptahedron (Fig. 7.6(d)). Consequently, the whole layer can be considered as a local mirror plane, just acting on the two neighboring layers. In the case of a *c*-layer, with the edge-sharing octahedra being related by inversion symmetry, the Ba AET is a distorted cuboctahedron. The inversion center is located in the middle of the edge shared by the inner square faces of the two octahedra. The edge shared by the two octahedra of a *c* link is also shared by two tetrahedra.

One description is by taking the Ba atoms as centers of disheptahedra or cuboctahedra, another one by decomposing these large AETs into six half-octahedra and six tetrahedra, i.e., to discuss the structure in terms of octahedra and tetrahedra only.

7.3   *cI2*-W based structures and their derivatives: Heusler and Hume-Rothery phases

A large number of crystal structures of intermetallics can be derived by substitution from the $cI2$-W structure type (229 $Im\bar{3}m$, W in $2a$ $0,0,0$) (Fig. 7.7(a)). Sticking to the same unit cell, just one derivative structure is possible: $cP2$-CsCl (221 $Pm\bar{3}m$, Cl in $1a$ $1/2, 1/2, 1/2$, Cs in $1b$ $0,0,0$) (Fig. 7.7(b)).

In the case of $(2 \times 2 \times 2)$-fold superstructures of the $cI2$-W type, four different stoichometries are possible—AB, AB$_3$, ABC$_2$, and ABCD: $cF16$-NaTl (227 $Fd\bar{3}m$, Tl in $8a$ $0,0,0$, Na in $8b$ $1/2, 1/2, 1/2$); $cF16$-BiF$_3$ (225 $Fm\bar{3}m$, Bi in $4a$ $0,0,0$, F in $4b$ $1/2, 1/2, 1/2$, F in $8c$ $1/4, 1/4, 1/4$); the Heusler phase

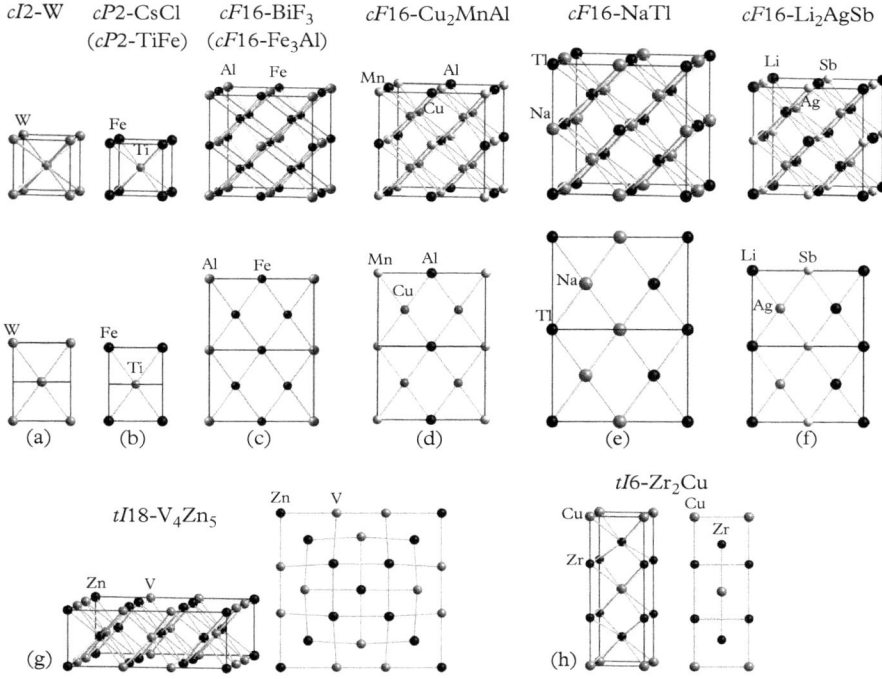

Fig. 7.7 *The cI2-W structure type together with several examples of its superstructures. In each case, the unit cell is depicted in perspective view as well as a (110) section through it. (a) cI2-W type; (b) cP2-TiFe (cP2-CsCl type); (c) cF16-Fe₃Al (cF16-BiF₃ type); (d) cF16-Cu₂MnAl type; (e) cF16-NaTl type; and (f) cF16-Li₂AgSb type. The tetragonal superstructures are shown with their unit cells and in projection along [010]: (g) tI18-V₄Zn₅ type; (h) tI6-Zr₂Cu.*

$cF16$-Cu$_2$MnAl (225 $Fm\bar{3}m$, Al in $4a$ $0,0,0$, Mn in $4b$ $1/2,1/2,1/2$, Cu in $8c$ $1/4,1/4,1/4$); $cF16$-Li$_2$CuSb (216 $F\bar{4}3m$, Li in $4a$ $0,0,0$ and in $4d$ $3/4,3/4,3/4$; Sb in $4b$ $1/2,1/2,1/2$; Ag in $4c$ $1/4,1/4,1/4$) (Fig. 7.7(c)–(f)); $cF16$-LiPdMgSn (216 $F\bar{4}3m$, Sn in $4a$ $0,0,0$ and Mg in $4b$ $1/2,1/2,1/2$; Cu in; Pd in $4c$ $1/4,1/4,1/4$; Li in $4d$ $3/4,3/4,3/4$) (Fig. 7.7).

The most common representative of $(3 \times 3 \times 3)$-fold superstructures is bcc γ-brass, $cI52$-Cu$_5$Zn$_8$ (217 $I\bar{4}3m$, Cu in $8c$ $0.828,0.828,0.828$; Zn in $8c$ $0.11,0.11,0.11$; Cu in $12e$ $0.355,0,0$; and Zn in $24g$ $0.313,0.313,0.036$) (Fig. 7.8).

Superstructures with other unit-cell sizes and symmetries are known as well. Two examples are depicted in Fig. 7.7(g)–(h): the $(3 \times 3 \times 1)$-fold superstructure $tI18$-V$_4$Zn$_5$ (139 $I4/mmm$, Zn in $2a$ $0,0,0$, and $8h$ $0.328,0.328,0$; V in $8i$ $0.348,0,0$), and the $(1 \times 1 \times 3)$-fold superstructure $tI6$-Zr$_2$Cu, 139 $I4/mmm$, Cu in $2a$ $0,0,0$; Zr in $4e$ $0,0,0.319$).

$cI52$-Cu_5Zn_8 (γ-brass)

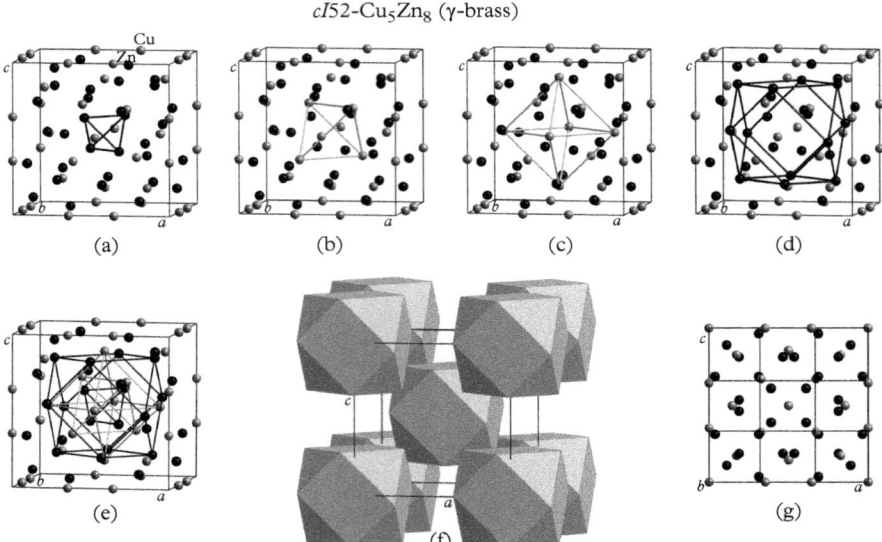

(a)   (b)   (c)   (d)

(e)

(f)

(g)

Fig. 7.8 *The I-cell γ-brass structure, $cI52$-Cu_5Zn_8, in different views (Cu. . . gray, Zn. . . black spheres). The structure is usually described by a bcc arrangement of 26-atom structural units (e, f), each one consisting of (a) an inner Zn tetrahedron (IT), surrounded first by (b) an outer tetrahedron (OT), then by (c) a Cu octahedron (OH) and, finally, (d)–(e) a Zn cuboctahedron (CO). All those polyhedra are distorted. The projection along [001] is shown in (g), with the (3 × 3 × 3) unit cells of the underlying $cI2$-W structure marked. The black spheres always correspond to the heavier atoms.*

In the case of $tI18$-V_4Zn_5, one can see that the central unit cell, decorated just by Zn, is larger than the surrounding ones. The melting temperature of this line compound is much higher than that of pure Zn, indicating strong attractive interactions between V and Zn, but is also much lower than that of elemental V. The Zn–Zn distances in this part with 2.701 Å are slightly shorter than the sum of their radii, those between V and V with 2.503 Å significantly shorter.

$tI6$-Cr_2Al has a broad compositional stability range and corresponds to an ordering state of the high-temperature solid solution (Al,Cr). The Cr–Cr distances with 2.439 Å are relatively short, while the shortest Al–Al distances are quite large (3.006 Å). The Al–Cr distances, 2.641 Å, are somewhere in-between.

7.3.1   Hume-Rothery phases and the system Cu–Zn

A significant number of intermetallic phases is assumed to be mainly electronically stabilized (see Table 7.1). What does this mean? In contrast to covalently bonded compounds or ionic crystals, where local interactions lead to stable structures, the electronic interaction typical for Hume-Rothery phases is of non-local

Table 7.1 Binary systems with Hume-Rothery phases (based on Cahn and Haasen (1996)). The number of valence electrons of an element is determined by the group number in the periodic table of elements; for the transition metals it was set to zero here.

Phases with cubic symmetry				Phases with hexagonal symmetry	
Disordered bcc structure β $1.36 \leq e/a \leq 1.59$ cI2-W	γ-brass structure γ $1.54 \leq e/a \leq 1.70$ cI52-Cu$_5$Zn$_8$	β-Mn structure μ $1.40 \leq e/a \leq 1.54$ cP20-Mn		c/a = 1.633 ζ $1.22 \leq e/a \leq 1.83$ hP2-Mg	c/a = 1.57 ε $1.65 \leq e/a \leq 1.89$ hP2-Mg
Cu-Be Au-Al	Cu-Zn	Mn-Zn	Cu-Si	Cu-Ga	Cu-Zn
Cu-Zn	Cu-Cd	Mn-In	Ag-Al	Cu-Si	Ag-Zn
Cu-Al	Cu-Hg	Fe-Zn	Au-Al	Cu-Ge	Ag-Cd
Cu-Ga	Cu-Al	Co-Zn	Co-Zn	Cu-As	Au-Sn
Cu-In	Cu-Ga	Ni-Zn		Cu-Sb	Au-Cd
Cu-Si	Cu-In	Ni-Cd		Ag-Cd	Li-Zn
Mn-Zn	Cu-Sn	Ni-In		Ag-Al	
	Ag-Li	Pd-Zn		Ag-Ga	
	Ag-Zn	Pt-Zn		Ag-In	
	Ag-Cd	Pt-Cd		Ag-Sn	
	Ag-Hg			Ag-As	
	Ag-In			Ag-Sb	
	Au-Zn			Au-Cd	
	Au-Cd			Au-Hg	
	Au-Ga			Au-In	
	Au-In			Au-Sn	
				Mn-Zn	

origin. The underlying mechanism is usually illustrated by the picture of Fermi-surface/Brillouin zone (FS/BZ) nesting leading to electron depletion (pseudo-gap) at the Fermi energy, lowering the energy of the occupied electron states in this way. In the case of a favorable valence-electron concentration (VEC), i.e., number of valence electrons per atom, e/a, the crystal structure modifies and adjusts itself in order to bring the almost spherical Fermi surface with radius $|\mathbf{k}|_F$ in close contact with the respective important Brillouin-zone planes, with diffraction vector $\mathbf{H}$. A BZ plane is important if it is strongly scattering, i.e., if the respective diffracted intensity $I(\mathbf{H})$ is strong. The interference of electron waves will lead to standing waves if the condition $2\mathbf{k}_F = \mathbf{H}$ is fulfilled. This is the case for electron waves with $\mathbf{k}_F = 1/\lambda$, and the wavelength $\lambda = 2d$, which are diffracted on atomic layers (Bragg planes) with distance $d = 1/|\mathbf{H}|$. FS/BZ nesting is especially efficient if the BZ is as spherical as possible, i.e., in high-symmetry structures. The closest to spherical symmetry of a BZ can be obtained in icosahedral quasicrystals, which are Hume-Rothery phases indeed.

In the following, we illustrate the role of the VEC on the binary system Cu–Zn with several electronically stabilized structures. The VEC increases with increasing Zn-content. The solid solution of *hcp* Zn in *ccp* Cu, called α-phase, has a very wide compositional stability field (up to 38.3 at.% Zn) with e/a varying in the range $1 \leq e/a \leq 1.4$. Around $e/a \approx 3/2$, the *bcc* β-phase follows, with an LT and an HT modification, where the LT structure (*cP2*-CsCl type) corresponds to an ordered variant of the substitutionally randomly disordered HT structure (*cI2*-W type).

The γ-phase (*cI52*-Cu$_5$Zn$_8$ type) appears in a broad range around $e/a \approx 21/13 \approx 1.62$ (21 valence electrons for 13 atoms). As mentioned in Section 7.3, its structure can be described as a $(3 \times 3 \times 3)$-fold superstructure of the *cI2*-W structure type, with the corner atoms as well as the central atom removed, and the other atoms relaxed (Fig. 7.8). Apart from this structure type with symmetry $I\bar{4}3m$ (217), called I-cell γ-brass, related structure types also exist. If the I-symmetry is broken by small differences in the structural units A and B, centered at the corners and the center of the unit cell, respectively, then the P-cell γ-brasses are obtained: *cP52*-Cu$_9$Al$_4$ type (215 $P\bar{4}3m$). This structure can be seen as cluster-decorated *cP2*-CsCl type. F-cell γ-brasses also exist (*cF416*-Cu$_{41}$Sn$_{11}$ type (216 $F\bar{4}3m$)), which can be described as a $(2 \times 2 \times 2)$-fold superstructures of the P-cell type, now with four slightly different 26-atom clusters A (in $4a$ $0, 0, 0$), B (in $4c$ $1/4, 1/4, 1/4$), C (in $4b$ $1/2, 1/2, 1/2$), and D (in $4d$ $3/4, 3/4, 3/4$). By this kind of ordering, energetically unfavorable close Sn-Sn contacts can be avoided, which would not be possible in P- or I-cells at this composition (Booth *et al.*, 1977). Finally, there are also pseudo-cubic R-cell brasses known with the *hR78*-Cr$_8$Al$_5$ structure type ($R3m$), featuring three symmetrically equivalent versions of the same 26-atom cluster per unit cell.

In the broad stability field of γ-brass (up to 57–68 at.% Zn), the Cu and Zn atoms show a specific site preference as a function of stoichiometry (Gourdon *et al.*, 2007). In all cases, the inner Cu$_4$Zn$_4$ tetrahedra star (i.e., the union of the inner and outer tetrahedra, IT+OT) remains unchanged (this includes the two sites sitting in an icosahedral environment), while the outer octahedron and cuboctahedron vary in their composition.

The ε-phase ($hP2$-Mg type) is observed around $e/a \approx 7/4$, and finally, the η-phase ($hP2$-Mg type) corresponds to a solid solution with a small compositional stability range (up to 2.8 at.%). The c/a-ratio of pure Zn equals 1.856, far from the ideal ratio 1.633 of a hcp sphere packing due to covalent bonding contributions within the hcp layers. At the boundary of the solid solution, with 2.8 at.% Cu in Zn, the c/a ratio decreases to 1.805. For the ε-phase, this ratio amounts to only 1.568, now much smaller than the ideal ratio 1.633.

7.3.2 Heusler phases

There are approximately 500 Heusler phases known so far, which are ternary or quaternary ($2 \times 2 \times 2$)-fold superstructures of the $cI2$-W structure type. One distinguishes between full and half-Heusler phases, which belong to the structure types $cF16$-Cu$_2$MnAl (225 $Fm\bar{3}m$, Al in $4a$ $0,0,0$, Mn in $4b$ $1/2,1/2,1/2$, Cu in $8c$ $1/4,1/4,1/4$), and $cF12$-LiAlSi (216 $F\bar{4}3m$, Al in $4a$ $0,0,0$, Li in $4b$ $1/2,1/2,1/2$, Si in $4c$ $1/4,1/4,1/4$), respectively. $cF16$-Cu$_2$MnAl can be considered to be a superstructure of the $cF16$-BiF$_3$ type (Fig. 7.7 (c)–(d)). Removing four of the eight Cu atoms in the eighth-cubes leads to the $cF12$-LiAlSi structure type (also called $cF12$-MgAgAs type) of the half-Heusler phases.

Heusler phases have the general composition A$_2$BC, with A and B transition elements. In some cases B can be a rare-earth or an alkali element, and C a main group element. Heusler phases are mostly semi-metallic. Their structure can also be described as four interpenetrating *fcc* sublattices, two of them occupied by the A atoms, and it has two magnetic sublattices. Half-Heusler compounds have composition ABC. With 8 or 18 valence electrons they are mostly semiconductors; with 27 valence electrons and if they are nonmagnetic, they can be superconducting. The structure can be described as consisting of two substructures, one of the $cF8$-ZnS type (mainly covalent bonding contributions), and one of the $cF8$-NaCl type (mainly ionic bonding contributions). They have only one magnetic sublattice, with the magnetic atom (RE or Mn) on the $cF8$-NaCl-like sublattice (Graf *et al.*, 2011).

Heusler and half-Heusler phases are still intensely studied because of the interesting (multifunctional) magnetic, thermoelectric, or superconducting properties of some of them. Their number of valence electrons allows us, to some extent, a prediction of the physical properties. Since most of their structures show great compositional flexibility, partial substitution of elements allows a fine tuning of the valence electron concentration and therewith of specific physical properties. For instance, the band gap of the semiconducting half-Heusler phases can be easily tuned in the range between 0 and 4 eV by varying the chemical composition. For a more general review see Graf *et al.* (2011), for instance, and for a review on the applications of half-Heusler phases see Casper *et al.* (2012).

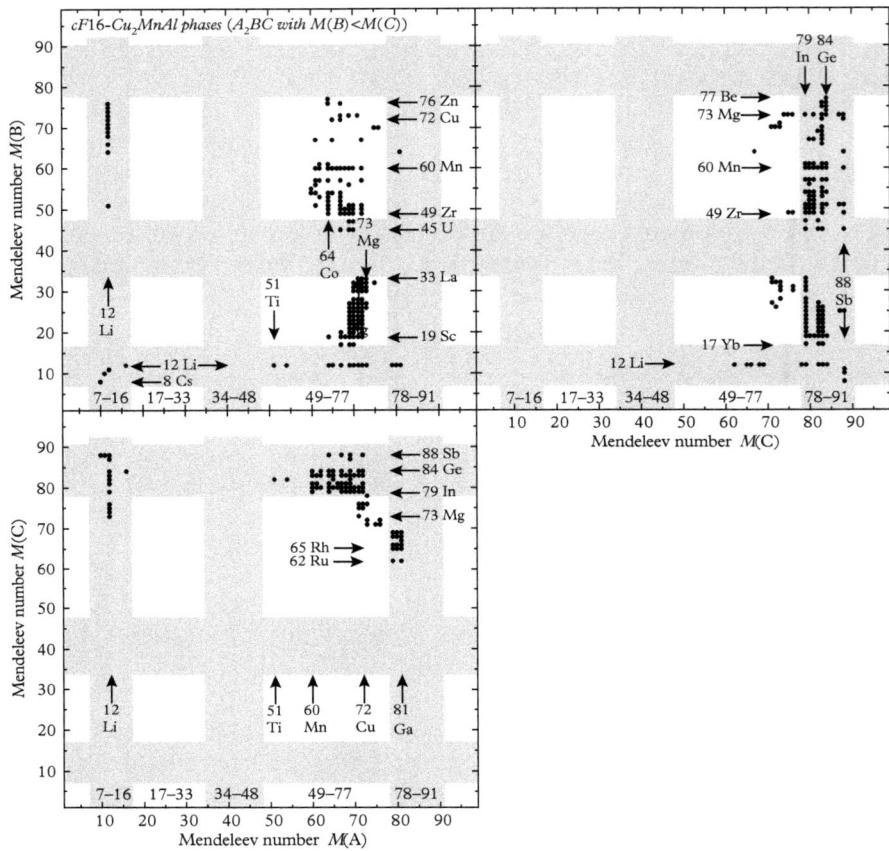

Fig. 7.9 *Occurrence of the Heusler phases, i.e, in structure type cF16-Cu₂MnAl. The constituting elements are given for the model formula* A₂BC *with* M(B) < M(C). *Only the 309 stoichiometric compounds are given.*

Among ternary intermetallics, 333 compounds are found to exhibit the structure type of the Heusler alloys $cF16$-Cu₂MnAl (Fig. 7.9) and 166 compounds that of half-Heusler alloys, $cF12$-MgAgAs (also called $cF12$-MgCuSb type).

Of the 333 Heusler phases with $cF16$-Cu₂MnAl-type structures, 309 are reported to be stoichiometric with 50% of element A with $M(A) = 10$–81 and 25% of elements B and C with $M(B) = 8$–77 and $M(C) = 62$–88 for $M(B) < M(C)$.

Of the 166 half-Heusler phases with $cF12$-MgAgAs-type structures, 161 are reported as stoichiometric with 33.3% of all three elements A, B, and C with $M(A) = 12$–72, $M(B) = 54$–80, and $M(C) = 80$–88 for $M(A) < M(B) < M(C)$ (Fig. 7.10).

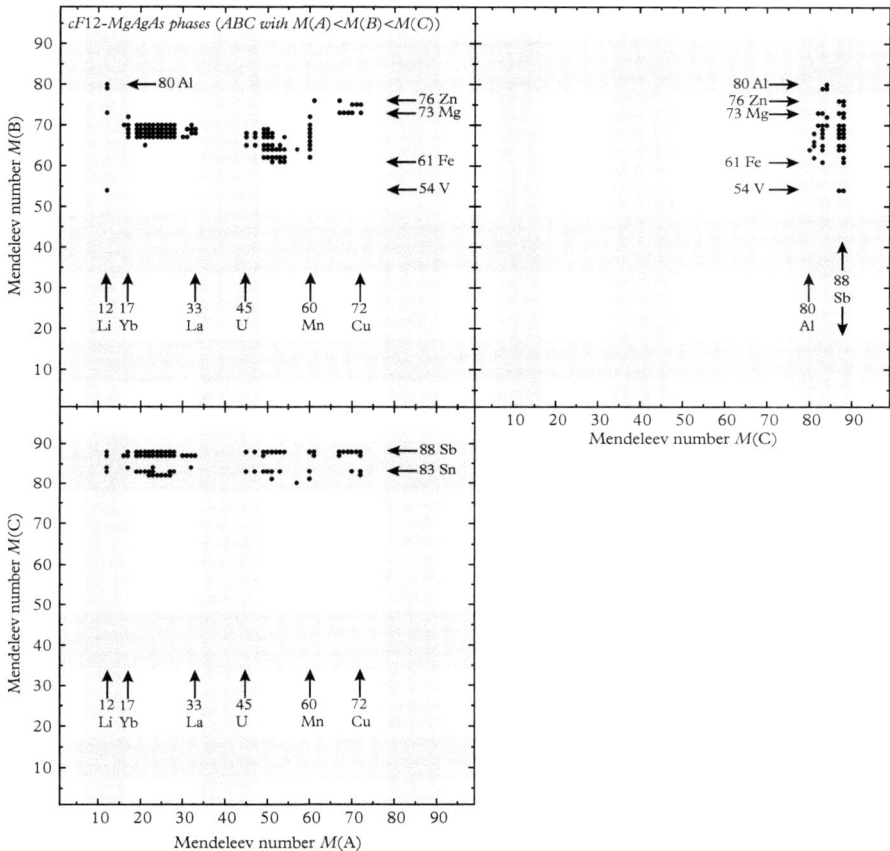

Fig. 7.10 *Occurrence of the half-Heusler phases, i.e., in structure type cF12-AgMgAs. The constituting elements are given for the model formula* ABC *with* $M(A) < M(B) < M(C)$. *Only the 161 stoichiometric compounds are given.*

7.4 Frank-Kasper phases: σ-, M-, P-, R-phases, and Laves phases

Frank-Kasper (FK)-phases are topologically close packed (*tcp*) intermetallic compounds with all of their atoms in 12-, 14-, 15-, or 16-coordination, respectively (Frank and Kasper, 1958; Frank and Kasper, 1959) (Fig. 7.11). *tcp* means that the structure has only tetrahedral interstices, i.e., that the structure can be described purely as a packing (3D tiling) of distorted tetrahedra. Consequently, the respective AETs, the four basic FK-polyhedra CN12, CN14, CN15, and CN16, all have triangular faces only, and 5- and 6-connected vertices such that no 6-connected vertices are linked by a common edge. Their duals are the pentagon dodecahedron and the only fullerenes with isolated hexagons, respectively (see Fig. 3.17 in Subsection 3.3.2).

7.4.1   Frank-Kasper polyhedra and their packings

This subsection is devoted to a more detailed discussion of the Frank-Kasper (FK) polyhedra, since they belong to the most important structural subunits in intermetallic phases. Their usual representation is shown in Fig. 7.11(a), (d), (g), and (j). However, since in most structures these coordination polyhedra (AETs) are partially overlapping, a different visualization may give a clearer picture. If the

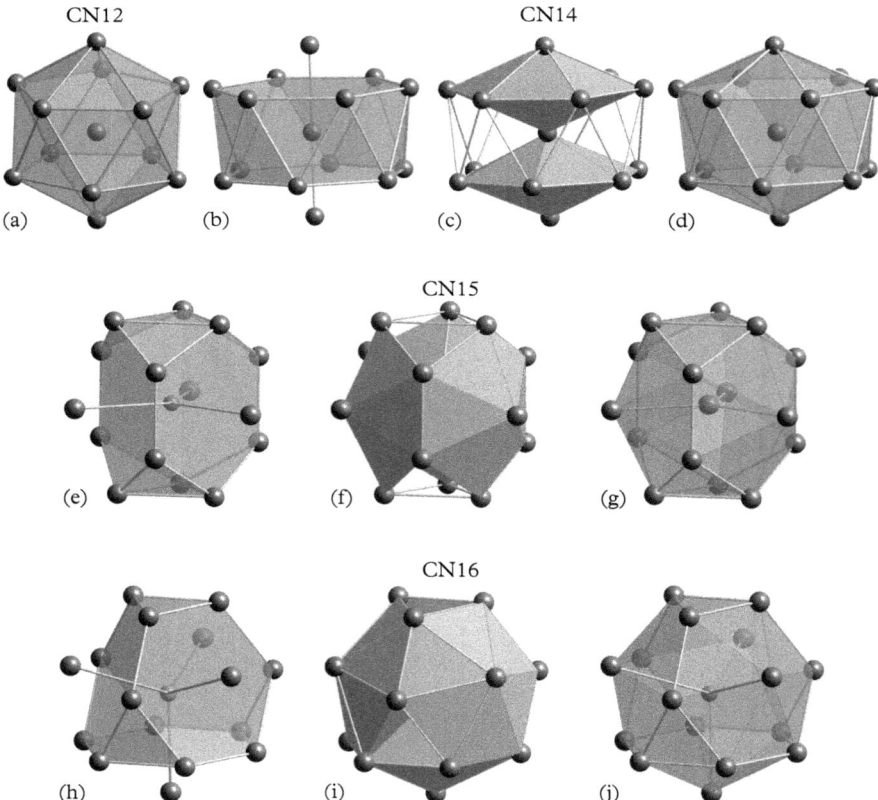

Fig. 7.11 *The Frank-Kasper (FK) polyhedra in different views: (a) CN12, with icosahedral shape; it can also be seen as pentagon-face-capped pentagonal antiprism; (b)–(d) CN14, a hexagon-face-capped hexagonal antiprism (b), which can also be visualized as two apex-sharing hexagonal bipyramids (hbps) (c);(e)–(g) CN15, the μ-phase polyhedron, a hexagon-face-capped vertex-truncated trigonal prism (e), which can also be seen as a union of three hbps (f); (h)–(j) CN16, also called Friauf polyhedron, a hexagon-face-capped vertex-truncated tetrahedron (h), which can also be seen a union of four hbps (i).*

hexagon-capping atom of one FK-polyhedron corresponds to the central atom of a neighboring one, then the unit tile representation as shown in Fig. 7.11(b), (e), and (h) may be more useful. Then the AETs do not overlap anymore, they are just sharing hexagon faces, constituting a 3D tiling. This representation also better shows the *major skeleton* of the structure, which connects all six-coordinated vertices via the *major ligand lines*. The visualizations in Fig. 7.11(c), (f), and (i) emphasize other subunits, apex-sharing hexagonal bipyramids (*hbps*). This representation as subunits, which are no AETs, however, because they are not centered, can give a clearer illustration of the building principles of complex intermetallics, as demonstrated later on.

Following Frank and Kasper (1958), we assume that the six-rings, decorated with atoms of type B, are plane regular hexagons with edge length 2, and define a_1 as the distance from the center of the FK polyhedron, decorated with an atom of type A, to the center of the hexagon; $(a_2 + 1)$ should be the distance from the center of the FK polyhedron to the vertices of the hexagon, and $2b$ the distance between neighboring vertices of different hexagons of a FK polyhedron. In a similar way, we take the distance between the vertices of the icosahedron to be 2, and define $(a_1 + 1)$ as their distance from the icosahedron center. The deviation of all these a- and b-values from unity is a measure of the degree of incompatibility of these arrangements with contact packing of hard spheres with radius:

- CN12, icosahedron built from 20 distorted tetrahedra, FK_{20}^{12}; dual polyhedron: pentagon dodecahedron F_{12}^{20}, with point group symmetry $m\bar{3}5$; $a_1 = [1/2(5 + \sqrt{5})]^{1/2} - 1 = 0.902$.

- CN14, hexagon-capped hexagonal antiprism built from 24 distorted tetrahedra, FK_{24}^{14}; dual polyhedron: fullerene F_{14}^{24}, with point group symmetry $\overline{12}2m$; $a_2 = (a_1^2 + 4)^{1/2} - 1$, $b = (a_1^2 + 2 - \sqrt{3})^{1/2}$. The central atom A can be greater or smaller than B, but $r_A = a_1$ is always smaller than a_2 and b. Consequently, for the central atom there is less space along the $\overline{12}$-axis than in the other directions. For $b = r_B = 1$, the distortion is rather large; it can be reduced by increasing b significantly. Some corresponding values are listed in Table 7.2.

- CN15, hexagon-capped vertex-truncated trigonal prism built from 26 distorted tetrahedra, FK_{26}^{15}; dual polyhedron: fullerene F_{15}^{26}, $\bar{6}2m$; $a_1 = 1$, $a_2 = \sqrt{5} - 1 = 1.236$, $b = 1/2\sqrt{3} = 0.866$. Consequently, b defines the size of the B atoms, yielding $r_B = b = 1/2\sqrt{3} = 0.866$, $r_A = a_1 = 1$, and the size ratio $r_A = 2/\sqrt{3} = 1.155r_B$. The distance between A and B atoms is with $d_{A\text{-}B} = \sqrt{5} = 2.236$ larger than $r_A + r_B = (2 + \sqrt{3})/2 = 1.866$.

- CN16, hexagon-capped vertex-truncated tetrahedron built from 28 distorted tetrahedra, FK_{28}^{16}; dual polyhedron: fullerene F_{16}^{28}, $\bar{4}3m$; $a_1 = \sqrt{3/2} = 1.225$, $a_2 = \sqrt{11/2} - 1 = 1.345$, b does not exist. Consequently, the size of the B atoms amounts to $r_B = 1$, $r_A = a_1 = \sqrt{3/2} = 1.225$, and the size ratio $r_A = \sqrt{3/2} = 1.225r_B$. The distance between A and B atoms is with $d_{A\text{-}B} = \sqrt{11/2} = 2.345$ larger than $r_A + r_B = \sqrt{3/2} + 1 = 2.225$.

Table 7.2 *The geometry of the CN14 FK-polyhedron: atomic radii $r_A = a_1$ and $r_B = b$ of the central-atom A and the coordinating atoms B, respectively, if all atoms should be in contact with one another.*

a_1	=	0.8	0.856	0.9	1.0	1.1	1.2
a_2	=	1.15	1.18	1.19	1.24	1.28	1.33
b	=	0.953	1	1.04	1.13	1.22	1.31

Table 7.3 *Examples of FK-phases with the relative frequencies of the structure constituting FK-polyhedra (based on Shoemaker and Shoemaker (1969)).*

Structure type	Name	Relative frequency of			
		CN12	CN14	CN15	CN16
$cP8$-Cr_3Si		0.25	0.75	0	0
$hP7$-Zr_4Al_3		0.44	0.28	0.28	0
$cF24$-$MgCu_2$	Laves phase	0.67	0	0	0.33
$hP12$-$MgZn_2$	Laves phase	0.67	0	0	0.33
$hP24$-$MgNi_2$	Laves phase	0.67	0	0.33	0
$tP30$-$Cr_{46}Fe_{54}$	σ-phase	0.33	0.54	0.13	0
$hR13$-W_7Fe_6	μ-phase	0.55	0.15	0.15	0. 15
$oP52$-$Nb_{10}Ni_9Al_3$	M phase	0.55	0.15	0.15	0.15
$oP56$-$Mo_{21}Cr_9Ni_{20}$	P phase	0.43	0.36	0.14	0.7
$hRP53$-$Mo_3Cr_2Co_5$	R phase	0.51	0.23	0.11	0.15
$cI162$-$Mg_{11}Al_6Zn_{11}$	Bergman phase	0.61	0.7	0.7	0.25

Examples of FK-phases with the relative frequencies of the their structure-constituting FK-polyhedra are listed in Table 7.3.

Structures that are exclusively composed of FK-polyhedra can be decomposed into flat atomic layers. The *primary layers* consist of triangles and pentagons (Laves phases, μ-phase, and M-phase), or of triangles and hexagons (σ-phase, $hP7$-Zr_4Al_3), or of triangles, pentagons, and hexagons (P phase). They are sandwiched by *secondary layers* of triangles and/or squares so that solely tetrahedral voids are formed. This constraint allows only capping of pentagons and hexagons of the primary layer by the vertices of the secondary layer, and not of triangles. Consequently, secondary layers are incomplete dual layers to the primary layers. While the primary layers are usually located on mirror planes, the secondary layers lie in-between. Pentagons/hexagons in subsequent primary layers form antiprisms centered by a vertex of the secondary layer in between.

Another property of structures where each atom is part of at least one FK-polyhedron is the existence of a *major skeleton*, the nodes of which, the *major sites*, are connected by *major ligands*. Any site in such a structure that is 12-coordinated is called *minor*, a $(12 + n)$-coordinated one, with $n \geq 1$, *major*. At a major site of coordination number Z, $(Z - 12)$ major ligands meet. Thus the central atoms of CN14, CN15, and CN16 FK-polyhedra are the meeting points of 2, 3, or 4 major ligands, which are (accurately or approximately) in line, 120° apart in a plane, or pointing to the vertices of a tetrahedron, respectively. The major skeleton uniquely defines an FK-structure.

The truncated tetrahedron has the vertex configuration 3.6^2 and corresponds to one of the 13 Archimedean solids (Fig. 3.14 in Subsection 3.3.1). As given in Table 3.6 and illustrated in Fig. 3.15 (both in Subsection 3.3.1), three different packings of uniform polyhedra with cubic symmetry are possible, where truncated tetrahedra are part of: (i) truncated tetrahedra plus tetrahedra (227 $Fd\bar{3}m$: $3.6^2 + 3^3$), (ii) truncated tetrahedra plus truncated octahedra plus cuboctahedra (225 $Fm\bar{3}m$: $3.6^2 + 4.6^2 + 3.4.3.4$), and (iii) truncated tetrahedra plus truncated cuboctahedra plus truncated cubes (225 $Fm\bar{3}m$: $3.6^2 + 4.6.8 + 3.8^2$).

There are also many ways FK-polyhedra can be packed in combination with other structural units. Examples are:

- $cF24$-MgCu$_2$ and $hP12$-MgZn$_2$ can be seen as cubic (sequence ABC) and hexagonal (sequence AB) packings, respectively, of flat layers of Friauf polyhedra with empty tetrahedral voids (Fig. 7.15).

- $hR39$-W$_6$Fe$_7$ (μ-phase): flat layers of densely packed W$_4$Fe$_{12}$ Friauf polyhedra with a sequence ABC (like the cubic Laves phase $cF24$-MgCu$_2$) with intercalated W-*hbp*s sharing atoms with the neighboring layers of Friauf polyhedra (Fig. 7.12). In a different view, the structure could be seen as a packing with sequence AaBbCc, where upper (lower) case letters symbolize properly shifted layers of CN16 (CN15) FK-polyhedra.

- Units of 1, 4, 34, and 146 Friauf polyhedra are found in the structures of $cF(5928 - x)$-Al$_{56.6}$Cu$_{3.9}$Ta$_{39.5}$ ($x = 20$, ACT-45) and $cF(23\,256 - x)$-Al$_{55.4}$Cu$_{5.4}$Ta$_{39.1}$ ($x = 122$, ACT-71) (see Subsection 7.4.4).

- Curved superclusters can be formed if the Friauf polyhedra are properly attached to each other. They can be arranged in a way that they form frameworks leaving pores, the inside of which correspond to fullerenes (see also Section 8.3).

- Spherical superclusters: The 104 atom Samson cluster is an example for the smallest spherical cluster that consists of 20 Friauf polyhedra. The outer shell corresponds to a 60-atom fullerene, the middle shell to a 32 atom triacontahedron, and the inner shell to a 12 atom icosahedron.

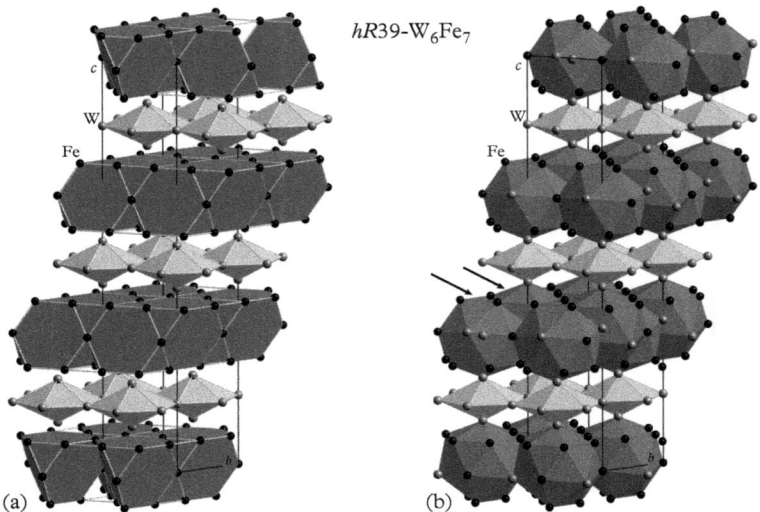

Fig. 7.12 *μ-phase hR39-W₆Fe₇ in two different descriptions: (a) W-centered Fe₁₂ truncated tetrahedra, W@Fe₁₂, alternating with layers of edge-sharing hexagonal W₈ bipyramids. (b) Layers of W-centered W₄Fe₁₂ Friauf polyhedra (CN16), W@W₄Fe₁₂, intercalated by W₈ hbps. The Friauf polyhedra are interpenetrating (one example marked by arrows) since each of the four W atoms at the vertices of a Friauf polyhedron sits at the same time at the center of another Friauf polyhedron.*

7.4.2   σ-, M-, P-, and R-phases

The stability of the σ-phase, $tP30\text{-}Cr_{46}Fe_{54}$ (136 $P4_2/mnm$), is assumed to be due to its favorable valence electron concentration, which is in the range $6.2 \leq e/a \leq 7$. Its structure can be described in different ways: (i) as a packing of hexagonal bipyramids (*hbp*s), which are vertex-connected along [001] (Fig. 7.13 (a)); in the (110)-layers, groups of four edge-connected *hbp*s are vertex-connected with each other, which is reflected in the section at $z = 0$ (Fig. 7.13 (e)); (ii) as a packing of partially interpenetrating CN15 FK-polyhedra (Fig. 7.13 (b)); (iii) as a packing of partially interpenetrating icosahedra (CN12 FK-polyhedra). So, the structure contains all but CN16 FK-polyhedra, because the CN14 FK-polyhedra can also be considered to contain pairs of *hbp*s (Fig. 7.13 (c)–(d)). This setting takes into account that the distances between atoms within each *hbp* are smaller than those between atoms of different *hbp*s along [001]. In the example shown ($tP30\text{-}Ta_{60}Al_{40}$), these distances amount to 2.634 Å between the apical atoms, to 2.508–2.873 Å between the equatorial atoms, and 2.849–3.241 Å between equatorial atoms between two different *hbp*s along [001]. The primary layers in $z = 0, 1/2$ are two-uniform hexagon/triangle tilings of the type ($6^2.3^2$; 6.3.6.3) (Fig. 7.13 (e)), while the secondary layers in $z = 1/4, 3/4$ are two-uniform hexagon/triangle tilings of the type $3^2.4.3.4$ (Fig. 7.13 (f)).

$tP30\text{-}Al_{40}Ta_{60}$ (σ–phase structure type)

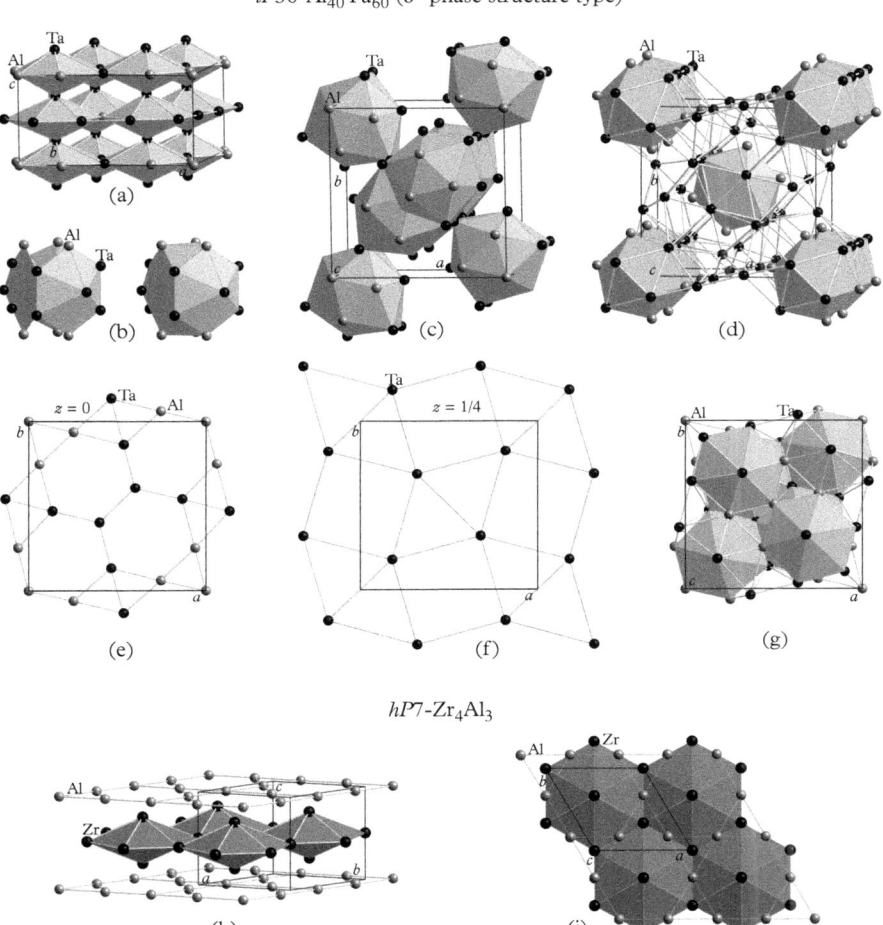

Fig. 7.13 *Structurally related FK phases in different descriptions: (a)–(g) The structure type of the σ-phase, $tP30\text{-}Cr_{46}Fe_{54}$, on the example of $tP30\text{-}Ta_{60}Al_{40}$ (a) as a packing of hexagonal bipyramids (hbps), (c) of CN15 FK-polyhedra, and (d) of icosahedra. The relationship between hbps and the CN15 FK-polyhedron is shown in (b). The primary layer (z = 0) of the structure is shown in (e), the secondary $3^3.4.3.4$ layer (z = 1/4) in (f), and in (g) the projection of the unit cell depicted in (a). The structure of $hP7\text{-}Zr_4Al_3$ in shown in (h)–(i). In our description, there are alternatingly stacked layers of edge-sharing Zr_8-hbps and Al-Kagomé layers.*

The P-phase, $oP56\text{-}Mo_{21}Cr_9Ni_{20}$, (62 *Pnma*), can be seen as a packing of edge-connected chains of *hbps* with icosahedra in between (Fig. 7.14 (a)–(b)), or from a different view, as a packing of CN12 and CN14 FK-polyhedra, respectively. The M-phase, $oP52\text{-}Nb_{10}Ni_9Al_3$, (62 *Pnma*), contains with *hbps* building blocks of the σ-phase and with double Friauf-polyhedra modules of the Laves phase (Fig. 7.14

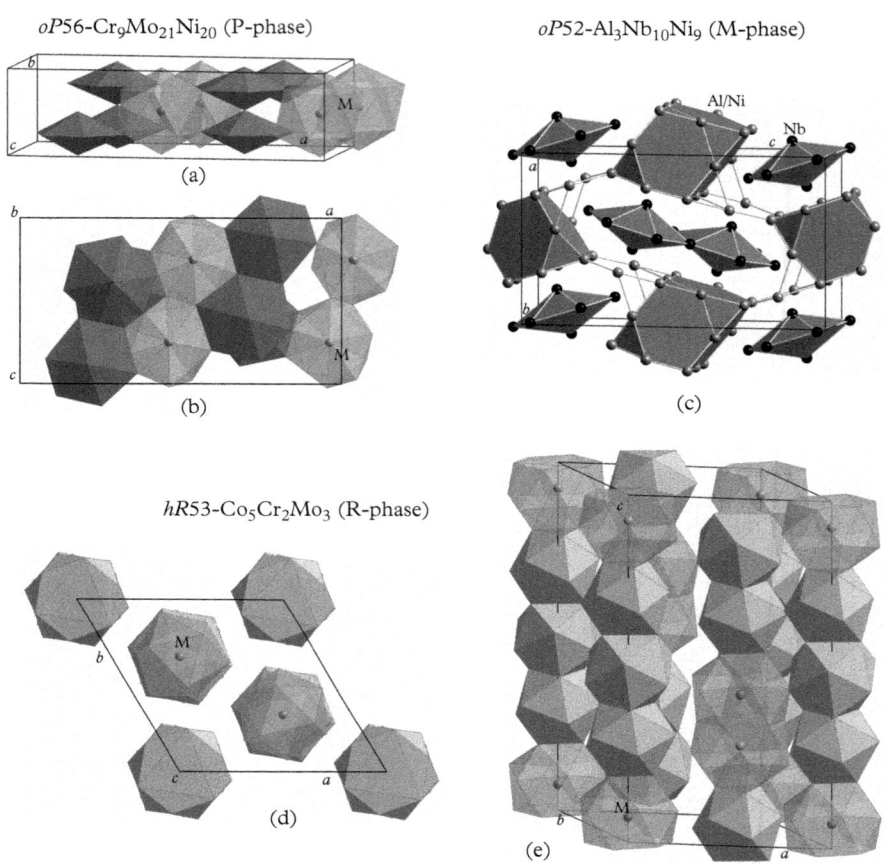

$oP56$-Cr$_9$Mo$_{21}$Ni$_{20}$ (P-phase)

$oP52$-Al$_3$Nb$_{10}$Ni$_9$ (M-phase)

(a)

(b)

(c)

$hR53$-Co$_5$Cr$_2$Mo$_3$ (R-phase)

(d)

(e)

Fig. 7.14 *Structurally related FK phases in different descriptions: the P-phase, $oP56$-Mo$_{21}$Cr$_9$Ni$_{20}$, is illustrated in (a)–(b), and the M-phase, $oP52$-Nb$_{10}$Ni$_9$Al$_3$, in (c). The R-phase, $hR159$-Mo$_3$Cr$_2$Co$_5$, in a projection along [001] (d) and in perspective view (e). The structure can be described as columnar stackings of CN16-(CN12)$_3$-CN16 FK-polyhedra, which are linked by tetrahedra.*

(c)). The R-phase, $hR159$-Mo$_3$Cr$_2$Co$_5$, (148 $R\bar{3}$), is built from close-packed chains of face-sharing FK-polyhedra with the sequence CN16-(CN12)$_3$-CN16 (Fig. 7.14 (d)–(e)). The R phase belongs to the FK-phases, which cannot be fully described by stackings of primary and secondary layers (Komura *et al.*, 1960).

7.4.3   Laves phases and related polytypes

Laves phases are binary or (pseudo)ternary intermetallic phases with more than 4000 entries and 1590 well-characterized structures in the PCD (Villars and Cenzual, 2011*a*). For a comprehensive review see Stein *et al.* (2004), Stein *et al.*

(2005). Their compositional stability fields range from very narrow (line compounds) to rather extended phase regions. The crucial structure-determining factors for these FK-phases are the atomic size ratios, the electronegativity differences, and the valence electron concentrations.

Laves phases can be described as stackings of layers of hexagonal-close-packed truncated tetrahedra (Fig. 7.15). Thereby, each truncated tetrahedron with one of its hexagon faces pointing downwards is surrounded by six others with opposite orientation. Neighboring truncated tetrahedra in adjacent layers can share these hexagon faces, only.

Depending on the kind of layer stacking, a similar number of polytypic structure types can result as is the case for close sphere packings. Consequently, we can use a similar notation for their classification, referring to the position of the centering atom of the truncated tetrahedra. Then, for the $cF24$-$MgCu_2$ structure type, the stacking sequence would correspond to ABC, and for $hP12$-$MgZn_2$ to AB. However, this notation does not take into account the relative rotation of the truncated tetrahedra, a problem that obviously does not exist in sphere packings: in addition to both pointing upwards with a triangular face and being displaced with respect to each other in an AB-manner, the layers are rotated against one another by 180°. This can be expressed by using a primed symbol like AB' for $hP12$-$MgZn_2$. Alternatively, the local symmetry relationship can be used to denote the stacking sequence, i.e., $\bar{1}\bar{1}\bar{1}$ for $cF24$-$MgCu_2$-type and mm for $hP12$-$MgZn_2$-type structures.

The second notation is easy to transfer to any number of stacking sequences, e.g., the slightly larger $hP24$-$MgNi_2$ structure type follows a $m\bar{1}m\bar{1}$ sequence. In the AB-notation, however, it is important to impose another rule to ensure consistency: the stacking runs "backward" within the primed planes. Then it corresponds to an AB'A'C sequence. It should be noted that the sequence of prime and non-prime layers have to be consistent. For reasons of clarity, the sequence of two m-related layers follows that of the first kind of layers, i.e., in alphabetical order in the case of AB', BC', and CA' and in reverse order in the case of A'C, B'A, and C'B. The much more complex $hR126$-$Mg(Ag_{0.1}Zn_{0.9})_2$ structure type can thus be described by either $m\bar{1}mm\bar{1}m\bar{1}m\bar{1}mm\bar{1}m\bar{1}m\bar{1}mm\bar{1}m\bar{1}$ or AB'A'CA'C'BCA'C'BC'A'CAB'A'CA'C'B. In the Jagodzinski notation, h corresponds to m, and c to $\bar{1}$. So, the layer sequence results to $(hchhchc)_3$.

Table 7.4 lists the above-discussed Laves phases, as well as a number of additional structures, which were identified to be Laves phase-type structures by inspection of their AETs in the PCD (Villars and Cenzual, 2011a). The compositions observed for binary intermetallic Laves phases in the most common structure types are illustrated by M/M-plots (Fig. 7.16).

In the following, the basic Laves phases and one large polytype are discussed in greater detail. The fundamental layers, consisting of hcp truncated tetrahedra (Friauf polyhedra) and tetrahedra, can be described as a Kagomé net (6.3.6.3) of the smaller B-atoms (Fig. 7.15(a)), followed by a triangle net (3^6) of the larger A atoms, which are centering half of the Friauf tetrahedra to be formed; then the

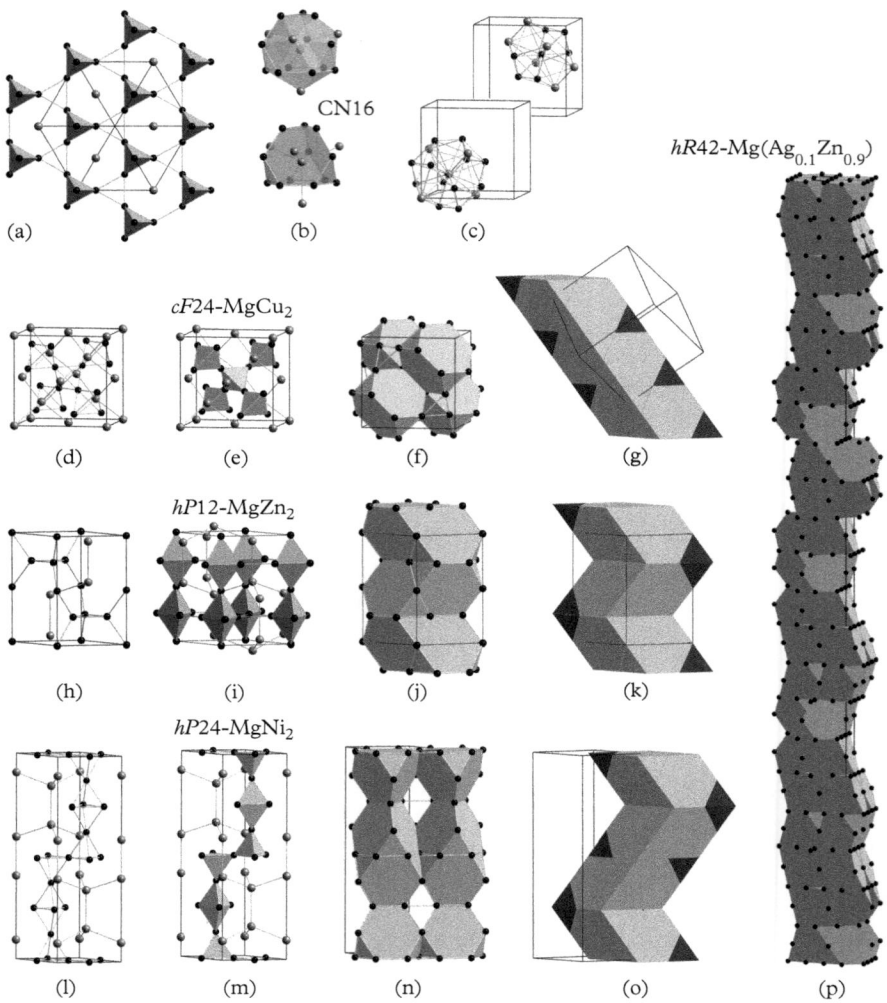

Fig. 7.15 *The structures of Laves phases in different representations: (a) The first three atomic layers are building one layer of Friauf tetrahedra: Kagomé net of TM atoms, followed by a triangle net of Mg atoms centering one half of the Friauf tetrahedra, and a triangle net of TM atoms forming small tetrahedra. (b) CN16 (Friauf) Frank-Kasper polyhedron of composition $TM_{12}Mg_4$ with truncated tetrahedron, TM_{12}, atoms shown below. (c) Unit cell of cF24-$MgCu_2$ with one CN16 and one CN12 polyhedron drawn in. (d)–(g) cF24-$MgCu_2$, (h)–(k) hP12-$MgZn_2$, and (l)–(o) hP24-$MgNi_2$ in different views. In (g), (k), and (o) the different stacking variants of double-Friauf polyhedra in these three Laves phases is shown schematically (Mg atoms omitted as in (f), (j), and (n)). (p) The packing of Friauf tetrahedra for the structure of hR42-$Mg(Ag_{0.1}Zn_{0.9})$. Mg. . . gray, TM. . . black. The scale is 3/4 of the usual scale.*

Table 7.4 *Basic Laves phases as well as their polytypes and derivative structures. The number of structure representatives refers to the PCD; the structural classification, where given, is either based on the stacking sequence, given in both the above discussed symbols and the Jagodzinski notation, or on superstructure parameters.*

Structure type	No. of representatives			Structural classification
	all	binary	ternary	
Basic Laves phases				
$cF24$-$MgCu_2$	806	223	523	ABC; $\bar{1}\bar{1}\bar{1}$; *ccc*
$hP12$-$MgZn_2$	456	154	265	AB'; *mm*; *hh*
$hP24$-$MgNi_2$	63	20	43	AB'A'C; $m\bar{1}m\bar{1}$; *hchc*
Superstructures of basic Laves phases				
$cF24$-$MgSnCu_4$	119	0	113	$(1 \times 1 \times 1)$-$MgCu_2$
$cF24$-Be_5Au	66	23	43	$(1 \times 1 \times 1)$-$MgCu_2$
$oF24$-$NdCo_2$	1	1	0	$(1 \times 1 \times 1)$-$MgCu_2$
$cP24$-$TmNi_2$	1	1	0	$(1 \times 1 \times 1)$-$MgCu_2$
$tP24$-$TmNi_2$	1	1	0	$(1 \times 1 \times 1)$-$MgCu_2$
$cF24$-$Li_{0.5}Ga_{0.5}Sn_{0.5}$	1	0	1	$(1 \times 1 \times 1)$-$MgCu_2$
$cF192$-$TmNi_2$	20	14	6	$(2 \times 2 \times 2)$-$MgCu_2$
$hP12$-$LuMn_5$	1	1	0	$(1 \times 1 \times 1)$-$MgZn_2$
$hP12$-Mg_2Cu_3Si	7	0	7	$(1 \times 1 \times 1)$-$MgZn_2$
$hP36$-$Nb_{6.4}Ir_4Al_{7.6}$	3	0	3	$(\sqrt{3} \times \sqrt{3} \times 1)$-$MgZn_2$
$hP24$-$Ca_2Mn_{0.32}Al_{3.68}$	1	0	1	$(1 \times 1 \times 1)$-$MgNi_2$
$hP96$-$Li_4Mg_8Zn_{12}$	1	0	1	$(2 \times 2 \times 1)$-$MgNi_2$
$oI12$-UMn_2	4	4	0	$MgCu_2$-derivative
$tI12$-YMn_2	3	3	0	$MgCu_2$-derivative
$mS12$-$Be_{2.3}FeAl_2$	1	0	1	$MgCu_2$-derivative
$hR18$-$TbFe_2$	8	7	1	$MgCu_2$-derivative
$hR18$-Mg_2Ni_3Si	5	0	5	$MgCu_2$-derivative
$oS24$-URe_2	5	2	3	$MgZn_2$-derivative
Laves phases with other stacking sequences				
$hP30$-$Mg(Cu_{0.5}Al_{0.5})_2$	1	0	1	AB'A'C'B'; $m\bar{1}\bar{1}\bar{1}m$; *hccch*
$hP36$-$Mg(Cu_{0.54}Al_{0.46})_2$	1	0	1	AB'A'CAB'; $m\bar{1}m\bar{1}mm$; *hchchh*
$hP36$-$Mg(Cu_{0.55}Ni_{0.45})_2$	1	0	1	ABC'B'A'C; $\bar{1}m\bar{1}\bar{1}m\bar{1}$; *chcchc*

continued

Table 7.4 *continued*

Structure type	No. of representatives			Structural classification
	all	binary	ternary	
$hP48$-Mg(Ag$_{0.03}$Zn$_{0.97}$)$_2$	3	0	3	AB'AB'A'CA'C; $mmm\bar{1}mmm\bar{1}$; $(hhhc)_2$
$hR54$-Mg(Cu$_{0.5}$Al$_{0.5}$)$_2$	4	0	4	AB'A'CA'C'BC'B';$m\bar{1}mm\bar{1}mm\bar{1}m$; $(hch)_3$
$hP60$-Mg(Ag$_{0.1}$Zn$_{0.9}$)$_2$	3	0	3	AB'A'CAB'ABC'B'; $m\bar{1}m\bar{1}mm\bar{1}m\bar{1}m$; $(hchch)_2$
$hP60$-Mg(Cu$_{0.5}$Al$_{0.5}$)$_2$	1	0	1	AB'A'C'B'ABCAB'; $m\bar{1}\bar{1}\bar{1}m\bar{1}\bar{1}\bar{1}mm$; $(hccc)_2hh$
$hP84$-LiMg$_7$Zn$_{13}$	1	0	1	AB'A'CA'CAB'ABC'BC'B'; $m\bar{1}mmm\bar{1}mm\bar{1}mmm\bar{1}m$; $(hchhch)_2$
$hP96$-Mg(Cu$_{0.54}$Al$_{0.46}$)$_2$	1	0	1	AB'ABC'B'AB'A'CA'C'BCA'C; $mm\bar{1}m\bar{1}mm\bar{1}mm\bar{1}m\bar{1}mm\bar{1}$; $(hhchchhc)_2$
$hR126$-Mg(Ag$_{0.1}$Zn$_{0.9}$)$_2$	1	0	1	AB'A'CA'C'ABC'B'CA'C'ABC'A'BC'A'B; $m\bar{1}mm\bar{1}m\bar{1}m\bar{1}mm\bar{1}m\bar{1}m\bar{1}mm\bar{1}m\bar{1}$; $(hchhchc)_3$
Total	1590	454	1033	*(plus 103 phases with 4+ components)*

A-atoms of a rotated triangle net cap all the triangles of the Kagomé net forming small tetrahedra; the subsequent A triangle net centers the other half of the Friauf tetrahedra; finally, the fundamental layer is finished by a rotated B-Kagomé net. The frequencies of the FK-polyhedra CN12 and CN16 are with 2/3 and 1/3, respectively, the same in the three types of basic Laves phases. Their prototype structures are characterized as follows:

$cF24$-**MgCu$_2$** 227 $Fd\bar{3}m$, Mg in 8a 0, 0, 0; Cu in 16d 5/8, 5/8, 5/8. The major skeleton is constituted from the larger Mg atoms occupying the sites of a diamond structure (lattice complex D), while the smaller Cu atoms constitute a 3D tetrahedral network (lattice complex T) around the Mg atoms, consisting of interpenetrating Kagomé nets connected via vertex-sharing tetrahedra (Fig. 7.15(a) and (d)–(g)). $cF24$-MgCu$_2$ can be described as a three-layer stacking variant ABC along the [111]-direction (Ramsdell: 3C, Jagodzinski: ccc). Ordered derivative structure types are: $cF24$-MgCu$_4$Sn and $cF24$-Be$_5$Au (216 $F\bar{4}3m$).

$hP12$-**MgZn$_2$** 194 $P6_3/mmc$, Mg in 4f 1/3, 2/3, 0.063; Zn in 2a 0, 0, 0, and 6h 0.830, 0.660, 1/4. The major skeleton is constituted from the larger Mg atoms occupying the sites of a lonsdaelite ("hexagonal diamond") structure, while the smaller Cu atoms constitute a 3D tetrahedral network around the Mg atoms, consisting of interpenetrating Kagomé nets connected via vertex-sharing double-tetrahedra (Fig. 7.15(h–k)). $hP12$-MgZn$_2$ can be described as two-layer stacking variant AB along the [001]-direction (Ramsdell: 2H, Jagodzinski: hh). Ordered derivative structure type: $hP12$-U$_2$OsAl$_3$.

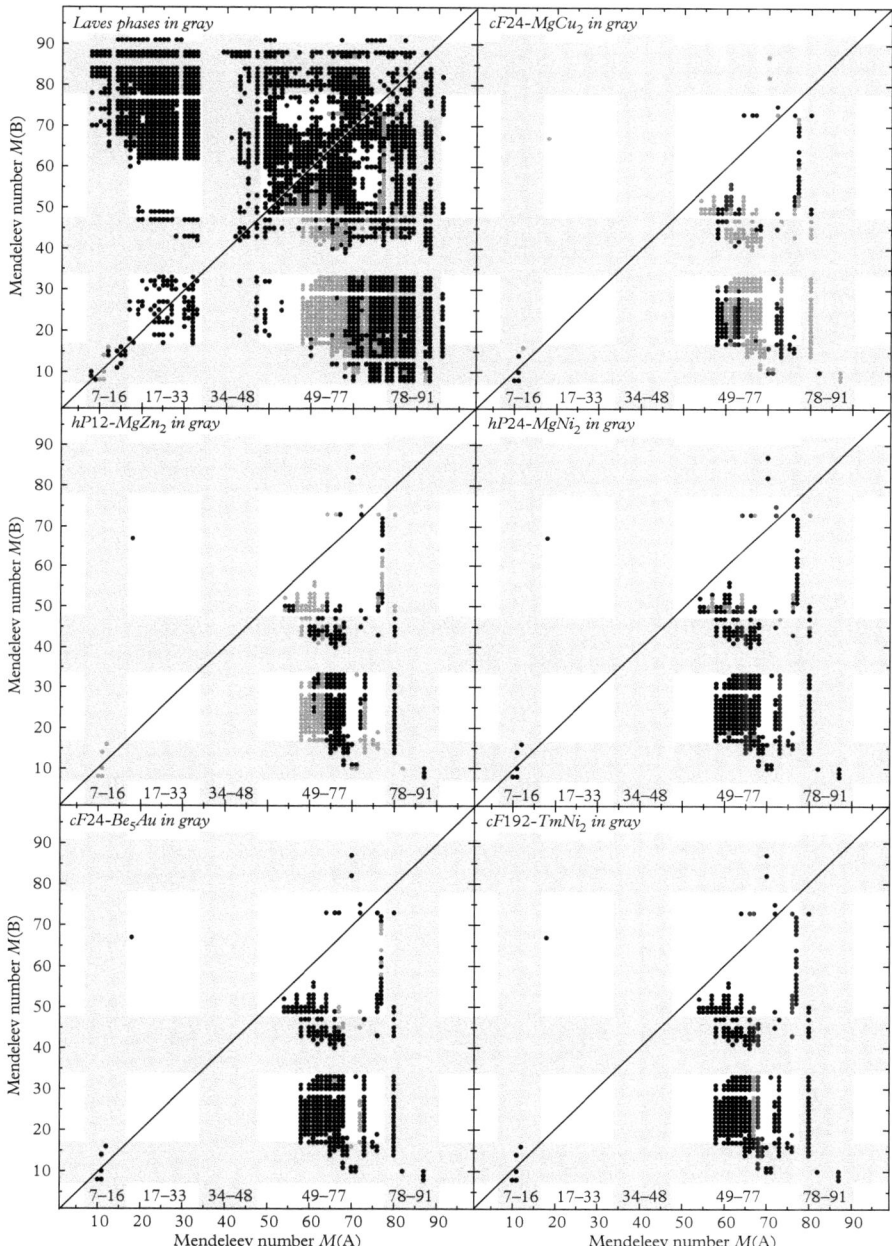

Fig. 7.16 *M/M composition plots of the binary intermetallic Laves phases. Depicted are all binary Laves phases (gray) in comparison with all binary intermetallic compounds (upper left subfigure), as well as the five most frequent structure types in comparison with all binary Laves phases (other five subfigures): cF24-MgCu$_2$, hP12-MgZn$_2$, hP24-MgNi$_2$, cF24-Be$_5$Au, and cF192-TmNi$_2$. M(A) and M(B) denote the Mendeleev numbers of the majority and minority elements, A and B, respectively. For AB-compounds, both element combinations are plotted.*

*hP*24-**MgNi₂** 194 *P*6₃/*mmc*, Mg in 4*e* 0, 0, 0.094, and 4*f* 1/3, 2/3, 0.094; Ni in 4*f* 1/3, 2/3, 0.844, 6*g* 1/2, 0, 0, and 6*h* 0.167, 0.334, 1/4. The major skeleton is constituted from the larger Mg atoms occupying the sites of an intergrowth of the diamond and lonsdaelite ("hexagonal diamond") structures, while the smaller Cu atoms constitute a 3D tetrahedral network around the Mg atoms, consisting of interpenetrating Kagomé nets connected via vertex-sharing single and double-tetrahedra (Fig. 7.15(l)–(o)), respectively. *hP*24-MgNi₂ can be described as a four-layer stacking variant ABAC along the [001]-direction (Ramsdell: 4H, Jagodzinski: *chch*).

*hR*42-**Mg(Ag₀.₁Zn₀.₉)** 166 *R3̄m*. (Fig. 7.15(l-o)), respectively. The structure of this example of a Laves polytype, ABC'B'ABC'BCA'C'BCA'CAB'A'CAB', can be described as a 21-layer stacking along the [001]-direction, with the primed letters indicating a layer rotated by π (Ramsdell: 21R, Jagodzinski: (*hchchch*)₃). There are also several other known polytypes (Komura and Kitano, 1977).

The crystal structure of the TM element influences whether a cubic or hexagonal Laves phase forms. If it is *fcc*, then the probability is high that the Laves phase is *fcc* as well. If phase transformations as a function of temperature are observed, then the LT structure is *fcc*, usually. There are several binary and ternary systems known such as Co–Nb or Al–Cr–Ti, where two or even all three types of basic Laves phases exist.

In the ideal prototype structures, only like atoms touch each other, and the two interpenetrating subsets of homogenous sphere packings constitute a heterogeneous sphere packing. In the case of *cF*24-MgCu₂, for instance, the Mg atoms of the *D* lattice complex form a homogeneous sphere packing with contact distance $d_1 = a/4\sqrt{3}$, with the lattice parameter *a*, and contact number $k = 4$. The Cu-atoms of the *T* lattice complex constitute another sphere packing with shortest distance $d_2 = a/4\sqrt{2}$ and contact number $k = 6$. The shortest Mg–Cu distance amounts to $d_3 = a/8\sqrt{11} > (d_1 + d_2)/2$ (Koch and Fischer, 1992). The ideal atomic size ratios amount to $r_{Mg}/r_{Cu} = \sqrt{3/2} = 1.225$. These size relationships are equally valid for the hexagonal Laves phases Stein *et al.* (2004). The experimentally observed radii ratios range between 1.05–1.70 Å, indicating that the size ratio is only one of the structure-determining factors. One should also keep in mind that atoms are no hard spheres and that the atomic radius also depends on the kind of bonding present in the respective compound.

Depending on their chemical composition, Laves phases can show a large variety of interesting physical properties:

- *Superconducting materials: cF*24-(Zr,Hf)V₂, *hP*12-ZrRe₂, *hP*24-HfMo₂, . . .
- *Magnetostrictive materials: cF*24-(Dy₁₋ₓTbₓ)Fe₂, . . .
- *Magnetocaloric materials: cF*24-RECo₂, *cF*24-RENi₂, *cF*24-REAl₂, . . .
- *Hydrogen storage materials: cF*24-Zr(V,Mn,Ni)₂, . . .

7.4.4 Complex cluster-based superstructures: the Al–Cu–Ta structure family

Cluster-based structures are based on recurrent, relatively densely packed structural units that are larger than AETs, i.e., consist of at least two coordination shells, and that allow a comparatively simple description. If the clusters are large and dense enough, then two different length scales can exist for electron waves as well as for phonons: the cluster diameters, and the period of the underlying lattice. This may lead to interesting physical properties.

In contrast to the building elements of modular structures, those in cluster-based structures are close to spherical, and the clusters may overlap. Clusters in intermetallic phases, however, are rather a convenient way to describe the constitution of crystal structures than crystal-chemically distinct objects. As we have seen in Subsection 7.3.1, γ-brasses can be described as packings of 26-atom clusters. In the following, we will use an alternative cluster description for the $(p \times p \times p) = p^3$-fold superstructures of the $cF16$-NaTl type, i.e., $(2p)^3$-fold superstructures of the basic $cI2$-W type. For the structures known so far, p can adopt the values 3, 4, 7, and 11 (Dshemuchadse *et al.*, 2011). We will discuss the building principles of cluster structures on the example of the structures in the system Al–Cu–Ta, based on the papers by Weber *et al.* (2009), Conrad *et al.* (2009), and Dshemuchadse *et al.* (2013).

The structures of AT-19, cF444-$Ta_{36.4}Al_{63.6}$, and ACT-h, hP386-$Ta_{39.0}Cu_{3.6}Al_{57.4}$

AT-19 is the simplest representative of the family of complex cluster-based structures in the system Al–Cu–Ta, with $p = 3$. The fundamental $Ta_{57}Al_{102}$ fullerene clusters occupy Wyckoff position $4d$ 3/4, 3/4, 3/4 resulting in a cubic close packing (Fig. 7.17). Each fullerene shell shares its pentagon faces with its 12 neighbors. All voids left in the packing are filled by either CN15 or CN16 FK-polyhedra, with their hexagon faces being part of the fullerene shells as well. The FK-polyhedra framework has the topology of a D net.

The electron localization function (ELF) shows maxima between the Al atoms forming the pentagons of the fullerene shell, i.e., the basal plane of the pentagonal bipyramids in the bifrusta, indicating covalent bonding contributions. The Al–Al distances are calculated to 2.517–2.562 Å in the fully relaxed structure, compared to 2.863 Å in the elemental structure. This gives the fullerene cluster some physical relevance beyond its suitability for the geometrical description of the structure.

ACT-h is also a structure that can be described as a close packing of fullerene clusters with the gaps filled by FK-polyhedra (Dshemuchadse, 2013) (Fig. 7.18). The main difference from ccp AT-19 is on one hand that ACT-h is hcp, and on the other hand that it consists of two kinds of fullerene shells, F_{32}^{60} and F_{39}^{74}, in contrast to AT-19 which shows F_{40}^{76} fullerene shells, only (the superscript gives the number of faces, the subscript that of vertices). The stacking sequence along [001] is

$cF444\text{-}Al_{63.6}Ta_{36.4}$

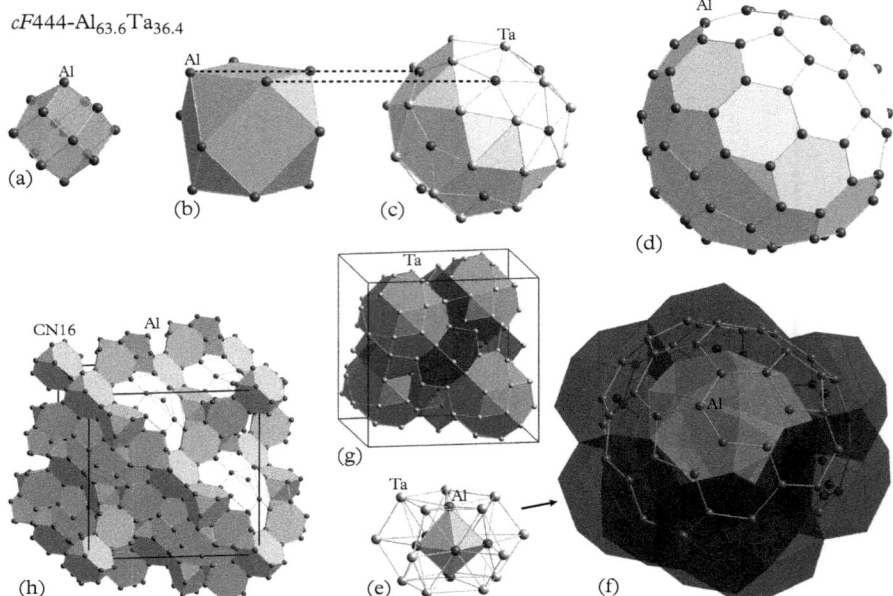

Fig. 7.17 *(a–d) Mutually dual shells of the endohedral* $Ta_{57}Al_{102}$ *fullerene cluster and (g–h) cluster structure of AT–19,* $cF444\text{-}Al_{63.6}Ta_{36.4}$. *(a)* $Ta@Al_{14}$ *rhombic dodecahedron; the distorted* Al_{12} *cuboctahedron shown in (b) merged with the* Ta_{28} *polyhedron depicted in (g) yields the* $Al_{12}Ta_{28}$ *polyhedron shown in (c); (d)* Al_{76} *fullerene shell. (e) Pentagonal* Ta_{15} *bifrustum around an empty pentagonal* Al_7 *bipyramid. (f) Partially open* $Ta_{57}Al_{102}$ *fullerene cluster showing the shell of bifrusta surrounding the cluster shell depicted in (c). (h) One unit cell of AT–19 illustrating the ccp packing of the endohedral fullerene clusters with the gaps filled with CN15 (not shown) and CN16 FK-polyhedra. (g) Different representation of the cluster packing based on the third fullerene cluster shells,* Ta_{28}; *these are surrounded by a shell of twelve bifrusta (e, f), shared with the other* Ta_{28} *polyhedra. The subfigures (g, h) are on half the usual scale.*

ABAC, with A corresponding to *hcp* layers of F_{32}^{60} polyhedra and B, C of F_{39}^{74} fullerene shells. The fullerenes share their twelve pentagon faces with each other.

In some cases, the structure of $hP386\text{-}Al_{57.4}Cu_{3.6}Ta_{39.0}$ shows some stacking disorder. Every other A layer is mirrored with respect to the ordered structure. As a consequence, the clusters in the B and C layers have to be adapted to this change. This leads to a different number and arrangement of atoms per cluster: the F_{74}^{39}-clusters turn into F_{76}^{40}-clusters, which are the kind of clusters building all three cubic Al-(Cu)-Ta compounds, AT-19, ACT-45, and ACT-71.

The structure of ACT-45, $cF(5928-x)\text{-}Ta_{39.5}Cu_{3.9}Al_{56.6}$

The structure of ACT-45 can be described as a packing of super-clusters consisting of a tetrahedral arrangement of pentagon-faces sharing

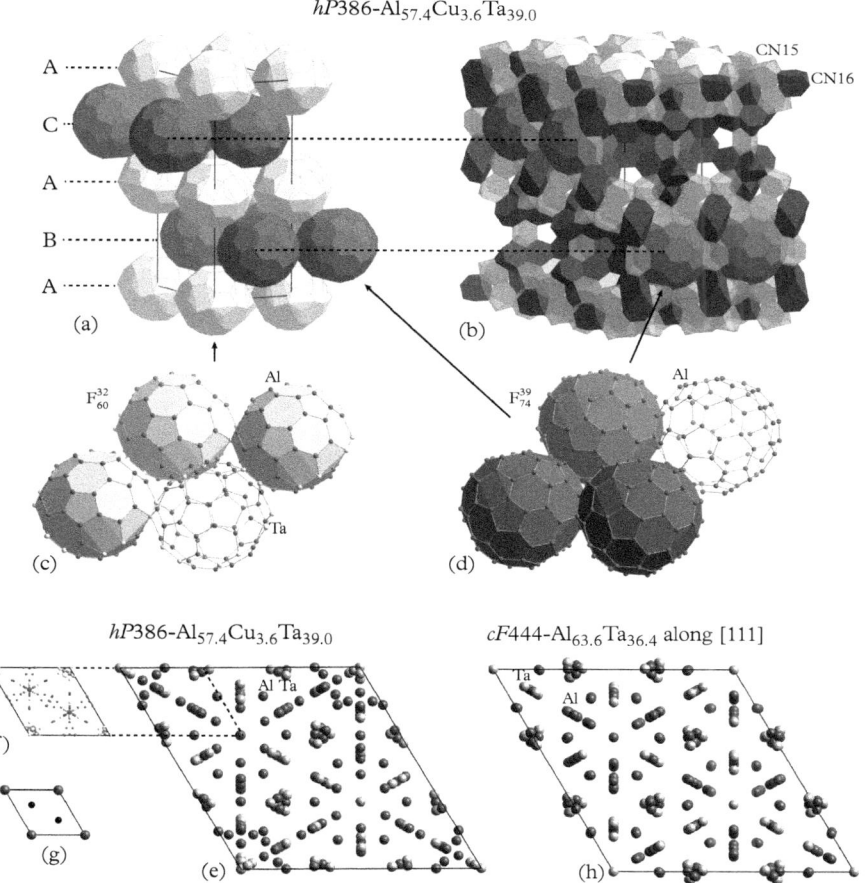

Fig. 7.18 *Cluster packing in hP386-Al$_{57.4}$Cu$_{3.6}$Ta$_{39.0}$ (ACT-h). The A-layers (c) consist of edge-sharing F_{60}^{32}-fullerene clusters (light gray), and the B- and C-layers (d) of pentagon-sharing F_{74}^{39}-clusters (dark gray). The structure results from an ABAC-stacking of these layers along [001]. The space in-between is filled with CN15 (FK$_{15}^{26}$) and CN16 (FK$_{16}^{28}$) FK-polyhedra. The close relationship between ACT-h and AT-19 is also reflected in the projections of the two structures along [001] (e) and [111] (h), respectively. The average structure of ACT-h (f) is related to the hP3-AlB$_2$ type as shown in (g). The figures (a, b) are on one third the usual scale, and (c, d) are on half the usual scale.*

fullerenes (4-supercluster), of Friauf polyhedra, pentagonal dodecahedra, bi-frusta, *h*-capped fullerenes, and Ta$_{81}$Al$_{36}$ superclusters. The tetrahedral void in the center of the 4-supercluster is occupied by a tetrahedral unit of four CN15 polyhedra (Fig. 7.19).

We can distinguish between a framework built from pentagonal dodecahedra and bifrusta, and the clusters filling the space in-between. This framework forms

$cF(5928-x)$-$Al_{56.6}Cu_{3.9}Ta_{39.5}$

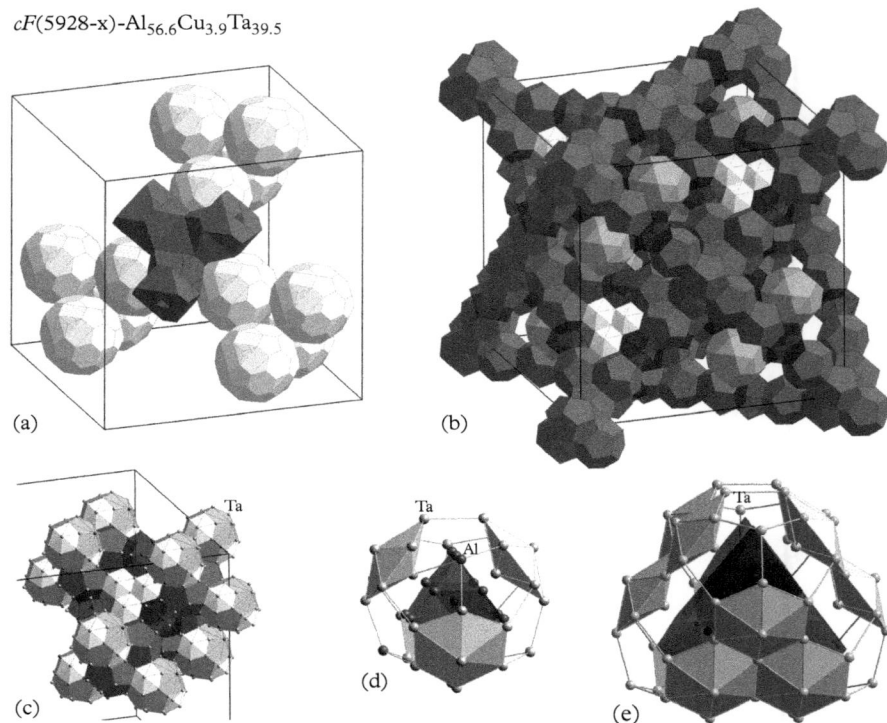

(a)   (b)

(c)   (d)   (e)

Fig. 7.19 *In* ACT-45, *($cF(5928-x)$-$Al_{56.6}Cu_{3.9}Ta_{39.5}$), the endohedral fullerene clusters form 4-superclusters, which share hexagon faces with the central 58-Laves block consisting of 58 Friauf polyhedra (a). All of them are enclosed in a framework of* $Ta_{21-x}Cu_xAl_{10}$ *dodecahedra and* $Ta_{15}Al_7$ *bifrusta (dark gray translucent). (c) The packing of dodecahedra, bifrusta, and* $Ta_{35}Al_{14}$ *h-bicapped fullerenes (d) in the upper right part of the unit cell. (e)* $Ta_{35}Al_{14}$ *supercluster of four interpenetrating* $Ta_{81}Al_{36}$ *h-bicapped fullerenes. The figures (a)–(c) are on one third the usual scale.*

two large interpenetrating tetrahedra related by inversion (which is not present in the total structure), equivalent to a stellated octahedron (stella octangula). Half of the outer star tetrahedra are occupied by $Ta_{81}Al_{36}$ superclusters, the other half by fullerene 4-superclusters. Each of the fullerenes of such a 4-supercluster centers one of the faces of the star tetrahedron and thereby also one face of the central octahedron. The center of the octahedron and the symmetrically equivalent mid-edge positions are occupied by a block of 58 Friauf polyhedra, later on called "58-Laves block", surrounded by four, hexagon-face sharing $Ta_{57}Al_{102}$ fullerenes.

The structure of ACT-71, $cF(23\,256-x)$-$Al_{55.4}Cu_{5.4}Ta_{39.1}$

In this compound, we have again a framework, but now of $Ta_{26}Al_2$ fullerene shells and bifrusta, and in the form of small- and medium-sized tetrahedra attached to large truncated tetrahedra. This arrangement corresponds to the network spanned by the Cu atoms of the cubic Laves phase $cF24$-$MgCu_2$. The space inside each of the eight small tetrahedra is filled by a 146-Laves block, each consisting of 146 Friauf polyhedra, centered at the sites $4b$ $1/2, 1/2, 1/2$ and $4d$ $3/4, 3/4, 3/4$ in the space group $F\bar{4}3m$. The central part of such a 146-Laves block consists of 30 Friauf polyhedra, tetrahedrally arranged like the cubic Laves phase. On top of each tetrahedron face of this unit there are 29 Friauf polyhedra assembled in two slabs like the hexagonal Laves phase. On the top faces of this 146-Laves block there are hexagonal bipyramids (*hbp*s) arranged as illustrated in Fig. 7.20 (b).

$cF(23\,256\text{-}x)$-$Al_{55.4}Cu_{5.4}Ta_{39.1}$

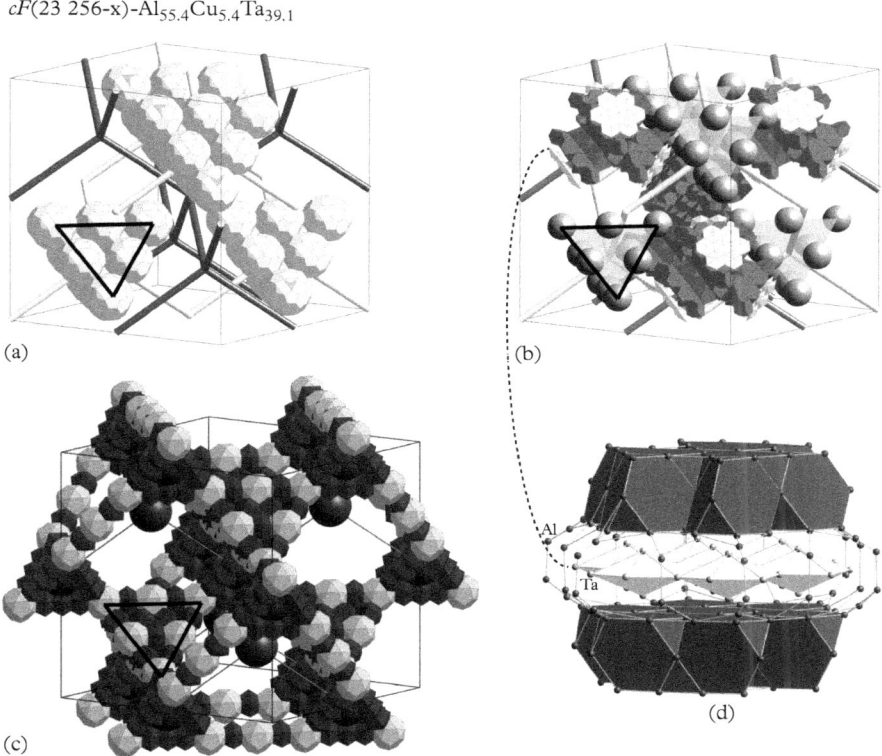

Fig. 7.20 *(a) One unit cell of* ACT-71, $cF(23\,256 - x)$-$Ta_{39.1}Cu_{5.4}Al_{55.4}$ *and the four fullerene 10-superclusters centered on the inner nodes of one D-net (light gray). (b) Distribution of the four 146-Laves blocks, with the hbps next to them, on the nodes of the other D-net (black). All nodes of the D-nets are occupied either by the 10-superclusters or the 146-Laves blocks. The Al_{76} fullerene shells share hexagon faces with the 146-Laves blocks. The positions of the fullerene cluster centers are marked by spheres. (c) Framework of $Ta_{28}Al_{12}$ fullerene shells (light gray) linked by bifrusta (dark gray). The triangles outlined in black in (a)–(c) mark the position of a 10-supercluster. (d) hbp-net sandwiched by slabs of Friauf polyhedra from adjacent 146-Laves blocks. The figures (a)–(c) are on one fifth the usual scale.*

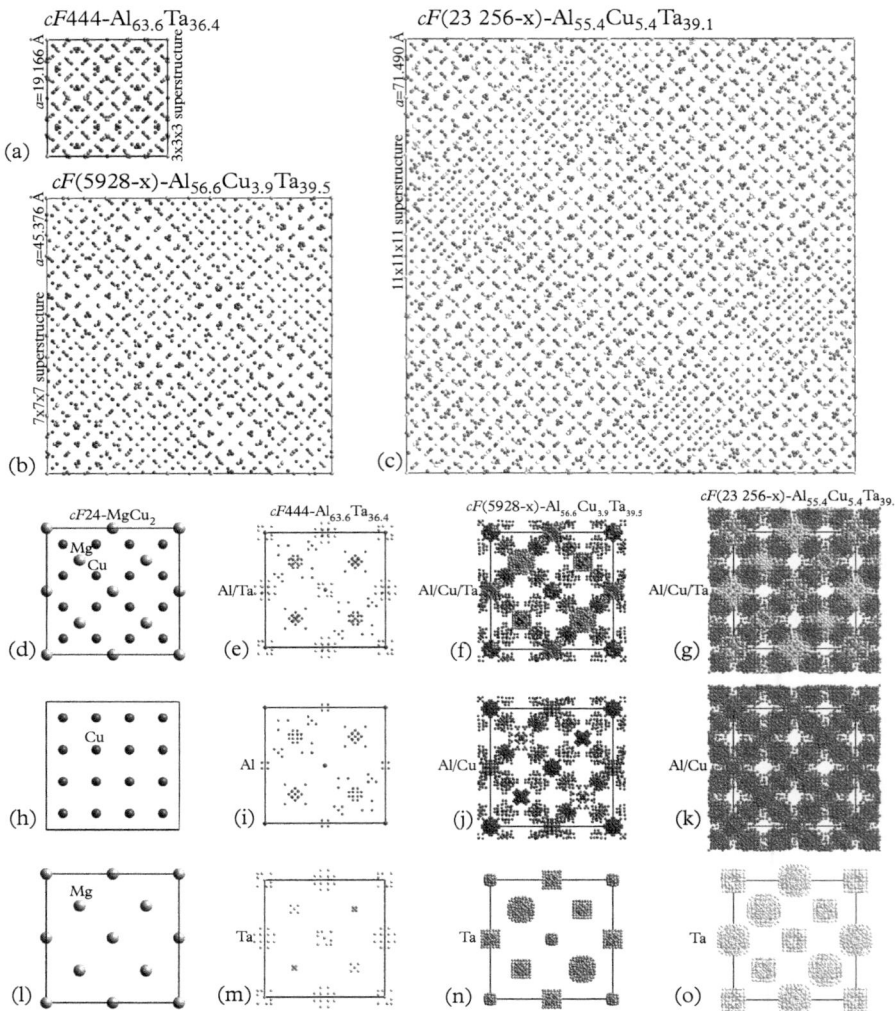

Fig. 7.21 *(a)–(c) The unit cells of the structures of* AT-19, ACT-45, *and* ACT-71 *in projections along [001]. A clear organization of atoms into subcells as well as atomic layers is obvious, as known for Frank-Kasper phases in general. The average structures are shown below ((e)–(g), (i)–(k), and (m)–(o)) in comparison with the structure of the cubic Laves phase* cF24-MgCu₂ *(d, h, and l). The first row depicts the projections of the full structures into one subcell each giving the average structures. In the second and third row, the partial structures are given as indicated next to each subfigure. For more information see Weber* et al.., *(2009). The figures (a)–(c) are on one third the usual scale.*

10-superclusters of ten fullerenes each are centered at $4a$ $0, 0, 0$ and $4c$ $1/4, 1/4, 1/4$, the sites of a double-diamond type structure ($cF16$-NaTl type), forming two interwoven D nets. The space inside the large truncated tetrahedra are occupied each by one 146-Laves block and two fullerene superclusters.

Projected and average structures

The close relationships between the structures AT-19, ACT-45, and ACT-71 can be illustrated nicely by comparing their projections and average structures. As mentioned above, they can be described as $(p \times p \times p) = p^3$-fold superstructures of the $cF24$-MgCu$_2$ or the $cF16$-NaTl types as basic structures. For the structures known so far, p can adopt the values 3, 4, 7, and 11 (Dshemuchadse *et al.*, 2011) (Fig. 7.21).

7.5 Zintl phases

Intermetallic compounds that are constituted from more-electropositive elements from groups 1 and 2 of the periodic table on the one hand, and of more-electronegative elements of groups 13–15, on the other hand, are called Zintl phases after their discoverer (Zintl and Dullenkopf, 1932). They combine properties of metals with those of polyanionic salts. In most cases, the polyanionic part of the structure consists of larger subunits such as clusters, layers, or 3D networks. Zintl and Dullenkopf (1932) introduced the original concept by means of the simple example of $cF16$-NaTl. In this compound, Na donates its electron to Tl, which then has an outermost shell isoelectronic to carbon. Consequently, the polyanionic substructure of the larger Tl⁻ anions should be that of diamond (D-net), while the smaller Na⁺ cations occupy the voids, forming another D-net. The *Zintl-Klemm concept*, a generalization of this example, says that the anions in such compounds form the same kind of substructures that the neutral main group elements with the same number of valence electrons form (*pseudoatom concept*).

The original Zintl concept has been modified and extended to also include transition metals (e.g., $cP2$-CsAu) or rare earth elements in the structure (Kauzlarich, 1996). This allows us to form large endohedral clusters, which can consist of many cluster shells. Due to the condition of charge balance, Zintl phases are stoichiometric line compounds. Skutterudites, AB$_3$ (A ... late TM, B ... P, As, Sb), with the $cI32$-CoAs$_3$ structure type, can be considered TM Zintl phases, for instance.

A more recent definition of Zintl phases says that they are compounds whose bonding and nonbonding states are completely occupied and separated from the antibonding states by no more than 2 eV (Miller, 1996). Accordingly, Zintl phases are semiconductors but they can also be metallic if they have a few extra electrons or holes relative to those needed for the 2-center-2-electron bonds. An example for such a metallic Zintl phase is $cP156$-K$_{29}$NaHg$_{48}$, which is almost isostructural to the true semiconducting Zintl phase $cP154$-K$_3$Na$_{26}$In$_{48}$, but is not electron-balanced since In has three and Hg only two valence electrons (Deiseroth and Biehl, 1999). The main structural building units are two Na-centered icosahedral closo clusters, $NaHg_{12}$ (n atoms on the n vertices of a polyhedron) and six K-centered hexagonal-antiprismatic arachno clusters, KHg_{12} ($n-2$ atoms on the n vertices of a polyhedron), embedded in a matrix of alkali metal atoms. The Hg atoms form a 3D network. Other examples of metallic Zintl phases are $hP8$-$BaSn_3$ and the series of polytypic superstructures in the homologous series $(BaSn_3)_m[Ba(Sn_yBi_{1-y})_3]_n$ (Fig. 7.6 in Section 7.2).

Closed-shell clusters in Zintl phases are assumed to have delocalized electrons. According to Wade's rule, $2n + 2$ bonding electrons are needed for closo deltahedral clusters (all n vertices occupied), $2n + 4$ for nido clusters (one vertex unoccupied), and $2n + 6$ for arachno clusters (two vertices unoccupied) (Wade, 1976). A deltahedron is a polyhedron with only equilateral triangles for faces, similar to a Greek capital letter delta, Δ. The clusters can be isolated or connected; the delocalized intracluster (endo-)bonds are longer than the 2-center-2-electron (exo-)bonds between clusters.

For reviews on Zintl phases see, for instance, Fässler and Hoffmann (1999) and Nesper (2014).

7.6   REME phases

REME phases are intermetallic compounds with composition RE:M:E = 1:1:1. RE means a rare earth metal (in most cases), an actinoid, or a group 1–4 element, M is a late transition metal from groups 8–12, and E is an element of groups 13–15 (Bojin and Hoffmann, 2003a; Bojin and Hoffmann, 2003b). In our discussion of intermetallic compounds we additionally exclude compounds containing one of the elements: B, C, N, Si, P, and As. In most cases, and depending on the actual structure type, a polyanionic substructure of the type $[ME]^{n-}$ exists balancing the charge of the counteraction RE^{n+}. Some of the REME phases are insulators, semiconductors, or semimetals, others have unusual magnetic and electronic properties. For instance, some representatives of the most common $oP12$-TiNiSi structure type are heavy-Fermion compounds. For a comprehensive review on the physical properties of REME phases see Gupta and Suresh (2015).

In their comprehensive review, Bojin and Hoffmann (2003a) listed a large number of structure types for REME phases with more than 2000 representatives. However, in our optimized database search based on the more restrictive criteria mentioned above, we found only 1074 intermetallic representatives among the 20 829 intermetallic compounds contained in the Table 7.5 lists the 20 most common structure types, which represent already 1013, i.e., 94% of all REME phases.

The chemical compositions that REME phases can adopt are shown in Fig. 7.22. Therein, the projected $M(RE)/M(M)$-, $M(RE)/M(E)$-, and

Table 7.5 *The twenty most common structure types of REME phases with the number of their representatives, representing 94% of all REME phases. Not listed are the 36 different structure types with only 3, 2, or 1 representatives, with altogether 61 representatives. Under "Comments", the rank of the space group of the respective structure type in the review article by Bojin and Hoffmann (2003a) is given, if applicable. If structure types are named differently than in this review, the corresponding structure types are given (= isopointal, ~ similar). The star, *, marks the 8 out of 20 structure types that belong to the hP3-AlB$_2$ derivative structures.*

Rank	Structure type	No. of reps.	No.	Space group	Wyckoff positions	Comments
1	$oP12$-TiNiSi	289	62	$Pnma$	$2ad\,6gh$	*1
2	$hP9$-ZrNiAl	268	189	$P\bar{6}2m$	$1a2d3fg$	2, $\sim hP9$-Fe$_2$P
3	$cF12$-MgAgAs	118	216	$F\bar{4}3m$	$4abc$	4
4	$oI12$-KHg$_2$	55	74	$Imma$	$4e\,8i$	*5, = $oI12$-CeCu$_2$
5	$hP12$-MgZn$_2$	50	194	$P6_3/mmc$	$2a\,4f\,6h$	3
6	$hP6$-CaIn$_2$	48	194	$P6_3/mmc$	$2b\,4f$	*3
7	$hP6$-LiGaGe	40	186	$P6_3mc$	$2ab^2$	*7
8	$hP6$-ZrBeSi	28	194	$P6_3/mmc$	$2acd$	*3, $\sim hP6$-Ni$_2$In
9	$hP6$-NdPtSb	22	186	$P6_3mc$	$2ab^2$	*7, $\sim hP6$-LiGaGe
10	$oP12$-HoNiGa	21	62	$Pnma$	$4c^3$	1, $\sim oP12$-Co$_2$Si
11	$hP18$-HfRhSn	11	190	$P\bar{6}2c$	$1s\,2d\,3fg$	–
12	$oP36$-AuYbGe	9	62	$Pnma$	$4c^9$	1
13	$oI36$-AuYbSn	9	44	$Imm2$	?	–
14	$tP6$-PbClF	9	129	$P4/nmm$	$2ac^2$	6, $\sim tP6$-Cu$_2$Sb
15	$hP3$-AlB$_2$	8	191	$P6/mmm$	$1a\,2d$	*8/9
16	$oP24$-YPdSi	8	59	$Pmmn$	$2ab\,4e^5$	*17, = $oP24$-GdPdGe
17	$oI36$-TiFeSi	7	46	$Ima2$	$4ab^4\,8c^2$	13
18	$hP3$-LiBaSi	5	187	$P\bar{6}m2$	$1ade$	–
19	$cF12$-CaF$_2$	4	225	$Fm\bar{3}m$	$4a\,8c$	19
20	$oP24$-LaNiAl	4	62	$Pnma$	$4c^6$	–
		1013				

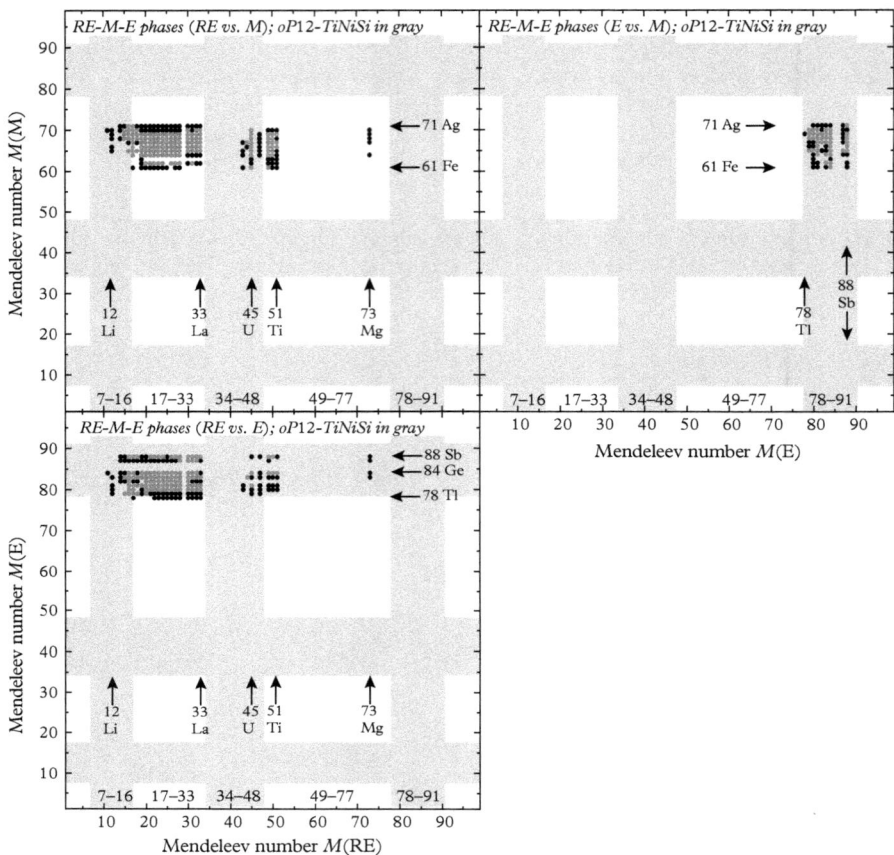

Fig. 7.22 *2D projections of the 3D $M(RE)/M(M)/M((E))$-plot of the intermetallic REME structures. The most common structure type, $oP12$-TiNiSi, is highlighted in gray.* $M(RE) = 73$ *corresponds to* Mg.

$M(M)/M(E)$-plots for all identified 1074 structures are shown, with the most common structure type, $oP12$-TiNiSi, highlighted.

$oI12$-TiNiSi, $oI12$-KHg$_2$, and $oI60$-EuAuSn

A large subclass of the REME phases belongs to the $hP3$-AlB$_2$ derivative structures, which are discussed in the next section. The corresponding structure types have been marked in Table 7.5 by an asterisk, *. Those REME structures can be described as a stackings of flat or puckered ME layers consisting of hexagonal rings. For instance, the structure of $oP12$-TiNiSi (Fig. 10.8(f)–(i)) can be derived by symmetry reduction ($I \rightarrow P$) from the $oI12$-KHg$_2$ structure type (Fig. 7.23(c)), which itself is an $hP3$-AlB$_2$ (Fig. 7.23(d)) derivative structure. The

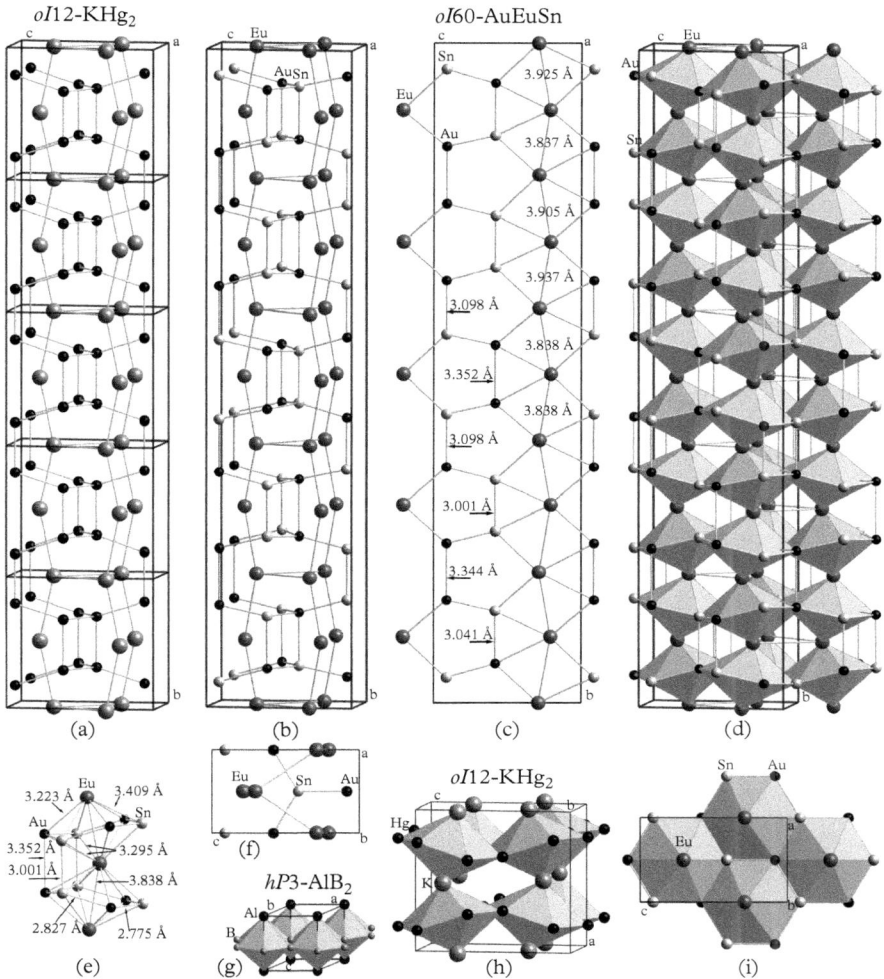

Fig. 7.23 *Structures of the REME phase o*I*60-AuEuSn (b)–(f), (i) and of the related phases (a, h) o*I*12-KHg$_2$ and (g) hP3-AlB$_2$ in different representations (Al, Au, Hg. . . black, B, Eu, K. . . gray, Sn. . . light gray). The structure of o*I*60-EuAuSn (b, d, f, i) can be seen as a fivefold superstructure (a) of the o*I*12-KHg$_2$ type (h). o*I*12-KHg$_2$ itself is a fourfold superstructure of the hP3-AlB$_2$ type (g). (d) The edge-connected Au$_3$Sn$_3$-rings, capped by Eu atoms, form puckered layers of hbps sharing the apical Eu atoms; (e) view of two hbps with interatomic distances. The along [010] projected unit cells depicted in (b) and (d) are shown in (f) and (i). The (011)-section of the structure for x = 0 (c) shows the distance modulation that is necessary to accommodate the large Eu cations, and which is also the reason for the fivefold superstructure.*

structure can be described by a [NiSi] polyanionic 3D four-connected substructure, wherein Ni is tetrahedrally coordinated by Si ($2.307 \text{ Å} \leq d_{\text{Ni-Si}} \leq 2.353 \text{ Å}$). According to Landrum *et al.* (1998), Ti carries a charge of +1.88, and Ni and Si almost evenly −0.91 and −0.97, respectively. The Ti–Si interactions ($2.571 \text{ Å} \leq d_{\text{Ti-Si}} \leq 2.633 \text{ Å}$) are quite strong, much stronger than the Ti–Ni ones ($2.749 \text{ Å} \leq d_{\text{Ti-Ni}} \leq 2.893 \text{ Å}$), while there are no indications of Ni–Ni bond formation despite the relatively short Ni–Ni distance of 2.659 Å.

As another example, the structure of *oI*60-EuAuSn is illustrated in Fig. 7.23. With the Eu^{2+} cations embedded in the polyanionic AuSn^{2-} framework, it can be seen as a fivefold superstructure of the *oI*12-KHg_2 type, which itself is a fourfold superstructure of the *hP*3-AlB_2 type. The symmetry reduction goes first via a *translationengleiche* transition of index two (t2) from space group 74 *Imma* of *oI*12-KHg_2 to 44 *Imm*2, and then via an *isomorphic* transition of index five (i5) to *oI*60-EuAuSn, with a five times larger *c* lattice parameter.

While the edge-connected planar 6-rings form flat layers in the case of *hP*3-AlB_2, they are puckered and form puckered layers in the other cases shown. The puckering leads to shorter distances between like atoms of neighboring rings (compare Figs. 7.23(a) and (h)). The Eu atoms have shorter distances (3.837 Å) to Eu atoms in [010] direction than to those in the (101) planes (4.224 Å, 4.791 Å, 5.296 Å) forming zigzag chains along [010] rather than a 3D framework. The distance 3.837 Å is shorter than two times the atomic radius, $2 \times r_{\text{Eu}} = 2 \times 1.995 = 3.990$ Å, but much larger than two times the ionic radius, $2 \times 1.176 = 2.352$ Å, although the divalent character of Eu was determined by the measurement of the magnetic susceptibility and by Mössbauer experiments (Pöttgen and Johrendt, 2000). If the distance between the Eu atoms is determined by the Eu atoms themselves, or by the Eu–Au and Eu–Sn distances under the constraint of shortest possible intra-*hbp* Sn–Au distances, then the tilt of the *hbp*s allows shorter inter-*hbp* Sn–Sn and Au–Au contacts. The fivefold superstructure along the [010] direction results from the Au/Sn ordering (see Fig. 7.23(i)).

The ideal size ratio of corner and center atoms decorating the vertices and the center of an isometric hexagonal prism with equal edge lengths and height is r_v : $r_c = 1 : (\sqrt{5} - 1) = 1 : 1.236$. Atomic and cationic radii of Eu and Eu^{2+} amount to 1.995 Å and 1.176 Å, respectively; the atomic radii of Au and Sn are 1.442 Å and 1.405 Å, respectively. Setting $r_v = r_{\text{Au}}$, we obtain $1.236 \times r_{\text{Au}} = 1.782$ Å, a value in-between the ionic and the atomic radius for Eu.

7.7   *hP*3-AlB_2 derivative structures

The following discussion is largely based on the review by Hoffmann and Pöttgen (2001), who identified 46 structure types with more than 1500 representatives as *hP*3-AlB_2 derivative structures. It is a good illustration of the variability of particular structure building principles in order to accommodate different realizations of chemical composition and bonding. The relationships between these 46 structure types have been described in detail by these authors using the *Bärnighausen tree* for the illustration of their group/subgroup relations (Fig. 7.24). The structure types studied in this paper have stoichiometries such as RT$_2$, RX$_2$, and RTX (R. . . alkaline earth, rare earth or actinoid metal; T. . . transition metal; X. . . main-group element). T and X elements substitute the B atoms forming a honeycomb network of 6-rings, while the R elements on the Al sites bicap them. Depending on the substituents, the 6-rings remain flat or get puckered and the structures show smaller or large distortions and lowering of the symmetry. For each space

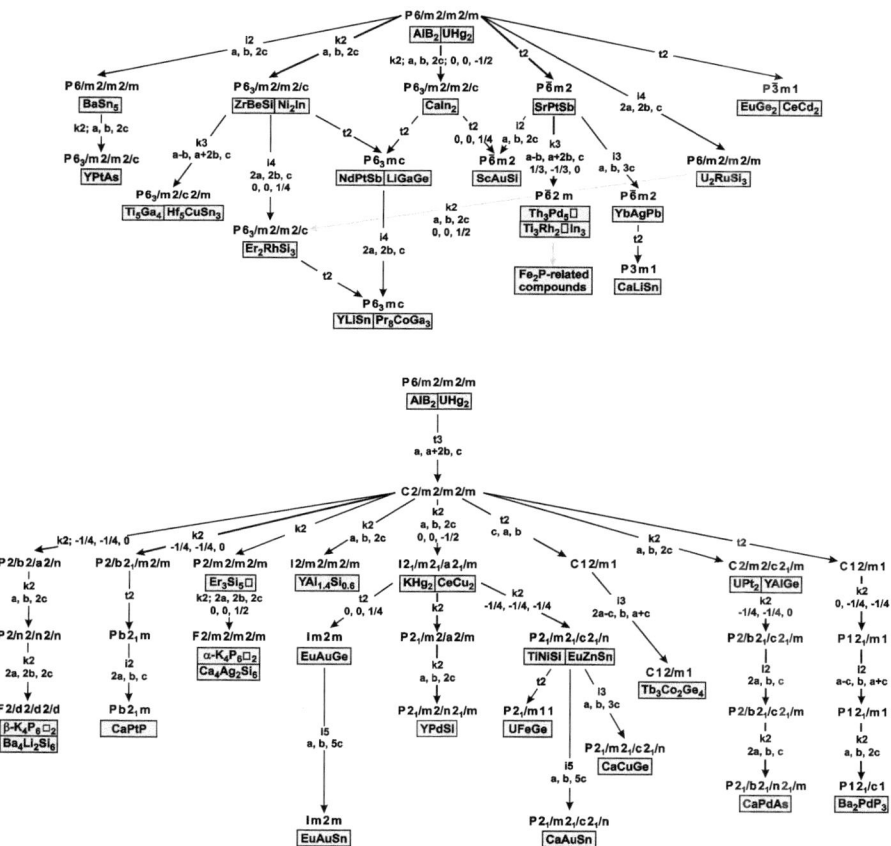

Fig. 7.24 Bärnighausen tree *of the group/subgroup relationships of the hP3-AlB*$_2$ *structure type and its distorted and/or ordered hexagonal/trigonal (upper part) and orthorhombic/monoclinic derivative structures. The indices of the* klassengleiche (k), *the* translationengleiche (t), *and the isomorphic* (i) *transitions, as well as the unit-cell transformations and origin shifts, are given. The numbers after* t, k, *and* i *indicate the index of the reduction in symmetry (from Hoffmann and Pöttgen (2001). Reprinted courtesy of Walter De Gruyter GmbH, Berlin, Germany.)*

group, there exists an infinite number of subgroups. Of course, only a few are realized for the $hP3$-AlB$_2$ derivative structures. We focus here on the two main branches, the hexagonal/trigonal one and the orthorhombic/monoclinc one, respectively (Table 7.6). The relationship between a hexagonal (h) lattice and an orthorhombic o lattice becomes clear if one uses its orthohexagonal (oh) representation ($\mathbf{a}_{oh} = \mathbf{a}_h$, $\mathbf{b}_{oh} = \mathbf{a}_h + 2\mathbf{b}_h$, $\mathbf{c}_{oh} = \mathbf{c}_h$; $b_{oh} = a_{oh}\sqrt{3}$). In this setting, $hP3$-AlB$_2$, 191 $P6/mmm$, would be described by $oC6$-AlB$_2$, $Cmmm$.

A *klassengleiche* (k) subgroup of a space group belongs to the same point group as the space group itself. For instance, the symmetry reduction may just be a de-centering operation such as $I \rightarrow P$ or $C \rightarrow P$. A *translationengleiche* (t) subgroup has the same lattice, and only the point group changes. An example would be the symmetry reduction from $6/mmm$ to $6_3/mmc$. *Isomorphic* (i) subgroups are a special case of the k-subgroups belonging to the same space group type; however, the asymmetric unit is increased. For instance, the symmetry group of $hP24$-Er$_2$RhSi$_3$ is an i-subgroup of index 4 of the symmetry group of $hP6$-ZrBeSi. The unit cell of $hP24$-Er$_2$RhSi$_3$ is $2 \times 2 \times 1$ larger than the one of $hP6$-ZrBeSi (Fig. 7.25). The index of a maximal subgroup is defined by the number of cosets of the subgroup in the minimal supergroup, and it is always a prime number or its square.

Isopointal $hP3$-AlB$_2$ and $hP3$-UHg$_2$

The aristotype structure $hP3$-AlB$_2$ can be described as a boron honeycomb 6^3-net, where each B-hexagon is bicapped by Al atoms. The distances between B atoms amount to $d_{\text{B-B}} = 1.732$ Å, between Al and B atoms $d_{\text{Al-B}} = 2.373$ Å, while with 3.245 Å they are much larger along [001], between the Al atoms and the B-nets, respectively. Consequently, the characteristic structural building blocks correspond to *hbp*s (see Fig. 7.23(d)). The AETs are trigonal Al prisms for the B atoms and hexagonal B prisms for the Al atoms.

In the case of isopointal $hP3$-UHg$_2$, the Hg atoms form the 6-rings that are bicapped now by the U atoms. The main difference between this structure and that of $hP3$-AlB$_2$ is the c/a ratio, which amounts to 1.080 for $hP3$-AlB$_2$ and to 0.647 for $hP3$-UHg$_2$. Consequently, in this case, the Hg–Hg distances in the (110) plane and along [001] with 2.873 Å and 3.218 Å, respectively, are closer to each other ($r_{\text{Hg}} = 1.503$ Å).

Isopointal $hP3$-EuGe$_2$ and $hP3$-CeCd$_2$; $hP3$-SrPtSb and $hP12$-U$_2$RuSi$_3$

The symmetry reduction leads to the trigonal *translationengleiche* subgroup of index 2, 162 $P\bar{3}m1$, which allows puckering of the honeycomb net. The c/a ratios of $hP3$-EuGe$_2$ and $hP3$-CeCd$_2$ with 1.218 and 0.680, respectively are again very different. In the case of $hP3$-EuGe$_2$, Eu is divalent and donates its two valence electrons to Ge, which become isoelectronic with As and form a puckered network as $hR2$-As does (valence electron concentration of the polyanion VEC = 5). While in $hP3$-EuGe$_2$ the Eu atoms are separated 4.995 Å ($r_{\text{Eu}} = 1.995$ Å), the Ce–Ce distances in $hP3$-CeCd$_2$ with only 3.450 Å are even shorter than the distance of 3.650 Å in elemental Ce.

Table 7.6 *hP3*-AlB$_2$-*derivative structures, with the space group and the number of representatives in the PCD (adapted from Hoffmann and Pöttgen (2001), with kind permission from Walter De Gruyter GmbH). The asterisk, *, indicates that only Si-, P-, or As-containing representatives are known,* ^{n} *means that the representatives are listed according to the binary structure type marked with the same number of atoms/unit cell.*

Hexagonal/trigonal structures				Orthorhombic/monoclinic structures			
Compound	**No.**	**Space group**	**N**	**Compound**	**No.**	**Space group**	**N**
$hP3$-AlB$_2$	191	$P6/mmm$	200	$oP8$-Er$_3$Si$_5$	47	$Pmmm$	1*
$hP3$-UHg$_2$	191	$P6/mmm$	48	$oI12$-YAl$_{1.4}$Si$_{0.6}$	71	$Immm$	?
$hP6$-BaSn$_5$	191	$P6/mmm$	1	$oI12$-KHg$_2$	74	$Imma$	252^2
$hP6$-ZrBeSi	194	$P6_3/mmc$	73	$oI12$-CeCu$_2$	74	$Imma$	0^2
$hP6$-Ni$_2$In	194	$P6_3/mmc$	89	$oC12$-UPt$_2$	63	$Cmcm$	2
$hP6$-CaIn$_2$	194	$P6_3/mmc$	215	$oC12$-YAlGe	63	$Cmcm$	10
$hP3$-SrPtSb	187	$P\bar{6}m2$	11	$oF40$-K$_4$P$_6$	69	$Fmmm$	5*
$hP3$-EuGe$_2$	162	$P\bar{3}m1$	7	$oF48$-Ca$_4$Ag$_2$Si$_6$	69	$Fmmm$	1*
$hP3$-CeCd$_2$	162	$P\bar{3}m1$	27	$oI12$-EuAuGe	44	$Imm2$	5
$hP12$-YPtAs	194	$P6_3/mmc$	34	$oP12$-TiNiSi	53	$Pmna$	403^3
$hP6$-NdPtSb	186	$P6_3mc$	34	$oP12$-EuZnSn	53	$Pmna$	0^3
$hP6$-LiGaGe	186	$P6_3mc$	78	$mC18$-Tb$_3$Co$_2$Ge$_4$	12	$C2/m$	10
$hP6$-ScAuSi	187	$P\bar{6}m2$	2	$oF80$-K$_4$P$_6$	70	$Fddd$	1*
$hP12$-U$_2$RuSi$_3$	191	$P6/mmm$	3*	$oF96$-Ba$_4$Li$_2$Si$_6$	70	$Fddd$	1
$hP18$-Ti$_5$Ga$_4$	193	$P6_3/mcm$	33	$oP12$-CaPtP	26	$Pmc2_1$	1*
$hP18$-Hf$_5$CuSn$_3$	194	$P6_3/mmc$	135	$oP24$-YPdSi	59	$Pmmn$	8
$hP8$-Th$_3$Pd$_5$	189	$P\bar{6}2m$	6^1	$mP12$-UFeGe	11	$P2_1/m$	1
$hP9$-Fe$_2$P	189	$P\bar{6}2m$	5	$oP36$-CaCuGe	62	$Pmna$	11
$hP8$-Ti$_3$Rh$_2$In$_3$	189	$P\bar{6}2m$	1	$oP48$-CaPdAs	62	$Pnma$	1*
$hP9$-YbAgPb	187	$P\bar{6}m2$	5	$mP24$-Ba$_2$PdP$_3$	14	$P2_1/c$	1*
$hP24$-Er$_2$RhSi$_3$	186	$P6_3mmc$	9*	$oI60$-EuAuSn	44	$Imm2$	1
$hP9$-CaLiSn	156	$P3m1$	4	$oP60$-CaAuSn	62	$Pmna$	1
$hP24$-YLiSn	186	$P6_3mc$	1				
$hP24$-Pr$_8$CoGa$_3$	186	$P6_3mc$	9				

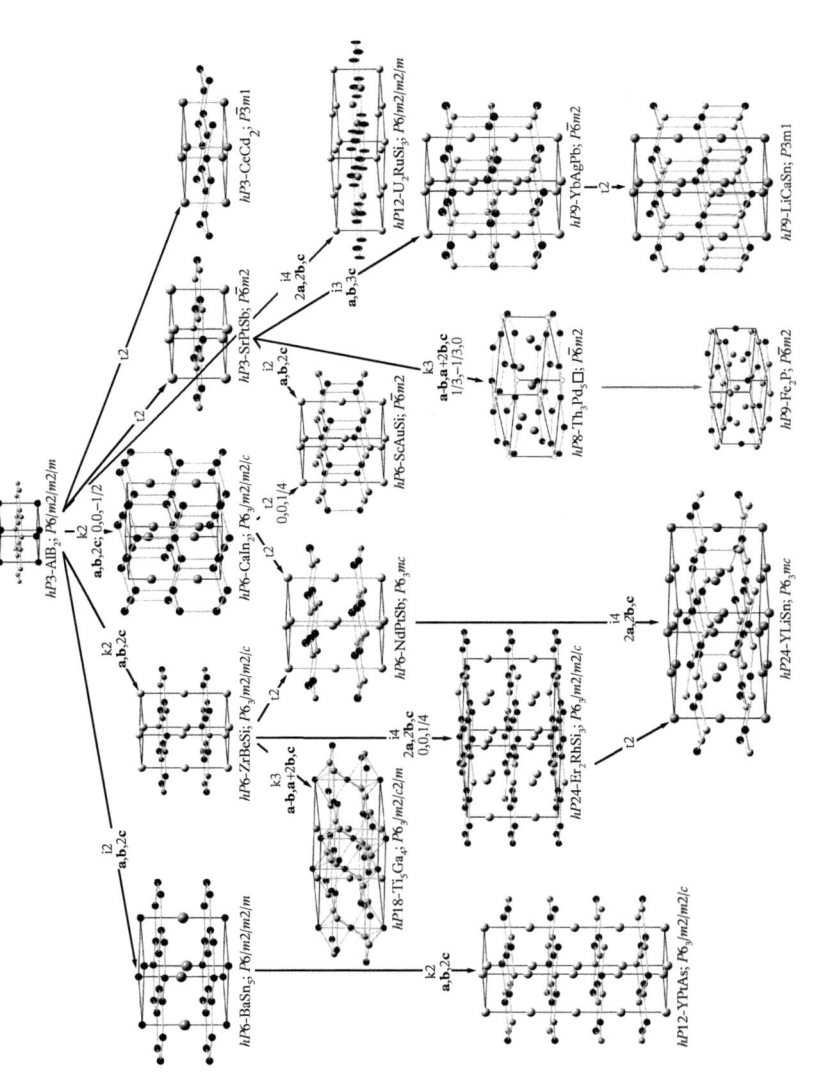

Fig. 7.25 *Group/subgroup relations of the hP3-AlB$_2$ structure type and its distorted and/or ordered hexagonal/trigonal derivative structures. The indices of the klassengleiche (k), the translationengleiche (t), and the isomorphic (i) transitions, as well as the unit-cell transformations and origin shifts, are given. The numbers after t, k, and i indicate the index of the reduction in symmetry (based on Hoffmann and Pöttgen (2001)).*

In the case of *hP3*-SrPtSb, Sr replaces Al, and Pt and Sb alternatingly occupy the B positions in the flat 6-rings. This lowers the symmetry to the *translationengleiche* subgroup of index 2, 187 $P\bar{6}m2$. *hP12*-U$_2$RuSi$_3$ can be described as a ($2 \times 2 \times 1$)-fold superstructure of *hP3*-AlB$_2$, and its symmetry relation as an *isomorphic* subgroup of index four. The structure consists of on average flat Si 6-rings, which are connected with each other by Ru atoms forming Ru$_2$Si$_4$ 6-rings again. All 6-rings are bicapped by U atoms. According to Pöttgen *et al.* (1994), Si could only be refined as split positions indicating a puckering of the Si 6-rings and a breaking of the mirror symmetry. The Si–Si distances $d_{\text{Si-Si}} = 2.342$ Å, $r_{\text{Si}} = 1.176$ Å correspond to the sum of atomic radii while those between Ru and Si with $d_{\text{Ru-Si}} = 2.3612$ Å, $r_{\text{Ru}} = 1.325$ Å are much shorter if puckering is not taken into account.

hP6-BaSn$_5$ and hP12-YPtAs

The *isomorphic* transition from *hP3*-AlB$_2$ to a subgroup of index two leads to the structure of *hP6*-BaSn$_5$ with a doubled lattice parameter *c*, where the Sn atoms not only decorate the flat 6-rings, but occupy alternatingly with Ba the capping sites along [001]. This gives two different Sn AETs. In the 6-rings, the distance $d_{\text{Sn-Sn}} = 3.102$ Å is shorter than that to the capping Sn atom with $d_{\text{Sn-Sn}} = 3.437$ Å. Due to the large diameters of the Ba atoms ($r_{\text{Ba}} = 2.174$ Å) the Sn-layers are shifted away from the *c*/4 positions leading to hexagonal Sn prisms of 2.959 Å height ($r_{\text{Sn}} = 1.405$ Å) around a central Sn. The height of the Sn-prism around the central Ba atom amounts to 4.138 Å. Of the same structure type as the superconductor *hP3*-MgB$_2$, *hP6*-BaSn$_5$ is also superconducting with a critical temperature of 4.4 K (Fässler *et al.*, 2001).

By a *klassengleiche* transition of index two the rotation axis, 6, is replaced by a screw axis, 6$_3$, leading to the doubled unit cell of the REME phase *hP12*-YPtAs. Now, all capping sites are occupied by Y, the slightly puckered 6-rings alternatingly with Pt and As. The short distance $d_{\text{Pt-As}} = 2.466$ Å indicates strong bonding in the layers ($r_{\text{Pt}} = 1.373$ Å, and $r_{\text{As}} = 1.245$ Å). The distances between the layers are also much larger than those between the Y ($d_{\text{Y-Y}} = 3.791$ Å, $r_{\text{Y}} = 1.776$ Å). However, due to the puckering the distances $d_{\text{Y-Pt}} = 2.997$ Å and $d_{\text{Y-As}} = 3.059$ Å can be rather short as well, at least to one Y atom in each case.

hP6-Ni$_2$In, hP6-ZrBeSi, and hP6-CaIn$_2$

Due to the *klassengleiche* symmetry reduction of index two to 186 $P6_3/mmc$, in all these cases the *c* lattice parameter is doubled compared to *hP3*-AlB$_2$. The honeycomb layers are flat in the case of *hP6*-Ni$_2$In and *hP6*-ZrBeSi, and puckered for *hP6*-CaIn$_2$. *hP6*-ZrBeSi is an ordered variant of the *hP6*-Ni$_2$In structure type. Zr and Be as well as Ni and In alternatingly occupy the 6-ring sites in their structure types, with Si and Ni, respectively, as capping atoms.

The Zintl phase $hP6$-CaIn$_2$ shows puckered 6-rings occupied by In atoms. The Ca valence electrons are transferred to the In atoms, which form a lonsdaleite-like "hexagonal diamond") 4-connected network (valence electron concentration of the polyanion VEC = 4).

Isopointal hP6-NdPtSb and hP6-LiGaGe; hP6-ScAuSi

All these compounds belong to the REME phases and result from a *translationengleiche* symmetry reduction of index two from the structure of $hP6$-CaIn$_2$. The puckered rings are alternatively occupied by Pt/Sb, Ga/Ge, and Au/Si, respectively, capped by the alkali and rare earth atoms.

In the Zintl phase $hP6$-LiGaGe, the strong puckering leads to an almost tetrahedral coordination of Ge by Ga ($3 \times d_{\text{Y-Pt}} = 2.542$ Å, $1 \times d_{\text{Y-Pt}} = 2.584$ Å). The polyanion [GaGe]$^-$ has a VEC = 4, and forms a lonsdaleite-like substructure. The distances $d_{\text{Li-Ga}} = 2.727$ Å and $d_{\text{Li-Ge}} = 2.743$ Å approximately correspond to the sums of the atomic radii ($r_{\text{Li}} = 1.52$ Å, $r_{\text{Ga}} = 1.221$ Å, and $r_{\text{Ge}} = 1.225$ Å). The isopointal phase $hP6$-NdPtSb shows less puckered 6-rings leading to a more 2D-like character of the AET around Sb ($3 \times d_{\text{Sb-Pt}} = 2.647$ Å, $1 \times d_{\text{Sb-Pt}} = 3.541$ Å).

$hP6$-ScAuSi also can be derived from $hP3$-SrPtSb by an i2 transformation, doubling the lattice parameter c and puckering the layers; Sr is replaced by Sc, Pt by Si, and Sb by Au. Due to the puckering Si becomes almost tetrahedrally coordinated forming a lonsdaleite-like substructure with rather short Au–Si distances ($3 \times d_{\text{Si-Au}} = 2.494$ Å, $1 \times d_{\text{Si-Si}} = 2.762$ Å; $d_{\text{Au-Au}} = 2.936$ Å; $r_{\text{Au}} = 1.442$ Å, $r_{\text{Si}} = 1.176$ Å).

hP9-YbAgPb, hP9-LiCaSn, hP8-Th₃Pd₅□

$hP9$-YbAgPb is a three-fold superstructure of $hP3$-SrPtSb and its symmetry is reduced by an *isomorphic* transition of index three. The structure contains one flat honeycomb layer sandwiched between two puckered ones, with all 6-rings decorated alternatingly by Ag and Pb and bicapped by Yb. The Ag–Pb distances are, with $d_{\text{Ag-Pb}} = 2.813$ Å, much smaller than the sum of the atomic radii, $r_{\text{Ag}} + r_{\text{Pb}} = 1.445 + 1.750 = 3.195$ Å, indicating strong bonding. The Pb–Pb distances between adjacent puckered layers with 3.354 Å are significantly shorter than $2 \times r_{\text{Pb}} = 3.500$ Å. The very short distance of 3.275 Å between the Yb, which is sandwiched between adjacent puckered layers, and the four nearest Pb atoms, is also a sign of the cationic character of Yb ($r_{\text{Yb}} = 1.940$ Å, $r_{\text{Yb}^{3+}} = 1.042$ Å).

A t2 symmetry reduction to a trigonal space group leads to the structure of $hP9$-LiCaSn. Due to the loss of the mirror planes perpendicular to the [001] direction, all three hexagon-layers can be puckered now. The shortest intralayer Li–Sn distances correspond with 2.939 Å approximately to the sum of the atomic radii, $r_{\text{Li}} + r_{\text{Sn}} = 1.520 + 1.405 = 2.925$ Å, while the interlayer distances are significantly larger. Ca-Li distances with 3.138 Å are shorter than the sum of radii, $r_{\text{Li}} + r_{\text{Ca}} = 1.520 + 1.974 = 3.494$ Å.

Another derivative structure of $hP3$-SrPtSb is $hP8$-Th₃Pd₅□, whose symmetry is reduced by a k3 transition and a transformation to a three times larger unit cell.

The □ marks an empty position on a 6-ring, which leads to a strongly distorted 5-ring. Each layer can be described by a covering of patches, which consist of a triangle edge- and vertex-connected with altogether six pentagons. All patches have the same orientation and decorate the vertices of a hexagonal lattice in a way that each patch overlaps with another patch by one pentagon. All Pd-pentagons are bicapped by Th atoms. The intralayer Pd–Pd distances are between 2.724 and 2.875 Å, close to the sum of the radii of 2.752 Å. The shortest Th–Pd distances with 2.933 Å are significantly shorter than the sum of radii, $r_{Th} + r_{Pd} = 1.798 + 1.376 = 3.174$ Å.

hP18-Ti$_5$Ga$_4$; hP24-Er$_2$RhSi$_3$ and hP24-YLiSn

The structure of $hP18$-Ti$_5$Ga$_4$ can be obtained from $hP6$-ZrBeSi by a k3 transition coupled with a cell transformation, replacement of Be by Ga, as well as Zr and Si by Ti and Ga, respectively. The 6-rings are planar but strongly trigonally distorted. According to Hoffmann and Pöttgen (2001), this relates this structure to the Nowotny chimney ladder phases (see, for instance, Fredrickson *et al.* (2004*a*); Fredrickson *et al.* (2004*b*)).

Reducing the symmetry of $hP6$-ZrBeSi by an i4 transition as well as doubling the lattice parameters b and c leads to the $hP24$-Er$_2$RhSi$_3$ structure type. Similar to the case of $hP12$-U$_2$RuSi$_3$, the slightly puckered polyanionic [RhSi$_3$] network consists of Si- decorated 6-rings that are connected with each other via Rh atoms. Consequently, each Si$_6$-ring is surrounded by six Rh$_2$Si$_4$-rings. Intralayer Si–Si distances are 2.345 Å, almost exactly those in the element (2.352 Å), those between Si and Rh with 2.351 Å are significantly shorter compared to the sum of radii $r_{Si} + r_{Rh} = 1.176 + 1.345 = 2.521$ Å. Er–Rh distances with 2.992 Å are also slightly shorter than the sum of radii, 3.079 Å.

The noncentrosymmetric structure of $hP24$-YLiSn can be obtained from that of centrosymmetric $hP24$-Er$_2$RhSi$_3$ by removal of the inversion center, a t2 symmetry reduction. The strongly puckered 6-rings are here alternatingly decorated by Li and Sn, and bicapped by Y. The shortest intralayer Li–Sn distances with 2.581 Å are definitely shorter than the sum of the atomic radii 2.925 Å listed above. The interlayer Li–Sn distances 2.990 Å are close to this value. The shortest Y–Sn distance amounts to 3.153 Å, close to the sum of atomic radii, 3.181 Å.

For a detailed discussion of the orthorhombic/monoclinic tree shown in Fig. 7.24, which allows a much stronger decoupling of atoms, see Hoffmann and Pöttgen (2001). The only examples discussed here are the structures of $oI12$-KHg$_2$ and $oI60$-AuEuSn, already shown before (Fig. 7.23).

7.8 Topological layer structures

Many crystal structures can be topologically decribed as layer structures, i.e., as a stacking of more or less flat atomic layers. However, this does not imply that these structures are layer structures in the crystal-chemical meaning of the word, with intra- and inter-layer chemical bonding differing from one another. The description of a structure in terms of layers can be the best way to visualize a complex structure and/or to reveal relationships to other structures. A few examples out of the many structure types that could be described in this way will be discussed.

Atomic layers may be considered to act as a kind of interface between different structural parts of complex structures. If these regions are symmetrically related, then only a few space groups can account for it. This has already been recognized by Samson (1964), who pointed out that every special Wyckoff position in the space groups 196 $F23$, 195 $P2_13$, 203 $Fd\bar{3}$, 205 $Pa\bar{3}$, 209 $F432$ (216 $F\bar{4}3m$), and 219 $F\bar{4}3c$, places a least one point on the set of $\{110\}$ planes. Since all these space groups belong to the cubic crystal system, which is characterized by its threefold symmetry along the space diagonals of the unit cell, the atomic layers interpenetrate each other forming 3D frameworks. This is illustrated on the example of the complex cubic structures in the system Al–Cu–Ta, which all crystallize in the space group $F\bar{4}3m$ (Fig. 7.21). In this space group, there are eight special positions, from 4 a 0, 0, 0 to 48 h x, x, z. However, in these complex intermetallics, the atoms on layers represent only a subset of all atoms. Therefore, they are structures with planar atomic layers as interfaces rather than layer structures.

In the following, we discuss a few examples of such topological layer structures. However, there are many more structures that could be classified in this way, but are described under different headings highlighting other characteristic features. Prominent examples are the close-packed structures and their derivatives (see Section 7.2), as well as the $cI2$-W derivative structures (see Section 7.3). Furthermore, many structure types of the previously discussed REME phases and $hP3$-AlB$_2$ derivative structures can be considered to be topological layer structures, besides many others not mentioned here.

The structure of $tI12$-CuAl$_2$ (140 $I4/mcm$; Cu in $4a$ 0, 0, 1/4; Al in $8h$ 0.1541, 0.6541, 0) can be described as a stacking of $3^2.4.3.4$ triangle/square nets located in $z = 0$ and 1/2, with the squares capped by Cu in $z = 1/4$ and 3/4 (Fig. 7.26(a)–(d)). Alternatively, it can be seen as a packing of edge-connected columns of face-sharing tetragonal antiprisms centered by Cu atoms, running along the [001] direction. The remaining space in-between has the shape of trigonal bipyramids.

Another stacking variant of $3^2.4.3.4$ triangle/square nets is realized in the $tP10$-U$_3$Si$_2$ structure type (100 $P4/mbm$; U in $2a$ 0, 0, 0 and $4h$ 0.181, 0.681, 1/2; Si in $4g$ 0.389, 0.889, 0). While in the case of $tI12$-CuAl$_2$ the nets are equally decorated with Al atoms and just related by the b-glide plane, the triangle/square nets are differently decorated in $tP10$-U$_3$Si$_2$. The net in $z = 0$ is decorated on its vertices by Si and in the centers of the squares by the much larger U atoms, while the one in $z = 1/2$ is decorated by U atoms only, which leads to almost regular triangles. The squares are regular anyway due to the fourfold rotation axis in both cases. From another point of view, the structure can be seen as consisting of a framework of vertex-sharing U-octahedra sandwiched between the dual triangle square nets.

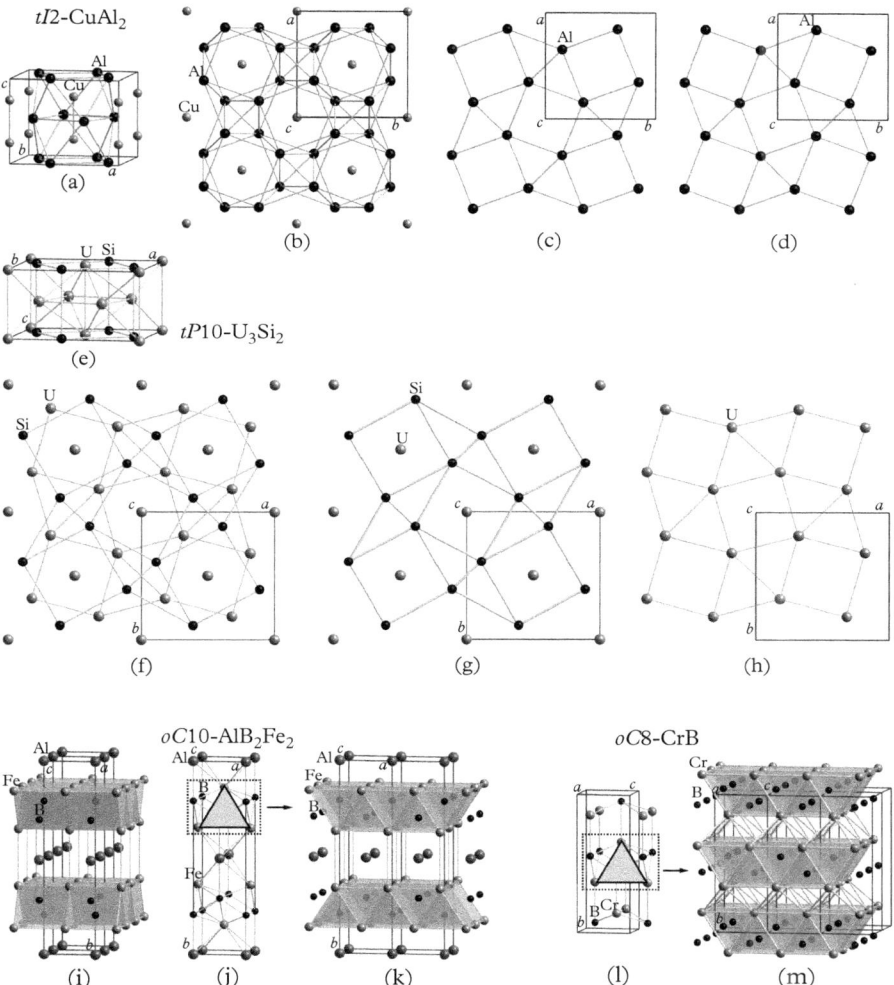

Fig. 7.26 *The structure of tI12-CuAl₂ in different views: (a) one unit cell (Cu. . . gray, Al. . . black), (b) projection along [001], sections at (c) z = 0 and (d) z = 1/2, respectively. The closely related structure of tP10-U₃Si₂ is shown in (e)–(h): (e) one unit cell (U. . . gray, Si. . . black), (f) projection along [001], sections in (g) z = 0 and (h) z = 1/2, respectively. (i)–(k) Structures of oC10-Fe₂AlB₂ (Fe. . . light gray, Al. . . medium gray, B. . . black) and (l)–(m) structures of oC8-CrB (Cr. . . gray, B. . . black) in perspective as well as in the polyhedral view. The polyhedral representations mark the slabs, which consist of trigonal prisms, and are the main structural building elements of (l) oC8-CrB. The alternating stackings of these slabs with those of the cP2-CsCl type are shown in (j).*

$oC10$-Fe_2AlB_2 (65 *Cmmm*; Al in 2*a* $0,0,0$; B in 4*i* $0,0,0.2071,0$; Fe in 4*j* $0,0.3540,1/2$) can be described in different ways: (i) as a stacking of 4^4 nets or (ii) as a prismatic stacking of $3^3.4^2$ triangle/square nets of Fe atoms, with B in the trigonal prismatic voids and Al centering the Fe cubes that are constituted from them (Fig. 7.26(i)–(k)). From another point of view, it can be seen as a combination of modules from the $oC8$-CrB and the $cP2$-FeAl ($cP2$-CsCl type) structures (Jeitschko, 1969), as a stacking of cubic and hexagonal slabs (Fig. 7.26(l)–(m)). In $oC10$-Fe_2AlB_2, the slabs of boron-centered trigonal prisms are arranged in a sequence ABBA; those in $oC8$-CrB show a sequence ACCA. B means a shift of the slab by half a triangle edge, C by half the trigonal prism height.

Layer structures like that of $aP20$-RESn$_3$ (RE = La-Nd, Sm) (Fornasini *et al.*, 2003), consist of one single layer that runs through the unit cell at an odd angle, so that it needs several unit cells until it enters another unit cell in the same way as the first one. Consequently, it has a period of several unit cells and such layers are stacked with shifted copies of themselves (also see Subsection 5.3.3).

7.9   Long-period (columnar) structures

We have already discussed several examples of long-period structures in Chapter 6 for the elements Rb, C, Sr, Ba, Bi, Sm, etc. In all these cases, the structures can be described either as modulated or as composite (host/guest) structures. In the case of intermetallic compounds, examples of long-period structures already discussed are the anti-phase domain structures in the system Au–Cu such as $oI40$-CuAu, or polytypic structures such as the Laves phase $hR42$-Mg(Ag$_{0.1}$Zn$_{0.9}$) (Fig. 7.15(l)–(o)), the Zintl compounds in the homologous series (BaSn$_3$)$_m$[Ba(Sn$_y$Bi$_{1-y}$)$_3$]$_n$ (Fig. 7.6) or the REME phase $oI60$-EuAuSn (Fig. 7.23).

The origin of the long period in one lattice direction, which is a multiple of the periods in the two other ones, can be a misfit between different parts of the structure and/or it can be electronically determined. In the case of host/guest structures, the dimensions of the host and the guest may be such that the periods of the two systems coincide only after m_h periods of the host and n_g of the guest partial structure, defining in this way the period of the total system to $c = m_h \times c_h = n_g h \times c_g$, with c_h and c_g the lattice parameter of the host and the guest partial structure, respectively, and c that of the combined structure in the direction of the long period. If the ratio c_h/c_g is irrational then the composite structure becomes incommensurate.

A modulated structure results from a modulation of a periodic basic structure (period c_b) with a displacive and/or substitutional modulation wave (wavelength λ), which can be commensurate or incommensurate depending on the ratio λ/c_b of the periods of the modulation wave and that of the basic structure. Commensurately modulated structures and superstructures differ only in the way in which they are described. If the deviation of the superstructure from the basic structure can be described more easily using a modulation wave, then this is the approach of choice. In all these cases, by a proper projection into the unit cell of the basic structure an average structure can be derived with $c_{av} = c_b$.

Nowotny chimney ladder structures, for instance, can be described as helical composite structures with composition-dependent periods (Fredrickson *et al.*, 2004*a*; Fredrickson *et al.*, 2004*b*; Sun and Lin, 2007) (Fig. 7.27). The general

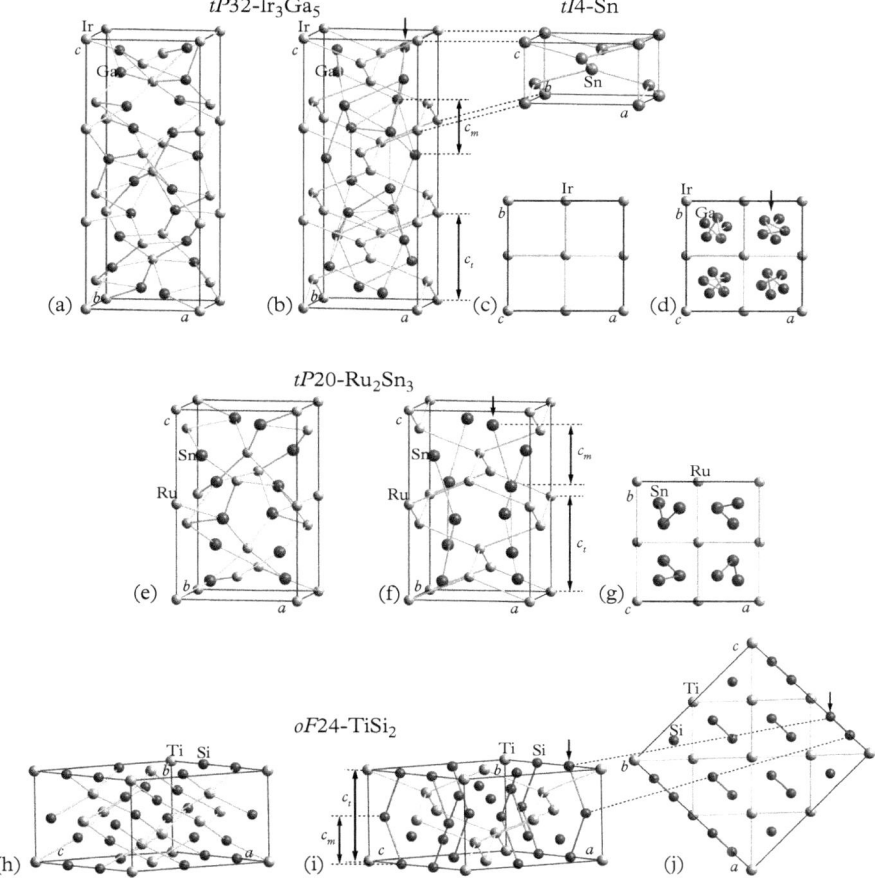

Fig. 7.27 *The structure of three Nowotny chimney ladder phases: (a, b, and d) tP32-Ir$_3$Ga$_5$ (t = 3, m = 5), (e–g) tP20-Ru$_2$Sn$_3$ (t = 2, m = 3), and (h–j) oF24-TiSi$_2$ (t = 12, m = 2). In (a), (e), and (h) stick bonds are shown between atoms in bonding distance (d$_{Ir-Ga}$ < 2.7Å; d$_{Ru-Sn}$ < 2.8Å; d$_{Ti-Si}$ < 2.6Å), while in (b, d), (f, g) and (i, j) they mark the host and guest substructures. In (d), (g), and (j), the structures are projected along the averaged guest-chain direction with period c$_m$, as indicated by the arrows. The period, c$_t$, of the host structure, and the period of the averaged guest structure, c$_m$, are indicated in (b), (f), and (i). In (c), two different projections of the structure of tI4-Sn are shown in relationship to that of a structure motif of tP32-Ir$_3$Ga$_5$.*

composition of this class of compounds is A_tB_m, with A a transition metal element from of groups 4–9, and B a main group element from groups 13–15, and t and m are integers with $1.25 \le m/t \le 2$. The host structure is formed by the A atoms, the guest structure by the B atoms. The actual lattice parameter results in $c = (2t - m)c_{av}$, and may reach more than 300 Å; it can even become incommensurate as observed in a few cases (Rohrer *et al.*, 2000; Rohrer *et al.*, 2001). An example is $(Mo,Rh)_{11}Ge_{18}$, a mutually modulated composite structure, which has satellite vectors $\mathbf{q_h} = 0.364c^*$ for the host and $\mathbf{q_g} = 0.389c^*$ for the guest substructure.

Examples of Nowotny chimney ladder phases are shown in Fig. 7.27: $tP32$-Ir_3Ga_5 (118 $P\bar{4}n2$), $tP20$-Ru_2Sn_3 (116 $P\bar{4}c2$), and $oF24$-$TiSi_2$ (70 $Fddd$) exhibit $t = 3, 2, 1$ host periods, c_t, and at the same time $m = 5, 3, 2$ guest periods, c_m, respectively, per unit cell. In Fig. 7.27(b), one sees the helical arrangement of the Ga atoms in the Ir diamond-like host framework. The different periodicities of the host/guest substructures are indicated. In each case, the host structure is formed by the TM atoms arranged in the form of tetra-helices, with the guest helices enclosed therein. The host substructure can be seen as constituting t modules of the $tI4$-Sn structure (see Fig. 7.27(b), (c)), while the guest appears as chains that are modulated by the interactions with the host and give rise to a superstructure of the host due to a size misfit.

Further examples are: $tP36$-Ir_4Ge_5 (116 $P\bar{4}c2$), $tP120$-$Mn_{11}Si_{19}$ (118 $P\bar{4}n2$), $tI56$-$Rh_{17}Ge_{22}$ (122 $I\bar{4}2d$), and $tP192$-$V_{17}Ge_{31}$ (118 $P\bar{4}n2$). The chemical composition and therewith the size of the unit cell is governed by the number of valence electrons per TM atom, which should be 14 (Fredrickson *et al.*, 2004a; Fredrickson *et al.*, 2004b). For instance, for $tP20$-Ru_2Sn_3, this reads $(2 \times 8 + 3 \times 4)/2 = 14$ or for $tP32$-Ir_3Ga_5, $(3 \times 9 + 5 \times 3)/3 = 14$.

An example of another group of long-period structures is $hP556$-$FeZn_{10}$ (194 $P6_3/mmc$) (Belin and Belin, 2000), a Hume-Rothery phase.

7.10   Hierarchical and modular structures

Formally, hierarchical structures can be derived from basic structures by replacing single atoms by groups of atoms. In a next step, the single atoms of the so-obtained structure can be substituted again by structural subunits (clusters) on a larger scale. Consequently, the same packing principles apply to the smaller and the larger scale(s). There are many examples known (see, e.g., Bodak *et al.* (2006), Deiseroth and Biehl (1999)):

- $cP8$-Cr_3Si → $cP156$-$K_{29}NaHg_{48}$ (see Fig. 7.28)
- $cI2$-W → $cI44$-$Ce_6Ni_6Si_2$
- $cP3$-$CaTiO_3$ → $cP39$-Mg_2Zn_{11}
- $cF16$-MCu_2Al → $cF116$-$Ce_3Pd_{20}Ge_6$
- $cF116$-Th_6Mn_{23} → $cF1124$-$Tb_{117}Fe_{52}Ge_{112}$
- quasiperiodic structures.

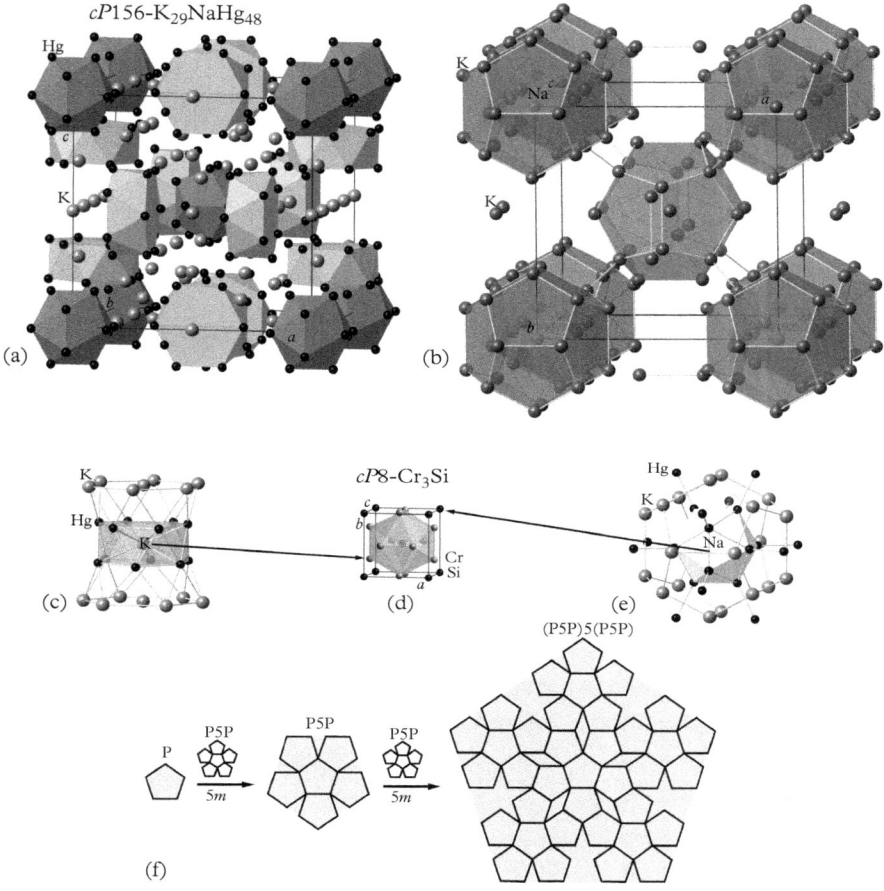

Fig. 7.28 *The structure of cP156-K$_{29}$NaHg$_{48}$ in relation to that of cP8-Cr$_3$Si. In (a), the icosahedral Na@Hg$_{12}$ clusters at the corners and the center of the unit cell are shown, while in (b) the second, pentagon-dodecahedral coordination shell consisting of 20 K atoms, is highlighted (Hg atoms are not shown). The third coordination shell is a Hg$_{12}$ icosahedron again. These clusters formally replace the Si atoms of the cP8-Cr$_3$Si unit cell, while the Cr atoms are substituted by hexagonal antiprismatic K@Hg$_{12}$ clusters. In (f), the fractal generation of pentagon tilings is shown: in each step, all pentagons are replaced by a patch of six pentagons, P5P, and rescaled to the original size of the pentagons. The scale of the structure images is 80% of the usual one.*

All self-similar structures, such as quasiperiodic or fractal ones, can be described as hierarchical structures as well. Due to self-similarity, particular structure motifs appear in the structure over and over on a larger and larger scale. In the case of the Penrose tiling, this can be realized by applying the substitution rule, illustrated in Subsection 3.2.3, Fig. 3.9. A fractal substitution is shown in

Fig. 7.28(f). There, each pentagon is replaced with a patch of six pentagons over and over. This can also be described as a process of local twinning. Each pentagon is reflected on each of its edges, arriving from P to P5P, and so on. The empty space remaining in such a generation process can be filled again with pentagons, yielding a structure model for a decagonal quasicrystal if done properly (Steurer, 2006a).

Modular structures are composed of parts (modules) of other structures. An example was discussed in Section 7.8 with the structure of $oC10$-Fe$_2$AlB$_2$. It can be seen as a combination of modules from the $oC8$-CrB and the $cP2$-FeAl ($cP2$-CsCl type) structures (Jeitschko, 1969), as a stacking of cubic and hexagonal slabs (Fig. 7.26(j)–(l)).

There are also many examples known for structures that can be described as hybrids between hierarchical structures and modules of other structures. We illustrate such a case using the example of $cP792$-V$_{11}$Cu$_9$Ga$_{46}$ (Lux et al., 2012) (see Fig. 7.29).

This structure can be described by replacing all the atoms in the $cP8$-Cr$_3$Si structure by so-called supercubes, Ga@V$_8$Ga$_{68}$. Such a supercube consists of a Ga-centered Ga-cuboctahedron, Ga@Ga$_{12}$, surrounded by triangle-face-sharing V@Ga10 centaur polyhedra. A centaur polyhedron is a hybrid between two different polyhedra, in our case by half a cube and half a icosahedron. The remaining empty space between the supercubes is filled with $cP2$-CsCl-like structural units, V@Ga$_8$ and Cu@Ga$_8$, respectively (Fig. 7.29 (a)–(c)). The lattice parameters of $cP792$-V$_{11}$Cu$_9$Ga$_{46}$ correspond to a $(8 \times 8 \times 8)$-fold superstructure of a simple bcc unit cell. However, such a 512-fold superstructure of $cI2$-W would have 1024 atoms compared to the 792 in our case. Consequently, our structure cannot be simply described as a superstructure.

The cuboctahedron, Ga@Ga$_{12}$, represents a structural subunit of the $cF4$-Cu structure, which is adopted by Ga under very high pressure. V has the $cI2$-W type structure. The phase diagrams of Cu–Ga and Ga–V show many intermetallic compounds, whereas that of Cu–V is empty. In the range between 19% and 27% Ga, Cu and Ga form a phase with the $cI2$-W type structure. In the system Ga–V a compound, $hR147$-V$_8$Ga$_{41}$, exists, which is closely related to $cP792$-V$_{11}$Cu$_9$Ga$_{46}$ (Girgis et al., 1975).

$hR147$-V$_8$Ga$_{41}$ (Girgis et al., 1975) is the simplest example of a structure built from these supercubes, Ga@V$_8$Ga$_{68}$. There, the supercubes occupy the Hg-sites of the $hR3$-Hg structure type, corresponding to a hierarchical structure. The supercubes are linked via vertices of the V@Ga$_{10}$ polyhedra. Adding Cu to this intermetallic phase, which does not form any compounds with V in the binary system, leads to the complex structure of $cP792$-V$_{11}$Cu$_9$Ga$_{46}$. Indeed, for atomic radii of Cu, Ga, and V of 1.28 Å, 1.22 Å, and 1.31 Å, respectively, the shortest atomic distance between Cu and V is with 2.69 Å rather large. In contrast, the shortest distances between Ga–V and Ga–Cu correspond with 2.50 Å and 2.53 Å more or less to the sum of the atomic radii.

cP792-V$_{11}$Cu$_9$Ga$_{46}$

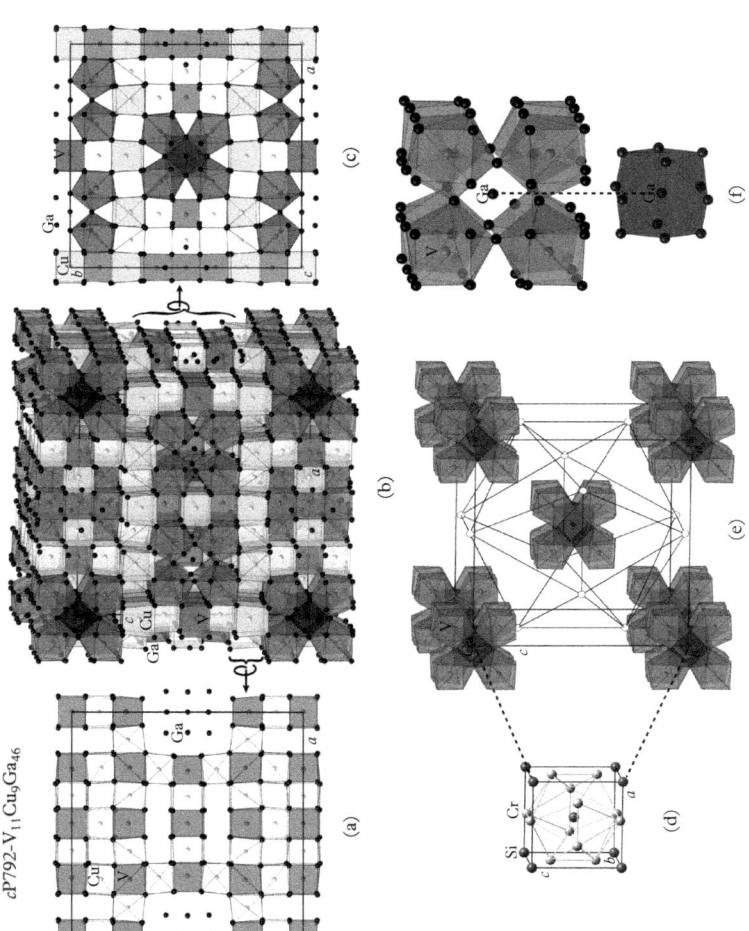

Fig. 7.29 The structure of cP792-V$_{11}$Cu$_9$Ga$_{46}$ (Lux et al., 2012) in different representations. The hierarchical structural part (e) consists of supercubes (f) on all the sites of the cP8-Cr$_3$Si structure (d). For clarity, the supercubes on the Cr sites are only marked by open circles. A supercube consists of a cuboctahedron centered by Ga, Ga@Ga$_{12}$, surrounded by triangle-face-sharing V@Ga10 centaur polyhedra, hybrids between cubes, and icosahedra (f). The space in-between the supercubes is filled with cP2-CsCl-like structural units, V@Ga$_8$ and Cu@Ga$_8$, respectively. Their arrangement is illustrated by sections of the structure shown in (b) at (a) $0.1 < z < 0.3$ and (c) $0.3 < z < 0.7$, respectively. Compared to our usual scale, the structures depicted in (a)–(c) and in (e) are shrunk by 50%.

7.11 Structures with one dominating element

In this section, structures will be discussed that contain at least 90% of one dominating element. How common are such structures, in which binary and ternary intermetallic systems do they occur? In the cases where each minority atom of type A is surrounded by two shells of majority atoms of type B, interesting properties may emerge. Such a case could be seen as an ordered distribution of atoms of type A in a matrix consisting just of B atoms, i.e., just as an ordered perturbation of the structure of B. No A atom would have another A atom as nearest or next-nearest neighbor. This is realized approximately in the structure of $cF184$-$ZrZn_{22}$, for instance (Fig. 7.30). Between two Zr atoms, there are, with four exceptions, two or more Zn atoms. The minimum Zr–Zr distance is > 6 Å.

Examples are:

- 95.7% Zn: $cF(420-18x)$-$MoZn_{22-x}$ (216 $F\bar{4}3m$) (Nasch and Jeitschko, 1999): despite the different decoration with atomic species, the structure is quite similar to those of $Mg_{44}Rh_7$, $Fe_{22}Zn_{78}$, Na_6Tl, and Mg_6Pd.

- 95.7% Zn: $cF184$-$ZrZn_{22}$ (227 $Fd\bar{3}m$): $MoBe_{22}$; $ReBe_{22}$; WBe_{22}.

- 94.1% Zn: $oC68$-$TiZn_{16}$ (63 $Cmcm$) (Chen *et al.*, 1995): the structure can be described as a stacking of layers A and B in the sequence ABAB along [100]. The layers of type A are flat and contain both Ti and Zn atoms, while those of type B are puckered and consist of Zn atoms only. The Ti atoms have exclusively Zn neighbors; the AETs corresponds to CN15 FK-polyhedra. Representative: $NbZn_{16}$.

- 93.1% Hg: $cI174$-Cs_2Hg_{27} (204 $Im\bar{3}$): One Cs–TM intermetallic with this structure type is known, with TM = Hg. The structure can be seen as a hierarchical substitution of the AgI structure type (Hoch and Simon, 2008).

- 92.9% Zn: $mC28$-$FeZn_{13}$ (12 $C2/m$) (Belin *et al.*, 2000; Liu *et al.*, 2008). The Fe atoms center Zn_{12} icosahedra, which form chains along the [001] direction by sharing vertices. Eight of such icosahedra surround Zn_2 dumbbells filling the space between them. The stability region of this phase reaches from the nominal Fe composition of 7.14% down to 5.9%, indicating that Fe atoms can be replaced by Zn to some extent but not vice versa.

- 92.9% Zn: $cF112$-$NaZn_{13}$ (226 $Fm\bar{3}c$): $LaCo_{13}$; $BaBe_{13}$; $BaCu_{13}$; $HfBe_{13}$; $MgBe_{13}$; $PuBe_{13}$; $Be_{13}Sb$; $ScBe_{13}$; $ThBe_{13}$; UBe_{13}; YBe_{13}; $ZrBe_{13}$; $CaZn_{13}$; $RbCd_{13}$; $EuZn_{13}$; $LaZn_{13}$; $Ba_{13}In$; $Ba_{13}Tl$; $BaZn_{13}$; $CaBe_{13}$.

- 92.3% Mn: $tI26$-$ThMn_{12}$ (139 $I4/mmm$) (Florio *et al.*, 1952) (Fig. 7.31): columns of square-face-sharing $ThMn_{20}$ polyhedra linked via vertices; it can be seen as intergrowth of $CaCu_5$- and Zr_4Al_3-blocks, or that in the $CaCu_5$ structure one half of the Ca atoms are replaced by dumbbells of TM atoms: $2RT_5 - R + 2T = RT_{12}$ (R. . . RE or Ca, T. . . TM). The only other binary representative is $SmZn_{12}$, which is strongly magnetic; the structure can be stabilized by ternary substitutions; however, this decreases the

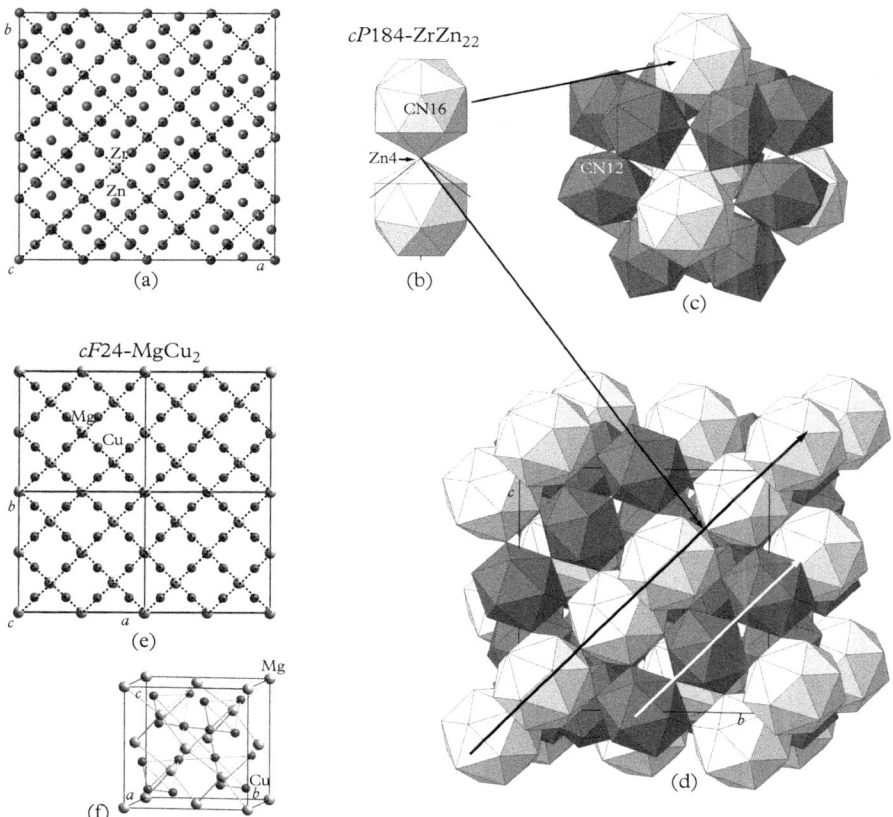

Fig. 7.30 *(a)–(d, (f) Different views of the structure of cF184-ZrZn$_{22}$ in comparison with the Laves phase cF24-MgCu$_2$ (e, f). The structure can be described as a vertex-connected packing of Zn-CN16 FK-polyhedra around the Zr atoms and Zn-CN12 icosahedra centered by the Zn3 atoms. The Zr atoms occupy the positions of Mg in the Laves phase, and the Zn3 atoms those of Cu. The projected structure of (a) cF184-ZrZn$_{22}$ is very similar to that of (e) 2 × 2 × 2 block of unit cells of (f) the cubic Laves phase.*

fraction of the major element to below 90%.: $R_{12-x}M_x$, with R a light RE, and M = Ti, V, Cr, Mo, W, or Al ($1 < x < 2$); CrBe$_{12}$; TiBe$_{12}$; WBe$_{12}$; CeMg$_{12}$; DyMn$_{12}$; DyZn$_{12}$; ErMn$_{12}$; ErZn$_{12}$; GdMn$_{12}$; GdZn$_{12}$; HoZn$_{12}$; LuZn$_{12}$; Mg$_{12}$Pr; TmMn$_{12}$; YMn$_{12}$; ScZn$_{12}$; SmZn$_{12}$; TbZn$_{12}$; TmZn$_{12}$; UZn$_{12}$; YZn$_{12}$; Mg$_{12}$Ce.

- 92.3% Al: $cI26$-WAl$_{12}$ (204 $Im\bar{3}$): MoAl$_{12}$; ReAl$_{12}$; NbBe$_{12}$; PdBe$_{12}$; PtBe$_{12}$; RuBe$_{12}$; TaBe$_{12}$; VBe$_{12}$.

- 91.7% Cd: $tI48$-BaCd$_{11}$ (141 $I4_1amd$) (Sanderson and Baenziger, 1953): packing of BaCd$_{22}$ polyhedra, which share square faces within layers and

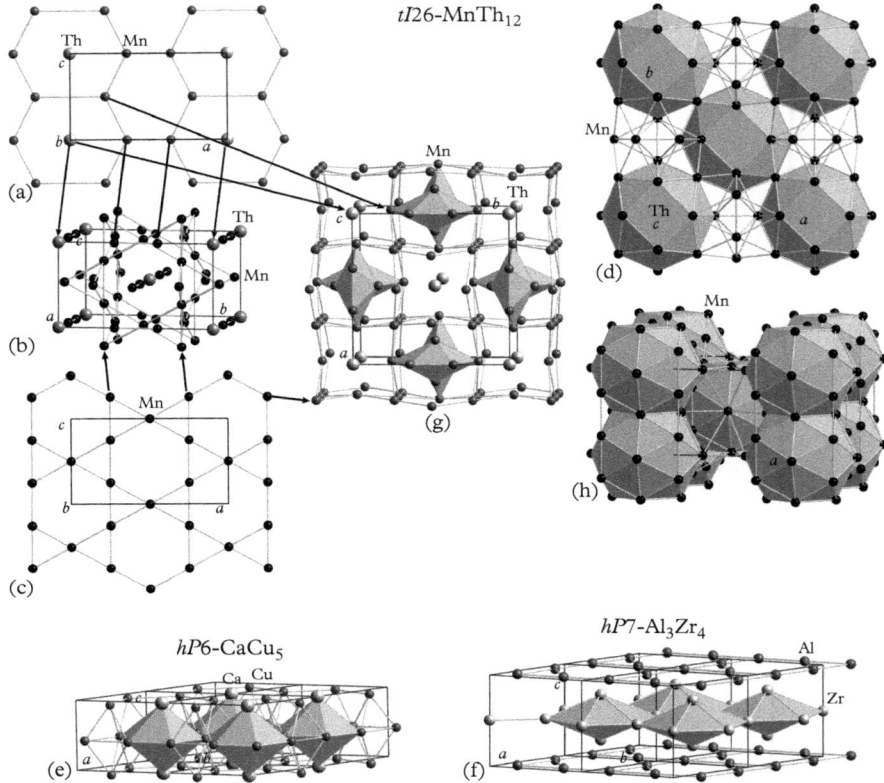

Fig. 7.31 *Different views of the structure of tI26-ThMn$_{12}$ in comparison with those of (e) hP6-CaCu$_5$ and (f) hP7-Zr$_4$Al$_3$. The large Th atoms sit in half of the hexagons in a honeycomb layer of Mn atoms (a), the other half of the hexagons is occupied by dumbbells of Mn atoms perpendicular to it, which can be visualized by hexagonal bipyramids (hbps) as illustrated in (g). The honeycomb layers are sandwiched between Mn Kagomé nets (c). This is similar to the structure of (f) hP7-Zr$_4$Al$_3$, and also bears some resemblance to the structure of (e) hP6-CaCu$_5$. From another perspective, the structure of tI26-ThMn$_{12}$ can also be described as a bcc packing of columns of square-face-sharing ThMn$_{20}$ polyhedra, which are linked to each other via their vertices (d, h).*

    vertices in [001] direction; representatives: SrCd$_{11}$, CaZn$_{11}$; REZn$_{11}$ with RE=La, Ce, Pr, Nd, Yb;

- 91.7% Hg: cP36-BaHg$_{11}$ (221 $Pm\bar{3}m$): CeCd$_{11}$; CaHg$_{11}$; EuCd$_{11}$; LaCd$_{11}$; NdCd$_{11}$; PrCd$_{11}$; PuCd$_{11}$; SmCd$_{11}$; SrCd$_{11}$; ThCd$_{11}$; UCd$_{11}$; KHg$_{11}$; RbHg$_{11}$; SrHg$_{11}$.

- 90.9% Al: cF176-VAl$_{10}$ (227 $Fd\bar{3}m$): The V-centered Al-icosahedra form a large tetrahedron with the icosahedra vertex-connected along the edges (four icosahedra/edge). Perpendicular to the threefold axes, the V atoms form layers with Kagomé structures. Another representative: $Ba_{10}Ga$.

- 90.7% Zn: $hP556$-$Fe_{13}Zn_{126}$ (194 $P6_3/mmc$) (Belin and Belin, 2000; Okamoto *et al.*, 2014). All 52 Fe atoms occupy exclusively the centers of ordered as well as disordered Zn_{12} icosahedra. Furthermore, there are also Zn-centered Zn_{12} icosahedra and Zn-centered Zn_{16} icosioctahedra, as well as a few "glue" Zn-atoms. The ordered (regular) Fe-centered icosahedra are connected with each other by sharing vertices and faces, respectively, forming slabs perpendicular to the [001] direction in this way. The slabs are linked to each other via face-sharing with Zn_{16} icosioctahedra. Around $z = 0$ and $z = 1/2$, the atoms form a slightly puckered 3.4.6.4 Archimedean small-rhombitrihexagonal tiling. The structure indicates that the chemical bonds between Fe and Zn are stronger than those between Zn and Zn, similar as is the case for the other complex phases, $cF408$-$Fe_{11}Zn_{40}$ and $mC28$-$CoZn_{13}$, in the system Fe–Zn.

It is remarkable that Zn-compounds are the most common representatives for this kind of structure. $cF184$-$ZrZn_{22}$ (227 $Fd\bar{3}m$, setting 2; Zr in $8a$ 1/8, 1/8, 1/8; Zn1 in $96g$ 0.0617, 0.0167, 0.3192; Zn2 in $48f$ 0.4867, 1/8, 1/8; Zn3 in $16d$ 1/2, 1/2, 1/2; Zn4 in $16c$ 0, 0, 0) can be described as a cluster-decorated cubic Laves phase (Fig. 7.30). The Zr atoms are surrounded by CN16 FK-polyhedra, which decorate the vertices of a 4-connected D-net, and are linked to each other via the Zn4 atoms. The Zn3-centered CN12 polyhedra (icosahedra) again form a 4-connected D-net, with smaller distances between the vertices, however. The icosahedra are linked via Zn2 atoms to each other and via Zn1 atoms to the CN16 polyhedra. A fraction of the Zn2 and Zn3 atoms is separated by more than 5 Å from the closest Zr atoms, and has at least one Zn atom between itself and Zr.

According to Ilyushin and Blatov (2009), two ternary superstructures to the $cF184$-$ZrZn_{22}$ structure type exist: the $cF184$-$CeCr_2Al_{20}$ structure type and the $cF184$-$Mg_3Ti_2Al_{18}$ structure type, with more than 100 representatives. Substituting the central atoms of the icosahedra, Zn3 = B, gives the composition AB_2X_{20} if Zr is given the label A. Consequently, in the case of $cF184$-$CeCr_2Al_{20}$, Ce occupies the centers of the CN16 polyhedra, and Cr those of the icosahedra. However, in the case of these superstructures, the majority element does not contribute more than 90% to the chemical composition anymore.

7.12 Alkali/alkaline earth metal compounds (groups 1 and 2 only)

The electronegativities and atomic radii of the group 1 and 2 elements range from $\chi = 0.7$ and $r = 2.66$ Å for Cs ($M = 8$) to 1.0 and 1.52 Å for Li ($M = 12$), respectively (Fig. 7.32). The electronegativities vary in a small range only, in contrast to

1	2	3	4	5	6	7	8	9	10	11	12	13	14	15	16
12 Li 1.0 1.52	77 Be 1.5 1.11														
11 Na 0.9 1.86	73 Mg 1.2 1.60											80 Al 1.5 1.43			
10 K 0.8 2.27	16 Ca 1.0 1.97	19 Sc 1.3 1.61	51 Ti 1.4 1.45	54 V 1.6 1.31	57 Cr 1.6 1.25	60 Mn 1.5 1.37	61 Fe 1.8 1.24	64 Co 1.8 1.25	67 Ni 1.8 1.25	72 Cu 1.9 1.28	76 Zn 1.6 1.34	81 Ga 1.6 1.22	84 Ge 1.8 1.23		
9 Rb 0.8 2.48	15 Sr 1.0 2.15	25 Y 1.2 1.78	49 Zr 1.4 1.59	53 Nb 1.6 1.43	56 Mo 1.8 1.36	59 Tc 1.9 1.35	62 Ru 2.2 1.33	65 Rh 2.2 1.35	69 Pd 2.2 1.38	71 Ag 1.9 1.45	75 Cd 1.7 1.49	79 In 1.7 1.63	83 Sn 1.8 1.41	88 Sb 1.9 1.45	
8 Cs 0.7 2.66	14 Ba 0.9 2.17	33* La 1.1 1.87	50 Hf 1.3 1.56	52 Ta 1.5 1.43	55 W 1.7 1.37	58 Re 1.9 1.37	63 Os 2.2 1.34	66 Ir 2.2 1.36	68 Pt 2.2 1.37	70 Au 2.4 1.44	74 Hg 1.9 1.50	78 Tl 1.8 1.70	82 Pb 1.8 1.75	87 Bi 1.9 1.55	91 Po 2.0 1.67
7 Fr 0.7	13 Ra 0.9 2.15	48+ Ac 1.1 1.88													

*Lanthanoids	32 Ce 1.1 1.87	31 Pr 1.1 1.82	30 Nd 1.1 1.81	29 Pm 1.1 1.63	28 Sm 1.2 1.62	18 Eu 1.2 2.00	27 Gd 1.2 1.79	26 Tb 1.1 1.76	25 Dy 1.2 1.75	24 Ho 1.2 1.74	23 Er 1.2 1.73	22 Tm 1.3 1.72	17 Yb 1.1 1.94	20 Lu 1.3 1.72
+Actinoids	47 Th 1.3 1.88	46 Pa 1.5 1.80	45 U 1.4 1.39	44 Np 1.4 1.30	43 Pu 1.3 1.51	42 Am 1.1	41 Cm 1.3	40 Bk 1.3	39 Cf 1.3	38 Es 1.3	37 Fm 1.3	36 Md 1.3	35 No 1.3	34 Lr 1.3

Fig. 7.32 *Elements constituting the compounds discussed in Section 7.12 are shaded in gray in the periodic table. Mendeleev numbers (top left in each box), Pauling electronegativities χ (relative to $\chi_F = 4.0$) (bottom left in each box), and atomic radii (half of the shortest distance between atoms in the crystal structure at ambient conditions) (bottom right in each box) of the metallic elements are given.*

the larger variation of the atomic radii. From their Mendeleev numbers, the group 2 elements Be ($\chi = 1.5$, $r = 1.11$ Å) and Mg ($\chi = 1.2$, $r = 1.60$ Å) with $M = 77$ and $M = 73$, respectively, are assigned to group 12. This is justified by their chemical properties, which differ from the other alkaline earth (AE) metals. In particular, Be, with its tendency to form covalent bonds, behaves quite differently. Within the group 1 and 2 elements, it just forms compounds of the $cF112$-NaZn$_{13}$ type. Be occupies the Zn position and forms AE-centered AEBe$_{24}$ snub cubes (AE = Ba, Sr, Ca, Mg) and Be-centered Be$_{12}$ icosahedra sharing triangle faces with the snub cubes in-between. The radii ratios of Be and the AE atoms range from 0.51 to 0.69. As is mentioned later on (Section 10.8), the structure type $cF112$-NaZn$_{13}$ is frequently found in hard-sphere self-assembled colloidal systems with size ratios around 0.49–0.63.

Among the 6441 binary and 13 026 ternary intermetallics, just 40 and 2 compounds, respectively, are known, which are built from elements of groups 1 and 2 only. Of the binary structures, twelve feature the simple sphere-packing structures:

- $cI2$-W: Cs$_{50}$Rb$_{50}$, Cs$_{50}$K$_{50}$, Li$_{90}$Mg$_{10}$, Ca$_{60}$Ba$_{40}$, Ba$_{50}$Sr$_{50}$, Ca$_{50}$Sr$_{50}$
- $cF4$-Cu: Li$_{81}$Mg$_{19}$, Ca$_{70}$Ba$_{30}$, Sr$_{78}$Ba$_{22}$, Ca$_{50}$Sr$_{50}$
- $hP2$-Mg: Mg$_{82}$Li$_{18}$
- $cP2$-CsCl: Sr$_{50}$Mg$_{50}$.

Nine intermetallics exhibit the Laves-phase structure type:

- $hP12$-$MgZn_2$: $K_{66.7}Cs_{33.3}$, $Na_{66.7}K_{33.3}$, $Na_{80}Ba_{20}$, $Na_{66.7}Cs_{33.3}$, $Li_{66.7}Ca_{33.3}$, $Mg_{66.7}Ba_{33.3}$, $Mg_{66.7}Sr_{33.3}$, $Mg_{66.7}Ca_{33.3}$
- $cF24$-$MgCu_2$: $Li_{66.7}Ca_{33.3}$.

Eleven more intermetallics have been reported with diverse other structure types such as:

- $tI16$-$Li_{90}Mg_{10}$ in structure type $tI16$-Li_3Mg_5
- $hR57$-$Mg_{89.5}Ba_{10.5}$ in structure type $hR57$-$Zn_{17}Th_2$
- $cF96$-$Ba_{50}Na_{50}$ in structure type $cF96$-$CdNi$
- $hP38$-$Mg_{89.5}Sr_{10.5}$ in structure type $hP38$-Th_2Ni_{17}
- $hP90$-$Mg_{80.9}Sr_{19.1}$ in structure type $hP90$-$SrMg_4$

as well as the eponymous compounds $tP20$-Li_2Sr_3, $hP26$-Cs_6K_7, $hP30$-Li_4Ba, $hP46$-$SrMg_{5.2}$, $hP94$-Sr_9Mg_{38}, and $tI252$-$Li_{44}Ba_{19}$. The latter complex structure shows Li_{19} polytetrahedral (anti-Mackay) clusters (Smetana *et al.*, 2007*a*).

Two more structure types are slightly more frequent than the others: the structure type $cF112$-$NaZn_{13}$ is featured in four compounds ($Be_{92.9}M_{7.1}$ with M = Ba, Sr, Ca, and Mg), and structure type $cF116$-Th_6Mn_{23} in three ($Li_{79.3}Sr_{20.7}$, $Mg_{79.3}Ba_{20.7}$, and $Mg_{79.3}Sr_{20.7}$).

The two ternary compounds are both quite complex: $hP(842$-$148)$-$Na_{11.3}Li_{49.0}Ba_{39.8}$ and $hR888$-$Li_{67.5}Ba_{26.5}Ca_{6.0}$. The main structural building units of the latter phase are endohedral fullerene-like clusters of the type $FK_{12}@F_{20}/FK_{32}@F_{60}$ (the subscripts give the number of atoms), which are located in Wyckoff position $6b$ 0 0 1/2 of the space group $R\bar{3}c$ (hexagonal setting) (Dshemuchadse and Steurer, 2014). Because Ba and Ca as well as Li and Na are immiscible, direct contacts between these atoms are avoided by the particular cluster shell arrangements in the structures. For instance, in the structure of $hP(842$-$148)$-$Na_{11.3}Li_{49.0}Ba_{39.8}$, Li_{26} cluster shells are formed, which are surrounded by a Ba_{28} cluster shell, in this way avoiding direct Li–Na contacts.

7.13 Alkali/alkaline earth metal compounds with TM elements

7.13.1 Compounds of Li, Na, K, Rb, or Cs with TM elements

The alkali metals have the largest atomic radii, ranging from $r_{Li} = 1.52$ Å to $r_{Cs} = 2.66$ Å and the smallest electronegativities ($\chi_{Cs} = 0.7 < \chi < \chi_{Li} = 1.0$). In contrast, the group 3–12 elements have much smaller atomic radii, ranging from $r_{Fe} = 1.24$ to $r_{La} = 1.87$ Å and much larger electronegativities

($\chi_{La} = 1.1 < \chi < \chi_{Au} = 2.4$), mainly caused by relativistic effects in the case of the heavier elements. Consequently, compounds of the alkali metals with late TM elements can show quite ionic bonding. Examples are platinides and aurides such as Cs^+Au^-, for instance. The large atomic size differences favor layer structures of the TM elements with the alkali metal atoms in-between. Examples for the layer structures are Kagomé or honeycomb nets, which can accommodate very large alkali metal atoms, or triangle/square nets for smaller ones. The most common structures, not already depicted elsewhere, are shown and discussed in the following.

Compounds of Li with TM elements

In the PCD, there are 35 binary and 2 ternary compounds of Li with TM elements listed. The five most common binary structure types of the binary Li–TM compounds are:

- $hP2$-LiRh (187 $P\bar{6}m2$): Four representatives with this structure type are known, with TM = Rh, Ir, Pt, and Pd. Ordering variant of the $hP2$-Mg type with a ratio of $c/a = 1.646$ for the LiRh compound.

- $cF32$-CuPt$_7$ (225 $Fm\bar{3}m$): Two representatives with this structure type are known, with TM = Pt and Pd. The structure corresponds to a $(2 \times 2 \times 2)$-fold superstructure of the $cF4$-Cu type.

- $oI4$-LiIr$_3$ (44 $Imm2$): Two representatives with this structure type are known, with TM = Rh and Ir. This structure type can be considered an ordered, orthorhombically distorted variant of the $hP2$-Mg type.

- $hP3$-Hg$_2$U (191 $P6/mmm$): Two representatives with this structure type are known, with TM = Pt and Pd. The structure is isopointal to $hP3$-AlB$_2$; however, the c/a-ratio is much smaller with 0.647 compared to 1.080 for $hP3$-AlB$_2$.

- $cF16$-NaTl (227 $Fd\bar{3}m$): Two representatives with this structure type are known, with TM = Cd and Zn. For the close packed structure of this classical Zintl phase, see Section 7.2.

To our knowledge, there is no structural information about ternary Li–TM–TM' phases available.

Compounds of Na with TM elements

In the PCD, there are 17 binary and no ternary compounds of Na with TM elements listed. Nine of them are Na–Hg compounds. The five most common binary structure types of the binary Na–TM compounds are:

- $cF24$-MgCu$_2$ (227 $Fd\bar{3}m$): Three representatives with this structure type are known, with TM = Pt, Au, and Ag.

- $hP18$-Na$_3$Hg (194 $P6_3/mmc$) (α-Na$_3$Hg): One representative with this structure type is known, with TM = Hg. This compound is the RT modification of Na_3Hg. This low-temperature structure could only be described with temperature-dependent triple-split positions of the Na atoms in the Hg_6 octahedral voids (Deiseroth and Rochnia, 1994).

- $hR12$-Na_3Hg (166 $R\bar{3}m$) (β-Na_3Hg): One representative with this structure type is known, with TM = Hg. The Hg atoms form a rhombohedrally distorted *ccp* packing with the Na atoms in the tetrahedral and octahedral voids. The Na atoms in the octahedral voids can only be described by multiple split positions over the whole stability range (Deiseroth and Rochnia, 1993).

- $hP3$-Hg_2U (191 $P6/mmm$): One representative with this structure type is known, with TM = Hg.

- $cF112$-$NaZn_{13}$ (226 $Fm\bar{3}c$): One representative with this structure type is known, with TM = Zn. The AET of the Na atoms corresponds to a snub cube, $Na@TM_{24}$. The TM atoms form $TM@TM_{12}$ icosahedra sharing faces with the snub cubes in-between, while the snub cubes are connected via their square faces (see Fig. 10.8).

Compounds of K with TM elements

In the PCD, there are 15 binary and no ternary compounds of K with TM elements listed. Seven of them are K-Hg compounds. The five most common binary structure types of the binary K–TM compounds are:

- $cF24$-$MgZn_2$ (194 $P6_3/mmc$): Two representatives with this structure type are known, with TM = Au and Ag.

- $cF112$-$NaZn_{13}$ (226 $Fm\bar{3}c$): Two representatives with this structure type are known, with TM = Cd and Zn.

- $oI10$-K_2Au_3 (71 $Immm$): One representative with this structure type is known, with TM = Au. The structure consists of flat hexagon/rhomb layers occupied by the Au atoms with short distances (2.68–2.80 Å, compared to 2.88 Å in the element). The K atoms are intercalated between these Au-layers. Although the K–K and Au–K distances are within the range of metallic interactions, due to the large difference in electronegativities it is assumed that K^+ cations and (Au_3^{2-}) polyanions are formed (Krieger-Beck, *et al.*, 1989).

- $oP48$-K_5Hg_7 (162 $P\bar{3}m$): One representative with this structure type is known, with TM = Hg. This structure can be derived from that of KHg_2 (a distorted $hP3$-AlB_2 structure) by replacing one-eighth of the Hg atoms by K atoms (Duwell and Baenziger, 1960).

- $aP8$-KHg (2 $P\bar{1}$): One representative with this structure type is known, with TM = Hg. The structure can be described by square-planar Hg_4 units stacked along [111]. The distances between these units (3.32 Å) are only slightly longer than those within (2.99 Å, compared to 3.01 Å in the element) (Biehl and Deiseroth, 1996).

Compounds of Rb with TM elements

In the PCD, there are 14 binary and no ternary compounds of Rb with TM elements listed. Six of them are Rb–Hg compounds. The five most common binary structure types of the binary Rb–TM compounds are:

- $cF112$-NaZn$_{13}$ (226 $Fm\bar{3}c$): Two representatives with this structure type are known, with TM = Cd and Zn.
- $oS20$-Rb$_3$Au$_7$ (65 $Cmmm$): One representative with this structure type is known, with TM = Au (Range *et al.*, 1994).
- $hP6$-CaCu$_5$ (191 $P6/mmm$): One representative with this structure type is known, with TM = Au.
- $cP46$-Rb$_3$Hg$_{20}$ (223 $Pm\bar{3}n$): One representative with this structure type is known, with TM = Hg. The structure is built of vertices sharing octacapped centered icosahedra Hg@Hg$_{12}$@Hg$_8$ (Todorov and Sevov, 2000)
- $tI48$-Rb$_5$Hg$_{19}$ (87 $I4/m$): One representative with this structure type is known, with TM = Hg. The structure can be described as an ordered defect variant of the $tI10$-BaAl$_4$ type (Biehl and Deiseroth, 1999).

Compounds of Cs with TM elements

In the PCD, there are eight binary and no ternary compounds of Cs with TM elements listed. Five of them are Cs–Hg compounds. The five most common binary structure types of the binary Cs–TM compounds are:

- $oI20$-Cs$_{0.34}$Zn$_4$ (74 $Imma$): One representative with this structure type is known, with TM = Zn (Wendorff and Röhr, 2006).
- $cI174$-Cs$_2$Hg$_{27}$ (204 $Im\bar{3}$): One representative with this structure type is known, with TM = Hg. The structure can be seen as a hierarchical substitution of the AgI structure type (Hoch and Simon, 2008).
- $aP8$-KHg (2 $P\bar{1}$): One representative with this structure type is known, with TM = Hg.
- $cF112$-NaZn$_{13}$ (226 $Fm\bar{3}c$): One representative with this structure type is known, with TM = Cd.
- $cP2$-CsCl (221 $Pm\bar{3}m$): One representative with this structure type is known, with TM = Au.

7.13.2   Compounds of Be, Mg, Ca, Sr, or Ba with TM elements

The structure-relevant chemical properties of the alkaline earth elements Be ($r_{Be} = 1.11$ Å, $\chi_{Be} = 1.5$) and Mg ($r_{Mg} = 1.60$ Å, $\chi_{Mg} = 1.2$) have more in common with the group 12 elements than with the other group 2 elements. This is accounted for by their Mendeleev numbers. The other alkaline earth metals behave similarly to the alkali metals. The most common structures, not already depicted elsewhere, are shown and discussed in the following.

Compounds of Be with TM elements

In the PCD, there are 94 binary and 24 ternary compounds of Be with TM elements listed. The five most common binary structure types of the binary Be–TM compounds are:

- $tI26$-ThMn$_{12}$ (139 $I4/mmm$): Fourteen representatives with this structure type are known, with TM = Ti, Ta, Nb, V, W, Mo, Cr, Mn, Fe, Co, Pt, Pd, Au, and Ag (see Fig. 7.31).

- $hP12$-MgZn$_2$ (194 $P6_3/mmc$): Eight representatives with this structure type are known, with TM = V, W, Mo, Cr, Re, Mn, Fe, and Ru.

- $hP19$-Be$_{15.34}$Rh$_{2.36}$ (187 $P\bar{6}m2$): Six representatives with this structure type are known, with TM = Fe, Ru, Os, Co, Rh, and Ir. The structure has some similarities with the $hP6$-CaCu$_5$ type.

- $cP2$-CsCl (221 $Pm\bar{3}m$): Seven representatives with this structure type are known, with TM = Ti, Fe, Co, Rh, Ni, Pd, and Cu.

- $cF24$-Be$_5$Au (216 $F\bar{4}3m$): Six representatives with this structure type are known, with TM = Re, Fe, Co, Pt, Pd, and Au. The structure is a superstructure of $cF24$-MgCu$_2$, derived by occupying the 8 Mg sites by 4 Au and 4 Be atoms. The 16 Cu sites are all filled with 16 Be atoms.

The only two ternary structure types of the 24 ternary Be–TM–TM' compounds are:

- $cF116$-Be$_{15}$Cu$_8$Ta$_6$ (225 $Fm\bar{3}m$): Thirteen representatives with this structure type are known for Be–Cu–(Zr, Hf, Ti, Ta, Nb), Be–Ni–(Zr, Hf, Ta, Nb), Be–Co–(Zr, Hf), and Be–Pd–(Zr, Hf). The structure (Fig. 7.33) can be derived from the $cF116$-Th$_6$Mn$_{23}$ type.

- $hR66$-Pr$_2$Mn$_{17}$C$_{1.77}$ (166 $R\bar{3}m$): One representative with this structure type is known, Pr$_2$Co$_{17}$Be$_{1.77}$. It can be regarded as a filled version of the Th$_2$Zn$_{17}$ type (Block and Jeitschko, 1986).

Compounds of Mg with TM elements

In the PCD, there are 95 binary and 34 ternary compounds of Mg with TM elements listed. The five most common binary structure types of the binary Mg–TM compounds are:

- $cP2$-CsCl (221 $Pm\bar{3}m$): Seven representatives with this structure type are known, with TM = Sc, Y, Rh, Pd, Au, Ag, and Hg.

$cF116$-$Be_{15}Cu_8Ta_6$

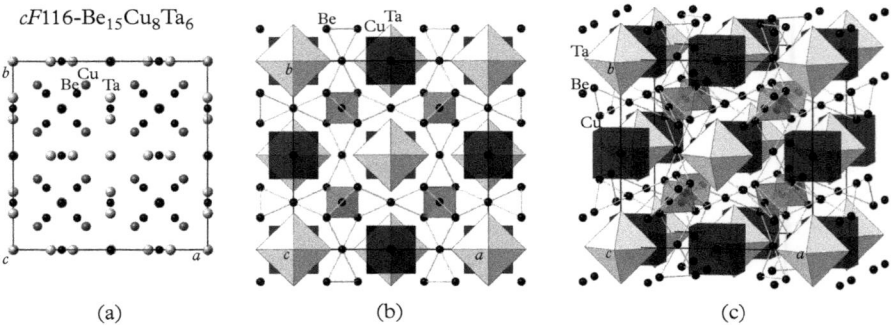

(a)   (b)   (c)

Fig. 7.33 *Different views of the structure of $cF116$-$Be_{15}Cu_8Ta_6$. (a) Projection along [010] of the atomic structure, (b) the polyhedra representation, and (c) in perspective view. The Ta atoms form octahedra, which are edge- and face-capped by Be atoms. Each vertex of the octahedra is connected to a Cu-cube, which is centered and edge-capped by Be, and face-capped by Ta. Be1 is coordinated by the aforementioned Cu-cube, Be2 and Be3 have distorted icosahedral coordination each, Be2@Be$_4$Cu$_4$Ta$_4$ and Be3@Be$_6$Cu$_3$Ta$_3$, respectively.*

- $hP24$-Cu_3P (163 $P\bar{3}c$): Five representatives with this structure type are known, with TM = Rh, Ir, Pt, Pd, and Au.

- $cF24$-$MgZn_2$ (194 $P6_3/mmc$): Four representatives with this structure type are known, with TM = Y, Co, Ir, and Zn.

- $hP8$-Na_3As (194 $P6_3/mmc$): Four representatives with this structure type are known, with TM = Ir, Pd, Au, and Hg.

- $cP4$-Cu_3Au (221 $Pm\bar{3}m$): Four representatives with this structure type are known, with TM = Ni, Pt, Pd, and Ag.

The five most common ternary structure types of the 34 ternary Mg–TM–TM' compounds are:

- $cF16$-Cu_2MnAl (225 $Fm\bar{3}m$): Three representatives with this structure type are known for Mg–Zn–(Y, Ag) and Mg–Cd–Ag.

- $cP208$-$Mg_{4.22}Zn_{21.08}Hf_{1.14}$ (200 $Pm\bar{3}$): Two representatives with this structure type are known for Mg–Zn–(Zr, Hf), a 1/1 approximant (Gómez *et al.*, 2008)

- $cP712$-$Mg_{2.5}Zn_{73.6}Sc_{11.18}$ (205 $Pa\bar{3}$): One representative with this structure type is known, $Mg_{2.5}Zn_{73.6}Sc_{11.18}$, a 2/1-approximant (Lin and Corbett, 2006).

- $hP36$-$Sc_3Ni_{11}Si_4$ (194 $P6_3/mmc$): One representative with this structure type is known, $Y_3Zn_{11}Mg_4$.

- $hP9$-$ZrNiAl$ (189 $P\bar{6}2m$): One representative with this structure type is known, MgYZn.

Compounds of Ca with TM elements

In the PCD, there are 68 binary and 31 ternary compounds of Ca with TM elements listed. The five most common binary structure types of the binary Ca–TM compounds are:

- $hP6$-CaCu$_5$ (191 $P6/mmm$): Five representatives with this structure type are known, with TM = Ni, Pt, Pd, Cu, and Zn.
- $cF24$-MgCu$_2$ (227 $Fd\bar{3}m$): Five representatives with this structure type are known, with TM = Rh, Ir, Ni, Pt, and Pd.
- $oI12$-KHg$_2$ (74 $Imma$): Four representatives with this structure type are known, with TM = Au, Ag, Cd, and Zn.
- $tI32$-Cr$_5$B$_3$ (140 $I4/mcm$): Four representatives with this structure type are known, with TM = Au, Ag, Hg, and Zn.
- $cP2$-CsCl (221 $Pm\bar{3}m$): Three representatives with this structure type are known, with TM = Pd, Hg, and Cd.

The five most common ternary structure types of the 31 ternary Ca–TM–TM' compounds are:

- $cF472$-Ca$_{21}$Zn$_{36}$Ni$_2$ (227 $Fd\bar{3}m$): One representative with this structure type is known, Ca$_{21}$Zn$_{36}$Ni$_2$.
- $oS68$-Ca$_4$Au$_{10}$In$_3$ (64 $Cmce$): One representative with this structure type is known, Ca$_4$Au$_{10}$Cd$_3$.
- $oP12$-HoNiGa (62 $Pnma$): One representative with this structure type is known, CaPdZn, which is related to the $hP3$-AlB$_2$ type.
- $oP12$-TiNiSi (62 $Pnma$): One representative, AuCaCd, with this structure type is known; it can be derived by symmetry reduction ($I \rightarrow P$) from the $oI12$-KHg$_2$ structure type, which itself is an $hP3$-AlB$_2$ derivative structure.
- $hP18$-YNi$_2$Al$_3$ (191 $P6/mmm$): One representative with this structure type is known, CaNi$_2$Zn$_3$.

Compounds of Sr with TM elements

In the PCD, there are 46 binary and 9 ternary compounds of Sr with TM elements listed. The five most common binary structure types of the binary Sr–TM compounds are:

- $oI12$-KHg$_2$ (74 $Imma$): Five representatives with this structure type are known, with TM = Au, Ag, Hg, Cd, and Zn.
- $cF24$-MgCu$_2$ (227 $Fd\bar{3}m$): Four representatives with this structure type are known, with TM = Rh, Ir, Pt, and Pd.

- *hR*45-Er$_3$Ni$_2$ (148 *R*$\bar{3}$): Three representatives with this structure type are known, with TM = Pt, Au, and Ag.

- *hP*20-Th$_7$Fe$_3$ (194 *P*6$_3$*mc*): Two representatives with this structure type are known, with TM = Au, and Ag. Its structure is shown in Fig. 7.39 (a)–(e).

- *tI*48-BaCd$_{11}$ (141 *I*4$_1$/*amd*): Two representatives with this structure type are known, with TM = Cd, and Zn.

There is just one ternary structure type among the nine ternary Sr–TM–TM' compounds:

- *oP*12-TiNiSi (62 *Pnma*): One representative, AuCdSr, with this structure type; it can be derived by symmetry reduction ($I \rightarrow P$) from the *oI*12-KHg$_2$ structure type, which itself is an *hP*3-AlB$_2$ derivative structure.

Compounds of Ba with TM elements

In the PCD, there are 36 binary and 1 ternary compound of Ba with TM elements listed. The five most common binary structure types of the binary Ba–TM compounds:

- *hP*6-CaCu$_5$ (191 *P*6/*mmm*): Four representatives with this structure type are known, with TM = Pt, Pd, Au, and Ag.

- *oI*12-KHg$_2$ (74 *Imma*): Four representatives with this structure type are known, with TM = Ag, Hg, Cd, and Zn.

- *cP*2-CsCl (221 *Pm*$\bar{3}$*m*): Three representatives with this structure type are known, with TM = Hg, Cd, and Zn.

- *tI*6-CuZr$_2$ (139 *I*4/*mmm*): Three representatives with this structure type are known, with TM = Hg, Cd, and Zn. See Fig. 7.3.

- *hR*45-Er$_3$Ni$_2$ (148 *R*$\bar{3}$): Three representatives with this structure type are known, with Ba-(Pt, Au, and Ag).

There is no ternary structure type of the single (pseudo-) ternary Ba–TM–TM' compounds.

7.14   Transition metal (TM) compounds (groups 3–12 only)

In the following, with the term "transition metal elements" (TM) we mean all elements from groups 3–12, not including the lanthanoids or any of the radioactive elements in the seventh period (Ac and following) (see Fig. 7.34). The Mendeleev numbers of these elements are M = 19, 25, 33, 49–72, and 74–76. The Pauling electronegativities, which generally decrease from periods three to five, increase

1	2	3	4	5	6	7	8	9	10	11	12	13	14	15	16
12 Li 1.0 1.52	77 Be 1.5 1.11														
11 Na 0.9 1.86	73 Mg 1.2 1.60											80 Al 1.5 1.43			
10 K 0.8 2.27	16 Ca 1.0 1.97	19 Sc 1.3 1.61	51 Ti 1.4 1.45	54 V 1.6 1.31	57 Cr 1.6 1.25	60 Mn 1.5 1.37	61 Fe 1.8 1.24	64 Co 1.8 1.25	67 Ni 1.8 1.25	72 Cu 1.9 1.28	76 Zn 1.6 1.34	81 Ga 1.6 1.22	84 Ge 1.8 1.23		
9 Rb 0.8 2.48	15 Sr 1.0 2.15	25 Y 1.2 1.78	49 Zr 1.4 1.59	53 Nb 1.6 1.43	56 Mo 1.8 1.36	59 Tc 1.9 1.35	62 Ru 2.2 1.33	65 Rh 2.2 1.35	69 Pd 2.2 1.38	71 Ag 1.9 1.45	75 Cd 1.7 1.49	79 In 1.7 1.63	83 Sn 1.8 1.41	88 Sb 1.9 1.45	
8 Cs 0.7 2.66	14 Ba 0.9 2.17	33* La 1.2 1.87	50 Hf 1.3 1.56	52 Ta 1.5 1.43	55 W 1.7 1.37	58 Re 1.9 1.37	63 Os 2.2 1.34	66 Ir 2.2 1.36	68 Pt 2.2 1.37	70 Au 2.4 1.44	74 Hg 1.9 1.50	78 Tl 1.8 1.70	82 Pb 1.8 1.75	87 Bi 1.9 1.55	91 Po 2.0 1.67
7 Fr 0.7	13 Ra 0.9 2.15	48^{+} Ac 1.1 1.88													

	* Lanthanoids													
	32 Ce 1.1 1.87	31 Pr 1.1 1.82	30 Nd 1.1 1.81	29 Pm 1.1 1.63	28 Sm 1.2 1.62	18 Eu 1.2 2.00	27 Gd 1.2 1.79	26 Tb 1.1 1.76	25 Dy 1.2 1.75	24 Ho 1.2 1.74	23 Er 1.2 1.73	22 Tm 1.3 1.72	17 Yb 1.1 1.94	20 Lu 1.3 1.72
+ Actinoids	47 Th 1.3 1.88	46 Pa 1.5 1.80	45 U 1.4 1.39	44 Np 1.4 1.30	43 Pu 1.3 1.51	42 Am 1.1	41 Cm 1.3	40 Bk 1.3	39 Cf 1.3	38 Es 1.3	37 Fm 1.3	36 Md 1.3	35 No 1.3	34 Lr 1.3

Fig. 7.34 *Elements constituting the compounds discussed in this section are shaded gray in the periodic table. Mendeleev numbers (top left in each box), Pauling electronegativities* χ *(relative to* $\chi_F = 4.0$*) (bottom left in each box), and atomic radii (half of the shortest distance between atoms in the crystal structure at ambient conditions) (bottom right in each box) of the metallic elements are given.*

from group 3 to 11, with La the least ($\chi = 1.1$) and Au the most ($\chi = 2.4$) electronegative element (see Fig. 6.2). The atomic radii, which increase from periods three to five, show a parabolic behavior. The atomic diameters have minimum values for each period for group 8 elements (Fe, Ru, Os) and maximum for group 3 elements.

Among the 6441 binary intermetallics, 1433 compounds (22.2%) are formed solely from transition metal elements, featuring 249 (26.4%) different structure types out of the 943 ones of binary intermetallics. The most common structure types are given in Table 7.7. Four out of the 18 most common structure types are just single-element ones. The most complex ones are the FK-phase $tP30$-$Cr_{0.49}Fe_{0.51}$ (σ-phase) and $cF96$-Ti_2Ni, which is constituted from icosahedral AETs and other AETs with locally fivefold symmetry (see Fig. 7.38).

The elements forming the binary transition metal intermetallics in the six most common structure types are illustrated in Fig. 7.35. In the top three structure types ($cF4$-Cu, $cI2$-W, and $hP2$-Mg), the major element constitutes up to 99.9% of the compounds. This does not mean that these phases are just solid solutions of element B in element A, keeping the structure of A. For instance, in the system Ag–Cd ($M = 71$ and $M = 75$, respectively), the HT-phase β-Ag_xCd_{1-x} ($0.4 \leq x \leq 0.55$) has a $cI2$-W-type structure. Another example is ε-Cr_xCo_{1-x} ($0.62 \leq x \leq 0.85$), which has a structure of type $hP2$-Mg, while Co crystallizes in the $cF4$-Cu type, and Cr in the $cI2$-W type.

Table 7.7 *Most common structure types of the 1433 binary intermetallic phases listed in the PCD, which are formed exclusively from transition metal elements. The top 18 structure types are given with the number of their representatives. Each structure type has at least 14 representatives and therefore represents at least 1.0% of all binary transition metal intermetallics. Only 14 out of the 18 structure types are binary, 4 are just unary.*

Rank	Structure type	No.	Space group	Wyckoff positions	No. of structures	% of all structures
1.	$cF4$-Cu	225	$Fm\bar{3}m$	$4a$	157	11.0%
2.	$cI2$-W	229	$Im\bar{3}m$	$2a$	125	8.7%
3.	$hP2$-Mg	194	$P6_3/mmc$	$2c$	114	8.0%
4.	$cP2$-CsCl	221	$Pm\bar{3}m$	$1ab$	79	5.5%
5.	$cP4$-Cu$_3$Au	221	$Pm\bar{3}m$	$1a\,3c$	61	4.3%
6.	$tP30$-Cr$_{0.49}$Fe$_{0.51}$	136	$P4_2/mnm$	$2a\,4f\,8i^2j$	47	3.3%
7.	$hP12$-MgZn$_2$	194	$P6_3/mmc$	$2a\,4f\,6h$	45	3.1%
8.	$cF24$-MgCu$_2$	227	$Fd\bar{3}m$	$8a\,16d$	43	3.0%
9.	$cP8$-Cr$_3$Si	223	$Pm\bar{3}n$	$2a\,6c$	37	2.6%
10.	$tP2$-CuAu	123	$P4/mmm$	$1ad$	29	2.0%
11.	$cF96$-Ti$_2$Ni	227	$Fd\bar{3}m$	$16c\,32e\,48f$	28	2.0%
12.	$tI6$-CuZr$_2$	139	$I4/mmm$	$2a\,4e$	19	1.3%
13.	$cI58$-Mn	217	$I\bar{4}3m$	$2a\,8c\,24g^2$	19	1.3%
14.	$tP2$-CuTi	123	$P4/mmm$	$1ad$	15	1.0%
15.	$hP24$-MgNi$_2$	194	$P6_3/mmc$	$4ef^2\,6gh$	15	1.0%
16.	$tI6$-MoSi$_2$	123	$I4/mmm$	$2a\,4e$	15	1.0%
17.	$hP6$-CaCu$_5$	191	$P6/mmm$	$1a\,2c\,3g$	14	1.0%
18.	$hP8$-Mg$_3$Cd	194	$P6_3/mmc$	$2d\,6h$	14	1.0%
					876	61.1%

The stability field of the $cF4$-Cu-type structures is essentially bounded by Co ($M = 64$) and Cu ($M = 72$) for the majority element A. All TM elements within these boundaries have the $cF4$-Cu structure type. The minority element B can be any of the transition elements. The adjacent $cI2$-W stability field is bounded by Zr ($M = 49$) and Fe ($M = 61$) for A elements. All TM elements within these boundaries have either the $hP2$-Mg or the $cI2$-W structure type. B can be any other TM element. Remarkable is the asymmetry for Re ($M = 58$), which can only form Re–TM compounds with Re as the minority element B. The majority

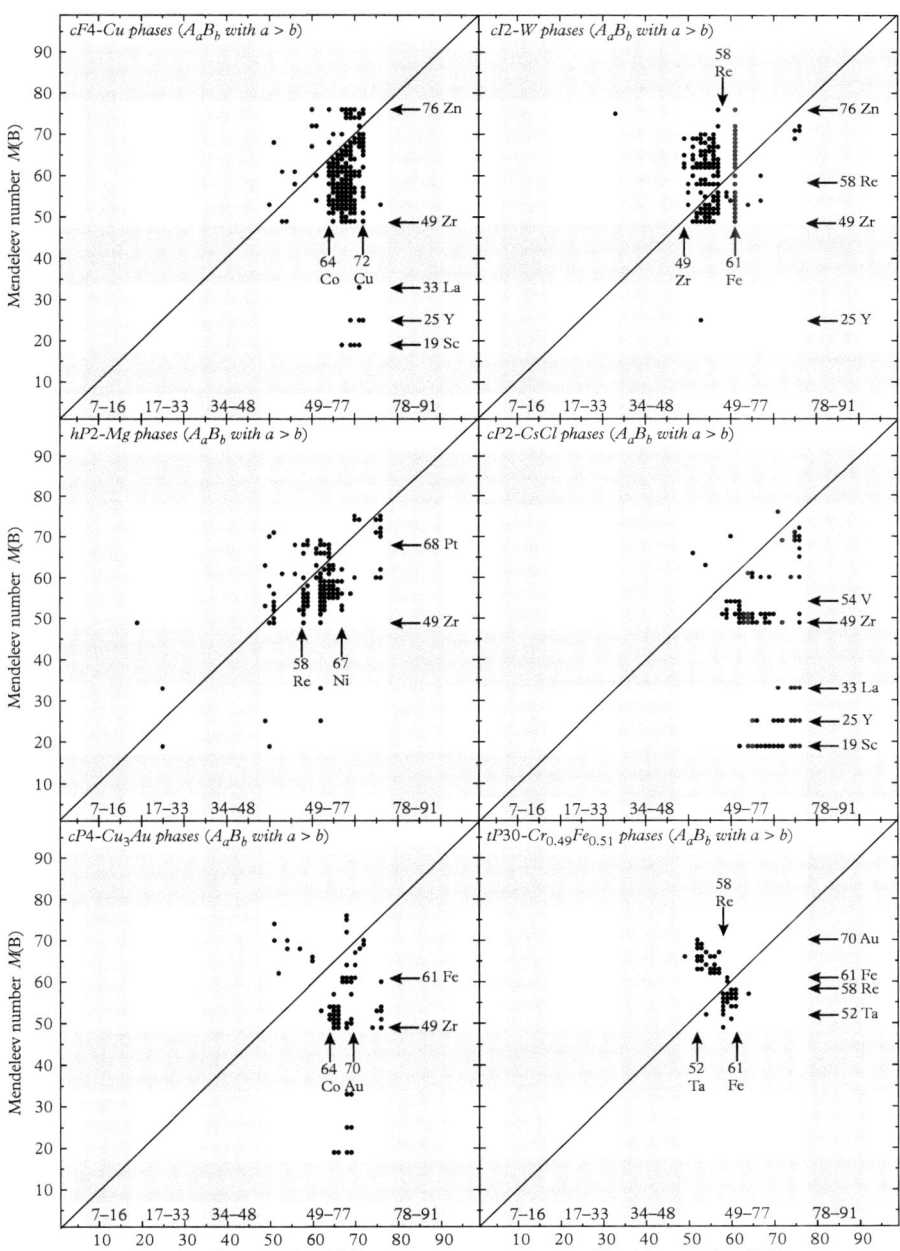

Fig. 7.35 *Occurrence of the six most common structure types among binary transition metal intermetallics. The Mendeleev numbers M of both elements are shown, with the major component* A *and the minor* B*. The depicted structure types are cF4-Cu, cI2-W, hP2-Mg, cP2-CsCl, cP4-Cu₃Au, and tP30-Cr₀.₄₉Fe₀.₅₁ (σ-phase).*

of phases with *hP2*-Mg-type structures are constituted from A elements from groups 7 and 8, and B atoms of group 4 to group 10 elements.

The compounds with *cP2*-CsCl-type structures contain 50–60 % of the majority element. This structure type is mainly adopted by late TM elements for A atoms and either the group three elements Sc ($M = 19$), Y ($M = 25$), and La ($M = 33$) or early TM elements for B atoms. In structure type *cP4*-Cu$_3$Au, the majority component makes up 66.7–82.5 % compared with 25% in the eponymous compound. The stability field encompasses groups 9 and 10 (all but one of the *cF4*-Cu type) for A atoms, and groups 4 (*hP2*-Mg) and 5 (*cI2*-W), as well as the *bcc* elements Mn ($M = 60$) and Fe ($M = 61$) for B atoms.

The structures of type *tP30*-Cr$_{0.49}$Fe$_{0.51}$ (σ-phase) on the other hand contain 50–80% of the majority component. There are two small stability fields. One comprises groups 5 and 6 (all *cI2*-W type) for A elements, and 8 to 11 (mostly *cF4*-Cu, and without Ag and Au) for B atoms. The other includes groups 7 and 8 (mostly *hP2*-Mg) for A atoms, and groups 5, 6 (all *cI2*-W type), and 7 (mostly *hP2*-Mg) for B elements.

The stoichiometries of all the 1433 binary transition metal intermetallics are visualized in the histogram depicted in Fig. 7.36. By far the most frequent composition is AB, followed by AB$_2$/A$_2$B and AB$_3$/A$_3$B. There are more than twenty different stoichiometries with at least ten representatives.

Among the 13 026 ternary intermetallics, 797 compounds are formed solely from transition metal elements, featuring 118 different structure types. The most common structure types are given in Table 7.8. The Laves phases *cF24*-MgCu$_2$ and *hP12*-MgZn$_2$ belong to the top three structure types. According to Stein *et al.* (2005), one distinguishes between pseudo-ternary Laves phases and true ternary Laves phases. If the stability of a binary Laves phase extends into the ternary region, then we can call this kind of ternary phase a solid solution, meaning it is just a pseudo-ternary Laves phase. In the case where no binary Laves phase exists in a ternary system, we call it a ternary Laves phase. However, this says nothing about the kind of structural ordering present.

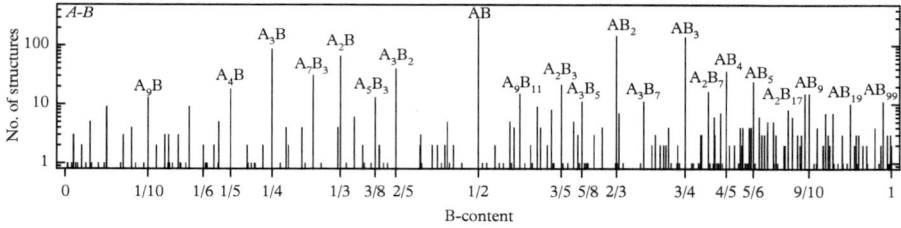

Fig. 7.36 *M/M-plots illustrating the occurrence of binary compositions* A$_m$B$_n$, *with* $M(A) < M(B)$ *of all the 1433 binary transition metal (TM–TM) intermetallics. The most frequent stoichiometries are labelled with the respective compositions. Note the logarithmic scale for the number of structures (compounds).*

Table 7.8 *Most common structure types of ternary intermetallic phases listed in the PCD, which are constituted from transition metal elements. The top 23 structure types are given, all of which have at least eight representative structures and therefore represent at least 1.0% of all ternary transition metal intermetallics. Only four of them are ternary.*

Rank	Structure type	No.	Space group	Wyckoff positions	No. of structures	% of all structures
1.	$cF24$-MgCu$_2$	227	$Fd\bar{3}m$	$8a\,16d$	87	10.9%
2.	$cF4$-Cu	225	$Fm\bar{3}m$	$4a$	63	7.9%
3.	$hP12$-MgZn$_2$	194	$P6_3/mmc$	$2a\,4f\,6h$	61	7.7%
4.	$cI2$-W	229	$Im\bar{3}m$	$2a$	59	7.4%
5.	$cP2$-CsCl	221	$Pm\bar{3}m$	$1ab$	38	4.8%
6.	$cP4$-Cu$_3$Au	221	$Pm\bar{3}m$	$1a\,3c$	33	4.1%
7.	$cP8$-Cr$_3$Si	223	$Pm\bar{3}n$	$2a\,6c$	25	3.1%
8.	$cF184$-CeCr$_2$Al$_{20}$	227	$Fd\bar{3}m$	$8a\,16cd\,48f\,96g$	22	2.8%
9.	$tP2$-CuAu	123	$P4/mmm$	$1ad$	22	2.8%
10.	$hP6$-CaCu$_5$	191	$P6/mmm$	$1a\,2c\,3g$	19	2.4%
11.	$tP30$-Cr$_{0.49}$Fe$_{0.51}$	136	$P4_2/mnm$	$2a\,4f\,8i^2j$	18	2.3%
12.	$hP2$-Mg	194	$P6_3/mmc$	$2c$	15	1.9%
13.	$cF96$-Ti$_2$Ni	227	$Fd\bar{3}m$	$16c\,32e\,48f$	13	1.6%
14.	$tI26$-ThMn$_{12}$	139	$I4/mmm$	$2a\,8fij$	12	1.5%
15.	$cF96$-Gd$_4$RhIn	216	$F\bar{4}3m$	$16e^3\,24fg$	11	1.4%
16.	$hR36$-PuNi$_3$	166	$R\bar{3}m$	$3ab\,6c^2\,18h$	11	1.4%
17.	$hR36$-BaPb$_3$	166	$R\bar{3}m$	$3a\,6c\,9e\,18h$	10	1.3%
18.	$tI12$-CuAl$_2$	140	$I4/mcm$	$4a\,8h$	10	1.3%
19.	$hP28$-Hf$_9$Mo$_4$B	194	$P6_3/mmc$	$2ac\,6h^2\,12k$	9	1.1%
20.	$hP16$-TiNi$_3$	194	$P6_3/mmc$	$2ad\,6gh$	9	1.1%
21.	$cP5$-CaTiO$_3$	221	$Pm\bar{3}m$	$4ac^2\,8d$	8	1.0%
22.	$oP8$-Cu$_3$Ti	59	$Pmmn$	$2ab\,4e$	8	1.0%
23.	$hR57$-Zn$_{17}$Th$_2$	166	$R\bar{3}m$	$6c^2\,9d\,18fh$	8	1.0%
					571	71.6%

In contrast to the also very frequent structure types $cF4$-Cu, $cI2$-W, and $hP2$-Mg, ternary Laves phases do not need to be chemically disordered. By lowering the symmetry from $Fd\bar{3}m$ to $F\bar{4}3m$, the Mg site, i.e., Wyckoff position $8a$ 0 0 0, splits into two sites $4a$ 0 0 0 and $4c$ 1/4 1/4 1/4. Consequently, they can be occupied by two different kinds of atoms. Although this derivative structure corresponds to the $cF24$-AuBe$_5$ type, one speaks of ternary Laves phases (see, e.g., Dogan and Pöttgen (2005), Tappe *et al.* (2012)). In the case of ordered ternary Laves phases of the type $hP12$-MgZn$_2$ (194 $P6_3/mmc$, Mg in $4f$ 1/3, 2/3, z; Zn in $2a$ 0, 0, 0 and $6h$ $x, 2x$, 1/4), such as U$_2$OsAl$_3$ (194 $P6_3/mmc$, Os in $2a$ 0, 0, 0; U in $4f$ 1/3, 2/3, z; Al in $6h$ $x, 2x$, 1/4), the symmetry remains unchanged, only the site occupations change.

In the case of the rather frequent $cP2$-CsCl structure type, and for the case of ternary compounds with stoichiometry 2:1:1, which have been studied just by X-ray powder diffraction methods, the actual structure type may be that of the Heusler phases, a derivative of the $cP2$-CsCl structure type (see, e.g., Dubenskyy *et al.* (2000)). This may also apply to other assignments of structure types.

Among the 23 structure types listed in Table 7.8, only four (with altogether 50 representatives) consist of three constituents (see Fig. 7.37), two of them are quite complex: the FK-phases $cF184$-CeCr$_2$Al$_{20}$ and $cF96$-Gd$_4$RhIn. $cF184$-CeCr$_2$Al$_{20}$ can be seen as superstructures of the $cF184$-ZrZn$_{22}$ structure type (see, e.g., Ilyushin and Blatov (2009)). The atomic layers typical for FK phases are nicely visible in the projection (Fig. 7.37 (a), (c)). The $cF96$-RE$_4$RhIn family of compounds show a 3D network of edge-connected Rh@RE$_6$ trigonal prisms, with the voids filled by further RE atoms and In$_4$ tetrahedra (Zaremba *et al.*, 2007). $hP28$-Hf$_9$Mo$_4$B can be described as a packing of tricapped trigonal prisms, B@Hf$_9$, which are vertex-sharing with columns of face-sharing icosahedra, Mo@Mo$_6$Hf$_6$, along [001]. There is also another representation possible, drawing the icosahedra, Mo@Mo$_4$Hf$_8$, around the Mo atoms in the other Wyckoff position.

7.14.1   Compounds of Sc, Y, or La with TM elements

There are many cases where the structures of ternary intermetallic phases are assigned to binary structure types, for instance to the $cF24$-MgCu$_2$ type. If the structure type was just chosen based on the comparison of Debye-Scherer photographs, which was quite common in the past, it is not clear whether the ternary phase was really partially disordered or rather of the $cF24$-MgCu$_4$Sn type (Fig. 7.55). The ordering of TM elements can be difficult to identify by diffraction experiments if the scattering powers of the constituting elements are too similar. Therefore, in the following we omit binary structure types of ternary phases even if they are the most frequent ones. The same applies to unary structure types assigned to binary phases. The most frequent structures, not already depicted elsewhere, are shown and discussed below.

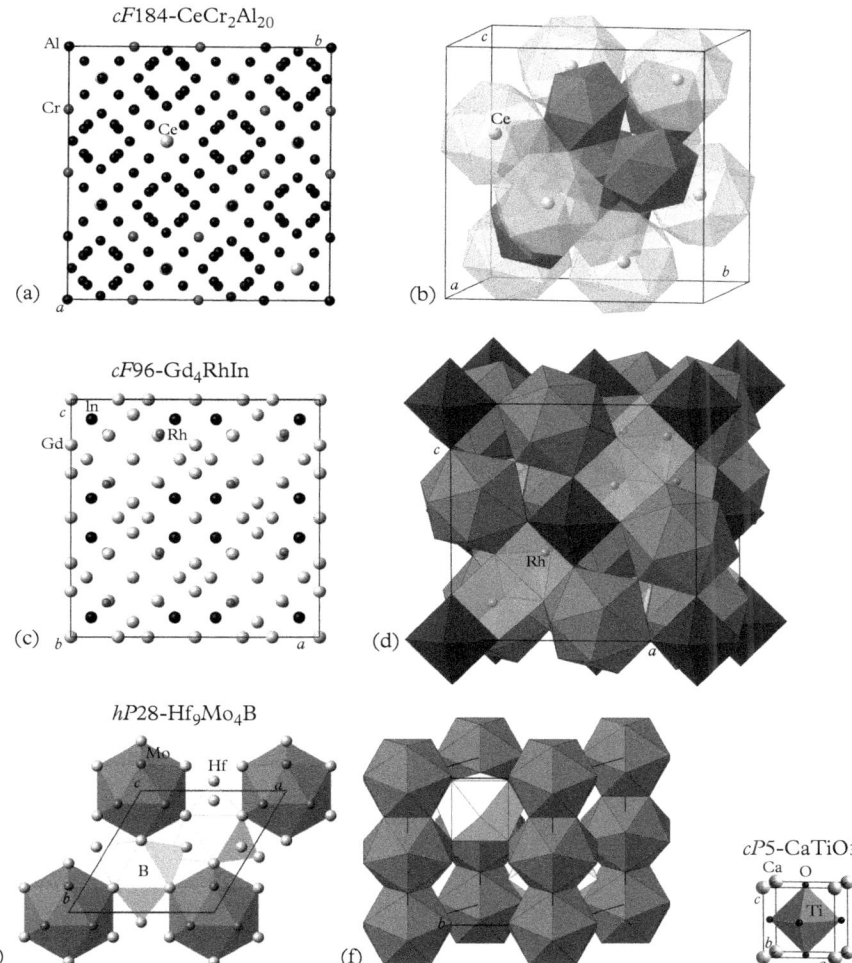

Fig. 7.37 *The four most common ternary structure types among ternary* TM–TM–TM' *intermetallics (see Table 7.8): (a)–(b) the structure of* cF184-CeCr$_2$Al$_{20}$ *in projection along [001] and in polyhedral representation.* Ce *occupies* CN16 FK-*polyhedra,* Ce@Al$_{16}$, *and* Cr-*centered icosahedra,* Cr@Al$_{12}$. *Each* CN16 *polyhedron shares its vertices with 12 icosahedra and 4 other* CN16 *polyhedra. Each icosahedron shares its vertices with six other icosahedra in trigonal antiprismatic arrangement and six* CN16 *polyhedra. (c)-(d) The structure of* cF96-Gd$_4$RhIn *in projection along [001] and in polyhedral representation. Four* Rh-*centered trigonal prisms share the triangle faces with a central empty tetrahedron. The* In-*centered icosahedra,* In@Gd$_9$In$_3$, *form* I4-*clusters, i.e., tetrahedral arrangements of four interpenetrating icosahedra, and share edges with the trigonal prisms,* Rh@Gd$_6$, *and the empty octahedra,* Gd$_6$. *The polyhedral structure of* hP28-Hf$_9$Mo$_4$B *is illustrated in the projection along [001] (e) and in perspective view (f). The* B *atoms center tricapped trigonal prisms,* B@Hf$_9$, *which are vertex-sharing with columns of face-sharing icosahedra,* Mo@Mo$_6$Hf$_6$, *along [001]. In (g) the perovskite structure,* cP5-CaTiO$_3$, *is shown with the* Ti@O$_6$ *octahedron marked.*

Compounds of Sc with TM elements

In the PCD, there are 78 binary and 20 ternary compounds listed with other TM elements. The five most common binary structure types of the binary Sc–TM compounds are:

- $cP2$-CsCl (221 $Pm\bar{3}m$): Thirteen representatives with this structure type are known, with TM = Ru, Co, Rh, Ir, Ni, Pt, Pd, Au, Ag, Cu, Hg, Cd, and Zn.

- $hP12$-MgZn$_2$ (194 $P6_3/mmc$): Seven representatives with this structure type are known, with TM = Re, Tc, Mn, Fe, Ru, Os, and Zn. Sc occupies the Mg site.

- $cF96$-Ti$_2$Ni (227 $Fd\bar{3}m$): Five representatives with this structure type are known, with TM = Ru, Co, Ir, Ni, and Pd. Sc occupies the Ni site. Its structure (Fig. 7.38) can be described as a packing of icosahedral and pentagonal prismatic structural subunits (Yurko *et al.*, 1959; Yurko *et al.*, 1962). It should be mentioned here that an icosahedral quasicrystal also exists in the system Ni–Ti–Zr. In $cF96$-Ti$_2$Ni, the Ni atoms ($32e$) form regular tetrahedra (edge length 2.82 Å), completely surrounded by Ti atoms, the Ti ($48f$) regular octahedra. The icosahedra around the Ni atoms, Ni@Ti$_9$Ni$_3$ (distances 2.48–2.89 Å), interpenetrate each other along 3 of the 12 five-fold axes, the shared volumes corresponding to pentagonal bipyramids in each case. The face-sharing icosahedra around Ti in $16c$, Ti@Ti$_6$Ni$_6$ (distances 2.48–2.90 Å), are arranged along the edges of large tetrahedra (see Fig. 7.38 (b)). In the remaining space between the icosahedra, their atoms form a framework of empty Ti octahedra and Ti and Ni tetrahedra, respectively (see Fig. 7.38 (c)). The AET around Ti in $48f$ is less regular, a distorted pentagonal prism, Ti@Ti$_{10}$Ni$_2$ (distance 2.61–3.03 Å), with two sides capped by Ni (distance 2.89 Å). In the case of isostructural $cF96$-Hf$_2$Ni, it was shown that this AET has covalent bonding contributions (Ivanovic *et al.*, 2006).

- $cF24$-MgCu$_4$Sn (216 $F\bar{4}3m$): Four representatives with this structure type are known, with TM = Fe, Co, Ir, and Ni. Sc occupies the Mg site. See Fig. 7.55.

- $cP140$-Rh$_{13}$Sc$_{57}$ (200 $Pm\bar{3}$): Four representatives with this structure type are known, with TM = Ru, Rh, Ir, and Pt. Sc occupies the Sc site. Its structure (Fig. 7.38) can be described as a packing of Rh@Sc$_{12}$ icosahedra and Rh@Sc$_{11}$ defective icosahedra (Cenzual *et al.*, 1985). Around a central almost regular icosahedron in the body center, twelve more vertex-sharing Rh@Sc$_{12}$ icosahedra are arranged icosahedrally, forming a kind of super-cluster arrangement. The icosahedra of neighboring superclusters share faces. The almost regular icosahedron in the origin is surrounded by vertex-sharing Rh@Sc$_{11}$ defective icosahedra. The Rh@Sc$_{11}$ defective icosahedra share edges with the Rh@Sc$_{12}$ icosahedra.

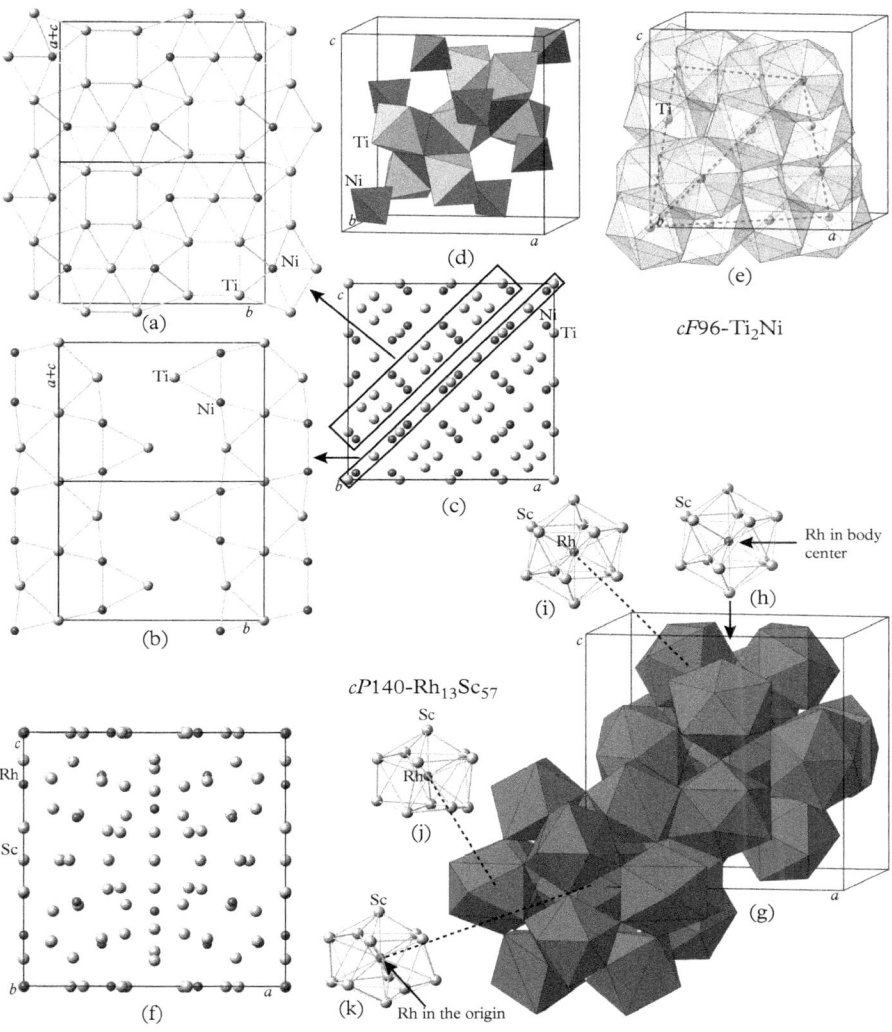

Fig. 7.38 *The structures of cF96-Ti₂Ni (a)–(e) and cP140-Rh₁₃Sc₅₇ (f)–(k) in different representations. The projected structure of cF96-Ti₂Ni (c) shows the atomic layer (a, b) arrangements typical for FK phases. The structure can be described as a packing of face-sharing icosahedra, Ti@Ti₆Ni6, as depicted in (e). The remaining space not covered by the polyhedra can be described as a framework of empty tetrahedra and octahedra (d). The structure of cP140-Rh₁₃Sc₅₇ (f)–(k) can be described as a packing of superclusters, each consisting of a regular Rh@Sc₁₂ icosahedron (h, k) surrounded by 12 more vertex–connected icosahedra (i). The icosahedra (j) surrounding the central icosahedron in the origin (k) have one missing vertex each.*

All the ternary structure types of the 20 ternary Sc–TM–TM' compounds are:

- $cF184$-CeCr$_2$Al$_{20}$ (227 $Fd\bar{3}m$): Compounds with this structure type are known for Sc–Zn–(Co, Ru). Sc occupies the Ce position, Zn the Cr site, and the other TM elements the Al sites. See Fig. 7.37.
- $cI192$-Cu$_{2.19}$Sc$_3$Zn$_{16}$ (204 $Im\bar{3}$): No additional compounds are known for this unique structure type.
- $cF24$-MgCu$_4$Sn (216 $F\bar{4}3m$): Compounds with this structure type are known for Sc–Ni–Au, only. Sc occupies the Mg position, Ni and Au the Cu and Sn sites, respectively. See Fig. 7.55.

Compounds of Y with TM elements

In the PCD, there are 122 binary and 113 ternary compounds listed with other TM elements. The five most common binary structure types of the binary Y–TM compounds are:

- $cP2$-CsCl (221 $Pm\bar{3}m$): Eight representatives with this structure type are known, with TM = Rh, Ir, Au, Ag, Cu, Hg, Cd, and Zn.
- $oP16$-Fe$_3$C (62 $Pnma$): Eight representatives with this structure type are known, with TM = Ru, Os, Co, Rh, Ir, Ni, Pt, and Pd. Y occupies the Fe position.
- $cF24$-MgCu$_2$ (227 $Fd\bar{3}m$): Seven representatives with this structure type are known, with TM = Mn, Fe, Co, Rh, Ir, Ni, and Pt. Y occupies the Mg site.
- $hP6$-CaCu$_5$ (191 $P6/mmm$): Six representatives with this structure type are known, with TM = Fe, Co, Rh, Ni, Cu, and Zn. Y occupies the Ca site.
- $hP12$-MgZn$_2$ (194 $P6_3/mmc$): Five representatives with this structure type are known, with TM = Re, Tc, Mn, Ru, and Os. Y occupies the Mg site.

The four most common ternary structure types of the 113 ternary Y–TM–TM' compounds are:

- $cF184$-CeCr$_2$Al$_{20}$ (227 $Fd\bar{3}m$): Six compounds with this structure type are known for Y–Zn–(Fe, Ru, Os, Co, Rh, Ir). Y occupies the Ce position, Zn the Al site and the other TM elements the Cr sites. See Fig. 7.37.
- $cF96$-Gd$_4$RhIn (216 $F\bar{4}3m$): Five compounds with this structure type are known for Y–Cd–(Ru, Ir, Ni, Pt, Pd). Y occupies the Gd position, Cd the In site and the other TM elements the Rh sites. See Fig. 7.37.
- $mS64$-Nd$_3$(Ti$_{0.21}$Fe$_{0.79}$)$_6$Fe$_{23}$ (12 $C2/m$): Five compounds with this structure type are known for Y–Fe–(Ti, Ta, V, Mo, Cr). Y occupies the Nd position, Fe the Fe sites and, together with the other TM elements, the (Ti$_{0.21}$Fe$_{0.79}$) sites.

- $tI26$-Nd$(Mo_{0.5}Fe_{0.5})_4$Fe$_8$ (139 $I4/mmm$): Three compounds with this structure type are known for Y–Fe–(V, Mo, Cr). Y occupies the Nd position, Fe the Fe sites and, together with the other TM elements, the $(Mo_{0.5}Fe_{0.5})$ sites.

Compounds of La with TM elements

In the PCD, there are 112 binary and 98 ternary compounds listed with other TM elements. The five most common binary structure types of the binary La–TM compounds are:

- $hP6$-CaCu$_5$ (191 $P6/mmm$): Seven representatives with this structure type are known, with TM = Co, Ir, Ni, Pt, Pd, Cu, and Zn. La occupies the Ca position.

- $cF24$-MgCu$_2$ (227 $Fd\bar{3}m$): Seven representatives with this structure type are known, with TM = Ru, Os, Co, Rh, Ir, Ni, and Pt. La occupies the Mg site.

- $hP20$-Th$_7$Fe$_3$ (194 $P6_3mc$): Five representatives with this structure type are known, with TM = Rh, Ir, Ni, Pt, and Pd. La occupies the Th site. Its structure is shown in Fig. 7.39 (a)–(e). It can be described as a packing of vertex-connected Fe@Th$_6$ trigonal prisms. Three of such prisms, sharing one vertex each (Fig. 7.39 (d)). The tetrahedral space between them and that to the next such unit forms a Th$_8$ *stella quadrangula* (Fig. 7.39 (e)). Each trigonal prism shares its base face with an octahdron, which forms columns along [001] (Fig. 7.39(b), (c)).

- $oS8$-TlI (63 $Cmcm$): Five representatives with this structure type are known, with TM = Rh, Ni, Pt, Pd, and Au.

- $oP16$-Fe$_3$C (62 $Pnma$): Five representatives with this structure type are known, with TM = Ru, Os, Co, Ir, and Ni. La occupies the Fe position.

The five most frequent ternary structure types of the 98 ternary La–TM–TM' compounds are:

- $cF96$-Gd$_4$RhIn (216 $F\bar{4}3m$): Six compounds with this structure type are known for La–Cd–(Ru, Co, Ir, Ni, Pt, Pd). La occupies the Gd position, Cd the In site and the other TM elements the Rh sites. See Fig. 7.37.

- $hP68$-Mg$_4$Pr$_{23}$Ir$_7$ (194 $P6_3mc$): Six compounds with this structure type are known for La–Cd–(Ru, Co, Rh, Ir, Ni, Pt). La occupies the Pr position, Cd the Mg site, and the other TM elements the Ir sites. In Fig. 7.39 (f)–(h), the structure of isostructural Mg$_4$Pr$_{23}$Ni$_7$ (Solokha *et al.*, 2009) is illustrated. In the projection along [001], one sees the close relationship to the above mentioned structure of $hP20$-Th$_7$Fe$_3$ (Fig. 7.39 (a)–(e)). We find the same kind of arrangement of three vertex-sharing Ni@Pr$_6$ trigonal prisms around a tetrahedron (T3-prism unit). There is also another one, sharing the faces with an octahedron (O3-prism unit) (Fig. 7.39 (g), (h)). One T3-prism

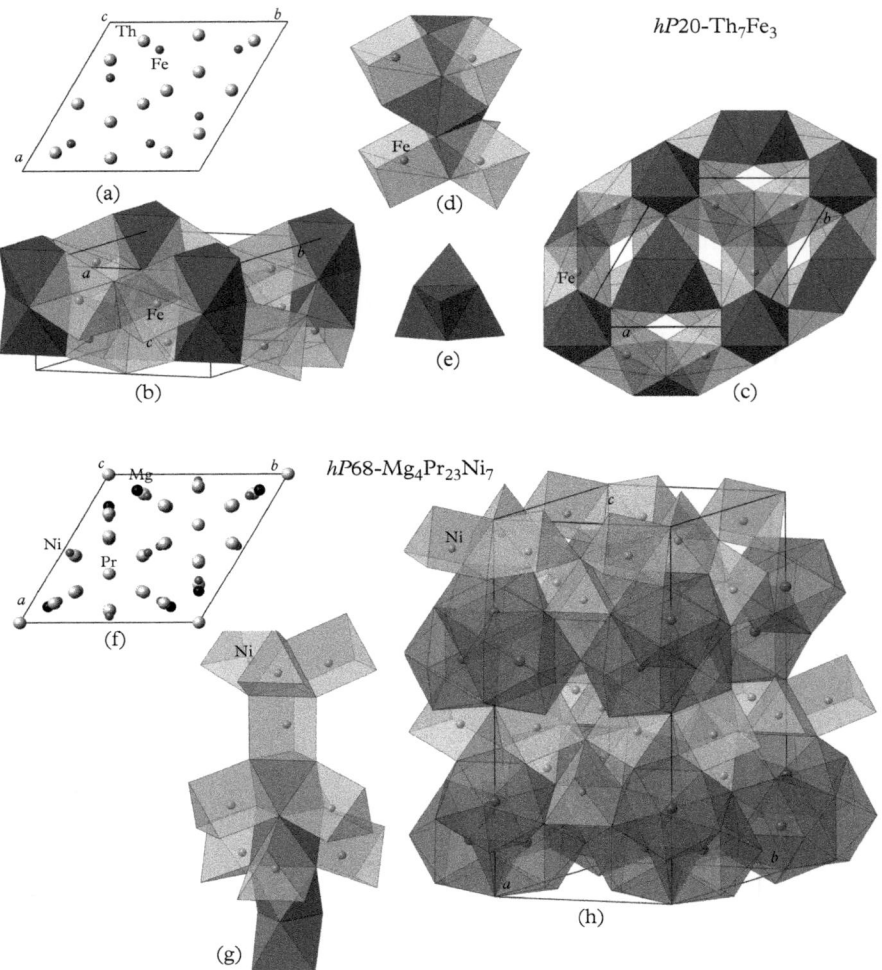

Fig. 7.39 *The structures of (a)–(e) hP20-Th₇Fe₃ and (f–h) hP68-Mg₄Pr₂₃Ni₇ (hP68-Mg₄Pr₂₃Ir₇ type). The close relationship between these two structures is obvious from the similarity of their [001]-projected structures (a, f) and of most of their structural units. Both structures show trigonal prismatic, tetrahedral, and octahedral subunits; however, in partly different arrangements. They differ mainly in the occurrence of* stellae quadrangulae, *typical for hP20-Th₇Fe₃, and the units of four interpenetrating icosahedra, which are found in hP68-Mg₄Pr₂₃Ni₇, only. In hP20-Th₇Fe₃, the columns of octahedra run along [001], and are centered in the origin of the unit cell (b, c). The structural units depicted in (d, e) are also lined up along [001], but they are centered in 1/3, 2/3, 0 and 2/3, 1/3, 0. In the case of hP68-Mg₄Pr₂₃Ni₇, the columns of octahedra, tetrahedra, and trigonal prisms (g) are also running along [001], but they are centered in 1/3, 2/3, 0 and 2/3, 1/3, 0, while the columns containing the icosahedra are centered at the origin of the unit cell (h).*

unit can be connected via a trigonal prism with an O3-prism unit (Fig. 7.39 (g)). Two centrosymmetrically related T3-prism units enclose a unit of four interpenetrating Mg@Th$_{12}$ icosahedra (h).

- *hP*9-ZrNiAl (189 $P\bar{6}2m$): Five compounds with this structure type are known for La–Cd–(Pd, Au), La–Hg–Pd, La–Zn–Ni, and La–Cu–Pd.

- *hR*57-Zn$_{14}$Gd$_2$Co$_3$ (166 $R\bar{3}m$): Five compounds with this structure type are known for La–Zn–(Fe, Co, Rh, Ni, Pd). La occupies the Gd position, Zn the Zn site, and the other TM elements the Co sites.

- *tP*10-Mo$_2$FeB$_2$ (127 $P4/mbm$): Four compounds with this structure type are known for La–Cd–(Rh, Ni, Pd, Au). La occupies the Mo position, Cd the Fe site, and the other TM elements the B sites.

7.14.2 Compounds of Ti, Zr, or Hf with TM elements

Compounds of Ti with TM elements

In the PCD, there are 132 binary and 143 ternary compounds listed with other TM elements. The five most common binary structure types of the binary Ti–TM compounds are:

- *cP*2-CsCl (221 $Pm\bar{3}m$): Thirteen representatives with this structure type are known, with TM = Re, Tc, Fe, Ru, Os, Co, Rh, Ir, Ni, Pt, Pd, Au, and Zn.

- *cP*4-Cu$_3$Au (221 $Pm\bar{3}m$): Seven representatives with this structure type are known, with TM = Co, Rh, Ir, Pd, Au, Hg, and Zn. Ti occupies the Au position except in the case of Au and Hg.

- *tI*6-CuZr$_2$ (139 $I4/mmm$): Six representatives with this structure type are known, with TM = Rh, Pd, Ag, Cu, Cd, and Ti. Ti occupies the Zr position. See Fig. 7.7. The structure can be described as a $(1 \times 1 \times 3)$-fold superstructure of the *cI*2-W type.

- *cP*8-Cr$_3$Si (223 $Pm\bar{3}n$): Five representatives with this structure type are known, with TM = Ir, Pt, Pd, Au, and Hg. Ti occupies the Cr position.

- *cF*96-Ti$_2$Ni (227 $Fd\bar{3}m$): Four representatives with this structure type are known, with TM = Fe, Co, Ni, and Cu. Ti occupies the Ti position. See Fig. 7.38.

All the ternary structure types of the 143 ternary Ti–TM–TM' compounds are:

- *cF*16-Li$_2$AgSb (216 $F\bar{4}3m$): Just one representative is known, Hg$_2$CuTi. The structure of this Zintl phase can be seen as a $(2 \times 2 \times 2)$-fold superstructure of the *cI*2-W type (Fig. 7.40).

- *cF*32-LiCa$_6$Ge (225 $Fm\bar{3}m$): Just one representative is known, CuPt$_6$Ti. The structure of this Zintl phase can be seen as a $(2 \times 2 \times 2)$-fold superstructure of the *cF*4-Cu type (Fig. 7.40).

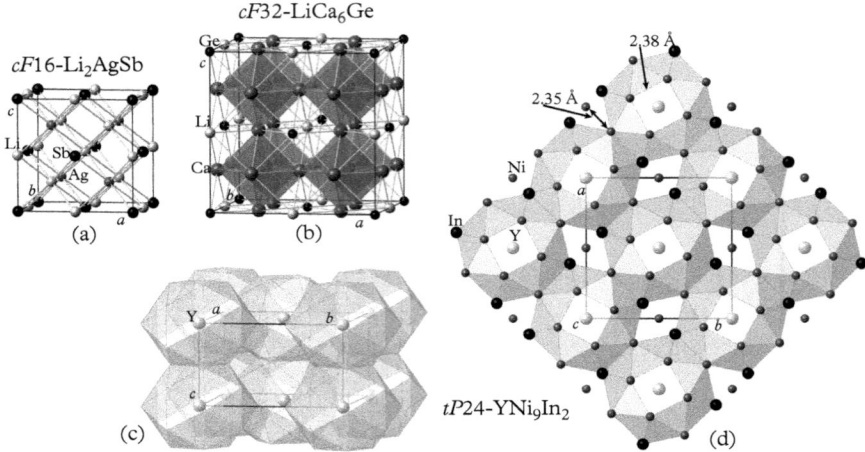

Fig. 7.40 *The structures of the Zintl phases (a) cF16-Li$_2$AgSb, (b) cF32-LiCa$_6$Ge, and the compound (c–d) tP24-YNi$_9$In$_2$. (a) cF16-Li$_2$AgSb can be seen as a (2 × 2 × 2)-fold superstructure of the cI2-W type. The indicated shortest atomic distances amount to 2.85 Å, which is shorter than the sum of atomic radii (r$_{Li}$ = 1.52 Å, r$_{Ag}$ = 1.45 Å, r$_{Sb}$ = 1.45 Å). (b) cF32-LiCa$_6$Ge can also be described as a (2 × 2 × 2)-fold superstructure, but now one of the cF4-Cu type. The indicated shortest atomic distance is 3.33 Å, which is much shorter than the sum of atomic radii of Ca, r$_{Ca}$ = 1.97 Å. The structure of tP24-YNi$_9$In$_2$ (c, d) can be described as a packing of Y@Ni$_{16}$In$_4$ polyhedra, which share their square faces along [001] and triangle faces along [110].*

- *mS64*-Nd$_3$(Ti$_{0.21}$Fe$_{0.79}$)$_6$Fe$_{23}$ (12 *C2/m*): Just one representative is known, Y$_3$(Ti$_{0.21}$Fe$_{0.79}$)$_6$Fe$_{23}$.

- *tP24*-YNi$_9$In$_2$ (127 *P4/mbm*): Just one representative is known, YFe$_9$Ti$_2$; however, several more indides contain lanthanoids. The structure can be described as a packing of Y@Ni$_{16}$In$_4$ polyhedra, which share their square Ni$_4$ faces ($d_{\text{Ni-Ni}}$ = 2.38 Å) along [001] and triangle faces along [110] (Fig. 7.40). The Ni atoms in the gaps between polyhedra are icosahedrally coordinated, Ni@Ni$_8$In$_4$, with Ni–Ni and Ni–In distances between 2.35 and 2.56 Å (r_{Ni} = 1.25 Å) and of 2.74 Å, respectively.

Compounds of Zr with TM elements

In the PCD, there are 164 binary and 158 ternary compounds listed with other TM elements. The five most common binary structure types of the binary Zr–TM compounds are:

- *cF24*-MgCu$_2$ (227 *Fd$\bar{3}$m*): Ten representatives with this structure type are known, with TM = V, W, Mo, Cr, Mn, Fe, Co, Ir, Ni, and Zn. Zr occupies the Mg site.

- $cP2$-CsCl (221 $Pm\bar{3}m$): Nine representatives with this structure type are known, with TM = Ru, Os, Co, Rh, Ir, Pt, Pd, Cu, and Zn.
- $hP12$-MgNi$_2$ (194 $P6_3/mmc$): Eight representatives with this structure type are known, with TM = V, Cr, Re, Tc, Mn, Fe, Ru, and Os. Zr occupies the Mg site.
- $tI6$-CuZr$_2$ (139 $I4/mmm$): Six representatives with this structure type are known, with TM = Pd, Au, Ag, Cu, Cd, and Zn. Zr occupies the Zr position. See Fig. 7.7.
- $tI2$-CuAl$_2$ (140 $I4/mcm$): Five representatives with this structure type are known, with TM = Fe, Co, Rh, Ir, Ni, and Hg. Zr occupies the Al position.

All the ternary structure types of the 158 ternary Zr–TM–TM' compounds are:

- $cF184$-CeCr$_2$Al$_{20}$ (227 $Fd\bar{3}m$): Six representatives with this structure type are known for Zr–Zn–(Mn, Fe, Ru, Co, Rh, Ni). Zr occupies the Ce position, Zn the Al site, and the other TM elements the Cr sites. See Fig. 7.37.
- $hP28$-Hf$_9$Mo$_4$B (194 $P6_3/mmc$): Three representatives with this structure type are known for Zr–Mo–(Fe, Co, Ni). Zr occupies the Hf position, Mo the Mo site, and the other TM elements the B sites. See Fig. 7.37.
- $cF16$-Cu$_2$MnAl (225 $Fm\bar{3}m$): Two compounds with this structure type are known, Cu$_2$ZrCd and Cu$_2$ZrZn.
- $cF116$-Mg$_6$Cu$_{16}$Si$_7$ (225 $Fm\bar{3}m$): Just one representative with this structure type is known, Zr$_6$Zn$_{16}$Cu$_7$.
- $cF24$-MgCu$_4$Sn (216 $F\bar{4}3m$): Just one representative with this structure type is known, ZrNi$_4$Zn. See Fig. 7.55.

Compounds of Hf with TM elements

In the PCD, there are 102 binary and 44 ternary compounds listed with other TM elements. The five most common binary structure types of the binary Hf–TM compounds are:

- $cF96$-Ti$_2$Ni (227 $Fd\bar{3}m$): Four representatives with this structure type are known, with TM = Mn, Fe, Os, Co, Rh, Ir, Ni, Pt, and Pd. Hf occupies the Ti position. See Fig. 7.38.
- $cF24$-MgCu$_2$ (227 $Fd\bar{3}m$): Eight representatives with this structure type are known, with TM = V, W, Mo, Cr, Fe, Co, Ni, and Zn. Hf occupies the Mg site.
- $tI6$-CuZr$_2$ (139 $I4/mmm$): Seven representatives with this structure type are known, with TM = Pd, Au, Ag, Cu, Hg, Cd, and Zn. Hf occupies the Zr position. See Fig. 7.7.

- *cP*2-CsCl (221 *Pm3̄m*): Six representatives with this structure type are known, with TM = Tc, Ru, Os, Co, Rh, and Pt.
- *hP*12-MgZn$_2$ (194 *P6$_3$/mmc*): Six representatives with this structure type are known, with TM = Cr, Re, Tc, Mn, Fe, and Os. Hf occupies the Mg site.

All the ternary structure types of the 44 ternary Hf–TM–TM' compounds are:

- *cF*184-CeCr$_2$Al$_{20}$ (227 *Fd3̄m*): Five representatives with this structure type are known for Hf–Zn–(Fe, Ru, Co, Rh, Ni). Hf occupies the Ce position, Zn the Al site, and the other TM elements the Cr sites. See Fig. 7.37.
- *hP*28-Hf$_9$Mo$_4$B (194 *P6$_3$/mmc*): Six representatives with this structure type are known for Zr–(Mo, W)–(Fe, Co, Ni). Zr occupies the Hf position, Mo and W, respectively, the Mo site, and the other TM elements the B sites. See Fig. 7.37.
- *cF*116-Mg$_6$Cu$_{16}$Si$_7$ (225 *Fm3̄m*): One representative with this structure type is known, Hf$_6$Zn$_{16}$Cu$_7$.

7.14.3   Compounds of V, Nb, or Ta with TM elements

Compounds of V with TM elements

In the PCD, there are 92 binary and 90 ternary compounds listed with other TM elements. The five most common binary structure types of the binary V–TM compounds are:

- *cP*8-Cr$_3$Si (223 *Pm3̄n*): Ten representatives with this structure type are known, with TM = Re, Os, Co, Rh, Ir, Ni, Pt, Pd, Au, and Cd. V occupies the Cr position except in the case of Re.
- *cP*2-CsCl (221 *Pm3̄m*): Five representatives with this structure type are known, with TM = Tc, Mn, Fe, Ru, and Os.
- *cP*4-Cu$_3$Au (221 *Pm3̄m*): Five representatives with this structure type are known, with TM = Co, Rh, Ir, Pt, Au, and Zn. V occupies the Au position except in the case of Pt and Au.
- *tP*2-CuAu (123 *P4/mmm*): Three representatives with this structure type are known, with TM = Rh, Ir, and Pt.
- *cF*24-MgCu$_2$ (227 *Fd3̄m*): Three representatives with this structure type are known, with TM = Zr, Hf, and Ta. Hf occupies the Cu site.

All the ternary structure types of the 90 ternary V–TM–TM' compounds are:

- *cF*32-LiCa$_6$Ge (225 *Fm3̄m*): Just one representative is known, CuPt$_6$V (Fig. 7.40).

- $tI26$-Nd$(Mo_{0.5}Fe_{0.5})_4$Fe$_8$ (139 $I4/mmm$): Just one representative is known, Y$(V_{0.5}Fe_{0.5})_4$Fe$_8$.
- $mS64$-Nd$_3(Ti_{0.21}Fe_{0.79})_6$Fe$_{23}$ (12 $C2/m$): One representative is known, Y$_3(V_{0.21}Fe_{0.79})_6$Fe$_{23}$.

Compounds of Nb with TM elements

In the PCD, there are 116 binary and 91 ternary compounds listed with other TM elements. The five most common binary structure types of the binary Nb–TM compounds are:

- $tP30$-Cr$_{46}$Fe$_{54}$ (σ phase) (136 $P4_2/mnm$): Six representatives with this structure type are known, with TM = Re, Os, Rh, Ir, Pt, and Pd.
- $cP8$-Cr$_3$Si (223 $Pm\bar{3}n$): Five representatives with this structure type are known, with TM = Os, Rh, Ir, Pt, and Au.
- $cP4$-Cu$_3$Au (221 $Pm\bar{3}m$): Four representatives with this structure type are known, with TM = Ru, Rh, Ir, Cd, and Zn.
- $hP12$-MgZn$_2$ (194 $P6_3/mmc$): Four representatives with this structure type are known, with TM = Cr, Mn, Fe, and Co.
- $hR39$-W$_6$Fe$_7$ (166 $R\bar{3}m$): Four representatives with this structure type are known, with TM = Fe, Co, Ni, and Zn.

The only ternary structure type among of the 91 ternary Nb–TM–TM' compounds is:

- $cF184$-CeCr2Al$_{20}$ (227 $Fd\bar{3}m$): Three representatives are known, for Nb–Zn–(Fe, Co, Ni).

Compounds of Ta with TM elements

In the PCD, there are 83 binary and 62 ternary compounds listed with other TM elements. The five most common binary structure types of the binary Ta–TM compounds are:

- $tP30$-Cr$_{46}$Fe$_{54}$ (σ phase) (136 $P4_2/mnm$): Eight representatives with this structure type are known, with TM = V, Re, Os, Rh, Ir, Pt, Pd, and Au.
- $hP12$-MgZn$_2$ (194 $P6_3/mmc$): Six representatives with this structure type are known, with TM = V, Cr, Mn, Fe, Co, and Zn.
- $hR39$-W$_6$Fe$_7$ (160 $R\bar{3}m$): Four representatives with this structure type are known, with TM = Fe, Co, Ni, and Zn.
- $cP4$-Cu$_3$Au (221 $Pm\bar{3}m$): Four representatives with this structure type are known, with TM = Ru, Co, Rh, and Ir.

- $cF24$-$MgCu_2$ (227 $Fd\bar{3}m$): Three representatives with this structure type are known, with TM = V, Cr, and Co.

The only ternary structure type among of the 62 ternary Ta–TM–TM' compounds is:

- $hP40$-$TaRhPd_2$ (194 $P6_3/mmc$): Three representatives are known, for Ta–Zn–(Fe, Co, Ni). It has a polytypic close-packed structure with stacking sequence ABCBCACBCB (Giessen and Grant, 1965).

7.14.4 Compounds of Cr, Mo, or W with TM elements

Compounds of Cr with TM elements

In the PCD, there are 70 binary and 97 ternary compounds listed with other TM elements. The five most common binary structure types of the binary Cr–TM compounds are:

- $tP30$-$Cr_{46}Fe_{54}$ (σ phase) (136 $P4_2/mnm$): Eight representatives with this structure type are known, with TM = Re, Tc, Mn, Fe, Ru Os, Co, and Ni.
- $cP8$-Cr_3Si (223 $Pm\bar{3}n$): Seven representatives with this structure type are known, with TM = Ru, Os, Co, Rh, Ir, Ni, and Pt.
- $cF24$-$MgCu_2$ (227 $Fd\bar{3}m$): Five representatives with this structure type are known, with TM = Zr, Hf, Ti, Ta, and Nb.
- $hP12$-$MgZn_2$ (194 $P6_3/mmc$): Five representatives with this structure type are known, with TM = Zr, Hf, Ti, Ta, and Nb.
- $hP24$-$MgNi_2$ (194 $P6_3/mmc$): Four representatives with this structure type are known, with TM = Zr, Hf, Ti, and Ta.

All the ternary structure types of the 97 ternary Cr–TM–TM' compounds are:

- $hR159$-$(Cr_{0.16}Mo_{0.38}Co_{0.46}$ (R phase) (148 $R\bar{3}$): Four representatives are known, for Cr-(Fe, Co)-(Mo, W).
- $cF32$-$LiCa_6Ge$ (225 $Fm\bar{3}m$): One representative is known, $CrCu_6Pt$ (Fig. 7.40).
- $tI26$-$NdFe_{10}Mo_2$ (139 $I4/mmm$): One representative is known, $YFe_{10}Cr_2$. Its structure can be described as ordering variant of the $tI26$-$ThMn_{12}$ type (see Fig. 7.31).
- $mS64$-$Nd_3(Ti_{0.21}Fe_{0.79})_6Fe_{23}$ (12 $C2/m$): One representative is known, $Y_3(Cr_{0.21}Fe_{0.79})_6Fe_{23}$. The structure can be described as a packing of polyhedra with 20 vertices, bounded by two squares and 40 triangles (Fig. 7.41).

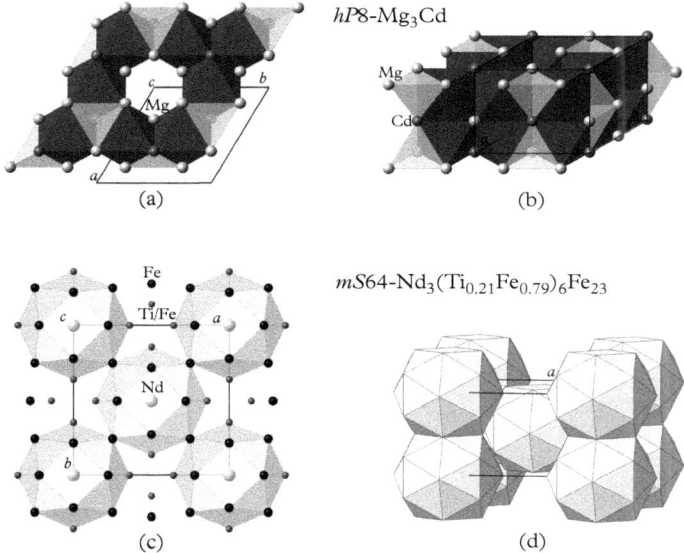

Fig. 7.41 *The structures of hP8-Mg₃Cd and mS64-Nd₃(Ti₀.₂₁ Fe₀.₇₉)₆Fe₂₃. (a)–(b) hP8-Mg₃Cd represents an ordered variant of the hP2-Mg structure type. The structure of mS64-Nd₃(Ti₀.₂₁Fe₀.₇₉)₆Fe₂₃ can be described as a packing of polyhedra with 20 vertices, bounded by 2 squares and 32 triangles.*

Compounds of Mo with TM elements

In the PCD, there are 80 binary and 83 ternary compounds listed with other TM elements. The five most common binary structure types of the binary Mo–TM compounds are:

- $tP30$-$Cr_{46}Fe_{54}$ (σ phase) (136 $P4_2/mnm$): Eight representatives with this structure type are known, with TM = Re, Tc, Mn, Fe, Ru, Os, Co, and Ir.

- $cP8$-Cr_3Si (223 $Pm\bar{3}n$): Six representatives with this structure type are known, with TM = Zr, Re, Tc, Os, Ir, and Pt.

- $hP8$-Mg_3Cd (194 $P6_3/mmc$): Four representatives with this structure type are known, with TM = Co, Rh, Ir, and Pt. The structure can be described as a $(2 \times 2 \times 1)$-fold superstructure of the $hP2$-Mg type (Fig. 7.41).

- $oP4$-AuCd (62 $Pmma$): Three representatives with this structure type are known, with TM = Rh, Ir, and Pt. The structure can be described as an orthorhombically distorted $hP2$-Mg type.

- $cF24$-$MgCu_2$ (227 $Fd\bar{3}m$): Two representatives with this structure type are known, with TM = Zr and Hf.

All the ternary structure types of the 83 ternary Mo–TM–TM' compounds are:

- $hP28$-Hf$_9$Mo$_4$B (194 $P6_3/mmc$): Six representatives with this structure type are known, for Mo–(Fe, Co, Ni)-(Zr, Hf). See Fig. 7.37.
- $hR159$-Cr$_{0.16}$Mo$_{0.38}$Co$_{0.46}$ (R phase) (148 $R\bar{3}$): Four representatives are known, for Mo–(Fe, Co)-(Cr, Mn). See Fig. 7.13.
- $oP56$-Cr$_{0.18}$Mo$_{0.42}$Ni$_{0.40}$ (P phase) (148 $R\bar{3}$): Two representatives are known, for Mo–(Ni)–(Cr, Fe). See Fig. 7.13.
- $tI26$-NdFe$_{10}$Mo$_2$ (139 $I4/mmm$): One ferromagnetic representative is known, YFe$_{10}$Mo$_2$. Its structure can be described as ordering variant of the $tI26$-ThMn$_{12}$ type (see Fig. 7.31).
- $mS64$-Nd$_3$(Ti$_{0.21}$Fe$_{0.79}$)$_6$Fe$_{23}$ (12 $C2/m$): One representative is known, Y$_3$(Mo$_{0.21}$Fe$_{0.79}$)$_6$Fe$_{23}$. The structure can be described as a packing of polyhedra with 20 vertices, bounded by 2 squares and 40 triangles (Fig. 7.41).

Compounds of W with TM elements

In the PCD, there are 54 binary and 33 ternary compounds listed with other TM elements, respectively. The five most common binary structure types of the binary W–TM compounds are:

- $tP30$-Cr$_{46}$Fe$_{54}$ (σ phase) (136 $P4_2/mnm$): Five representatives with this structure type are known, with TM = Re, Tc, Ru Os, and Ir.
- $hP8$-Mg$_3$Cd (194 $P6_3/mmc$): Three representatives with this structure type are known, with TM = Co, Rh, and Ir.
- $cF24$-MgCu$_2$ (227 $Fd\bar{3}m$): Two representatives with this structure type are known, with TM = Zr and Hf.
- $hR39$-W$_6$Fe$_7$ (166 $R\bar{3}m$): Two representatives with this structure type are known, with TM = Fe and Co.
- $tI32$-W$_5$Si$_3$ (140 $I4/mcm$): One representative with this structure type is known, W$_5$Zr$_3$.

All the ternary structure types of the 83 ternary W–TM–TM' compounds are:

- $hP28$-Hf$_9$Mo$_4$B (194 $P6_3/mmc$): Three intermetallics with this structure type are known for W–Hf–(Fe, Co, Ni). See Fig. 7.37.
- $hR159$-Cr$_{0.16}$Mo$_{0.38}$Co$_{0.46}$ (R phase) (148 $R\bar{3}$): Two representatives are known for W–Cr–(Fe, Co).

7.14.5   Compounds of Mn, Tc, or Re with TM elements

Compounds of Mn with TM elements

In the PCD, there are 99 binary and 102 ternary compounds listed with other TM elements. The five most common binary structure types of the binary Mn–TM compounds are:

- $cP2$-CsCl (221 $Pm\bar{3}m$): Seven representatives with this structure type are known, with TM = V, Rh, Ni, Pd, Au, Hg, and Zn.
- $hP12$-MgZn$_2$ (194 $P6_3/mmc$): Seven representatives with this structure type are known, with TM = Sc, Y, Zr, Hf, Ti, Ta, and Nb.
- $tP30$-Cr$_{46}$Fe$_{54}$ (σ phase) (136 $P4_2/mnm$): Six representatives with this structure type are known, with TM = Ti, V, Mo, Cr, Re, and Tc.
- $cP4$-Cu$_3$Au (221 $Pm\bar{3}m$): Six representatives with this structure type are known, with TM = Rh, Ir, Ni, Pt, Pd, and Zn.
- $tP2$-CuAu (123 $P4/mmm$): Five representatives with this structure type are known, with TM = Rh, Ir, Ni, Pt, and Pd.

All the ternary structure types of the 102 ternary Mn–TM–TM' compounds are:

- $hR159$-Cr$_{0.16}$Mo$_{0.38}$Co$_{0.46}$ (R phase) (148 $R\bar{3}$): Two representatives are known for Mn–Mo–(Fe, Co).
- $tP24$-Ce(Mn$_{0.55}$Ni$_{0.45}$)$_{11}$ (127 $P4/mbm$): Two representatives are known for Mn–La–(Ni, Cu).
- $oP4$-MnHg$_{0.5}$Au$_{0.25}$ (62 $Pmma$): One representative is known, MnHg$_{0.5}$Au$_{0.25}$.
- $cF184$-CeCr$_2$Al$_{20}$ (227 $Fd\bar{3}m$): One representative is known, ZrMn$_2$Zn$_{20}$
- $cF32$-LiCa$_6$Ge (225 $Fm\bar{3}m$): One representative is known, MnPt$_6$Cu (Fig. 7.40).

Compounds of Tc with TM elements

In the PCD, there are 42 binary and no ternary compounds listed with other TM elements. The five most common binary structure types of the binary Tc–TM compounds are:

- $tP30$-Cr$_{46}$Fe$_{54}$ (σ phase) (136 $P4_2/mnm$): Five representatives with this structure type are known, with TM = W, Mo, Cr, Mn, and Fe.
- $cP2$-CsCl (221 $Pm\bar{3}m$): Four representatives with this structure type are known, with TM = Hf, Ti, Ta, and V.
- $hP12$-MgZn$_2$ (194 $P6_3/mmc$): Four representatives with this structure type are known, with TM = Sc, Y, Zr, and Hf.
- $cP8$-Cr$_3$Si (223 $Pm\bar{3}n$): One representative with this structure type is known, Tc$_3$Mo.

- $cF32$-CuPt$_7$ (225 $Fm\bar{3}m$): One representative with this structure type is known, TcZn$_7$.

Compounds of Re with TM elements

In the PCD, there are 63 binary and 17 ternary compounds listed with other TM elements. The five most common binary structure types of the binary Re–TM compounds are:

- $tP30$-Cr$_{46}$Fe$_{54}$ (σ phase) (136 $P4_2/mnm$): Nine representatives with this structure type are known, with TM = Zr, Ta, Nb, V, W, Mo, Cr, Mn, and Fe.

- $cI58$-Ti$_5$Re$_{24}$ (217 $I\bar{4}3m$) (χ phase) (Fig. 7.42): Five representatives with this structure type are known, with TM = Sc, Zr, Hf, Ti, and Nb. This structure corresponds to an ordering type of the $cI58$-Mn structure type. It can also be described as a packing of face-sharing CN16 FK-polyhedra, Ti1@Re$_{15}$Ti2, centered on the Ti1 atoms in the 16f Wyckoff position. The CN16 FK-polyhedra, Ti2@Re$_{12}$Ti1$_4$, around the Ti2 in 2a, share faces with half of the other set of CN16 FK-polyhedra, and hexagonal bipyramids with the other half. However, the structure of $cI58$-Ti$_5$Re$_{24}$ can be fully described by just a bcc packing of a supercluster of four face-sharing CN16 FK-polyhedra, Ti1@Re$_{15}$Ti2, in a tetrahedral arrangement.

- $hP12$-MgZn$_2$ (194 $P6_3/mmc$): Four representatives with this structure type are known, with TM = Sc, Y, Zr, and Hf.

- $cP8$-Cr$_3$Si (223 $Pm\bar{3}n$): Three representatives with this structure type are known, with TM = V, W, and Mo.

- $hR276$-Zr$_{21}$Re$_{25}$ (165 $R\bar{3}c$) (Fig. 7.42): Two representatives with this structure type are known, with TM = Zr and Hf. All the Re atoms on six different sites show icosahedral coordination, one of the four Zr atoms has a CN16 FK coordination, Zr@Re$_{12}$Zr$_4$ (Cenzual and Parthé, 1986). Units of three face-sharing CN16 FK-polyhedra form columns along [001] around 1/3, 2/3, 0 and 2/3, 1/3, 0.

There is not a single ternary structure type among the 17 ternary Re–TM–TM' compounds in the PCD.

7.14.6   Compounds of Fe, Ru, or Os with TM elements

Compounds of Fe with TM elements

In the PCD, there are 113 binary and 182 ternary compounds listed with other TM elements. The five most common binary structure types of the binary Fe–TM compounds are:

- $hP12$-MgZn$_2$ (194 $P6_3/mmc$): Eight representatives with this structure type are known, with TM = Sc, Zr, Hf, Ti, Ta, Nb, W, and Mo.

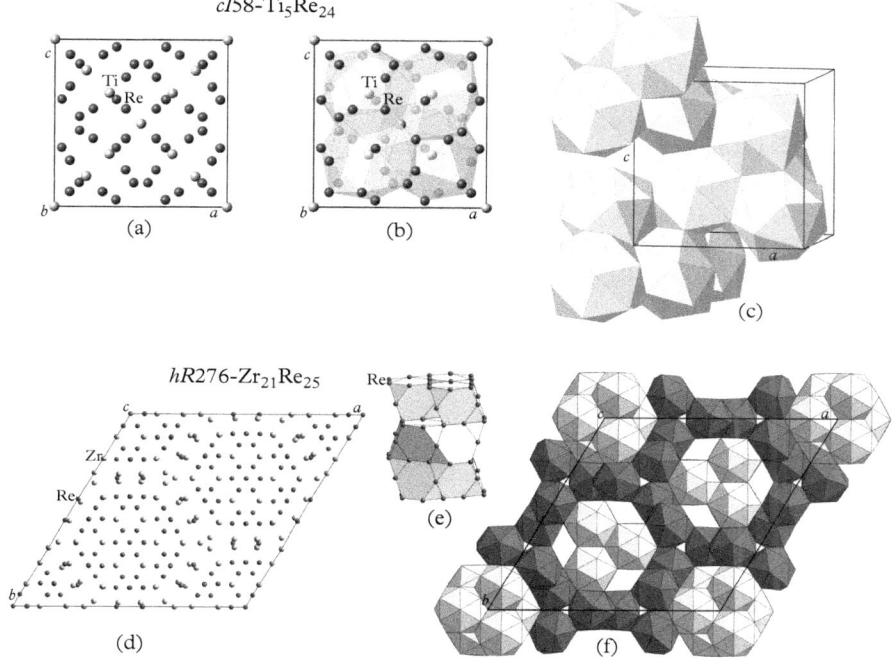

Fig. 7.42 *The structures of cI58-Ti₅Re₂₄ (a)–(c) and hR276-Zr₂₁Re₂₅ (d)–(f). In (a)–(b) and (d, f) the projected structures are shown in different representations. The structure of cI58-Ti₅Re₂₄ can be described as a bcc packing of superclusters, which each consist of four CN16 FK-polyhedra (b)–(c). The structure of hR276-Zr₂₁Re₂₅ consists of columns of CN16 FK-polyhedra surrounded by distorted icosahedral structural units centered on Re atoms (f). The skeleton of such columns (e), face-sharing truncated tetrahedra, are structural units of the hexagonal Laves phase hP12-MgZn₂. The subfigures (d)–(f) are on half the usual scale.*

- $tP30$-$Cr_{46}Fe_{54}$ (σ phase) (136 $P4_2/mnm$): Five representatives with this structure type are known, with TM = V, Mo, Cr, Re, and Tc.
- $cP2$-CsCl (221 $Pm\bar{3}m$): Four representatives with this structure type are known, with TM = Ti, V, Co, and Rh.
- $cP4$-Cu₃Au (221 $Pm\bar{3}m$): Four representatives with this structure type are known, with TM = Ni, Pt, Pd, and Au.
- $cF24$-MgCu₂ (227 $Fd\bar{3}m$): Two representatives with this structure type are known, with TM = Sc, Y, Zr, and Hf.

The ternary structure types of the 182 ternary Fe–TM–TM' compounds are:

- $mS64$-Nd₃(Ti₀.₂₁Fe₀.₇₉)₆Fe₂₃ (12 $C2/m$): Five representatives are known for Fe–Y–(Ti, Ta, V, Mo, Cr).

- $cF184$-$CeCr_2Al_{20}$ (227 $Fd\bar{3}m$): Four compounds are known, for Fe–Zn–(Y, Zr, Hf, Nb).
- $hR159$-$Cr_{0.16}Mo_{0.38}Co_{0.46}$ (R phase) (148 $R\bar{3}$): Three representatives are known for Fe–Mo–(Cr, Mn) and Fe–Cr–W.
- $hP28$-Hf_9Mo_4B (194 $P6_3/mmc$): Three representatives with this structure type are known for Fe–Mo–(Zr, Hf) and Fe–W–Hf. See Fig. 7.37.
- $tI26$-$NdFe_{10}Mo_2$ (139 $I4/mmm$): Three representatives with this structure type are known for Fe–Y–(V, Mo, Cr). Its structure can be described as ordering variant of the $tI26$-$ThMn_{12}$ type (see Fig. 7.31).

Compounds of Ru with TM elements

In the PCD, there are 68 binary and 27 ternary compounds listed with other TM elements. The five most common binary structure types of the binary Ru–TM compounds are:

- $cP2$-CsCl (221 $Pm\bar{3}m$): Seven representatives with this structure type are known, with TM = Sc, Zr, Hf, Ti, Ta, Nb, and V.
- $tP30$-$Cr_{46}Fe_{54}$ (σ phase) (136 $P4_2/mnm$): Three representatives with this structure type are known, with TM = W, Mo, and Cr.
- $tP2$-CuTi (123 $P4/mmm$): Three representatives with this structure type are known, with TM = Ta, Nb, and V.
- $cP4$-Cu_3Au (221 $Pm\bar{3}m$): Two representatives with this structure type are known, with TM = Ta and Nb.
- $hP12$-$MgZn_2$ (194 $P6_3/mmc$): Two representatives with this structure type are known, with TM = Sc and Zr.

All the ternary structure types of the 27 Ru–TM–TM' compounds are:

- $cF96$-Gd_4RhIn (216 $F\bar{4}3m$): One representative with this structure type is known, La_4RuCd. See Fig. 7.37.
- $hP68$-$Mg_4Pr_{23}Ir_7$ (186 $P6_3mc$) (Fig. 7.39): One representative with this structure type is known, $Cd_4La_{23}Ru_7$.

Compounds of Os with TM elements

In the PCD, there are 59 binary and 12 ternary compounds listed with other TM elements. The five most common binary structure types of the binary Os–TM compounds are:

- $tP30$-$Cr_{46}Fe_{54}$ (σ phase) (136 $P4_2/mnm$): Five representatives with this structure type are known, with TM = Ta, Nb, W, Mo, and Cr.

- $hP12$-MgZn$_2$ (194 $P6_3/mmc$): Five representatives with this structure type are known, with TM = Sc,Y, Zr, Hf, and La.
- $cP8$-Cr$_3$Si (223 $Pm\bar{3}n$): Four representatives with this structure type are known, with TM = Nb, V, Mo, and Cr.
- $cP2$-CsCl (221 $Pm\bar{3}m$): Four representatives with this structure type are known, with TM = Sc, Zr, Hf, Ti, and V.
- $oI142$-Hf$_{54}$Os$_{17}$ (71 $Immm$): Two representatives with this structure type are known, with TM = Zr and Hf. This structure is closely related to the $cP140$-Sc$_{57}$Rh$_{13}$ type discussed above (Cenzual *et al.*, 1985).

There is just one ternary structure type among the the 12 ternary Os–TM–TM' compounds are:

- $cF184$-CeCr$_2$Al$_{20}$ (227 $Fd\bar{3}m$): One representative is known, YOs$_2$Zn$_{20}$

7.14.7 Compounds of Co, Rh, or Ir with TM elements

Compounds of Co with TM elements

In the PCD, there are 116 binary and 180 ternary compounds listed with other TM elements. The five most common binary structure types of the binary Co–TM compounds are:

- $cF24$-MgCu$_2$ (227 $Fd\bar{3}m$): Eight representatives with this structure type are known, with TM = Sc, Y, Zr, Hf, Ti, Ta, Nb, and La.
- $cP2$-CsCl (221 $Pm\bar{3}m$): Five representatives with this structure type are known, with TM = Sc, Zr, Hf, Ti, and Fe.
- $cP4$-Cu$_3$Au (221 $Pm\bar{3}m$): Five representatives with this structure type are known, with TM = Ti, Ta, V, Pt, and Au.
- $cF96$-Ti$_2$Ni (227 $Fd\bar{3}m$): Four representatives with this structure type are known, with TM = Sc, Zr, Hf, and Ti. See Fig. 7.38.
- $tP30$-Cr$_{46}$Fe$_{54}$ (σ phase) (136 $P4_2/mnm$): Three representatives with this structure type are known, with TM = V, Mo, and Cr.

The five most common ternary structure types of the 180 ternary Co–TM–TM' compounds are:

- $cF184$-CeCr$_2$Al$_{20}$ (227 $Fd\bar{3}m$): Five representatives are known for Co–Zn–(Sc, Y, Zr, Hf, Nb).
- $hR159$-Cr$_{0.16}$Mo$_{0.38}$Co$_{0.46}$ (R phase) (148 $R\bar{3}$): Three representatives are known for Co–Mo–(Cr, Mn) and Co–Cr–W.
- $hP28$-Hf$_9$Mo$_4$B (194 $P6_3/mmc$): Three representatives with this structure type are known for Co–Mo–(Zr, Hf) and Co–W–Hf. See Fig. 7.37.

- $hR69$-$Y_{1.8}(Fe_{0.7}Co_{0.3})_{17.4}$ (166 $R\bar{3}m$): Unique structure type.
- $cF96$-Gd_4RhIn (216 $F\bar{4}3m$): One representative with this structure type is known, La_4CoCd. See Fig. 7.37.

Compounds of Rh with TM elements

In the PCD, there are 107 binary and 57 ternary compounds listed with other TM elements. The five most common binary structure types of the binary Rh–TM compounds are:

- $cP4$-Cu_3Au (221 $Pm\bar{3}m$): Nine representatives with this structure type are known, with TM = Sc, Zr, Hf, Ti, Ta, Nb, V, Cr, and Mn.
- $cP2$-$CsCl$ (221 $Pm\bar{3}m$): Eight representatives with this structure type are known, with TM = Sc, Y, Zr, Hf, Ti, Mn, Fe, and Zn.
- $tP2$-$CuAu$ (123 $P4/mmm$): Three representatives with this structure type are known, with TM = Nb, V, and Mn.
- $tP30$-$Cr_{46}Fe_{54}$ (σ phase) (136 $P4_2/mnm$): Two representatives with this structure type are known, with TM = Ta and Nb.
- $oP4$-$AuCd$ (62 $Pmma$): Two representatives with this structure type are known, with TM = Nb and Mo.

The five most common ternary structure types of the 57 ternary Rh–TM–TM' compounds are:

- $cF184$-$CeCr_2Al_{20}$ (227 $Fd\bar{3}m$): Three representatives are known for Rh–Zn–(Y, Zr, Hf).
- $oP24$-$LaNiAl$ (62 $Pnma$): One representative is known, $LaRhZn$.
- $hP40$-$Ta(Rh_{0.33}Pd_{0.67})_3$ (194 $P6_3/mmc$): One representative is known, $Ta(Rh_{0.33}Pd_{0.67})_3$.
- $oP12$-$TiNiSi$ (62 $Pnma$): One compound is known, $YRhZn$.
- $hP68$-$Mg_4Pr_{23}Ir_7$ (186 $P6_3mc$): One representative with this structure type is known, $Cd_4La_{23}Rh_7$.

Compounds of Ir with TM elements

In the PCD, there are 104 binary and 23 ternary compounds listed with other TM elements. The five most common binary structure types of the binary Ir–TM compounds are:

- $cP4$-Cu_3Au (221 $Pm\bar{3}m$): Eight representatives with this structure type are known, with TM = Sc, Zr, Hf, Ti, Ta, Nb, V, and Mn.
- $tP30$-$Cr_{46}Fe_{54}$ (σ phase) (136 $P4_2/mnm$): Five representatives with this structure type are known, with TM = Zr, Ta, Nb, W, and Mo.

- *cP8*-Cr₃Si (223 *Pm3̄n*): Five representatives with this structure type are known, with TM = Ti, Nb, V, Mo, and Cr.
- *tP2*-CuAu (123 *P4/mmm*): Five representatives with this structure type are known, with TM = Ti, Ta, Nb, V, and Mn.
- *cP2*-CsCl (221 *Pm3̄m*): Four representatives with this structure type are known, with TM = Sc, Y, Zr, and T.

All the ternary structure types of the 23 ternary Ir–TM–TM' compounds are:

- *cF96*-Gd₄RhIn (216 *F4̄3m*): Three representatives are known for Ir–Cd–(Y, La). See Fig. 7.37.
- *cF184*-CeCr₂Al₂₀ (227 *Fd3̄m*): One representative is known, YIr₂Zn₂₀.
- *hP68*-Mg₄Pr₂₃Ir₇ (186 *P6₃mc*): One representative with this structure type is known, Cd₄La₂₃Ir₇.

7.14.8   Compounds of Ni, Pd, or Pt with TM elements

Compounds of Ni with TM elements

In the PCD, there are 149 binary and 261 ternary compounds listed with other TM elements, respectively. The five most common binary structure types of the binary Ni–TM compounds are:

- *cF24*-MgCu₂ (227 *Fd3̄m*): Five representatives with this structure type are known, with TM = Sc, Y, Zr, Hf, and La.
- *cP2*-CsCl (221 *Pm3̄m*): Four representatives with this structure type are known, with TM = Sc, Ti, Mn, and Zn.
- *hP36*-Ce₂Ni₇ (194 *P6₃/mmc*): Three representatives with this structure type are known, with TM = Sc, Y, and La. The structure can be described by a face-sharing packing of the two different Ce-AETs. The one is a truncated tetrahedron Ce1@Ni₁₂, the other a double hexagonal antiprism, Ce2@Ni₁₈ (Fig. 7.43).
- *cP4*-Cu₃Au (221 *Pm3̄m*): Three representatives with this structure type are known, with TM = Mn, Fe, and Au.
- *oP8*-Cu₃Ti (59 *Pmmn*): Three representatives with this structure type are known, with TM = Ta, Nb, and Mo.

The five most common ternary structure types of the 261 ternary Ni–TM–TM' compounds are:

- *cF184*-CeCr₂Al₂₀ (227 *Fd3̄m*): Three representatives are known, for Ni–Zn–(Zr, Hf, Nb).
- *hP28*-Hf₉Mo₄B (194 *P6₃/mmc*): Three representatives with this structure type are known for Ni–Mo–(Zr, Hf) and Ni–W–Hf. See Fig. 7.37.

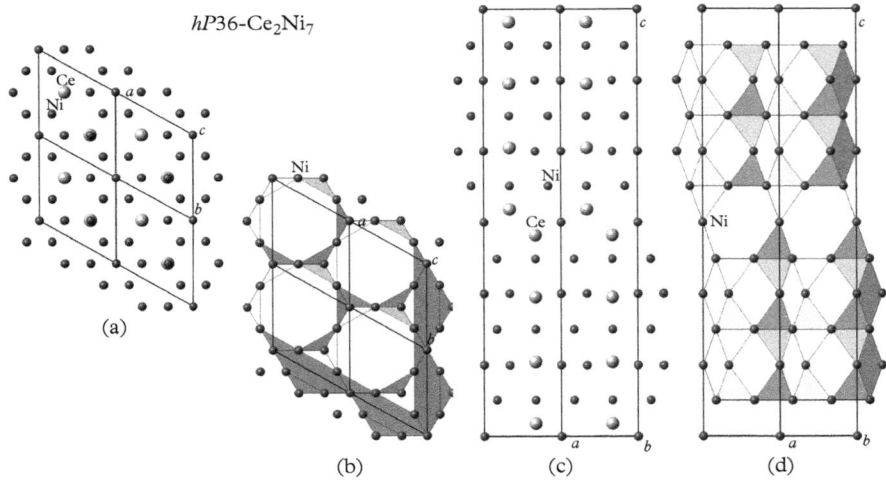

Fig. 7.43 *The structure of hP36-Ce₂Ni₇ in different representations. It can be described by a face-sharing packing along [001] of the two different Ce-AETs. One is a truncated tetrahedron* Ce1@Ni₁₂ *(centered at z = 0, 1/2), the other a double hexagonal antiprism,* Ce2@Ni₁₈ *(centered at z = ±1/6, ±2/6).*

- $cF24$-MgCu$_4$Sn (216 $F\bar{4}3m$): Three representatives with this structure type are known for Ni–Au–(Sc, Y) and Ni–Zn–Zr. Ni occupies the Cu site. See Fig. 7.55.
- $oP56$-Cr$_{0.18}$Mo$_{0.42}$Ni$_{0.40}$ (P phase) (148 $R\bar{3}$): Two representatives are known, for Ni–Mo–(Cr, Fe).
- $cF96$-Gd$_4$RhIn (216 $F\bar{4}3m$): Two representatives are known for Ni–Cd–(Y, La). See Fig. 7.37.

Compounds of Pd with TM elements

In the PCD, there are 128 binary and 62 ternary compounds listed with other TM elements. The five most common binary structure types of the binary Pd–TM compounds are:

- $cP4$-Cu$_3$Au (221 $Pm\bar{3}m$): Nine representatives with this structure type are known, with TM = Sc, Y, Hf, Ti, Cr, Mn, Fe, Cu, and La.
- $cP2$-CsCl (221 $Pm\bar{3}m$): Seven representatives with this structure type are known, with TM = Sc, Zr, Ti, Mn, Cu, Cd, and Zn.
- $tP2$-CuAu (123 $P4/mmm$): Five representatives with this structure type are known, with TM = Mn, Fe, Cu, Hg, and Zn.

- *oI*6-MoPt$_2$ (71 *Immm*): Four representatives with this structure type are known, with TM = Ta, Nb, V, and Mo. This is an othorhombically distorted $(3 \times 1 \times 1)$-fold superstructure of the *cI*2-W type.

- *tI*6-CuZr$_2$ (139 *I*4/*mmm*): Three representatives with this structure type are known, with TM = Zr, Hf, and Ti. See Fig. 7.7.

The five most common ternary structure types of the 62 ternary Pd–TM–TM' compounds are:

- *cF*96-Gd$_4$RhIn (216 *F*43*m*): Two representatives are known for Pd–Cd–(Y, La). See Fig. 7.37.

- *hP*9-ZrNiAl (59 *Pmmn*): Five compounds with this structure type are known for Pd–La–(Cd, Cu, Hg).

- *oS*16-MgCuAl$_2$ (63 *Cmcm*): One representative with this structure type is known, LaPdCd$_2$.

- *tP*10-Mo$_2$FeB$_2$ (127 *P*4/*mbm*): One representative with this structure type is known, Pd$_2$CdLa$_2$

- *hP*40-Ta(Rh$_{0.33}$Pd$_{0.67}$) (194 *P*6$_3$/*mmc*): One representative with this structure type is known, Ta(Rh$_{0.33}$Pd$_{0.67}$).

Compounds of Pt with TM elements

In the PCD, there are 129 binary and 74 ternary compounds listed with other TM elements. The five most common binary structure types of the binary Pt–TM compounds are:

- *cP*4-Cu$_3$Au (221 *Pm*3*m*): Thirteen representatives with this structure type are known, with TM = Y, Zr, Hf, V, Cr, Mn, Fe, Co, Ag, Cu, Cd, Zn, and La.

- *tP*2-CuAu (123 *P*4/*mmm*): Nine representatives with this structure type are known, with TM = V, Cr, Mn, Fe, Co, Ni, Hg, Cd, and Zn.

- *cP*8-Cr$_3$Si (223 *Pm*3*n*): Five representatives with this structure type are known, with TM = Ti, Nb, V, Mo, and Cr.

- *oP*4-AuCd (62 *Pmma*): Four representatives with this structure type are known, with TM = Ti, Nb, and V.

- *oI*6-MoPt$_2$ (71 *Immm*): Four representatives with this structure type are known, with TM = Nb, V, W, and Mo.

The five most frequent ternary structure types of the 74 ternary Pt–TM–TM' compounds are:

- *cF*32-LiCa$_6$Ge (225 *Fm*3*m*): Six representatives with this structure type are known, with Pt–Cu–(Ti, V, Cr, Mn, Fe, Ni) (Fig. 7.40).

- $cF96$-Gd_4RhIn (216 $F\bar{4}3m$): Two representatives are known for Pt–Cd–(Y, La). See Fig. 7.37.
- $hP28$-$Ti(Ni_{0.11}Pt_{0.89})_3$ (164 $P\bar{3}m$): Two representatives are known for Pt–Ti–(Ni, Pd).
- $hP68$-$Mg_4Pr_{23}Ir_7$ (186 $P6_3mc$): One representative with this structure type is known, $Cd_4La_{23}Pt_7$.
- $tP4$-$ZnCdPt_2$ (123 $P4/mmm$): One representative with this structure type is known, $tP4$-$ZnCdPt_2$.

7.14.9 Compounds of Cu, Ag, or Au with TM elements

Compounds of Cu with TM elements

In the PCD, there are 86 binary and 133 ternary compounds listed with other TM elements, respectively. The five most common binary structure types of the binary Cu–TM compounds are:

- $cP2$-$CsCl$ (221 $Pm\bar{3}m$): Five representatives with this structure type are known, with TM = Sc, Y, Zr, Pd, and Zn.
- $cP4$-Cu_3Au (221 $Pm\bar{3}m$): Three representatives with this structure type are known, with TM = Pt, Pd, and Au.
- $hP68$-$Ag_{51}Gd_{14}$ (175 $P6/m$): Three representatives with this structure type are known, with TM = Y, Zr, and Hf.
- $tI6$-$CuZr_2$ (139 $I4/mmm$): Three representatives with this structure type are known, with TM = Zr, Hf, and Ti. See Fig. 7.7.
- $hP6$-$CaCu_5$ (191 $P6/mmm$): Two representatives with this structure type are known, with TM = Y and La.

The five most common ternary structure types of the 133 ternary Cu–TM–TM' compounds are:

- $cF32$-$LiCa_6Ge$ (225 $Fm\bar{3}m$): Six representatives are known for Cu–Pt–(Ti, V, Cr, Mn, Fe, Ni) (Fig. 7.40).
- $cF16$-Cu_2MnAl (225 $Fm\bar{3}m$): Three representatives with this structure type are known, Cu–Zr–(Cd, Zn) and Cu–Au–Zn.
- $tP24$-$Ce(Mn_{0.55}Ni_{0.45})_{11}$ (127 $P4/mbm$): Two representatives are known for Cu–La–(Mn, Cd).
- $cF16$-Li_2AgSb (216 $F\bar{4}3m$): Two representatives are known for Cu–Hg–Ti and Cu–Cd–La.
- $cF116$-$Mg_6Cu_{16}Si_7$ (225 $Fm\bar{3}m$): Two representatives are known for Cu–Zn–(Zr, Hf).

Compounds of Ag with TM elements

In the PCD, there are 48 binary and 26 ternary compounds listed with other TM elements. The five most common binary structure types of the binary Ag–TM compounds are:

- $cP2$-CsCl (221 $Pm\bar{3}m$): Five representatives with this structure type are known, with TM = Sc, Y, Cd, Zn, and La.
- $tI6$-CuZr$_2$ (139 $I4/mmm$): Three representatives with this structure type are known, with TM = Zr, Hf, and Ti. See Fig. 7.7.
- $hP68$-Ag$_{51}$Gd$_{14}$ (175 $P6/m$): Two representatives with this structure type are known, with TM = La and Hg.
- $cI52$-Cu$_5$Zn$_8$ (217 $I\bar{4}3m$): Two representatives with this structure type are known, with TM = Cd and Zn. This compound is known as γ-brass.
- $tP4$-CuTi (128 $P4/mnc$): Two representatives with this structure type are known, with TM = Zr and Hf.

The only known ternary structure type among the 26 ternary Ag–TM–TM' compounds are:

- $cF16$-Cu$_2$MnAl (225 $Fm\bar{3}m$): Two representatives with this structure type are known, Ag–Au–(Cd, Zn).

Compounds of Au with TM elements

In the PCD, there are 111 binary and 42 ternary compounds listed with other TM elements. The five most common binary structure types of the binary Au–TM compounds are:

- $tI10$-MoNi$_4$ (87 $I4/m$): Six representatives with this structure type are known, with TM = Hf, Ti, V, Cr, and Mn. This structure is an ordered superstructure of the $cF4$-Cu type (see Fig. 7.3).
- $tI6$-MoSi$_2$ (139 $I4/mmm$): Six representatives with this structure type are known, with TM = Sc, Y, Zr, Hf, Ti, and Mn.
- $cP2$-CsCl (221 $Pm\bar{3}m$): Six representatives with this structure type are known, with TM = Sc, Y, Ti, Mn, Cd, and Zn.
- $cP4$-Cu$_3$Au (221 $Pm\bar{3}m$): Six representatives with this structure type are known, with TM = Ti, V, Fe, Co, Ni, and Cu.
- $oP8$-Cu$_3$Ti (59 $Pmmn$): Four representatives with this structure type are known, with TM = Sc, Y, Zr, and Hf.

All the ternary structure types of the 42 ternary Au–TM–TM' compounds are:

- $cF16$-Cu$_2$MnAl (225 $Fm\bar{3}m$): Three representatives with this structure type are known, Au–Ag–(Cd, Zn) and Au–Cu–Zn.

- $cF24$-$MgCu_4Sn$ (216 $F\bar{4}3m$): Two representatives with this structure type are known for Au–Ni–(Sc, Y). See Fig. 7.55.

- $hP9$-$ZrNiAl$ (59 $Pmmn$): Two representatives with this structure type are known for Au–Cd–(Y, La).

- $tP10$-Mo_2FeB_2 (127 $P4/mbm$) (127 $P4/mbm$): One representative with this structure type is known, Au_2CdLa_2

7.14.10   Compounds of Zn, Cd, or Hg with TM elements

Compounds of Zn with TM elements

In the PCD, there are 136 binary and 134 ternary compounds listed with other TM elements. The five most common binary structure types of the binary Zn–TM compounds are:

- $cP2$-$CsCl$ (221 $Pm\bar{3}m$): Twelve representatives with this structure type are known, with TM = Sc, Y, Zr, Ti, Mn, Rh, Ni, Pd, Au, Ag, Cu, and La.

- $cP4$-Cu_3Au (221 $Pm\bar{3}m$): Six representatives with this structure type are known, with TM = Zr, Ti, Nb, V, Mn, and Pt.

- $cI52$-Cu_5Zn_8 (217 $I\bar{4}3m$): Seven representatives with this structure type are known, with TM = Fe, Rh, Ir, Ni, Pd, and Ag.

- $mS28$-$Zn_{13}Co$ (12 $C2/m$): Four representatives with this structure type are known, with TM = Mn, Fe, Co, and Rh. See Section 7.11.

- $tI6$-$CuZr_2$ (139 $I4/mmm$): Three representatives with this structure type are known, with TM = Zr, Hf, and Ti. See Fig. 7.7.

The five most common ternary structure types of the 134 ternary Zn–TM–TM' compounds are:

- $cF184$-$CeCr_2Al_{20}$ (227 $Fd\bar{3}m$): Twenty-two representatives with this structure type are known for Zn with Fe–(Y, Zr, Hf, Nb), Ru–(Sc, Y, Zr, Hf), Co–(Sc, Y, Zr, Hf, Nb), Rh–(Y, Zr, Hf), Y–(Os, Ir), Ni–(Zr, Hf, Nb), and Zr–Mn. See Fig. 7.37.

- $cF16$-Cu_2MnAl (225 $Fm\bar{3}m$): Five representatives with this structure type are known, for Zn with Ag–(Cd, Zn) and Cu–Zn.

- $cF24$-$MgCu_4Sn$ (216 $F\bar{4}3m$): Three representatives with this structure type are known for Zn with Au–(Ag, Cu) and Cu–Zr. See Fig. 7.55.

- $hR57$-$Ce_2Co_{15}Al_2$ (166 $R\bar{3}m$): Two representatives with this structure type are known for Zn–La–(Co, Ni).

- $cF116$-$Mg_6Cu_{16}Si_7$ (225 $Fm\bar{3}m$): Two representatives are known for Zn–Cu–(Zr, Hf).

Compounds of Cd with TM elements

In the PCD, there are 70 binary and 40 ternary compounds listed with other TM elements. The five most common binary structure types of the binary Cd–TM compounds are:

- $cP2$-CsCl (221 $Pm\bar{3}m$): Six representatives with this structure type are known, with TM = Sc, Y, Pd, Au, Ag, and La.
- $cI52$-Cu$_5$Zn$_8$ (217 $I\bar{4}3m$): Four representatives with this structure type are known, with TM = Pd, Au, Ag, and Cu.
- $tI6$-CuZr$_2$ (139 $I4/mmm$): Three representatives with this structure type are known, with TM = Zr, Hf, and Ti. See Fig. 7.7.
- $tP4$-CdTi (129 $P4/nmm$): Three representatives with this structure type are known, with TM = Zr, Hf, and Ti.
- $tP2$-CuTi (123 $P4/mmm$): Three representatives with this structure type are known, with TM = Pd, Au, and La.

The five most common ternary structure types of the 40 ternary Cd–TM–TM' compounds are:

- $cF96$-Gd$_4$RhIn (216 $F\bar{4}3m$): Eleven representatives are known for Cd–Y–(Ru, Ir, Ni, Pt, Pd) and Cd–La–(Co, Ir, Ni, Pd, Pt, Ru). See Fig. 7.37.
- $hP68$-Mg$_4$Pr$_{23}$Ir$_7$ (186 $P6_3mc$): Six representatives with this structure type are known, Cd–La–(Co, Ir, Ni, Pt, Rh, Ru).
- $tP10$-Mo$_2$FeB$_2$ (127 $P4/mbm$): Five representatives with this structure type are known, Cd–La–(Au, Ni, Pd, Rh) and Cd–Cu–Y.
- $hP9$-ZrNiAl (59 $Pmmn$): Three representatives with this structure type are known for Cd–Au–(Y, La) and Cd–La–Pd.
- $cF16$-Cu$_2$MnAl (225 $Fm\bar{3}m$): Two representatives with this structure type are known, Cd–Ag–Au and Cd–Cu–Zr.

Compounds of Hg with TM elements

In the PCD, there are 45 binary and 10 ternary compounds listed with other TM elements. The five most common binary structure types of the binary Hg–TM compounds are:

- $cP2$-CsCl (221 $Pm\bar{3}m$): Four representatives with this structure type are known, with TM = Sc, Y, Mn, and La.
- $tP2$-CuAu (123 $P4/mmm$): Four representatives with this structure type are known, with TM = Zr, Ti, Pt, and Pd.

- *h*P8-Mg$_3$Cd (194 *P*6$_3$/*mmc*): Three representatives with this structure type are known, with TM = Sc, Y, and La.
- *c*P8-Cr$_3$Si (223 *Pm*$\bar{3}$*n*): Two representatives with this structure type are known, with TM = Zr and Ti.
- *c*P4-Cu$_3$Au (221 *Pm*$\bar{3}$*m*): Six representatives with this structure type are known, with TM = Zr and Ti.

All the ternary structure types of the 10 ternary Hg–TM–TM compounds are:

- *c*F16-Li$_2$AgSb (216 *F*$\bar{4}$3*m*): One representative is known, Hg$_2$CuTi.
- *h*P9-ZrNiAl (59 *Pmmn*): One representative is known, LaPdHg.

7.15   Intermetallic compounds with at least one (semi)metallic element from groups 13–16

In the following, additional information is given for intermetallics with at least one constituent from groups 13–16 (Fig. 7.44). In Table 7.9, the distribution is shown separately for binary and ternary compounds. Remarkably, the ratio of the total number of structure types to that of unique structure types ranges from 1.5 to 2.1 for this class of intermetallic compounds, and is therefore quite close to

1	2	3	4	5	6	7	8	9	10	11	12	13	14	15	16
12 Li 1.0 1.52	77 Be 1.5 1.11														
11 Na 0.9 1.86	73 Mg 1.2 1.60											80 Al 1.5 1.43			
10 K 0.8 2.27	16 Ca 1.0 1.97	19 Sc 1.3 1.61	51 Ti 1.4 1.45	54 V 1.6 1.31	57 Cr 1.6 1.25	60 Mn 1.5 1.37	61 Fe 1.8 1.24	64 Co 1.8 1.25	67 Ni 1.8 1.25	72 Cu 1.9 1.28	76 Zn 1.6 1.34	81 Ga 1.6 1.22	84 Ge 1.8 1.23		
9 Rb 0.8 2.48	15 Sr 1.0 2.15	25 Y 1.2 1.78	49 Zr 1.4 1.59	53 Nb 1.6 1.43	56 Mo 1.8 1.36	59 Tc 1.9 1.35	62 Ru 2.2 1.33	65 Rh 2.2 1.35	69 Pd 2.2 1.38	71 Ag 2.4 1.45	75 Cd 1.7 1.49	79 In 1.7 1.63	83 Sn 1.8 1.41	88 Sb 1.9 1.45	
8 Cs 0.7 2.66	14 Ba 0.9 2.17	33* La 1.1 1.87	50 Hf 1.3 1.56	52 Ta 1.5 1.43	55 W 1.7 1.37	58 Re 1.9 1.37	63 Os 2.2 1.34	66 Ir 2.2 1.36	68 Pt 2.2 1.37	70 Au 2.4 1.44	74 Hg 1.9 1.50	78 Tl 1.8 1.70	82 Pb 1.8 1.75	87 Bi 1.9 1.55	91 Po 2.0 1.67
7 Fr 0.7	13 Ra 0.9 2.15	48$^+$ Ac 1.1 1.88													

* Lanthanoids	32 Ce 1.1 1.87	31 Pr 1.1 1.82	30 Nd 1.1 1.81	29 Pm 1.1 1.63	28 Sm 1.2 1.62	18 Eu 1.2 2.00	27 Gd 1.2 1.79	26 Tb 1.1 1.76	25 Dy 1.2 1.75	24 Ho 1.2 1.74	23 Er 1.2 1.73	22 Tm 1.3 1.72	17 Yb 1.1 1.94	20 Lu 1.3 1.72
+ Actinoids	47 Th 1.3 1.88	46 Pa 1.5 1.80	45 U 1.4 1.39	44 Np 1.4 1.30	43 Pu 1.3 1.51	42 Am 1.1	41 Cm 1.3	40 Bk 1.3	39 Cf 1.3	38 Es 1.3	37 Fm 1.3	36 Md 1.3	35 No 1.3	34 Lr 1.3

Fig. 7.44 *Elements constituting the compounds discussed in Section 7.15 at least partly are shaded gray in the periodic table. Mendeleev numbers (top left in each box), Pauling electronegativities χ (relative to χ$_F$ = 4.0) (bottom left in each box), and atomic radii (half of the shortest distance between atoms in the crystal structure at ambient conditions) (bottom right in each box) of the metallic elements are given.*

Table 7.9 *The number of intermetallic compounds with at least one component out of groups 13–16. For each subclass, the ratio of the total number of non-unique structure types to the number of unique structure types is also given.*

Group 13–16 element	Binaries: No. of		Ternaries: No. of	
	compounds	non-unique/unique structure types	compounds	non-unique/unique structure types
Al	423	174/110 = 1.6	2449	384/204 = 1.9
Ga	397	164/96 = 1.7	1842	374/197 = 1.9
In	279	106/63 = 1.7	1260	255/127 = 2.0
Tl	167	47/24 = 2.0	127	50/34 = 1.5
Ge	425	171/109 = 1.6	2434	420/198 = 2.1
Sn	381	144/89 = 1.6	1470	328/161 = 2.0
Pb	168	64/36 = 1.8	344	80/41 = 2.0
Sb	343	130/75 = 1.7	1176	257/124 = 2.1
Bi	187	74/44 = 1.7	365	89/45 = 2.0
Po	24	3/0	0	0/0

the ratio derived for all intermetallics, $2166/1087 = 1.99$. The electronegativities of the (semi)metallic elements of groups 13–16 range between 1.5 and 2.0, which is comparable to those of the intermediate TM elements, while their atomic radii show a larger variation (Fig. 7.44).

That the number of compounds decreases drastically going from top to bottom of a group does not necessarily reflect a lower compound formation probability, but the fact that fewer phase diagram studies have been performed to date. One of the reasons for this may be the lack of technological applications due to the toxicity (Po, Pb, and Tl), radioactivity (Po), or scarcity (Po) of these elements.

How strongly do the preferred structure types differ within a group and between groups? This is shown in Table 7.10 for binary and ternary structure types.

7.15.1 Aluminides

The distribution of ternary aluminides in the ternary concentration diagram of all intermetallics is depicted in Fig. 7.45 together with the frequencies of the subset of binary aluminides as a function of their stoichiometries. Since the constituting elements A, B, and C stand for metallic elements with increasing Mendeleev numbers, $M(A) < M(B) < M(C)$, Al with $M(Al) = 80$, in most intermetallics $A_xB_yC_z$ will be represented by the letter C. Consequently, most compounds with Al as the majority element will agglomerate in the concentration triangle close to the upper corner marked C. With a few exceptions, the distribution of aluminides does not differ much from the general one. Perhaps the most significant difference can be seen along the otherwise densely occupied tieline connecting AC_2 with ABC_2. This means that compounds of the type $A_xB_yAl_z$ are rather rare for $x+y+z = 1$, $x = z/2$, and $y = 1 - 3z/2$, i.e., $A_{z/2}B_{1-3z/2}Al_z$ and $z \geq 0.5$. In contrast,

Table 7.10 *Structure types of binary and ternary intermetallics, respectively, with one constituent M out of groups 13–16 in comparison. The percentage of structures belonging to each particular structure type is given. Condition to be listed here is a minimum occurrence in three columns. Binary and ternary structure types are listed above and below the dashed line, respectively.*

Structure type	No.	Space group	Sites occupied	M=Al	Ga	In	Tl	Ge	Sn	Pb	Sb	Bi	Po
$hP16$-Mn$_5$Si$_3$	193	$P6_3/mcm$	$4d\,6g^2$	–	3.0	1.8	5.4	5.6	5.0	10.1	5.2	7.5	–
$cP4$-Cu$_3$Au	221	$Pm\bar{3}m$	$1a\,3c$	5.4	4.8	9.3	12	1.4	6.0	14.3	–	–	–
$cP2$-CsCl	221	$Pm\bar{3}m$	$1ab$	4.3	2	6.5	12.6	–	1	–	2.0	2.1	–
$oP12$-Co$_2$Si	62	$Pnma$	$4c^3$	2.8	–	1.4	–	1.4	1.6	3.0	–	–	–
$cP8$-Cr$_3$Si	223	$Pm\bar{3}n$	$2a\,6c$	–	1.0	–	1.8	–	1.3	1.8	1.2	–	–
$hP6$-Co$_{1.75}$Ge	194	$P6_3/mmc$	$2acd$	6.8	3.6	–	2.4	–	1.5	1.6	–	–	–
$tI32$-W$_5$Si$_3$	140	$I4/mcm$	$4ab\,8h\,16k$	–	1.5	1.4	3.6	1.6	–	1.8	–	–	–
$hP4$-NiAs	194	$P6_3/mmc$	$2ac$	–	–	–	–	–	1.3	1.8	3.2	2.1	25
$tI84$-Ho$_{11}$Ge$_{10}$	139	$I4/mmm$	$4de^2\,8h^2j\,16mm^2$	–	–	–	–	2.4	2.6	–	1.2	2.1	–
$tI12$-CuAl$_2$	140	$I4/mcm$	$4a\,8h$	1.2	–	–	2.4	–	1.3	2.4	–	–	–
$hP8$-Mg$_3$Cd	194	$P6_3/mmc$	$2d\,6h$	2.1	1.3	1.8	–	–	1.6	–	–	–	–
$tI32$-Cr$_5$B$_3$	140	$I4/mcm$	$4ac\,8h\,16l$	–	2.0	3.0	1.4	1.0	–	–	–	–	–
$tP2$-CuAu	123	$P4/mmm$	$1ad$	–	–	2.9	1.8	–	–	2.4	–	–	–
$oS32$-Pu$_3$Pd$_5$	63	$Cmcm$	$4c^2\,8efg$	–	–	4.3	8.4	–	1.8	–	–	–	–
$hP6$-CaIn$_2$	194	$P6_3/mmc$	$2b\,4f$	–	1.0	1.4	1.8	–	–	–	–	–	–

$tI10$-BaAl$_4$	139	$I4/mmm$	$2a\,4de$	2.1	1.5	1.1	–	–	–	–	–	–	–
$oP36$-Sm$_5$Ge$_4$	62	$Pnma$	$4c^3\,8d^3$	–	–	–	–	3.8	2.6	7.1	–	–	–
$cF8$-NaCl	225	$Fm\bar{3}m$	$4ab$	–	–	–	–	–	–	–	7.6	10.2	62.5
$hP9$-ZrNiAl	189	$P\bar{6}2m$	$1a2d3fg$	2.0	2.2	8.5	19.7	2.1	5.1	5.2	–	1.6	–
$oP12$-TiNiSi	62	$Pnma$	$4c^3$	1.4	2.6	1.0	–	5.5	4	4.7	3.7	2.2	–
$cF16$-Cu$_2$MnAl	225	$Fm\bar{3}m$	$4abcd$	2.3	–	6.6	–	1.1	–	4.9	1.9	–	–
$cF12$-MgAgAs	216	$F\bar{4}3m$	$4abc$	–	–	–	–	–	2.2	2.0	6.0	12.3	–
$hP6$-LiGaGe	186	$P6_3mc$	$2ab^2$	–	–	–	–	–	1.6	6.7	1.0	1.4	–
$hP18$-CuHf$_5$Sn$_3$	193	$P6_3/mcm$	$2b\,4d\,6g^2$	–	–	–	–	–	1.4	12.2	3.9	3.8	–
$tI10$-CeAl$_2$Ga$_2$	139	$I4/mmm$	$2a\,4de$	1.3	3.1	–	–	7.6	–	–	–	–	–
$tP10$-Mo$_2$FeB$_2$	127	$P4/mbm$	$2a\,4gh$	–	–	6.0	–	–	1.2	4.7	–	–	–
$tI80$-La$_6$Co$_{11}$Ga$_3$	140	$I4/mcm$	$4ad\,8f\,16kl^3$	–	–	–	3.1	–	–	1.5	–	1.1	–
$tP10$-CaBe$_2$Ge$_2$	129	$P4/nmm$	$2abc^3$	–	–	–	–	1.3	–	–	2.4	2.7	–

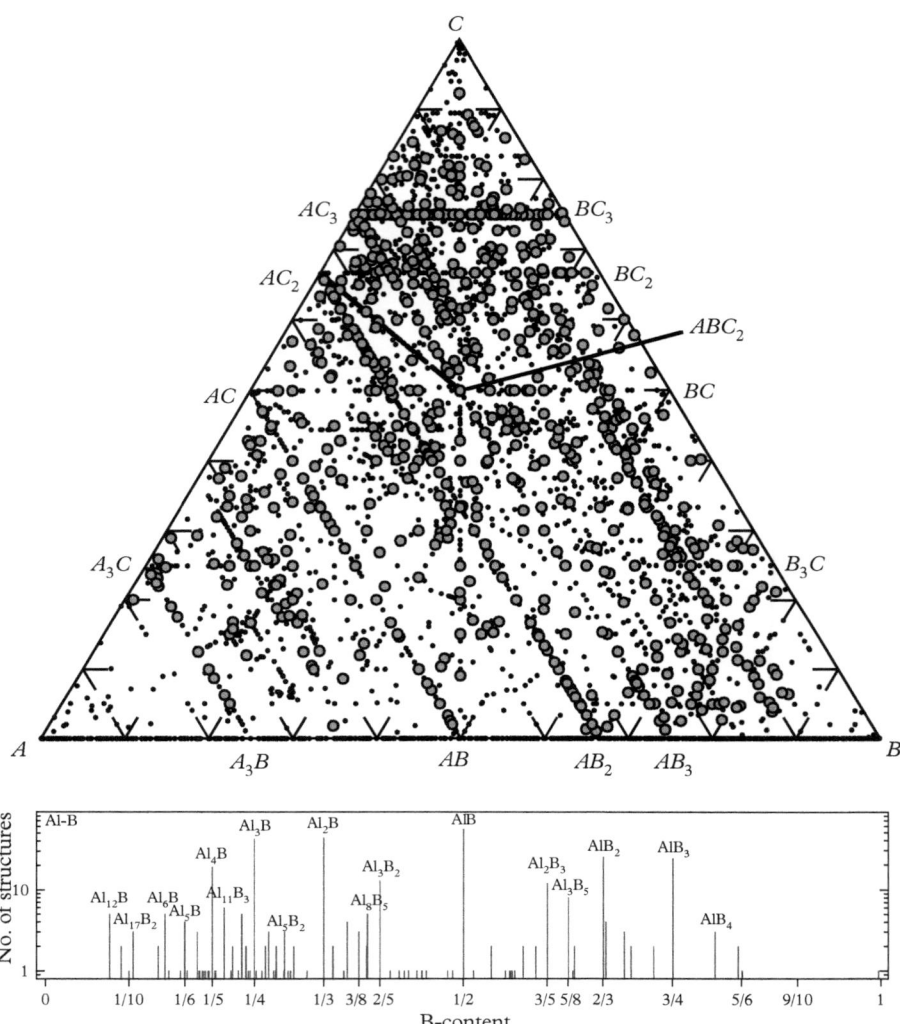

Fig. 7.45 *(top) Concentration diagram of the 20 829 intermetallic compounds (black dots) contained in* Pearson's Crystal Data *(PCD) (Villars and Cenzual, 2011a). The aluminides are marked by larger gray circles. A, B, and C stand for metallic elements with increasing Mendeleev numbers, $M(A) < M(B) < M(C)$. (bottom) Frequencies of the binary Al–B compounds as a function of stoichiometry (note the logarithmic scale).*

the lines $AC–A(B_yC_{1-y})–AB$, $AC_2–A(B_yC_{1-y})_2–AB_2$, $AC_3–A(B_yC_{1-y})_3–AB_3$, and $AC_2–(A_{1-y}B_y)_2–BC_2$, $AC_3–(A_{1-y}B_y)_3–BC_3$, all with $0 \leq y \leq 1$, are rather densely populated.

Furthermore, the distribution of aluminides on lines with the concentration being constant for one element is also not homogeneous. The $A_{const.}$ and, for high

C-concentration, $C_{const.} = Al_{const.}$ lines are much more densely populated than the $B_{const.}$ ones. Along $B_{const.}$, the A–C stoichiometry varies, i.e., the concentration of a low-M element vs. the high-M element aluminum. This appears to be less favorable than the variation of the B/C ratio with the concentration of the low-M element A being constant.

Binary aluminides

As mentioned previously, among the 20 829 IMs, there are 13 026 ternary and 6441 binary IMs. Among the binary IMs, 423 aluminides (6.6%) are listed in the PCD, featuring 174 (8.0%) out of the altogether 2166 structure types. This gives ≈ 2.4 representatives per structure type. The compositions of all the 423 binary Al-intermetallics are marked in Fig. 7.46, and their most common structure types are listed in Table 7.11.

In contrast to the other triels forming binary compounds with all alkali metals, the only known binary alkali-metal aluminides are those with Li ($M = 12$) as a constituent. Their structures can be seen as derivatives of the $cI2$-W type, the structure of which is also adopted by $cI2$-Li itself. These structures are that of the Zintl phase $cF16$-LiAl ($cF16$-NaTl type), with a 3D network of Al atoms, of $hR15$-Li$_3$Al$_2$ with puckered honeycomb layers of Al, and of $mS26$-Li$_9$Al$_4$ with zigzag-chains of Al (Tebbe *et al.*, 2007).

The alkaline earth elements Mg ($M = 73$), Ca ($M = 14$), Sr ($M = 15$), and Ba ($M = 16$) form binary aluminides in contrast to Be ($M = 77$) and Ra ($M = 13$).

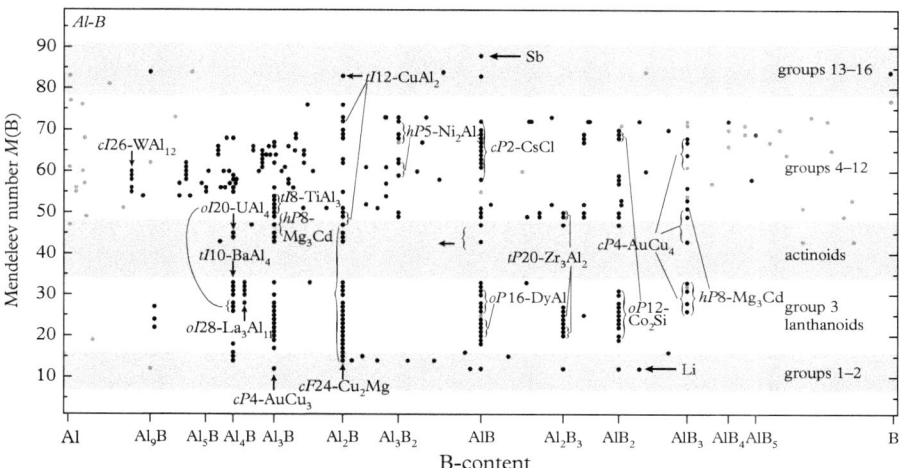

Fig. 7.46 *Stoichiometries of the 423 binary aluminides, Al$_a$B$_b$ vs. the Mendeleev number M(B) of the other element in the compound. The 47 disordered binary phases adopting unary structure types cF4-Cu, cI2-W, and hP2-Mg are marked by gray dots. The chemical compositions of the 14 remaining most common structure types (see Table 7.11) with 147 structures are marked. There remain 234 structures adopting structure types with less than five representatives.*

Table 7.11 *Most common structure types of the 423 binary aluminides listed in the PCD. The top 17 structure types out of the 174 are given, all of which have at least five representative structures and therefore represent more than 1% of all binary Al-intermetallics each. The chemical compositions of their representatives are shown in Fig. 7.47.*

Rank	Structure type	No.	Space group	Wyckoff positions	No. of reps.	% of all reps.
1.	$cF4$-Cu	225	$Fm\bar{3}m$	$4a$	28	6.6%
2.	$cP4$-Cu$_3$Au	221	$Pm\bar{3}m$	$1a\,3c$	23	5.4%
3.	$cF24$-MgCu$_2$	227	$Fd\bar{3}m$	$8a\,16d$	23	5.4%
4.	$cP2$-CsCl	221	$Pm\bar{3}m$	$1ab$	18*	4.3%
5.	$oP12$-Co$_2$Si	62	$Pnma$	$4c^3$	12	2.8%
6.	$cI2$-W	229	$Im\bar{3}m$	$2a$	11	2.6%
7.	$oP16$-DyAl	57	$Pbcm$	$4cd^3$	10*	2.4%
8.	$tI10$-BaAl$_4$	139	$I4/mmm$	$2a\,4de$	9	2.1%
9.	$hP8$-Mg$_3$Cd	194	$P6_3/mmc$	$2d\,6h$	9	2.1%
	($hP8$-Ni$_3$Sn)					
10.	$hP2$-Mg	194	$P6_3/mmc$	$2c$	8	1.9%
11.	$hP5$-Ni$_2$Al$_3$	164	$P\bar{3}m$	$1a\,2d^2$	6	1.4%
12.	$oI20$-UAl$_4$	74	$Imma$	$4ae^2\,8h$	6	1.4%
13.	$tP20$-Zr$_3$Al$_2$	136	$P4_2/mnm$	$4dfg\,8j$	6	1.4%
14.	$tI12$-CuAl$_2$	140	$I4/mcm$	$4a\,8h$	5	1.2%
15.	$oI28$-La$_3$Al$_{11}$	71	$Immm$	$2ad\,4hi\,8j^2$	5	1.2%
16.	$tI8$-TiAl$_3$	139	$I4/mmm$	$2ab\,4d$	5	1.2%
17.	$cI26$-WAl$_{12}$	204	$Im\bar{3}$	$2a\,24g$	5	1.2%
					189	44.6%

* The number of representatives of the $cP2$-CsCl type is actually 12 instead of 18, since the aluminides of the lanthanoids have been erroneously assigned to this instead to the $oP16$-DyAl type in older publications, and the PCD contains both assignments.

The latter, however, may have just not been studied because of its radioactivity. Be–Al alloys are technologically important light-weight materials for the aerospace industry. There are a couple of rather complex cluster-based magnesium aluminides such as the Samson phase $cF1168$-Mg$_2$Al$_3$, and its high-temperature modification $hR293$-Mg$_2$Al$_3$ (Feuerbacher *et al.*, 2007). Their structures can be described as polytetrahedral with a preference of FK-polyhedra. Their large amount of partially occupied sites has been interpreted as necessary to adjust the valence electron concentration.

There are four calcium aluminides known, the Al-richest one with the $tI10$-$BaAl_4$ structure type, one with the cubic Laves phase, $cF24$-$MgCu_2$, type, and two low symmetric ones: the metallic Zintl phases $mS54$-$Ca_{13}Al_{14}$ and $aP22$-Ca_8Al_3. $mS54$-$Ca_{13}Al_{14}$ contains a 2D Al network of 6-, 4-, and 3-membered rings with Ca in-between, while $aP22$-Ca_8Al_3 ($aP22$-Ca_8In_3 structure type) shows isolated Ca and Al atoms. $aP22$-Ca_8Al_3 can be described as a derivative of the $cF16$-BiF_3 (or $cF16$-$AlFe_3$) structure type with a cation deficiency (Huang and Corbett, 1998).

For strontium, three aluminides exist at ambient conditions: $tI10$-Sr-Al_4, $cP64$-$SrAl$, $oI12$-$SrAl_2$, which transforms at high pressure to $cF24$-$SrAl_2$, and $cP60$-Sr_8Al_7. The latter structure shows isolated tetrahedral and triangular Al clusters centered close to the Na and Cl sites in the $cF8$-NaCl structure type, while Sr forms CN16 and CN13 polyhedra.

In the case of barium, there are the eponymous compounds $tI10$-$BaAl_4$, $hP20$-$Ba_{21}Al_{40}$ ($hP61$-$Ba_{21}Al_{40}$), and $hP18$-Ba_4Al_5. The latter two are structurally closely related (Fornasini, 1975). The structures can be derived from the Laves phase $hP24$-$MgNi_2$. In summary, $tI10$-$BaAl_4$ is the only structure type that Ca, Sr, and Ba have in common.

Most of the aluminides of the group 3 elements and of the lanthanoids form compounds at the same stoichiometries, but not always with the same structure types:

- Al_4B: Ce, Eu, La, Nd, Pr, and Sm on the Ba sites of the $tI10$-$BaAl_4$ type
- Al_3B: Ce, Gd, La, Nd, Pr, and Sm on the Cd sites of $hP8$-Mg_3Cd type
- Al_2B: Ce, Dy, Er, Eu, Gd, Ho, La, Lu, Nd, Pr, Sc, Sm, Tb, Tm, Y, and Yb on the Mg sites of the $cF24$-$MgCu_2$ type
- AlB: Ce, Gd, La, and Pr on the Ce sites of the $oS16$-$CeAl$ type, and Dy, Er, Gd, Ho, Lu, Nd, Pr, Sm, Tb, and Tm on the Dy sites of $oP16$-$DyAl$
- Al_2B_3: Dy, Gd, Ho, and Tb on the Gd sites of the $tP20$-Gd_3Al_2 type
- AlB_2: Dy, Er, Gd, Ho, Lu, Nd, Pr, Sm, Tb, and Y on the Co sites of the $oP12$-Co_2Si type
- AlB_3: Ce, Dy, Er, La, Lu, Pr, Sc, Sm, Tb, Tm, Y, and Yb on the Cu sites of the $cP4$-Cu_3Au type, and Dy, Er, Ho, and Tb on the Ho sites of the $hR16$-$HoAl_3$ type.

In the case of the aluminides of the few actinoids studied so far, a similar distribution of stoichiometries has also been observed, for example:

- Al_3B: Th on the Cd sites of $hP8$-Mg_3Cd type, and Pu, U on the Au sites of the $cP4$-Cu_3Au type
- Al_2B: Th on the Al sites of the $hP3$-AlB_2 type, and Pu, U on the Mg site of the $cF24$-$MgCu_2$ type.

In the case of binary TM aluminides, the stoichometries show a more scattered distribution, and there are more structure types for a given stoichiometry. This is illustrated in Fig. 7.46. The structures of some binary TM aluminides are quite complex; here we list the few of those with more than 100 atoms per primitive unit cell:

- Ta ($M = 52$): $cF444$-$Ta_{39}Al_{69}$
- Cr ($M = 57$): $mS732$-$Cr_{23}Al_{77}$, $hP574$-$Mn_{55}Al_{226}$
- Mn ($M = 60$): $cP564$-Mn_8Al_{39}, $hP574$-$Mn_{55}Al_{226}$
- Co ($M = 64$): $oP240$-$Co_{12.5}Al_{37}$
- Rh ($M = 65$): $oS884$-$Rh_{99}Al_{343}$
- Ir ($M = 66$): $oP236$-$Ir_{13}Al_{45}$, $hP236$-Ir_9Al_{28}
- Pt ($M = 68$): $cF416$-$Li_{21}Si_5$.

All of them are unique structure types except for $hP574$-$Mn_{55}Al_{226}$, which has two representatives. Finally, there are no ordered aluminides known of the (semi)metallic main group elements with the exception of $cF8$-AlSb with the $cF8$-ZnS structure type.

In the following, we discuss some characteristic features of the five most frequent binary structure types of binary aluminides listed in Table 7.11. Their chemical compositions are shown in Fig. 7.47, and their structures are depicted in Fig. 7.48. The assignment of the unary structure types $cF4$-Cu, $cI2$-W, and $hP2$-Mg to binary intermetallic phases indicates either disordered structures or constituting elements, which cannot be distinguished by the diffraction method used for structure analysis. Nearly all of the remaining structure types occur at their defined compositions. In some of them, Al always plays the role of the majority element ($cF24$-MgCu$_2$, $tI10$-BaAl$_4$, $hP5$-Ni$_2$Al$_3$, $oI20$-UAl$_4$, $oI28$-La$_3$Al$_{11}$, $tI8$-TiAl$_3$, and $cI26$-WAl$_{12}$) or of the minority element ($oP12$-Co$_2$Si, and $tP20$-Zr$_3$Al$_2$). In others, it appears to switch between those roles ($cP4$-Cu$_3$Au, $hP8$-Mg$_3$Cd, and $tI12$-CuAl$_2$). Obviously, no such statement can be made for 1:1-stoichiometries, such as $cP2$-CsCl or $oP16$-DyAl. An overview over the respective structure types and the compositions of their representatives is shown in Fig. 7.47.

- $cP4$-Cu$_3$Au (221 $Pm\bar{3}m$): This structure can be considered as an ordering variant of the $cF4$-Cu structure type (see also Section 7.2 and Fig. 7.3).
- $cF24$-MgCu$_2$ (227 $Fd\bar{3}m$): The structure type of this cubic Laves phase can be described as a cubic packing (ABC) of flat layers of face-sharing Friauf polyhedra with empty tetrahedral voids (see also Subsection 7.4.3 and Fig. 7.15).
- $cP2$-CsCl (221 $Pm\bar{3}m$): This structure corresponds to an ordering variant of the $cP2$-W structure type (see also Section 7.3 and Fig. 7.7).

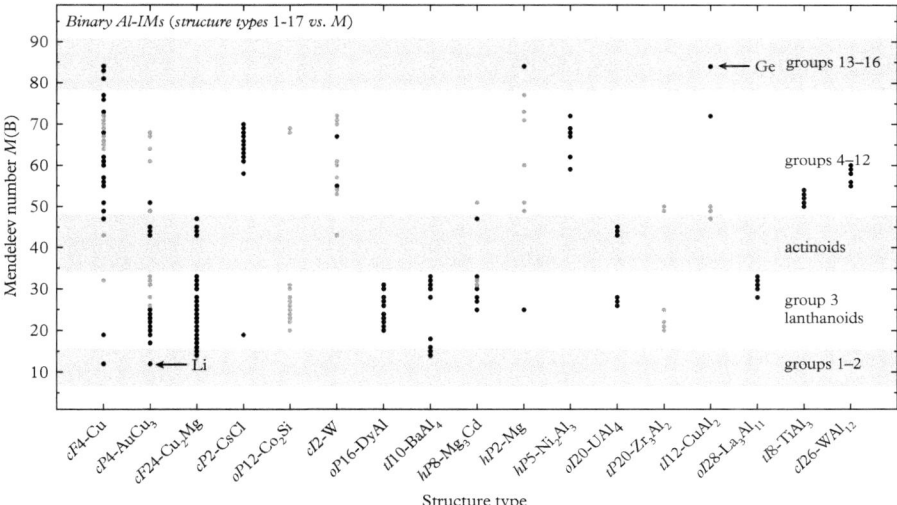

Fig. 7.47 *Occurrence of the 17 most common structure types among binary aluminides. The Mendeleev number M of the other element in the compound is shown over the rank of the structure type, as given in Table 7.11. Structures with Al as their major component (≥ 50%) are marked in black, those with minor Al-content are shown in gray color.*

- *oP*12-Co₂Si (62 *Pnma*): This structure type consists of symmetrically equivalent flat atomic layers (distorted square/triangle net) stacked along [010] in a way to form three different types of octahedra between the layers, with the space between them filled by tetrahedra. The face-sharing octahedra of type I form isolated chains running through the corners and center of the unit cell along [010]. The type II octahedra form flat (110) layers by sharing two opposite edges each and their apical vertices along [010]. Octahedra of type I and II share faces with type I octahedra. Type III octahedra share part of their edges with each other and form double-chains running along [100] and connecting the layers of type II octahedra with each other. The remaining channels are filled by the chains of type I octahedra. The structure is also related to the *hP*6-Ni₂In structure type, which itself is a derivative of the *hP*6-AlB₃ type (see Section 7.7).

- *oP*16-DyAl (57 *Pbcm*): There are flat pentagon/triangle nets decorated by Dy and Al in $z = 0$ and $z = 1/4$. In-between, part of the Al atoms form chains running along [001], centering columns of interpenetrating slightly distorted icosahedra, which share edges with in each case four neighboring other columns. Four edge-connected columns of icosahedra enclose an open space in the shape of a column of face-sharing octahedra running along [001].

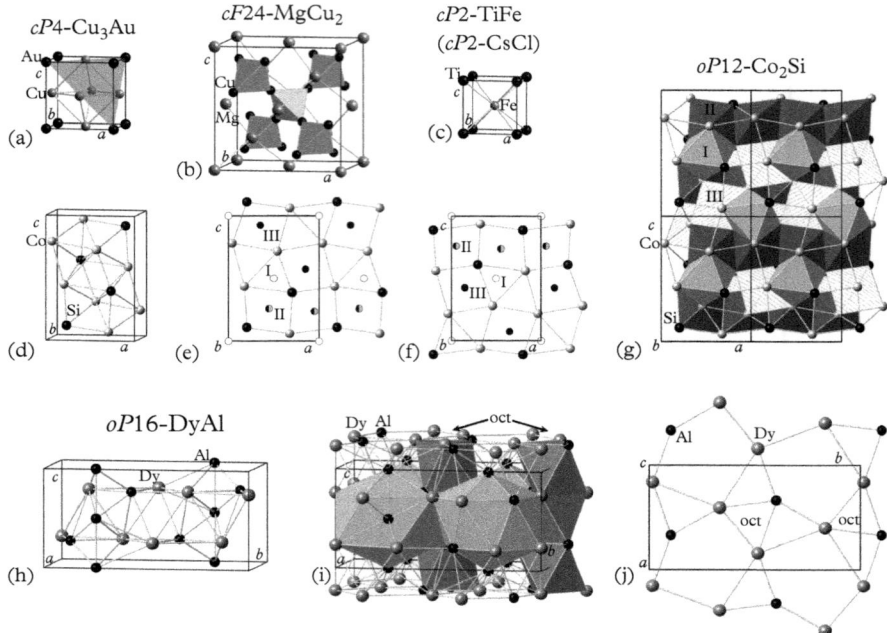

Fig. 7.48 *The five most common binary structure types among binary aluminides (see Table 7.11): one unit cell each of the structures of (a) cP4-Cu₃Au with the (111) plane shaded gray, (b) cF24-MgCu₂, (c) cP2-TiFe (cP2-CsCl structure type), (d) oP12-Co₂Si, and (h) oP16-DyAl, respectively. In (e)–(f), the two flat layers in y = 1/4, 3/4 are given, which constitute the structure of oP12-Co₂Si. The number I, II, and III mark the type of the octahedra centered at the respective symbols, and the packing of which is shown in (g). In (i) the edge-connected distorted icosahedra are highlighted as well as the columns of face-sharing octahedra, and in (j) the constituting pentagon/triangle layer in z = 1/4 is shown.*

Ternary aluminides

The 2449 ternary Al-containing IMs feature 384 different structure types, resulting in ≈ 6.4 representatives per structure type. This is much more than in the case of binary aluminides with only ≈ 2.4 representatives per structure type. It should be mentioned here that the most complex structure type known so far, $cF(23\,256 - x)$-Ta$_{39.1}$Cu$_{5.4}$Al$_{55.4}$ as well as a large class of both decagonal and icosahedral quasicrystals and their approximants are based on Al as their main constituent. The most common structure types of ternary aluminides are listed in Table 7.12.

In the following, we discuss some characteristic features of the five most frequent ternary structure types of the ternary aluminides listed in Table 7.12. Their chemical compositions are shown in Fig. 7.50, and their structures are depicted in Fig. 7.49. Among the 28 structure types listed, there are 12 ternary, 14 binary, and 2 unary ones.

Table 7.12 *Most common structure types of ternary aluminides. The top 28 structure types out of a total of 384 are given, all of which have at least 24 representatives. In total, the 1388 representatives of these 28 structure types make up 56.5% of all 2449 ternary Al-intermetallics.*

Rank	Structure type	No.	Space group	Wyckoff positions	No. of reps.	% of all reps.
1.	$cF24$-$MgCu_2$	227	$Fd\bar{3}m$	$8a\,16d$	179	7.3%
2.	$hP12$-$MgZn_2$	194	$P6_3/mmc$	$2a\,4f\,6h$	104	4.2%
3.	$cP4$-Cu_3Au	221	$Pm\bar{3}m$	$1a\,3c$	98	4.0%
4.	$hP106$-$Ho_6Mo_4Al_{43}$	193	$P6_3/mcm$	$2b\,6g^2\,8h\,12ijk^3\,24l$	88	3.6%
5.	$tI26$-$CeMn_4Al_8$	139	$I4/mmm$	$2a\,8fij$	73	3.0%
6.	$cF184$-$CeCr_2Al_{20}$	227	$Fd\bar{3}m$	$8a\,16cd\,48f\,96g$	72	2.9%
7.	$cP2$-$CsCl$	221	$Pm\bar{3}m$	$1ab$	70	2.9%
8.	$cF16$-Cu_2MnAl	225	$Fm\bar{3}m$	$4ab\,8c$	56	2.3%
9.	$hP9$-$ZrNiAl$	189	$P\bar{6}2m$	$1a2d3fg$	49	2.0%
10.	$hR57$-$Zn_{17}Th_2$	166	$R\bar{3}m$	$6c^2\,9d\,18fh$	46	1.9%
11.	$hP6$-$CaCu_5$	191	$P6/mmm$	$1a\,2c\,3g$	40	1.6%
12.	$oS56$-$Y_2Co_3Ga_9$	63	$Cmcm$	$4ac\,8efg^2\,16h$	39	1.6%
13.	$hP8$-Mg_3Cd	194	$P6_3/mmc$	$2d\,6h$	37	1.5%
14.	$cF116$-$Mg_6Cu_{16}Si_7$	225	$Fm\bar{3}m$	$4a\,24de\,32f^2$	35	1.4%
15.	$oP12$-$TiNiSi$	62	$Pnma$	$2ad\,6gh$	34	1.4%
16.	$tI10$-$CeAl_2Ga_2$	139	$I4/mmm$	$2a\,4de$	33	1.3%
17.	$oI12$-KHg_2	74	$Imma$	$4e\,8i$	33	1.3%
18.	$hP6$-$PrNi_2Al_3$	191	$P6/mmm$	$1a\,2c\,3g$	33	1.3%
19.	$tI26$-$ThMn_{12}$	139	$I4/mmm$	$2a\,8fij$	33	1.3%
20.	$cF4$-Cu	225	$Fm\bar{3}m$	$4a$	30	1.2%
21.	$oS52$-$YbFe_2Al_{10}$	63	$Cmcm$	$4c\,8def^2g^2$	30	1.2%
22.	$hP3$-AlB_2	191	$P6/mmm$	$1a\,2d$	27	1.1%
23.	$hR36$-$BaPb_3$	166	$R\bar{3}m$	$3a\,6c\,9e\,18h$	26	1.1%
24.	$hP16$-$TiNi_3$	194	$P6_3/mmc$	$2ad\,6gh$	26	1.1%
25.	$hP38$-$Gd_3Ru_4Al_{12}$	194	$P6_3/mmc$	$2ab\,4f\,6gh^2\,12k$	25	1.0%
26.	$oI28$-La_3Al_{11}	71	$Immm$	$2ad\,4hi\,8l^2$	24	1.0%
27.	$hP38$-Th_2Ni_{17}	194	$P6_3/mmc$	$2bc\,4f\,6g\,12jk$	24	1.0%
28.	$cI2$-W	229	$Im\bar{3}m$	$2a$	24	1.0%
					1388	56.5%

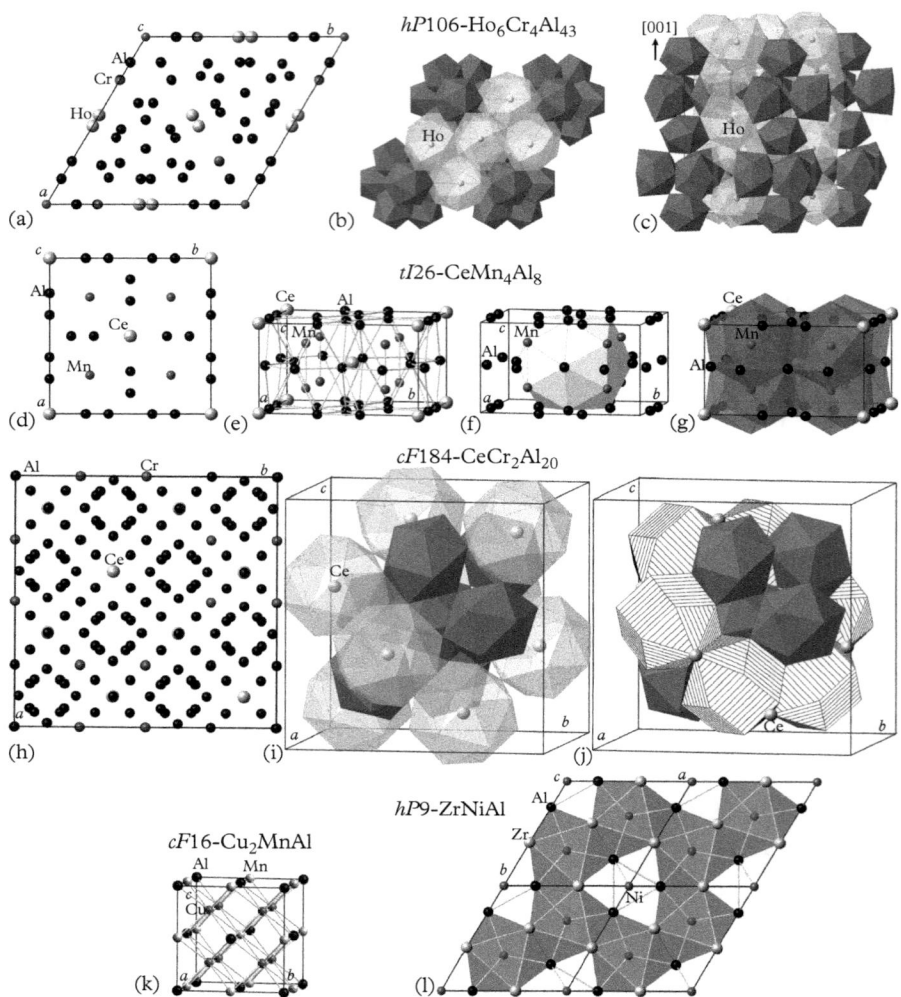

Fig. 7.49 *The structures of the five most common ternary structure types of ternary aluminides (see Table 7.12). The structure and polyhedral representation of hP106-Ho₆Cr₄Al₄₃ is shown (a)–(b) in projection along [001], and (c) in a projection perpendicular to it (the scale of the polyhedral representations is only 50%). The packing of the AETs of Ho and Cr is illustrated. The structure of tI26-CeMn₄Al₈ is shown in projection along [001] and in perspective unit cell view in (d)–(g). The complex structure of cF184-CeCr₂Al₂₀ is shown in projection along [001] in (h); the packing of Ce and Cr AETs, respectively, is illustrated in (i), and that of Al CN16 polyhedra and Ce AETs in (j). The structures of (k) cF16-Cu₂MnAl and (l) hP9-ZrNiAl have already been discussed and shown before and are depicted here for convenience.*

- $hP106$-$Ho_6Mo_4Al_{43}$ (193 $P6_3/mcm$) represented by $hP106$-$Ho_6Cr_4Al_{43}$: Cr1 and Cr2 in Al/Ho and Al icosahedra, respectively; Al either in distorted icosahedra or bicapped pentagonal prisms; Ho in irregular CN17 polyhedron (15 Al + Cr + Ho). The CN17 polyhedra form columns along [001] by sharing triangle faces on one side and interpenetrating on the opposite side. The icosahedra around Cr are vertex-connected in a way to form columns along [001] with a sequence of 3-1-3-1 per period (3 denotes a cluster of three vertex-connected icosahedra).

- $tI26$-$CeMn_4Al_8$ (139 $I4/mmm$): This structure type can be described as an ordered $tI26$-$ThMn_{12}$ type structure. Mn is coordinated by an icosahedron that is distorted by two Ce atoms. These polyhedra interpenetrate each other forming columns along [001]. Ce centers truncated octahedra that are Al-capped on the hexagon faces, and form by square-face-sharing columns along [001]. These columns are vertex-linked laterally.

- $cF184$-$CeCr_2Al_{20}$ (227 $Fd\bar{3}m$): Complex structure type with Ce in CN16 FK-polyhedra of Al, Cr in Al-icosahedra, Al1 in hexagonal Al-prism, with the hexagon faces capped by Ce, Al2 and Al3 in pentagonal Al-prisms, with the pentagon faces capped by Cr and Cr/Ce, respectively. The icosahedra are vertex-connected and form 6-rings centered by a vertex-sharing pair of Al-CN16 FK-polyhedra each. See Fig. 7.49.

- $cF16$-Cu_2MnAl (225 $Fm\bar{3}m$): This structure type represents a $(2 \times 2 \times 2)$-fold superstructure of the $cI2$-W type. Al forms an fcc unit cell, Mn occupies the edge centers and the body center, and Cu centers the eight-cubes (see also Fig. 7.9 in Section 7.3.2).

- $hP9$-$ZrNiAl$ (189 $P\bar{6}2m$): Ni atoms center trigonal Zr-prisms, which form by edge-sharing trigonally distorted 6-rings. Three Al atoms cap the Zr-prisms, thereby forming Ni-centered trigonal prisms (see also Fig. 10.8 in Section 10.8).

The 14 ternary aluminides adopting the binary structure types listed in Table 7.12 have broad stability ranges and are inherently, at least partially, disordered: for $cF24$-$MgCu_2$, $hP12$-$MgZn_2$, $cP2$-CsCl, and $cP4$-Cu_3Au the Al-content varies in the ranges 1.7–66.7%, 6.7–66.7%, 5.0–56.0%, and 2.5–75%, respectively. More pseudo-binary structure types with varying amounts of Al among their representatives are $hP8$-Mg_3Cd (2.5–75.0%), $oI12$-KHg_2 (2.0–63.3%; 18 structures with 1:1:1 composition), $tI26$-$ThMn_{12}$ (6.5–76.9%), $hP3$-AlB_2 (6.0–58.3%), $hR36$-$BaPb_3$ (67.5–75%; 24 structures with 75% Al and disorder on Ba-site with $M \in [17, 28]$; 2 structures with Ge and Al together adding up to 75%), $hP16$-$TiNi_3$ (24 structures with 75% Al and disorder on the Ti-site with $M \in [17, 28]$; two structures with Ni as the major component and Al mixed with Ti or Ta amounting to 25%), $oI28$-La_3Al_{11} (39.3–78.6%), and $hP38$-Th_2Ni_{17} (10.5–57.9%). The two most common unary structure types have very large ranges for their Al-content, as well: $cF4$-Cu (0.1–99.5%), $cI2$-W (4.0–34.0%).

In contrast, the 88 $hP106$-$Ho_6Mo_4Al_{43}$-type structures, for instance, are line compounds with well-defined stoichiometry (with two representatives being subject to minor element exchange between the Al-deficit and excess of the Mo-equivalent component: $Ho_6Mo_4Al_{43}$ and $Yb_6Cr_4Al_{43}$). The structures adopting the $cF184$-$CeCr_2Al_{20}$ structure type are for the most part stoichiometric with four compounds exhibiting disorder on the respective sites of at most 2.3 *at.%*. Most $cF16$-Cu_2MnAl-type representatives (49 out of 56) contain 25% Al, whereas five contain 50% Al. The 49 $hP9$-$ZrNiAl$-type structures are all stoichiometric, as are the $oS56$-$Y_2Co_3Ga_9$-type structures with Al taking the place of Ga. The compositions are depicted in Fig. 7.50 with M/M-plots of the two non-Al-elements.

The stability regions of all the six structure types shown in Fig. 7.50 appear well defined. Representatives of the $hP106$-$Ho_6Mo_4Al_{43}$ structure type can be found for most lanthanoids in combination with early TM elements. The missing dot in the stability field refers to the not-observed compound $hP106$-$Y_6V_4Al_{43}$. Furthermore, there are also some U-containing representatives of this type. In the case of the structure type $tI26$-$CeMn_4Al_8$, almost all lanthanoids and Th can be constituents as well as the TM elements of the first row of the periodic table, Cr, Fe, and Cu. This structure type can be seen as an ordered $tI26$-$ThMn_{12}$ type.

The $cF184$-$CeCr_2Al_{20}$ structure type is adopted by Sr, Ca and, mainly, the early lanthanoids on one hand, and most of the early TM elements on the other hand. The structure type of the Heusler phase, $cF16$-Cu_2MnAl, has representatives constituted from one of the early and one of the late TM elements in addition to aluminum.

In the case of the structure type $hP9$-$ZrNiAl$, aluminides are known that contain Ca, most of the lanthanoids, Pu, U, Th Zr, or Hf as one component and Ni, Pd, or Cu as the other one. Finally, representatives of the $oS56$-$Y_2Co_3Ga_9$ structure type can have as one constituent Ca, most of the lanthanoids or U, and as the other one Co, Rh, Ir, or Pd.

The compounds adopting structure type $cF116$-$Mg_6Cu_{16}Si_7$ mostly exhibit Al in the role of Cu in the eponymous compounds (21 structures with exact stoichiometry and 8 slightly off-stoichiometric), but it also adopts the Si-role in six representative compounds. As the previous 1:1:1-compounds, $oP12$-$TiNiSi$-type structures all appear to be fully stoichiometric. Nearly half of the $tI10$-$CeAl_2Ga_2$-type structures contain 40% of Al (16 structures), while the remaining 17 structures contain 34.0–76.0% of Al. In the $hP6$-$PrNi_2Al_3$ structure type, 14 representatives have Al adopt the eponymous Al-role, 10 appear to have it in the role of Ni, and the remaining 9 display disorder between the three elements in the structure formula. All $oS52$-$YbFe_2Al_{10}$-type structures have Al in its eponymous role, as is also the case for the $hP38$-$Gd_3Ru_4Al_{12}$-type structures (with minor Co-Al-disorder in the $U_3Co_4Al_{12}$-compound).

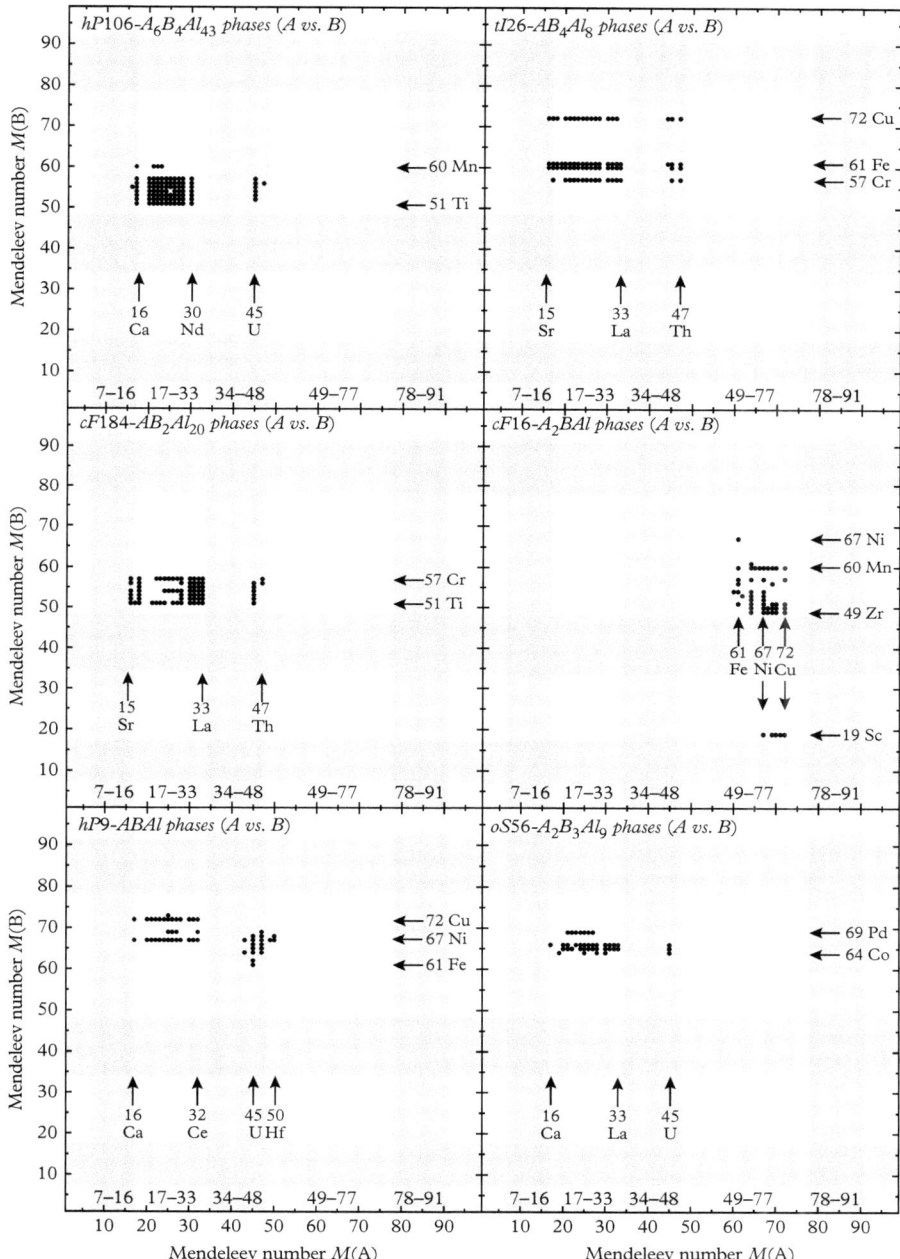

Fig. 7.50 *Chemical compositions of the six most common ternary structure types of ternary aluminides. The Mendeleev numbers M of the other two elements in the compound are shown. The depicted structure types are hP106-Ho₆Mo₄Al₄₃, tI26-CeMn₄Al₈, cF184-CeCr₂Al₂₀, cF16-Cu₂MnAl, hP9-ZrNiAl, and oS56-Y₂Co₃Ga₉ (ranks 4, 5, 6, 8, 9, and 12, respectively, in Table 7.12). Only seven structures were omitted for structure type cF16-Cu₂MnAl, where the chemical decoration of the structure changes.*

7.15.2   Gallides

Gallium, *o*C8-Ga, has a smaller atomic radius than aluminum, *c*F4-Al, (r_{Al} = 1.43 Å, r_{Ga} = 1.22 Å) and a higher electronegativity (χ_{Al} = 1.5, χ_{Ga} = 1.8). The distribution of gallides in the ternary concentration diagram of intermetallics is depicted in Fig. 7.51 together with the frequencies of the subset of binary

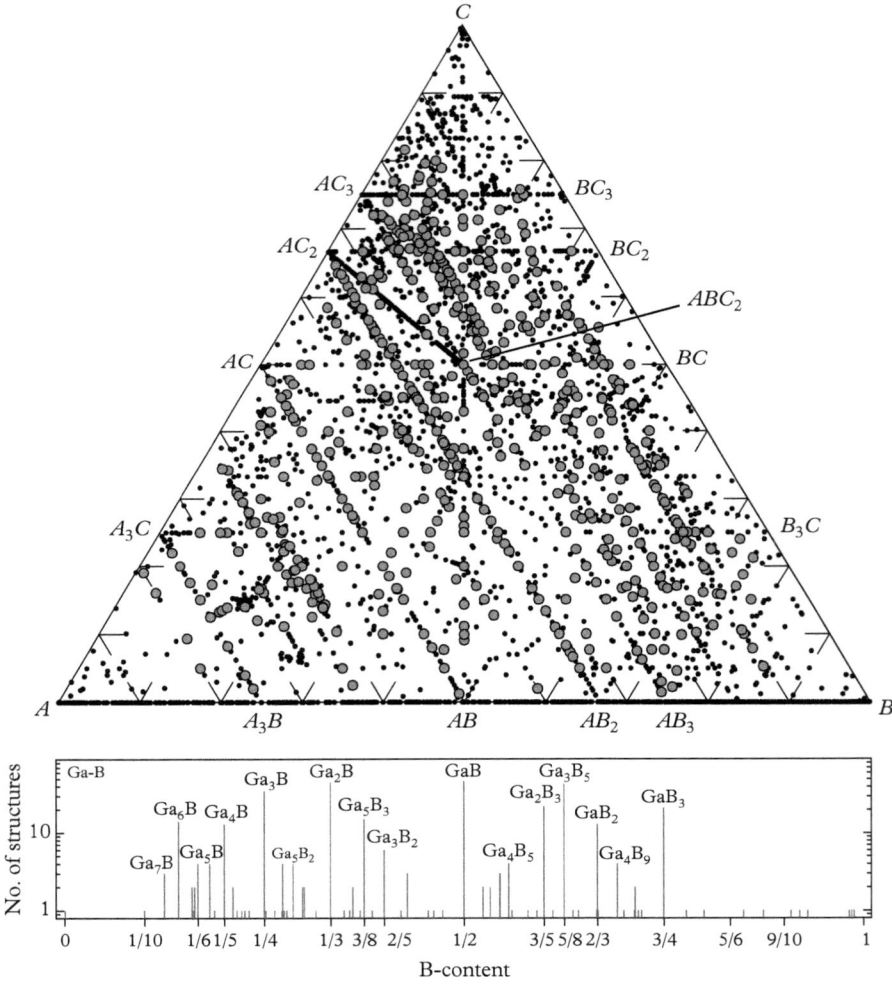

Fig. 7.51 *(top) Concentration diagram of the 20 829 intermetallic compounds (black dots) contained in* Pearson's Crystal Data *(PCD) (Villars and Cenzual, 2011a). The Ga-containing compounds are marked by large gray circles. A, B, and C stand for metallic elements with increasing Mendeleev numbers, M(A) < M(B) < M(C). (bottom) Frequencies of the binary Ga–B compounds as a function of stoichiometry (note the logarithmic scale).*

gallides. Since the constituting elements A, B, and C stand for metallic elements with increasing Mendeleev numbers, $M(A) < M(B) < M(C)$, Ga with $M(Ga) = 81$, will in most intermetallics $A_xB_yC_z$ be represented by the letter C. Consequently, most compounds with Ga as majority element will agglomerate in the concentration triangle close to the top corner marked C.

The distribution of ternary gallides in the concentration triangle (Fig. 7.51) is less dense but otherwise similar to that of the aluminides (Fig. 7.45). The main difference is along the tieline AC_3–BC_3, where gallides with 8 representatives are rare in contrast to aluminides with 158. Taking a closer look at the structure types featured at this tieline, one finds in the case of the ternary aluminides that only 9 structures belong to ternary structure types, the remaining 149 adopt binary ones such as $cP4$-Cu_3Au (39), $hP16$-$TiNi_3$ (34), $hR36$-$BaPb_3$ (24), $hP8$-Mg_3Cd (24), $hR60$-$HoAl_3$ (22), etc. In the case of the ternary gallides, these binary structure types are not represented among the ternary gallides (only among the binary ones), and five out of the eight compounds on this tieline adopt ternary structure types. The scarcity of solid solutions in ternary gallides of the type $A_xB_{1-x}Ga_3$ may originate from covalent bonding contributions of Ga. The frequency distribution of binary gallides A_xB_y shows some differences to that of the aluminides, mainly for large values of x or of y.

Binary gallides

As mentioned previously, among the 20 829 IMs there are 13 026 ternary and 6441 binary IMs. Among the binary IMs, 397 gallides can be found in the PCD, featuring 164 structure types this means ≈ 2.4 representatives per structure type, similar to aluminides (≈ 2.4). The compositions of the 397 binary Ga-intermetallics are shown in Fig. 7.52, and their most common structure types are listed in Table 7.13. In contrast to the situation for the aluminides, a significant number of alkali gallides exist for all alkali metals but radioactive Fr.

The compositions of the $cF4$-Cu-, $hP2$-Mg-, and $cI2$-W-type structures are rather diverse (containing 5–25%, 2–60%, and 3–50% Ga, respectively). Nearly all of the remaining structure types occur at their defined compositions. In some of them, Ga always plays the role of the majority element ($hP3$-AlB_2, $tP14$-$PuGa_6$, $oP32$-Tm_3Ga_5, $tP16$-$IrIn_3$, $tI10$-$BaAl_4$, $hP6$-$CaIn_2$, and $oI12$-KHg_2) or of the minority element ($hP16$-Mn_5Si_3, $tI80$-Gd_3Ga_2, $tP32$-Ba_5Si_3, $tI32$-Cr_5B_3, $tI32$-W_5Si_3, $mS32$-Y_5Ga_3, $cP8$-Cr_3Si, and $tP10$-U_3Si_2). In particular, for the compositions A_5B_3 and A_3B_5 there is no significant number of aluminides known in contrast to gallides. In others, it appears to switch between those roles ($cP4$-Cu_3Au, $tI8$-$TiAl_3$, and $hP8$-Mg_3Cd). Obviously, no such statement can be made for 1:1-stoichiometries, such as $oS8$-TlI, and only small deviations are found for $cP2$-$CsCl$. An overview of the respective intermetallic systems is shown in Fig. 7.53.

The most common binary structure types of binary aluminides and gallides differ considerably. With the exception of $cP4$-Cu_3Au, the second most frequent

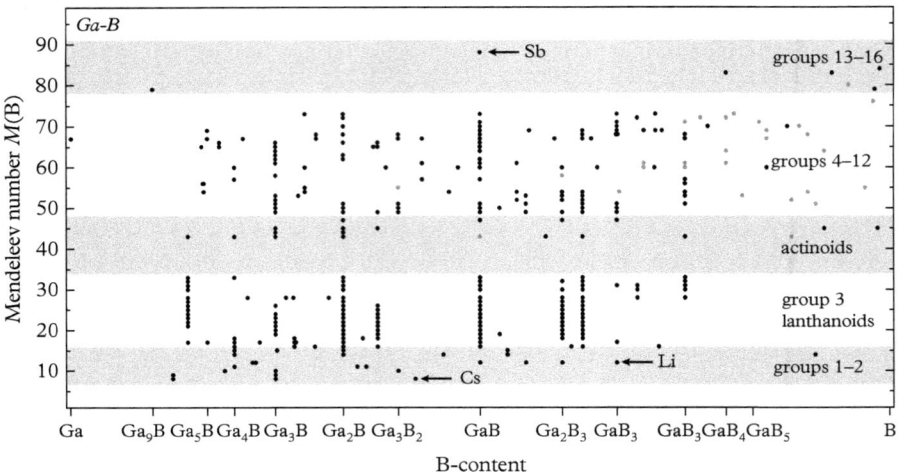

Fig. 7.52 *Stoichiometries of the 397 binary gallides, Ga_aB_b vs. the Mendeleev number M(B) of the other element in the compound. The 28 binary phases adopting unary structure types cF4-Cu, cI2-W, and hP2-Mg are marked by gray dots.*

structure type in both cases (short 2, 2), $cP2$-CsCl (4, 11), $tI10$-BaAl$_4$ (8, 14), $hP8$-Mg$_3$Cd (9, 19), and $tI8$-TiAl$_3$ (16, 15), there are no common binary structure types with frequencies $\geq 1\%$.

In the following, we discuss some characteristic features of the five most frequent binary gallide structure types listed in Table 7.13. Their structures are shown in Fig. 7.54.

- $hP3$-AlB$_2$ (191 $P6/mmm$): For a detailed discussion of this structure type see Section 7.7. The structure can be considered as a stacking of honeycomb nets with Al at the vertices. B sits in the center of the hexagonal prisms constituted from the Al atoms of adjacent layers. From groups 1 and 2, only Ba, Sr, and Ca adopt this structure type, however, almost all lanthanoids and the actinoids Np, Pu, and U do. There are no TM or main group element gallides with this structure type known so far. The only known aluminides with this structure type contain La and Th, respectively.

- $cP4$-Cu$_3$Au (221 $Pm\bar{3}m$): For a detailed discussion of this structure type, a substitutional derivative of the $cF4$-Cu type, see Section 7.2. Almost all lanthanoids form gallides with this structure type, the early ones with stoichometry RE$_3$Ga, the late ones with REGa$_3$. This structure type is also known for the gallides of the actinoids Np, Pu, and U. Out of the groups from 4 to 16, only Ti, Fe, Ni, and Pt adopt this structure type. This distribution does not differ too much from that of the aluminides.

Table 7.13 *Most common structure types of Ga-containing binary intermetallics. The top 23 structure types are given, all of which have at least four representative structures and therefore represent more than 1% of all binary Ga-intermetallics each.*

Rank	Structure type	No.	Space group	Wyckoff positions	No. of reps.	% of all reps.
1.	$hP3$-AlB_2	191	$P6/mmm$	$1a\,2d$	19	4.8%
2.	$cP4$-Cu_3Au	221	$Pm\bar{3}m$	$1a\,3c$	19	4.8%
3.	$oS8$-TlI	63	$Cmcm$	$4c^2$	15	3.8%
4.	$tP14$-$PuGa_6$	125	$P4/nbm$	$2c\,4g\,8m$	14	3.5%
5.	$hP16$-Mn_5Si_3	193	$P6_3/mcm$	$4d\,6g^2$	12	3.0%
6.	$cF4$-Cu	225	$Fm\bar{3}m$	$4a$	11	2.8%
7.	$tI80$-Gd_3Ga_2	140	$I4/mcm$	$4ac\,8gh^2\,16l\,32m$	11	2.8%
8.	$hP2$-Mg	194	$P6_3/mmc$	$2c$	11	2.8%
9.	$tP32$-Ba_5Si_3	130	$P4/ncc$	$4c^2\,8f\,16g$	8	2.0%
10.	$tI32$-Cr_5B_3	140	$I4/mcm$	$4ac\,8h\,16l$	8	2.0%
11.	$cP2$-CsCl	221	$Pm\bar{3}m$	$1ab$	8	2.0%
12.	$oP32$-Tm_3Ga_5	62	$Pnma$	$4c^4\,8d^2$	8	2.0%
13.	$tP16$-$IrIn_3$	136	$P4_2/mnm$	$4cf\,8j$	7	1.8%
14.	$tI10$-$BaAl_4$	139	$I4/mmm$	$2a\,4de$	6	1.5%
15.	$tI8$-$TiAl_3$	139	$I4/mmm$	$2ab\,4d$	6	1.5%
16.	$cI2$-W	229	$Im\bar{3}m$	$2a$	6	1.5%
17.	$tI32$-W_5Si_3	140	$I4/mcm$	$4ab\,8h\,16k$	6	1.5%
18.	$mS32$-Y_5Ga_3	12	$C2/m$	$4d\,6g^2$	6	1.5%
19.	$hP8$-Mg_3Cd	194	$P6_3/mmc$	$2d\,6h$	5	1.3%
20.	$hP6$-$CaIn_2$	194	$P6_3/mmc$	$2b\,4f$	4	1.0%
21.	$cP8$-Cr_3Si	223	$Pm\bar{3}n$	$2a\,6c$	4	1.0%
22.	$oI12$-KHg_2	74	$Imma$	$4e\,8i$	4	1.0%
23.	$tP10$-U_3Si_2	127	$P4/mbm$	$2a\,4gh$	4	1.0%
					202	50.9%

- $oS8$-TlI (63 $Cmcm$): This structure can be seen as a stacking along [001] of two symmetrically equivalent distorted honeycomb layers, which are shifted against each other; thereby, layers are formed of face-sharing octahedra alternating with layers of face-sharing trional prisms. Gallides with this structure type are known with Ca and all lanthanoids but Yb and Eu,

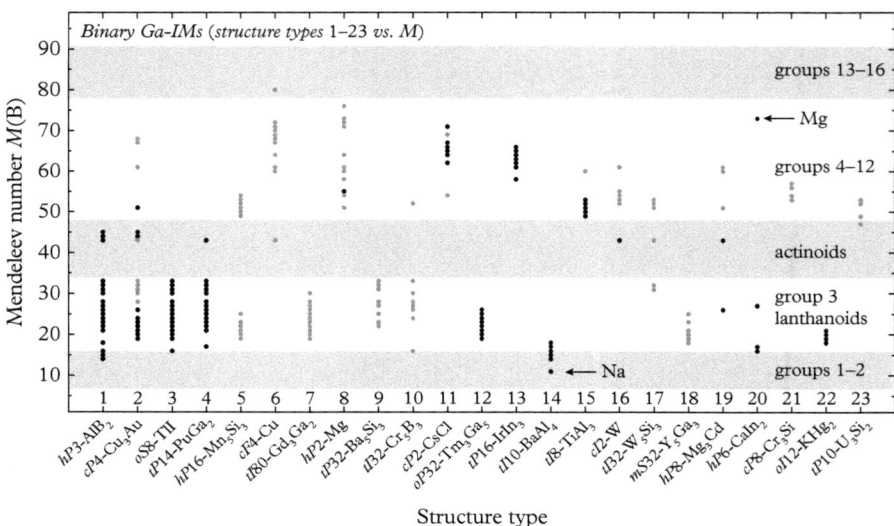

Fig. 7.53 *Occurrence of the 23 most common structure types among binary Ga-intermetallics. The Mendeleev number M of the other element in the compound is shown over the rank of the structure type, as given in Table 7.13. Structures with Ga as their majority component (≥ 50%) are marked in black, and those with minor Ga-content are shown in gray.*

which have atomic radii of 1.94 and 2.00 Å, respectively, by far the largest among the lanthanoids. For comparison, there are four aluminides known with this structure type containing Y, Th, Zr, and Hf, respectively.

- $tP14$-PuGa$_6$ (125 $P4/nbm$): The structure can be described as a packing in layers of the face-sharing Pu-AETs, where the layers are connected via short Ga–Ga bonds (2.52 Å). This structure type is known for gallides of all lanthanoids but Eu, Sc, and Lu, as well as for Pu as the only actinoid. There are no aluminides known with this structure type.

- $hP16$-Mn$_5$Si$_3$ (193 $P6_3/mcm$): This structure type, one of the Nowotny phases, can be described as a packing of face-sharing CN16 FK-polyhedra, leaving empty spaces corresponding to columns of face-sharing octahedra running along [001] (see also Subsection 7.15.9). In the case of this structure type, Ga is always the minority component (sitting on the Si site). Majority constituents can be Sc, Lu, Tm, Ho, and Y on one hand, and Zr, Hf, Ti, Ta, Nb, and V on the other hand. For comparison, there are four aluminides known with this structure type containing Y, Zr, Hf, and Ta, respectively.

Finally, an interesting binary transition metal gallide, TM$_4$Ga$_5$ (TM = Ta, Nb, Ta/Mo), shall be discussed briefly (Fredrickson *et al.*, 2015). It is characterized by quite an unusual homoatomic clustering of the TM elements and Ga,

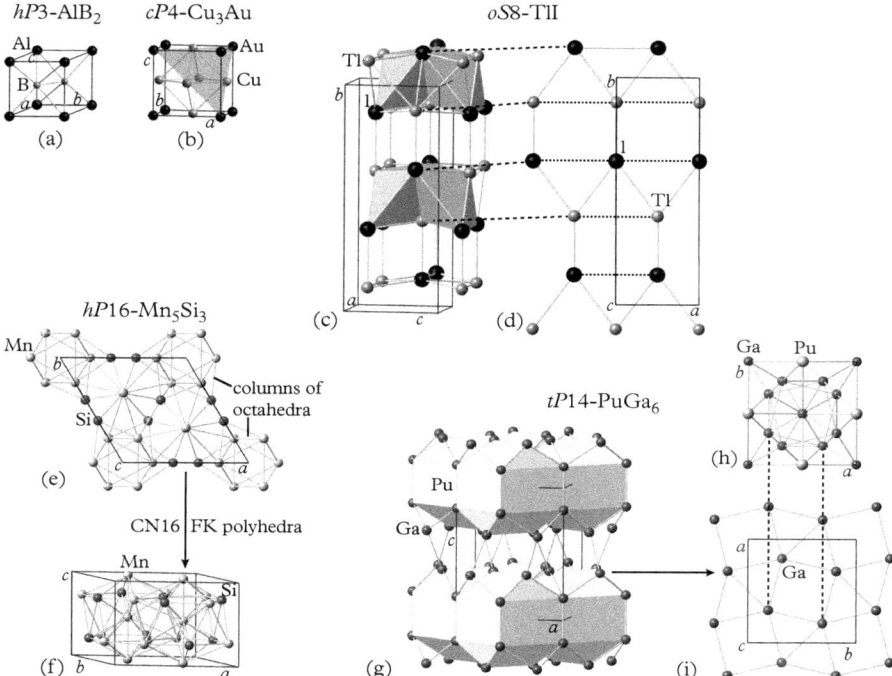

Fig. 7.54 *The structures of the five most frequent binary structure types of binary gallides. For a detailed discussion of the structures of (a) hP3-AlB$_2$ and (b) cP4-Cu$_3$Au see Sections 7.7 and 7.2, respectively. In (c) the structure of oS8-TlI is shown with the layers of face-sharing octahedra highlighted. The structure of one of the two symmetrically equivalent layers stacked along [001] is depicted in (d). The elongated hexagons can also be seen to consist of squashed triangles and squares (dotted lines) as is the case in the $3^2.4.3.4$ triangle square tiling representing one of the constituting layers (i) of the structure of (g)–(h) tP14-PuGa$_6$. (g) The packing in layers is illustrated of the face-sharing Pu-AETs located at the edge-centers of the unit cell. The layers are connected by short Ga–Ga bonds (2.52 Å). In (e, f) the unit cell of hP16-Mn$_5$Si$_3$ is shown containing two face-sharing CN16 FK-polyhedra. Around the corners of the unit cell, columns of face-sharing octahedra are constituted from the surrounding FK-polyhedra.*

respectively. The TM elements form *bcc* cubes, TM@TM$_8$, which are linked through face-capping Ga dumbbells, forming a primitive cubic framework in this way. The resulting empty spaces are alternatingly filled by distorted TM-pentagonal dodecahedra and dimers of *bcc* fragments. Ga tetrahedra and icosahedral units fill the rest of the empty space. The compounds show a deep pseudogap at the Fermi energy mainly caused by strong TM–TM and TM–Ga bonding, which provides a local 18 electron configuration to the TM atoms, despite the electron concentration being only 8.75 electrons per TM atom. The Ga atoms at the periphery of these clusters have a supportive role in stabilizing them.

Ternary gallides

The 1842 ternary Ga-containing IMs feature 374 different structure types, i.e., ≈ 4.9 representatives per structure type. This is significantly less than in the case of ternary aluminides (≈ 6.4). In Table 7.13, the most common structure types among binary Ga-IMs are given. In Table 7.14, the most common structure types among ternary Ga-IMs are given, and the distribution of the chemical compositions of six of the most common ones are shown in Fig. 7.56.

The most common ternary structure types (with frequencies $\geq 1\%$) of ternary aluminides and gallides differ considerably. There are only six ternary structure types common to both of them: $tI10$-CeAl$_2$Ga$_2$ (16, 3), $oP12$-TiNiSi (15, 6), $oS56$-Y$_2$Co$_3$Ga$_9$ (12, 7), $hP9$-ZrNiAl (9, 10), $cF16$-Cu$_2$MnAl (8, 12), and $cF116$-Mg$_6$Cu$_{16}$Si$_7$ (14, 23).

In the following, we discuss some characteristic features of the five most frequent ternary structure types of ternary gallides listed in Table 7.14. Their structures are shown in Fig. 7.55.

- $tI10$-CeAl$_2$Ga$_2$ (139 $I4/mmm$): This structure type is an ordered derivative of the $tI10$-BaAl$_4$ type (see also Fig. 10.8 in Section 10.8). Ce sits in 4-capped hexagonal prisms (CeAl$_8$Ga$_8$), while Ga centers tricapped trigonal prisms (GaAl$_4$Ce$_4$).

- $oP12$-TiNiSi (62 $Pnma$): The structure contains a 3D 4-connected network of Ti atoms ($d_{\text{Ti-Ti}} = 3.15\text{–}3.22$ Å), where the large channels are filled with zigzag-bands of edge-sharing Ni$_2$Si$_2$ rhomb units. It can be also described as a distorted $hP6$-Ni$_2$In type structure.

- $oS56$-Y$_2$Co$_3$Ga$_9$ (64 $Cmca$): The Co and Ga atoms have distorted icosahedral or defective icosahedral AETs, respectively; Y centers a strongly distorted hexagonal prism of Co/Ga atoms, with three more Ga atoms coordinating Y in the equatorial plane. The structure can be described as a stacking along [001] of puckered Ga/Co-triangle nets and flat Ga/Y-triangle/pentagon nets.

- $hP9$-ZrNiAl (189 $P\bar{6}2m$): The structure can be subdivided into face-sharing tricapped trigonal Ni@Zr$_6$Al$_3$ prisms (see also Fig. 10.8 in Section 10.8).

- $cF16$-Cu$_2$MnAl (225 $Fm\bar{3}m$): This structure type of the Heusler phases can be described as $(2 \times 2 \times 2)$-fold superstructure of the $cI2$-W type, and is discussed in Subsection 7.3.2.

The ternary gallides of the binary structure types $oI12$-KHg$_2$ and $hP6$-CaIn$_2$ have varying Ga-contents (23.3–58.7% and 16.7–66.7%, respectively). Similarly, the $tI10$-CeAl$_2$Ga$_2$-type structures exhibit a high degree of disorder between the elements on the Al- and Ga-positions of the prototype: while the Ga-content of 16.0–75.0% is connected with the content in a second element in these compounds ($M = 64\text{–}80$ and content 5.0–64.0%), the third element always makes

Table 7.14 *Most common structure types of Ga-containing ternary intermetallics. The top 24 structure types are given, all of which have at least 18 representative structures and therefore represent more than 1% of all ternary Ga-intermetallics.*

Rank	Structure type	No.	Space group	Wyckoff positions	No. of reps.	% of all reps.
1.	$oI12$-KHg$_2$	74	*Imma*	$4e\,8i$	78	4.2%
2.	$hP6$-CaIn$_2$	194	*P6$_3$/mmc*	$2b\,4f$	61	3.3%
3.	$tI10$-CeAl$_2$Ga$_2$	139	*I4/mmm*	$2a\,4de$	58	3.1%
4.	$oI28$-La$_3$Al$_{11}$	71	*Immm*	$2ad\,4hi\,8j^2$	57	3.1%
5.	$cP4$-Cu$_3$Au	221	*Pm$\bar{3}$m*	$1a\,3c$	54	2.9%
6.	$oP12$-TiNiSi	62	*Pnma*	$4ccc$	48	2.6%
7.	$oS56$-Y$_2$Co$_3$Ga$_9$	64	*Cmca*	$4ac\,8efg^2\,16h$	48	2.6%
8.	$tI26$-ThMn$_{12}$	139	*I4/mmm*	$2a\,8fij$	41	2.2%
9.	$hR57$-Zn$_{17}$Th$_2$	166	*R$\bar{3}$m*	$6c^2\,9d\,18fh$	41	2.2%
10.	$hP9$-ZrNiAl	189	*P$\bar{6}$2m*	$1a\,2d\,3fg$	40	2.2%
11.	$hP3$-AlB$_2$	191	*P6/mmm*	$1a\,2d$	39	2.1%
12.	$cF16$-Cu$_2$MnAl	225	*Fm$\bar{3}$m*	$4abcd$	37	2.0%
13.	$tP7$-HoCoGa$_5$	123	*P4/mmm*	$1abc\,4i$	32	1.7%
14.	$cF24$-MgCu$_2$	227	*Fd$\bar{3}$m*	$8a\,16d$	32	1.7%
15.	$hP6$-CaCu$_5$	191	*P6/mmm*	$1a\,2c\,3g$	31	1.7%
16.	$hP12$-MgZn$_2$	194	*P6$_3$/mmc*	$2a\,4f\,6h$	31	1.7%
17.	$cI34$-Y$_4$PdGa$_{12}$	229	*Im$\bar{3}$m*	$2a\,8c\,12de$	29	1.6%
18.	$oI10$-W$_2$CoB$_2$	71	*Immm*	$2a\,4hj$	23	1.2%
19.	$cP8$-Cr$_3$Si	223	*Pm$\bar{3}$n*	$2a\,6c$	22	1.2%
20.	$hP38$-Th$_2$Ni$_{17}$	194	*P6$_3$/mmc*	$2bc\,4f\,6g\,12jk$	22	1.2%
21.	$hP18$-YNi$_2$Al$_3$	191	*P6/mmm*	$1a\,2d\,3f\,6kl$	22	1.2%
22.	$tI10$-BaAl$_4$	139	*I4/mmm*	$2a\,4de$	21	1.1%
23.	$cF116$-Mg$_6$Cu$_{16}$Si$_7$	225	*Fm$\bar{3}$m*	$4a\,24de\,32f^2$	21	1.1%
24.	$oI26$-ScFe$_6$Ga$_6$	71	*Immm*	$2a\,4ghij\,8k$	18	1.0%
					906	49.2%

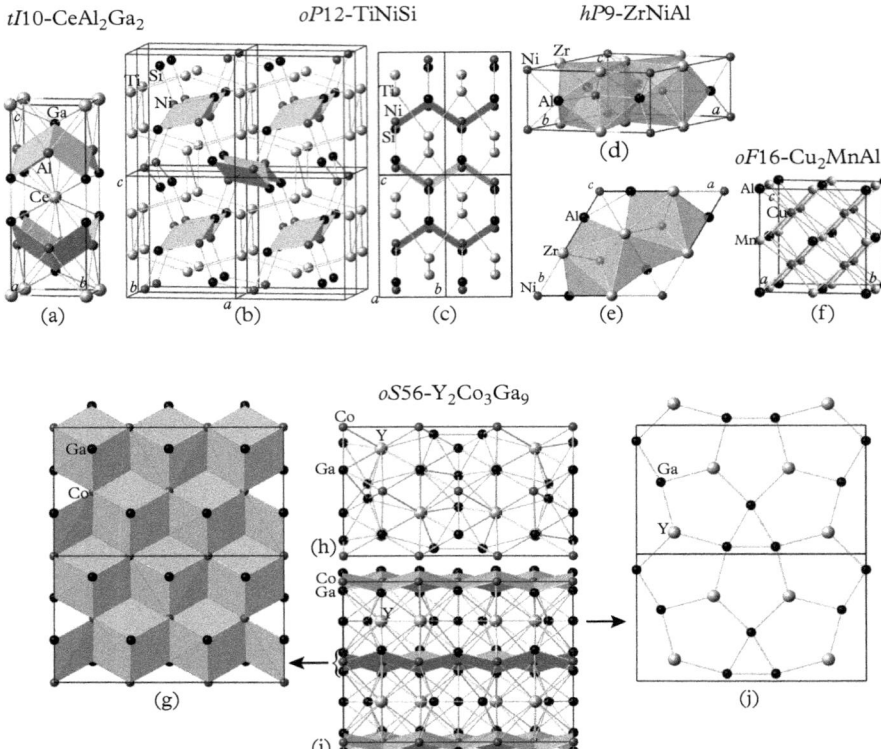

Fig. 7.55 *The structures of the five most frequent ternary structure types of ternary gallides. (a) tI10-CeAl$_2$Ga$_2$, with the edge-sharing Al$_2$Ga$_2$ rhomb units shaded gray. The structure of oP12-TiNiSi is shown in two different projections (b)–(c). The bands of edge-sharing Ni$_2$Si$_2$ rhomb units are shaded gray. One unit cell of hP9-ZrNiAl in perspective view and projection along [001] (d)–(e). The structure type of the Heusler phases cF16-Cu$_2$MnAl is depicted in (f). The structure of oS56-Y$_2$Co$_3$Ga$_9$ (g)–(j) can be composed of two kinds of layers, a puckered Ga/Co-triangle net (Co$_2$Ga$_2$ rhomb units shaded gray) and a flat Ga/Y-triangle/pentagon net.*

up 20.0% of the formula and has M-values 10–33. While the oI28-La$_3$Al$_{11}$-type ternary gallides always have an element with the smallest M-value (17–27) that can be assigned the role of La in the prototype compound (with a content of 21.4%), the remaining elements are disordered (M(B) = 60–76) with Ga-contents 30.0–68.6%.

cP4-Cu$_3$Au-type compounds again have varying Ga-contents (1.5–68.0%). Only one oP12-TiNiSi-structure, on the other hand, deviates from the strict 1:1:1-composition with a slight exchange between Ga and the B-element (M = 64–73), with undisturbed A-elements in all cases (M = 16–45).

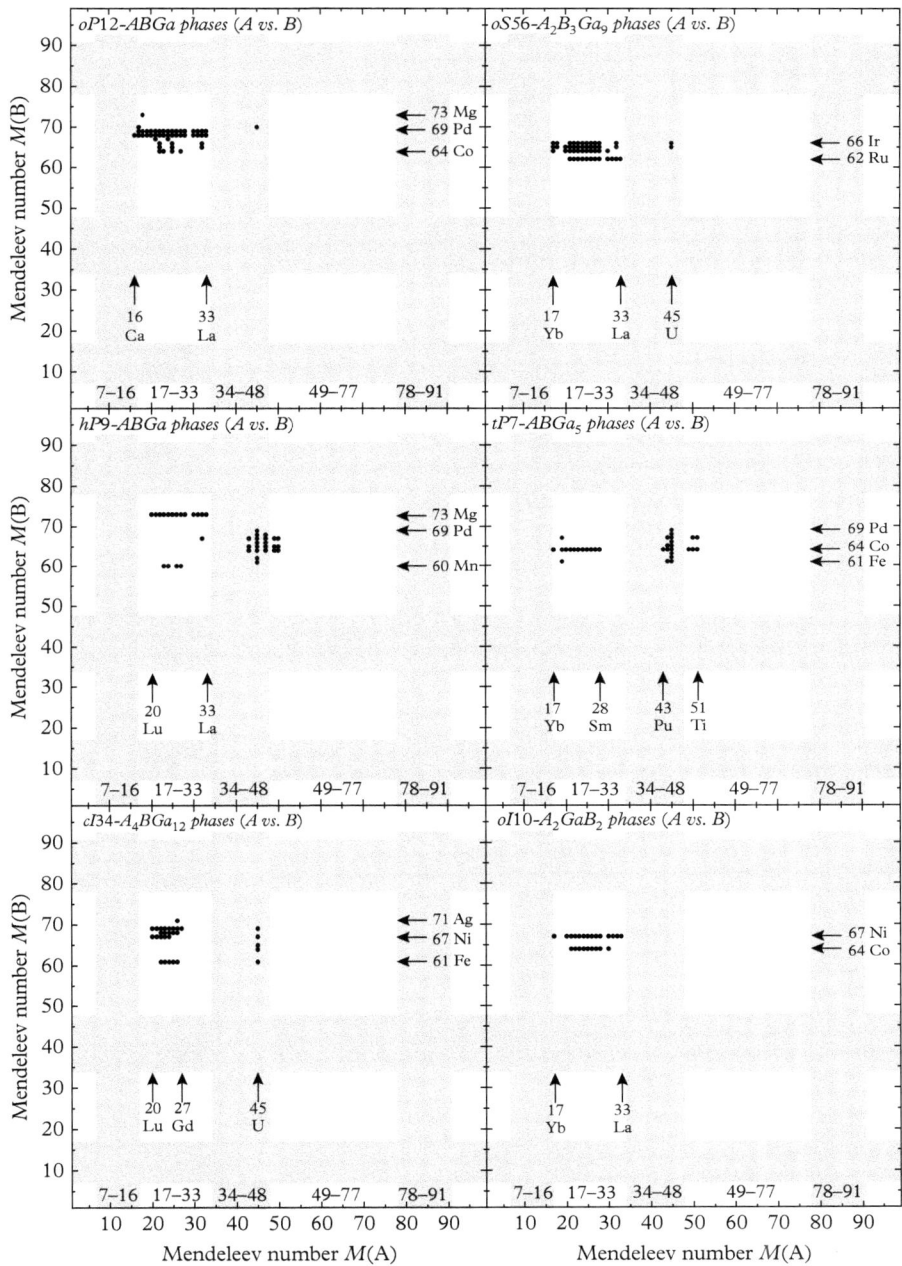

Fig. 7.56 *Occurrence of six of the most common structure types among ternary Ga-intermetallics. The Mendeleev numbers M of the other two elements in the compound are shown. The depicted structure types are oP12-TiNiSi, oS56-Y$_2$Co$_3$Ga$_9$, hP9-ZrNiAl, tP7-HoCoGa$_5$, cI34-Y$_4$PdGa$_{12}$, and oI10-W$_2$CoB$_2$ (ranks 6, 7, 10, 13, 17, and 18, respectively, in Table 7.14). Where* A *and* B *have equal shares in the chemical formula, they are assigned so that* M(A) < M(B). *These are the most common—inherently ternary—structure types, whose sites are occupied in an ordered manner for the most part.*

All $oS56$-Y$_2$Co$_3$Ga$_9$-type representatives among ternary Ga-intermetallics have undisturbed compositions and are composed of elements $M(A) = 17$–45, $M(B) = 62$–66, containing 64.3% of Ga. The $tI26$-ThMn$_{12}$-type compounds have Ga-components of 35.4–56.2%, which share the Mn-role with the B-element ($M(B) = 60$–72), while the A-element on the Th-position remains stoichiometric at 7.7% ($M(A) = 17$–50). Similarly, the $hR57$-Zn$_{17}$Th$_2$-representatives have Ga at contents 10.5–51.1% share the role of Zn in the prototype with the B-element ($M(B) = 60$–72), while the Th-role is fulfilled by elements with $M(A) = 17$–49 and only in one case deviates from the stoichiometric 10.5%. The 1:1:1-structure type $hP9$-ZrNiAl, again, is always reported with the correct stoichiometry and contains elements $M(A) = 20$–50 and $M(B) = 60$–73 in addition to Ga.

The $hP3$-AlB$_2$-type structures exhibit large variations in their stoichiometries with Ga-contents 9.9–66.7%. Most $cF16$-Cu$_2$MnAl-type structures contain 25% of Ga (29 out of 37 compounds), others contain 50% (6 compounds), and the rest deviates from the basic stoichiometry. All $tP7$-HoCoGa$_5$-type structures are stoichiometric with Ga in its eponymous role and other elements $M(A) = 17$–51 and $M(B) = 61$–69. Both, the $cF24$-MgCu$_2$- and $hP12$-MgZn$_2$-type structures have varying stoichiometries and the Ga-content ranges between 5.0–33.3% and 10–42.1%, respectively. The structures of type $hP6$-CaCu$_5$ also have varying Ga-contents (12.5–43.4%), but together with the B-elements with $M(B) = 64$–72, these add up to the Cu-content in the prototype composition, while all A-elements have a 16.7%-content and have M-values 17–45. The $cI34$-Y$_4$PdGa$_{12}$-type structures all have the strict composition of the prototype compound (with one minor deviation in the Tb$_4$AgGa$_{12}$-compound with a small Ga-excess and Ag-deficit) with 23.5% of elements $M(A) = 20$–45, 5.9% of $M(B) = 61$–71, and 70.6% of Ga.

The $oI10$-W$_2$CoB$_2$-type structures have 20% Ga and 40% each of elements $M(A) = 17$–33 and $M(B) = 64$–67. Minor Ga-deficiencies in RE$_2$GaNi$_2$ (RE = La, Ce) are compensated by increased Ni-contents (45% and 43%, respectively). The ternary $cP8$-Cr$_3$Si-type representatives contain 5.0–25.0% Ga. Of the 22 compounds, 15 can be directly identified with the 3:1-composition with either the Cr- or the Si-positions being occupied by a mixture of two elements, with $M = 51$–72 and 80–88 (incl. Ga), respectively. Three more structures have reversed occupancies with 25% V/Nb and 75% mixed Ga-Co/Pt, while the remaining four structures have stoichiometries that are incompatible with a direct identification with the Cr- and Si-sites. The $hP38$-Th$_2$Ni$_{17}$ structures contain 5.3–31.6% Ga, apparently adopting the role of Ni together with elements $M(B) = 60$–67, while the Th-content of 10.5% is contributed by $M(A) = 17$–32. Of the 22 $hP18$-YNi$_2$Al$_3$-representatives, only one is not fully stoichiometric, with a small B-Ga-exchange, while generally the compounds consist of 16.7% of $M(A) = 22$–33, 50.0% of $M(B) = 64$–67, and 33.3% of Ga. Ga-contents in $tI10$-BaAl$_4$-type structures range between 40.0% and 75.0% and always amount to 80% of the entire formula, when adding component B with $M(B) = 68$–83 (or, as in one case $M(B) = 12$). The Ba-role is played by $M(A) = 14$–33.

Apart from one representative, all $cF116$-$Mg_6Cu_{16}Si_7$-type compounds are fully stoichiometric and composed of 20.7% $M(A) = 19$–51 and 24.1% $M(B) = 62$–69; only $Zr_6Ga_{16}Ni_7$ actually contains 27.6% of Ni and only 51.7% of Ga. The $oI26$-$ScFe_6Ga_6$-type structures contain 46.2–53.8% Ga and 38.5–46.2% of the B-element with $M = 61$–64, so that the Sc-role is apparently filled by $M(A) = 17$–50, always with the stoichiometric content of 7.7%.

7.15.3 Indides

Indium, $tI2$-In, has a larger atomic radius than aluminum, $cF4$-Al, ($r_{Al} = 1.43$ Å, $r_{In} = 1.63$ Å) and the same electronegativity ($\chi_{Al} = 1.5$, $\chi_{In} = 1.5$). The distribution of indides in the ternary concentration diagram of intermetallics is depicted in Fig. 7.57 together with the frequencies of the subset of binary indides. Since the constituting elements A, B, and C stand for metallic elements with increasing Mendeleev numbers, $M(A) < M(B) < M(C)$, In with $M(In) = 79$, will in most intermetallics $A_xB_yC_z$ be represented by the letter C. Consequently, most compounds with In as majority element will agglomerate in the concentration triangle close to the corner marked C. The distribution of ternary indides in the concentration triangle (Fig. 7.57) is much sparser but otherwise similar to that of the aluminides (Fig. 7.45). The main difference is in the In-rich part of the concentration triangle. The frequency distribution of binary indides A_xB_y shows many differences to that of the aluminides, mainly for large values of x.

Binary indides

Among the binary IMs, 279 In-containing IMs are listed in the PCD, featuring 106 structure types; these are ≈ 2.6 representatives per structure type, similar to aluminides (≈ 2.4). The compositions of the 279 binary In-intermetallics are shown in Fig. 7.58, and their most common structure types are listed in Table 7.15. In contrast to the situation for the aluminides, there exist a significant number of alkali indides for all alkali metals but radioactive Fr. The compositions of the $cF4$-Cu-, $cI2$-W-, and $hP2$-Mg-type structures are rather diverse (containing 0.1–90%, 2–62%, and 2–25% In, respectively). The same holds true for $tI2$-In-, $hP1$-$Hg_{0.1}Sn_{0.9}$-, and $tI4$-Sn-type structures (with 25–98.2%, 20–70%, 5–50% In, respectively). Nearly all of the remaining structure types occur at their defined compositions. In some of them, In always plays the role of the majority element ($oS32$-Pu_3Pd_5, $hP6$-$CaIn_2$, $tP16$-$IrIn_3$, $tI10$-$BaAl_4$, $oI12$-KHg_2, $cI40$-Ru_3Sn_7, and $oP32$-Tm_3Ga_5) or of the minority element ($hP16$-Mn_5Si_3, $oP12$-Co_2Si, $cP52$-Cu_9Al_4, $tI32$-W_5Si_3, and $tP4$-$SrPb_3$). In others, it appears to switch between those roles ($cP4$-Cu_3Au, $hP6$-$Co_{1.75}Ge$, $tP2$-CuAu, $hP8$-Mg_3Cd, and $hP5$-Ni_2Al_3). Obviously, no such statement can be made for 1:1-stoichiometries, as in $cP2$-CsCl. An overview over the respective intermetallic systems is shown in Fig. 7.59. A lot of these structure types, however, have representatives whose compositions differ from the prototypes and therefore have to

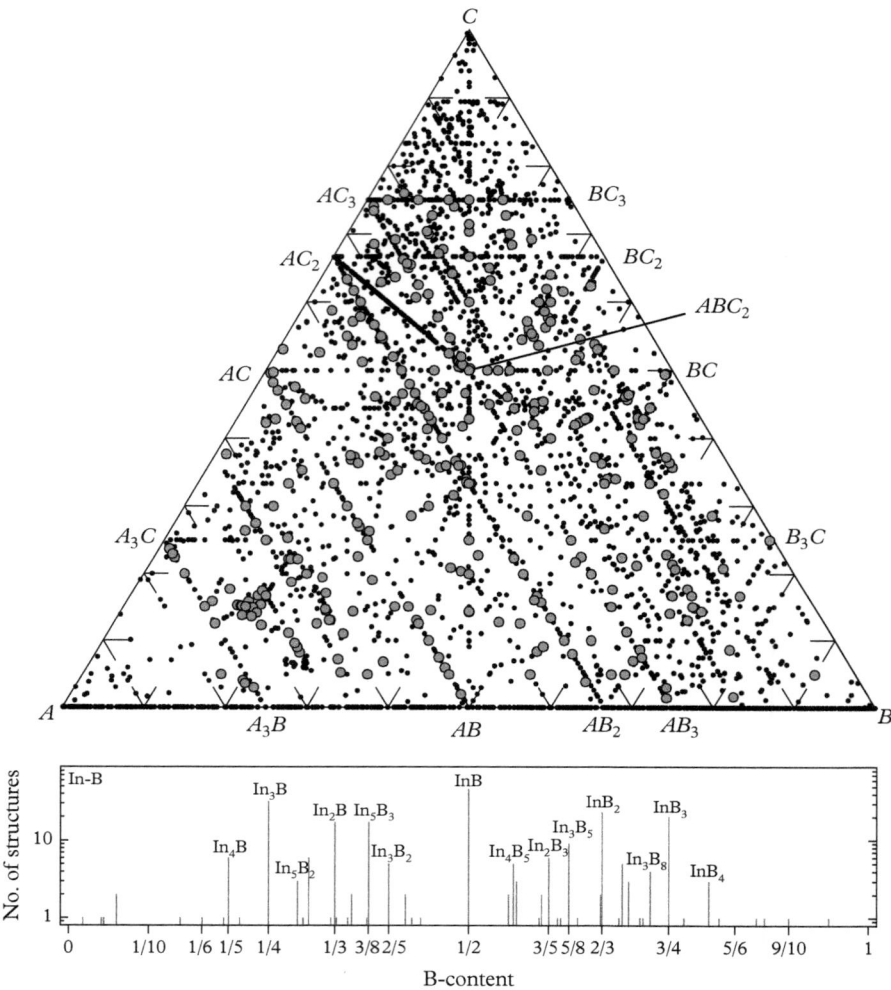

Fig. 7.57 *(top) Concentration diagram of the 20 829 intermetallic compounds (black dots) contained in* Pearson's Crystal Data *(PCD) (Villars and Cenzual, 2011a). The In-containing compounds are marked by large gray circles. A, B, and C stand for metallic elements with increasing Mendeleev numbers, $M(A) < M(B) < M(C)$. (bottom) Frequencies of the binary In–B compounds as a function of stoichiometry.*

exhibit some disorder. This is true for 11 out of the 24 most common structure types.

The most common binary structure types of binary aluminides and indides differ considerably. With the exception of six structure types, $cP4$-Cu_3Au, the second most frequent structure type in the case of aluminides and the most frequent

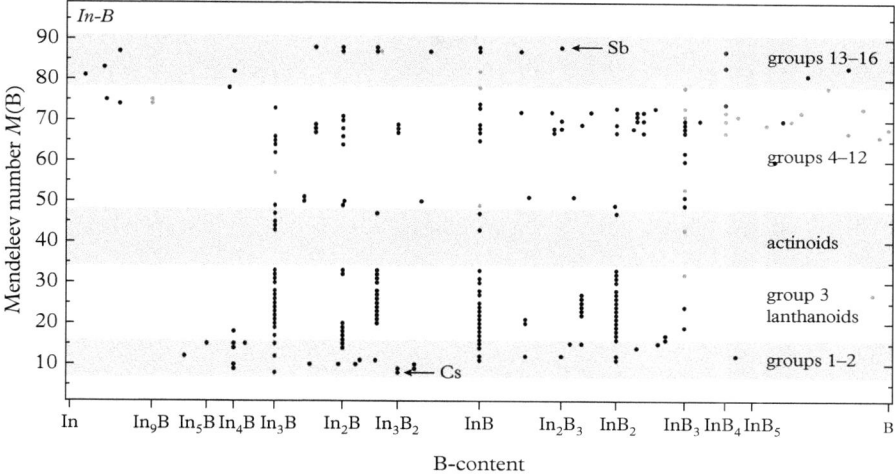

Fig. 7.58 *Stoichiometries of the 279 binary indides, In_aB_b vs. the Mendeleev number $M(B)$ of the other element in the compound. The 27 binary phases adopting unary structure types cF4-Cu, cI2-W, and hP2-Mg are marked by gray dots.*

one among the indides (short 2, 1), $cP2$-CsCl (4, 3), $hP8$-Mg$_3$Cd (9, 10), $oP12$-Co$_2$Si (5, 13), $hP5$-Ni$_2$Al$_3$ (11, 16), and $tI10$-BaAl$_4$ (8, 19), there are no common binary structure types with frequencies $\geq 1\%$.

With ten common structure types, there are significantly more similarities between the frequencies of binary structure types of gallides and indides: $cP4$-Cu$_3$Au (2, 1), $cP2$-CsCl (11, 3), $hP8$-Mg$_3$Cd (19, 10), $hP16$-Mn$_5$Si$_3$ (5, 11), $hP6$-CaIn$_2$ (20, 12), $tP16$-IrIn$_3$ (13, 15), $tI32$-W$_5$Si$_3$ (17, 17), $tI10$-BaAl$_4$ (14, 19), $oI12$-KHg$_2$ (22, 20), and $oP32$-Tm$_3$Ga$_5$ (12, 24).

In the following, we discuss some characteristic features of the five most frequent binary structure types listed in Table 7.15. Their structures are shown in Fig. 7.60.

- $cP4$-Cu$_3$Au (221 $Pm\bar{3}m$): This structure type is an ordered derivative of the $cF4$-Cu type; see Section 7.2 and Fig. 7.3. Li and Mg are the only elements of groups 1 and 2 adopting this structure type. All lanthanoids except Eu and Pm form structures of this type, as well as the actinoids Pu, Np, U, and Th. While all these elements occupy the Au site in the structure type, the transition elements Zr, Ti, Ni, Pt, and Ag sit on the Cu sites.

- $hP6$-Co$_{1.75}$Ge (194 $P6_3/mmc$): The structure is a deficient variant of the $hP6$-Ni$_2$In type, which itself can be seen as a superstructure of the $hP3$-AlB$_2$ type or as a filled derivative of the $hP4$-NiAs type. The positions of Ge in $2c$ 1/3, 2/3, 1/4 and Co1 in $2a$ 0, 0, 0 are fully occupied, while that of Co2 in $2d$ 1/3, 2/3, 3/4 is only partially filled. Co2, together with In,

Table 7.15 *Most common structure types of In-containing binary intermetallics. The top 24 structure types are given, all of which have at least three representative structures and therefore represent more than 1% of all binary In-intermetallics each.*

Rank	Structure type	No.	Space group	Wyckoff positions	No. of reps.	% of all reps.
1.	$cP4$-Cu_3Au	221	$Pm\bar{3}m$	$1a\,3c$	26	9.3%
2.	$hP6$-$Co_{1.75}Ge$	194	$P6_3/mmc$	$2acd$	19	6.8%
3.	$cP2$-$CsCl$	221	$Pm\bar{3}m$	$1ab$	18	6.5%
4.	$cF4$-Cu	225	$Fm\bar{3}m$	$4a$	16	5.7%
5.	$tI2$-In	139	$I4/mmm$	$2a$	13	4.7%
6.	$oS32$-Pu_3Pd_5	63	$Cmcm$	$4c^2\,8efg$	12	4.3%
7.	$tP2$-$CuAu$	123	$P4/mmm$	$1ad$	8	2.9%
8.	$cI2$-W	229	$Im\bar{3}m$	$2a$	6	2.2%
9.	$hP2$-Mg	194	$P6_3/mmc$	$2c$	5	1.8%
10.	$hP8$-Mg_3Cd	194	$P6_3/mmc$	$2d\,6h$	5	1.8%
11.	$hP16$-Mn_5Si_3	193	$P6_3/mcm$	$4d\,6g^2$	5	1.8%
12.	$hP6$-$CaIn_2$	194	$P6_3/mmc$	$2b\,4f$	4	1.4%
13.	$oP12$-Co_2Si	62	$Pnma$	$4c^3$	4	1.4%
14.	$cP52$-Cu_9Al_4	215	$P\bar{4}3m$	$4e^4\,6fg\,12i^2$	4	1.4%
15.	$tP16$-$IrIn_3$	136	$P4_2/mnm$	$4cf\,8j$	4	1.4%
16.	$hP5$-Ni_2Al_3	162	$P\bar{3}m$	$1a\,2d^2$	4	1.4%
17.	$tI32$-W_5Si_3	140	$I4/mcm$	$4ab\,8h\,16k$	4	1.4%
18.	$hP1$-$Hg_{0.1}Sn_{0.9}$	191	$P6/mmm$	$1a$	3	1.1%
19.	$tI10$-$BaAl_4$	139	$I4/mmm$	$2a\,4de$	3	1.1%
20.	$oI12$-KHg_2	74	$Imma$	$4e\,8i$	3	1.1%
21.	$cI40$-Ru_3Sn_7	229	$Im\bar{3}m$	$12de\,16f$	3	1.1%
22.	$tI4$-Sn	141	$I4_1/amd$	$4a$	3	1.1%
23.	$tP4$-$SrPb_3$	123	$P4/mmm$	$1ac\,2e$	3	1.1%
24.	$oP32$-Tm_3Ga_5	62	$Pnma$	$4c^4\,8d^2$	3	1.1%
					178	63.9%

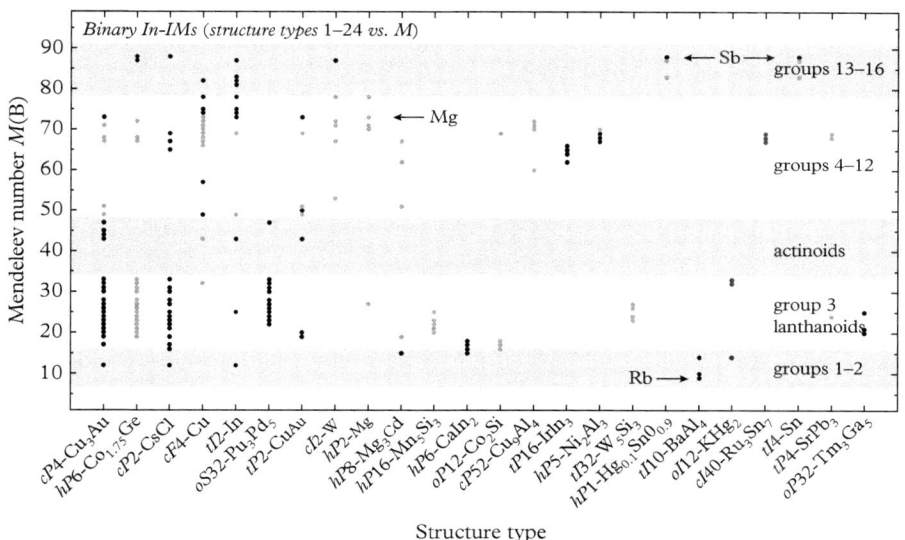

Fig. 7.59 *Occurrence of the 24 most common structure types among binary In-intermetallics. The Mendeleev number M of the other element in the compound is shown over the rank of the structure type, as given in Table 7.15. Structures with In as their major component (≥ 50%) are marked in black, those with minor In-content are shown in gray.*

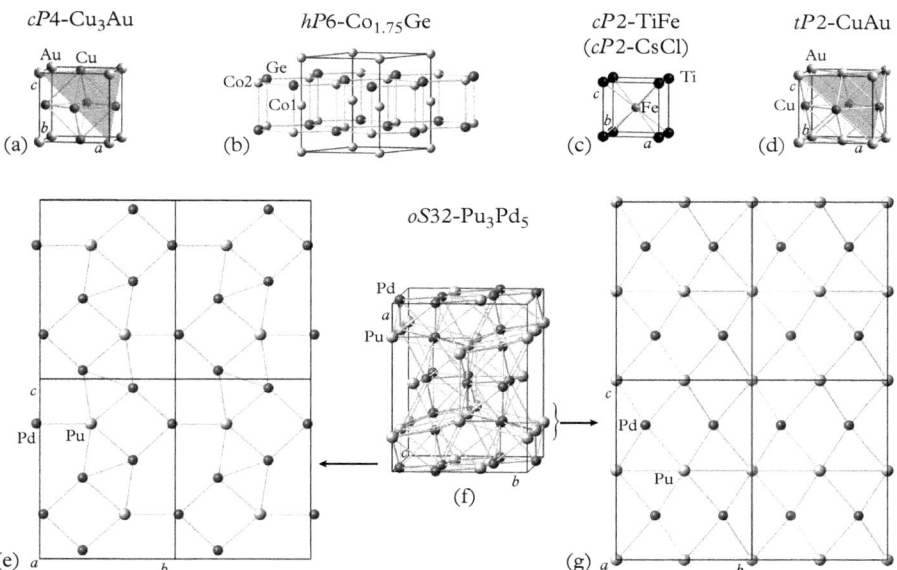

Fig. 7.60 *The structures of the five most frequent binary structure types of binary indides. (a) cP4-Cu₃Au (also see Section 7.3); (b) hP6-Co₁.₇₅Ge, a superstructure of the hP3-AlB₂ type; (c) cP2-FeTi (cP2-CsCl type) (also see Section 7.3); (d) tP2-CuAu (also see Section 7.2). (e)–(g) The structure of oS32-Pu₃Pd₅ in different representations: (f) one unit cell showing the layer structure; (e) the flat layer, and (g) the puckered layer.*

forms honeycomb nets with Co1 centering the hexagonal prisms formed. All lanthanoids except Yb, Pm, and Eu form structures of this type as well as the transition elements Ni, Pt, and Cu, and sit on the Ge sites, in contrast to the main group elements Bi and Sb, which occupy the Co site.

- *cP*2-CsCl (example *cP*2-FeTi) (221 *Pm$\bar{3}$m*): This structure type is an ordered derivative of the *cI*2-W type; see Section 7.3 and Fig. 7.7. Li and Ca are the only elements of groups 1 and 2 adopting this structure type. All lanthanoids except Ce, Eu, and Pm form structures of this type as well as the transition elements Rh, Ni, and Pd, and the main group element Bi.

- *oS*32-Pu$_3$Pd$_5$ (63 *Cmcm*): This structure can be decomposed into layers stacked along [100]. A puckered layer, with non-bonding distances ($d_{\text{Pu-Pd}} = 2.97–3.13$ Å) between the atoms, is sandwiched between two symmetrically equivalent flat layers. The puckering is caused by fitting this layer in the best way between the flat triangle/square/hexagon layers. Here, the Pu–Pd and Pd–Pd distances are with 2.87–2.97 Å and 2.82–2.85 Å significantly shorter. This structure is only adopted by the lanthanoids from Er to La except for Pm, as well as by the actinoid Th.

- *tP*2-CuAu (123 *P4/mmm*): This structure type is an ordered derivative of the *cF*4-Cu type; see Section 7.2 and Fig. 7.3. The indides crystallizing in this structure type are constituted from Sc, Lu, Pu, Zr, Hf, Ti , Pd, and Mg.

Ternary indides

The 1260 ternary In-containing IMs feature 255 different structure types, i.e., ≈ 4.9 representatives per structure type. This is significantly less than in the case of ternary aluminides (≈ 6.4). In Table 7.16, the most common structure types among ternary In-IMs are given, and the distribution of the chemical compositions of six of the most common ones are shown in Fig. 7.62. The most common ternary structure types (with frequencies $\geq 1\%$) of ternary aluminides, gallides, and indides differ considerably. There are only three ternary structure types common to all three of them: *hP*9-ZrNiAl (9, 10, 1), *cF*16-Cu$_2$MnAl (8, 12, 2), and *oP*12-TiNiSi (15, 6, 21), and one structure type common to aluminides and indides, only: *tI*26-CeMn$_4$Al$_8$ (5, 22).

In the following, we discuss some characteristic features of the five most frequent ternary structure types of ternary indides listed in Table 7.16. Their structures are shown in Fig. 7.61.

- *hP*9-ZrNiAl (189 *P$\bar{6}$2m*): The structure can be subdivided into face-sharing tricapped trigonal Ni@Zr$_6$Al$_3$ prisms (see also Fig. 10.8 in Section 10.8).

- *cF*16-Cu$_2$MnAl (225 *Fm$\bar{3}$m*): This structure type of the Heusler phases can be described as $(2 \times 2 \times 2)$-fold superstructure of the *cI*2-W type, and is discussed in Subsection 7.3.2.

Table 7.16 *Most common structure types of In-containing ternary intermetallics. The top 22 structure types are given, all of which have at least 12 representative structures and therefore represent at least 1% of all ternary In-intermetallics.*

Rank	Structure type	No.	Space group	Wyckoff positions	No. of reps.	% of all reps.
1.	$hP9$-ZrNiAl	189	$P\bar{6}2m$	$1a2d3fg$	107	8.5%
2.	$cF16$-Cu$_2$MnAl	225	$Fm\bar{3}m$	$4ab\,8c$	83	6.6%
3.	$tP10$-Mo$_2$FeB$_2$	127	$P4/mbm$	$2a\,4gh$	75	6.0%
4.	$cF24$-MgCu$_4$Sn	216	$F\bar{4}3m$	$4ac\,16e$	47	3.7%
5.	$oS16$-MgCuAl$_2$	63	$Cmcm$	$4c^2\,8f$	40	3.2%
6.	$hP6$-CaIn$_2$	194	$P6_3/mmc$	$2b\,4f$	39	3.1%
7.	$cP4$-Cu$_3$Au	221	$Pm\bar{3}m$	$1a\,3c$	36	2.9%
8.	$cP2$-CsCl	221	$Pm\bar{3}m$	$1ab$	34	2.7%
9.	$tP7$-HoCoGa$_5$	123	$P4/mmm$	$1abc\,4i$	31	2.5%
10.	$oP22$-Lu$_5$Ni$_2$In$_4$	55	$Pbam$	$2a\,4g^2h^3$	30	2.4%
11.	$tP80$-Gd$_{14}$Co$_3$In$_{2.7}$	137	$P4_2/nmc$	$4c^2d^2\,8fg^5\,16h$	28	2.2%
12.	$hP3$-AlB$_2$	191	$P6/mmm$	$1a\,2d$	23	1.8%
13.	$tP11$-Ho$_2$CoGa$_8$	123	$P4/mmm$	$1a\,2egh\,4i$	20	1.6%
14.	$tP24$-YNi$_9$In$_2$	127	$P4/mbm$	$2ac\,4g\,8jk$	19	1.5%
15.	$oS24$-YNiAl$_4$	63	$Cmcm$	$4ac^3\,8f$	17	1.3%
16.	$hP3$-Hg$_2$U	191	$P6/mmm$	$1a\,2d$	16	1.3%
17.	$tI26$-ThMn$_{12}$	139	$I4/mmm$	$2a\,8fij$	15	1.2%
18.	$tP20$-U$_2$Pt$_2$Sn	136	$P4_2/mnm$	$4dfg\,8j$	14	1.1%
19.	$cF96$-Gd$_4$RhIn	216	$F\bar{4}3m$	$16e^3\,24fg$	13	1.0%
20.	$oS48$-Nd$_{11}$Pd$_4$In$_9$	65	$Cmmm$	$2ac\,4gi^2\,8pq^3$	13	1.0%
21.	$oP12$-TiNiSi	62	$Pnma$	$2ad\,6gh$	13	1.0%
22.	$tI26$-CeMn$_4$Al$_8$	139	$I4/mmm$	$2a\,8fij$	12	1.0%
					725	57.6%

- $tP10$-Mo$_2$FeB$_2$ (127 $P4/mbm$): The structure consists of two nets stacked along [001]: one $3^2.4.3.4$ net (snub square tiling) decorated just by Mo atoms, and the dual net (Cairo pentagonal tiling) with its vertices occupied by Fe and B. Consequently, B occupies the centers of trigonal Mo-prisms, and Fe centers Mo-cubes that are at four sides capped by B; Mo centers all-side-capped pentagonal prisms.

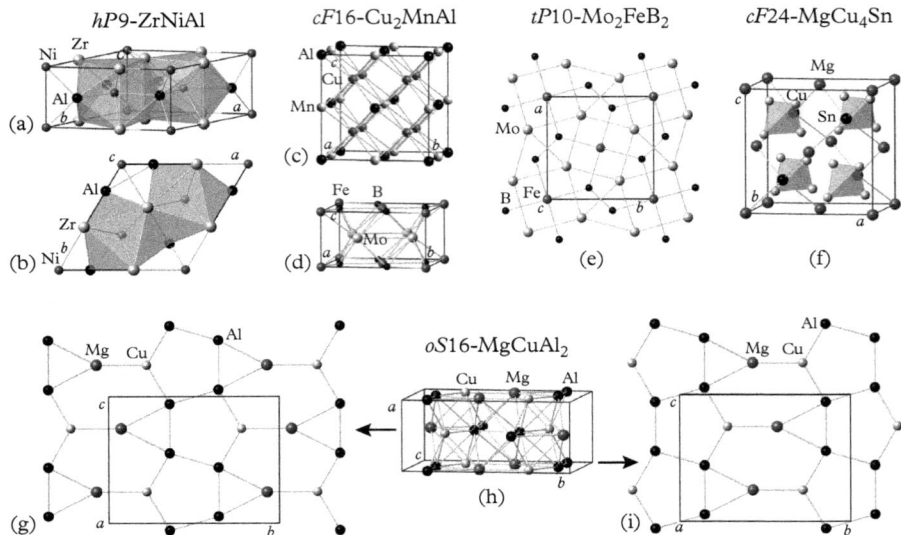

Fig. 7.61 *The structures of the five most frequent ternary structure types of ternary indides.*
(a)–(b) The structure of hP9-ZrNiAl in different projections with the tricapped trigonal AETs of
Ni shaded. The structure of the Heusler phase cF16-Cu₂MnAl is depicted in (c), and of
tP10-Mo₂FeB₂ (127 P4/mbm) (d)–(e). The projection along [001] shows the superposition of
the Mo-decorated snub square tiling and its dual, the Cairo pentagonal tiling with its vertices
occupied by Fe and B. cF24-MgCu₄Sn (f) corresponds to an ordered variant of the cubic Laves
phase cF24-MgCu₂. oS16-MgCuAl₂ (g)–(i) can be described as a stacking along [100] of
symmetrically equivalent copies of a triangle/pentagon/hexagon layer.

- $cF24$-MgCu$_4$Sn (216 $F\bar{4}3m$): Ordering structure of the cubic Laves phase $cF24$-MgCu$_2$. See Fig. 7.55.

- $oS16$-MgCuAl$_2$ (63 $Cmcm$): Ordered variant of the $oS16$-Re$_3$B structure type. Mg in a partially side-capped pentagonal prism; Cu in a tricapped trigonal prism; Al in a distorted cuboctahedron.

Nearly all $hP9$-ZrNiAl-type representatives among ternary In-intermetallics have perfect 1:1:1-stoichiometry (with the exception of ScPdIn, which is slightly Pd-deficient—32.9%). The A- and B-elements have values $M(A) = 17–47$ and $M(B) = 65–73$. The $cF16$-Cu$_2$MnAl-type structures mostly contain 25% of In, with five representatives containing 50% and one outlier with 10%, where it seems to be complemented with 40% of Mg. Also the $tP10$-Mo$_2$FeB$_2$ (127 $P4/mbm$)-type structures are quite consistent with 20% In, except for four slightly deviating compositions (in the (Gd/Dy/Er/Tm)$_2$InNi$_2$-compounds—with 20.4–20.9% In). All $cF24$-MgCu$_4$Sn-type structures are perfectly stoichiometric with 16.7% In and $M(A) = 16–49$ or 73 and $M(B) = 64–72$. Similarly,

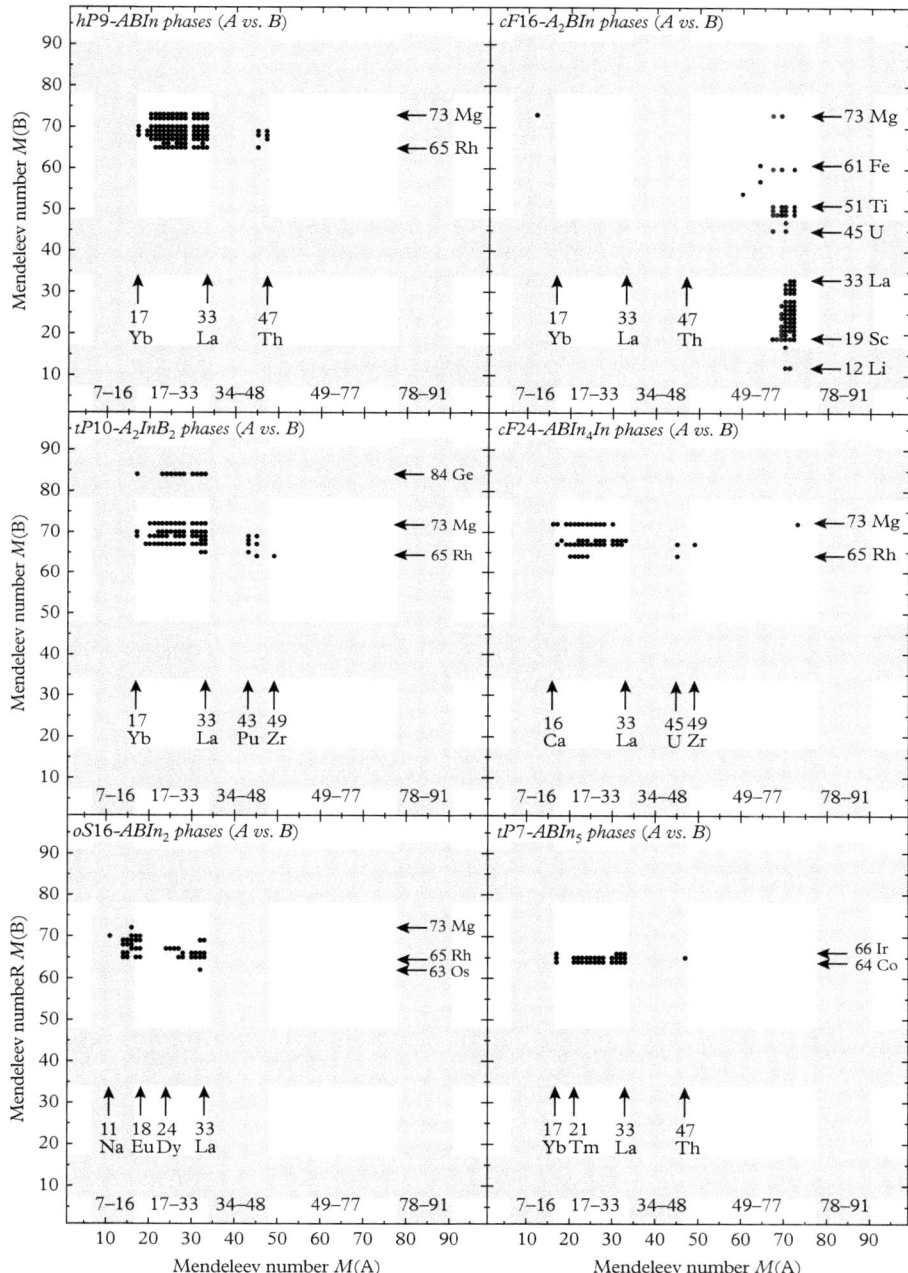

Fig. 7.62 *Occurrence of six of the most common structure types among ternary In-intermetallics. The Mendeleev numbers M of the other two elements in the compound are shown. The depicted structure types are hP9-ZrNiAl, cF16-Cu$_2$MnAl, tP10-Mo$_2$FeB$_2$ (127 P4/mbm), cF24-MgCu$_4$Sn, oS16-MgCuAl$_2$, and tP7-HoCoGa$_5$ (ranks 1–5, and 9, respectively, in Table 7.16). Where A and B have equal shares in the chemical formula, they are assigned so that M(A) < M(B). These are the most common—inherently ternary—structure types. Only six structures were omitted for structure type cF16-Cu$_2$MnAl, where the chemical decoration of the structure changes.*

the $oS16$-MgCuAl$_2$-type structures all have 1:1:2-stoichiometry with 50% In, containing elements $M(A) = 11$–33 and $M(B) = 62$–72.

The $hP6$-CaIn$_2$-type structures have various stoichiometries: the majority of 21 compounds have 1:1:1 compositions (all A–Zn/Cd–In-compounds, plus Ca–Ga–In), 10 more contain 50% In, 16.7% Cu, and 33.3% with $M(A) = 22$–32 (or, in one case, Sr–Ga–In), and the remaining 8 have diverse ratios of the B-element: In with always 33.3% of the A-element ($M(A) = 15$–33). $cP4$-Cu$_3$Au-type structures contain 1.3–75% In: 22 contain 25% A-element ($M(A) = 21$–49) and 8 other ones contain 75% In, both with varying subdivisions of the remaining formula among the two other elements. Of the six additional structures, three have completely different stoichiometries and three contain 75% A. The $cP2$-CsCl-type representatives contain 5.0–45.0% In and consist of largely varying elements—$M(A) = 11$–71 and $M(B) = 65$–83—with various stoichiometries (19 with 50% B and 13 with 50% A). In contrast, all $tP7$-HoCoGa$_5$-type structures are fully stoichiometric and are built from elements with $M(A) = 17$–33 or 47 and $M(B) = 64$–66 (i.e., Co, Rh, Ir). Also $oP22$-Lu$_5$Ni$_2$In$_4$-type structures exhibit the eponymous stoichiometry and contain elements $M(A) = 19$–33 or 49/50 and $M(B) = 65$–69.

$tP80$-Gd$_{14}$Co$_3$In$_{2.7}$-type structures have roughly the same stoichiometries, containing 12.3–17.4% In and additionally 66.8–71.1% with $M(A) = 19$–27 and 13.9–18.3% with $M(B) = 64$–69. The structures of type $hP3$-AlB$_2$ contain 10.0–56.7% In and the complementary 10.0–56.7% of the B-element with $M(B) = 60$–72 or 81, with a consistent A-content of 33.3% with $M(A) = 14$–33. All $tP11$-Ho$_2$CoGa$_8$ have the optimal stoichiometry with 72.7% In, as well as 18.2% of element A ($M(A) = 17$–33) and 9.1% of element B ($M(B) = 64$–69). Also $tP24$-YNi$_9$In$_2$-type structures are perfectly stoichiometric with 16.7% In, as well as 8.3% of A ($M(A) = 18$–33) and 75% of B with $M(B) = 67$ or 72 (i.e., Ni or Cu, respectively). The $oS24$-YNiAl$_4$-type structures contain In in the role of Al in the prototype (66.7%) and are built by A- and B-elements with $M(A) = 15$–33 or and $M(B) = 67$–72.

In $hP3$-Hg$_2$U-type structures, In shares the Hg-role with B-elements $M(B) = 60$–76 with contents of 36.7–55.0% and 11.7–30.0%, respectively, complemented with 33.3% of the A-element ($M(A) = 17$–32). Similarly, the $tI26$-ThMn$_{12}$-type structures contain a constant 7.7% of A with $M(A) = 16$–33, as well as 43.1–53.8% In, with B-elements Ag or Cu making up the rest of the structure ($M(B) = 71$ and 72, respectively). The $tP20$-U$_2$Pt$_2$Sn-type structures always contain 20% of In and respectively 40% of elements A ($M(A) = 19$–21 or 49–51) and B ($M(B) = 67$–72). The $cF96$-Gd$_4$RhIn-type structures contain 16.7% In, as well as the same amount of element B ($M(B) = 65$ or 66, i.e., Rh or Ir, respectively) and 66.7% of A ($M(A) = 20$–27). $oS48$-Nd$_{11}$Pd$_4$In$_9$-type structures contain 37.5% In in the eponymous role, as well as 45.8% of A ($M(A) = 23$–33) and 16.7% of B ($M(B) = 64$–49). In the perfect 1:1:1-compositions of the $oP12$-TiNiSi-type structures, elements A and B have values $M(A) = 15$–20 and $M(B) = 65$–70. In the $tI26$-CeMn$_4$Al$_8$-type structures, the A-content is constant at 7.7% with elements $M(A) = 17$–32, while 46.2–53.8% In and the complementary amount of Cu $(M(B) = 72)$ share the combined roles of Mn and Al in the prototype structure.

7.15.4 Thallides

Thallium, $hP2$-Mg, has a larger atomic radius than aluminum, $cF4$-Al, $(r_{Al} = 1.43$ Å, $r_{Tl} = 1.70$ Å) and a slightly lower electronegativity $(\chi_{Al} = 1.5, \chi_{In} = 1.4)$. The distribution of thallides in the ternary concentration diagram of intermetallics is depicted in Fig. 7.66 together with the frequencies of the subset of binary thallides. Since the constituting elements A, B, and C stand for metallic elements with increasing Mendeleev numbers, $M(A) < M(B) < M(C)$, Tl with $M(Tl) = 78$, will in most intermetallics $A_xB_yC_z$ be represented by the letter C. Consequently, most compounds with Tl as majority element will agglomerate in the concentration triangle close to the corner marked C.

Binary thallides

Among the binary IMs, 167 Tl-containing IMs can be found, featuring 47 structure types, i.e., ≈ 3.6 representatives per structure type, significantly more than in the case of the aluminides (≈ 2.4). In Table 7.17, the most common structure types among binary Tl-IMs are given. The compositions of the 167 binary thallides are shown in Fig. 7.63, and their most common structure types are listed in Table 7.17. In contrast to the situation for the aluminides, a significant number of alkali thallides are known for all alkali metals but radioactive Fr.

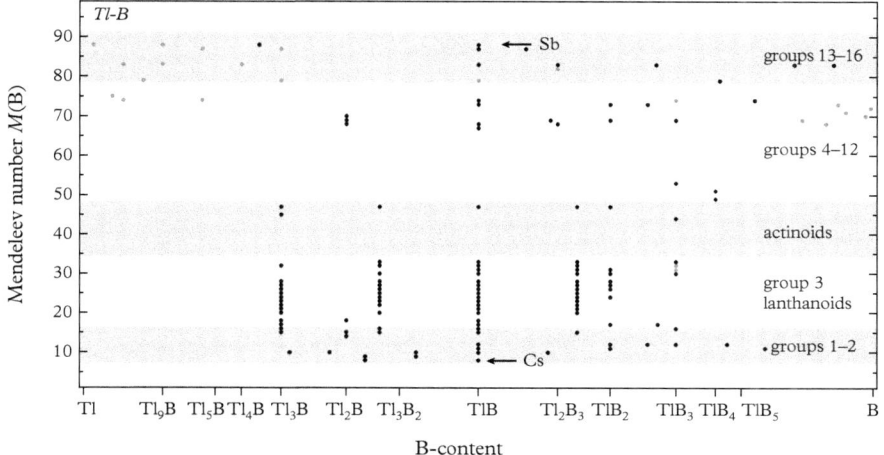

Fig. 7.63 *Stoichiometries of the 167 binary thallides,* Tl_aB_b *vs. the Mendeleev number M(B) of the other element in the compound. The 23 binary phases adopting unary structure types cF4-Cu, cI2-W, and hP2-Mg are marked by gray dots.*

Table 7.17 *Most common structure types of Tl-containing binary intermetallics. The top 16 structure types are given, all of which have at least three representative structures and therefore represent more than 1.5% of all binary Tl-intermetallics each. In sum, the 129 representatives of these structure types make up 77.2% of all binary Tl-intermetallics.*

Rank	Structure type	No.	Space group	Wyckoff positions	No. of reps.	% of all reps.
1.	$cP2$-CsCl	221	$Pm\bar{3}m$	$1ab$	21	12.6%
2.	$cP4$-Cu$_3$Au	221	$Pm\bar{3}m$	$1a\,3c$	20	12.0%
3.	$oS32$-Pu$_3$Pd$_5$	63	$Cmcm$	$4c^2\,8efg$	14	8.4%
4.	$cF4$-Cu	225	$Fm\bar{3}m$	$4a$	13	7.8%
5.	$tP2$-CuTi	123	$P4/mmm$	$1ad$	9	5.4%
6.	$hP16$-Mn$_5$Si$_3$	193	$P6_3/mcm$	$4d\,6g^2$	9	5.4%
7.	$hP2$-Mg	194	$P6_3/mmc$	$2c$	7	4.2%
8.	$hP6$-Co$_{1.75}$Ge	194	$P6_3/mmc$	$2acd$	6	3.6%
9.	$tI32$-W$_5$Si$_3$	140	$I4/mcm$	$4ab\,8h\,16k$	6	3.6%
10.	$tI32$-Cr$_5$B$_3$	140	$I4/mcm$	$4ac\,8h\,16l$	5	3.0%
11.	$tI12$-CuAl$_2$	194	$P6_3/mmc$	$4a\,8h$	4	2.4%
12.	$hP6$-CaIn$_2$	194	$P6_3/mmc$	$2b\,4f$	3	1.8%
13.	$cP8$-Cr$_3$Si	223	$Pm\bar{3}n$	$2a\,6c$	3	1.8%
14.	$tP2$-CuAu	123	$P4/mmm$	$1ad$	3	1.8%
15.	$tI2$-In	139	$I4/mmm$	$2a$	3	1.8%
16.	$cI2$-W	229	$Im\bar{3}m$	$2a$	3	1.8%
					129	77.2%

The compositions of the $cF4$-Cu-, $hP2$-Mg-, and $cI2$-W-type structures are rather diverse (containing 0.3–90%, 4.5–98.8%, and 75–90% Tl, respectively). The same holds true for $tI2$-In-type structures (with 15–40% Tl). Nearly all of the remaining structure types occur at their defined compositions. In some of them, Tl always plays the role of the majority element ($oS32$-Pu$_3$Pd$_5$ and $hP6$-CaIn$_2$) or of the minority element ($hP16$-Mn$_5$Si$_3$, $hP6$-Co$_{1.75}$Ge, $tI32$-W$_5$Si$_3$, $tI32$-Cr$_5$B$_3$, and $cP8$-Cr$_3$Si). In $cP4$-Cu$_3$Au and $tI12$-CuAl$_2$, it appears to switch between those roles.

Obviously, no such statement can be made for 1:1-stoichiometries, as in $cP2$-CsCl, $tP2$-CuTi, and $tP2$-CuAu. An overview over the respective intermetallic systems is shown in Fig. 7.64. Apart from the four unary structure types mentioned above, nearly all structures crystallize in ideal stoichiometries: only

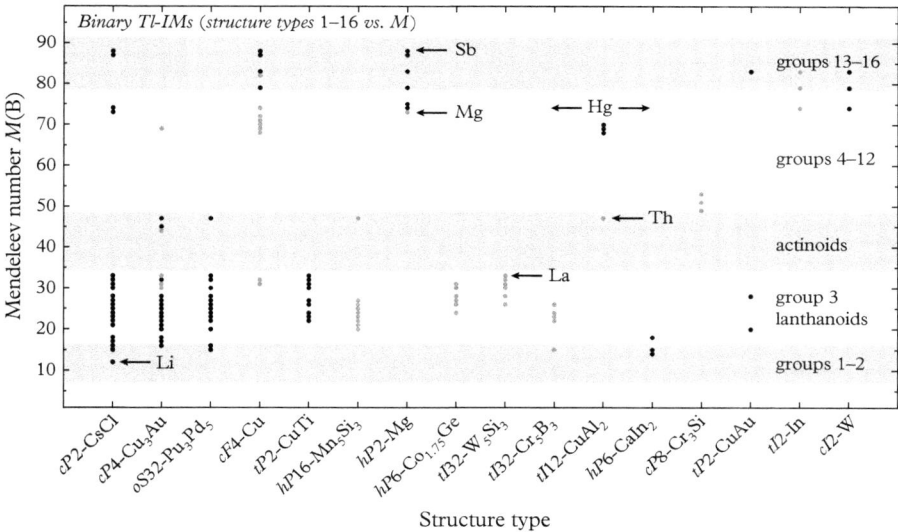

Fig. 7.64 *Occurrence of the 16 most common structure types among binary Tl-intermetallics. The Mendeleev number M of the other element in the compound is shown over the rank of the structure type, as given in Table 7.17. Structures with Tl as their major component (≥ 50%) are marked in black, those with minor Tl-content are shown in gray.*

one $cP8$-Cr_3Si-compound is slightly off, while all $hP6$-$Co_{1.75}$Ge-structures have 2:1-compositions.

In the following, we discuss some characteristic features of the five most common ternary structure types of binary thallides listed in Table 7.17. Their structures are shown in Fig. 7.65.

- $cP2$-CsCl (example $cP2$-FeTi) (221 $Pm\bar{3}m$): This structure type is an ordered derivative of the $cI2$-W type (see also Section 7.3 and Fig. 7.7). Representatives of this structure type are known for Li, Sr, Ca, and Mg of groups 1 and 2, all lanthanoids except Lu and Pm, as well as Hg, Bi, and Sb.

- $cP4$-Cu_3Au (221 $Pm\bar{3}m$): This structure type is an ordered derivative of the $cF4$-Cu type (see also Section 7.2 and Fig. 7.3). It is adopted by thallides of Ca, all lanthanoids except Lu and Pm, the actinoids Np, U, and Th, as well as Pd. Only in the case of Nd, Pr, La, Np, and Pd is Tl a minority element occupying the Au position.

- $oS32$-Pu_3Pd_5 (63 $Cmcm$): This structure can be decomposed into layers stacked along [100]. A puckered layer, with non-bonding distances ($d_{Pu–Pd} = 2.97$–3.13 Å) between the atoms, is sandwiched between two symmetrically equivalent flat layers. The puckering is caused by fitting this layer

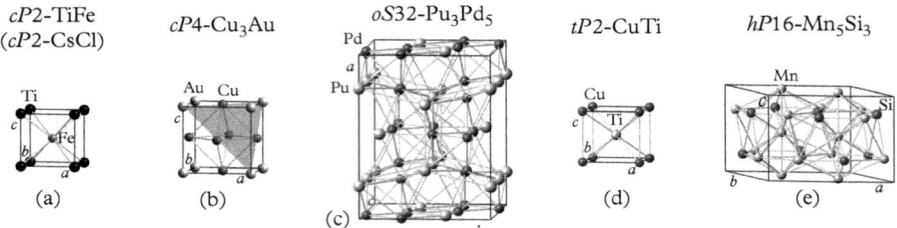

Fig. 7.65 *The structures of the five most frequent binary structure types of binary thallides. (a) cP2-CsCl (see also Section 7.3 and Fig. 7.7); (b) cP4-Cu₃Au (see also Section 7.2 and Fig. 7.3). The structure of oS32-Pu₃Pd₅ (c) can be described by layers stacked along [100], where a puckered layer is sandwiched between two symmetrically equivalent flat layers (for more details see Fig. 7.54). The structure of tP2-CuTi (d) corresponds to a tetragonally distorted cP2-CsCl type. The unit cell of hP16-Mn₅Si₃ (e) contains two face-sharing CN16 FK-polyhedra (see also Fig. 7.54).*

in the best way between the flat triangle/square/hexagon layers. Here, the Pu–Pd and Pd–Pd distances are with 2.87–2.97 Å and 2.82–2.85 Å significantly shorter (for more details see Fig. 7.54). This structure type is only found for thallides containing Sr, Ca, the lanthanoids except Yb, Eu, Tm, Pm, and Pr, as well as Th, all on the Pu sites.

- *tP2*-CuTi (123 *P4/mmm*): This structure can be described as a tetragonally distorted *cP2*-CsCl type or chemically ordered *tP2*-Pa type. For the change of the AETs with the c/a-ratio see Section 3.1. It is adopted exclusively by the lanthanoids Er, Ho, Dy, Tb, Gd, Nd, Pr, Ce, and La.

- *hP16*-Mn₅Si₃ (193 *P6₃/mcm*): This structure type, one of the Nowotny phases, can be described as a packing of face-sharing CN16 FK-polyhedra, leaving empty spaces corresponding to columns of face-sharing octahedra running along [001] (see also Subsection 7.15.9 and Fig. 7.54). Thallides are known for the lanthanoids Er, Ho, Dy, Tb, Gd, Nd, Pr, Ce, La, and Th, which all occupy the Mn sites.

Ternary thallides

The 127 ternary Tl-containing IMs feature 50 different structure types, i.e., ≈ 2.5 representatives per structure type. The distribution of ternary thallides in the concentration triangle (Fig. 7.57) is very sparse and hardly comparable to that of the aluminides (Fig. 7.45). In Table 7.18, the most common structure types among ternary Tl-IMs are given, and the distribution of the chemical compositions of six of the most common ones are shown in Fig. 7.67. In the ternary compounds, Tl is most often the element with the highest M-value (78) and will therefore be assigned the C-position, while the A-elements will strictly have smaller M-values than B-elements in all formulae $A_aB_bTl_c - M(A) < M(B)$.

Table 7.18 *Most common structure types of Tl-containing ternary intermetallics. The top nine structure types are given, all of which have at least three representative structures and therefore represent at least 2% of all ternary Tl-intermetallics.*

Rank	Structure type	No.	Space group	Wyckoff positions	No. of reps.	% of all reps.
1.	$hP9$-ZrNiAl	189	$P\bar{6}2m$	$1a\,2d\,3fg$	25	19.7%
2.	$hP6$-CaIn$_2$	194	$P6_3/mmc$	$2b\,4f$	17	13.4%
3.	$cP4$-Cu$_3$Au	221	$Pm\bar{3}m$	$1a\,3c$	11	8.7%
4.	$cP2$-CsCl	221	$Pm\bar{3}m$	$1ab$	9	7.1%
5.	$tI80$-La$_6$Co$_{11}$Ga$_3$	140	$I4/mcm$	$4ad\,8f\,16kl^3$	4	3.1%
6.	$tP2$-CuTi	123	$P4/mmm$	$1ad$	4	3.1%
7.	$cF16$-NaTl	227	$Fd\bar{3}m$	$8ab$	3	2.4%
8.	$oS16$-MgCuAl$_2$	63	$Cmcm$	$4c^2\,8f$	3	2.4%
9.	$cI46$-K$_9$NaTl$_{13}$	204	$Im\bar{3}$	$2a\,8c\,12e\,24g$	3	2.4%
					79	62.2%

The $hP9$-ZrNiAl-type structures among ternary Tl-intermetallics all have perfect 1:1:1-stoichiometry and are composed of elements $M(A) = 17$–33 and $M(B) = 69$ or 73 (i.e., Pd and Mg, respectively). Also all $hP6$-CaIn$_2$-type representatives have 1:1:1-stoichiometry, while here the constituting elements are $M(A) = 22$–33 and $M(B) = 72$, 75, or 76 (i.e., Cu, Cd, and Zn, respectively). In seven of the $cP4$-Cu$_3$Au-type structures, the A-element makes up 25% of the formula and in the remaining ones 75%, with $M(A) = 16$–33 and $M(A) = 68$ or 69, respectively. In the first case, Tl shares the Cu-role equally with the B-elements $M(B) = 79$–83, while in the latter one the sharing is mostly equal, too (with $M(B) = 87$ or 88), and in one case asymmetric (with $M(B) = 82$, i.e., Pb, where the Tl-content is 20%).

The $cP2$-CsCl-type structures are more diverse; however, one component always makes up for 50% of the structure and the other two have rather similar M-values. These are, for 50%-Tl-compounds, 25&27, as well as 27&33, while in the case of 50% A ($M(A) = 12$–27), the B-element has M-values of 71–79, compared with $M(Tl) = 78$. Only one structure contains 50% of B with $M(A) = 60$ and 25% Tl. All $tI80$-La$_6$Co$_{11}$Ga$_3$-type representatives contain 5% of Tl, assuming the role of the last Ga in the original notation of the structure type – $tI80$-La$_6$Co$_9$(Co$_{0.5}$Ga$_{0.5}$)$_4$Ga; the remaining elements have $M(A) = 28$–33 and $M(B) = 61$ or 64 (i.e., Fe and Co, respectively). Both, $tP2$-CuTi- and $cF16$-NaTl-type structures have rather diverse compositions, with Tl-contents of 25–50% and 12.5–25%, respectively. In the $oS16$-MgCuAl$_2$-type structures, 50% of Tl are combined with 25% each of $M(A) = 15$ or 18 and $M(B) = 68$ or 69 (Sr, Eu, and Pt, Pd, respectively). The $cI46$-K$_9$NaTl$_{13}$-type structures again all have

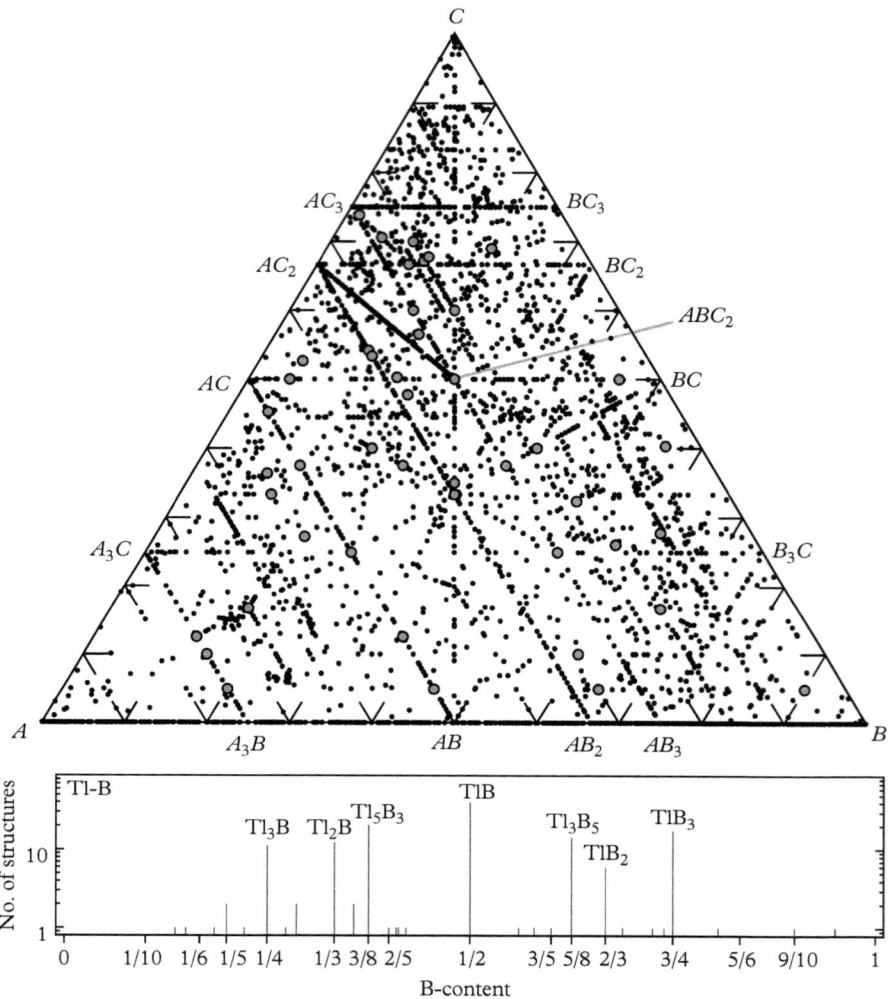

Fig. 7.66 *(top) Concentration diagram of the 20 829 intermetallic compounds (black dots) contained in* Pearson's Crystal Data *(PCD) (Villars and Cenzual, 2011a). The Tl-containing compounds are marked by large gray circles. A, B, and C stand for metallic elements with increasing Mendeleev numbers, $M(A) < M(B) < M(C)$. (bottom) Frequencies of the binary Tl–B compounds as a function of stoichiometry.*

the same stoichiometry with 56.5% Tl, as well as 26.1% and 17.3% of elements A and B ($M(A) = 8$–10, i.e., Cs, Rb, K, and $M(B) = 11$, i.e., Na).

In the following, we discuss some characteristic features of the four most common ternary structure types of ternary indides listed in Table 7.18. Their structures are shown in Fig. 7.68.

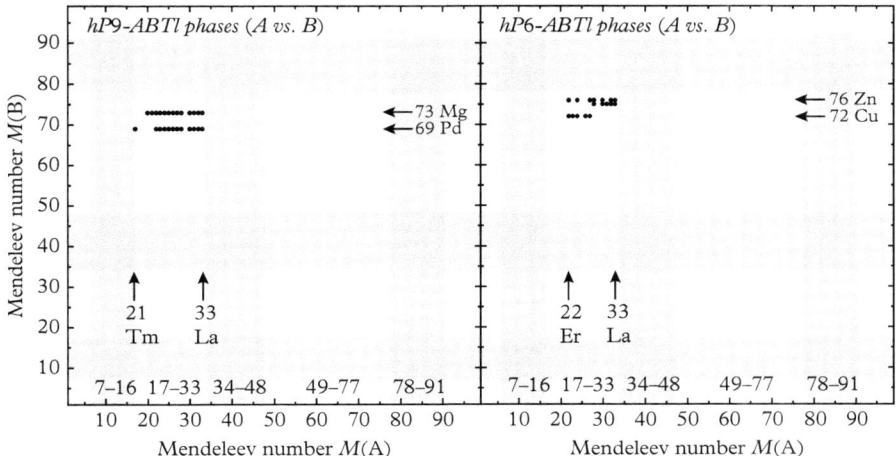

Fig. 7.67 *Occurrence of the two most common structure types among ternary Tl-intermetallics, hP9-ZrNiAl and hP6-CaIn₂. The Mendeleev numbers M of the other two elements in the compound are shown. Since* A *and* B *have equal shares in the chemical formulae, they are assigned so that* $M(A) < M(B)$.

- $hP9$-ZrNiAl (189 $P\bar{6}2m$): The structure can be subdivided into face-sharing tricapped trigonal Ni@Zr₆Al₃ prisms (see also Fig. 10.8 in Section 10.8).

- $tI80$-La₆Co₁₁Ga₃ (140 $I4/mcm$): The structure can be described as a stacking along [001] of layers of edge-sharing bicapped square antiprisms, Ga@La₁₀ , and layers of isolated Ga/Co@Co₁₂ icosahedra. The layers forming the La-square antiprisms can be described as a snub square tiling, $3^2.4.3.4$.

- $oS16$-MgCuAl₂ (63 $Cmcm$): Ordered variant of the $oS16$-Re₃B structure type. Mg in a partially side-capped pentagonal prism; Cu in a tricapped trigonal prism; Al in a distorted cuboctahedron (see also Fig. 7.55).

- $cI46$-K₉NaTl₁₃ (Cordier and Müller, 1994), which may have the actual composition $cI46$-K₆Na₄Tl₁₃ (Dong and Corbett, 1995) (204 $Im\bar{3}$): Zintl phase with naked polyanionic centered icosahedral Tl_{13}^{10-} and Tl_{13}^{11-} clusters in a *bcc* arrangement. The icosahedra are linked via K-dumbbells along the main axes, and via Na atoms, capping eight triangle faces, along the [110] directions.

7.15.5   Germanides

The structural chemistry of germanides is very rich, and also contains a variety of quaternary structure types with many representatives. We will show exemplarily representatives that can be derived from the $tP10$-RE₂InGe₂ structure type.

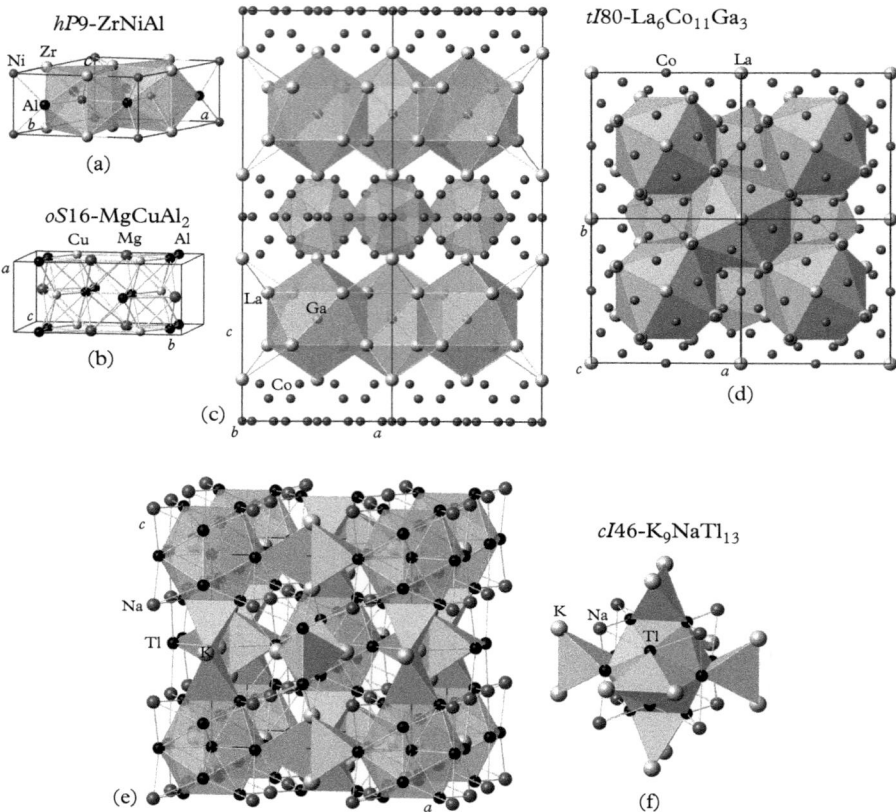

Fig. 7.68 *The structures of the four most frequent ternary structure types of ternary thallides. (a) hP9-ZrNiAl: The structure can be subdivided into face-sharing tricapped trigonal Ni@Zr$_6$Al$_3$ prisms (see also Fig. 10.8 in Section 10.8). oS16-MgCuAl$_2$ (b) can be described as a stacking along [100] of symmetrically equivalent copies of a triangle/pentagon/hexagon layer. The structure of (c)–(d) tI80-La$_6$Co$_{11}$Ga$_3$ can be described as a stacking of layers of edge-sharing bicapped square antiprisms, Ga@La$_{10}$, and layers of isolated Ga/Co@Co$_{12}$ icosahedra. The structure of cI46-K$_9$NaTl$_{13}$ (e)–(f) consists of a bcc arrangement of Tl-centered Tl-icosahedra, which are linked via K-dumbbells along the main axes. Eight triangle faces are capped by Na atoms connecting icosahedra along the [110] directions.*

Binary germanides

Among the binary IMs, 425 Ge-containing phases can be found, featuring 171 structure types; these are ≈ 2.5 representatives per structure type. Their compositions can be found in Fig. 7.69, while in Table 7.19 the most common structure types among binary Ge-IMs are given.

The compositions of the *cF*4-Cu- and *hP*2-Mg-type structures are rather diverse (containing 2.5–15.0% and 15–70% Ge, respectively). The same holds true

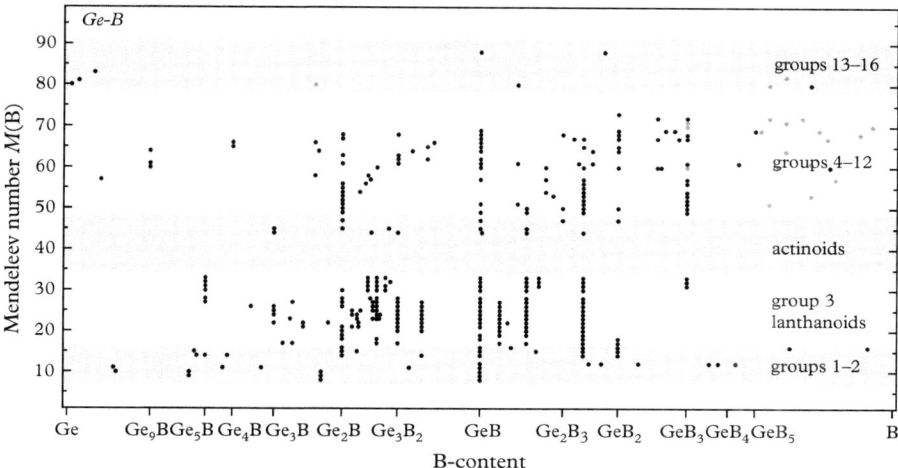

Fig. 7.69 *Stoichiometries of the 425 binary germanides, Ge_aB_b vs. the Mendeleev number $M(B)$ of the other element in the compound. The 20 binary phases adopting unary structure types cF4-Cu, cI2-W, and hP2-Mg are marked by gray dots.*

for $cF8$-C-type structures with 90.0–99.6% Ge. It should be mentioned here that $cF8$-Ge has the diamond structure, and it forms binary compounds with the $cF4$-Cu structure type with the following elements: Ag, Al, Au, Cu, Ni, Pb, Pd, and Pt, all of them with the $cF4$-Cu structure type themselves. These intermetallic phases can be considered as solid solutions since the atomic radii are also within the appropriate range, 1.25–1.44 Å, compared to $r_{Ge} = 1.22$. In contrast, the electronegativity of Ge, $\chi_{Ge} = 2.0$, is significantly outside of the range, 1.4–1.8, of most of the other constituents. The elements $hP2$-Co and $cI58$-Mn also form with $cF8$-Ge binary compounds with the $cF4$-Cu structure type, which cannot be classified as solid solutions in the same way, because none of the element structures are of the $cF4$-Cu type. Both elements, however, have a HT modification of this type in contrast to Ge. In the system Cu–Ge, for instance, the Cu-rich phase of the $cF4$-Cu type can be considered as a solid solution, but not the somewhat Ge-richer phase, which crystallizes in the $hP2$-Mg type. Whether this is really the case or just due to the used qualitative structure identification method (Debye-Scherrer camera, (Zhou, 1991)) remains an open question, and not only for this particular case.

Nearly all of the remaining structure types occur at their defined compositions. In some of them, Ge always plays the role of the majority element ($hP3$-AlB$_2$, $tI12$-ThSi$_2$, $oF64$-Y$_3$Ge$_5$, $oS28$-Er$_3$Ge$_4$, $oS32$-Gd$_3$Ge$_4$, $oS12$-ZrSi$_2$, $oS16$-DyGe$_3$, $oI12$-LaGe$_5$, and $oS44$-Nd$_4$Ge$_7$) or of the minority element ($hP16$-Mn$_5$Si$_3$, $oP36$-Sm$_5$Ge$_4$, $tI84$-Ho$_{11}$Ge$_{10}$, $tP32$-Ti$_3$P, $tI32$-W$_5$Si$_3$, $oP12$-Co$_2$Si, and $tI32$-Cr$_5$B$_3$). In $cP4$-Cu$_3$Au and $oP12$-PbCl$_2$, it appears to switch between those

Table 7.19 *Most common structure types of Ge-containing binary intermetallics. The top 24 structure types are given, all of which have at least five representative structures and therefore represent at least 1% of all binary Ge-intermetallics each.*

Rank	Structure type	No.	Space group	Wyckoff positions	No. of reps.	% of all reps.
1.	$hP16$-Mn_5Si_3	193	$P6_3/mcm$	$4d\,6g^2$	24	5.6%
2.	$oP36$-Sm_5Ge_4	62	$Pnma$	$4c^3\,8d^3$	16	3.8%
3.	$oS8$-TlI	63	$Cmcm$	$4c^2$	15	3.5%
4.	$hP3$-AlB_2	191	$P6/mmm$	$1a\,2d$	13	3.1%
5.	$tI12$-$ThSi_2$	141	$I4_1/amd$	$1a2d$	13	3.1%
6.	$cF4$-Cu	225	$Fm\bar{3}m$	$4a$	10	2.4%
7.	$tI84$-$Ho_{11}Ge_{10}$	139	$I4/mmm$	$4de^2\,8h^2j\,16mn^2$	10	2.4%
8.	$oF64$-Y_3Ge_5	43	$Fdd2$	$8a^2\,16b^3$	10	2.4%
9.	$oS28$-Er_3Ge_4	63	$Cmcm$	$4c^2\,8f^3$	8	1.9%
10.	$cF8$-C	227	$Fd\bar{3}m$	$8a$	7	1.6%
11.	$oS32$-Gd_3Ge_4	71	$Immm$	$4ac^2\,8f^2$	7	1.6%
12.	$tP32$-Ti_3P	86	$P4_2/n$	$8g^4$	7	1.6%
13.	$tI32$-W_5Si_3	140	$I4/mcm$	$4ab\,8h\,16k$	7	1.6%
14.	$oS12$-$ZrSi_2$	63	$Cmcm$	$4c^3$	7	1.6%
15.	$oP12$-Co_2Si	62	$Pnma$	$4c^3$	6	1.4%
16.	$tI32$-Cr_5B_3	140	$I4/mcm$	$4ac\,8h\,16l$	6	1.4%
17.	$cP4$-Cu_3Au	221	$Pm\bar{3}m$	$1a\,3c$	6	1.4%
18.	$oS16$-$DyGe_3$	63	$Cmcm$	$4c^4$	6	1.4%
19.	$cP8$-FeSi	198	$P2_13$	$4a^2$	6	1.4%
20.	$oI12$-$LaGe_5$	71	$Immm$	$2ad\,8l$	6	1.4%
21.	$hP2$-Mg	194	$P6_3/mmc$	$2c$	6	1.4%
22.	$oP8$-FeAs	62	$Pnma$	$4c^2$	5	1.2%
23.	$oS44$-Nd_4Ge_7	30	$C222_1$	$4ab^2\,8c^4$	5	1.2%
24.	$oP12$-$PbCl_2$	62	$Pnma$	$4c^3$	5	1.2%
					211	49.6%

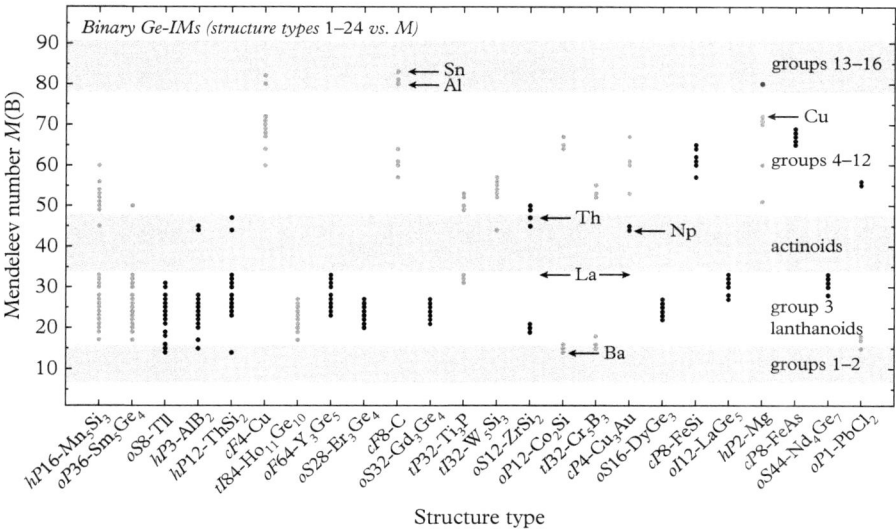

Fig. 7.70 *Occurrence of the 24 most common structure types among binary* Ge-*intermetallics. The Mendeleev number M of the other element in the compound is shown over the rank of the structure type, as given in Table 7.19. Structures with* Ge *as their major component (≥ 50%) are marked in black, those with minor* Ge-*content are shown in gray.*

roles. Obviously, no such statement can be made for 1:1-stoichiometries, as in $oS8$-TlI, $cP8$-FeSi, and $oP8$-FeAs. An overview over the respective intermetallic systems is shown in Fig. 7.70.

In the following, we discuss some characteristic features of the five most frequent binary structure types of binary germanides listed in Table 7.19. Their structures are shown in Fig. 7.71.

- $hP16$-Mn_5Si_3 (193 $P6_3/mcm$): This structure type, one of the Nowotny phases, can be described as a packing of face-sharing CN16 FK-polyhedra, leaving empty spaces corresponding to columns of face-sharing octahedra running along [001] (also see Subsection 7.15.9 and Fig. 7.54). Ge always occupies the Si positions. As constituting elements all lanthanoids, except Eu and Pm, have been identified, as well as the actinoid U, and also the group 4 and 5 elements Zr, Hf, Ti, Ta, Nb, and V, as well as Mo and Mn for germanides of this type.

- $oP36$-Sm_5Ge_4 (62 $Pnma$): The pseudo-tetragonal structure can be described as a stacking of layers of face-sharing, Ge-capped, body-centered Sm-cubes, where the squares are part of a snub-square tiling, $3^2.4.3.4$. The shortest Ge–Ge distances between Ge atoms within the layers amount to 2.657 Å (2.450 Å in the element structure). These polyhedra-layers are

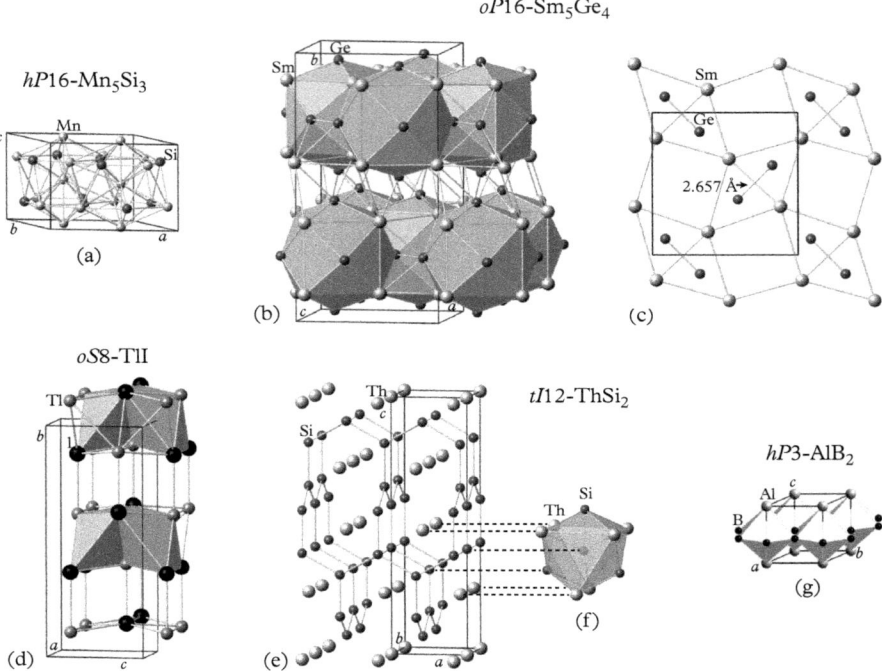

Fig. 7.71 *The structures of the four most frequent binary structure types of binary germanides. (a) The unit cell of hP16-Mn$_5$Si$_3$ contains two face-sharing CN16 FK-polyhedra (also see Fig. 7.54). (b) The structure of oP36-Sm$_5$Ge$_4$ can be described as a stacking of layers of face-sharing, Ge-capped, body-centered Sm-cubes, where the squares are part of a snub-square tiling, $3^2.4.3.4$ (c). In (d) the structure of oS8-Tll is shown with the layers of face-sharing octahedra highlighted (also see Fig. 7.54). The structure of tl12-ThSi$_2$ is depicted in (e), with the fundamental structural unit, a Si-tricapped trigonal prism of Th atoms, shown in polyhedral form (f). For a detailed discussion of the structure of hP3-AlB$_2$ (g) see Section 7.7.*

shifted against each other and stacked along [010], with Sm–Ge distances between 2.9 and 3.0 Å. Representatives of this structure type are known for the lanthanoids (except Eu and Pm) and Hf.

- oS8-Tll (63 *Cmcm*): This structure can be seen as a stacking along [001] of two symmetrically equivalent distorted honeycomb layers that are shifted against each other; thereby, layers are formed of face-sharing octahedra alternating with layers of face-sharing trigonal prisms (also see Fig. 7.54). This structure type is known for germanides based on the alkaline earth element Ba, Sr, and Ca, as well as on all lanthanoids except Yb, Lu, Pm, Ce, and La.

- hP3-AlB$_2$ (191 *P6/mmm*): For a detailed discussion see Section 7.7. The structure can be considered as a stacking of honeycomb nets with Al at the vertices. B sits in the center of the hexagonal prisms formed by the Al atoms of adjacent layers. This structure type is known to be adopted by Sr, Ca, and all lanthanoids but Yb, Lu, Pm, Ce, and La.

- $tI12$-ThSi$_2$ (141 $I4_1/amd$): The Si atoms form a 3D 3-connected network with the large Th atoms in the large holes, and the Si atoms in trigonal Th prisms. The Si–Si distances are between 2.382 Å and 2.396 Å (2.352 in the element), that of Si–Th with 3.155 Å significantly larger than the sum of their atomic radii. Germanides with this structure type are known to constitute Ba, Ho, Dy, Y, Tb, Gd, Sm, Nd, Pr, Ce, and La, as well as Np and Th.

Ternary germanides

The 2434 ternary Ge-containing IMs feature 420 different structure types, i.e., ≈ 5.8 representatives per structure type. In Table 7.20, the most common structure types among ternary Ge-IMs are given. In the ternary compounds, Ge is most often the element with the highest M-value (84) and will therefore be assigned the C-position, while the A-elements will strictly have smaller M-values than B-elements in all formulae $A_a B_b Ge_c$—$M(A) < M(B)$.

Nearly all ternary Ge-intermetallics with $tI10$-CeAl$_2$Ga$_2$-type structures are stoichiometric with 40% of Ge, as well as 20% of an element A with $M(A) = 14$–47 and 40% of element B with $M(B) = 57$–80. Most $oP12$-TiNiSi-type structures have 1:1:1-stoichiometry and the A- and B-elements have values of $M(A) = 15$–60 and $M(B) = 56$–75; three more structures are found in the Li/Ba/Sr–Ca–Ge systems ($M(A) = 12, 14, 15$, and $M(B) = 16$) and two structures with a significantly lower Ge-content ($\approx 7\%$) are found in the Zr/Hf–Sb–Sn systems ($M(A) = 49, 50$ and $M(B) = 88$). All $oS16$-CeNiSi$_2$-type structures have Ge-contents in the range 50–65.6%, as well as A-contents of 25–33.3% and B-contents of 3.8–25% with elements $M(A) = 15$–47 and $M(B) = 57$–80. The $hP3$-AlB$_2$-type structures contain 12.7–60% Ge, as well as 9.9–40% (although mostly 33%) of element A with $M(A) = 14$–47 and 4.8–54% of element B with $M(B) = 22, 24, 33, 51$, or 61–81. Most $oS18$-CuCe$_2$Ge$_6$-type structures have ideal stoichiometries with 66.7% Ge, as well as 22.2% of element A ($M(A) = 17$–33) and 11.1% of element B ($M(A) = 60$–72).

Most $hP9$-ZrNiAl-type structures are 1:1:1-stoichiometric and contain elements with values $M(A) = 12$–60 and $M(B) = 17$–76. Nearly all $cP40$-Yb$_3$Rh$_4$Sn$_{13}$-type representatives contain 65% Sn in addition to 15% of element A ($M(A) = 16$–32) and 20% of element B ($M(B) = 62$–66). The $tP6$-PbClF-type structures all have 1:1:1 stoichiometry and feature elements $M(A) = 11$–60 and $M(B) = 51$–88. All representatives of the $oI32$-YIrGe$_2$ structure type contain 50% Ge, while the A- and B-elements are both featured with 25% and have values of $M(A) = 16$–45 and $M(B) = 65$–69. Most of the $oI12$-KHg$_2$-type structures have 1:1:1 stoichiometry and contain elements with $M(A) = 15$–45 and $M(B) = 62$–76. All $hP13$-MgFe$_6$Ge$_6$-type structures contain 46.2 Ge in addition to 7.7% of element A ($M(A) = 12$–64) and 46.2% of element B ($M(B) = 57$–73).

Table 7.20 *Most common structure types of Ge-containing ternary intermetallics. The top 22 structure types are given, all of which have at least 24 representative structures and therefore represent at least 1% of all binary Ge-intermetallics. In sum, the 1109 representatives of these structure types make up 45.6% of all ternary Ge-intermetallics.*

Rank	Structure type	No.	Space group	Wyckoff positions	No. of reps.	% of all reps.
1.	$tI10$-CeAl$_2$Ga$_2$	139	$I4/mmm$	$2a\,4de$	186	7.6%
2.	$oP12$-TiNiSi	62	$Pnma$	$2ad\,6gh$	133	5.5%
3.	$oS16$-CeNiSi$_2$	63	$Cmcm$	$4c^4$	97	4.0%
4.	$hP3$-AlB$_2$	191	$P6/mmm$	$1a\,2d$	82	3.4%
5.	$oS18$-CuCe$_2$Ge$_6$	38	$Amm2$	$2a^5b^4$	73	3.0%
6.	$hP9$-ZrNiAl	189	$P\bar{6}2m$	$1a2d3fg$	52	2.1%
7.	$cP40$-Yb$_3$Rh$_4$Sn$_{13}$	223	$Pm\bar{3}n$	$2a\,6c\,8e\,24k$	47	1.9%
8.	$tP6$-PbClF	129	$P4/nmm$	$2ac^2$	35	1.4%
9.	$oI32$-YIrGe$_2$	71	$Immm$	$4gi^2j\,8l^2$	35	1.4%
10.	$oI12$-KHg$_2$	74	$Imma$	$4e\,8i$	32	1.3%
11.	$hP13$-MgFe$_6$Ge$_6$	191	$P6/mmm$	$1a\,2cde\,6i$	32	1.3%
12.	$hP16$-Mn$_5$Si$_3$	193	$P6_3/mcm$	$4d\,6g^2$	32	1.3%
13.	$cI34$-U$_4$Re$_7$Si$_6$	229	$Im\bar{3}m$	$2a\,8c\,12de$	32	1.3%
14.	$oI40$-U$_2$Co$_3$Si$_5$	72	$Ibam$	$4ab, 8gj^3$	31	1.3%
15.	$oI22$-Cu$_4$Gd$_3$Ge$_4$	71	$Immm$	$2a\,4hij\,8l$	30	1.2%
16.	$tI10$-BaNiSn$_3$	107	$I4mm$	$2a^3\,4b$	28	1.2%
17.	$cF16$-Cu$_2$MnAl	225	$Fm\bar{3}m$	$4abcd$	26	1.1%
18.	$mS20$-Sc$_2$CoSi$_2$	12	$C2/m$	$4i^5$	26	1.1%
19.	$tP38$-Sc$_5$Co$_4$Si$_{10}$	127	$P4/mbm$	$2a\,4gh^2\,8i^2j$	26	1.1%
20.	$hP8$-Y$_{0.5}$Co$_3$Ge$_3$	191	$P6/mmm$	$1a\,2ce\,3g$	25	1.0%
21.	$oP48$-ZrCrSi$_2$	55	$Pbam$	$4e^4h^2\,8i^3$	25	1.0%
22.	$cP54$-Na$_4$Si$_{23}$	223	$Pm\bar{3}n$	$2a\,6cd\,16i\,24k$	24	1.0%
					1109	45.6%

The $hP16$-Mn$_5$Si$_3$-type structures contain 5.0–37.5% Ge and are formed with additional elements A (M(A) = 18–60) and B (M(B) = 24–88) with contributions of 12.5–62.5% and 2.5–50.0%, respectively. The $cI34$-U$_4$Re$_7$Si$_6$-type structures again are all stoichiometric with 35.4% Ge, as well as 23.5% A (M(A) = 17–60) and 41.2% B (M(B) = 59–66).

The structures of type $oI40$-$U_2Co_3Si_5$ mostly contain 50% Ge, 20% of element A, and 30% of element B, with values $M(A) = 21$–33 and $M(B) = 62$–69. The representatives of the $oI22$-$Cu_4Gd_3Ge_4$ structure type all contain 36.4% Ge, as well as mostly 27.3% of element A ($M(A) = 21$–33) and again 36.4% of element B ($M(B) = 60$–72). All $tI10$-$BaNiSn_3$-type structures are stoichiometric with 60% Ge and 20% each of element A and B (with $M(A) = 14$–33 and $M(B) = 61$–69). The structures of type $cF16$-Cu_2MnAl all contain 25% Ge, but the division of the remaining formula into A- and B-elements varies. Only one compound is non-stoichiometric (Fe–Co–Ge system), seven contain 50% of element A and 25% of element B (with $M(A) = 12$, 64, or 67 and $M(B) = 73$–77), and 18 contain 25% of element A and 50% of element B (with $M(A) = 12$, 19, or 51–61 and $M(B) = 16$, 60–72, or 80).

The $mS20$-Sc_2CoSi_2-type compounds all contain 40% of both, Ge and element B ($M(B) = 62$–66), as well as 20% of element A with $M(A) = 17$–33. The $tP38$-$Sc_5Co_4Si_{10}$-type structures contain 52.6% of Ge, 26.3% of element A with $M(A) = 17$–27 and 21.1% of element B with $M(B) = 63$–66. The $hP8$-$Y_{0.5}Co_3Ge_3$-type structures contain 46.2–47.3% Ge and most are composed of 6.5–7.7% of element A ($M(A) = 17$–31) and 45.8–46.7% of element B ($M(B) = 60$–64), while one compound has reversed roles with 7.6% Mg ($M(B) = 73$) and 46.2% of Ni ($M(A) = 67$). All $oP48$-$ZrCrSi_2$-type structures have the prototype stoichiometry with 50% Ge and 25% of elements A and B, each, with $M(A) = 15$–51 and $M(B) = 57$–65. The $cP54$-Na_4Si_{23}-type representatives contain 42.6–83.4% Ge, as well as 14.8–19.3% of element A ($M(A) = 9$–15) and 0.9–42.6% of element B ($M(B) = 60$–81).

In the following, we discuss some characteristic features of the five most frequent ternary structure types of the ternary germanides listed in Table 7.20. Remarkably, three of the five structure types are Ce-compounds. Their structures are shown in Fig. 7.72.

- $tI10$-$CeAl_2Ga_2$ (139 $I4/mmm$): This structure type is an ordered derivative of the $tI10$-$BaAl_4$ type (see also Fig. 10.8 in Section 10.8). Ce sits in 4-capped hexagonal prisms ($CeAl_8Ga_8$), while Ga centers tricapped trigonal prisms ($GaAl_4Ce_4$).

- $oP12$-$TiNiSi$ (62 $Pnma$): The structure contains a 3D 4-connected network of Ti atoms ($d_{Ti-Ti} = 3.15$–3.22 Å), where the large channels are filled with zigzag-bands of edge-sharing Ni_2Si_2 rhomb units. It can be also described as a distorted $hP6$-Ni_2In type structure.

- $oS16$-$CeNiSi_2$ (63 $Cmcm$): This structure type and that of $oS18$-$CuCe_2Ge_6$ have some similarities such as their layer structure. The Si–Ni bonds with 2.31 Å are rather short indicating stronger bonding. Si–Si distances in the zigzag chains with 2.48 Å are longer than in the element (2.35 Å).

- $oS18$-$CuCe_2Ge_6$ (38 $Amm2$): In contrast to the previous structure, this structure type shows Ge–Ge double layers. One forms empty square prisms

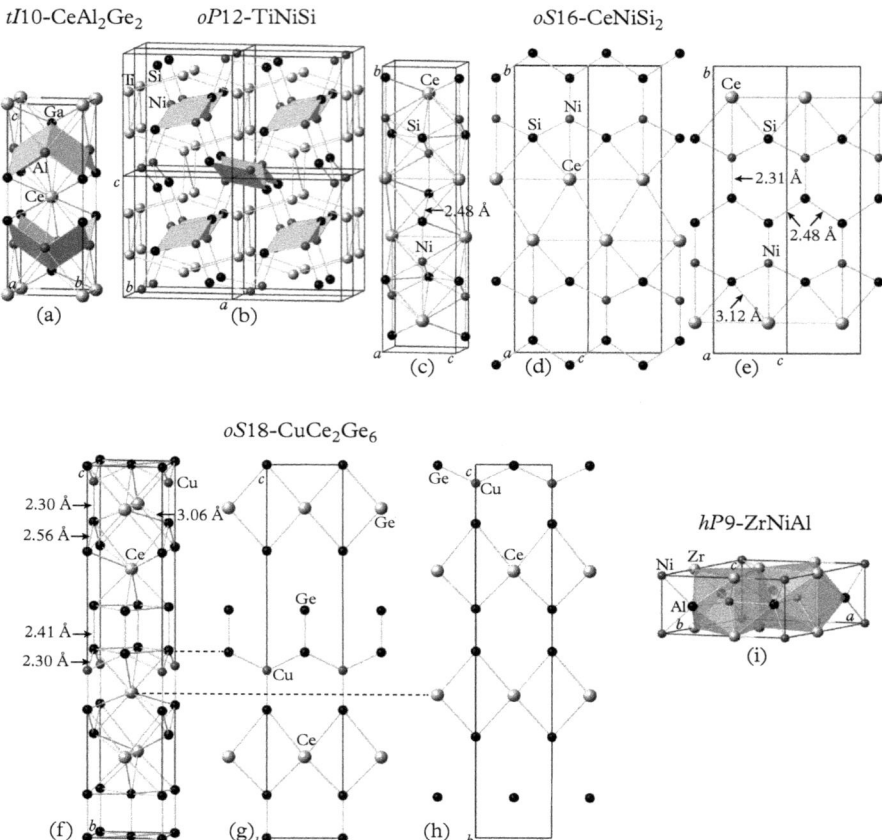

Fig. 7.72 *The structures of the five most frequent ternary structure types of ternary germanides. (a) tI10-CeAl₂Ga₂, with the edge-sharing Al₂Ga₂ rhomb units shaded gray. The structure of oP12-TiNiSi is shown in two different projections (b)–(c). The bands of edge-sharing Ni₂Si₂ rhomb units are shaded gray. One unit cell (c) of the structure of oS16-CeNiSi₂, and the constituting layers in x = 0 and x = 1/2, respectively. The structure of oS18-CuCe₂Ge₆ in perspective view (f) and the constituting layers in y = 0 and y = 1/2, respectively. (i) One unit cell of hP9-ZrNiAl in perspective view.*

with 2.93 Å edge length of the squares and 2.41–2.48 Å height; the other one is symmetrically equivalent under the base-centering symmetry operation.

- *hP*9-ZrNiAl (189 $P\bar{6}2m$): The structure can be subdivided into face-sharing tricapped trigonal Ni@Zr₆Al₃ prisms (also see Fig. 10.8 in Section 10.8).

Quaternary germanides

There is a significant number of known quaternary germanides. We present here examples of two related quaternary rare-earth germanides, $mS22$-RE₄M₂InGe₄

Table 7.21 *Occurence of quaternary germanides of the type* $RE_4M_2InGe_4$ *and* $RE_4RhInGe_4$ *(Oliynyk et al., 2015)*

Compound	La	Ce	Pr	Nd	Sm	Gd	Tb	Dy	Ho	Er	Tm	Lu
$RE_4Mn_2InGe_4$	+	+	+	+	+	+	+	+	+	+	+	+
$RE_4Fe_2InGe_4$	–	+	+	+	+	+	+	+	+	+	+	+
$RE_4Ru_2InGe_4$	–	+	+	+	+	+	+	+	+	+	+	+
$RE_4Co_2InGe_4$	–	+	+	+	+	+	+	+	+	+	+	+
$RE_4Rh_2InGe_4$	–	+	+	+	+	+	–	–	–	–	–	–
$RE_4Ir_2InGe_4$	+	+	+	+	–	–	–	–	–	–	–	–
$RE_4Ni_2InGe_4$	–	–	–	–	+	+	+	+	+	+	+	–
$RE_4RhInGe_4$	–	–	–	–	–	–	+	+	+	+	–	–

(M = Fe, Co, Ni, Ru, Rh, and Ir), $mS22$-$Ho_4Ni_2InGe_4$ structure type, and $mS40$-$RE_4RhInGe_4$ (own type), because they have, with more than sixty, a comparably large number of representatives (see Table 7.21), and nicely illustrate the well-ordered structural assembly of four types of elements (Oliynyk *et al.*, 2015).

The structures of $mS22$-$Ho_4Ni_2InGe_4$ and $mS40$-$Tb_4RhInGe_4$ as well as their relationship to the structure of $tP10$-Mo_2FeB_2 (127 $P4/mbm$) (see also Subsubsection 7.15.3 and Fig. 7.55) are shown in Fig. 7.73. In sits in cuboctahedra, which are slightly elongated along [010].

In both germanides, the constituting atomic layers are built up by alternating strips of a square/triangle tiling and of a Cairo pentagonal tiling. The square/triangle tiling can be cut out of the $3^2.4.3.4$ tiling characteristic for the structure of $tP10$-Mo_2FeB_2 (127 $P4/mbm$). Depending on the direction of the cut, either the structure motifs needed for $mS22$-$Ho_4Ni_2InGe_4$ or $mS40$-$Tb_4RhInGe_4$ are obtained. The Cairo pentagonal tiling is the dual tiling to the $3^2.4.3.4$ square/triangle tiling.

- $mS22$-$Ho_4Ni_2InGe_4$ (12 $C2/m$): Within the layers, In has only Ge neighbors, with distances In–Ge ranging from 2.84 to 3.08 Å, which is close to the sum of radii (2.85 Å), although Ge and In themselves are immiscible and do not form any compounds. However, the studies by Oliynyk *et al.* (2015) show weak covalent bonding between Ge atoms. This coordination is further stabilized via the Ho atoms in the squares below and above In, which form a cuboctahedron together with the Ge atoms. The distances In–Ho range from 3.32 to 3.38 Å (3.37 Å), those of Ge–Ho range from 2.89 to 3.12 Å (2.97 Å). The Ho–Ho distances range from 3.61 to 3.75 Å, significantly larger than the sum of radii, 3.49 Å. Along [010], i.e., between the layers, the shortest Ho–Ho distance with 4.21 Å is even larger. Within the layers, there are also dumbbells of Ge atoms with the shortest Ge–Ge distance in this

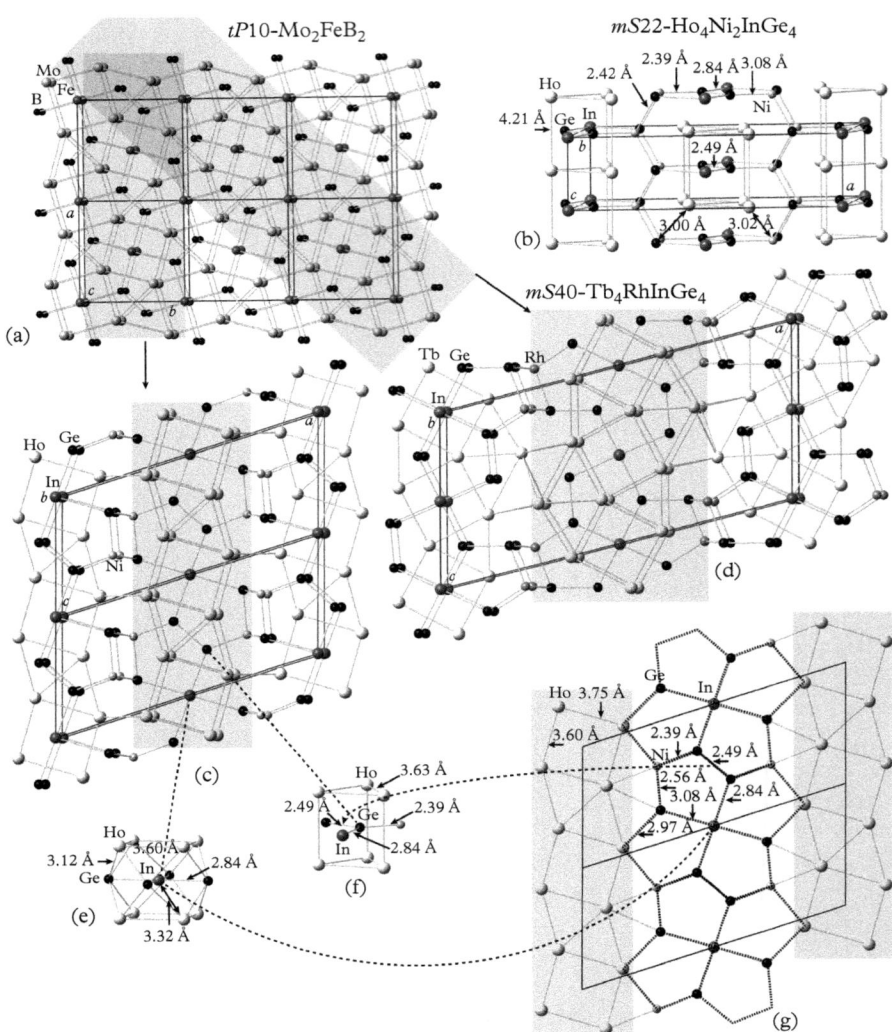

Fig. 7.73 *Structure motifs (marked by gray-shaded rectangles) of the (a) tP10-Mo$_2$FeB$_2$ (127 P4/mbm) structure type can be found in the structures of (b, c, g) mS22-Ho$_4$Ni$_2$InGe$_4$ and (d) mS40-Tb$_4$RhInGe$_4$. The In atoms are coordinated by distorted Ho$_8$Ge$_4$ cuboctahedra (e), and half of the Ge atoms by tricapped trigonal prisms, Ho$_6$GeNiIn, (f). The shortest Ge–Ge distance of 2.49 Å can be found between the Ge atoms centering neighboring trigonal prisms. In (g), one of the two symmetrically equivalent flat atomic layers is shown, which constitute the structure of mS22-Ho$_4$Ni$_2$InGe$_4$.*

compound, 2.49 Å (sum of radii 2.45 Å) resulting from covalent bonding contributions. Ni has two Ge and two Ho atoms as neighbors, one Ge with the very short distance of 2.39 Å (2.47 Å), which indicates a strong covalent bonding contribution, and one with a longer one of 2.56 Å. The distances to the Ho atoms correspond with 2.97–3.03 Å to the sum of radii, 2.99 Å.

* $mS40$-Tb$_4$RhInGe$_4$ (12 $C2/m$): In the structure of this compound, the square/triangle and Cairo pentagon tiling strips are connected in a different way. This also leads to different atomic distances. The shortest Ge–Ge distances in this compound are with 2.60–2.62 Å (sum of radii 2.45 Å) significantly larger than in $mS22$-Ho$_4$Ni$_2$InGe$_4$. This may be caused by the larger atomic radii of Rh (1.35 Å) and Tb (1.76 Å) compared to Ni (1.25 Å) and Ho (1.74 Å). Like Ni, Rh has also two Ge and two Ho atoms as neighbors, both Ge with the short distances of 2.54–2.55 Å (2.57 Å). The distances to the Tb atoms with 3.08 and 3.21 Å are close to the sum of radii, 3.11 Å. The shortest Tb–Tb distance amounts to 3.54 Å (3.52 Å).

7.15.6 Stannides

Binary stannides

Among the binary IMs, 381 containing Sn can be found, featuring 144 structure types, i.e., ≈ 2.7 representatives per structure type. This is slightly more than in the case of binary germanides (≈ 2.5). Their compositions can be found in Fig. 7.74, while in Table 7.22, the most common structure types among binary Sn-IMs are given.

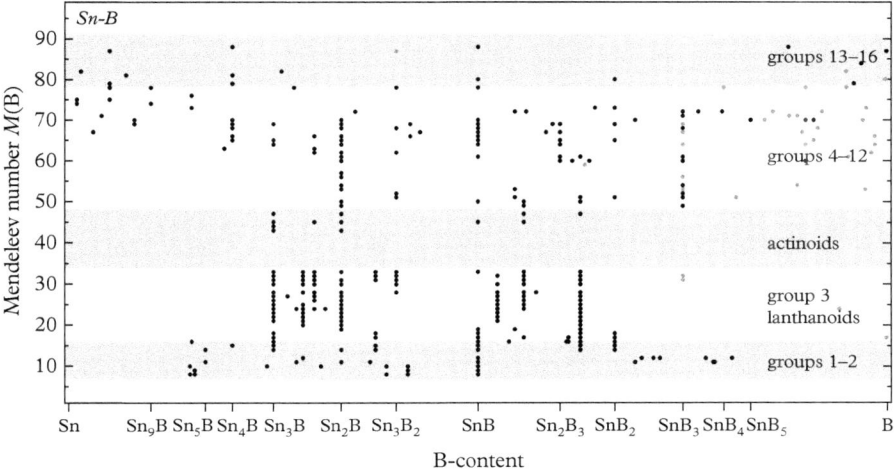

Fig. 7.74 *Stoichiometries of the 381 binary stannides,* Sn$_a$B$_b$ *vs. the Mendeleev number* M(B) *of the other element in the compound. The 34 binary phases adopting unary structure types cF4-Cu, cI2-W, and hP2-Mg are marked by gray dots.*

Table 7.22 *Most common structure types of* Sn-*containing binary intermetallics. The top 30 structure types are given, all of which have at least four representative structures and therefore represent at least 1% of all binary* Sn-*intermetallics each.*

Rank	Structure type	No.	Space group	Wyckoff positions	No. of reps.	% of all reps.
1.	$cP4$-Cu_3Au	221	$Pm\bar{3}m$	$1a\,3c$	23	6.0%
2.	$hP16$-Mn_5Si_3	193	$P6_3/mcm$	$4d\,6g^2$	19	5.0%
3.	$cF4$-Cu	225	$Fm\bar{3}m$	$4a$	16	4.2%
4.	$hP1$-$Hg_{0.1}Sn_{0.9}$	191	$P6/mmm$	$1a$	14	3.7%
5.	$tI84$-$Ho_{11}Ge_{10}$	139	$I4/mmm$	$4de^2\,8h^2j\,16mn^2$	10	2.6%
6.	$hP2$-Mg	194	$P6_3/mmc$	$2c$	10	2.6%
7.	$oP36$-Sm_5Ge_4	62	$Pnma$	$4c^3\,8d^3$	10	2.6%
8.	$hP6$-$Co_{1.75}Ge$	194	$P6_3/mmc$	$2acd$	9	2.4%
9.	$tI4$-Sn	141	$I4_1/amd$	$4a$	9	2.4%
10.	$oS12$-$ZrSi_2$	63	$Cmcm$	$4c^3$	9	2.4%
11.	$cI2$-W	229	$Im\bar{3}m$	$2a$	8	2.1%
12.	$oS16$-$GdSn_{2.75}$	38	$Amm2$	$2a^4b^4$	7	1.8%
13.	$oS32$-Pu_3Pd_5	63	$Cmcm$	$4c^2\,8efg$	7	1.8%
14.	$oP12$-Co_2Si	62	$Pnma$	$4c^3$	6	1.6%
15.	$oP14$-Er_2Ge_5	59	$Pmmn$	$2a^4b^3$	6	1.6%
16.	$hP8$-Mg_3Cd	194	$P6_3/mmc$	$2d\,6h$	6	1.6%
17.	$oS28$-Tb_3Sn_7	65	$Cmmm$	$2ac\,4i^4j^2$	6	1.6%
18.	$oS28$-Ce_2Sn_5	65	$Cmmm$	$2ac\,4i^3j^3$	5	1.3%
19.	$cP8$-Cr_3Si	223	$Pm\bar{3}n$	$2a\,6c$	5	1.3%
20.	$tI12$-$CuAl_2$	69	$Fmmm$	$4a\,8h$	5	1.3%
21.	$aP20$-Nd_2Sn_3	147	$P\bar{1}$	$2i^{10}$	5	1.3%
22.	$hP4$-$NiAs$	194	$P6_3/mmc$	$2ac$	5	1.3%
23.	$oS8$-TlI	63	$Cmcm$	$4c^2$	5	1.3%
24.	$oS12$-$ZrGa_2$	65	$Cmmm$	$2ac\,4ij$	5	1.3%
25.	$cF12$-CaF_2	225	$Fm\bar{3}m$	$4a\,8c$	4	1.0%
26.	$tI32$-Cr_5B_3	140	$I4/mcm$	$4ac\,8h\,16l$	4	1.0%
27.	$cP2$-$CsCl$	221	$Pm\bar{3}m$	$1ab$	4	1.0%
28.	$oF48$-Mg_2Cu	70	$Fddd$	$16fg^2$	4	1.0%
29.	$tI64$-$NaPb$	142	$I4_1/acd$	$16ef\,32g$	4	1.0%
30.	$hP18$-Ti_5Ga_4	193	$P6_3/mcm$	$2b\,4d\,6g^2$	4	1.0%
					234	61.4%

The compositions of the $cF4$-Cu-, $hP2$-Mg-, and $cI2$-W-type structures are rather diverse (containing 0.1–25.0%, 2.0–37.0%, and 2.7–60.0% Sn, respectively). The same holds true for $hP1$-$Hg_{0.1}Sn_{0.9}$- and $tI4$-Sn-type structures with 50.0–96.0% and 80.0–99.0% Sn, respectively. Nearly all of the remaining structure types occur at their defined compositions. In some of them, Sn always plays the role of the majority element ($oS12$-$ZrSi_2$, $oS16$-$GdSn_{2.75}$, $oS32$-Pu_3Pd_5, $oP14$-Er_2Ge_5, $oS28$-Tb_3Sn_7, $oS28$-Ce_2Sn_5, $tI12$-$CuAl_2$, $aP20$-Nd_2Sn_3, $oS12$-$ZrGa_2$, and $oF48$-Mg_2Cu) or of the minority element ($hP16$-Mn_5Si_3, $tI84$-$Ho_{11}Ge_{10}$, $oP36$-Sm_5Ge_4, $hP6$-$Co_{1.75}Ge$, $oP12$-Co_2Si, $cP8$-Cr_3Si, $tI32$-Cr_5B_3, and $hP18$-Ti_5Ga_4). In $cP4$-Cu_3Au, $hP8$-Mg_3Cd, and $cF12$-CaF_2, it appears to switch between those roles. Obviously, no such statement can be made for 1:1-stoichiometries, as in $hP4$-NiAs, $oS8$-TlI, $cP2$-CsCl, and $tI64$-NaPb. An overview over the respective intermetallic systems is shown in Fig. 7.75.

In the following, we discuss some characteristic features of the five most frequent binary structure types of binary stannnides listed in Table 7.22. Their structures are shown in Fig. 7.76.

- $cP4$-Cu_3Au (221 $Pm\bar{3}m$): This structure type is an ordered derivative of the $cF4$-Cu type (see also Section 7.2 and Fig. 7.3). This basic structure type is adopted by stannides of the minority elements Ca, Y, and the lanthanoids except Lu, Tm, and Pm, the actinoids Pu, Np, U, and Th, as well as the TM

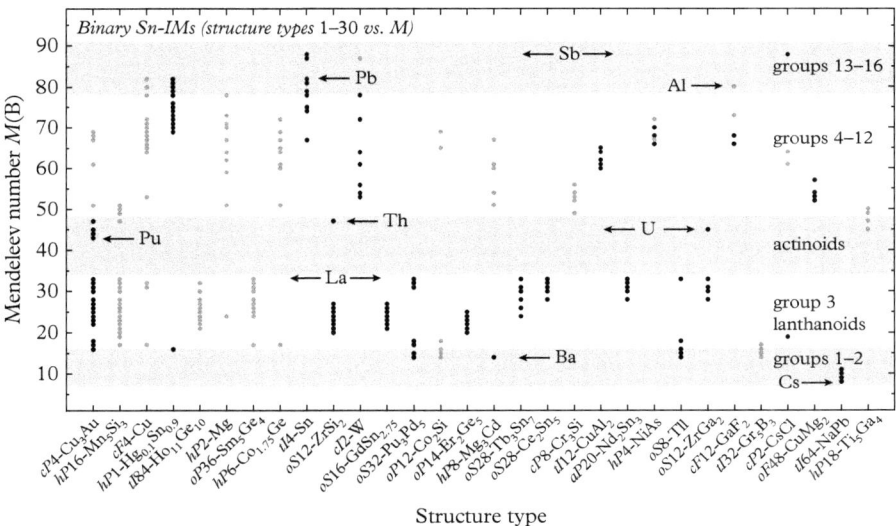

Fig. 7.75 *Occurrence of the 30 most common structure types among binary* Sn-*intermetallics. The Mendeleev number M of the other element in the compound is displayed over the rank of the structure type, as given in Table 7.22. Structures with Sn as their major component ($\geq$ 50%) are marked in black, those with minor Sn-content are shown in gray.*

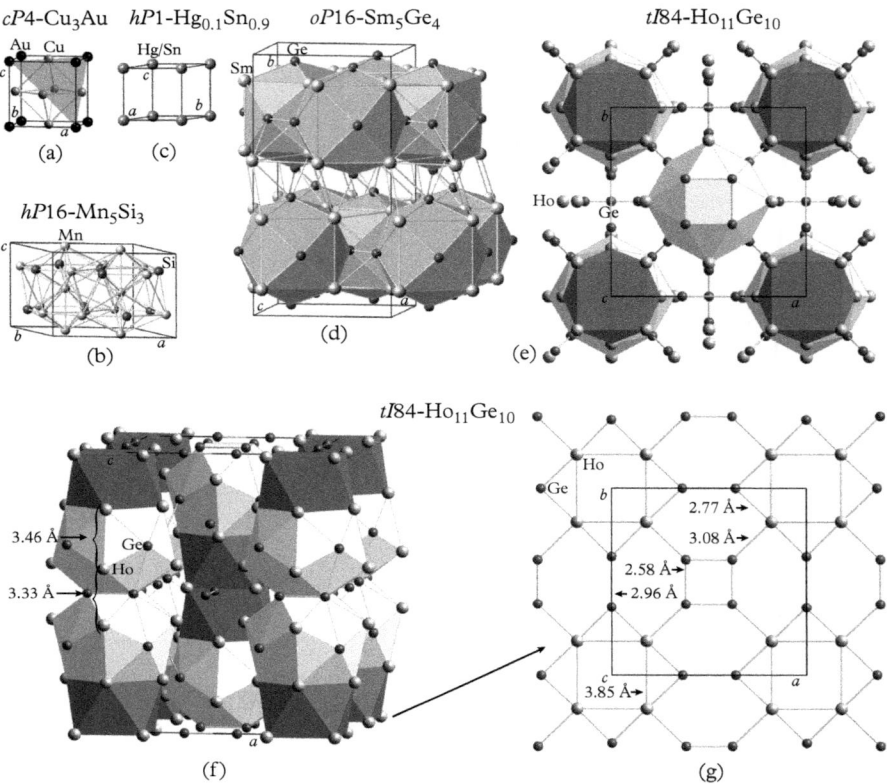

Fig. 7.76 *The structures of the five most frequent binary structure types of binary stannides.*
(a) cP4-Cu₃Au (also see Section 7.2 and Fig. 7.3). (b) The unit cell of hP16-Mn₅Si₃ contains
two face-sharing CN16 FK-polyhedra (also see Fig. 7.54). (c) The structure of
hP1-Hg₀.₁Sn₀.₉ (hP1-HgSn₉) is the simplest hexagonal phase possible with statistical Hg/Sn
distribution. The structure of oP36-Sm₅Ge₄ (d) can be described as a stacking of layers of
face-sharing, Ge-capped, body-centered Sm-cubes, where the squares are part of a snub-square
tiling, 3².4.3.4. (e)–(g) tI84-Ho₁₁Ge₁₀ in different representations. The atomic layer shown in
(g) is located in z = 0.

elements Ti, Fe, Ni, Pt, and Pd, which correspond to the majority elements
in this structure type.

- *hP16-Mn₅Si₃* (193 *P6₃/mcm*): This structure type, one of the Nowotny
  phases, can be described as a packing of face-sharing CN16 FK-polyhedra,
  leaving empty spaces corresponding to columns of face-sharing octahedra
  running along [001] (also see Subsection 7.15.9 and Fig. 7.54). The stannides of this structure type are known to contain Y and the lanthanoids
  except Eu and Pm, Th Zr, Hf, and Ti. All of them can occupy the Mn
  positions, only.

- $hP1$-$Hg_{0.1}Sn_{0.9}$ ($hP1$-$HgSn_9$) (191 $P6/mmm$): Alloys of Cd and Hg, with In and Sn that have this structure type are stable for valence electron concentrations between 3.80 and 3.95 (Che *et al.*, 1991). The axial ratio c/a decreases with increasing electron concentration. The structure differs from both that of $hR1$-Hg and of $tI4$-Sn. This structure can be formed from stannides of the elements Ca, Pd, Au, Ag, Cu, Mg, Hg, Cd, Zn, Tl, In, Al, Ga, and Pb, which are located at the Hg sites.

- $tI84$-$Ho_{11}Ge_{10}$ (139 $I4/mmm$): The rather complex structure contains Ge squares with a 2.58 Å edge length and Ge dumbbells with a rather large 2.96 Å distance. The Ho atoms form square antiprisms with a 3.85 Å edge length of the squares and 3.72 Å of the other edges. Together with some larger Ho and Ge squares, columnar structure units can be defined running along [001]. The shortest Ho–Ho distance belong with 3.33 Å, compared to the sum of atomic radii 3.48 Å, to the shortest ones known. This structure type is known to be formed from the lanthanoids Y, Tm, Er, Ho, Dy, Tb, Gd, Sm, Nd, and Ce, all occupying the Ho site in this structure type.

- $oP36$-Sm_5Ge_4 (62 $Pnma$): The pseudo-tetragonal structure can be described as a stacking of layers of face-sharing, Ge-capped, body-centered Sm-cubes, where the squares are part of a snub-square tiling, $3^2.4.3.4$. For a more detailed description see Subsection 7.15.5 and Fig. 7.71. Stannides of this structure type are known to form with Y and the lanthanoids Yb, Dy, Tb, Gd, Sm, Nd, Pr, Ce, and La, all of them sitting on the Sm site.

As another example, we briefly dicuss the triclinic structure type of $aP20$-Nd_2Sn_3, ranked 21 in Table 7.22, because it has such a low symmetry and shows some interesting features (Fig. 7.77). Topologically, it can be described as a layer structure based on a single type of atomic layer. It is oriented with respect to the conventional crystallographic unit cell such that it has a periodicity of $1a \times 2b \times 3c$. It exhibits inversion centers in every square of the band running through the decagons (Fig. 7.77 (e)). The individual layers are also related by inversion symmetry with these symmetry elements located in the origin and the body center of the triclinic unit cell as well as on the edge and face centers.

The structure of the layer can be described as triangle/square tiling, where every other triangle originating from the band of Sn squares, i.e., those consisting of Sn atoms only, is opened up forming squashed pentagons. This creates the band of overlapping decagons shown in (Fig. 7.77 (e)). The Sn-square in the center of the decagons forms with Sn atoms from the two neighboring layers of octahedra, the faces of which are capped by Nd (Fig. 7.77 (c)). This arrangement corresponds to one unit cell of the $cP4$-Cu_3Au structure type (see Fig. 7.3). The framework of Sn–Sn connected pseudo-cubes also constitutes the structures of $oP20$-Ce_3Sn_7 and $oP28$-Ce_2Sn_5 (Boucherle *et al.*, 1988). The distorted pentagon/triangle band as well as that of the mixed-atom squares is part of the structure type $oC12$-$ZrSi_2$, which is also adopted by a series of rare earth stannides (Fornasini *et al.*, 2003).

$aP20\text{-Nd}_2\text{Sn}_3$

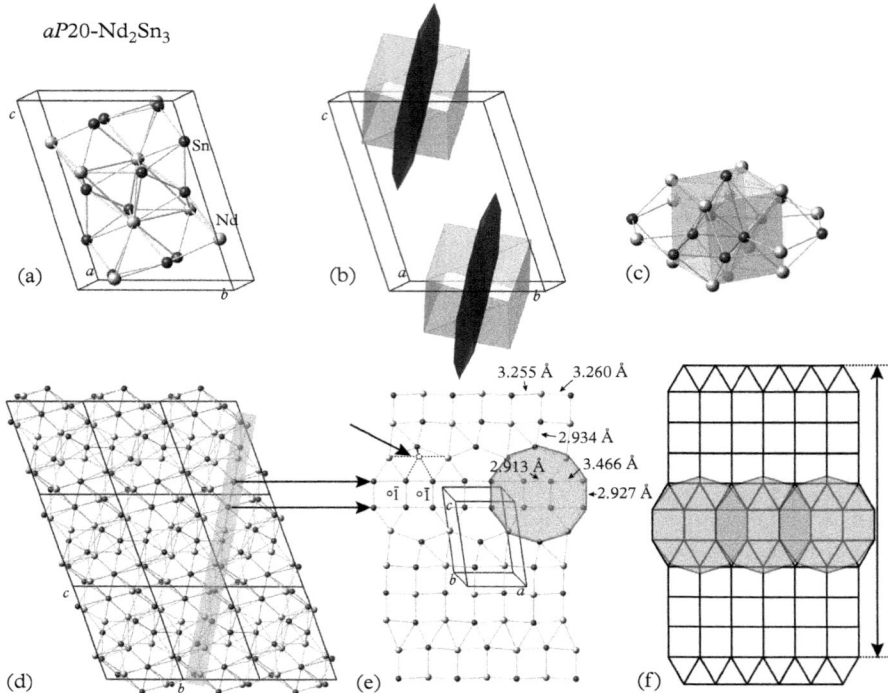

Fig. 7.77 *The structure of aP20-Nd$_2$Sn$_3$ in different views. (a) One unit cell as a ball-and-stick model and (b) with pseudo-cubic unit cells and decagonal structural units, respectively, in polyhedral representation. (c) Single Nd-pseudo-cube centered by a Sn-octahedron, which is located within the decagonal structure unit. Along [100] projected structure (d) with one layer shaded gray, which is shown in (e). The inversion centers within the layer are located in every square inside the decagons. (f) Schematical representation of one period of the layer.*

Ternary stannides

The 1470 ternary Sn-containing IMs feature 328 different structure types, i.e., ≈ 4.5 representatives per structure type. In Table 7.23, the most common structure types among ternary Sn-IMs are given, and the compositions of the six most common ones can be found in Fig. 7.78. In the ternary compounds, Sn is most often the element with the highest M-value (83) and will therefore be assigned the C-position, while the A-elements will strictly have smaller M-values than B-elements in all formulae $A_aB_bSn_c$, $M(A) < M(B)$. In the following, we discuss some characteristic features of the five most frequent ternary structure types of ternary stannides listed in Table 7.23. Their structures are shown in Fig. 7.79.

- $oS16\text{-CeNiSi}_2$ (63 *Cmcm*): This structure type can be described as a layer structure. The Si–Ni bonds with 2.31 Å are rather short indicating stronger

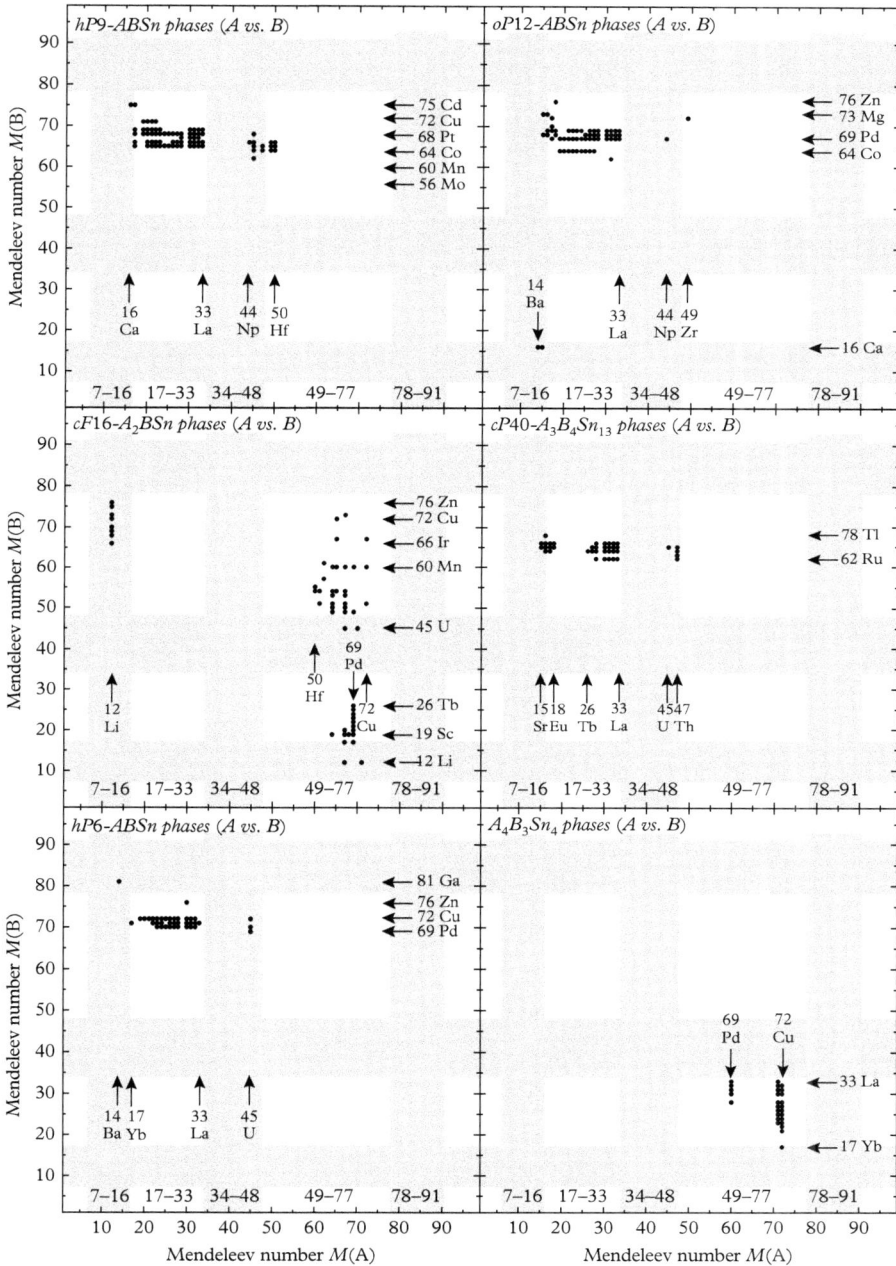

Fig. 7.78 *Occurrence of six of the most common structure types among ternary Sn-intermetallics. The Mendeleev numbers M of the other two elements in the compound are shown. The depicted structure types are hP9-ZrNiAl, oP12-TiNiSi, cF16-Cu$_2$MnAl, cP40-Yb$_3$Rh$_4$Sn$_{13}$, hP6-CaIn$_2$, and oI22-Cu$_4$Gd$_3$Ge$_4$ (ranks 2–6, and 9, respectively, in Table 7.23). Where A and B have equal shares in the chemical formula, they are assigned so that M(A) < M(B). These are the most common structure types, whose sites are occupied in an ordered manner for the most part.*

Table 7.23 *Most common structure types of Sn-containing ternary intermetallics. The top 22 structure types are given, all of which have at least 16 representative structures and therefore represent at least 1% of all ternary Sn-intermetallics.*

Rank	Structure type	No.	Space group	Wyckoff positions	No. of reps.	% of all reps.
1.	$oS16$-CeNiSi$_2$	63	$Cmcm$	$4c^4$	88	6.0%
2.	$hP9$-ZrNiAl	189	$P\bar{6}2m$	$1a\,2d\,3fg$	75	5.1%
3.	$oP12$-TiNiSi	62	$Pnma$	$4c^3$	59	4.0%
4.	$cF16$-Cu$_2$MnAl	225	$Fm\bar{3}m$	$4abcd$	57	3.9%
5.	$cP40$-Yb$_3$Rh$_4$Sn$_{13}$	223	$Pm\bar{3}n$	$2a\,6c\,8e\,24k$	39	2.7%
6.	$hP6$-CaIn$_2$	194	$P6_3/mmc$	$2b\,4f$	38	2.6%
7.	$cP4$-Cu$_3$Au	221	$Pm\bar{3}m$	$1a\,3c$	38	2.6%
8.	$cF12$-MgAgAs	216	$F\bar{4}3m$	$4abc$	32	2.2%
9.	$oI22$-Cu$_4$Gd$_3$Ge$_4$	71	$Immm$	$2a\,4hij\,8l$	27	1.8%
10.	$hP6$-LiGaGe	186	$P6_3mc$	$2ab^2$	24	1.6%
11.	$hP18$-CuHf$_5$Sn$_3$	193	$P6_3/mcm$	$2b\,4d\,6g^2$	21	1.4%
12.	$tP10$-CaBe$_2$Ge$_2$	129	$P4/nmm$	$2abc^3$	19	1.3%
13.	$oI36$-Ho$_6$Co$_2$Ga	71	$Immm$	$2ac\,4gj\,8lmn$	18	1.2%
14.	$hP13$-MgFe$_6$Ge$_6$	191	$P6/mmm$	$1a\,2cde\,6i$	18	1.2%
15.	$cP8$-Cr$_3$Si	223	$Pm\bar{3}n$	$2a\,6c$	17	1.2%
16.	$hP30$-Lu$_3$Co$_{7.77}$Sn$_4$	186	$P6_3mc$	$2ab^2\,6c^4$	17	1.2%
17.	$tP10$-Mo$_2$FeB$_2$	127	$P4/mbm$	$2a\,4gh$	17	1.2%
18.	$hP28$-CeNi$_5$Sn	194	$P6_3/mmc$	$2abcd\,4f^2\,12k$	16	1.1%
19.	$oP28$-Cu$_5$AuCe	62	$Pnma$	$4c^5\,8d$	16	1.1%
20.	$hP6$-NdPtSb	186	$P6_3mc$	$2ab^2$	16	1.1%
21.	$cF116$-Tb$_5$Rh$_6$Sn$_{17}$	216	$F\bar{4}3m$	$4acd\,16e^2\,24f\,48h$	16	1.1%
22.	$hP12$-YPtAs	194	$P6_3/mmc$	$2ab\,4f^2$	16	1.1%
					684	46.5%

bonding. Si–Si distances in the zigzag chains with 2.48 Å are longer than in the element (2.35 Å) (also see Fig. 7.72).

- $hP9$-ZrNiAl (189 $P\bar{6}2m$): The structure can be subdivided into face-sharing tricapped trigonal Ni@Zr$_6$Al$_3$ prisms (also see Fig. 10.8 in Section 10.8).

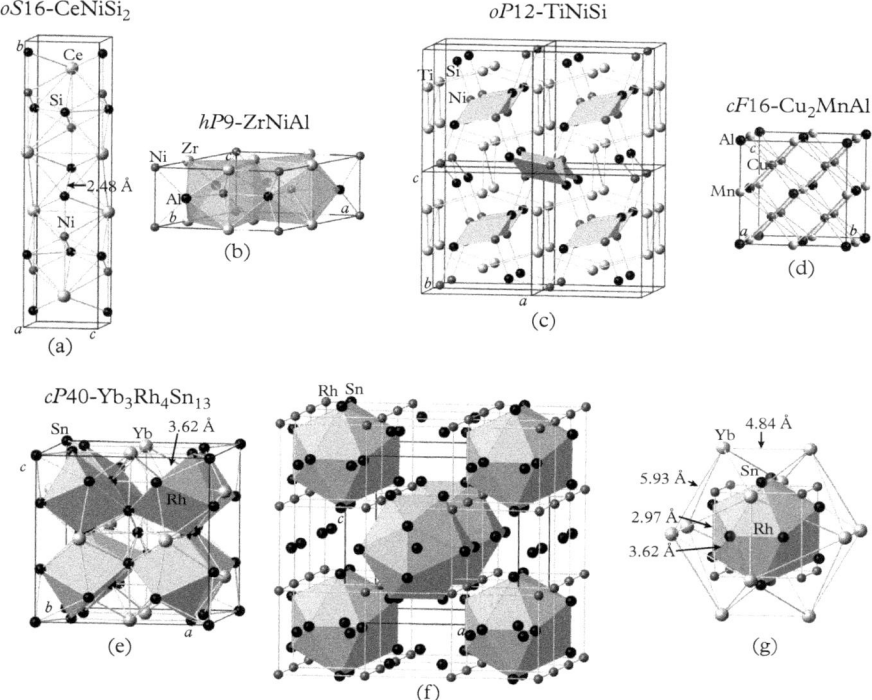

oS16-CeNiSi$_2$

hP9-ZrNiAl

oP12-TiNiSi

cF16-Cu$_2$MnAl

cP40-Yb$_3$Rh$_4$Sn$_{13}$

Fig. 7.79 *The structures of the five most frequent ternary structure types of ternary stannides. (a) The structure of oS16-CeNiSi$_2$ can be described as a layer structure. The Si–Ni bonds with 2.31 Å are rather short indicating stronger bonding. Si–Si distances in the zigzag chains with 2.48 Å are longer than in the element (2.35 Å). (b) The unit cell of hP9-ZrNiAl with face-sharing tricapped trigonal Ni@Zr$_6$Al$_3$ prisms shaded. (c) The structure of oP12-TiNiSi contains a 3D 4-connected network of Ti atoms, with zigzag-bands of edge-sharing Ni$_2$Si$_2$ rhomb units in the large channels. The structure of cF16-Cu$_2$MnAl (d) can be described as a (2 × 2 × 2)-fold superstructure of the cI2-W type. The unit cell of cP40-Yb$_3$Rh$_4$Sn$_{13}$ is shown in (e), with the vertex-sharing trigonal Sn-prisms marked. The prisms are tricapped by Yb atoms. The arrangement of Sn-icosahedra centered at the Sn atoms in 0,0,0 and 1/2,1/2,1/2 and of the Rh-cubes is depicted in (f). A single structural subunit Sn@Sn$_{12}$Rh$_8$Yb$_{12}$ is shown in (g).*

- oP12-TiNiSi (62 *Pnma*): The structure contains a 3D 4-connected network of Ti atoms ($d_{\text{Ti-Ti}} = 3.15$–3.22 Å), where the large channels are filled with zigzag-bands of edge-sharing Ni$_2$Si$_2$ rhomb units. It can be also described as a distorted hP6-Ni$_2$In type structure (see also Fig. 7.72).

- cF16-Cu$_2$MnAl (225 *Fm$\bar{3}$m*): This structure type of the Heusler phases can be described as a (2 × 2 × 2)-fold superstructure of the cI2-W type, and is discussed in Subsection 7.3.2.

- $cP40$-$Yb_3Rh_4Sn_{13}$ (223 $Pm\bar{3}n$): The Yb atoms are coordinated by distorted Sn-cuboctahedra (capped with four Rh and two Yb atoms), and the Rh atoms by tricapped trigonal prisms Sn_6Yb_3. The Yb atoms form a distorted, Sn-centered, icosahedron ($d_{Yb\text{-}Yb} = 4.84$–5.93 Å, $d_{Sn\text{-}Yb} = 5.41$ Å), with the Rh atoms, in a cube-arrangement ($d_{Sn\text{-}Rh} = 5.41$ Å), centering eight triangle faces. A smaller Sn-icosahedron ($d_{Sn\text{-}Sn} = 2.97$–3.62 Å) is nested inside the Yb-icosahedron, capping the remaining 12 triangle faces from the inside ($d_{Sn\text{-}Yb} = 3.40$ Å).

The representatives of the $oS16$-$CeNiSi_2$-type among ternary Sn-intermetallics contain 50–66.2 % Ga, whereas only 19 out of 88 compounds exhibit the correct 1:1:2 stoichiometry. Of these, 12 are $LiRESn_2$ with rare earth elements $M(B) = 20$–33, wheres all other compounds are composed of A- and B-elements with $M(A) = 14$–47 and $M(B) = 60$–72 and contents of 25–66.2 % and 0.7–25%, respectively. All $hP9$-$ZrNiAl$-representatives, on the other hand, have 1:1:1-stoichiometry and contain elements $M(A) = 16$–50 and $M(B) = 62$–75. The same holds true for $oP12$-$TiNiSi$-type structures with $M(A) = 14$–49 and $M(B) = 62$–76, as well as two compounds in the Sr/Ba–Ca–Sn-systems with $M(B) = 16$. Among the 57 $cF16$-Cu_2MnAl-type structures, only three are not stoichiometric (in systems Li–Ti/Sb–Sn, as well as Ti–Ir–Sn), the rest of which all contain 25% Sn. The rest of the formula mostly contains 25% of A with $M(A) = 12$–61 or 67 and 50% of B with $M(B) = 60$–72. Eight compounds are formed with 50% of Li as the A-element and 25% of $M(B) = 66$–76 and three more contain 50% of $M(A) = 65$ or 67 (i.e., Rh and Ni, respectively) and 25% of $M(B) = 67$–73. All $cP40$-$Yb_3Rh_4Sn_{13}$-type structures are stoichiometric, containing 65% Sn, as well as 15% $M(A) = 15$–47 and 20% $M(B) = 62$–78.

The ternary $hP6$-$CaIn_2$-type Sn-compounds all have 1:1:1 stoichiometry and contain elements with $M(A) = 14$–45 and $M(B) = 69$–81. The $cP4$-Cu_3Au-type structures contain 3.0–75% Sn; it is found mostly in the role of Cu (34 out of 38 compounds), often shared with the B-element $M(B) = 69$–82 and opposite either only element A, or both, A and B, with values $M = 16$–47. Four compounds contain 75% of either element Ni or combined Pt and Pd, while then Sn makes up for 25% together with Al, Ga, or Ge ($M(B) = 80$–84), or by itself. Most $cF12$-$MgAgAs$-type structures are 1:1:1-stoichiometric, with three exceptions containing 25–40% Sn, and are forming with elements $M(A) = 12$–72 and $M(B) = 61$–79. The structures with $oI22$-$Cu_4Gd_3Ge_4$-type all have an ideal stoichiometry with 27.3% of element A ($M(A) = 14$–45) and 36.4% of both, element B ($M(B) = 60$–72) and Sn. The $hP6$-$LiGaGe$-type compounds, again, all have stoichiometry 1:1:1 and—in addition to Sn—contain metals with $M(A) = 16$–51 and $M(B) = 69$–76. $hP18$-$CuHf_5Sn_3$-type structures are mostly stoichiometric with 33.3% of Sn, as well as 55.6% of element A with $M(A) = 17$–51 and 11.1% of B with $M(B) = 51$, 60–72, or 82. In one case, Sn switched roles with the A-element U (33.3%) and makes up for 55.6% of the formula, in addition to the B-element Ti with 11.1%. Most $tP10$-$CaBe_2Ge_2$-type compounds contain 20%

of element A (M(A) = 14–33) and 40% each of element B (M(B) = 65–76) and Sn. One case with slightly shifted stoichiometry also contains U (M = 45) as element A and Co as B (M = 64).

The oI36-Ho$_6$Co$_2$Ga-type structures all have ideal stoichiometry, constituted from 66.7% of element A with M(A) = 20–30, 22.2% of Co or Ni (M(B) = 64 and 67, respectively), and 11.1% Sn. hP13-MgFe$_6$Ge$_6$-type compounds contain 46.2% of Sn in addition to 7.7% of element A (M(A) = 12–50) and 46.2% of element B (M(B) = 53, 60, or 61, i.e., Nb, Mn, and Fe, respectively), with one structure with reversed A- and B-roles with elements 46.2% Mn and Mg (with M(B) = 73 and 7.7% content). The structures found to form type cP8-Cr$_3$Si contain either 75% of Nb or V (M = 53 and 54, respectively) as element A or the same portion divided between elements A and B with M = 49–61. The remaining 25% of the formula are made up of either element B and Sn combined, with M(B) = 79–88, or of Sn only. The Sn-content varies in the range of 12.5–25%. The hP30-Lu$_3$Co$_{7.77}$Sn$_4$-type structures mostly have the correct stoichiometry and therefore contain 26.7% of Sn, as well as 20% of element A (M(A) = 16–31) and 53.3% of element B (M(B) = 64, 67, or 72, i.e., Co, Ni, or Cu). The structures in type tP10-Mo$_2$FeB$_2$ (127 $P4/mbm$) contain 20% of Sn, as well as 40% of each elements A and B with M(A) = 17–50 and M(B) = 61–70. The Sn-content in hP28-CeNi$_5$Sn-type structures is 14.3–20.8%, while elements A (M(A) = 17–45) and B (M(B) = 67 or 72, i.e., Ni and Cu, respectively) make up for 13.0–14.3% and 66.0–71.4% of the formula, respectively. The oP28-Cu$_5$AuCe-type representatives contain 14.3% of both, Sn and element A (M(A) = 17–47) and 71.4% of element B, which is mostly Ni but in two instances also Fe and Ni (M(B) = 72, 61, and 67, respectively). The structures of type hP6-NdPtSb have 1:1:1-stoichiometry and contain elements with M(A) = 17–32 and M(B) = 70–76 in addition to Sn. The Sn-content in structures of type cF116-Tb$_5$Rh$_6$Sn$_{17}$ lies between 51.0–63.4%; elements A (M(A) = 17–27) and B (M(B) = 63–66) have contents of 15.9–21.3% and 19.3–29.4%, respectively. Lastly, the hP12-YPtAs-type structures again all have 1:1:1-stoichiometry and are built from elements M(A) = 15–33 and Zn or Ga (M(B) = 76 or 81, respectively).

7.15.7 Plumbides

Binary plumbides

Among the binary IMs, 168 Pb-containing IMs can be found, featuring 64 structure types, i.e., ≈ 2.6 representatives per structure type, i.e., comparable to the case of germanides (≈ 2.5). Their compositions can be found in Fig. 7.80, while in Table 7.24, the most common structure types among binary Pb-IMs are given.

The compositions of the cF4-Cu-, hP2-Mg-, and cI2-W-type structures are rather diverse (containing 2–99.9%, 6–80%, and 3–50% Pb, respectively). Nearly all of the remaining structure types occur at their defined compositions. In some of them, Pb always plays the role of the majority element (tI12-CuAl$_2$ and tI6-CuZr$_2$) or of the minority element (hP16-Mn$_5$Si$_3$, oP36-Sm$_5$Ge$_4$,

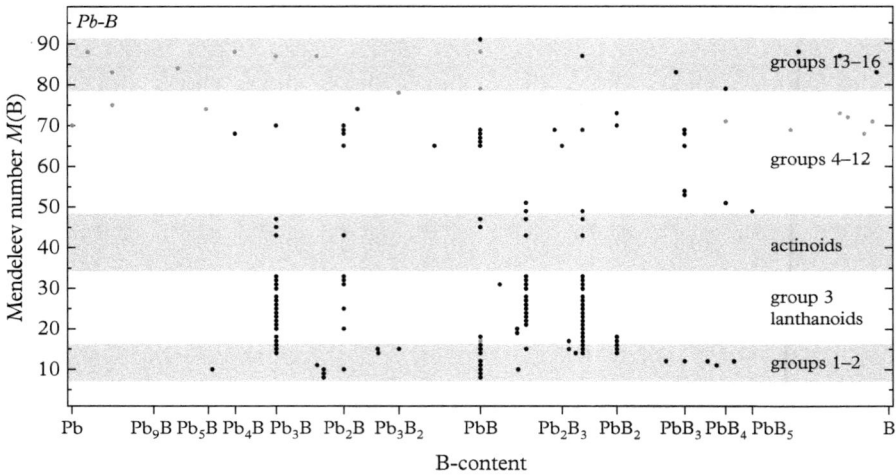

Fig. 7.80 *Stoichiometries of the 168 binary plumbides, Pb_aB_b vs. the Mendeleev number M(B) of the other element in the compound. The 20 binary phases adopting unary structure types cF4-Cu, cI2-W, and hP2-Mg are marked by gray dots.*

Table 7.24 *Most common structure types of Pb-containing binary intermetallics. The top 15 structure types are given, all of which have at least three representative structures and therefore represent more than 1.5% of all binary Pb-intermetallics each.*

Rank	Structure type	No.	Space group	Wyckoff positions	No. of reps.	% of all reps.
1.	$cP4$-Cu_3Au	221	$Pm\bar{3}m$	$1a\,3c$	24	14.3%
2.	$hP16$-Mn_5Si_3	193	$P6_3/mcm$	$4d\,6g^2$	17	10.1%
3.	$cF4$-Cu	225	$Fm\bar{3}m$	$4a$	13	7.7%
4.	$oP36$-Sm_5Ge_4	62	$Pnma$	$4c^3\,8d^3$	12	7.1%
5.	$oP12$-Co_2Si	62	$Pnma$	$4c^3$	5	3.0%
6.	$tI12$-$CuAl_2$	69	$Fmmm$	$4a\,8h$	4	2.4%
7.	$tP2$-CuAu	123	$P4/mmm$	$1ad$	4	2.4%
8.	$tI6$-$CuZr_2$	139	$I4/mmm$	$2a\,4e$	4	2.4%
9.	$hP2$-Mg	194	$P6_3/mmc$	$2c$	4	2.4%
10.	$tI64$-NaPb	142	$I4_1/acd$	$16ef\,32g$	4	2.4%
11.	$cP8$-Cr_3Si	223	$Pm\bar{3}n$	$2a\,6c$	3	1.8%
12.	$hP4$-NiAs	194	$P6_3/mmc$	$2ac$	3	1.8%
13.	$hP18$-Ti_5Ga_4	193	$P6_3/mcm$	$2b\,4d\,6g^2$	3	1.8%
14.	$cI2$-W	229	$Im\bar{3}m$	$2a$	3	1.8%
15.	$tI32$-W_5Si_3	140	$I4/mcm$	$4ab\,8h\,16k$	3	1.8%
					106	63.1%

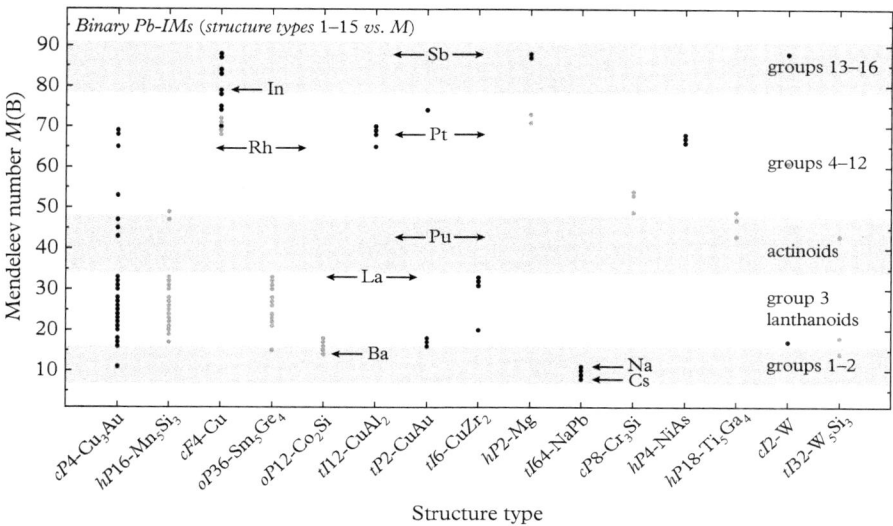

Fig. 7.81 *Occurrence of the 15 most common structure types among binary* Pb-*intermetallics. The Mendeleev number M of the other element in the compound is shown over the rank of the structure type, as given in Table 7.24. Structures with* Pb *as their major component (≥ 50%) are marked in black, those with minor* Pb-*content are shown in gray color.*

$oP12$-Co_2Si, $cP8$-Cr_3Si, $hP18$-Ti_5Ga_4, and $tI32$-W_5Si_3). In others, it appears to switch between those roles ($cP4$-Cu_3Au). Obviously, this is different for 1:1-stoichiometries, as in $tP2$-CuAu (one structure containing 65% Pb vs. 35% Hg), $tI64$-NaPb, or $hP4$-NiAs. An overview of the respective intermetallic systems is shown in Fig. 7.81.

In the following, we discuss some characteristic features of the five most common binary structure types of binary plumbides listed in Table 7.24. Their structures are shown in Fig. 7.82.

- $cP4$-Cu_3Au (221 $Pm\bar{3}m$): This structure type is an ordered derivative of the $cF4$-Cu type (also see Section 7.2 and Fig. 7.3). This close packed structure type is adopted by plumbides with the elements Na, Ca, and Y, all lanthanoids but Pm, the actinoids Pu, U, and Th, as well as Nb, Rh, Pt, and Pd, all of them occupying the Au position.

- $hP16$-Mn_5Si_3 (193 $P6_3/mcm$): This structure type, one of the Nowotny phases, can be described as a packing of face-sharing CN16 FK-polyhedra, leaving empty spaces corresponding to columns of face-sharing octahedra running along [001] (also see Subsection 7.15.9 and Fig. 7.54). The Mn site of this structure type can be occupied in stannides by Sc, Y, Yb, Lu, Tm, Er, Ho, Dy, Tb, Gd, Sm, Nd, Pr, Ce, and La.

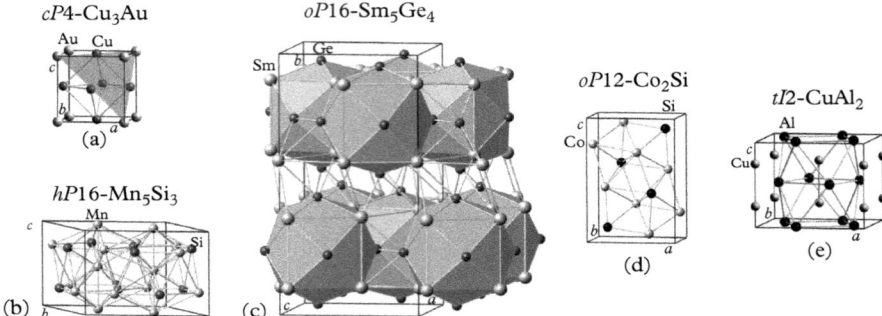

Fig. 7.82 *The structures of the four most frequent binary structure types of binary plumbides. (a) cP4-Cu₃Au (also see Section 7.2 and Fig. 7.3). (b) The unit cell of hP16-Mn₅Si₃ contains two face-sharing CN16 FK-polyhedra (also see Fig. 7.54). (c) The structure of oP36-Sm₅Ge₄ can be described as a stacking of layers of face-sharing, Ge-capped, body-centered Sm-cubes, where the squares are part of a snub-square tiling, 3².4.3.4. (d) oP12-Co₂S: This structure type consists of one type of flat atomic layers (distorted square/triangle net) stacked along [010] in a way to form three different types of octahedra between the layers, with the space in-between filled by tetrahedra (see Fig. 7.48). (e) The structure of tI12-CuAl₂ can be described as a stacking of 3².4.3.4 triangle/square nets located in z = 0 and 1/2, with the squares capped by Cu in z = 1/4 and 3/4.*

- *oP36-Sm₅Ge₄* (62 *Pnma*): The pseudo-tetragonal structure can be described as a stacking of layers of face-sharing, Ge-capped, body-centered Sm-cubes, where the squares are part of a snub-square tiling, 3².4.3.4. For a more detailed description see Subsection 7.15.5 and Fig. 7.71. In plumbides of this structure type, the Sm sites can be occupied by Sr, Tm, Er, Ho, Dy, Tb, Gd, Sm, Nd, Pr, Ce, and La.

- *oP12-Co₂Si* (62 *Pnma*): This structure type consists of one type of flat atomic layers (distorted square/triangle net) stacked along [010] in a way to form three different types of octahedra between the layers, with the space in-between filled by tetrahedra (also see Fig. 7.48). The face-sharing octahedra of type I form isolated chains running through the corners and center of the unit cell along [010]. The type II octahedra form flat (110) layers by sharing two opposite edges each and their apical vertices along [010]. Octahedra of type I and II share faces with type I octahedra. Type III octahedra share part of their edges with each other and form double-chains running along [100] and connecting the type II octahedra layers with each other. The remaining channels are filled by the chains of type I octahedra. Only five representatives of this structure type are known among the stannides. These contain the elements Ba, Sr, Ca, Yb, and Eu, all of them occupying the Co sites.

- $tI12$-CuAl$_2$ (69 *Fmmm*): The structure of $tI12$-CuAl$_2$ can be described as a stacking of $3^2.4.3.4$ triangle/square nets located in $z = 0$ and $1/2$, with the squares capped by Cu in $z = 1/4$ and $3/4$ (see Fig. 7.26(a)–(d)). There are only four stannides known with this structure type: Rh, Pt, Pd, and Au, all occupying the Cu sites.

Ternary plumbides

The 344 ternary Pb-containing IMs feature 80 different structure types, i.e., ≈ 4.3 representatives per structure type, this is slightly less than in the case of germanides (≈ 5.8). In Table 7.25, the most common structure types among ternary Pb-IMs are given, and the distribution of the chemical compositions of six of the most common ones are shown in Fig. 7.84. In the ternary compounds, Pb is most often the element with the highest M-value (82) and will therefore be assigned the C-position, while the A-elements will strictly have smaller M-values than B-elements in all formulae A$_a$B$_b$Pb$_c$, $M(A) < M(B)$.

In the following, we discuss some characteristic features of the five most frequent ternary structure types of ternary plumbides listed in Table 7.25. Their structures are shown in Fig. 7.83.

- $hP18$-CuHf$_5$Sn$_3$ (193 *P6$_3$/mcm*): The structure can be described as consisting of columns of face-sharing, Cu-centered regular Hf-octahedra, surrounded each by rings of six edge-sharing hexagonal bipyramids, which themselves are in a columnar arrangement. In the related $hP16$-Mn$_5$Si$_3$ structure type, the Cu-site is empty.

- $hP6$-LiGaGe (186 *P6$_3$mc*): The structure of this Zintl phase can be described as a superstructure of the $hP6$-CaIn$_2$ type, which itself is a twofold superstructure of the $hP3$-AlB$_2$ type (also see Section 7.7).

- $hP9$-ZrNiAl (189 *P$\bar{6}$2m*): The structure can be subdivided into face-sharing tricapped trigonal Ni@Zr$_6$Al$_3$ prisms (also see Fig. 10.8 in Section 10.8).

- $cF16$-Cu$_2$MnAl (225 *Fm$\bar{3}$m*): This structure type of the Heusler phases can be described as a $(2 \times 2 \times 2)$-fold superstructure of the $cI2$-W type, and is discussed in Subsection 7.3.2.

- $oI36$-Ho$_6$Co$_2$Ga (71 *Immm*): The structure can be described as a packing of Ga@Ho$_{12}$ icosahedra, which are centered at one half of the unit-cell faces as well as edge centers and share eight faces each with Ho$_6$ octahedra. The Ga atoms on the remaining face and edge centers of the unit cell are coordinated by eight Ho atoms forming cubes in this way. Three cubes are bicapped by a part of the Co atoms; the other Co atoms cap octahedra faces. The Ho–Ho distances range between 3.42 and 3.53 Å (in the element 3.486 Å), the shortest Co–Co distances amount to 2.51 Å (in the element 2.506 Å), and the distance Ga–Ho with 2.94 Å is slightly shorter than the sum of radii (2.964 Å) as well.

Table 7.25 *Most common structure types of* Pb-*containing ternary intermetallics. The top 19 structure types are given, all of which have at least five representative structures and therefore represent at least 1% of all ternary* Pb-*intermetallics.*

Rank	Structure type	No.	Space group	Wyckoff positions	No. of reps.	% of all reps.
1.	$hP18$-CuHf$_5$Sn$_3$	193	$P6_3/mcm$	$2b\,4d\,6g^2$	42	12.2%
2.	$hP6$-CaIn$_2$	194	$P6_3/mmc$	$2b\,4f$	26	7.6%
3.	$hP6$-LiGaGe	186	$P6_3mc$	$2ab^2$	23	6.7%
4.	$hP9$-ZrNiAl	189	$P\bar{6}2m$	$1a\,2d\,3fg$	18	5.2%
5.	$cF16$-Cu$_2$MnAl	225	$Fm\bar{3}m$	$4abcd$	17	4.9%
6.	$oI36$-Ho$_6$Co$_2$Ga	71	$Immm$	$2ac\,4gj\,8lmn$	16	4.7%
7.	$tP10$-Mo$_2$FeB$_2$	127	$P4/mbm$	$2a\,4gh$	16	4.7%
8.	$oP12$-TiNiSi	62	$Pnma$	$4c^3$	16	4.7%
9.	$cI38$-Sm$_{12}$Ni$_6$In	204	$Im\bar{3}$	$2a\,12e\,24g$	15	4.4%
10.	$cP4$-Cu$_3$Au	221	$Pm\bar{3}m$	$1a\,3c$	14	4.1%
11.	$oS10$-Mn$_2$AlB$_2$	191	$P6/mmm$	$2a\,4ij$	9	2.6%
12.	$cF24$-Be$_5$Au	216	$F\bar{4}3m$	$4ac\,16e$	8	2.3%
13.	$cF16$-Li$_2$AgSb	216	$F\bar{4}3m$	$4abcd$	7	2.0%
14.	$cF12$-MgAgAs	216	$F\bar{4}3m$	$4abc$	7	2.0%
15.	$oI12$-KHg$_2$	74	$Imma$	$4e\,8i$	6	1.7%
16.	$cF144$-Ca$_{11}$Ga$_7$	225	$Fm\bar{3}m$	$4ab\,24e\,32f^2\,48h$	5	1.5%
17.	$cI58$-La$_4$Re$_6$O$_{19}$	197	$I23$	$2a\,8c\,12de\,24f$	5	1.5%
18.	$tI80$-La$_6$Co$_{11}$Ga$_3$	140	$I4/mcm$	$4ad\,8f\,16kl^3$	5	1.5%
19.	$cP40$-Yb$_3$Rh$_4$Sn$_{13}$	223	$Pm\bar{3}n$	$2a\,6c\,8e\,24k$	5	1.5%
					260	75.6%

The $hP18$-CuHf$_5$Sn$_3$-type structures (Nowotny phases) among ternary Pb-intermetallics all contain 33.3% Pb, as well as 55.6% of element A ($M(A) = 16$–33) and 11.1% of element B ($M(B) = 54$–76 or 88). $hP18$-CuHf$_5$Sn$_3$ (Fig. 7.83) can be considered a filled $hP16$-Mn$_5$Si$_3$ type. In addition to the Wyckoff positions $4d$ and $6g$ in space group 193 $P6_3/mcm$, $2b$ is also occupied. Cu is surrounded by Hf-octahedra, which share one face along the [001] direction. The irregular AET of Hf(1) consists of 2 Cu, 4 Hf, and 5 Sn; Hf(2) centers a bicapped hexagonal

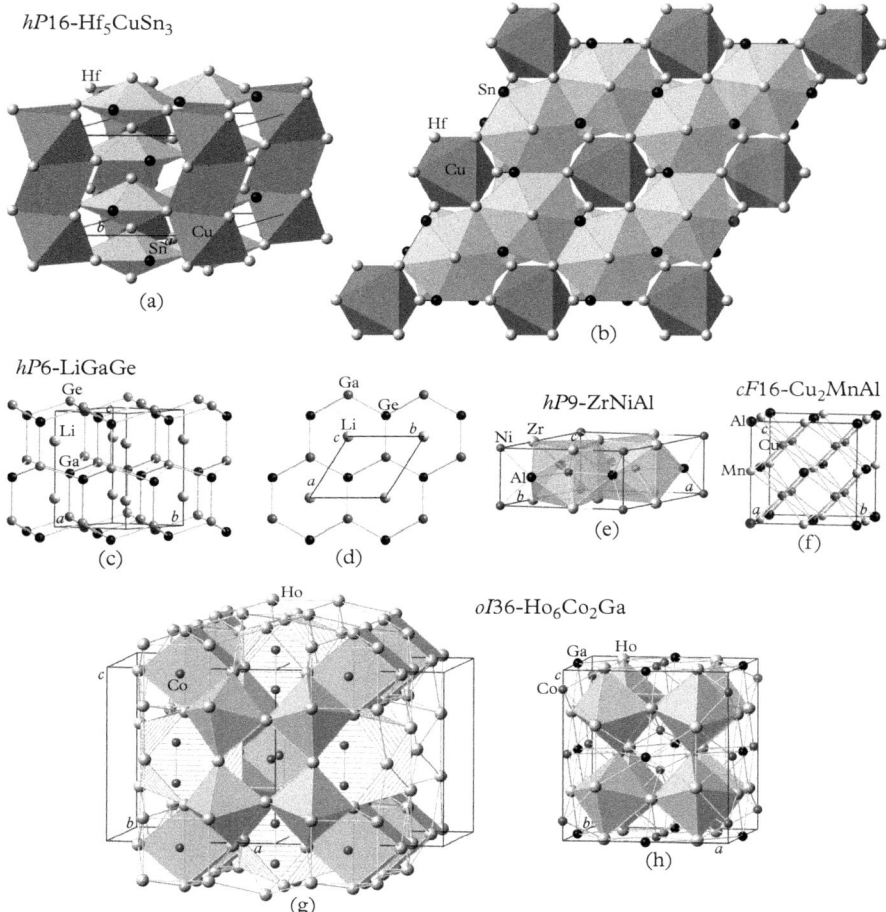

Fig. 7.83 *The structures of the five most frequent ternary structure types of ternary plumbides. The hP16-Hf$_5$CuSn$_3$ structure type: (a) One unit cell with the columns of Cu-centered Hf-octahedra and of hexagonal bipyramids, Hf$_5$Sn$_3$, shaded; (b) projection along [001]. hP6-LiGaGe (c) can be described as a superstructure of the hP6-CaIn$_2$ type with a 3D 4-connected network. (d) The unit cell of hP9-ZrNiAl with face-sharing tricapped trigonal Ni@Zr$_6$Al$_3$ prisms shaded. The structure of cF16-Cu$_2$MnAl (e) can be described as a (2 × 2 × 2)-fold superstructure of the cI2-W type. The structure of oI36-Ho$_6$Co$_2$Ga (g) can be described as a packing of Ga@Ho$_{12}$ icosahedra, Ga@Ho$_8$ cubes and Ho$_6$ octahedra.*

antiprism of 6 Sn and 8 Hf atoms, Sn by 9 Hf atoms. The Cu atom sits in Wyckoff position $2b$, which is empty in the $hP16$-Mn$_5$Si$_3$ type.

With three exceptions, all $hP6$-CaIn$_2$-type structures have 1:1:1-stoichiometry between Pb, A (M(A) = 17–33), and B (M(B) = 70–72 or 76); only

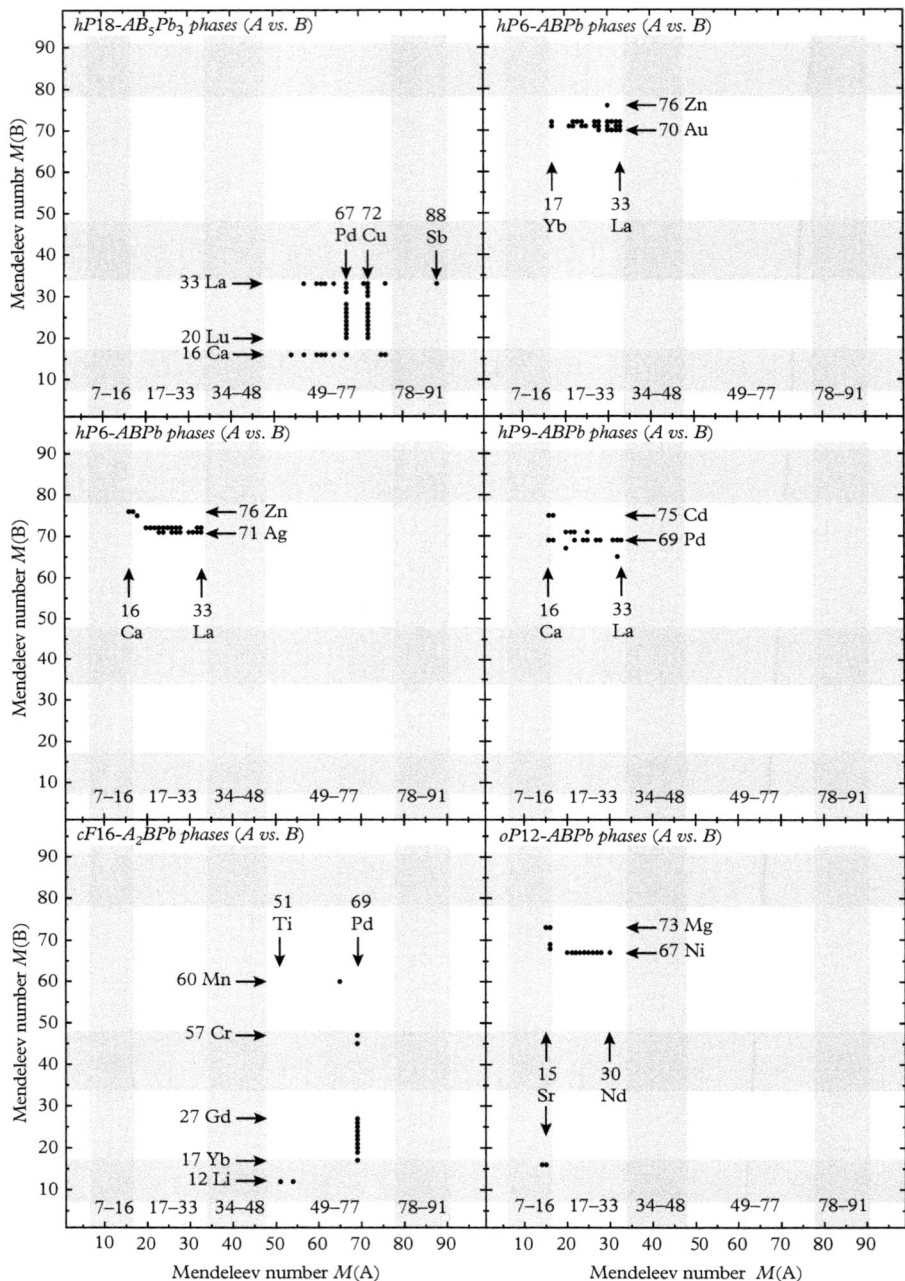

Fig. 7.84 *Occurrence of six of the most common structure types among ternary Pb-intermetallics. The Mendeleev numbers M of the other two elements in the compound are shown. The depicted structure types are hP18-CuHf$_5$Sn$_3$, hP6-CaIn$_2$, hP6-LiGaGe, hP9-ZrNiAl, cF16-Cu$_2$MnAl, and oP12-TiNiSi (ranks 1–5, and 8, respectively, in Table 7.25). Where A and B have equal shares in the chemical formula, they are assigned so that M(A) < M(B). Only two structures were omitted for structure type cF16-Cu$_2$MnAl, where the chemical decoration of the structure changes.*

Y/Er/Tm–Ag–Pb have small Ag–Pb-exchanges with Pb-contents 40.7–46.0%. Representatives of structure types $hP6$-LiGaGe and $hP9$-ZrNiAl also exhibit 1:1:1-stoichiometries and are formed with elements $M(A) = 16$–33, as well as $M(B) = 71$–76 and $M(B) = 65$–75, respectively. The $cF16$-Cu$_2$MnAl-type structures all contain 25% Pb and mostly have exact stoichiometries with 25% A ($M(A) = 12$–47 or 60) and 50% B ($M(B) = 51$–69), except for one compound with reversed A- and B-identities (50% with $M(A) = 12$ and 25% with $M(B) = 69$), as well as one with 37.5% of each, A and B ($M = 12$ and 73, respectively). The $oI36$-Ho$_6$Co$_2$Ga-type representatives contain 5.6–7.2% Pb, as well as 66.6–69.3% A with $M(A) = 20$–27 and 23.9–27.8% B with $M(B) = 64$ or 67 (i.e., Co and Ni, respectively); although the assignment of the element type is unambiguous, the stoichiometries are somewhat washed out.

In nearly all $tP10$-Mo$_2$FeB$_2$ (127 $P4/mbm$)-type structures, 20% Pb are present in addition to 40% of both, A- and B-elements with $M(A) = 16$–33 and $M(B) = 69$ or 70 (Pd or Au). One of two exceptions contains 40% Pb and only 20% of the B-element (Li–V–Pb), while the other is somewhat Pb-deficient (Ce–Rh–Pb). For structure type $oP12$-TiNiSi, again only 1:1:1-stoichiometries are found, combining Pb with elements A with $M(A) = 14$–30 and B with $M(B) = 16$ or 67–73. The $cI38$-Sm$_{12}$Ni$_6$In-type structures all contain 5.3% Pb, in addition to 63.2% A ($M(A) = 23$–33) and 31.2% B ($M(B) = 64$ or 67, i.e., Co and Ni, respectively). $cP4$-Cu$_3$Au-type structures contain 5–75% Pb and have diverse combinations of elements sharing either the Cu- or Au-role. All $oS10$-Mn$_2$AlB$_2$-type structures contain 20% Pb and 40% of each, A- and B-elements with $M(A) = 20$–28 and $M(B) = 67$, i.e., Ni. The Pb-content in the $cF24$-Be$_5$Au-type structures varies in the range 6.7–11.0%, while it shares the Be-role with the B-element Cu ($M(B) = 72$), and the A-elements ($M(A) = 20$–27) make up for the 16.7% in the Au-role.

Most $cF16$-Li$_2$AgSb-type structures contain 25% of Pb, as well as 50% of Na or Li ($M(A) = 11$ and 12, respectively) and 25% of the B-element with $M(B) = 70$–75. Another two compounds in the Na–Zn/In–Pb-systems have differing stoichiometries with 24% and 8.3% Pb, respectively. The stoichiometries of the $cF12$-MgAgAs-type structures, on the other hand, are all 1:1:1, involving elements $M(A) = 22$–27 and $M(B) = 70$ (Au), as well as a compound in the Cu–Mg–Pb-system. Also the $oI12$-KHg$_2$-type representatives stick to the 1:1:1-stoichiometry and are constituted from elements $M(A) = 14$–18 and $M(B) = 71$–76. The $cF144$-Ca$_{11}$Ga$_7$-type structures contain 10.3–19.4% Pb, as well as 58.3–60.6% of element A ($M(A) = 15$ or 16, i.e., Sr and Ca, respectively) and 22.2% or 29.1% of element B ($M(B) = 79$–81). The $cI58$-La$_4$Re$_6$O$_{19}$-type structures, again, are stoichiometric with 65.5% Pb, 13.8% A ($M(A) = 16$–30), and 20.7% Rh (i.e., $M(B) = 65$). $tI80$-La$_6$Co$_{11}$Ga$_3$-type structures contain 5% Pb and 30% and 65% of elements A and B ($M(A) = 28$–33 and $M(B) = 61$ or 64, i.e., Fe and Co, respectively). The structures of type $cP40$-Yb$_3$Rh$_4$Sn$_{13}$, finally contain 65% Pb, in addition to 15% A with values $M(A) = 15$–33 and 20% Rh.

7.15.8   Antimonides

Binary antimonides

Among the binary IMs, 343 Sb-containing IMs can be found, featuring 130 structure types, i.e., ≈ 2.6 representatives per structure type, i.e., comparable to the case of germanides (≈ 2.5). Their compositions can be found in Fig. 7.85, while in Table 7.26, the most common structure types among binary Sb-IMs are given.

The compositions of the $cF4$-Cu- and $hP2$-Mg-type structures are rather diverse (containing 0.1–50.0% and 1.2–25.0% Sb, respectively). The same holds true for $cP1$-Po-type structures with 60.0–90.0% Sb. Nearly all of the remaining structure types occur at their defined compositions. In some of them, Sb always plays the role of the majority element ($oS24$-SmSb$_2$, $oS6$-HoSb$_2$, $oP6$-FeAs$_2$, $mP28$-Dy$_2$Sb$_5$, and $mP6$-CaSb$_2$) or of the minority element ($hP16$-Mn$_5$Si$_3$, $tI12$-La$_2$Sb, $oP32$-Yb$_5$Sb$_3$, $oP32$-Y$_5$Bi$_3$, $tP56$-Ca$_{16}$Sb$_{11}$, $hP6$-Co$_{1.75}$Ge, $cF16$-BiF$_3$, $cP8$-Cr$_3$Si, $tI84$-Ho$_{11}$Ge$_{10}$, and $hP8$-Na$_3$As). In $cI28$-Th$_3$P$_4$ it appears to switch between those roles. Obviously, no such statement can be made for 1:1-stoichiometries, as in $cF8$-NaCl, $hP4$-NiAs, $cP2$-CsCl, $tP2$-CuTi, $cF8$-ZnS, and mostly also $tI4$-Sn. An overview over the respective intermetallic systems is shown in Fig. 7.86.

In the following, we discuss some characteristic features of the five most common binary structure types of binary antimonides listed in Table 7.26. Their structures are shown in Fig. 7.87.

- $cF8$-NaCl (225 $Fm\bar{3}m$): This structure type can be described as *ccp* packing of the larger atoms with the smaller atoms in the octahedral voids (also see

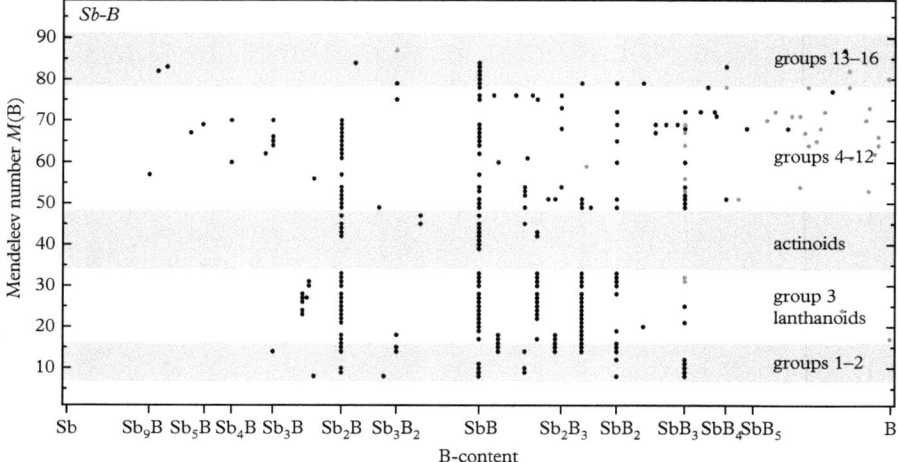

Fig. 7.85 *Stoichiometries of the 343 binary antimonides, Sb$_a$B$_b$ vs. the Mendeleev number M(B) of the other element in the compound. The 17 binary phases adopting unary structure types cF4-Cu, cI2-W, and hP2-Mg are marked by gray dots.*

Table 7.26 *Most common structure types of Sb-containing binary intermetallics. The top 25 structure types are given, all of which have at least four representative structures and therefore represent at least 1% of all binary Sb-intermetallics each. In sum, the 193 representatives of these structure types make up 56.3% of all binary Sb-intermetallics.*

Rank	Structure type	No.	Space group	Wyckoff positions	No. of reps.	% of all reps.
1.	$cF8$-NaCl	225	$Fm\bar{3}m$	$4ab$	26	7.6%
2.	$hP16$-Mn$_5$Si$_3$	193	$P6_3/mcm$	$4d\,6g^2$	18	5.2%
3.	$cI28$-Th$_3$P$_4$	220	$I\bar{4}3d$	$12a\,16c$	16	4.7%
4.	$hP4$-NiAs	194	$P6_3/mmc$	$2ac$	11	3.2%
5.	$cF4$-Cu	225	$Fm\bar{3}m$	$4a$	10	2.9%
6.	$oS24$-SmSb$_2$	64	$Cmca$	$8ef^2$	10	2.9%
7.	$tI12$-La$_2$Sb	139	$I4/mmm$	$4ce^2$	9	2.6%
8.	$oP32$-Yb$_5$Sb$_3$	62	$Pnma$	$4c^4\,8d^2$	9	2.6%
9.	$cP2$-CsCl	221	$Pm\bar{3}m$	$1ab$	7	2.0%
10.	$oS6$-HoSb$_2$	21	$C222$	$2a\,4k$	7	2.0%
11.	$tP2$-CuTi	123	$P4/mmm$	$1ad$	6	1.7%
12.	$oP6$-FeAs$_2$	58	$Pnnm$	$2a\,4g$	6	1.7%
13.	$oP32$-Y$_5$Bi$_3$	62	$Pnma$	$4c^4\,8d^2$	6	1.7%
14.	$tP56$-Ca$_{16}$Sb$_{11}$	113	$P\bar{4}2_1m$	$2c^2\,4d^2\,e^5\,8f^3$	5	1.5%
15.	$hP6$-Co$_{1.75}$Ge	194	$P6_3/mmc$	$2acd$	5	1.5%
16.	$mP28$-Dy$_2$Sb$_5$	11	$P2_1/m$	$2e^{14}$	5	1.5%
17.	$cP1$-Po	221	$Pm\bar{3}m$	$1a$	5	1.5%
18.	$cF16$-BiF$_3$	225	$Fm\bar{3}m$	$4c^2\,8d$	4	1.2%
19.	$mP6$-CaSb$_2$	11	$P2_1/m$	$2e^3$	4	1.2%
20.	$cP8$-Cr$_3$Si	223	$Pm\bar{3}n$	$2a\,6c$	4	1.2%
21.	$tI84$-Ho$_{11}$Ge$_{10}$	139	$I4/mmm$	$4de^2\,8h^2j\,16mn^2$	4	1.2%
22.	$hP2$-Mg	194	$P6_3/mmc$	$2c$	4	1.2%
23.	$hP8$-Na$_3$As	194	$P6_3/mmc$	$2bc\,4f$	4	1.2%
24.	$tI4$-Sn	141	$I4_1/amd$	$4a$	4	1.2%
25.	$cF8$-ZnS	216	$F\bar{4}3m$	$4ac$	4	1.2%
					193	56.3%

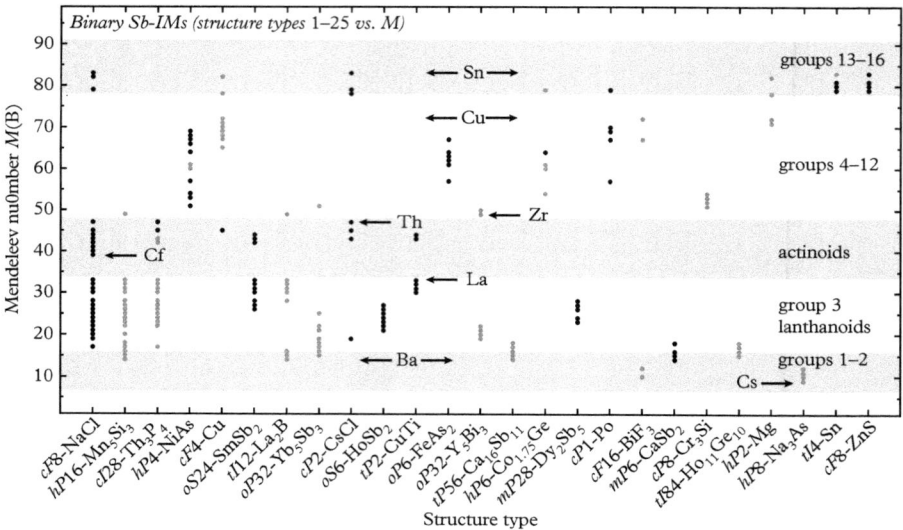

Fig. 7.86 *Occurrence of the 25 most common structure types among binary Sb-intermetallics. The Mendeleev number M of the other element in the compound is shown over the rank of the structure type, as given in Table 7.26. Structures with Sb as their major component ($\geq 50\%$) are marked in black, those with minor Sb-content are shown in gray.*

the respective paragraph in Subsection 7.15.9). This structure type is adopted for antimonides of the elements: Y and the lanthanoids except Eu and Pm, the actinoids Cf, Bk, Cm, Am, Pu, Np, U, and Th, as well as the main group elements In, Pb, and Sn.

- $hP16$-Mn_5Si_3 (193 $P6_3/mcm$): In this structure type, Si and Mn form distorted CN16 FK-polyhedra, which share two triangular faces. Thereby, the Mn1 atoms are arranged in linear chains in 1/3, 2/3, z, parallel to [001], with short Mn–Mn distances of 2.407 Å ($r_{Mn} = 1.367$ Å). The Mn2 atoms form face-sharing columns of octahedra in 0, 0, z, also parallel to [001]. There, the Mn–Mn distances amount to 2.822 Å and 2.907 Å, respectively (also see the respective paragraph in Subsection 7.15.9). Antimonides of this structure type are known for the elements Ba, Sr, Ca, Y, and the lanthanoids except Tm and Pm, as well as Zr. Sb always occupies the Si site.

- $cI28$-Th_3P_4 (220 $I\bar{4}3d$): The basic structural unit can be seen as a trigonal P-centered Th antiprism, with three of the six sides plus the top and bottom capped by P atoms. These Th antiprisms are stacked in a similar way along [111] as the Mn2 octahedra in $hP16$-Mn_5Si_3 (see also the respective paragraph in Subsection 7.15.9). The P site in this structure type can be occupied by Y, the lanthanoids Yb, Er, Ho, Dy, Tb, Gd, Sm, Nd, Pr, Ce, and La, as well as the actinoids Am and Pu; in contrast, U and Th occupy the Th sites.

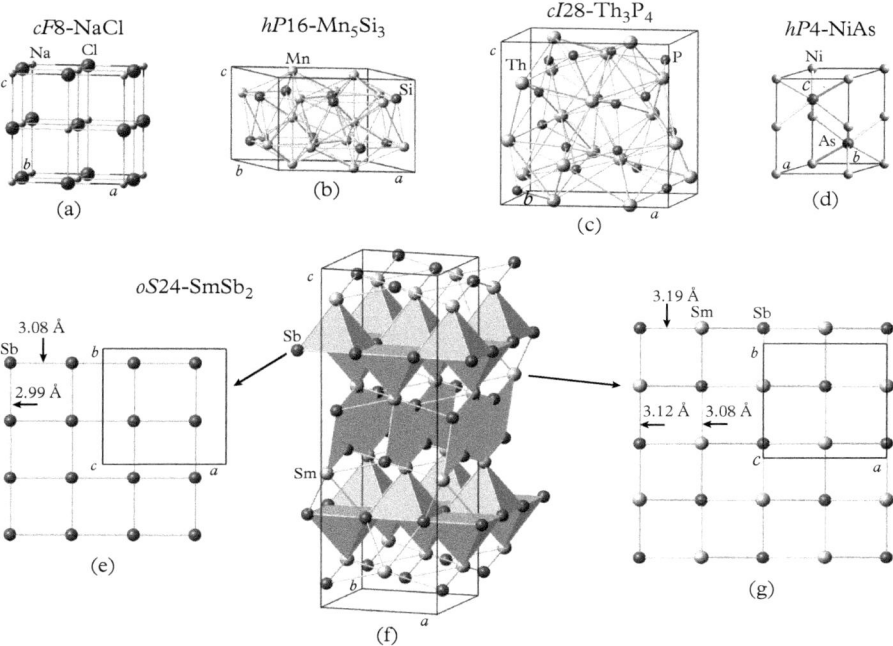

Fig. 7.87 *The structures of the five most frequent binary structure types of binary antimonides in different projections: (a) cF8-NaCl, (b) hP16-Mn₅Si₃, (c) cI28-Th₃P₄, (d) hP4-NiAs, and (e)–(g) oS24-SmSb₂. In (e), the slabs of alternatingly up-and-down Sm-capped square nets of Sb atoms are shaded, and the zigzag bands constituted from rhombs connecting the slabs are shown as well. In (e) the flat Sb-layer is depicted and in (g) the puckered Sb/Sm layer is shown.*

- *hP4-NiAs* (194 $P6_3/mmc$): This structure type can be described as an *hcp* packing of the larger atoms with the smaller atoms in the octahedral voids. Antimonides with this structure type are only formed with TM elements: Ti, Nb, V, Cr, Mn, Fe, Co, Ir, Ni, Pt, and Pd.

- *oS24-SmSb₂* (64 *Cmca*): The structure can be decomposed into slabs and bands. The slabs consist of alternatingly up-and-down Sm-capped square nets of Sb atoms with Sb–Sb distances between 2.99 and 3.08 Å. The flat zigzag bands are constituted from edge-sharing Sb₂Sm₂ rhombs connecting the slabs with each other. The rhombs running along [100] all have the same shape, but form a slightly puckered band. The flat zigzag band running along [010] consists of two types of rhombs, a skinny one and a fat one. The short diagonal of the skinny rhomb with 2.78 Å gives the very short distance between two Sb atoms (the distance in the element amounts to 2.90 Å). These antimonides form with Sb always on the Sb sites and the elments Tb, Gd, Sm, Nd, Pr, Ce, La, Am, Pu, and Np.

Ternary antimonides

The 1176 ternary Sb-containing IMs feature 257 different structure types, i.e., ≈ 4.6 representatives per structure type, i.e., slightly less than in the case of germanides (≈ 5.8). In Table 7.27, the most common structure types among ternary Sb-IMs are given, and the distribution of the chemical compositions of six of the most common ones are shown in Fig. 7.90. In the ternary compounds, Sb is most often the element with the highest M-value (88) and will therefore be assigned the C-position, while the A-elements will strictly have smaller M-values than B-elements in all formulae $A_a B_b Sb_c$ ($M(A) < M(B)$).

In the following, we discuss some characteristic features of the five most frequent ternary structure types of ternary antimonides listed in Table 7.27, and shown in Fig. 7.88.

- $tP8$-CuHfSi$_2$ (129 $P4/nmm$): This structure type is equivalent to that of $tP8$-CuZrSi$_2$. The structure consists of a flat 4^4 net of Si, followed by a puckered rotated 4^4 net of Si/Hf, followed by a flat 4^4 net of Cu (also see the section on binary bismuthides and Fig. 7.95).

- $cF12$-MgAgAs (216 $F\bar{4}3m$): This structure type is also known as $cF12$-LiAlSi, in which the Al (Ag and other TM) atoms form a ccp packing, the Si (As and other main group) elements occupy half of the tetrahedral voids, and Li (Mg and RE) atoms are sitting in the octahedral voids (see Fig. 7.2).

- $hP18$-CuHf$_5$Sn$_3$ (193 $P6_3/mcm$): The structure consists of columns of face-sharing, Cu-centered regular Hf-octahedra, each surrounded by rings of six edge-sharing hexagonal bipyramids, which themselves are in a columnar arrangement (also see Fig. 7.83).

- $oP12$-TiNiSi (62 $Pnma$): The structure contains a 3D 4-connected network of Ti atoms ($d_{\text{Ti-Ti}} = 3.15$–3.22 Å), where the large channels are filled with zigzag-bands of edge-sharing Ni$_2$Si$_2$ rhomb units. It can be also described as a distorted $hP6$-Ni$_2$In type structure (also see Fig. 7.72).

- $cI34$-LaFe$_4$P$_{12}$ (204 $Im\bar{3}$): One representative of this structure type (filled skutterudite) is $cI34$-LaFe$_4$Sb$_{12}$. It can be described as a bcc arrangement of distorted La@Sb$_{12}$ icosahedra that are connected via face-sharing Fe@Sb$_6$ octahedra. The shortest La–Sb and Fe–Sb distances are 3.41 Å (sum of radii equals 3.32 Å) and 2.55 Å (sum of radii equals 2.69 Å), respectively. The shortes Sb–Sb distances range from 2.93 Å to 2.98 Å.

The structure of $oP20$-CeCrSb$_3$ (rank 15 in Table 7.27), which is related to the $tI10$-BaAl$_4$ structure type, is illustrated in Fig. 7.89. It can be described as chains of along [001] face-sharing Cr-centered Sb-octahedra (Cr–Sb distances: 2.696–2.736 Å), which are edge-connected with equivalent chains along [010]. One vertex of each octahedron forms a flat pyramid with the Ce atoms, which themselves form a square net of 4.314 Å edge length (Ce–Sb distances:

Table 7.27 *Most common structure types of* Sb-*containing ternary intermetallics. The top 20 structure types are given, all of which have at least 12 representative structures and therefore represent at least 1% of all ternary* Sb-*intermetallics each.*

Rank	Structure type	No.	Space group	Wyckoff positions	No. of reps.	% of all reps.
1.	$tP8$-CuHfSi$_2$	129	$P4/nmm$	$2abc^2$	87	7.4%
2.	$cF12$-MgAgAs	216	$F\bar{4}3m$	$4abc$	71	6.0%
3.	$hP18$-CuHf$_5$Sn$_3$	193	$P6_3/mcm$	$2b\,4d\,6g^2$	46	3.9%
4.	$oP12$-TiNiSi	62	$Pnma$	$4c^3$	44	3.7%
5.	$cI34$-LaFe$_4$P$_{12}$	204	$Im\bar{3}$	$2a\,8c\,24g$	39	3.3%
6.	$cI40$-Au$_3$Y$_3$Sb$_4$	220	$I\bar{4}3d$	$12ab\,16c$	37	3.1%
7.	$tI32$-Mo$_5$SiB$_2$	140	$I4/mcm$	$4ac\,8h\,16l$	30	2.6%
8.	$tP10$-CaBe$_2$Ge$_2$	129	$P4/nmm$	$2abc^3$	28	2.4%
9.	$hP6$-BeZrSi	194	$P6_3/mmc$	$2acd$	24	2.0%
10.	$cF16$-Cu$_2$MnAl	225	$Fm\bar{3}m$	$4abcd$	22	1.9%
11.	$tI32$-Nb$_5$Sn$_2$Si	140	$I4/mcm$	$4ab\,8h\,16k$	21	1.8%
12.	$hP4$-NiAs	194	$P6_3/mmc$	$2ac$	21	1.8%
13.	$hP9$-K$_2$UF$_6$	189	$P\bar{6}2m$	$1a\,2d\,3fg$	19	1.6%
14.	$hP5$-Ce$_2$SO$_2$	162	$P\bar{3}m$	$1a\,2d^2$	18	1.5%
15.	$oP20$-CeCrSb$_3$	57	$Pbcm$	$4c^2d^3$	17	1.4%
16.	$hP6$-Co$_{1.75}$Ge	194	$P6_3/mmc$	$2acd$	17	1.4%
17.	$cI28$-Th$_3$P$_4$	220	$I\bar{4}3d$	$12a\,16c$	14	1.2%
18.	$tI32$-W$_5$Si$_3$	140	$I4/mcm$	$4ab\,8h\,16k$	14	1.2%
19.	$hP6$-LiGaGe	186	$P6_3mc$	$2ab^2$	12	1.0%
20.	$oP12$-PbCl$_2$	62	$Pnma$	$4c^3$	12	1.0%
					593	50.4%

3.259–3.283 Å). At approximately half way up the unit cell, an almost square net of Sb atoms (edge lengths: 3.039 and 3.092 Å) forms the bases of alternatingly up- and down-pointing pyramids with Ce atoms at the vertices (Ce–Sb distances: 3.309–3.318 Å). The structure can also be viewed as a layer structure. At temperatures below 115 K, this compound shows two ferromagnetic ordering transitions.

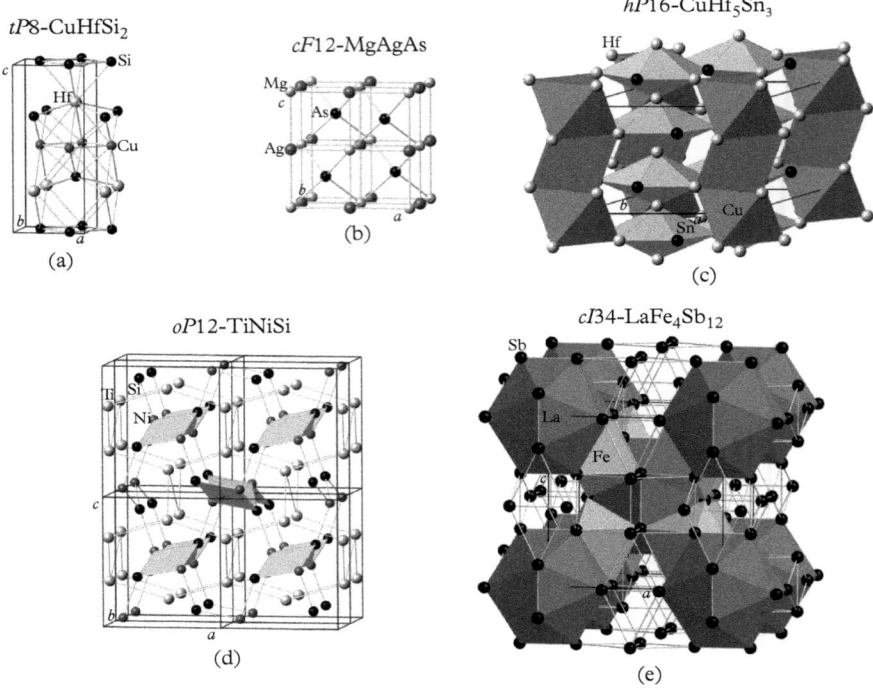

Fig. 7.88 *The structures of the five most frequent ternary structure types of ternary antimonides in different projections: (a) the structure of tP8-CuHfSi$_2$ can be described as a stacking of flat and puckered 4^4 nets decorated with Si, Si/Hf, and Cu, respectively. (b) cF12-MgAgAs: the Al atoms form a ccp packing, with Si occupying half of the tetrahedral voids and Mg atoms all the octahedral voids. (c) The hP16-CuHf$_5$Sn$_3$ structure type: one unit cell with the columns of Cu-centered Hf-octahedra and of hexagonal bipyramids, Hf$_5$Sn$_3$, shaded dark and light gray, respectively. The structure of oP12-TiNiSi (d) contains a 3D 4-connected network of Ti atoms, with zigzag-bands of edge-sharing Ni$_2$Si$_2$ rhomb units in the large channels. (e) The structure of cI34-LaFe$_4$Sb$_{12}$ can be described as a bcc arrangement of distorted La@Sb$_{12}$ icosahedra that are connected via face-sharing Fe@Sb$_6$ octahedra.*

The oP20-CeCrSb$_3$ structure type contains structural units that are also found in tI10-ThCr$_2$Si$_2$ (also the structure type of BaFe$_2$As$_2$ and related pnictogen superconductors), tP10-CaBe$_2$Ge$_2$, and tP8-HfCuSi$_2$ type structures (Brylak and Jeitschko, 1995). oP60-CeNiSb$_3$ can be seen as a threefold superstructure along [001], caused by a different stacking order of the Ni-centered Sb-octahedra. The compound shows a steep decrease in the electrical conductivity below 6 K (Macaluso *et al.*, 2004). The structure of oP40-LaPdSb$_3$ is as twofold superstructure along [001], intermediately between oP20-CeCrSb$_3$ and oP60-CeNiSb$_3$ (Thomas *et al.*, 2006a).

$oP20$-CeCrSb$_3$

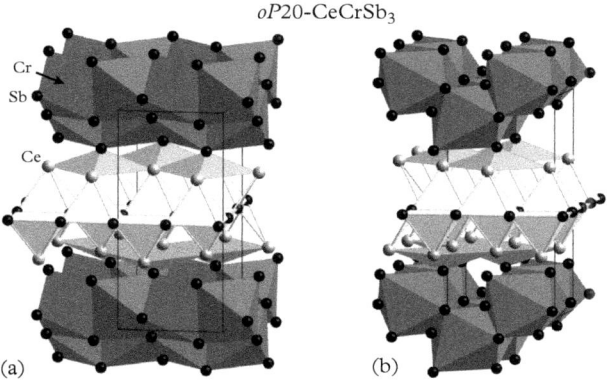

(a)   (b)

Fig. 7.89 *The oP20-CeCrSb$_3$ structure type. Structural characteristics in a projection along (a) [001] and (b) [010].*

The ternary Sb-intermetallics crystallizing in structure type $tP8$-CuHfSi$_2$ contain 35.8–60.8% Sb; most of them (68 out of 87) are stoichiometric with 50% Sb, as well as 25% of elements A and B, each. All compounds feature components with $M(A) = 17$–51 and $M(B) = 60$–76. Nearly all $cF12$-MgAgAs-type structures have 1:1:1-stoichiometry, featuring—apart from Sb—elements A and B with $M(A) = 12$–72 and $M(B) = 54$–76. Among the $hP18$-CuHf$_5$Sn$_3$-type structures, diverse behaviors can be found: a large group of 27 compounds contains 55.6% of Sb, 33.3% of A ($M(A) = 28$–45), and 11.1% of B ($M(B) = 49$–60). One compound has reversed A- and B-roles with contents of 11.1% and 33.3%, respectively (Sc–U–Sb system with $M(A) = 19$ and $M(B) = 45$). Another group of 16 structures contains around 33.3% of Sb, around 55.6% of element A, and 11.1% of element B with $M(A) = 49$–51 and $M(B) = 61$–84, while the last two structures contain only 11% of Sb and 56% and 33.3% of elements A and B, respectively—these are found in the La–Ge/Pb–Sb-systems. Although most $oP12$-TiNiSi-type structures are reported to have exact 1:1:1-stoichiometry, the Sb-content varies overall in the range 33.3–60.0%, while the elements A and B have values $M(A) = 11$–75 and $M(B) = 15$–18 and 61–84. The majority of the $cI34$-LaFe$_4$P$_{12}$-type structure is stoichiometric with 70.6% of Sb, as well as 5.9% of element A with $M(A) = 10$–33 and 23.5% of element B ($M(B) = 61$–63, i.e., Fe, Ru, and Os). The remaining structures have higher $M(B)$-values: for $M(B) = 64$ and 65 the stoichiometry deviates only slightly from these values ($M(A) = 16$–33); for $M(B) = 78$–83 the A- and B-roles are reversed ($M(A) = 61$ or 64) and also scattered around above values, while the compound in the Ni–Sn–Sb-system ($M(A) = 67$ and $M(B) = 83$) has a much smaller Sb-content, compensated by the B-element.

All $cI40$-Au$_3$Y$_3$Sb$_4$-type structures are stoichiometric with 40% Sb, as well as 20% of A and B, each, with elements $M(A) = 17$–50 and $M(B) = 64$–72. The

$tI32$-Mo$_5$SiB$_2$-type structures are all stoichiometric, too, and contain 12.5% of Sb, 62.5% of element A ($M(A)$ = 20–31), and 25.0% of B ($M(B)$ = 61–70). The structures of type $tP10$-CaBe$_2$Ge$_2$ contain 40–50% of Sb and are constituted from elements A and B with $M(A)$ = 12–33 and $M(B)$ = 30–32 and 67–72. Nearly all $hP6$-BeZrSi-type structures are 1:1:1-stoichiometric; they contain metals with values $M(A)$ = 10–64 and $M(B)$ = 14 and 60–77. The $cF16$-Cu$_2$MnAl-type structures mostly contain 25% of Sb, but the disordered compounds contain up to 42.6% Sb. The forming elements are rather diverse with $M(A)$ = 8–69 and $M(B)$ = 10–83, while the roles within the structure switch between A- and B-elements repeatedly. The structures of type $tI32$-Nb$_5$Sn$_2$Si contain 25.0–32.5% Sb and an equivalent 5.0–12.5% of element B with $M(B)$ = 54–84. The content in A is always 62.5% and only elements Zr, Hf, and Ti ($M(A)$ = 49–51, respectively) are featured. The Sb-content in $hP4$-NiAs-type structures lies between 15% and 50% and the forming elements have values $M(A)$ = 49–70 and $M(B)$ = 60–87.

All $hP9$-K$_2$UF$_6$-type structures contain 22.2% Sb, as well as 66.7% of element A and 11.1% of element B, with $M(A)$ = 19–50 and $M(B)$ = 56 and 60–68. The structures of type $hP5$-Ce$_2$SO$_2$ are also nearly all stoichiometric with 40% of each, Sb and element B, as well as 20% of A, with metals $M(A)$ = 14–18 and $M(B)$ = 60 or 73–76; one compound with a different stoichiometry is formed with Mn ($M(A)$ = 60). The $oP20$-CeCrSb$_3$-type structures all have the stoichiometry of the prototype, containing 60% Sb and 20% of each, element A and B, with $M(A)$ = 17–47 and $M(B)$ = 54 or 57 (i.e., V and Cr, respectively). The stoichiometries of the $hP6$-Co$_{1.75}$Ge-type representatives are very diverse with Sb-contents in the range of 14.7–47.6%; elements A and B have M-values $M(A)$ = 51–67 and $M(B)$ = 60–83. Among the $cI28$-Th$_3$P$_4$-representatives, a first group of structures contains 42.9% Sb and 42.9% and 14.3% of elements A and B or vice versa ($M(A)$ = 17–30 and $M(B)$ = 24–33). The remaining compounds contain 21.4–39.3% Sb, always 57.1% of element A ($M(A)$ = 26, 27 or 33, i.e., Tb, Gd, and La, respectively) and 3.6–21.4% of element B with $M(B)$ = 82–87. The $tI32$-W$_5$Si$_3$-type structures contain 25.0–31.9% Sb and the complementary amount of 5.6–12.5% of element B ($M(B)$ = 57–81), as well as the constant amount of 62.5% of either Zr, Hf, or Ti ($M(A)$ = 49–51). All $hP6$-LiGaGe-type structures have composition 1:1:1 and are built from elements with $M(A)$ = 12–33 and $M(B)$ = 69–77. Nearly all $oP12$-PbCl$_2$-type structures contain 66.7% Sb, 11.0% of element A ($M(A)$ = 21–32), and 22.3% of U ($M(B)$ = 45); one compound has a different stoichiometry and was found in the Zr–Sn–Sb ($M(A)$ = 49 and $M(B)$ = 83).

7.15.9   Bismuthides

Binary bismuthides

Among the binary IMs 187 contain Bi, featuring 74 structure types, resulting in ≈ 2.5 representatives per structure type, i.e., comparable to the case of germanides

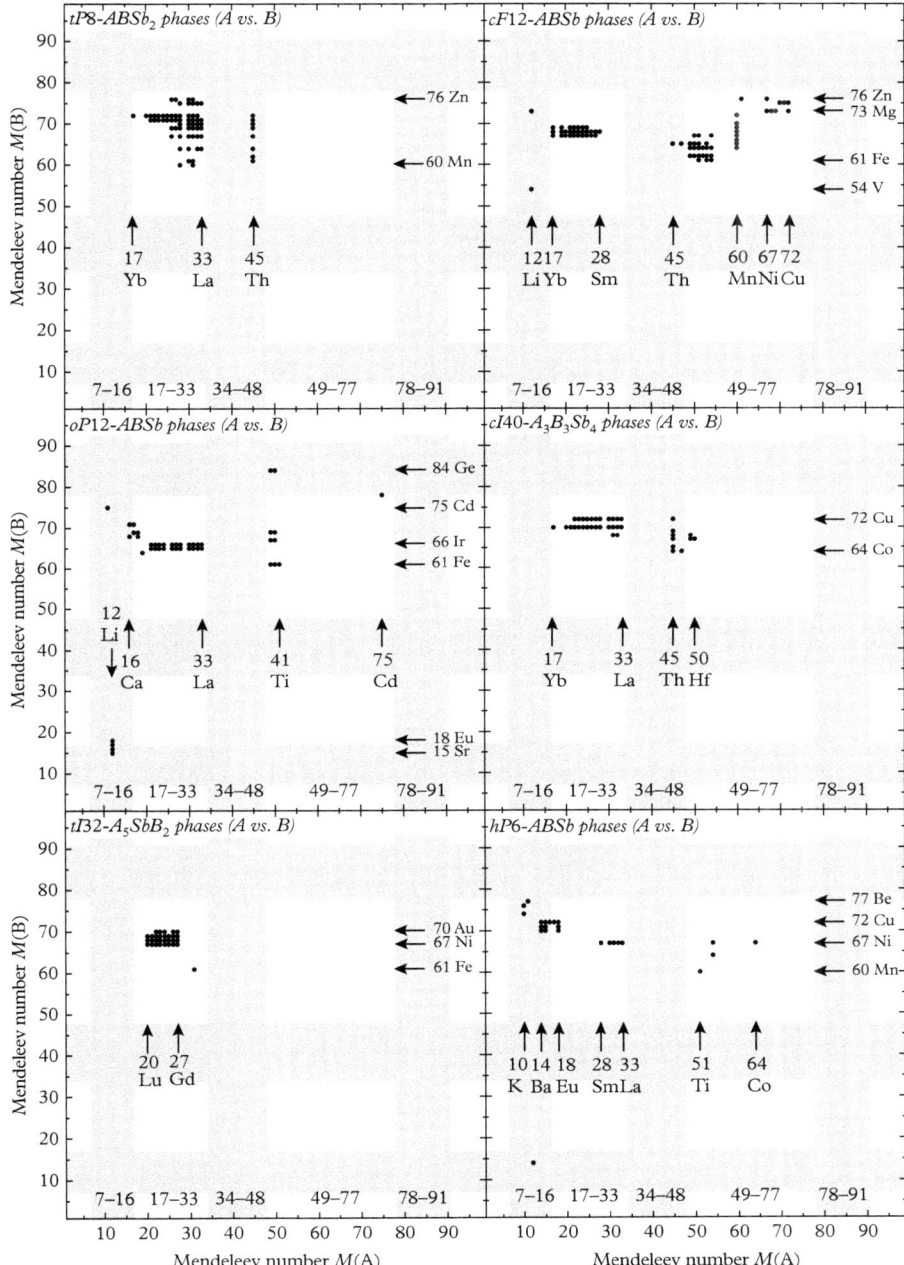

Fig. 7.90 *Occurrence of six of the most common structure types among ternary Sb-intermetallics. The Mendeleev numbers M of the other two elements in the compound are shown. The depicted structure types are tP8-CuHfSi$_2$, cF12-MgAgAs, oP12-TiNiSi, cI40-Au$_3$Y$_3$Sb$_4$, tI32-Mo$_5$SiB$_2$, and hP6-BeZrSi (ranks 1, 2, 4, 6, 7, and 9, respectively, in Table 7.27). Only 19 structures were omitted for structure type tP8-CuHfSi$_2$, where the chemical decoration of the structure changes. Where A and B have equal shares in the chemical formula, they are assigned so that M(A) < M(B). These are the most common structure types, whose sites are occupied in an ordered manner for the most part.*

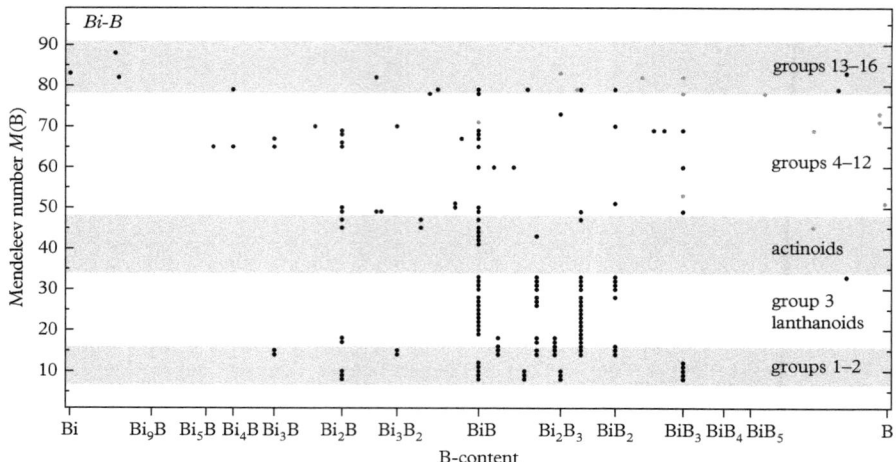

Fig. 7.91 *Stoichiometries of the 187 binary bismuthides, Bi_aB_b vs. the Mendeleev number M(B) of the other element in the compound. The 13 binary phases adopting unary structure types cF4-Cu, cI2-W, and hP2-Mg are marked by gray dots.*

(≈ 2.5). Their compositions are illustrated in Fig. 7.91. In Table 7.28, the twenty most common structure types among binary Bi-IMs are listed, four of them just unary ones. The compositions of the unary structure types $cF4$-Cu-, $hP2$-Mg-, and $cI2$-W-type structures are rather diverse (containing 1.0–25.0%, 0.4–50.0%, and 25.0–40.0% Bi, respectively). The same holds true for $hR6$-As-type structures with 94.0–99.9% Bi. Nearly all of the remaining structure types occur at their defined compositions. Only $cI2$-W is a structure type of a Bi-allotrope (Bi-V) that is stable at high pressure.

In some of the structure types, Bi always plays the role of the minority element ($hP16$-Mn$_5$Si$_3$, $tI12$-La$_2$Sb, $oP32$-Y$_5$Bi$_3$, $oP32$-Yb$_5$Sb$_3$, $tP56$-Ca$_{16}$Sb$_{11}$, $cF16$-BiF$_3$, $tI84$-Ho$_{11}$Ge$_{10}$, $hP6$-Co$_{1.75}$Ge, $mS20$-Eu$_3$Ga$_2$, and $hP8$-Na$_3$As). In $cI28$-Th$_3$P$_4$ and $cF24$-MgCu$_2$ it appears to switch between the roles of minority and majority element. Obviously, no such statement can be made for 1:1-stoichiometries, as in $cF8$-NaCl, $cP2$-CsCl, $hP4$-NiAs, and $mP32$-CsSb. No structure types consistently feature Bi as the majority component. The six most common structure types are mainly adopted by Ln- and An-bismuthides, respectively. An overview of the systems with bismuthides crystallizing in the twenty topmost structure types is shown in Fig. 7.92. The most common structure types of the binary bismuthides (ranks 1–5 in Table 7.28) are illustrated in Fig. 7.93, and will be discussed in the following.

- $cF8$-NaCl (225 $Fm\bar{3}m$): All RE elements except Yb, Pm, as well as the actinoids Cm, Am, Pu, Np, and U are known to adopt bismuthides in this structure type at equiatomic composition. This structure type can be

Table 7.28 *Most common structure types of* Bi-*containing binary intermetallics. The top 20 (out of 74) structure types are given, all of which have at least three representative structures and therefore represent at least 1.5% of all binary* Bi-*intermetallics each.*

Rank	Structure type	No.	Space group	Wyckoff positions	No. of reps.	% of all reps.
1.	$cF8$-NaCl	225	$Fm\bar{3}m$	$4ab$	19	10.2%
2.	$hP16$-Mn$_5$Si$_3$	193	$P6_3/mcm$	$4d\,6g^2$	14	7.5%
3.	$cI28$-Th$_3$P$_4$	220	$I\bar{4}3d$	$12a\,16c$	14	7.5%
4.	$tI12$-La$_2$Sb	139	$I4/mmm$	$4ce^2$	8	4.3%
5.	$oP32$-Y$_5$Bi$_3$	62	$Pnma$	$4c^4\,8d^2$	8	4.3%
6.	$oP32$-Yb$_5$Sb$_3$	62	$Pnma$	$4c^4\,8d^2$	7	3.7%
7.	$tP56$-Ca$_{16}$Sb$_{11}$	113	$P\bar{4}2_1m$	$2c^2\,4d^2e^5\,8f^3$	5	2.7%
8.	$cF4$-Cu	225	$Fm\bar{3}m$	$4a$	5	2.7%
9.	$hP2$-Mg	194	$P6_3/mmc$	$2c$	5	2.7%
10.	$cF16$-BiF$_3$	225	$Fm\bar{3}m$	$4c^2\,8d$	4	2.1%
11.	$cP2$-CsCl	221	$Pm\bar{3}m$	$1ab$	4	2.1%
12.	$tI84$-Ho$_{11}$Ge$_{10}$	139	$I4/mmm$	$4de^2\,8h^2j\,16mn^2$	4	2.1%
13.	$cF24$-MgCu$_2$	227	$Fd\bar{3}m$	$8a\,16d$	4	2.1%
14.	$hP4$-NiAs	194	$P6_3/mmc$	$2ac$	4	2.1%
15.	$hR6$-As	166	$R\bar{3}m$	$6c$	3	1.6%
16.	$hP6$-Co$_{1.75}$Ge	194	$P6_3/mmc$	$2acd$	3	1.6%
17.	$mP32$-CsSb	11	$P2_1/c$	$4e^8$	3	1.6%
18.	$mS20$-Eu$_3$Ga$_2$	15	$C2/c$	$4e\,8f^2$	3	1.6%
19.	$hP8$-Na$_3$As	194	$P6_3/mmc$	$2bc\,4f$	3	1.6%
20.	$cI2$-W	229	$Im\bar{3}m$	$2a$	3	1.6%
					123	65.8%

described as a *ccp* packing of the larger atoms with the smaller atoms in the octahedral voids. Since the RE elements, with radii between 1.62 to 1.995 Å, are larger than that of Bi with 1.545 Å, Bi can be considered formally to occupy the octahedral voids. Therefore, the structure can be described as a packing of Bi-centered edge-sharing RE octahedra and empty RE tetrahedra. In the case of the actinoids, the opposite is true.

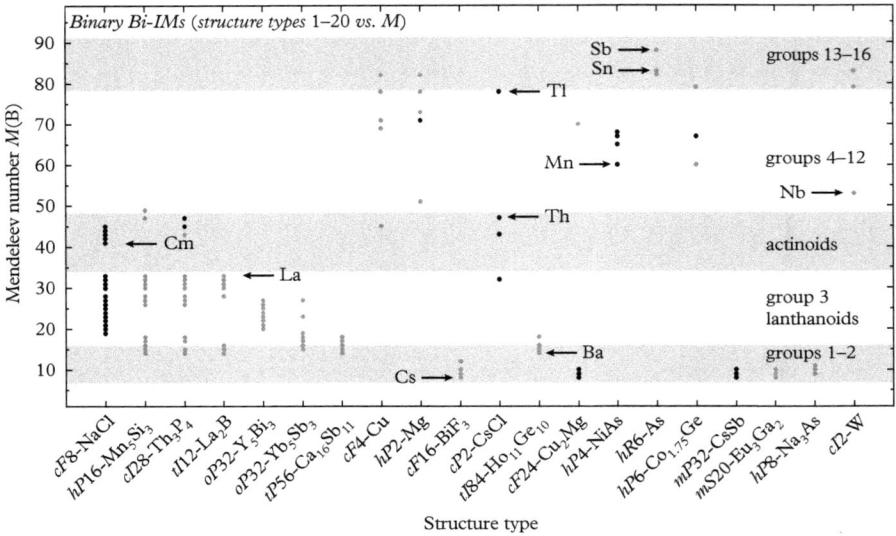

Fig. 7.92 *Occurrence of the 20 most common structure types among binary Bi-intermetallics. The Mendeleev number M of the other element in the compound is shown over the rank of the structure type, as given in Table 7.28. Structures with Bi as their major component ($\geq 50\%$) are marked in black; those with minor Bi-content are shown in gray.*

- $hP16$-Mn_5Si_3 (193 $P6_3/mcm$): This structure type is adopted by bismuthides of divalent Ba, Sr, Yb, and Eu, as well as by the trivalent RE metals Tb, Gd, Sm, Nd, Pr, Ce, La, and Th, as well as Zr. In this structure type, Si and Mn form distorted CN16 FK-polyhedra, which share two triangular faces. Thereby, the Mn1 atoms are arranged in linear chains in $1/3, 2/3, z$, parallel to [001], with short Mn–Mn distances of 2.407 Å ($r_{Mn} = 1.367$ Å). The Mn2 atoms form face-sharing columns of octahedra in $0, 0$, and z, also parallel to [001]. There, the Mn–Mn distances amount to 2.822 Å and 2.907 Å, respectively. In the bismuthides, Bi takes over the Si sites.

- $cI28$-Th_3P_4 (220 $I\bar{4}3d$): This structure type is adopted by bismuthides of divalent Ba, Sr, Ca, Yb, and Eu, as well as by the trivalent RE metals Tb, Gd, Sm, Nd, Pr, Ce, La, and Pu, U, Th, as well as Zr. The basic structural unit can be seen as a trigonal P-centered Th antiprism, with three of the six sides plus the top and bottom capped by P atoms. These Th antiprisms are stacked in a similar way along [111] as the Mn2 octahedra in $hP16$-Mn_5Si_3. As shown in Fig. 7.93 (c), (f), motifs of the projected stuctures (highlighted by gray triangles) are quite similar. In these bismuthides, Bi takes over the Th sites, except in the case of Th and Zr, as constituting elements.

- $tI12$-La_2Sb (139 $I4/mmm$): This structure type is adopted by bismuthides of divalent Ba, Sr, and Ca as well as by the trivalent RE metals Sm, Nd,

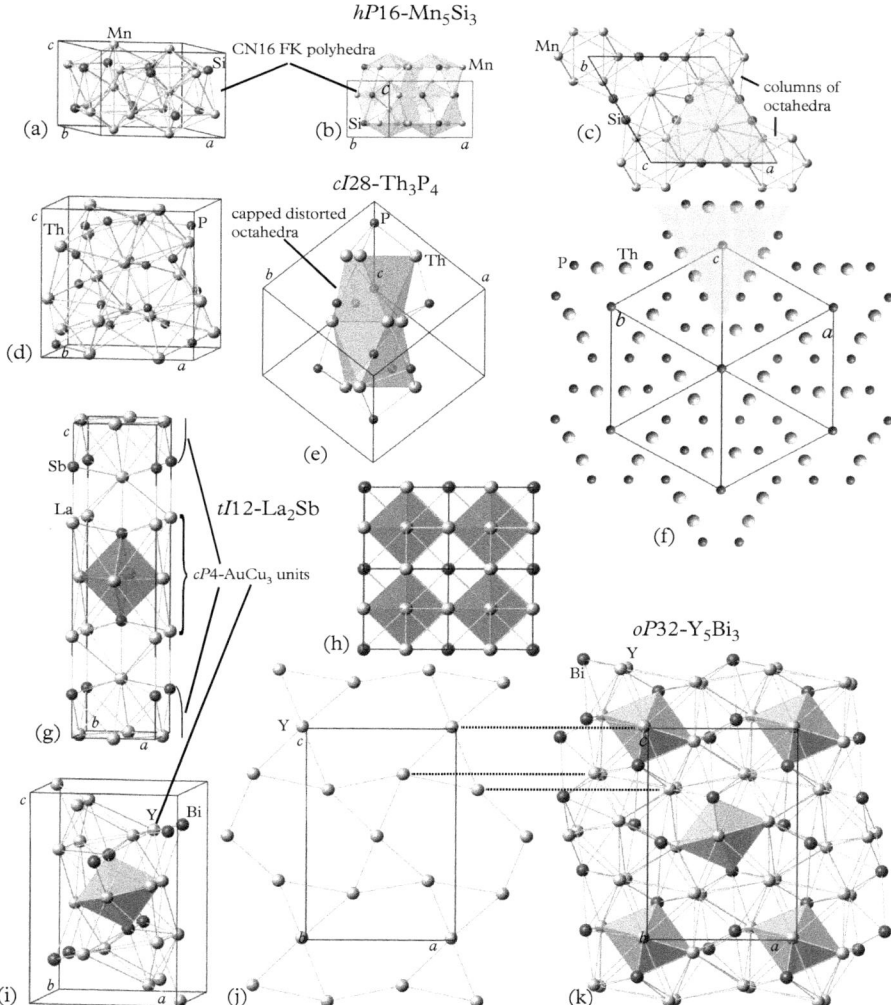

Fig. 7.93 *Structures of the most common structure types of binary bismuthides (ranks 2–5 in Table 7.28) in different projections: (a)–(c) hP16-Mn$_5$Si$_3$, (d)–(f) cI28-Th$_3$P$_4$, (g)–(h) tI12-La$_2$Sb, and (i)–(k) oP32-Y$_5$Bi$_3$. Some characteristic structural features such as FK-polyhedra and octahedra are marked and described in the text. One structural unit common to all these structure types are empty or centered octahedra. In (c) the face-sharing octahedra form columns along [001], and in (e) along [111]. In (g) they share all four vertices within the (110) planes, and in (k) they form chains along [010] linked via two opposite vertices.*

Pr, Ce, and La. This structure can be seen to contain LaSb$_3$ units ($cP4$-Cu$_3$Au structure type), which are alternatingly shifted by $(1/2, 1/2, z)$. In these bismuthides, Bi takes over the Sb sites.

- $oP32$-Y$_5$Bi$_3$ (62 *Pnma*): This structure type is adopted by bismuthides of the trivalent RE metals Lu, Tm, Er, Ho, Dy, Y, Tb, and Gd, only. The structure shows similar pseudo-cubic units, centered by octahedra, as $tI12$-La$_2$Sb. Along [010], in the corners of the unit cell, there are columns of trigonal Y-prisms centered by Bi atoms, which share edges forming deformed hexagonal channels. The octahedra share two of their vertices forming chains along [010], at the same time the pseudo-cubes, which enclose them, share faces. According to Wang *et al.* (1976), the $hP16$-Mn$_5$Si$_3$ structure type is stable in the Ln$_5$Bi$_3$ series with a radius ratio Ln/Bi $\gtrsim$ 1.05, then due to the finite compressibility of the Ln ion core, the $oP32$-Y$_5$Bi$_3$ structure type becomes more stable. The Bi–Gd distances in $hP16$-Gd$_5$Bi$_3$ and $oP32$-Gd$_5$Bi$_3$, for instance, differ only slightly with 3.31 to 3.28 Å, respectively, because of the strong metallic interaction and the low CN in the trigonal prismatic AET. The authors note also a close similarity of the structure type $oP32$-Y$_5$Bi$_3$ to $oP32$-Yb$_5$Sb$_3$, which has rank 6 in Table 7.28.

The alkali and alkaline earth metals as well as the lanthanoids and actinoids all have congruently melting compounds between 25% and 50% Bi composition. Their melting temperatures exceed those of the elements considerably. For instance, $cF16$-BiLi$_3$ ($cF16$-BiF$_3$ type) melts at 2093 K compared to the melting temperatures of Bi and Li of 544.4 K and 453.6 K.

Ternary bismuthides

The 365 ternary Bi-containing IMs feature 89 different structure types, with ≈ 4.1 representatives per structure type, i.e., slightly less than in the case of germanides (≈ 5.8). In Table 7.29, the most common structure types among ternary bismuthides are given, and the distribution of the chemical compositions of four of the most common ones are shown in Fig. 7.94. In the ternary compounds, Bi is most often the element with the highest Mendeleev number (87), and will therefore be assigned to the C-position, while the A-elements will always have smaller M-values than B-elements in all formulae of the type A$_a$B$_b$Bi$_c$ ($M(A) < M(B)$).

The ternary Bi-intermetallics crystallizing in structure type $cF12$-MgAgAs all have 1:1:1-stoichiometry; they are formed with the additional elements A, with $M(A) = 12$–72 and B with $M(B) = 54$–76. With only one slightly deviating compound, the $tI32$-Mo$_5$SiB$_2$-type structures contain 12.5% of Bi, as well as 62.5% of element A with $M(A) = 20$–27 and 25.0% of element B with $M(B) = 64$–70. All $hP9$-K$_2$UF$_6$-type structures contain 22.2% Bi, as well as 66.7% and 11.1% of elements A and B, respectively, with $M(A) = 20$–27, 49, or 50, and $M(B) = 60$–67. The structures of type $cI40$-Au$_3$Y$_3$Sb$_4$ all contain 40% Bi and 30%, each, of elements A and B ($M(A) = 28$–33 or 45, and $M(B) = 65$–72). The $tP8$-CuHfSi$_2$-type structures contain 50.0–57.1% Bi, as well as 25.0–28.6% of element A ($M(A) = 12$–33) and 14.3–25.0% of element B ($M(B) = 33$, 60 or 67–76).

Table 7.29 *Most common structure types of* Bi-*containing ternary intermetallics. The top 27 (out of 89) structure types are given, all of which have at least four representative structures and therefore represent at least 1% of all ternary* Bi-*intermetallics.*

Rank	Structure type	No.	Space group	Wyckoff positions	No. of reps.	% of all reps.
1.	$cF12$-MgAgAs	216	$F\bar{4}3m$	$4abc$	45	12.3%
2.	$tI32$-Mo$_5$SiB$_2$	140	$I4/mcm$	$4ac\,8h\,16l$	31	8.5%
3.	$hP9$-K$_2$UF$_6$	189	$P\bar{6}2m$	$1a\,2d\,3fg$	24	6.6%
4.	$cI40$-Au$_3$Y$_3$Sb$_4$	220	$I\bar{4}3d$	$12ab\,16c$	17	4.7%
5.	$tP8$-CuHfSi$_2$	129	$P4/nmm$	$2abc^2$	16	4.4%
6.	$hP18$-CuHf$_5$Sn$_3$	193	$P6_3/mcm$	$2b\,4d\,6g^2$	14	3.8%
7.	$hP6$-BeZrSi	194	$P6_3/mmc$	$2acd$	10	2.7%
8.	$tP10$-CaBe$_2$Ge$_2$	129	$P4/nmm$	$2abc^3$	10	2.7%
9.	$hP4$-NiAs	194	$P6_3/mmc$	$2ac$	8	2.2%
10.	$oP12$-TiNiSi	62	$Pnma$	$4c^3$	8	2.2%
11.	$cP4$-Cu$_3$Au	221	$Pm\bar{3}m$	$1a\,3c$	7	1.9%
12.	$oI36$-Ho$_{12}$Co$_5$Bi	71	$Immm$	$2ac\,4ej\,8lmn$	7	1.9%
13.	$tP6$-PbClF	129	$P4/nmm$	$2ac^2$	7	1.9%
14.	$tI16$-SrZnBi$_2$	139	$I4/mmm$	$4cde^2$	7	1.9%
15.	$oP32$-Y$_2$HfS$_5$	62	$Pnma$	$4c^4\,8d^2$	7	1.9%
16.	$cF16$-BiF$_3$	225	$Fm\bar{3}m$	$4c^2\,8d$	6	1.6%
17.	$tI208$-Ca$_{14}$AlSb$_{11}$	142	$I4_1/acd$	$8ab\,16ef\,32g^5$	6	1.6%
18.	$cI28$-Th$_3$P$_4$	220	$I\bar{4}3d$	$12a\,16c$	6	1.6%
19.	$oP48$-Zn$_{4.23}$Yb$_9$Sb$_9$	55	$Pbam$	$2ac\,4g^5\,h^6$	6	1.6%
20.	$hP9$-ZrNiAl	189	$P\bar{6}2m$	$1a2d3fg$	6	1.6%
21.	$hP6$-LiGaGe	186	$P6_3mc$	$2abb$	5	1.4%
22.	$tI28$-V$_4$SiSb$_2$	140	$I4/mcm$	$4a\,8h\,16k$	5	1.4%
23.	$oP44$-Ca$_9$Mn$_4$Bi$_9$	55	$Pbam$	$2ac\,4g^5\,h^5$	4	1.1%
24.	$hP5$-Ce$_2$SO$_2$	162	$P\bar{3}m$	$1a\,2d^2$	4	1.1%
25.	$cF4$-Cu	225	$Fm\bar{3}m$	$4a$	4	1.1%
26.	$cF88$-Cu$_4$Mn$_3$Bi$_4$	225	$Fm\bar{3}m$	$8c\,24de\,32f$	4	1.1%
27.	$tI80$-La$_6$Co$_{11}$Ga$_3$	140	$I4/mcm$	$4ad\,8f\,16kl^3$	4	1.1%
					278	76.2%

The structures of type $hP18$-CuHf$_5$Sn$_3$ contain either 55.6% Bi or 33.3% (with one slightly off stoichiometry—34.9% in system Ti–Zn–Bi). Mostly, element B (M(B) = 60, 72, 73, or 76) has content 11.1%, while element A then takes on the role complementary to Bi: 33.3% of M(A) = 30–33 with M(B) = 60 or 73, i.e., Mn or Mg, respectively, as well as 11.1% of M(A) = 26–33 or 51 with M(B) = 72 (or 76—see above), i.e., Cu. In one structure, A- and B-contents are reversed (Sc–La–Bi; 11.1% with M(A) = 19 and 33.3% with M(B) = 33). The $hP6$-BeZrSi-type structures all have 1:1:1-stoichiometry and contain elements A and B with M(A) = 14–18 and M(B) = 70–72. Of the $tP10$-CaBe$_2$Ge$_2$-type structures, only two have 1:2:2-stoichiometry with 40% Bi and 20% of each, A and B, in this case Sr/Eu and Pd, respectively (M(A) = 15 or 18 and M(B) = 69). The remaining structures contain 44.4% Bi, as well as 22.2% of element A with M(A) = 24–33 and 33.3% Ni (M(B) = 67). The $hP4$-NiAs-type structures contain 10–50% Bi, 20–50% of element A (M(A) = 30 or 60–69), and 15–40% of element B (M(B) = 60–88). All $oP12$-TiNiSi-type structures have stoichiometry 1:1:1; they are either built with Li (M(A) = 12) and an element B with M(B) = 15–18, or elements A with values M(A) = 28–32 and Rh (M(B) = 65).

The representatives of structure type $cP4$-Cu$_3$Au mostly contain 75% of Pt or Pd (M(A) = 68 or 69, respectively) with a largely equal division of the remaining 25% between Bi and element B (M(B) = 78–82); one compound is formed with 25% of Ba (M(A) = 14) and a mixture of Pb (M(B) = 82) and Bi for the remaining 75%. All $oI36$-Ho$_{12}$Co$_5$Bi-type structures contain 5.6% Bi and 27.8% Co (M(B) = 64), as well as 66.7% of element A with M(A) = 21–27. The 1:1:1-compounds of type $tP6$-PbClF are built either with elements M(A) = 8–11 and M(B) = 16, 60, or 73, or—in one case—in the U–Sb–Bi-system (M(A) = 45 and M(B) = 88). The $tI16$-SrZnBi$_2$-type structures contain 50% Bi, as well as 25%, each, of elements A (M(A) = 14, 15, or 32) and B (M(B) = 60, 75, 76, or 84). Most $oP32$-Y$_2$HfS$_5$-type structures contain 25% Bi, 62.5% of element A (M(A) = 22–27) and 12.5% of Co, Ni, or Cu (M(B) = 64, 67, or 72, respectively), while the latter one is slightly Cu-deficient, compensated by a higher Bi-content of 28.8%. The $cF16$-BiF$_3$-type structures contain 12.5–27.9% Bi, combined with elements A and B with M(A) = 12 or 69 and M(B) = 49–54, 80, or 84.

All $tI208$-Ca$_{14}$AlSb$_{11}$-type structures are stoichiometric with 42.3% Bi, 53.8% of element A (M(A) = 14–18), and 3.8% of Mn or In (M(B) = 60 or 79). The $cI28$-Th$_3$P$_4$-type representatives mostly contain 57.1% of element A with M(A) = 26–33, the remaining formula being made up by Bi and elements B with values M(B) = 82–88; the only compound where A and B combine to 57.1%, is built from elements with M-values are 24 and 27, i.e., Dy and Gd, respectively. $oP48$-Zn$_{4.23}$Yb$_9$Sb$_9$-type structures contain 40.4–40.9% Bi, as well as similar amounts of element A (M(A) = 15–18) and 18.2–19.2% of element B, i.e., Cd or Zn (M(B) = 75 and 76, respectively). All $hP9$-ZrNiAl-type structures have stoichiometry 1:1:1 and contain elements with M(A) = 11 or

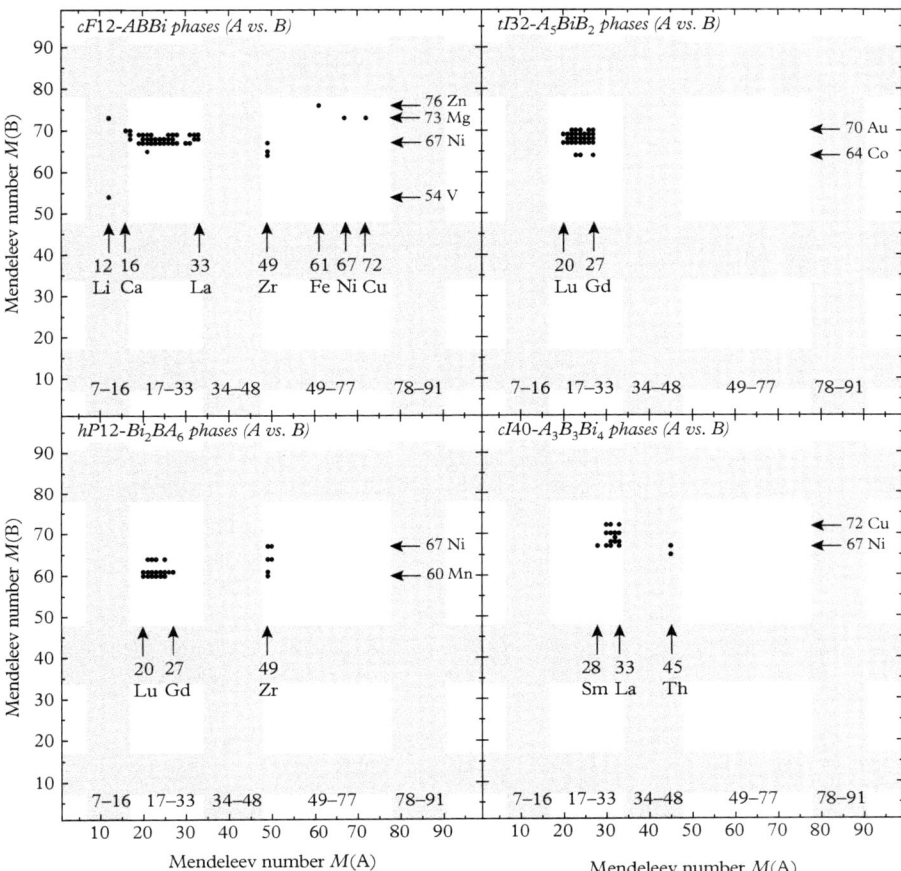

Fig. 7.94 *Occurrence of the four most common structure types among ternary Bi-intermetallics. The Mendeleev number M of the other two elements in the compound are shown. The depicted structure types are cF12-MgAgAs, tI32-Mo$_5$SiB$_2$, hP9-K$_2$UF$_6$, and cI40-Au$_3$Y$_3$Sb$_4$ (ranks 1–4, respectively, in Table 7.29). Where A and B have equal shares in the chemical formula, they are assigned so that M(A) < M(B).*

22–27 and $M(B) = 14$ or 65. Also the $hP6$-LiGaGe-type structures have 1:1:1-stoichiometries, while being built with elements A and B with $M(A) = 16$–18 and $M(B) = 70$–72. The $tI28$-V$_4$SiSb$_2$-type structures are stoichiometric with 28.6% of Bi, 57.1% of Ti ($M(A) = 51$), and 14.3% of element B with $M(B) = 57$–67. $oP44$-Ca$_9$Mn$_4$Bi$_9$-type representatives contain 40.9% Bi, the same amount of element A (Sr or Ca with $M(A) = 15$ and 16, respectively), and 18.2% of element B with values $M(B) = 60$, 75, or 76. The structures of type $hP5$-Ce$_2$SO$_2$

contain 40%, each, of Bi and Mn or Mg ($M(B) = 60$ or 73), and 20% of element A ($M(A) = 14$–16). Four $cF4$-Cu-type structures contain 0.2–15% Bi, in addition to a mixture of elements $M(A) = 71$–78 and $M(B) = 72$–82. All $cF88$-Cu$_4$Mn$_3$Bi$_4$-type structures contain 36.4% Bi; three contain 45.5% Mn ($M(A) = 60$) and 18.2% of element B with $M(B) = 65$–69, while the eponymous compound was reported to be built from 27.3% Mn and 36.4% Cu ($M(B) = 72$). The $tI80$-La$_6$Co$_{11}$Ga$_3$-type structures all contain 5% Bi in addition to 30% of element A ($M(A) = 28$–33) and 65% of element B ($M(B) = 61$ or 64, i.e., Fe and Co, respectively).

In the following, the structures of the topmost five structure types of ternary bismuthides (see Table 7.29) are discussed and then illustrated in Fig. 7.95.

- $cF12$-MgAgAs (216 $F\bar{4}3m$): This structure type is also known as $cF12$-LiAlSi, in which the Al (Ag and other TM) atoms form a *ccp* packing, the Si (As and other main group) elements occupy half of the tetrahedral voids, and Li (Mg and RE) atoms are sitting in the octahedral voids (see Fig. 7.2).

- $tI32$-Mo$_5$SiB$_2$ (Rawn *et al.*, 2001) (140 $I4/mcm$): The structure type can be seen as an ordered variant of the $tI32$-Cr$_5$B$_3$ type. In bismuthides, the Mo site can be occupied by one of the RE elements Lu, Tm, Er, Ho, Dy, Tb, and Gd, the Si site by Co, Ni, Pt, Pd, and Au, and the B site by Bi. The structure can be seen as being constituted from Mo-centered Mo cubes, which are edge-connected so that their base and top squares form a $3^2.4.3.4$ triangle/square tiling. The B atoms cap the side squares of the cubes, and the Si atoms center the square antiprisms constituted from the top and bottom squares of along [001] neighboring Mo cubes.

- $hP9$-K$_2$UF$_6$ (Brunton, 1969) (189 $P\bar{6}2m$): This structure type, which can be seen as an ordered $hP9$-Fe$_2$P version, is also known as the $hP9$-Zr$_6$CoAs$_2$ structure type. The structure can be geometrically seen as a packing of large hexagonal K-prisms centered by U atoms. U is coordinated by nine F atoms, three within the K honeycomb layers, and three above and below them in the U layers. In bismuthides, Bi sits in the K-position, RE elements (Lu, Tm, Er, Ho, Dy, Y, Tb, and Gd) plus Zr and Hf in the F-site, and TM atoms (Mn, Fe, Ru, Os, Co, Rh, Ir, and Ni) at the U-site.

- $cI40$-Au$_3$Y$_3$Sb$_4$ (220 $I\bar{4}3d$): Bi is located at the Sb-site, RE elements at the Au site, and TM atoms at the Au position. The Au atoms center elongated Sb-tetrahedra, which are vertex-connected, while Y atoms center the space in-between.

- $tP8$-CuHfSi$_2$ (129 $P4/nmm$): This structure type is equivalent to that of $tP8$-CuZrSi$_2$. In the respective bismuthides, Bi occupies the Si-site, Li, Ba, Sr, Ca and the lanthanoids replace Hf, and La, Mn, Ni, Pd, Pt, Au, Ag, Cu, Hg, Cd, and Zn are located on the Cu-site. The structure consists of a flat 4^4 net of Si, followed by a puckered rotated 4^4 net of Si/Hf, followed by a flat 4^4 net of Cu.

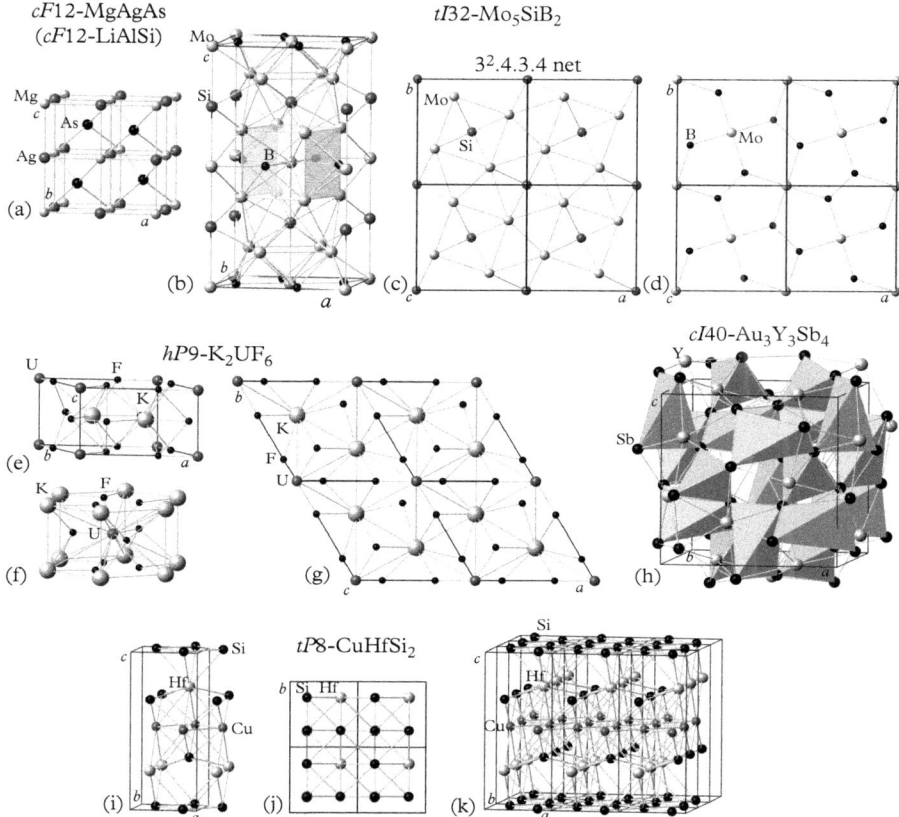

Fig. 7.95 *The topmost five structure types of ternary bismuthides: (a) cF12-MgAgAs (cF12-LiAlSi), (b)–(d) tI32-Mo₅SiB₂, (e)–(g) hP9-K₂UF₆ (hP9-Zr₆CoAs₂), (h) cI40-Au₃Y₃Sb₄, and (i)–(k) tP8-CuHfSi₂ (tP8-CuZrSi₂). In (b), the main structural unit, a body-centered Mo cube, is shaded gray. The base and top faces of the edge-connected cubes form a 3².4.3.4 triangle/square tiling each, with the squares capped by Si (c). The B/Mo pentagon tiling in-between—(d)—contains the body center of the Mo-cube. (f) The basic structural building unit of hP9-K₂UF₆ is a hexagonal K-prism, centered by a U atom, which is 9-coordinated by F. (h) The structure of cI40-Au₃Y₃Sb₄ can be seen as consisting of vertex-connected elongated Sb-tetrahedra centered by Au, with the large Y distributed in the spaces between. (j) The structure of tP8-CuHfSi₂ can be described by a stacking of flat and puckered 4⁴ nets decorated with Si, Si/Hf, and Cu, respectively.*

7.15.10   Polonides

Among the binary IMs, only 24 Po-containing intermetallics are listed in the PCD, featuring just three structure types, i.e., 8.0 representatives per structure type. In contrast, no ternary polonides can be found there. The comparatively small

Table 7.30 *The structure types of binary polonides.*

Rank	Structure type	No.	Space group	Wyckoff positions	No. of structures	% of all structures
1.	*cF*8-NaCl	225	$Fm\bar{3}m$	4ab	15	62.5%
2.	*hP*4-NiAs	194	$P6_3/mmc$	2ac	6	25.0%
3.	*cF*8-ZnS	216	$F\bar{4}3m$	4ac	3	12.5%
					24	100.0%

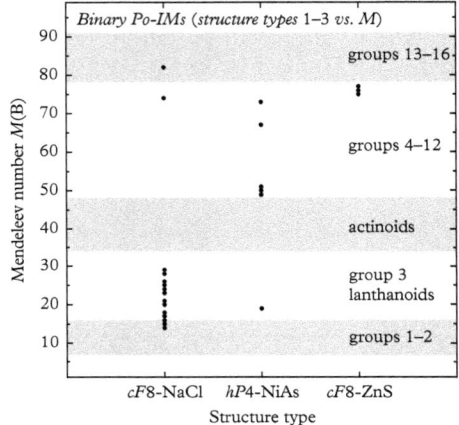

Fig. 7.96 *Occurrence of the three different structure types among binary polonides. Po has the Mendeleev number $M(Po) = 91$. The Mendeleev number $M(B)$ of the other element in the compound is marked by a dot over the respective structure type.*

number of polonides known so far has to be attributed to the toxicity and radio-activity of this element as well as to its scarcity. In Table 7.30, the structure types of the binary polonides are given, and in Fig. 7.96 their constituting elements are shown.

All of these 24 Po-intermetallics have 1:1-compositions and simple structures. The *cF*8-NaCl structure type is almost exclusively adopted by RE polonides (plus Ba, Ca, Hg, and Pb), which have metallic character. The *hP*4-NiAs type by TM compounds (plus Sc and Mg), and the few *cF*8-ZnS type structures with the group 12 elements Cd, Zn, and Be, have a more ionic character. Po has with

$\chi = 2.0$ the highest electronegativity of the metallic main group elements, and can be expected to be negatively charged (-2) in all these compounds except in those with Hg.

7.16 Lanthanoid/lanthanoid and actinoid/actinoid compounds

Because of their close chemical similarity (for electronegativities and atomic radii see Fig. 7.97), no ordered binary compounds are known so far between two lanthanoids (Ln) or two actinoids (An). What is observed, however, are solid solutions of one component in one of the allotropes of the other, depending on temperature. "Mischmetall" (mixed metal) is a solid solution of usually 50% Ce, 25% Ln, and small amounts of Nd and Pr, as it can be obtained from the ore monazite. Another solid solution is the five-component high-entropy alloy (HEA), HoDyYTbGd, which has been studied recently (Feuerbacher *et al.*, 2015). In most cases, the structures of the different allotropes stable at different temperatures are of the types $hP2$-Mg, $cI2$-W, $cF4$-Cu, or $hP4$-La. In several cases, a pseudo-binary LT-phase (δ) has been identified in the range between 20% and 80%, with a structure of the $hR9$-Sm type. Binary phases assigned to unary structure types with only one crystallographic site occupied must be disordered,

1	2	3	4	5	6	7	8	9	10	11	12	13	14	15	16
12 Li 1.0 1.52	77 Be 1.5 1.11														
11 Na 0.9 1.86	73 Mg 1.2 1.60											80 Al 1.5 1.43			
10 K 0.8 2.27	16 Ca 1.0 1.97	19 Sc 1.3 1.61	51 Ti 1.4 1.45	54 V 1.6 1.31	57 Cr 1.6 1.25	60 Mn 1.5 1.37	61 Fe 1.8 1.24	64 Co 1.8 1.25	67 Ni 1.8 1.25	72 Cu 1.9 1.28	76 Zn 1.6 1.34	81 Ga 1.6 1.22	84 Ge 1.8 1.23		
9 Rb 0.8 2.48	15 Sr 1.0 2.15	49 Y 1.2 1.78	53 Zr 1.4 1.59	56 Nb 1.6 1.43	59 Mo 1.8 1.36	62 Tc 1.9 1.35	65 Ru 2.2 1.33	69 Rh 2.2 1.35	71 Pd 2.2 1.38	75 Ag 2.4 1.45	79 Cd 1.7 1.49	83 In 1.7 1.63	84 Sn 1.8 1.41	88 Sb 1.9 1.45	
8 Cs 0.7 2.66	14 Ba 0.9 2.17	33* La 1.1 1.87	50 Hf 1.3 1.56	52 Ta 1.5 1.43	55 W 1.7 1.37	58 Re 1.9 1.37	63 Os 2.2 1.34	66 Ir 2.2 1.36	68 Pt 2.2 1.37	70 Au 2.4 1.44	74 Hg 1.9 1.50	78 Tl 1.8 1.70	82 Pb 1.8 1.75	87 Bi 1.9 1.55	91 Po 2.0 1.67
7 Fr 0.7	13 Ra 0.9 2.15	48+ Ac 1.1 1.88													

* Lanthanoids													
32 Ce 1.1 1.87	31 Pr 1.1 1.82	30 Nd 1.1 1.81	29 Pm 1.1 1.63	28 Sm 1.2 1.62	18 Eu 1.2 2.00	27 Gd 1.2 1.79	26 Tb 1.1 1.76	25 Dy 1.2 1.75	24 Ho 1.2 1.74	23 Er 1.2 1.73	22 Tm 1.3 1.72	17 Yb 1.1 1.94	20 Lu 1.3 1.72
+ Actinoids													
47 Th 1.3 1.88	46 Pa 1.5 1.80	45 U 1.4 1.39	44 Np 1.4 1.30	43 Pu 1.3 1.51	42 Am 1.1	41 Cm 1.3	40 Bk 1.3	39 Cf 1.3	38 Es 1.3	37 Fm 1.3	36 Md 1.3	35 No 1.3	34 Lr 1.3

Fig. 7.97 *Elements constituting the compounds discussed in this section are shaded gray in the periodic table. Mendeleev numbers (top left in each box), Pauling electronegativities χ (relative to $\chi_F = 4.0$) (bottom left in each box), and atomic radii (half of the shortest distance between atoms in the crystal structure at ambient conditions) (bottom right in each box) of the metallic elements are given.*

supposing they have been determined reliably in spite of the weak contrast for X-ray diffraction.

In the PCD, there are 73 entries for binary Ln–Ln phases with the following structure types: $cF4$-Cu, $cI2$-W, $hP2$-Mg, $hP4$-La, and $hR9$-Sm. All these structure types are adopted by one or the other rare-earth element. Most of them show the $hP2$-Mg structure at ambient conditions and the $cI2$-W type at elevated temperatures, while at higher pressures the $hR9$-Sm type is the most common one. There is just one entry of a ternary Ln–Ln–Ln phase, which is a solid solution, however.

In the following, all binary Ln–Ln phases are given, whose structure type differs from the ones of both constituting elements at ambient conditions. However, at high pressures at least one of the two constituents has the structure type of the resulting binary phase. This could be understood in the way that the second component stabilizes a high-pressure phase at ambient conditions.

- $cF4$-Cu: $cI2$-Eu/$hP2$-Yb and $hP2$-Gd/$hP2$-Yb. Gd and Yb both have high-pressure allotropes with the $cF4$-Cu structure type.
- $hP4$-La: $cF4$-Ce–($hP2$-Ho, $hP2$-Tb, $hP2$-Gd, and $hR9$-Sm). Ho, Tb, Gd, and Sm all have high-pressure allotropes with the $hP4$-La structure type.
- $hR9$-Sm: $hP4$-La–($hP2$-Gd, $hP2$-Dy), $cF4$-Ce–($hP2$-Er, $hP2$-Ho, $hP2$-Tb, $hP2$-Gd), $hP4$-Pr–($hP2$-Ho, $hP2$-Tb, $hP2$-Gd), $hP4$-Nd–($hP2$-Tm, $hP2$-Ho, $hP2$-Dy, and $hP2$-Gd). Gd, Dy, Er, Ho, Tb, Tm, and Dy all have high-pressure allotropes with the $hR9$-Sm structure type.

A similar situation applies to the An–An phases. There are 17 binary phases in the PCD, all of them but one crystallizing in unary structure types: $cF4$-Cu, $cI2$-W, $oP8$-Np, $tP4$-Np, $mP16$-Pu, $oF8$-Pu, $mS34$-Pu, $oS4$-U, and $tP30$-U. There is no entry for a ternary An–An–An phase. In the following, all binary An–An phases are given, whose structure type differs from the ones of both constituting elements at ambient conditions. However, at high temperatures at least one of the two constituents has the structure type of the resulting binary phase. This could be understood in such a way that the second component stabilizes a HT-phase at ambient conditions.

- $cF4$-Cu: $mP16$-Pu–($hP4$-Am, $oC4$-U). Am has a high-pressure allotrope with the $cF4$-Cu structure type.
- $cI2$-W: $mP16$-Pu–($oP8$-Np, $oC4$-U). All three actinoids have HT-phases with the $cI2$-W structure type.

From the few phase diagrams explored so far, a (pseudo)-binary phase (δ) has also been experimentally observed but its structure has not been determined yet. In the case of Np–Pu it was identified as orthorhombic (Sheldon and Peterson, 1985a), for Np–U as $cP58$ (Sheldon and Peterson,

$hR147\text{-}U_{0.4}Pu_{0.6}$

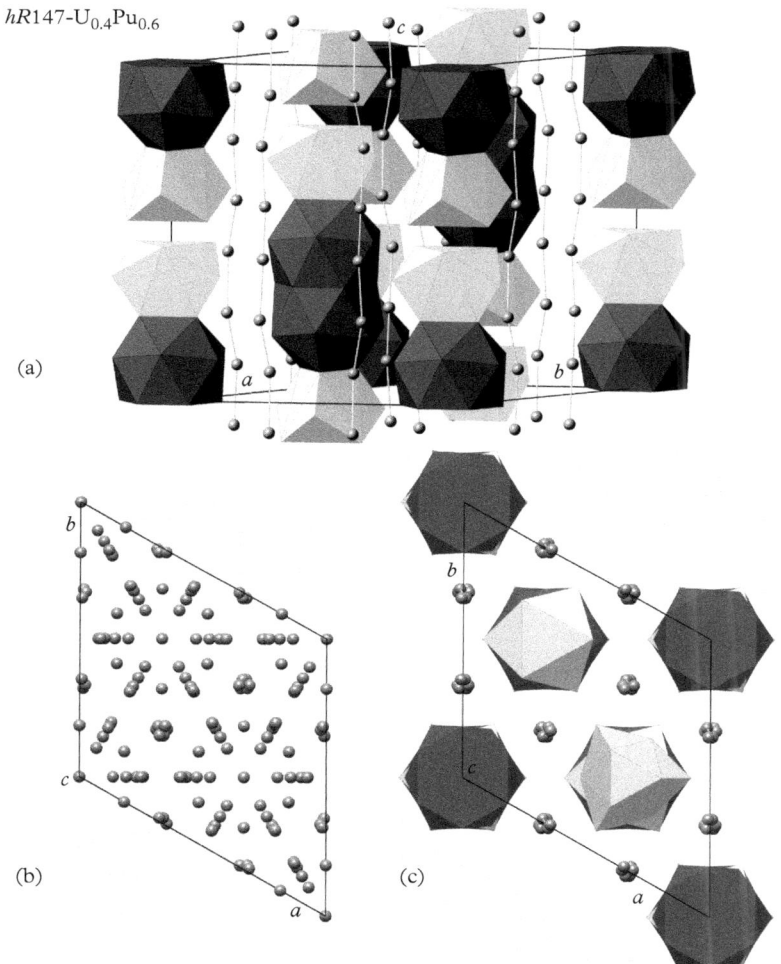

Fig. 7.98 *The structure of hR147-$U_{0.4}Pu_{0.6}$ in different projections: (a, c) Polyhedron representations with a quite regular CN16 Frank-Kasper polyhedron (dark gray), and a distorted CN15 one (light gray) stacked along [001]. In-between are Pu zigzag chains with alternating distances of 3.349 Å and 2.913 Å. (b) Projection of the structure along [001].*

1985b). In the system Pu–U even two binary phases have been observed, and de-nominated $U_{0.4}Pu_{0.6}$-ht (η) and $hR147\text{-}U_{0.4}Pu_{0.6}$ (ζ) (Lawson *et al.*, 1996). The latter phase is shown in Fig. 7.98. Pu and U atoms could not be distinguished in the structure analysis due to their very similar X-ray atomic scattering factors. Although their atomic radii differ by more than ten percent, $r_{Pu} = 1.561$ Å and

$r_U = 1.385$ Å, their atomic volumes in the RT structures, $V_{at}^{Pu} = 20.43$ Å^3 and $V_{at}^U = 20.75$ Å^3, are very similar to that of $hR147\text{-}U_{0.4}Pu_{0.6}$ with $V_{at}^U = 21.17$ Å^3.

In terms of polyhedra packing, the structure can be fully described by the packing of three more or less distorted Frank-Kasper polyhedra. Pu1 and Pu9 are coordinated with distorted CN15 and quite regular CN16 FK-polyhedra, respectively, Pu10 with distorted CN14 FK-polyhedra. There are some similarities between the AETs and their way of stacking between this compound and $cI58$-Mn.

7.17   High-pressure phases of selected intermetallic compounds

As has been shown before in Chapter 6, a majority of the chemical elements undergoes structural phase transformations as a function of pressure. Under compression, the interatomic distances are decreased, the atomic orbitals overlap to a larger extent, the electronic band structure gets modified, i.e., the bands are usually broadened. For some elements under pressure, a differentiation of atoms at different sites in the crystal structure can take place, leading to host/guest structures, for instance, a behavior that is known for particular binary or ternary compounds mostly. Under pressure, covalent bonding contributions can also increase. Frequently the structures of the high-pressure modifications of the lighter elements are closely related to those of the heavier homologues due to their stronger atomic interactions.

Intermetallic phases behave similarly but in a more complex manner than the elements. Unfortunately, the data base is small since only the structures of relatively few intermetallics have been studied so far as a function of pressure. In the case of intermetallic compounds, we can additionally define a "chemical pressure" induced by atoms that are "too large" to fit smoothly in their AETs. In the case of compounds consisting of "soft" and "hard" atoms behaving as pure elements quite differently under pressure, the structural changes may be more pronounced than in the case of intermetallics constituted from "hard atoms", only.

In the following, we shortly discuss the structural changes that selected intermetallics undergo under hydrostatic compression based on the review by Demchyna *et al.* (2006). These are Zintl phases, AlB$_2$ type structures, and Laves phases. We are not going to discuss another kind of intermetallics that are unstable or metastable under ambient conditions, but form stable compounds if synthesized under high pressure.

Zintl phases can be considered to be polar intermetallics, which under pressure can create a pseudo-gap due to the reduction of the electronic density of states at the Fermi level. In Table 7.31, a selection of 1:1 Zintl-phases is shown with the structure type their structure belongs to under ambient conditions (LP), the pressure under which the structural transformation takes place, as well as the high-pressure (HP) structure type. The LP-structures listed are, with a few exceptions,

Table 7.31 *Structural data for selected low-pressure* (LP) *and high-pressure* (HP) *Zintl-phases with composition* AB. *Listed are compound formula, structure type of the LP phase, transformation pressure* p *at ambient temperature, structure type of the HP phase, and references.*

AB	LP-structure type	P [GPa]	HP-structure type	References
LiIn	$cF16$-NaTl	11	$cP2$-CsCl	Schwarz *et al.* (1998*b*)
KTl	$oC48$-KTl	2	$cF16$-NaTl	Evers and Oehlinger (2000)
LaSb	$cF8$-NaCl	11	$tP2$-MnHg	Leger *et al.* (1984)
CeSb	$cF8$-NaCl	10	$tP2$-MnHg	Leger *et al.* (1984)
PrSb	$cF8$-NaCl	13	$tP2$-MnHg	Hayashi *et al.* (2000)
NdSb	$cF8$-NaCl	15	$tP2$-MnHg	Hayashi *et al.* (2000)
SmSb	$cF8$-NaCl	19	$tP2$-MnHg	Hayashi *et al.* (2000)
GdSb	$cF8$-NaCl	22	$tP2$-MnHg	Hayashi *et al.* (2000)
TbSb	$cF8$-NaCl	21	$tP2$-MnHg	Hayashi *et al.* (2000)
DySb	$cF8$-NaCl	22	$cP2$-CsCl	Shirotani *et al.* (2001)
HoSb	$cF8$-NaCl	22	$cP2$-CsCl	Shirotani *et al.* (2001)
ErSb	$cF8$-NaCl	25	$cP2$-CsCl	Shirotani *et al.* (2001)
TmSb	$cF8$-NaCl	22	$cP2$-CsCl	Shirotani *et al.* (2001)
YbSb	$cF8$-NaCl	13	$cP2$-CsCl	Hayashi *et al.* (2004)
LuSb	$cF8$-NaCl	24	$cP2$-CsCl	Shirotani *et al.* (2001)
NpSb	$cF8$-NaCl	12	$tP2$-MnHg	Méresse *et al.* (1999)
PuSb	$cF8$-NaCl	18	$cP2$-CsCl	Méresse *et al.* (1999)
	$cP2$-CsCl	42	$tP2$-MnHg	Méresse *et al.* (1999)
CeBi	$cF8$-NaCl	13	$cP2$-CsCl	Leger *et al.* (1985)
		13	$tP2$-MnHg	Leger *et al.* (1985)
UBi	$cF8$-NaCl	5	$cP2$-CsCl	Méresse *et al.* (1999)
NpBi	$cF8$-NaCl	8	$cP2$-CsCl	Méresse *et al.* (1999)
PuBi	$cF8$-NaCl	10	$tP2$-MnHg	Méresse *et al.* (1999)
	$cF8$-NaCl	42	$cP2$-CsCl	Méresse *et al.* (1999)
AmBi	$cF8$-NaCl	14	$tP2$-MnHg	Méresse *et al.* (1999)
CmBi	$cF8$-NaCl	12	$cP2$-CsCl	Méresse *et al.* (1999)
	$cP2$-CsCl	20	$tP2$-MnHg	Méresse *et al.* (1999)

continued

Table 7.31 *continued*

AB	LP-structure type	P [GPa]	HP-structure type	References
GdCu	cP2-CsCl	12.8	oP4-AuCd	Degtyareva *et al.* (1997)
LaAg	cP2-CsCl	5	oP4-AuCd	Degtyareva *et al.* (1997)
NdAg	cP2-CsCl	3.4	oP4-AuCd	Degtyareva *et al.* (1997)
AgZn	hP9-AgZn	3.1	cP2-CsCl	Iwasaki *et al.* (1985)
NdZn	cP2-CsCl	4.2	oP4-AuCd	Degtyareva *et al.* (1997)
CeZn	cP2-CsCl	2	oP4-AuCd	Degtyareva *et al.* (1997)
LiCd	cF16-NaTl	11	cP2-CsCl	Schwarz *et al.* (1998*b*)

all of either the cF8-NaCl or the cP2-CsCl type, and they transform in most cases to derivative structures of the cP2-CsCl type or the tP2-MnHg type. This means that they can be described by different decorations (atoms/vacancies) of undistorted or distorted cubic primitive lattices.

The classical Zintl phase LiIn undergoes a transition from the cF16-NaTl structure type to the cP2-CsCl type at 11 GPa. Thereby the more covalent In–In bonding in the polyanionic 3D 4-connected diamond net is largely lost (Schwarz *et al.*, 1998*b*). KTl crystallizes under high pressure only, with the cF16-NaTl type structure. At ambient conditions, the K$^+$ ion is too large for the interstices in the polyanionic 3D 4-connected diamond net constituted from the Tl$^-$ ions: $r_{Na}/r_{Tl} = 1.843/1.588 = 1.16$ vs. $r_K/r_{Tl} = 2.296/1.588 = 1.45$ (Evers and Oehlinger, 2000). Under pressure, the more compressible K thus can be squeezed into the voids of the polyanionic network, where the 3D 4-connected diamond net provides a higher coordination of the K$^+$ ions than the oC48-KTl structure.

The structure of tP2-MnHg can be seen as a slightly tetragonally distorted cP2-CsCl-type structure, with a c/a ratio slightly larger than 1. In the HP-structures of RESb (RE = La, Ce, Pr, and Nd), however, it drops to 0.82 after the transition, which shows a volume collapse of 10–11%. The coordination number of the RE atoms increases from 6 Sb to 8 Sb + 2 RE, all at a distance $1/2(2a^2 + c^2)^{1/2}$. The rather small atomic distances indicate a metallic or covalent Sb–Sb bonding; there are no indications of a 4f electronic transition (Leger *et al.*, 1984). In the case of PrSb and NdSb the atomic distance Pr–Sb indicates covalent bonding (Hayashi *et al.*, 2000). The heavier RESb show a transition cF8-NaCl to cP2-CsCl with a much smaller volume collapse of 1–3%, reflecting the increased covalent character of chemical bonding in these compounds. The transition pressures increase with, due to the lanthanide contraction, decreasing lattice parameters of the LP-phases, with the exception of YbSb (Shirotani *et al.*, 2001; Hayashi *et al.*, 2004).

In the case of lanthanoid and actinoid compounds, the application of pressure may influence the 4f and 5f electrons. This does not seem to be the case for the lanthanoid compounds discussed above, but it applies for compounds of the heavier actinoids. The structure of PuBi, for instance, first transforms from the $cF8$-NaCl type at 10 GPa, with a volume collapse of $\approx 12\%$ to the $tP2$-MnHg type and, thereafter, at 42 GPa without a volume discontinuity to the $cP2$-CsCl type (Méresse *et al.*, 1999). In contrast, UBi and NpBi transform directly to the $cP2$-CsCl type, and AmBi to the $tP2$-MnHg, only, CmBi transforms first to the $cP2$-CsCl type and then to the $tP2$-MnHg just as PuSb does. In the case of PuBi, the formation of covalent Sb–Sb bonds seems to be the driving force for the formation of the tetragonal structure, while for the higher actinoids, interactions of An 5f and Bi 6p electrons are believed to be decisive (Méresse *et al.*, 1999).

The structures of LaAg, NdAg, NdZn, and CeZn all transform at rather low pressures between 2 and 5 GPa from the $cP2$-CsCl type to the $oP4$-AuCd type (Degtyareva *et al.*, 1997), which can be described as an orthorhombically distorted superstructure of $hP2$-Mg, with its [100] direction parallel to the six-fold axis. In the prototype structure, Au is surrounded by a disheptahedron of eight Cd and four Au atoms, Cd by the same AET but now decorated with eight Au and four Cd atoms.

The AlB$_2$ derivative phases shown in Table 7.32 transform into other AlB$_2$ derivative structures, mostly with lower symmetry or just smaller c/a ratios. CeGa$_2$, crystallizing in the $hP3$-AlB$_2$ structure type, with $c/a = 0.931$ at 24 GPa, transforms into the isopointal $hP3$-UHg$_2$ type with $c/a = 0.817$ at 24 GPa. The structure types $hP3$-UHg$_2$ and $hP3$-AlB$_2$ differ just by their c/a ratios which are either significantly smaller than or close to 1, respectively. While the coordination of Al

Table 7.32 *Structural data for selected low-pressure* (LP) *and high-pressure* (HP) AlB$_2$ *derivative phases, and Laves phases (below the dashed line). Listed are compound formula, structure type of the* LP *phase, transformation pressure* p *at ambient temperature, structure type of the* HP *phase, and references.*

AB$_2$	LP-structure type	P [GPa]	HP-structure type	References
CeGa$_2$	$hP3$-AlB$_2$	16	$hP3$-UHg$_2$	Shekar *et al.* (2004)
HoGa$_2$	$hP3$-AlB$_2$	4	$hP3$-UHg$_2$	Schwarz *et al.* (1998a)
TmGa$_2$	$oI12$-KHg$_2$	22	$hP3$-UHg$_2$	Schwarz *et al.* (1996)
YbGa$_2$	$hP6$-CaIn$_2$	22	$hP3$-UHg$_2$	Schwarz *et al.* (2001b)
GdGa$_2$	$hP3$-AlB$_2$	7.7	$hP3$-UHg$_2$	Schwarz *et al.* (2001a)
LaCu$_2$	$hP3$-AlB$_2$	1.6	$oI12$-CeCu$_2$	Lindbaum *et al.* (2000a)
KHg$_2$	$oI12$-KHg$_2$	2.5	$hP3$-UHg$_2$	Beister *et al.* (1993)
ThAl$_2$	$hP3$-AlB$_2$	0.3	$cF24$-MgCu$_2$	Godwal *et al.* (1986)
UAl$_2$	$cF24$-MgCu$_2$	0.3	$hP24$-MgNi$_2$	Sahu *et al.* (1995)
UMn$_2$	$cF24$-MgCu$_2$	3	orthorhombic	Lindbaum *et al.* (2000b)

equals 9, that for U in the latter amounts to 11. So it is more likely to form under pressure (Shekar *et al..*, 2004). In the case of $HoGa_2$, $c/a = 0.813$, and for $GdGa_2$ it equals 0.809 at 8 GPa, in KHg_2 even 0.584 at 11.7 GPa (Beister *et al.*, 1993).

$ThAl_2$ is an intermetallic compound that transforms from $hP3$-AlB_2 into the cubic Laves phase structure type $cF24$-$MgCu_2$. The radii ratio of Th and Al is with $r_{Th}/r_{Al} = 1.798/1.432 = 1.256$ already at ambient conditions close to the ideal value for Laves phases. Since Th is much more compressible than Al, its atomic volume can be compressed to $\approx 50\%$ at ≈ 70 GPa while ≈ 200 GPa is needed for Al; high pressure can drive the phase transition to the Laves phase.

UAl_2, a classic spin-fluctuation system, just transforms from the cubic Laves phase to the hexagonal double-period Laves phase $hP24$-$MgNi_2$. The transition has been interpreted to be caused by the pressure-induced increase in the f electron delocalization changing the e/a ratio from 1.66 to 1.8, the stability range of the hexagonal Laves phase (Sahu *et al.*, 1995).

7.18   High-entropy alloys

7.18.1   Introduction and definitions

High-entropy alloys (HEAs) were introduced under this name at the beginning of this century (Yeh, 2002; Yeh *et al.*, 2004*a*; Yeh *et al.*, 2004*b*). Actually, single phase HEAs would be better called solid solutions and not alloys, because the term "alloys" is commonly used for multiphase materials. As mentioned by Cantor (2014), these kinds of multicomponent alloys have been studied for the first time by Vincent (1981), followed up by Knight (1995) as well as Ranganathan (2003).

HEAs were originally defined as single-phase alloys consisting of $n = 5$–11 principal elements with concentrations of each component approximately equal to $100/n$ at.%. This definition allows, at least theoretically, the preparation of a large number of HEAs with promising physical properties. The underlying assumption is that the mixing entropy ΔS_{mix} would be large enough to prevent the formation of intermetallics and phase separation. However, this is true for only a few of them, depending on the equilibration temperature. So far, the maximum number of constituents found forming single-phase HEAs amounts to $n = 6$ (He *et al.*, 2014). According to the "confusion principle" (Greer, 1993), the more constituting elements are contained in an alloy, the lower is the chance to arrange them in a low-energy crystal structure, and the higher is the probability of glass formation, followed by phase separation if thermally equilibrated. To some extent, HEAs can be considered to be the thermodynamically stable counterparts to bulk metallic glasses (BMGs). Their deformed but still existing simple average lattices allow for a larger ductility than that of BMGs.

For the formation of a thermodynamically stable substitutional solid solution, the condition $\Delta G_{mix} = \Delta H_{mix} - T\Delta S_{mix} < 0$ must hold. This is always the case if $|\Delta H_{mix}| < T\Delta S_{mix}$ and/or $\Delta H_{mix} < 0$. ΔH_{mix} can be seen as a measure for the degree of structural ordering, which may range from phase separation ($\Delta H_{mix} \gg 0$) via short-range order (clustering) to long-range order ($\Delta H_{mix} \ll 0$), i.e., the formation of intermetallic compounds. Although the mixing entropy increases logarithmically with the number n of constituents, the number of potentially coexisting intermetallics increases as well, even linearly with n. For instance, doubling the number of constituents from three to six increases the mixing entropy just by a factor of 0.69, i.e., from 1.1 to 1.79 in units of the gas constant R. Consequently, increasing the number of constituents does not automatically mean that the formation of intermetallics can be suppressed. It should be kept in mind as well that HEAs can only be stable above a particular threshold temperature $T_{th} < T < T_m$ (with T_m the melting temperature), defined by $|\Delta H_{mix}| = T_{th}\Delta S_{mix}$. This has implications for the applications of HEAs at temperatures below T_{th}, where significant diffusion is still possible.

The configurational mixing entropy of an HEA can be described as $\Delta S_{conf} = -R\sum_{i=1}^{n} x_i ln(x_i)$, with x_i the fraction of the i-th element, and R the universal gas constant. ΔS_{conf} can be seen as a measure for the degree of randomness (disorder) of the structure. In the case of HEAs consisting of five elements, the mixing enthalpy was estimated to be in the range $-15 < \Delta H_{mix} < 5$ kJ/mol, and the configurational mixing entropy $\Delta S_{conf} \geq 1.61\ R$ (Zhang and Zhou, 2007). Thereby, the configurational mixing entropy was calculated assuming that each atom is statistically distributed in the same way without any short-range ordering and specific atomic interactions. For the prediction of HEAs see, for instance, the recent paper by Troparevsky *et al.* (2015) and references therein.

In the decade after the publication of the first papers on HEAs (Yeh *et al.*, 2004*a*; Yeh *et al.*, 2004*b*), more than 400 articles on HEAs have appeared. Some recent reviews are: Kozak *et al.* (2015), Tsai and Yeh (2014), and Zhang *et al.* (2014*b*). Unfortunately, most samples studied were neither quenched from a thermodynamic equilibrium state nor single-phase, causing some confusion in this field. It seems that the definition of HEAs has been extended, now also including multiphase alloys (Yeh, 2015). This, however, raises the question about the difference between classical alloys and HEAs. However, in the following we essentially stick to the original definition, and only discuss HEAs that are approximately equiatomic, single-phase, substitutional solid solutions in thermal equilibrium. All of them (known so far) have simple average crystal structures of either the *cF*4-Cu or the *cI*2-W type. The only exception so far are HEAs such as HoDyYGdTb, made up of *hcp* rare earth elements, which adapt an *hcp* structure as well.

7.18.2   Stability regions

Until recently, more than twenty multinary systems have been searched for HEAs, based on more than one hundred combinations of four to nine elements out of twenty-one (Ho, Dy, Y, Gd, Tb, Ti, Zr, Hf, V, Nb, Ta, Cr, Mo, W, Mn, Fe, Co, Ni, Pd, Cu, and Al). Only a few compositions lead to single-phase, close to equiatomic HEAs (at least in the as-cast state) (Kozak *et al.*, 2015): CrFeCoNi

(Lucas *et al.*, 2012), CrFeCoNiAl$_{0.3}$ (Ma *et al.*, 2013), CrFeCoNiMn (Cantor *et al.*, 2004; Otto *et al.*, 2013), CrFeCoNiMnAl$_{0.2}$ (He *et al.*, 2014), CrFeCoNiPd (Lucas *et al.*, 2011), NbMoTaW (Senkov *et al.*, 2010), VNbMoTaW (Senkov *et al.*, 2010), TiZrHfNbTa (Senkov *et al.*, 2011), and HoDyYGdTb (Feuerbacher *et al.*, 2015).

In the case of the Al-based HEAs, $cF4$-Cu type average structures were observed for the systems CrFeCoNiAl$_x$ (Kao *et al.*, 2009; Wang *et al.*, 2012a) and CrFeCoNiMnAl$_x$ (He *et al.*, 2014), with $0 < x < 0.3$, while $cI2$-W and $cP2$-CsCl type average structures, respectively, result at higher Al contents ($x \geq 1$). At ambient conditions, Cr and Fe are *bcc*, Co *hcp*, Ni and Al *ccp*, while Mn has a complex superstructure of the $cI2$-W type. At elevated temperatures, Fe and Co transform to *ccp*, and at high pressure Mn to *bcc*. Al has a larger atomic radius than the TM elements and can consequently cause a high internal local pressure.

HEAs containing light TM only, such as CrFeCoNi and CrFeCoNiMn, adopt the $cF4$-Cu structure type, while for those containing refractory elements such as NbMoTaW, VNbMoTaW, and TiZrHfNbTa the $cI2$-W structure type is observed. V, Nb, Mo, Ta, and W all have structures of the $cI2$-W type; Ti, Zr, and Hf are all *hcp* at ambient conditions and *bcc* at high temperatures. The atomic radii of Ti, Nb, and Ta are $\approx 5\%$ larger than those of V, Mo, and W, and $\approx 10\%$ smaller than those of Zr and Hf.

While all of the above-listed HEAs have either *bcc* or *fcc* average structures, not surprisingly HoDyYGdTb is *hcp* as the pure elements are. The lattice parameters follow the rule of mixtures (the atomic radii are in the range 1.743–1.787 Å). The authors assume that also *hcp* HEAs that are constituted from the other RE should also form although the atomic radii span a much larger range from 1.724 to 1.995 Å (Feuerbacher *et al.*, 2015). RE HEAs might be of interest for studying magnetism or heavy-fermion behavior in disordered systems.

The probability of the formation of HEAs is higher in multinary systems, the binary boundary systems of which show extended solid solutions (Bei, 2013; Otto *et al.*, 2013; Zhang *et al.*, 2014a). This is the case for the *bcc* HEAs NbTaMoW, VNbTaMoW, ZrHfNbMo, and TiZrHfNbMo, which do not show significant deviations from a random solid solution when annealed at temperatures ≥ 2000 K (Maiti, 2015). All 28 binary subsystems except six, (Zr,Hf)–(V,Mo,W), show full solid solubility above a given threshold temperature (see Table 7.33). However, V, Mo, and W can solve up to 25, 4, and 11% of Hf, and 9.5, 5, and 3.5% of Zr at sufficiently high temperatures.

Some more quantitative prerequisites that are known to favor HEA formation are (Broer and Pettifor, 1988; Guo *et al.*, 2011; Zhang *et al.*, 2012):

- Atomic size difference $\delta \leq 6.6\%$, with $\delta = \sqrt{\sum_{i=1}^{n} c_i (1 - \frac{r_i}{\bar{r}})^2}$, c_i and r_i the concentration and the atomic radius of the i-th element, respectively, and $\bar{r}$ the average atomic radius.

Table 7.33 *Solid solubility of the binary subsystems of refractory HEAs. All constituting elements as well as the resulting HEAs are of the cI2-W structure type, at least above a threshold temperature [K], which is given for a 1:1 composition in each case. If no temperature but a plus sign is given, this means that for the solid solutions no phase transformation was observed down to ambient temperature. A minus sign indicates no solid solution at all.*

	Ti	Zr	Hf	V	Nb	Ta	Mo	W
Ti		≈ 880	≈ 1270	≈ 880	≈ 870	≈ 800	≈ 1070	≈ 1300
Zr	≈ 880		≈ 1650	–	≈ 1240	≈ 2000	–	–
Hf	≈ 1270	≈ 1650		–	≈ 1510	≈ 1350	–	–
V	≈ 880	–	–		+	≈ 1550	+	+
Nb	≈ 870	≈ 1240	≈ 1510	+		+	+	+
Ta	≈ 800	≈ 2000	≈ 1350	≈ 1550	+		+	+
Mo	≈ 1070	–	–	+	+	+		+
W	≈ 1300	–	–	+	+	+	+	

- Valence electron concentration $VEC \geq 8$ for *fcc* solid solutions and $VEC < 6.87$ for *bcc* ones, where $VEC = \sum_{i=1}^{n} c_i (VEC)_i$.

- Parameter $\Omega \geq 1.1$, with $\Omega = \frac{T_m \Delta S_{mix}}{|\Delta H_{mix}|}$, $T_m = \sum_{i=1}^{n} c_i T_m^i$ the melting temperature of the solid solution, and T_m^i the melting temperature of the *i*-th element. ΔH_{mix} and ΔS_{mix} are the mixing enthalpy and entropy, respectively, calculated based on Miedema's approach (Miedema, 1976).

7.18.3 Structures and properties

The average structures of HEAs are quite simple, mainly of the *cI2*-W or the *cF4*-Cu type. Not much is known, however, about the short-range ordering, and the lattice distortion resulting therefrom (Fig. 7.99; also compare with Fig. 7.7 (h), for instance). This knowledge is important for understanding the physical properties of these materials, in particular their mechanical strength, plasticity, and deformation characteristics. See, for instance, the recent papers by Zou *et al.* (2014), Diao *et al.* (2015), and references therein.

HEAs are potentially very valuable materials with tunable properties. For instance, the refractory-metal-based HEAs possess high-temperature strength and thermal resistance comparable to superalloys, with a rather weak decrease of yield strength up to 1900 K (Senkov *et al.*, 2012). This effect has been assiged to the sluggish diffusion in HEAs caused by lattice distortions (Tsai *et al.*, 2013). When reducing the dimensions of HEA samples, their strength and ductility can be significantly increased (see, e.g., Zou *et al.* (2014)).

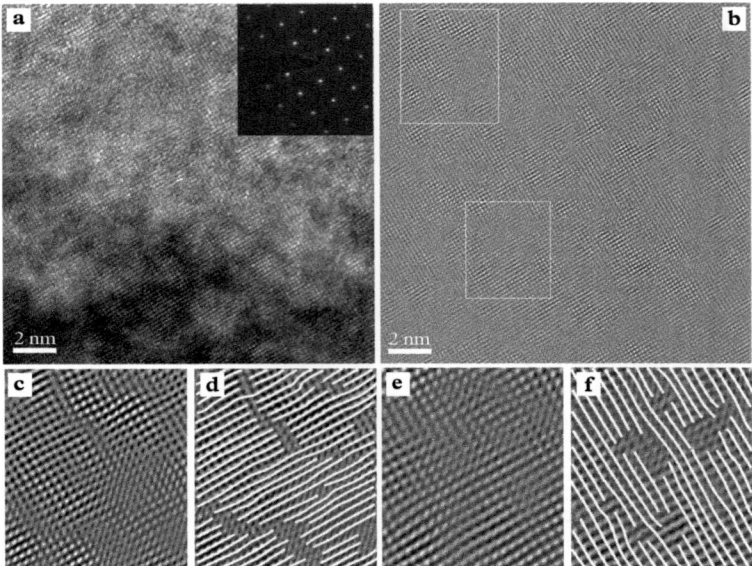

Fig. 7.99 *High-resolution TEM images of the* Nb–Mo–Ta–W *HEA homogenized at 1800° C for seven days: (a) a bright-field TEM image oriented along the [100] zone axis and the corresponding electron-diffraction pattern; (b) an inverse fast Fourier transform image of the area (a); (c) and (e) are enlarged images of the indicated boxes in (b); (d) and (f) show lattice fringes that are traced for (c) and (f), respectively, to indicate the regions with lattice distortions. Reprinted from Acta Materialia (Zou* et al., *2014), with permission from Elsevier.*

Some of the HEAs studied so far possess both high strength and ductility (to some extent). This is an interesting property for low-temperature applications among others. It was shown by first-principles calculations that the stacking-fault energy decreases with the number of constituents (Zaddach et al., 2013). For instance, it amounts to ≈ 100 mJ/m^2 and 7.7 mJ/m^2 for NiAl and $Cr_{26}Fe_{20}Co_{20}Ni_{14}Mn_{20}$, respectively. This is even lower than commercial austenitic stainless steels, with 18 mJ/m^2 for AISI 304L, for instance.

Remarkably, HEAs can also show type-II superconductivity, as has been demonstrated on the example of bcc $Ta_{34}Nb_{33}Hf_8Zr_{14}Ti_{11}$ (Koželj et al., 2014). The transition temperature $T_c = 7.27$ K is higher than that expected from the weighted average of the elements (Ta:4.47 K, Nb: 9.25 K, Hf: 0.128 K, Zr: 0.61 K, Ti: 0.40 K): $T_c = 4.71$ K. It has already been shown by Corsan and Cook (1970) that binary alloys of the type Nb_xTa_y have transition temperatures more or less corresponding to the rule of mixture in contrast to V_xNb_y and V_xTa_y, which show a minimum in T_c below the T_c of the constituting elements.

8

Complex intermetallics (CIMs)

We discussed the meaning of complexity for intermetallics in Section 3.4. It is simpler to say which of several structures is more or less complex than to define on an absolute scale which IM can be called a CIM, i.e., what makes a structure complex. The term "complex" is commonly used intuitively for big structures with distinct structural subunits and correspondingly different length scales. Thus, we apply qualitative rather than quantitative criteria for classifying whether a structure is complex. However, for a statistical analysis of a big database, one needs well-defined parameters for a classification of structures into complex and non-complex ones. The purpose of such a classification is the identification of intermetallic compounds that have structures with multiple structural length scales, that on the scale of atoms and that on the scale of endohedral clusters or other structural subunits. These different length scales can be the origin of interesting physical properties, which may lead to useful applications.

A well-defined and feasible approach is to base our classification of complexity simply on the number of atoms per primitive unit cell, equating complex structures with big structures. This information is easily accessible from the database *Pearson's Crystal Data* (PCD) (Villars and Cenzual, 2011*a*), since it is already contained in the Pearson symbol. It results from the number of atoms per unit cell given there, divided by the multiplicity of the centering type ($P \ldots 1$, $A, B, C, S, I \ldots 2$, $R \ldots 3$, and $F \ldots 4$). Due to the lack of a 3D unit cell, this approach cannot be used for non-3D-periodic CIMs such as quasicrystals or amorphous intermetallic phases.

It is obvious that only structures above a given threshold can show several hierarchical levels (see Section 7.10) and different structural length scales, ranging from the atomic scale to that of structural subunits such as endohedral clusters or large polyanions. Accordingly, CIMs will have structures with hundreds or thousands of atoms per unit cell, and the problem now is to quantify the criterion, to define the threshold value for classifying a structure as complex. Here, we will use the same approach that is employed for the determination of coordination polyhedra (AETs), i.e., the maximum-gap method (Brunner and Schwarzenbach, 1971). However, one should keep in mind that such a threshold value is a quite arbitrary number, although it may be useful to get an idea at which compositions compounds with larger structures are preferentially formed.

Intermetallics: Structures, Properties, and Statistics. First Edition. Walter Steurer and Julia Dshemuchadse.
© Walter Steurer and Julia Dshemuchadse 2016. Published in 2016 by Oxford University Press.

If we have a look at the general unit cell size distributions shown in Figs. 5.7 and 5.8(a), we can identify gaps at around 84 and 170 atoms per primitive unit cell. Since CIMs may not be equally distributed over all symmetries, it may be beneficial to have a closer look at the histograms for each of the 14 Bravais lattice types separately (Fig. 8.1), in order to use different threshold values for structures of different symmetries, if necessary.

We start with the largest group of CIMs, which is found in the highest-symmetric Bravais lattice (*cF*). The respective unit cell size distribution histogram shows multiple pronounced gaps: most notably at around 15, 50, and 90 atoms per primitive unit cell. Beyond that, there are even wider gaps between less and less common unit-cell sizes. Right above 100 atoms per primitive unit cell, there seems to be a considerable number of *cF*-structures. This threshold is also indicated around approximately 100 atoms per (primitive) unit cell in most of the other histograms shown in Fig. 8.1. That for *cI* IMs has a large gap above the threshold, while in case of *cP* IMs, the threshold falls right in the middle of a significant gap, and for *tI* IMs right below 100. In the histogram for *tP* IMs, there is a subtle gap right below 100, but it could also convincingly be argued that it has a large gap above this value. The gaps are also less clear in the case of the trigonal and hexagonal lattices: for *hR* IMs there is a gap a little below the threshold, whereas for *hP* IMs there seems to be a blurry gap a little above that value. There are only a few *oF* structures in the respective histogram, and only four of them have larger unit cell sizes than 100—a small gap can be found at that value, but also a slightly wider one below and a much larger one at higher unit cell sizes. The histogram for *oI* IMs has a wide gap at 100 and only a handful of structures above this threshold. That for *oS* IMs is less convincing, exhibiting a smaller gap right above 100. The distribution of *oP* IMs could be interpreted as having either a small gap below or a slightly larger gap a little above 100, while *mS* IMs are a clear case again with a wide gap around 100, but also the distribution of *mP* IMs features a sizeable gap above 100. No triclinic structures with 100 or more atoms per unit cell are known so far. The gaps "separating" complex and non-complex intermetallics, are specified in Table 8.1. Given are the (near-)empty ranges of unit cell sizes, as well as the number of complex intermetallics within each Bravais lattice and the fraction that complex intermetallics make up within that respective lattice.

Our definition of CIMs as periodic compounds with primitive unit cells comprising 100 or more atoms applies to approximately 2% of all periodic intermetallics. In their distribution, there does not seem to be a significant trend towards specific (high- or low-symmetry) Bravais lattice types, point group symmetries, or space group types (Dshemuchadse and Steurer, 2015). In contrast, the distribution of their chemical compositions seems to be more distinct from that of IMs in general. Specifically, the lanthanoids and transition metal elements appear to be underrepresented, while alkali and alkaline earth metals as well as metalloid elements, both kinds together forming Zintl phases, are overrepresented as majority elements in CIMs.

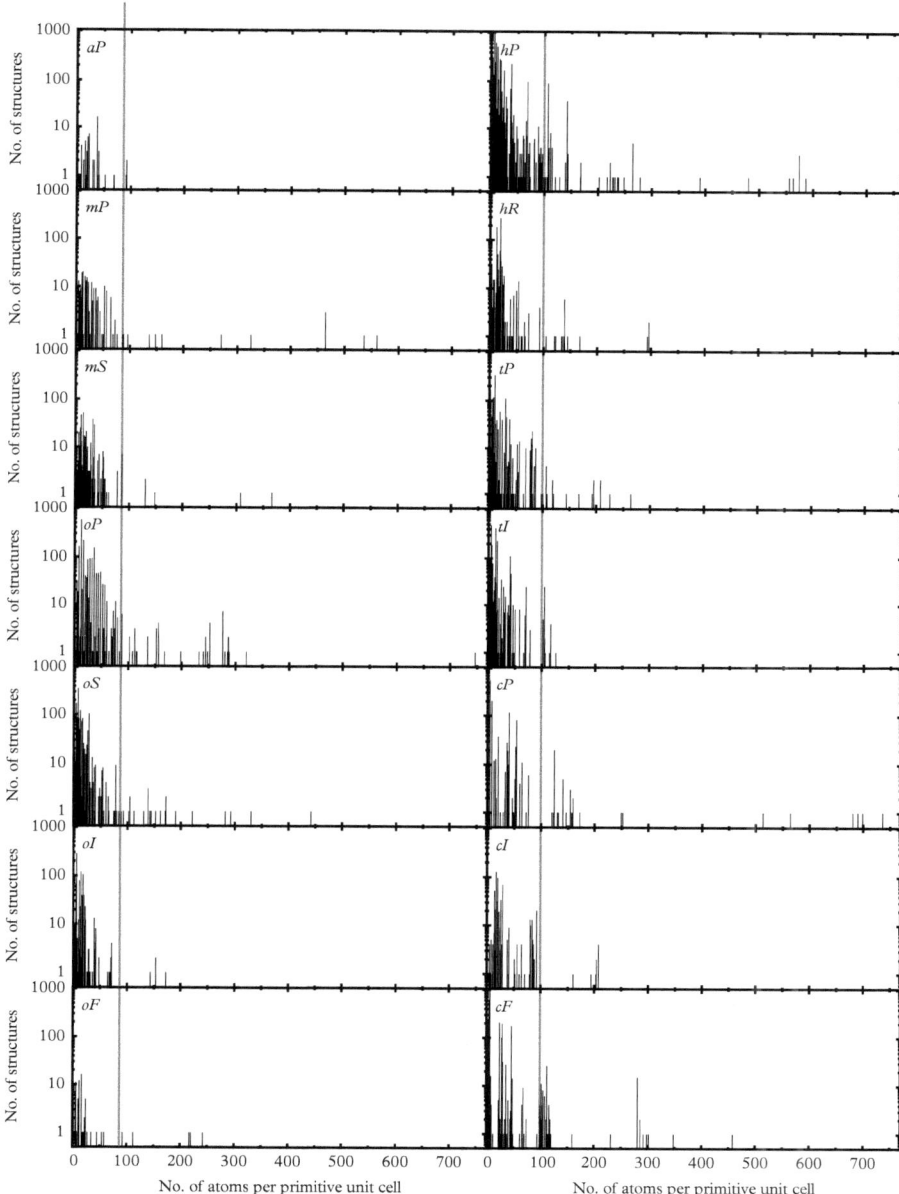

Fig. 8.1 *Distribution of the unit cell sizes (number of atoms per primitive unit cell) of intermetallics for the 14 Bravais lattice types. The vertical gray lines at 100 atoms per primitive unit cell run through small histogram gaps in most cases, and separate IMs from CIMs according to our definition.*

Table 8.1 *Gaps in the histograms shown in Fig. 8.1 for the definition of CIMs, and their distribution over the Bravais lattices (Dshemuchadse, 2013).*

Bravais lattice	Gap between IMs and CIMs	No. of CIMs	Fraction of CIMs
aP	92-...	0	0.0%
mP	96-136	10	4.3%
mS	86-130	5	1.1%
oP	88-102	44	2.4%
oS	92-102	23	1.6%
oI	71-143	4	0.4%
oF	92-112	4	5.1%
tP	86-97	22	1.5%
tI	78-102	38	1.8%
hP	128-139	74	1.4%
hR	73-92	22	2.3%
cP	76-120	47	2.7%
cI	92-160	9	0.9%
cF	74-94	104	3.0%
all	≈ 100	406	1.9%

Surprisingly, the fraction of CIMs of all intermetallics with a given number of constituents appears to be higher in binary than in ternary compounds (Table 8.2). However, this may be biased by the small relative number of ternary intermetallics studied so far. The combinations of CIM-forming elements also seem to show some specific tendencies. They are given in the M/M-plots for all binary and ternary intermetallics depicted in Fig. 8.2, again with the CIMs highlighted within the set of IMs in general. Despite the generally very wide distribution of intermetallics in such M/M-plots, we find that transition metal elements are underrepresented, and the majority elements for CIMs mostly have M-values in the range 70–85. This trend can be seen for both binaries and ternaries.

More specifically: elements from groups 12–14 in the periodic table of elements tend to account for major components in most complex intermetallics, whereas transition metals from groups 4–11 are underrepresented. Especially the prevalence of majority elements Al, Ga, Zn, or Cd hints at the close relationship between CIMs and quasicrystals, which also tend to occur in these systems. Large quasicrystal approximant structures by definition belong into the category of CIMs, but other intermetallic systems that have not been proven yet to feature quasiperiodic structures also tend to form complex phases in related intermetallic systems.

Table 8.2 *Distribution of the 20 829 intermetallic compounds, as well as the 406 complex intermetallics among them, according to the number of constituting elements (Dshemuchadse, 2013).*

No. of constituents n	No. of IMs with n components	No. of CIMs with n components	Fraction of CIMs
1	277	0	0.0%
2	6441	201	3.1%
3	13 026	184	1.4%
4	973	21	2.2%
5	65	0	0.0%
6	14	0	0.0%
7	13	0	0.0%
8	8	0	0.0%
9	2	0	0.0%
all	20 829	406	100%

Additionally, the region in the plots that has low M-values—lanthanoids ($M \approx 10$–35)—on both axes is very sparse in CIMs. The same trend can be seen with both plots over the majority element in ternaries; however, the third ternary M/M-plot that shows how the occurrence of specific elements in the two minor components correlates, shows no such tendency. This means that only one out of two components in binary intermetallics, and only two out of three in ternary ones, tend to have values in this region. This largely seems to be true for all intermetallics, but appears even more pronounced in the distribution of the complex compounds.

The stoichiometries of all binary and ternary compounds can be illustrated in one- and two-dimensional plots, respectively. For binaries, it is obvious that the most common stoichiometries among intermetallics are ratios of small integer numbers, whereas complex compounds tend to occur mostly in-between those values, i.e., at ratios of larger numbers that are enabled by large unit cells (see Fig. 8.3). In the case of ternaries, the plot in Fig. 8.4 illustrates the occurrence of ternary compositions. Ternary intermetallics in general seem to occur most frequently on lines connecting simple binary compositions, most prominently AC_3–BC_3, AC_2–BC_2, AC–BC (all: left – right), A_3C–A_3B, AC–AB, AC_2–AB_2 (all: top left – bottom right), AC_2–B, and AC–B (both: left – bottom right)[1].

---

[1] It is important to note that the enforced order of elements A, B, and C by $M(A) < M(B) < M(C)$ might skew how clear these motifs are in different sextants of the triangular diagram; however, it should not affect the general appearance of these features.

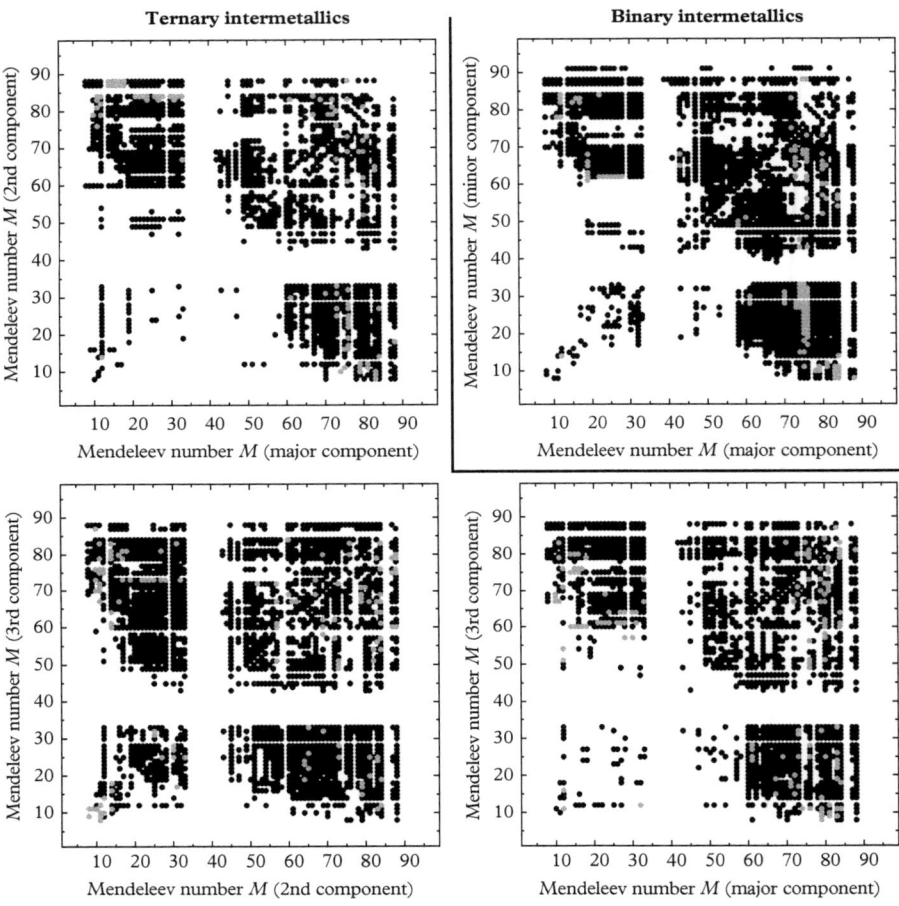

Fig. 8.2 *M/M-plots showing the distribution of CIMs in binary and ternary intermetallic compounds. CIMs are marked by gray dots. Binary compounds are shown with the minority component vs. the majority component (top right), whereas three plots specify the combinations of elements in ternary intermetallics: the component with the second-highest content vs. the majority component (top left), the minority component vs. the second component (bottom left), and the minority component vs. the majority component (bottom right).*

However, CIMs rarely seem to coincide with these pseudo-binary compositions. Another trend that is noticeable in this representation is that the element with the highest Mendeleev number $M(C)$ seems to be often the majority element in ternary CIMs. This can be recognized by the accumulation of a large number of points indicating CIMs in the top corner of the ternary composition diagram. The same is true for the binaries in Fig. 8.3, where the bulk of intermetallics can be found on the B-rich part of the composition plot.

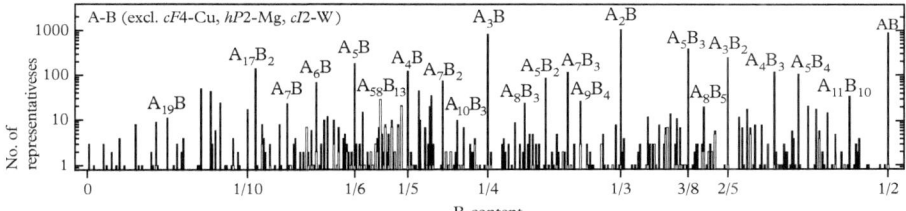

Fig. 8.3 *Stoichiometries of binary intermetallics with complex compounds highlighted (black-outlined bars). The constituting elements of each compound are sorted according to their Mendeleev numbers, $M(A) < M(B)$.*

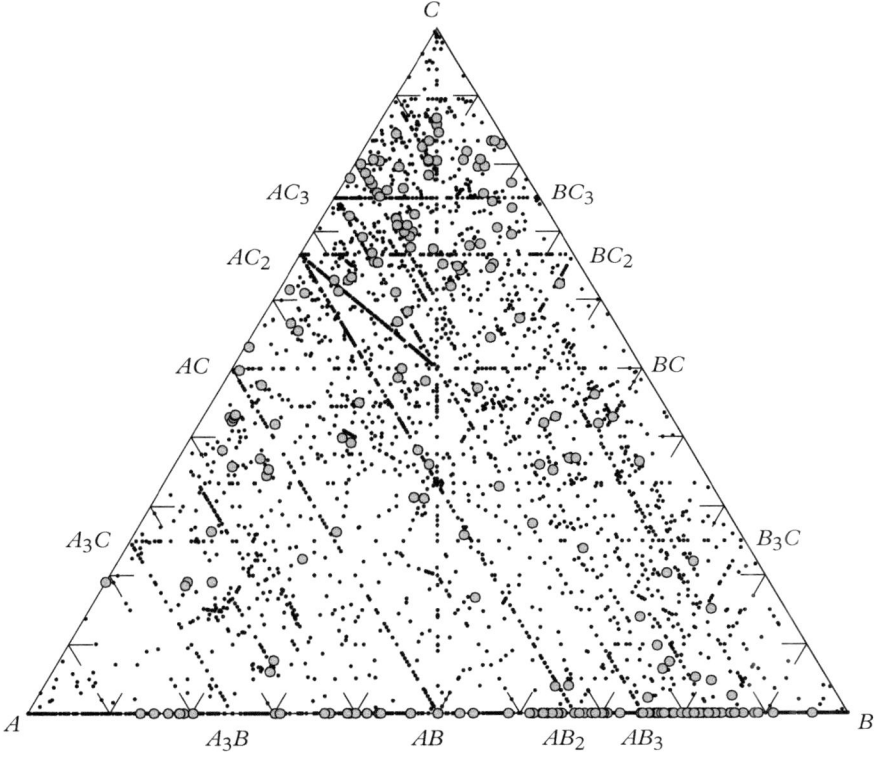

Fig. 8.4 *Occurrence of stoichiometries of ternary intermetallics with CIMs highlighted (large gray circle). The constituting elements of each compound are sorted according to their Mendeleev numbers, $M(A) < M(B) < M(C)$.*

In the following, we will discuss structures of cluster-based CIMs, the largest group of CIMs. Some other CIMs have been discussed before, most of them can be described as a kind of superstructure, such as the long-period structures of $tP120$-$Mn_{11}Si_{19}$ or $tP192$-$V_{17}Ge_{31}$ (see Section 7.9), for instance. Other examples are hierarchical and modular structures such as $cP156$-$K_{29}NaHg_{48}$ or $cP792$-$V_{11}Cu_9Ga_{46}$ (see 7.10), to name just a few.

We already discussed, in Section 3.1, what we mean by cluster, and that the chemical bonding between atoms within a cluster and to atoms outside a cluster may but does not need to differ. The driving forces behind the formation of clusters can be quite diverse. In case of complex stoichiometries, the formation of close to spherical structural subunits can be beneficial for the maximization of the packing density (*ccp* or *hcp* cluster packings). Furthermore, clusters can be stabilized by particular valence electron concentrations (see Zintl-Klemm concept and Wade-Mingos rules in Chapter 2). Also, covalent bonding contributions and hybridization effects leading to pseudo-gaps at the Fermi edge can play a cluster-stabilizing role as we know from quasicrystals and their approximants. The clusters constituting the structures to be discussed in the following are mainly endohedral (nested) fullerene-like clusters. The basics of their internal structure, the nesting of cluster shells, have already been discussed in Subsection 3.3.2.

8.1   Cluster structures of face-centered cubic CIMs

The Bravais lattice type featuring the largest group of CIMs is *fcc*/*cF* (see Table 8.1). This is also the symmetry of the intermetallic structure with the highest number of atoms per unit cell known so far, $cF(23\,256 - 122)$-$Al_{55.4}Ta_{39.1}Cu_{5.4}$. The 56 *fcc* CIMs published until 2011 were the subject of a comprehensive comparative study (Dshemuchadse *et al.*, 2011), which is the basis of the following discussion in this section.

Fig. 8.5 shows the distribution of the *fcc* intermetallic compounds according to their unit cell size up to 1600 atoms. There are several significant gaps, with one of them right below our general threshold for CIMs, 100 atoms per primitive unit cell which corresponds to 400 atoms per *fcc* unit cell. Another wide gap ranges from around 500 to 1100 atoms per *fcc* unit cell. This indicates two main groups of *fcc* CIMs. The first group with $cF \approx 400$ contains 43 structures, 36 with space group symmetry 216 $F\bar{4}3m$ and 7 with 227 $Fd\bar{3}m$. The second group comprises 11 structures with $cF \approx 1100$ in four different space groups: 1 in 216 $F\bar{4}3m$, 2 in 227 $Fd\bar{3}m$, 1 in 203 $Fd\bar{3}$, 4 in 225 $Fm\bar{3}m$, and 3 in 226 $Fm\bar{3}c$. Two structures with symmetry 216 $F\bar{4}3m$ stand out from these two groups: $cF(5928-20)$-$Al_{56.6}Ta_{39.5}Cu_{3.9}$ and $cF(23\,256-122)$-$Al_{55.4}Ta_{39.1}Cu_{5.4}$. Both have exceptionally large unit cells, the latter one not having been matched in size by any other intermetallic structure thus far. The first group obviously stands out due to its apparent uniformity, all of these structures with very similar unit cell sizes having the same space group symmetry, non-centrosymmetric 216 $F\bar{4}3m$, or the same group with added inversion symmetry, 227 $Fd\bar{3}m$. All structures have a set of Wyckoff sites in common and only differ by successive replacement of another set of Wyckoff sites. Therefore, their cluster structures are closely related and can be described as a series of subtypes of a common structural aristotype.

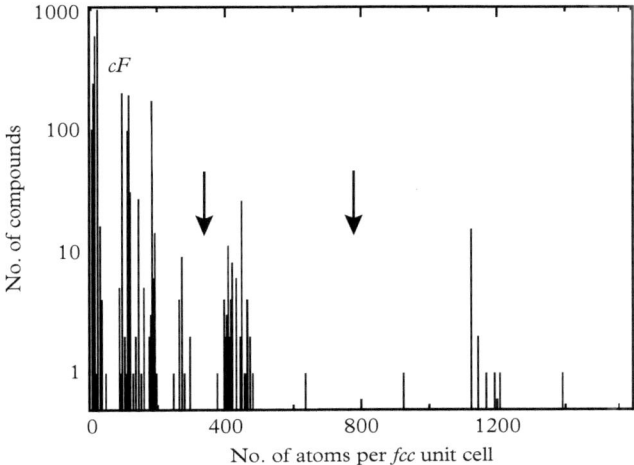

Fig. 8.5 *Distribution of the fcc unit cell sizes of intermetallic compounds within the Bravais lattice type cF. The two gaps mentioned above are marked by arrows.*

8.1.1 Face-centered cubic CIMs as superstructures

All *fcc* CIMs can be regarded as $(p \times p \times p)$-fold superstructures (with $p = 3$, 4, 7, 11) of a simple basic structure (with $p = 1$). The subunit cell is well-defined in reciprocal space, and can be determined from the positions of a subset of the strongest reflections (see Fig. 8.6). The simulated $hk0$-reciprocal-space planes of *fcc* CIMs depicted there indicate $(3 \times 3 \times 3)$-, $(4 \times 4 \times 4)$-, $(7 \times 7 \times 7)$-, and $(11 \times 11 \times 11)$-fold superstructures. These superstructures correspond to $cF444$-$Al_{63.6}Ta_{36.4}$ and $cF1192$-$Al_{53.6}Mg_{46.4}$, both representatives of the groups of structures with approximately 400 and approximately 1200 atoms per *fcc* unit cell, as well as the two unique giant-unit-cell structures $cF5928$-$Al_{56.6}Ta_{39.5}Cu_{3.9}$ and $cF23\,256$-$Al_{55.4}Ta_{39.1}Cu_{5.4}$.

The periodicities of the basic structures of all these superstructures appear to be the same, because their respective highest-intensity reflections, $6\,6\,0$, $8\,8\,0$, $14\,14\,0$, and $22\,22\,0$, respectively, coincide in reciprocal space. Considering that only reflections with all-odd or all-even indices are allowed for face-centered lattices, $2\,2\,0$ should be one of the most intense reflections in the $hk0$-plane. Therefore, the order of the superstructures, should the basic structures be face-centered as well, can be obtained from the factor between above-listed reflections and $2\,2\,0$, which are indeed 3, 4, 7, and 11.

With the unit cell dimensions of the basic structure known, the atomic positions of the supercells can be projected into the basic unit cell. The resulting "average structures" of the superstructures show the projected atoms scattered

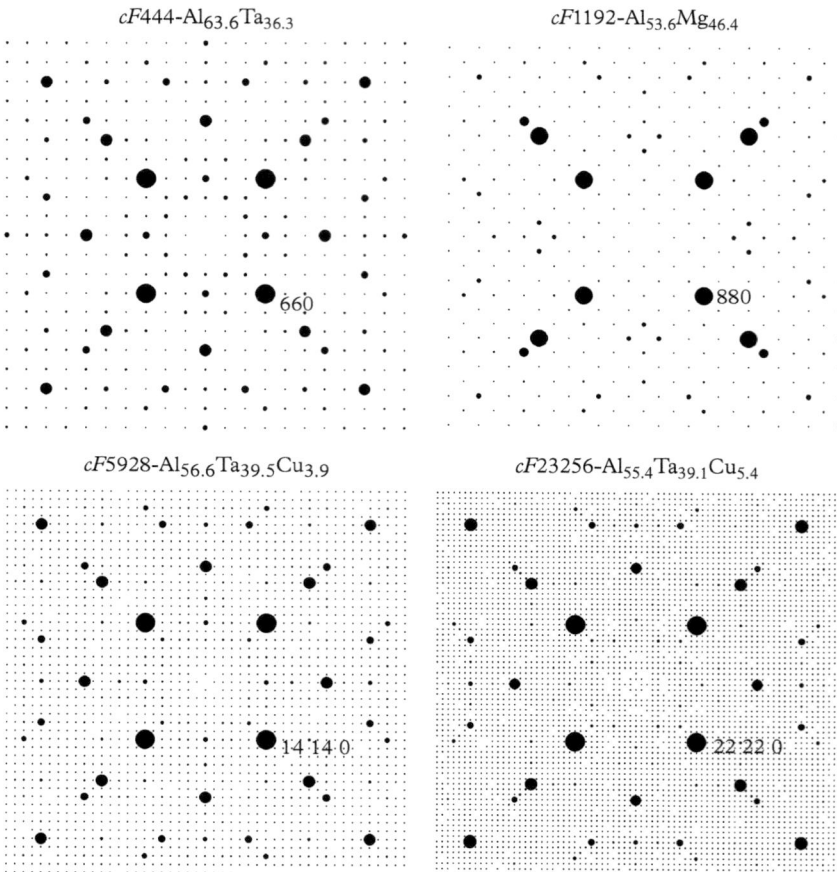

Fig. 8.6 *Simulated diffraction patterns of representative fcc CIMs that can be described as $(3 \times 3 \times 3)$-, $(4 \times 4 \times 4)$-, $(7 \times 7 \times 7)$-, and $(11 \times 11 \times 11)$- superstructures. Shown are the hk0-reciprocal-space planes of cF444-Al$_{63.6}$Ta$_{36.4}$, cF1192-Al$_{53.6}$Mg$_{46.4}$, cF5928-Al$_{56.6}$Ta$_{39.5}$Cu$_{3.9}$, and cF23 256-Al$_{55.4}$Ta$_{39.1}$Cu$_{5.4}$. The most intense reflections, which all coincide in their absolute coordinates in reciprocal space, are labeled with their indices.*

around special Wyckoff positions in the subunit cell. The "basic structure" can then be derived by taking these special Wyckoff positions as its atomic positions.

The common basic structure for the different *fcc* CIMs is of the *cF*16-NaTl structure type. This structure itself can be described as a $(2 \times 2 \times 2)$-fold superstructure of the *cP*2-CsCl structure type. However, if the *fcc* lattice is to be conserved, the 16-atom unit cell of the *cF*16-NaTl structure type is the smallest possible unit cell of a basic structure. In previous studies, $cF \approx 400$-structures were frequently described as $(2 \times 2 \times 2)$-fold superstructures of the γ-brass structure type $cI52$-Cu_5Zn_8 (Johansson and Westman, 1970; Arnberg *et al.*, 1976; Booth *et al.*, 1977; Fornasini *et al.*, 1978; Lidin *et al.*, 1994; Thimmaiah *et al.*, 2003; Berger *et al.*, 2008). The γ-brass structure, in turn, is a $(3 \times 3 \times 3)$-fold superstructure of $cP2$-CsCl, but does not preserve the face-centering of the lattice.

Fig. 8.7 shows the projections of the unit cell of $cF444$-$Al_{63.6}Ta_{36.4}$ as well as of all three possible average structures, corresponding to its possible descriptions as a (b) $(2 \times 2 \times 2)$-fold, (c) $(3 \times 3 \times 3)$-fold, and (d) $(6 \times 6 \times 6)$-fold superstructure, respectively. The corresponding space group symmetries are 216 $F\bar{4}3m$, in the case of the supercell, and 215 $P\bar{4}3m$, 216 $F\bar{4}3m$, and 221 $Pm\bar{3}m$ for the three average structures, respectively. Note that the $cF16$-NaTl structure type has centrosymmetric space group symmetry 227 $Fd\bar{3}m$, and that the space group of the here-discussed average structure is the non-centrosymmetric subgroup 216 $F\bar{4}3m$. The indices of the group-subgroup relationships between the supercell and all three basic structures are 2, 27, and 108, respectively (Dshemuchadse *et al.*, 2011). The description as a $(3 \times 3 \times 3)$-fold superstructure seems to be appropriate, as it preserves the symmetry and also enables the description of all other *fcc* CIMs as superstructures of the same basic unit cell. The derivation of the other periodicities and space groups is equally possible for all *fcc* CIMs (for more details, see Dshemuchadse *et al.* (2011) and the respective supplementary information).

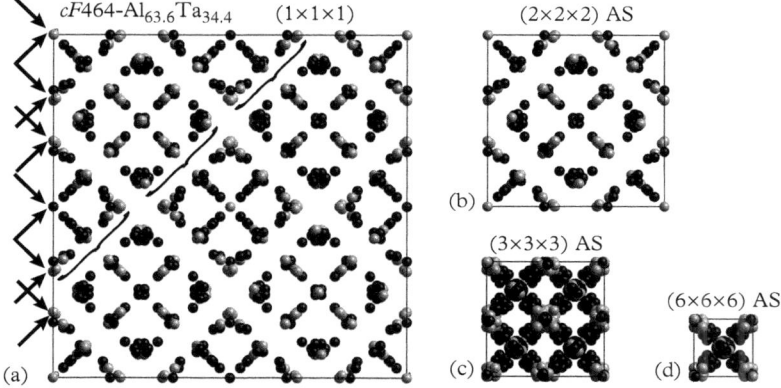

Fig. 8.7 *Projections along* [100] *of the unit cells of (a) the structure of* $cF444$-$Al_{63.6}Ta_{36.4}$ *(216 $F\bar{4}3m$) and its different average structures according to the description of* $cF444$-$Al_{63.6}Ta_{36.4}$ *as a (b) $(2 \times 2 \times 2)$-fold, (c) $(3 \times 3 \times 3)$-fold, and (d) $(6 \times 6 \times 6)$-fold superstructure. The unit cells have lattice parameters of 19.2 Å, 9.6 Å, 6.4 Å, and 3.2 Å, respectively. The structure types underlying the average structures in (b)–(d) are those of* $cP52$-Al_4Cu_9 *(215 $P\bar{4}3m$),* $cF16$-NaTl *(227 $Fd\bar{3}m$), and* $cP2$-CsCl *(221 $Pm\bar{3}m$), respectively. The arrows mark two sets of atomic layers parallel to the* $\langle110\rangle$ *directions. The braces show the positions of one set of the three-layer stacks.*

The structures of *fcc* CIMs can also be described as constituted from layers that are stacked along all ⟨110⟩-directions (see Fig. 8.7(a) and Dshemuchadse *et al.* (2011)). The {110}-layers in space group 216 $F\bar{4}3m$ contain at least one representative of every special Wyckoff position, i.e., of all symmetrically independent atomic sites in the unit cell, except for those occupying general positions (Samson, 1964). The same is true for space groups 196 $F23$, 203 $Fd\bar{3}$, 209 $F432$, 216 $F\bar{4}3m$, and 219 $F\bar{4}3c$, and with minor restrictions also for 210 $F4_132$, 227 $Fd\bar{3}m$, and 228 $Fd\bar{3}c$, as well as for the primitive space groups 198 $P2_13$, 205 $Pa\bar{3}$, and with minor restrictions for 212 $P4_332$ and 210 $P4_132$.

If the *fcc* CIMs are projected along their main lattice directions, the layer structure becomes clearly visible. For the $cF \approx 400$-structures, the period of the layer stacking is $p = 3$: one flat layer that coincides with a mirror plane is sandwiched between two puckered layers (Fig. 8.7(a)). Each adjacent pair of such three-layer stacks (3S) is related by a glide plane, yielding the stacking sequence 3S-3S'-3S-3S' along [110], with 3S' representing the three-layer stack symmetrically equivalent to 3S. More generally speaking, the stacking periodicity is $2p$, always with one flat layer and $(p-1)$ puckered ones.

In the cubic structures discussed here, the network of {110}-layers consists of six mutually intersecting, symmetrically equivalent stacks. This leaves only a few degrees of freedom for the decoration of these layers with atoms. Apparently, either the layer formation is the driving force automatically leading to the observed endohedral clusters or the other way round. The dimensions of the unit cells of the respective structures depend on the number of layers contained in each stack, which is in turn connected with the size of the endohedral clusters, as well as with their packing. The number of layers in each stack, p, is identical to the order of the superstructures, i.e., 3, 4, 7, and 11, for the structures discussed here. The arrangement of the approximately equidistant flat and puckered layers also explains the strong $2p\,2p\,0$ reflections in the diffraction patterns.

8.1.2   CIMs with space group symmetry 216 $F\bar{4}3m$

Of the 39 CIMs with space group symmetry 216 $F\bar{4}3m$, 36 can be regarded as $(3 \times 3 \times 3)$-fold superstructures of a $cF16$-NaTl-like basic structure. The other three structures, $cF1124$-$Cu_{56.9}Cd_{43.1}$, $cF(5928-20)$-$Al_{56.6}Ta_{39.5}Cu_{3.9}$, and $cF(23\,256-122)$-$Al_{55.4}Ta_{39.1}Cu_{5.4}$, can be described as $(4 \times 4 \times 4)$-fold, $(7 \times 7 \times 7)$-fold, and $(11 \times 11 \times 11)$-fold superstructures of the same basic structure, respectively.

All 36 of the smallest CIMs with space group symmetry 216 $F\bar{4}3m$ share a considerable amount of atomic sites and thus can be grouped into four different subtypes of one aristotype. They are formed by the successive substitution of specific sites, leading to ideal numbers of atoms per unit cell of 432, 440, 448, and 456, respectively. A varying number of sites can be left unoccupied and an additional element of flexibility is the occurrence of disordered positions that are not fully occupied, resulting in non-integer average numbers of atoms per unit cell.

We usually express these by, e.g., $cF(432-x)$, which refers to a structure belonging to subtype I, with an average number of x unoccupied positions per unit cell[2].

The sites that are actually occupied in $cF \approx 400$-structures are given in Table 8.3 with their ideal coordinates. The upper part of Table 8.3 contains the sites that are occupied in all of the structures, while the lower part of the table lists the sites that are exchanged between subtypes. The coordinates given for sites with free parameters are idealized values and correspond to the above-mentioned basic structures.

Following from the large number of atomic positions that these structures have in common, the cluster structures that can be interpreted into their atomic arrangements vary only slightly. A typical example of these structures is $cF444$-$Al_{63.6}Ta_{36.4}$, which can be described as a cubic close-packing of three-shell clusters of the fullerene-like type F_{76}^{40} (with 76 vertices and 40 faces) (Conrad *et al.*, 2009) (see Fig. 7.17 in Subsection 7.4.4 for the cluster structure). The main cluster in these structures is located at one of the highest-symmetry positions: $4a\ 0, 0, 0$, $4b$ $1/2, 1/2, 1/2$, $4c\ 1/4, 1/4, 1/4$, or $4d\ 3/4, 3/4, 3/4$, all with site symmetry $\bar{4}3m$. Origin shifts of $+(1/4, 1/4, 1/4)$ and multiples of it are permitted in structures in space group $F\bar{4}3m$.

The immediate effect of swapping one of the doublets of a 16e-position (16e6-8) and a 24f/g-position (24f2, 24g1/2) with one of the 48h-positions (48h5-7) becomes apparent when regarding the first-shell clusters around the points of highest symmetry, i.e., positions 4a-d in space group $F\bar{4}3m$. One such exchange, defined as the "first" one in these structures, is 16e7 & 24f2 $\rightarrow$ 48h6. It describes the transition between subtypes I and II. The "second" transition— 16e8 & 24g2 $\rightarrow$ 48h7—describes the transition from subtype II to II, and the "third" one—16e6 & 24g1 $\rightarrow$ 48h5—describes the transition from subtype III to IV. The sequence of these swaps is determined by the choice of origin and can be changed by origin shifts and thus interchanging sites 4a-d, as well as the sites within each group of equivalent positions demarcated by horizontal lines in Table 8.3.

The four subtypes all feature rhombic dodecahedra (rd) and/or Friauf polyhedra FK_{16}^{24} (Fp) around sites 4a-d. The sequence along the body diagonal of the face-centered cubic unit cell, i.e., on $4a\ 0, 0, 0$, $4c\ 1/4, 1/4, 1/4$, $4b\ 1/2, 1/2, 1/2$, $4d\ 3/4, 3/4, 3/4$, are rd-rd-rd-rd (subtype I), rd-rd-Fp-rd (subtype II), rd-rd-Fp-Fp (subtype III), and rd-Fp-Fp-Fp (subtype IV). Some structures also exhibit incomplete rd-clusters, i.e., partially capped octahedra or cubes.

The second cluster shell is only constituted of atoms occupying positions listed in the upper part of Table 8.3 and therefore it is the same around all positions 4a-d in all subtypes I-IV: a Frank-Kasper polyhedron FK_{40}^{76}.

---

[2] An additional margin of error resulting from intermediate structure types—e.g., I/II—or slight variations of the original four—e.g., II', III'. They, too, can have numbers of atoms per unit cell differing further from the values listed above.

Table 8.3 *Atomic sites of the here-discussed fcc CIMs with space group 216 F43m. The Wyckoff position is included in the site label. The subtypes, which feature the respective sites, are given. In the upper part of the table, the sites are given that are occupied in all of the structures, while the lower part of the table lists the sites that are exchanged between subtypes. '(All)' means that the sites can be but do not need to be occupied in all subtypes. The general expressions of the sites with free parameters are: $16e\ x,x,x;\ 24f\ x,0,0;\ 48h\ x,x,z$.*

Site	Symmetry	Idealized coordinates	Subtypes
16e1	.3m	1/6, 1/6, 1/6	*All*
16e2	.3m	5/12, 5/12, 5/12	*All*
16e3	.3m	2/3, 2/3, 2/3	*All*
16e4	.3m	11/12, 11/12, 11/12	*All*
48h1	..m	1/6, 1/6, 1/48	*All*
48h2	..m	1/12, 1/12, 13/48	*All*
48h3	..m	1/6, 1/6, 25/48	*All*
48h4	..m	1/12, 1/12, 37/48	*All*
4a	$\bar{4}3m$	0, 0, 0	*(All)*
4b	$\bar{4}3m$	1/2, 1/2, 1/2	*(All)*
4c	$\bar{4}3m$	1/4, 1/4, 1/4	*(All)*
4d	$\bar{4}3m$	3/4, 3/4, 3/4	*(All)*
16e5	.3m	3/48, 3/48, 3/48	I, II, III, IV
16e6	.3m	15/48, 15/48, 15/48	I, II, III
16e7	.3m	27/48, 27/48, 27/48	I
16e8	.3m	39/48, 39/48, 39/48	I, (II)
24f1	2.mm	1/6, 0, 0	I, II, III, IV
24f2	2.mm	1/3, 0, 0	I
24g1	2.mm	1/12, 1/4, 1/4	I, II, III
24g2	2.mm	7/12, 1/4, 1/4	I, II
48h5	..m	5/24, 5/24, 19/48	IV
48h6	..m	1/24, 1/24, 31/48	II, III, IV
48h7	..m	5/24, 5/24, 43/48	III, IV

The third cluster shell of all the clusters located at these positions contain atomic positions of neighboring first-shell clusters and consequently vary between positions ($4a$-d) and subtypes. The most conspicuous of these shapes is indeed the fullerene-like F_{76}^{40}-shape, which is also dual to the second-shell FK_{16}^{24}-polyhedron. It can be found in subtypes III and IV around site $4c$.

In addition to the $cF\approx400$-structures, three more CIMs are found to exhibit space group 216 $F\bar{4}3m$. $cF1124$-$Cu_{56.9}Cd_{43.1}$ was first reported by Samson (1967) and was determined based on the specific properties of the $\{110\}$-planes in this space group, as mentioned above. Two interpenetrating three-dimensional frameworks, one composed of Friauf polyhedra and the other of icosahedra, were found to be arranged in a diamond-type net, the Friauf polyhedra sharing hexagonal faces and the icosahedra fitting into cavities and sharing vertices with the truncated-tetrahedron basis of the Friauf polyhedra. Alternative descriptions published since then describe the same structure in terms of octahedra, tetrahedra, and Friauf polyhedra (Andersson, 1980; Hellner and Pearson, 1987), and as a relative to a metastable icosahedral quasicrystal in the Cd–Cu system, also including the $I3$-cluster concept (Kreiner and Schäpers, 1997).

Reinvestigating $cF1124$-$Cu_{56.9}Cd_{43.1}$ with respect to multishell endohedral clusters (see Fig. 8.8), we found a description of the unit cell involving two clusters at two of the four highest-symmetry sites. The one at $4b$ $1/2, 1/2, 1/2$ consists of a tetrahedron (which is regarded rather as a cluster center than a shell by itself), surrounded by a 22-atom polyhedron with triangular faces that exhibits tetrahedral symmetry, and then a fullerene-like second shell F_{40}^{22}-polyhedron (with partially capped faces). The site $4c$ $1/4, 1/4, 1/4$ is occupied by a Cd-atom, which is then surrounded by a Friauf polyhedron FK_{16}^{28}, a fullerene-like F_{28}^{16}-cluster (with all 12 pentagonal faces capped by one atom each, as well as all four hexagonal faces capped by a triangle of atoms each), and a 70-atom polyhedron (consisting of 12 triangles, 36 pentagons, and 4 hexagons), which in its capped form yields the 110-atom Frank-Kasper-like polyhedron FK_{110}^{266}. The F_{40}^{22}-polyhedra at positions $4b$ and the 70-atom polyhedra at positions $4c$ form a three-dimensional network by sharing hexagonal faces.

The two giant-unit cell structures $cF(5928 - 20)$-$Al_{56.6}Cu_{3.9}Ta_{39.5}$ and $cF(23\,256 - 122)$-$Al_{55.4}Cu_{5.4}Ta_{39.1}$ are related to $cF444$-$Al_{63.6}Ta_{36.4}$ in that they all occur in the same intermetallic system Al–(Cu)–Ta, exhibit all the same space group symmetry 216 $F\bar{4}3m$, and are $(3 \times 3 \times 3)$-, $(7 \times 7 \times 7)$-, and $(11 \times 11 \times 11)$-fold superstructures, respectively, of the same basic structure type, $cF16$-NaTl. The dominant feature in the cluster structure of $cF444$-$Al_{63.6}Ta_{36.4}$, described above, is the fullerene-like, F_{76}^{40} three-shell cluster, which is packed densely, and is sharing all of its pentagonal faces with its 12 closest cluster neighbors. In the medium and large structure that occur in the ternary extension of this system, Al–Cu–Ta, superclusters constituted from four or ten of these fullerene-like clusters occur. For a detailed discussion see Conrad *et al.* (2009) or Section 7.4.4.

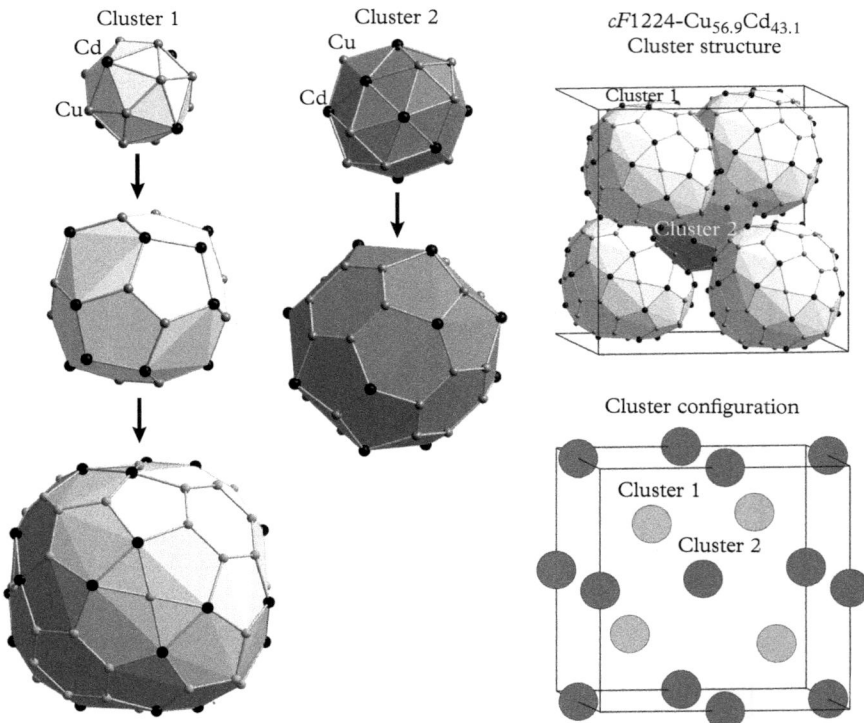

Fig. 8.8 *Typical cluster structure of the only $cF \approx 1200$-compound with space group symmetry 216 $F\bar{4}3m$: $cF1124$-$Cu_{56.9}Cd_{43.1}$. The unit cell is on half the usual scale.*

8.1.3   CIMs with space group symmetry 227 $Fd\bar{3}m$

Similar to the CIMs with space group symmetry 216 $F\bar{4}3m$, those with space group 227 $Fd\bar{3}m$ include a number of $cF \approx 400$-structures as well as two compounds of the type $cF \approx 1200$. The seven smaller structures can again be described as $(3 \times 3 \times 3)$-fold superstructures, whereas the two larger ones correspond to $(4 \times 4 \times 4)$-fold superstructures, with the basic structure of the $cF16$-NaTl type in all cases.

The atomic positions of the $cF \approx 400$-structures with space group 227 $Fd\bar{3}m$ are very similar to those with space group symmetry 216 $F\bar{4}3m$, although they obviously have double multiplicity due to the additional center of inversion at $1/8, 1/8, 1/8$ and symmetrically equivalent positions. The cluster structure, illustrated on an averaged $cF464$ structure (see Fig. 8.9), features a three-shell cluster at $8a$ $1/4, 1/4, 1/4$ that consists of a Friauf polyhedron FK_{16}^{28} in the center, surrounded by a fullerene-like F_{28}^{16}-cluster (also termed a triakis tetrahedron; with all faces capped), and a fullerene-like F_{84}^{44}-cluster, which is dual to the capped version of the

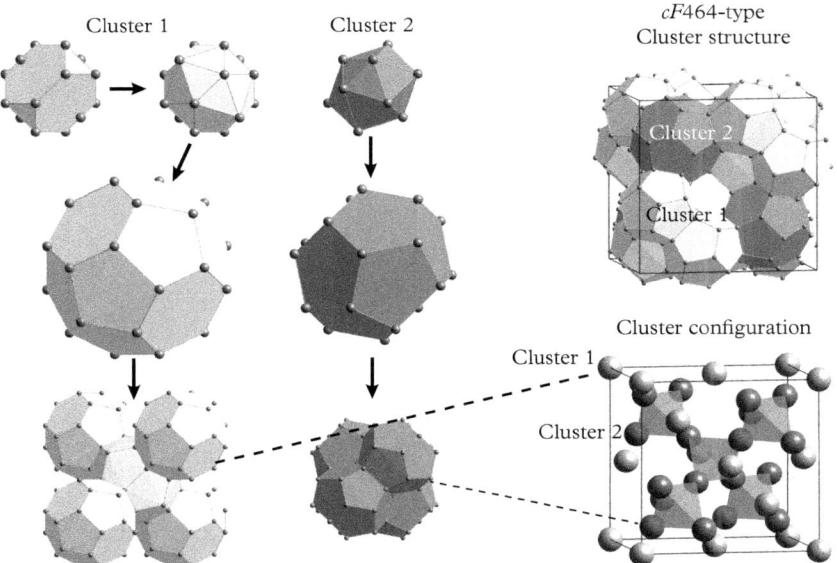

Fig. 8.9 *Cluster structure of an averaged CIM of type cF464 with space group 227 Fd3̄m. Shown are the two-shell clusters that constitute the seven cF ≈ 400-compounds and two cF ≈ 1200-compounds, as described in the text. The superclusters and the unit cells are on half the usual scale.*

second-shell F_{28}^{16}-cluster. The outermost cluster shells intersect one another, while the second cluster shells share hexagonal faces, resulting in each F_{28}^{16}-polyhedron being coordinated tetrahedrally by four F_{28}^{16}-polyhedra. A second type of cluster completes this network: the empty $16d$ $5/8, 5/8, 5/8$ positions are surrounded by an icosahedron, which in turn is the center of a pentagonal dodecahedron F_{20}^{12}. The centers of four of these two-shell dodecahedra are located on Wyckoff position $8b$ $1/2, 1/2, 1/2$, which, together with the two-shell F_{28}^{16}-polyhedra, fill space completely.

The first *fcc* CIM that was reported ever to contain more than 1000 atoms per unit cell was $cF(1192 - 40)$-$Cd_{66.7}Na_{33.3}$ (Samson, 1962). This structure is built almost entirely from Friauf polyhedra, however, not in a layered manner as is the case in Laves phases, but *via* the formation of spherical clusters (Samson clusters) from these building blocks. The compound was studied later on by quantum-mechanical calculations (Fredrickson *et al.*, 2007; Lee *et al.*, 2007), complementing the complex geometric picture with an analysis of the chemical bonding, resulting in a subdivision of the structure into electron-rich and electron-poor regions. The basically isostructural compound $cF(1192 - 23)$-$Al_{53.6}Mg_{46.4}$, known as 'Samson phase', exhibits a striking degree of structural disorder (Samson, 1965). Recent reinvestigations confirmed Samson's pioneering work (Feuerbacher *et al.*, 2007), and added a new view at this highly complex structure (Sikora *et al.*, 2008; Wolny *et al.*, 2008).

$cF(1192 - 40)$-$Cd_{66.7}Na_{33.3}$ and $cF(1192 - 23)$-$Al_{53.6}Mg_{46.4}$ can be described based on multishell clusters on the two highest-symmetric Wyckoff positions in the unit cell. At $8a$ $0,0,0$, a tetrahedron is centering an 22-atom cluster with triangular faces, which is surrounded by a fullerene-like F_{40}^{22}-polyhedron (that is partially capped), similar to the one found in $cF1124$-$Cu_{56.9}Cd_{43.1}$ at $4b$. The second cluster, centered at $8b$ $1/2, 1/2, 1/2$, consists of a central atom, surrounded by a Friauf polyhedron FK_{16}^{28}, which is surrounded by a fullerene-like F_{28}^{16}-cluster. Thus far, it is identical to the cluster on site $4c$ in $cF1124$-$Cu_{56.9}Cd_{43.1}$, however, its outer shells turn out to be of higher symmetry: all faces of the second-shell cluster are capped, resulting in a FK_{44}^{84}-polyhedron that is in turn surrounded by a third-shell fullerene-like F_{76}^{40}-cluster (with all faces capped, yielding a FK_{16}^{228}-polyhedron). The second-shell cluster at $8a$ and the third-shell cluster around $8b$ contain all atomic positions of these structures and therefore cover space completely while overlapping with one another.

8.1.4 Some more face-centered cubic CIMs

The CIMs with *fcc* space group types can mostly be assigned to two groups that are defined by their approximate unit cell size: $cF \approx 400$-structures and those with $cF \approx 1200$ (see Fig. 8.5). The former group only contains structures with space group symmetries 216 $F\bar{4}3m$ and 227 $Fd\bar{3}m$, discussed above. The latter group exhibits those as well, but also contains compounds that have space group symmetries 225 $Fm\bar{3}m$, 226 $Fm\bar{3}c$, and 203 $Fd\bar{3}$.

- 225 $Fm\bar{3}m$: The structures of $cF1124$-$Tb_{41.6}Ge_{39.9}Fe_{18.5}$ (Pecharskii *et al.*, 1987) as well as the isotypic structures in the systems Pr–Sn–Co and Gd–Ge–Fe, and of $cF(1208 - 64)$-$Dy_{40.9}Sn_{39.2}Co_{19.9}$ (Salamakha *et al.*, 2001) were found to be closely related to one another and to have the same cluster structure. A single atom at $4a$ $0,0,0$ is surrounded by a disordered rhombic dodecahedron, which is enclosed by a capped rhombicuboctahedron or a deltoidal icositetrahedron; the third cluster shell is a 80-atom polyhedron with 6 quandrangular and 144 triangular faces. At $4b$ $1/2, 1/2, 1/2$, an atom is surrounded by a cube, a capped rhombicuboctahedron or deltoidal icositetrahedron, and a 48-atom polyhedron with 8 triangular and 42 quadrangular faces. At Wyckoff position $8c$ $1/4, 1/4, 1/4$ (which also includes position $3/4, 3/4, 3/4$), a 22-atom cluster with only triangular faces, is located, which is centered by an atom, similar to $cF \approx 1200$-type compounds with symmetries 216 $F\bar{4}3m$ ($4b$) and 227 $Fd\bar{3}m$ ($8a$). The second cluster shell is the same kind of 28-atom cluster that was found in the $cF \approx 400$-structures: a tetrahedrally truncated rhombic dodecahedron with capped pentagonal faces, FK_{40}^{76}. The two three-shell clusters at $4a$ and $4b$ together with the two-shell cluster at $8c$ fully describe the structures with symmetry 225 $Fm\bar{3}m$, packing densely by sharing quadrangular and triangular faces.

- 226 $Fm\bar{3}c$: The structure of $cF(944-22)$-$Zn_{67.1}Sn_{20.8}Mo_{12.1}$ exhibits an unusual unit cell size, not exactly fitting into the previously mentioned $cF\approx400$- and $cF\approx1200$-categories (Hillebrecht *et al.*, 1997). Its cluster structure is bult from two clusters. The first one consists of an atom at $8a$ $1/4,1/4,1/4$ centering a rhombic dodecahedron, enclosed by a snub cube, surrounded by a 60-atom third-shell cluster (with 24 pentagonal, 6 quadrangular, and 24 triangular faces). The second cluster is located at $8b$ $0,0,0$ and consists of a central atom within an icosahedron, surrounded by another icosahedron (with atoms capping all faces), which is enclosed by a small rhombicosidodecahedron. Both three-shell clusters overlap with their symmetrically equivalent clusters and cover all sites within the unit cell. Two packings of clusters can be formed by choosing either the two-shell cluster at $8a$ and the three-shell cluster at $8b$ or *vice versa*, and either of these packings fully describe the structure. Equivalent structures were found in the systems Zn–Ru–Sb and Zn–Ru–Sb (Xiong *et al.*, 2010).

- 203 $Fd\bar{3}$: The structure of $cF1392$-$Cd_{86.2}Eu_{13.8}$ can be seen as an approximant for the icosahedral (Tsai-type) quasicrystals in the related Ca–Cd and Yb–Cd systems (Gómez and Lidin, 2004). Originally it was described based on two symmetrically inequivalent triacontahedra, each centered by a disordered tetrahedron within a dodecahedron, which is enclosed by an icosahedron surrounded by an icosidodecahedron. The cluster can be also decomposed in a different way. Then, the disordered tetrahedron constitutes still the cluster center, and the dodecahedron forms the first cluster shell. The icosidodecahedron can be regarded as the second cluster shell (with its pentagonal faces capped by atoms that had previously been described as a separate cluster shell in the shape of an icosahedron). The third cluster shell is formed by the same atoms that were formerly regarded as making up the triacontahedron, but can also be viewed differently: as being arranged in a fullerene-like shell F_{80}^{42}. Two three-shell clusters formed by these polyhedra are located at the positions $8a$ $0,0,0$ and $8b$ $1/2,1/2,1/2$ (these sites being arranged in a double-diamond lattice). The third cluster shells overlap with one another. All Wyckoff sites contained in the three-shell cluster around $8a$, together with the two-shell cluster around $8b$, cover all atomic positions within the unit cell.

8.2 Cluster structures of hexagonal CIMs

As can be seen in Table 8.1, hexagonal plus trigonal CIMs form the second largest group of CIMs, following right after the *fcc* CIMs discussed in the previous section. Their cluster structures have so far only been reviewed by Dshemuchadse (2013), with numerous different packings of Frank-Kasper polyhedra and fullerene-like multishell clusters being observed. However, a somewhat universal description as was found for the *fcc* CIMs (see Section 8.1) was not possible.

Already the distributions of the unit cell sizes of structures with Bravais lattice types hR and hP (see Fig. 8.1) are much less discrete, and cannot be separated into clear subgroups. At closer inspection, almost no hP-CIMs with unit cell sizes of around 200 atoms are reported, with another, narrower gap being located at around 130 atoms. The latter one was chosen as a lower bound for the definition of hexagonal CIMs, so as to keep the number of investigated structures with 33 within a feasible range. For hR-CIMs, obvious gaps in the histogram can be found at $hR \approx 200$ and $hR \approx 300$. The centered equivalent of a 100-atom-containing primitive cell lies at $hR300$, which was chosen as a threshold for CIMs, resulting in the selection of 15 hR-structures. For a complete list of all hP- and hR-CIMs discussed here see Table 4.1 in Dshemuchadse (2013).

The 230 space groups contain 25 trigonal and 27 hexagonal groups, but only 38 of these 52 are featured among intermetallics in the PCD. Classified by Bravais lattice, 31 of these belong to hP and 7 to hR (out of a total 45 and 7, respectively). The hP-CIMs can be grouped into 18 structure types, the hR-CIMs into 7 different ones (an important criterion is a similar c/a-ratio for all equivalent structures with hexagonal unit cells). Five of the former and five of the latter were found to exhibit a structure built with multishell endohedral clusters similar to the cF-CIMs discussed above. These were also discussed in Dshemuchadse and Steurer (2014), among others. More details on these structures, as well as the ones that were not found to exhibit a multi-shell cluster structure as discussed in this chapter, can be found in Dshemuchadse (2013). In the following, some characteristic examples are discussed in greater detail.

- $hP139$-$Ga_{47.5}Mg_{26.4}Cu_{15.5}Li_{10.6}$ (187 $P\bar{6}m2$) (Lin and Corbett, 2008): this is a unique structure type constituted from three multishell clusters: a FK_{15}^{26}-polyhedron centered by an atom is located at Wyckoff position $1f$; it is surrounded by a fullerene-like F_{26}^{15}-shell (with capped faces: FK_{41}^{78}), which is then enclosed by the fullerene-like polyhedron F_{78}^{41}. These three-shell clusters are packed closely along three lattice directions, sharing hexagonal faces along [001] and pentagonal faces in the (110)-plane. A network of Frank-Kasper (FK) polyhedra between these F_{78}^{41}-shells consists of mostly FK_{16}^{28}- as well as fewer FK_{15}^{26}- and FK_{12}^{20}-polyhedra. On site $1d$, a similar cluster arrangement can be found. The first and second cluster shells are equivalent to the ones around $1f$, but the third shell is distorted in a way that transforms 6 of the 29 hexagonal faces of F_{78}^{41} into pentagons, resulting in a 75-atom polyhedron which is not quite fullerene-like. The empty FK_{12}^{20}-polyhedra (icosahedra) mentioned in the cluster packing above center other multishell clusters. The icosahedra are surrounded by a pentagon dodecahedron, F_{20}^{12}, with capped faces, FK_{32}^{60}, which is in turn surrounded by a C_{60}-fullerene-like (F_{60}^{32}) cluster.
- $hP(218-20)$-$Ga_{62.6}Li_{29.3}Cd_{8.1}$ (163 $P\bar{3}1c$) (Tillard-Charbonnel *et al.*, 1994): this structure constituted from endohedral clusters in the following way:

at Wyckoff position $2b$, an empty icosahedron FK_{12}^{20} is surrounded by a second-shell (incomplete) pentagon dodecahedron F_{20}^{12} (with capped faces: FK_{32}^{60}), which is in turn surrounded by a F_{60}^{32}-cluster shell. An *hcp* layer of these three-shell clusters is stacked along the $\bar{c}$-direction of the structure and a network of FK-polyhedra can be found in the interstices. An alternative description of the fullerene-like F_{60}^{32}-polyhedron is as a packing of 20 FK_{16}^{28} Friauf-polyhedra around the central icosahedron. A second fullerene-like structure is located at a position of slightly lower symmetry ($4f$) and consists of a central atom within a FK_{16}^{28} Friauf-polyhedron, within a defective fullerene-like F_{28}^{16}-shell (with capped faces: approximately FK_{44}^{82}), within a larger shell of the fullerene-like F_{44}^{84}-shape. These large three-shell clusters overlap in all three lattice directions, forming a two-dimensional hexagonal arrangement that is stacked in an *AABB*-manner, and contain all atomic positions that are occupied in these structure, hence describing it completely.

- Three *hP*198-*hP*238-compounds: The two structures of *hP*(224 − 2.2)-In$_{55.0}$K$_{23.4}$Tl$_{16.1}$Na$_{5.4}$ (Flot *et al.*, 1997*a*) and *hP*238-In$_{67.2}$K$_{32.8}$ (Li and Corbett, 2003) were described in space group 164 *P*$\bar{3}$*m*1. This space group has a common supergroup, 194 *P*6$_3$/*mmc*, with the space group 163 *P*$\bar{3}$1*c* of the Ga–Li–Cd-compound described above, and their cluster structures are closely related. Due to its lower symmetry and the corresponding splitting of Wyckoff sites into multiple symmetrically inequivalent ones, both of the multishell clusters found in the Ga–Li–Cd-structure are present in these two compounds in almost identical form, but each of them in two slightly different versions. The FK_{12}^{20}@F_{20}^{12} (capped: FK_{32}^{60}) @F_{60}^{32}-cluster is present in the same arrangement on site 1*a*, and in a slightly altered version (with six heptagonal faces replacing six pentagonal ones, as well as four quadrangular faces in addition) on site 1*b*. Both these sites in space group 164 *P*$\bar{3}$*m*1 are directly linked to the 2*b* site in 163 *P*$\bar{3}$1*c via* the group-supergroup-subgroup relationships mentioned above. The central site of the other fullerene-like cluster—4*f* in 163 *P*$\bar{3}$1*c*—is also split into two inequivalent positions of almost the same cluster, both centered around different 2*d* sites. The sequence of a central atom within a FK_{16}^{28}-polyhedron surrounded by a F_{28}^{16}-shell (with capped faces: FK_{44}^{82}) enclosed by a F_{44}^{84}-cluster is the same for both of them, in one case with similar defects as reported for the Ga–Li–Cd-structure, in the other in the ideal geometry. The packing of these clusters is unchanged with respect to the structure with symmetry 163 *P*$\bar{3}$1*c*, and the network of FK-polyhedra between the F_{60}^{32} clusters is only slightly altered compared with the structure of *hP*(218 − 20)-Ga$_{62.6}$Li$_{29.3}$Cd$_{8.1}$, with one additional cluster shell occurring, FK_{19}^{34}, probably in direct connection with the small changes to some of the fullerene-like clusters.

- *hP*272-Al$_{55.3}$Li$_{26.9}$Mg$_{12.0}$Cu$_{5.7}$ (*P*6$_3$/*mmc*) (Le Bail *et al.*, 1991): This structure exhibits two three-shell and one two-shell cluster worth describing. At site 4*f*, a central atom is surrounded by an FK_{15}^{26}-polyhedron, which is surrounded by a fullerene-like F_{26}^{15}-shell (with capped faces: FK_{41}^{78}). The third cluster shell with roughly the shape of F_{78}^{41} in this structure is defective in that one hexagonal face is replaced by a triangular one. The clusters themselves are the same that were found in $hP139$-$Ga_{47.5}Mg_{26.4}Cu_{15.5}Li_{10.6}$, but do not overlap with one another and instead pack together by sharing faces. As was the case in multiple cases above, this cluster can alternatively be described as a packing of FK-polyhedra. The second kind of three-shell cluster is found on a $6h$-site and consists of an empty icosahedron FK_{12}^{20}, surrounded by a pentagonal dodecahedron F_{20}^{12} (only partially capped to an incomplete version of FK_{32}^{60}), surrounded by a slightly defective F_{60}^{32} fullerene-like shell. If only the second shells of both above-mentioned clusters are taken into account, i.e., F_{26}^{15} and F_{20}^{12}, a covering of the entire unit cell can be achieved by adding only one more two-shell cluster: around an atom sitting on a $4e$-site, a disordered Friauf polyhedron FK_{16}^{28} within another fullerene-like shell of F_{24}^{14} can be described.

- $hP386$-$Al_{57.4}Ta_{39.0}Cu_{3.6}$ (194 $P6_3/mmc$) (Dshemuchadse *et al.*, 2013): One three-shell cluster is located around the empty site $2a$, consisting of a rhombohedron (adopting one of at least six different orientations), surrounded by a pentagon dodecahedron F_{20}^{12} (with capped faces resulting in a rhombic triacontahedron FK_{32}^{60}), which is in turn surrounded by the fullerene-like shell of F_{60}^{32}. The other three-shell cluster is centered at a central atom at position $2c$, which is surrounded by a disordered 11-atom cluster shell within a 27-atom-polyhedron (consisting of 12 pentagonal and 14 triangular faces; with capped pentagons resulting in FK_{39}^{74}), which is located within the fullerene-like F_{74}^{39}-cluster shell. Each of these two kinds of clusters is packed densely on a hexagonal lattice within layers perpendicular to the $\vec{c}$-lattice direction. These layers are then stacked in an $ABAC$-stacking sequence with B- and C-layers being composed of the same cluster and both A-layers of another one, resulting effectively in an $AB'AC'$-stacking. The interstices between the fullerene-like clusters are filled with different FK-polyhedra, mostly FK_{15}^{26} and FK_{16}^{28}. An alternative description is suggested in the same manuscript (Dshemuchadse *et al.*, 2013) and makes use of the second-shell clusters mentioned above, as well as additional polyhedra. In addition to F_{20}^{12} at $2a$ and the 27-vertex shape at $2c$, a F_{28}^{16} fullerene-like second-shell cluster found at a $4f$-site and a set of pentagonal bifrusta are necessary to describe the entire unit cell of the structure.

- $hP390$-$In_{49.8}Na_{49.2}Ni_{1.0}$ (194 $P6_3/mmc$) (Sevov and Corbett, 1993*a*): This structure is similar to that of $hP386$-$Al_{57.4}Cu_{3.6}Ta_{39.0}$ discussed before. There are only a few differences in the description of its disorder motifs (Dshemuchadse *et al.*, 2013).

- $hP694$-$Li_{49.0}Ba_{39.8}Na_{11.3}$ ($P\bar{3}$) (Smetana *et al.*, 2007*a*): The AETs found in this structure are mostly FK-polyhedra: FK_{16}^{28} and some FK_{15}^{26} around the larger Ba-atoms, FK_{12}^{20} around the smaller Na- and Li-atoms. The structure exhibits a high degree of disorder thus rendering its cluster description ambiguous and incomplete. The same three-shell cluster can be found around four atomic positions—$1a$, $1b$, and two $2d$-sites: in each case, the central atom is surrounded by an icosahedron FK_{12}^{20}, which is located within a pentagon dodecahedron F_{20}^{12} (with capping atoms: FK_{32}^{60}), which is then surrounded by the fullerene-like third shell in the shape of F_{60}^{32}. Within each layer, these clusters form a hexagonal lattice with an overall stacking motif of *ABBACC*. They do not, however, form a close packing or describe the entire structure.

- $hR360$-$Ga_{65.0}Na_{35.0}$ (Frank-Cordier *et al.*, 1982*a*) and $hR366$-$In_{65.0}K_{35.0}$ (Cordier and Müller, 1992) ($R\bar{3}m$): Both structures are very similar. The constituting clusters are the same as in $hP238$-$In_{67.2}K_{32.8}$ and related structures, but they are packed in a different manner. An empty icosahedron FK_{12}^{20} is located at the $9e$ position, surrounded by F_{20}^{12} (with capping atoms: FK_{32}^{60}), which is then surrounded by a F_{60}^{32}-shell. These clusters overlap and each one contains all atomic positions found in the structure. Another fullerene-like three-shell cluster, also observed in the $hP238$-compound, is featured in these structures, as well: a central atom enclosed by an FK_{19}^{34}-polyhedron, which is located within a F_{28}^{16}-shell (with capping atoms a modified version of FK_{44}^{84}), which is in turn surrounded by the third-shell F_{84}^{44}-cluster. This kind of cluster also overlaps with its equivalents within the unit cell, but one atomic position is not contained within. It can also, alternatively, be built from FK-polyhedra FK_{16}^{28} and FK_{19}^{34} and, together with a set of FK_{12}^{20}-icosahedra, these make up the unit cell almost entirely.

- Eight isostructural $hR387$-$hR417$-compounds (166 $R\bar{3}m$): $hR(414-27.3)$-$Ga_{73.6}Na_{20.2}K_{6.2}$ (Belin and Charbonnel, 1986; Flot *et al.*, 1998), $hR(417-14.9)$-$Ga_{61.1}Na_{15.7}Cd_{12.8}K_{10.4}$ (Flot *et al.*, 1997*b*), $hR417$-$Ga_{66.9}Na_{24.5}Cu_{8.7}$ (Tillard-Charbonnel *et al.*, 1992*a*), $hR417$-$Ga_{66.2}Na_{24.5}Cu_{5.0}Cd_{4.3}$ (Chahine *et al.*, 1994), $hR417$-$Ga_{58.4}Na_{24.5}Zn_{17.2}$ (Tillard-Charbonnel *et al.*, 1992*b*), $hR417$-$In_{69.2}K_{24.5}Au_{6.3}$ (Li and Corbett, 2006), $hR417$-$In_{65.5}K_{24.5}Mg_{10.0}$ (Li and Corbett, 2006), and $hR(420-9.0)$-$In_{65.7}K_{24.8}Zn_{9.5}$ (Li and Corbett, 2006; Cordier and Müller, 1995). A four-shell cluster is located at site $3b$: an icosahedron FK_{12}^{20} is surrounded by a pentagon dodecahedron F_{20}^{12} (with capping atoms: FK_{32}^{60}), which is enclosed in a fullerene-like F_{60}^{32}-shell, followed directly by another cluster shell of fullerene-like shape: F_{80}^{42}. Apart from one atomic position, all independent sites are contained within this large cluster. Another cluster, this one located at $9e$, features the same first and second shells: FK_{12}^{20} within F_{20}^{12} (with capping atoms: FK_{32}^{60}), which are followed by an elongated version of the F_{60}^{32}-polyhedron through the insertion of ten additional vertices. Most sites are again surrounded by FK-type coordination polyhedra (FK_{16}^{28}, FK_{19}^{34}, and FK_{12}^{20})

- $hR(441-5.9)$-$Cd_{40.2}Na_{33.8}Sn_{26.0}$ (166 $R\bar{3}m$) (Todorov and Sevov, 1997): Another familiar cluster was found here as well: an empty icosahedron

Table 8.4 *CIMs with fullerene-like multishell clusters with orthorhombic lattice symmetry (primitive—oP, base-centered—oS, body-centered—oI, and face-centered oF).*

Compound	Space group	Site	Cluster shells	Ref.
$oP244$-Ga$_{63.9}$Na$_{36.1}$	53 *Pmna*	4a	$FK_{12}^{20}@\,F_{20}^{12}/FK_{32}^{60}@\,F_{60}^{32}$	[1]
				[2]
$oP244$-Na$_{36.1}$Zn$_{33.1}$Sn$_{30.8}$	"	"	"	[3]
$oP248$-In$_{62.9}$K$_{29.3}$Na$_{7.8}$	"	"	"	[4]
$oP(756-14)$-In$_{53.1}$Na$_{46.3}$Ni$_{0.5}$	59 *Pmmn*	2a	$X_{10}@\,X_{25}/FK_{37}^{70}@\,F_{70}^{37}$	[5]
		2a	$X_{18}@\,X_{29}/FK_{41}^{78}@\,F_{78}^{41}$	
		2b	$X_9@\,F_{20}^{12}/FK_{32}^{60}@\,F_{60}^{32}$	
		4e	$FK_{15}^{26}@\,F_{26}^{15}/X_{41}@\,X_{78}$	
		4f	$FK_{12}^{20}@\,F_{20}^{12}/X_{32}@\,F_{60}^{32}$	
$oS(288-5)$-Ga$_{77.4}$Li$_{22.6}$	63 *Cmcm*	4b	$FK_{12}^{20}@\,X_{20}/X_{32}@\,F_{60}^{32}$	[4]
		4c	$FK_{16}^{28}@\,X_{32}/X_{44}@\,X_{84}$	
$oS(344-4)$-In$_{64.7}$Na$_{35.3}$	63 *Cmcm*	4a	$FK_{12}^{20}@\,X_{20}/X_{32}@\,F_{60}^{32}$	[6]
		4c	$FK_{16}^{28}@\,X_{32}/FK_{44}^{84}@\,X_{84}$	[7]
		8g	$FK_{12}^{20}@\,F_{20}^{12}/X_{32}@\,F_{60}^{32}$	
$oS(348-25)$-Ga$_{71.5}$Li$_{21.1}$K$_{7.4}$	63 *Cmcm*	4a	$FK_{12}^{20}@\,F_{20}^{12}/FK_{32}^{60}@\,F_{60}^{32}$	[8]
$oI344$-Na$_{37.2}$Zn$_{31.5}$Sn$_{31.3}$	72 *Ibam*	4b	$X_{16}@\,X_{30}/FK_{46}^{88}@\,F_{84}^{44}$	[3]
$oF968$-Ga$_{56.8}$Na$_{26.4}$Au$_{16.8}$	69 *Fmmm*	4b	$FK_{12}^{20}@\,F_{20}^{12}/FK_{32}^{60}@\,F_{60}^{32}$	[9]
$oF(920-39)$-Ga$_{71.0}$Na$_{18.2}$Li$_{10.9}$	"	"	"	[10]
$oF(904-34)$-Ga$_{74.8}$Na$_{23.0}$Rb$_{2.2}$	"	"	"	[11]

[1] Ling and Belin (1982), [2] Frank-Cordier *et al.* (1982b), [3] Kim and Fässler (2009), [4] Carrillo-Cabrera *et al.* (1994), [5] Sevov and Corbett (1996), [6] Sevov and Corbett (1993b), [7] Cordier and Müller (1993), [8] Belin (1983), [9] Tillard-Charbonnel *et al.* (1993), [10] Charbonnel and Belin (1984), and [11] Charbonnel and Belin (1987).

FK_{12}^{20} on site $9d$ within F_{20}^{12} (with capping atoms: FK_{32}^{60}) within a F_{60}^{32}-polyhedron. These clusters overlap and cover all but four atomic positions in the structure, describing slabs of the structure completely, but omitting the atoms between these slabs. Another cluster is located at site $3a$, where a central atom is surrounded by a FK_{18}^{32}-polyhedron, which is surrounded by an elongated version of the pentagonal dodecahedron consisting of 26 vertices in total (with capping atoms: 44-vertex polyhedron), with the third cluster shell having the fullerene-like shape FK_{44}^{84}. This three-shell cluster forms a hexagonal close-packing perpendicular to the $\vec{c}$-direction and is ABC-stacked along $\vec{c}$. The remaining empty space between these polyhedra can be described entirely with FK_{16}^{28}- and FK_{12}^{20}-polyhedra.

- $hR879$-$Al_{63.1}Mg_{36.9}$ (166 $R3m$) (Feuerbacher *et al.*, 2007): The structure is closely related to that of $cF(1192-23)$-$Al_{53.6}Mg_{46.4}$ discussed above (Section 8.1.3). Atoms in this structure also mostly have FK-type coordination shells, specifically FK_{16}^{28}, FK_{15}^{26}, and FK_{12}^{20}. Around two of these, located at two different $3a$-sites, rather irregular three-shell clusters are located, derived from the fullerene-like shapes of F_{76}^{40} and F_{78}^{41}. The third shells of two of these different clusters overlap and form pairs stacked along the $\vec{c}$-direction. Compatible with the rhombohedral symmetry, three more of the "other kind" of three-shell cluster can be found to surround each one, forming an overall tetrahedral coordination. These arrangements, together with a three-dimensional network of FK_{16}^{28} Friauf-polyhedra, almost completely tile the unit cell of this structure.

- $hR888$-$Li_{67.5}Ba_{26.5}Ca_{6.0}$ (167 $R\bar{3}c$) (Smetana *et al.*, 2007a): The relatively prevalent three-shell cluster of an icosahedron FK_{12}^{20} within a pentagon dodecahedron F_{20}^{12} (with capping atoms: FK_{32}^{60}) within a C_{60}-like F_{60}^{32}-polyhedron is located at a $6b$-site. However, due to a failure to explain the remaining sites as similar FK-type or fullerene-like clusters, these remain isolated and quite a few atomic positions remain unexplained.

Table 8.5 *CIMs with fullerene-like multishell clusters with tetragonal lattice symmetry (primitive – tP and body-centered – tI).*

Compound	Space group	Site	Cluster shells	Ref.
$tP(228-3)$-$In_{62.7}Na_{37.3}$	137 $P4_2/nmc$	$2a$	$FK_{16}^{28}@X_{32}/FK_{44}^{84}@X_{84}$	[1]
$tP(906-28)$-$Al_{59.0}Li_{28.9}Cu_{6.0}Zn_{6.0}$	"	$2c$	$FK_{12}^{20}@F_{20}^{12}/FK_{32}^{60}@F_{60}^{32}$	[2]
		$2f$	$FK_{16}^{28}@X_{28}/\,X_{44}\,@X_{84}$	
		$4i$	$FK_{12}^{20}@F_{20}^{12}/FK_{32}^{60}@F_{60}^{32}$	
		$4i$	$FK_{16}^{28}@X_{28}/\,X_{44}\,@X_{80}$	
		$8o$	$X_{12}\,@X_{20}/\,X_{32}\,@F_{60}^{32}$	
		$8p$	$FK_{16}^{28}@F_{28}^{16}/FK_{44}^{84}@X_{84}$	
$tI232$-$Sn_{64.6}Rh_{20.7}Er_{14.9}$	142 $I4_1/acd$	$8b$	$X_{18}\,@X_{32}/FK_{44}^{84}@F_{84}^{44}$	[3]
$tI252$-$Li_{69.8}Ba_{30.2}$	122 $I\bar{4}2d$	$16e$	$FK_{12}^{20}@F_{20}^{12}/FK_{32}^{60}@X_{60}$	[4]

[1] Sevov and Corbett (1992), [2] Leblanc *et al.* (1991), [3] Hodeau *et al.* (1984), and [4] Smetana *et al.* (2007b).

Table 8.6 *CIMs with fullerene-like multishell clusters with cubic primitive lattice symmetry (cP).*

Compound	Space group	Site	Cluster shells	Ref.
$cP140$-Sc$_{81.4}$Rh$_{18.6}$	200 $Pm\bar{3}$	$1b$	FK^{20}_{12} @ X_{30} / FK^{80}_{42} @ X_{80}	[1]
$cP140$-Sc$_{81.4}$Ir$_{18.6}$	"	"	"	[1]
$cP140$-Sc$_{81.4}$Pt$_{18.6}$	"	"	"	[1]
$cP140$-Sc$_{81.4}$Ru$_{18.6}$	"	"	"	[1]
$cP140$-Sc$_{82.9}$Fe$_{17.1}$	"	"	"	[2]
$cP146$-Mg$_{41.1}$Ag$_{26.0}$Al$_{32.9}$	"	($1a$ & $1b$)	"	[3]
$cP154$-In$_{62.3}$Na$_{33.8}$K$_{3.9}$	223 $Pm\bar{3}n$	$2a$	FK^{20}_{12} @ F^{12}_{20} / FK^{60}_{32} @ X_{60}	[4]
		$6c$	FK^{24}_{14} @ F^{14}_{24} / X_{38} @ X_{72}	[5]
$cP154$-In$_{62.3}$Na$_{33.8}$Rb$_{3.9}$	"	"	"	[4]
$cP154$-In$_{62.3}$Na$_{33.8}$Cs$_{3.9}$	"	"	"	[4]
$cP156$-Hg$_{61.5}$K$_{37.2}$Na$_{1.3}$	"	($2a$ & $6d$)	"	[6]
$cP157$-Tl$_{68.8}$K$_{31.2}$	200 $Pm\bar{3}$	$1a$	FK^{20}_{12} @ F^{12}_{20} / FK^{60}_{32} @ F^{32}_{60}	[7]
		$1b$	FK^{20}_{12} @ F^{12}_{20} / FK^{60}_{32} @ X_{60}	
		$6f$	FK^{24}_{14} @ F^{14}_{24} / X_{38} @ X_{72}	
$cP(163-5)$-Zn$_{79.7}$Mg$_{15.2}$Ti$_{5.2}$	200 $Pm\bar{3}$	$1a$	FK^{20}_{12} @ F^{12}_{20} / FK^{60}_{32} @ F^{32}_{60}	[8]
		$1b$	FK^{20}_{12} @ F^{12}_{20} / FK^{60}_{32} @ F^{32}_{60}	
		$6g$	FK^{26}_{15} @ X_{27} / X_{42} @ X_{82}	
$cP(166-7)$-Zn$_{79.8}$Mg$_{15.9}$Hf$_{4.3}$	"	"	"	[8]
$cP(166-7)$-Zn$_{79.8}$Mg$_{15.9}$Zr$_{4.3}$	"	"	"	[8]
$cP172$-Cd$_{86.0}$Ce$_{14.0}$	201 $Pn\bar{3}$	$2a$	FK^{4}_{4} @ F^{12}_{20} / FK^{60}_{32} @ X_{34} @ F^{42}_{80};	[9]
$cP(561-48)$-Al$_{69.6}$Pd$_{24.3}$Mn$_{6.1}$	200 $Pm\bar{3}$	$1a$	FK^{20}_{12} @ F^{12}_{20} / FK^{60}_{32} @ F^{32}_{60}	[10]
$cP680$-Zn$_{68.2}$Mg$_{21.2}$Y$_{10.6}$	205 $Pa\bar{3}$	$8c$	FK^{20}_{12} @ F^{12}_{20} / FK^{60}_{32} @ F^{32}_{60}	[11]
		$8c$	FK^{28}_{16} @ F^{16}_{28} / X_{44} @ X_{80}	
$cP(704-5)$-Zn$_{84.3}$Sc$_{12.8}$Mg$_{2.9}$	205 $Pa\bar{3}$	$8c$	FK^{4}_{4} @ F^{12}_{20} / FK^{60}_{32} @ X_{30} @ F^{42}_{80};	[12]
		$8c$	FK^{20}_{12} @ X^{32} @ F^{32}_{60}	
$cP(704-14)$-In$_{44.5}$Ag$_{40.4}$Yb$_{15.1}$	"	"	"	[13]
$cP736$-In$_{42.9}$Ag$_{42.9}$Eu$_{14.1}$	205 $Pa\bar{3}$	$8c$	FK^{4}_{4} + FK^{8}_{6} @ F^{12}_{20} / FK^{60}_{32} @ X_{30} @ F^{42}_{80}	
		$8c$	FK^{20}_{12} @ X^{32} @ X_{60}	[14]

[1] Cenzual *et al.* (1985), [2] Andrusyak and Kotur (1991), [3] Kreiner and Spiekermann (1997), [4] Sevov and Corbett (1993c), [5] Carrillo-Cabrera *et al.* (1993), [6] Deiseroth and Biehl (1999), [7] Cordier *et al.* (1993), [8] Gómez *et al.* (2008), [9] Armbrüster and Lidin (2000), [10] Sugiyama *et al.* (1998), [11] Brühne *et al.* (2005), [12] Lin and Corbett (2006), [13] Li *et al.* (2008), and [14] Gómez *et al.* (2009).

8.3   Fullerene-like three-shell clusters

In addition to the more detailed analysis of the *cF*-, as well as the *hP*- and *hR*-CIMs discussed above, a search for fullerene-like multishell clusters in CIMs of all lattice symmetries rendered a wealth of additional findings (Dshemuchadse and Steurer, 2014). Fullerene-like three-shell clusters were found in CIMs with seven other Bravais lattices: *oP*, *oS*, *oI*, *oF*, *tP*, *tI*, *cP*, while none were found in *mP*-, *mS*-, or *cI*-CIMs. No *aP*-intermetallics reported thus far have large enough unit cells to be characterized as CIMs. In the following, three tables orthorhombic (Table 8.4), tetragonal (Table 8.5), and cubic (Table 8.6) CIMs are listed with the structure of their clusters. These clusters mostly have *F*- and *FK*-shapes; derivative cluster shapes are denoted as *X*-clusters.

9

Quasicrystals (QCs)

The discovery of quasicrystals (QCs) (Shechtman *et al.*, 1984) opened up a completely new world in the field of intermetallic phases. At that time, quite a few rather complex intermetallic phases were already known, the famous Samson phases $cF1192$-$NaCd_2$ and $cF1168$-Mg_2Al_3 belonging to the largest ones (Samson, 1964; Samson, 1965). However, the novel kind of long-range order, i.e., quasiperiodic order, was a big surprise for everybody, and even unbelievable for the famous chemist and double Nobel laureate Linus Pauling (Pauling, 1985; Pauling, 1989). It took a while until the structure of QCs, their stabilizing mechanism and the driving forces for the evolution of quasiperiodic long-range order were basically understood (for a short review see Steurer (2011*b*)). It should be mentioned here that quasiperiodic ordering is not restricted to intermetallic phases; it is also quite common on the mesoscale for particles (colloids, nanoparticles, etc.) with specific interaction potentials (Dotera, 2011; Dotera, 2012). Even more, quasiperiodic ordering has also been identified in particular packings of polyhedra, forced by two different characteristic length scales (Haji-Akbari *et al.*, 2009).

It should be emphasized here that the intermetallic phases classified as QCs do not necessarily have quasiperiodic structures in the strict mathematical meaning of the word. To clarify, this has been one of the the the main goals of QC structure analysis so far. Strictly quasiperiodic structures are purely point diffractive (Bragg reflections only), and can be described by the higher-dimensional (nD) approach. Real QCs, of course, will show continuous contributions in their Fourier spectrum originating from thermal diffuse scattering, defects, and structural disorder if any, i.e., in the same way as periodic intermetallic phases.

There are two main classes of intermetallic QCs known so far: decagonal QCs (DQCs) and icosahedral QCs (IQCs). The terminology "decagonal" and "icosahedral" refers to the Laue symmetry ($10/m$, $10/mmm$ and $m\bar{3}\bar{5}$, respectively) of their diffraction patterns (intensity weighted reciprocal lattice) or, equivalently, to the symmetry of the interatomic vector map (auto-correlation function or Patterson map). It also refers to the "bond-orientational order" of a QC structure, which is equivalent to its vector map. The full space-group symmetry of a quasiperiodic structure can best be described in the framework of the nD approach. However, an equivalent description is also possible in 3D reciprocal space based on the symmetry relationships between the complex structure factors $F(\mathbf{H})$

Intermetallics: Structures, Properties, and Statistics. First Edition. Walter Steurer and Julia Dshemuchadse.
© Walter Steurer and Julia Dshemuchadse 2016. Published in 2016 by Oxford University Press.

(Rabson *et al.*, 1991). It has to be kept in mind that within the 3D direct-space description of QCs (e.g., decorated tilings or coverings), they do not show any singular point of global tenfold or icosahedral symmetry, generally. However, these symmetries can be found locally everywhere in the respective QC structures.

Apart from the relatively common DQCs and IQCs, a few intermetallic QCs with octagonal and dodecagonal symmetries have been experimentally observed as well (see, for instance, Iwami and Ishimasa (2015)). However, they have been found to be either metastable or of very poor quality, and will not be discussed here. In contrast, dodecagonal symmetry is the most common one in the case of mesoscopic QCs (Barkan *et al.*, 2011). For more information on these QCs as well as on the quasiperiodic self-assembly on the meso- and macro-scale see Steurer and Deloudi (2009), for instance.

The search for stable intermetallic QCs has been mostly based on two parameters: the valence electron concentration and the atomic diameter ratio (Tsai, 2003; Tsai, 2013). Furthermore, the existence of rational approximants drove the search in specific binary or ternary intermetallic systems, which was quite successful in many cases.

9.1   Quasicrystal structure analysis

QC structure analysis is fundamentally different from the determination of periodic crystal structures. In the case of periodic crystals, the goal is the determination of the distribution of atoms in the unit cell, while the ordering of the unit cells is known *a priori*, because they form a lattice. In contrast, if we describe a QC as a covering then both the structure of the covering clusters and their arrangement have to be determined. While there are 14 Bravais lattice types, only, an infinite number of different tilings are possible. In terms of the *n*D description, we again have a quite limited number of *n*D Bravais lattice types. However, the shapes of the $(n-3)$D occupation domains (atomic surfaces), which also contain the information on the long-range order of the covering clusters, are only restricted by the closeness condition.

In the best case, the fundamental clusters and the way they are allowed to overlap can be first studied on periodic approximants, giving a sound basis for the subsequent determination of the QC structure itself. There are two kinds of periodic approximants to QCs known: general approximants and rational approximants. General approximants are only in some way related to QCs, they just show similar structure motifs (clusters). In contrast, the structures of rational approximants can be generated based on the higher-dimensional approach, just by shearing the hypercrystal structure properly. They not only show the same kind of clusters as the QCs but also the way they can overlap. Fortunately, rational approximants are available for most IQCs and for some DQCs.

A full structure analysis of a QC should be performed employing both electron microscopy and diffraction methods. X-ray and neutron diffraction can give an accurate picture of a QC structure; however, it will be averaged modulo one nD unit cell if just Bragg scattering is taken into account. One has to keep in mind that already the indexing of the reflection data set with n reciprocal basis vectors implies the validity and applicability of the nD approach. It has been demonstrated in a simulation that refining an orientationally fivefold twinned approximant crystal mimicking a DQC, for instance, can lead to excellent R-factors (R... reliability factor of a model structure refined against observed diffraction data). Thus, without further experimental information one would accept such a sample as a DQC.

Electron microscopy, in particular if microscopes corrected for spherical aberration (C_s-corrected TEMs) are used, can give a rather accurate picture of the local arrangement of atoms (clusters) averaged over the sample thickness of ≈ 10 nm. However, laterally (up to several hundred nm), the structures are not averaged. Surface structure analysis by scanning tunneling microscopy (STM) on terraced samples, can also give valuable structural information with atomic resolution across areas of up to several 100 nm. The dynamic structure (phonon dispersion spectrum) can be explored by inelastic neutron scattering giving information about chemical bonding in the same way as for periodic crystals. The main difference is that QCs have no Brillouin zone due to lack of periodicity. However, one can define a Jones zone based on a set of strong Bragg reflections, which also define a periodic average structure (PAS). Quantitative quantum mechanical (first principles) calculations are only possible on approximants due to the lack of periodic boundary conditions for QCs. They can give insight in the electronic structure, the character of chemical bonding, and the existence of (pseudo)gaps at the Fermi level, if any. Consequently, one can extrapolate the findings for the case of QCs; however, they cannot be quantified beyond that point yet.

QC structures should be described in both 3D par-space sections and by illustrating the content of the nD unit cell. The 3D par-space sections show the characteristic building elements (clusters), and how they are arranged on the underlying tiling. Due to the irrational slope of the par-space relative to the nD lattice, nD sections with different perp-space coordinates are not fully congruent to each other. The structural ordering principles, however, i.e., the structure of the clusters and their mutual arrangements, are equivalent in all perp-space sections.

The nD unit cell description allows a complete representation of a QC structure in closed form, and enables the comparison of structures of different QCs. One has to keep in mind, however, that in contrast to the limited number of 3D lattices (14 Bravais lattices), there exists an infinite number of different 2D tilings ("quasilattices") underlying DQCs. In the case of IQCs, the number of different 3D quasiperiodic tilings is restricted to the 3D Ammann tiling. Based on an always limited experimental data set, the determination of the quasilattice of a DQC is always an approximate one, only. In the following, we will focus on the 3D par-space description, and on the comparison with the structure of rational approximants, where available.

Caveat I: It cannot be emphasized often enough that in the QC-community the term 'cluster' is used just as a synonym for 'structural subunit' or 'repeat unit'. It can but does not need to be an energetically favorable arrangement of atoms existing in both QCs and approximants. The cluster approach allows the description of a quasiperiodic structure basically by a single type of partially over-lapping clusters, resulting in a covering. These overlapping clusters correspond (if properly chosen) to a projected cutout of an nD lattice. Consequently, if one cluster is related to the proper projection of an nD unit cell onto 3D par-space, then it may be called a "quasi-unit cell" (Steinhardt *et al.*, 1998). Otherwise, two or more clusters and 'glue' atoms might be necessary for constituting a QC structure.

The centers of the properly arranged covering clusters form a tiling. The atomic decoration of the clusters and the resulting covering can be mapped on this tiling. Consequently, the cluster and tiling decoration approaches are fully equivalent. Depending on the experimental method used, one or the other de-scription seems to be more natural. For C_s-corrected HAADF- and ABF-STEM images favoring one or the other case compare, for instance, Hiraga and Yasuhara (2013*a*) and Hiraga and Yasuhara (2013*b*). Focusing on the tiling description and connecting, for instance, the TM atoms, shows the major skeleton of the structure as well as the bond-orientational order. In the cluster approach, the focus is on the kind of packing and how it looks in the case of QCs and their approximants.

Caveat II: Although the experimental data resulting from X-ray diffraction (XRD) and electron microscopy (EM, in particular transmission electron micros-copy, TEM) of the intermetallic phases classified as "quasicrystals" appear to be in very good agreement with quasiperiodic structure models, there is some un-certainty left as to how closely these structure models might approach their true structures. Particularly, in the case of DQCs, there is a continously varying se-quence of generalized PTs (Pavlovich and Klèman, 1987) as well as of Masakova (Masakova *et al.*, 2005) tilings, which could serve as quasilattices. Neither XRD nor EM could properly distinguish many of them. This situation is completely different from that of periodic crystal structures with their clearly derivable 14 Bravais lattice types.

9.2 Decagonal quasicrystals (DQCs)

Decagonal phases belong to the class of axial QCs. This means, in the special par-space section containing the origin of the nD lattice, they have a unique ten-fold rotation axis (like tetragonal or hexagonal crystal structures have fourfold or sixfold axes, respectively), which is perpendicular to the quasiperiodic layers. From a purely geometrical point of view, DQCs can be seen as periodic stackings of flat or slightly puckered quasiperiodic layers. Alternatively, DQC structures can be described geometrically as packings of partially, systematically overlapping decaprismatic clusters. In both cases, the quasiperiodic arrangement of structural building units can be illustrated as decorations of quasiperiodic tilings. Amazingly, the tilings underlying the structures of the DQCs known so far are all closely related to the 2D PT: pentagon PT, rhomb PT, hexagon–boat–star (HBS) tiling, and Masakova tilings (Masakova *et al.*, 2005; Deloudi *et al.*, 2011).

All stable DQCs observed so far are ternary intermetallic compounds of the type A–B–C (Fig. 9.1). Most of them are line compounds without an extended compositional stability range. Some Al-based DQCs have compositionally quite extended stability fields, because two of their three constituents can replace one another in the structure to a large extent. The majority (> 50 at%) element A corresponds to either Al or Zn. The concentration of the minority element C can be as low as ≈ 2 at% in the case of C being one of the rare earth elements (RE). All stable DQCs can be assigned to either the Al–TM(1)–TM(2) or the Zn–Mg–RE class. In each of the systems Al–Pd–Mn, Al–Pd–Re, and Zn–Mg–Dy, both stable DQCs and IQCs have been identified at slightly different stoichiometries.

A commonly used classification scheme of DQCs is based on their translation period along the tenfold axis, which is always a multiple *n* of stacks of two

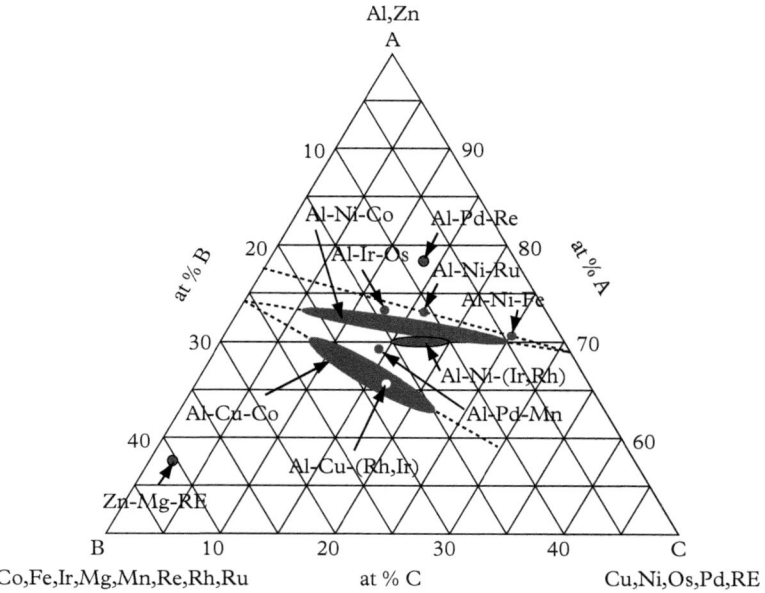

Fig. 9.1 *Schematic representation of the compositional stability fields of DQCs (adapted from Steurer and Deloudi (2009)). RE denotes the rare earth metals Y, Dy, Ho, Er, Tm, and Lu in the case of d-Zn–Mg–RE. Note the different ranges of the coordinates [at.%]: 50 ≤ A ≤ 100, 0 ≤ B ≤ 50, 0 ≤ C ≤ 50.*

quasiperiodic atomic layers. In the following those d(decagonal)-phases are listed, the structures of which have already been reliably quantitatively determined (only the best known d-phase is given, if in a single ternary system more than one DQC has been studied):

2-layer periodicity ($n = 1$)

d-Al$_{65.0}$Co$_{14.6}$Cu$_{20.4}$ (Kuczera *et al.*, 2012)

d-Al$_{61.9}$Rh$_{19.6}$Cu$_{18.5}$ (Kuczera *et al.*, 2012)

d-Al$_{57.6}$Ir$_{16.5}$Cu$_{25.9}$ (Kuczera *et al.*, 2012)

d-Al$_{70.6}$Co$_{6.7}$Ni$_{22.7}$ (Cervellino *et al.*, 2002)

d-Zn$_{59.8}$Mg$_{38.5}$Dy$_{1.7}$ (Oers *et al.*, 2014)

4-layer periodicity ($n = 2$)

d-Al$_{72.5}$Co$_{18.5}$Ni$_{9.0}$ (Strutz *et al.*, 2010)

6-layer periodicity ($n = 3$)

d-Al$_{70}$Mn$_{17}$Pd$_{13}$ (Weber and Yamamoto, 1998)

d-Al$_{70}$Mn$_{17}$Pd$_{13}$ (Yamamoto *et al.*, 2004)

8-layer periodicity ($n = 4$)

d-Al$_{75}$Os$_{10}$Pd$_{15}$ (Cervellino, 2002)

d-Al$_{73}$Os$_{12.5}$Ir$_{14.5}$ (Katrych *et al.*, 2007)

The structures of the 4-layer DQCs can be seen as twofold superstructures of the 2-layer DQCs. In most cases these superstructures are laterally disordered, in contrast to the average 2-layer DQCs. This means that the true period in the columnar clusters of these DQCs corresponds to four layers; however, the lateral correlation between different clusters is limited. In contrast, DQCs with 6- and 8-layer periodicity, respectively, have been described using different types of clusters. Furthermore, the 6-layer DQCs are related to IQCs as reflected in their diffraction patterns showing icosahedral pseudo-symmetry.

9.2.1   Example: DQC ($n = 1$) in the system Al–Cu–Rh

The structures of the 2-layer DQCs are all rather similar. d-Al–Cu–Co and d-Al–Co–Ni (at some compositions) show some structured diffuse scattering indicating an actual 4-layer periodicity of the columnar clusters, which are laterally only short-range-ordered, however. In contrast, for d-Al–Cu–Rh and d-Al–Cu–Ir there are no such indications of superstructure ordering or any other correlated disorder. Therefore, we will discuss just the example of d-Al$_{61.9}$Cu$_{18.5}$Rh$_{19.6}$, the structure of which has been determined employing the 3D tiling-decoration approach (Kuczera *et al.*, 2012) (Fig. 9.2). Also in the same study the structures of the closely related DQCs in the systems Al-Cu-Co and Al-Cu-Ir were determined for comparison.

The structure of d-Al$_{61.9}$Cu$_{18.5}$Rh$_{19.6}$ in the 5D-description is illustrated in Fig. 9.3. The 5D space group is $P10_5/mmc$. In the center of this figure, a 2D section through the 5D unit cell projected along [00100] is shown. x_1, x_2, and x_3 are coordinates in the physical par-space, x_4 and x_5 in perp-space. x_3 runs parallel to the tenfold screw axis, 10_5. The section is spanned by the physical space coordinate x_1 and the perpendicular space coordinate x_5, so that the long body diagonal of the 4D rhombohedral subunit cell is depicted. It runs from 00000 to $1\bar{1}0\bar{1}\bar{1}$. In the left, one 5D unit cell projected on the 2D perp-space, spanned by x_4 and x_5, is shown. In the right subfigure, the closeness and nearest-neighbor conditions, respectively, between the atomic surfaces A, B, C, and D are shown.

The basic structural subunits can be described as decaprismatic columnar clusters with ≈ 33 Å diameter and ≈ 4.2 Å period, which decorate the vertices of a PPT with ≈ 20 Å edge length (Fig. 9.2). The 5D space group is $P10_5/mmc$. The atomic layers are located on mirror planes perpendicular to the 10_5 screw axis at $x_3 = 1/4$ and 3/4, and are symmetrically related by the 10_5 screw axis and the c glide plane. The period along the tenfold axis amounts to 4.278(5) Å.

The inner atomic arrangement of such a columnar cluster (Fig. 9.4) can be described as a column of apex-sharing pentagonal bipyramids of Al atoms, with the apical Al atoms pentagonally coordinated by TM atoms. This innermost columnar cluster shell can also be described as a cylindrically wound-up *hcp* layer of Al and TM atoms. The outer columnar cluster shell can be characterized in the same way. By the way, the description as cylindrically wound-up *hcp* layer applies to any structure created by the iterative action of an $N_{N/2}$ screw axis (with $N > 4$ and an even number) on an atom. Both shells are linked by Al atoms. The clusters are arranged in a way that puckered atomic layers are formed running through the structure like lattice planes in a periodic crystal structure.

A HT study of d-Al–Cu–Rh up to 1223 K did not reveal any significant structural changes with temperature (Kuczera *et al.*, 2014). The best on-average quasiperiodic order was found to exist between 1083 K and 1153 K. This may indicate that this DQC is a HT phase, stabilized by entropy. However, the main source of this entropic contribution might not be related to phason flips, but rather

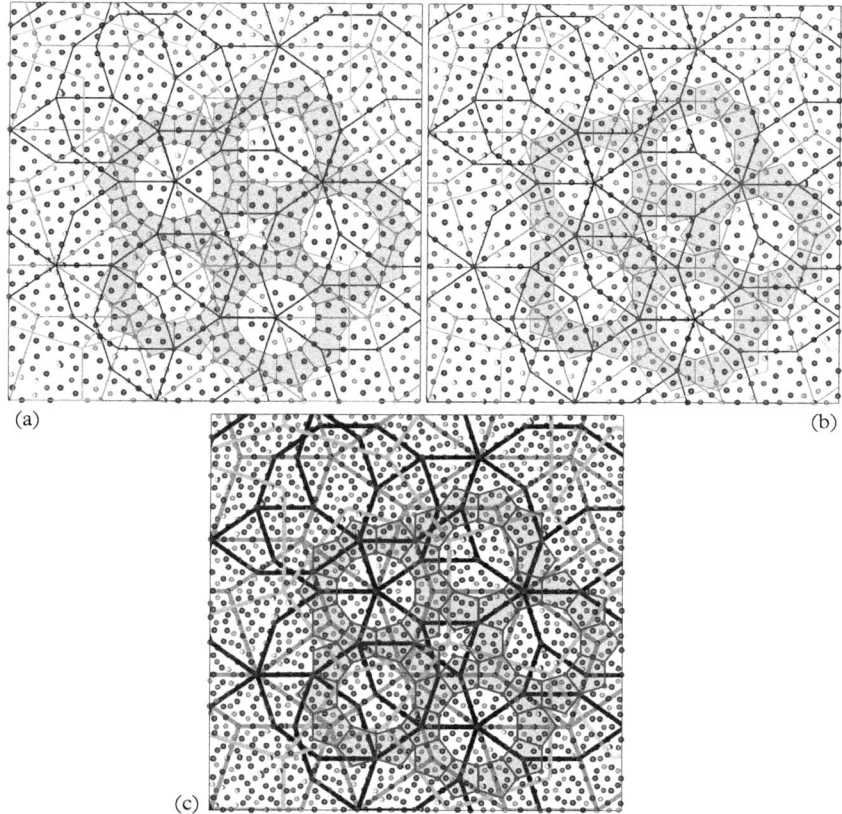

Fig. 9.2 78×78 Å^2 *sections of the atomic layers of* d-Al$_{61.9}$Cu$_{18.5}$Rh$_{19.6}$ *in (a) $x_3 = 1/4$, (b) $x_3 = 3/4$, and (c) projected along x_3. The shaded edge-sharing pentagons highlight pentagonal bipyramidal TM structure motifs, which are also prominent structure motifs in the case of the approximants in the systems Al–Co, Al–Ir, and Al–Rh. The edge length of the gray Penrose rhombs amounts to ≈ 17 Å, of the light-gray large pentagons (forming a PPT) to ≈ 20 Å, and the diameter of the black outlined decagons ≈ 33 Å. The corners of the shaded pentagons in (a) and (c) are all occupied by TM atoms (after Kuczera et al. (2012)). Reproduced with permission of the International Union of Crystallography.*

to lattice vibrations, occupational and chemical disorder. Perhaps, one should mention that in the systems Al–Co, Al–Ir, and Al–Rh metastable binary DQCs were observed and stable approximants are known. In contrast, no approximants have been observed in the system Al–Cu. The role of Cu seems to be to stabilize the DQC by adjusting the valence electron concentration and allowing for more kinds of chemical disorder.

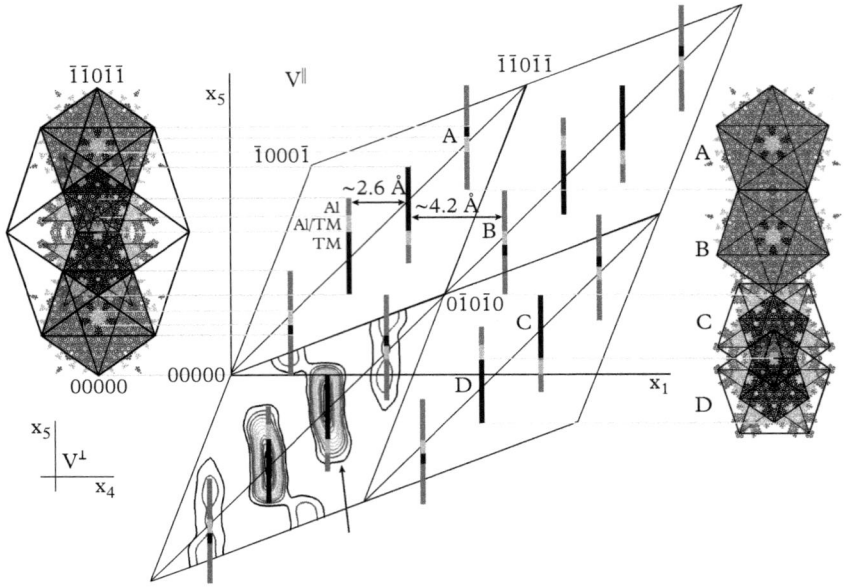

Fig. 9.3 *Characteristic projections and sections of the 5D unit cell of* d-Al$_{61.9}$Cu$_{18.5}$ Rh$_{19.6}$. *Left: perp-space projection of one 5D unit cell; middle: (10001) section through four, along [00100] projected, 5D unit cells. The lower-left unit cell section is underlaid with the corresponding Fourier map; right: projection of the atomic surfaces (occupation domains) marked A, B, C, and D in the middle in order to illustrate the closeness condition. Each dot corresponds to an atom lifted from 3D to 5D space. TM dots and strokes are black, Al/TM mixed positions are light gray, and Al dark gray. The black pentagrams should serve as guides to the eyes (from Kuczera* et al., *(2012)). Reproduced with permission of the International Union of Crystallography.*

9.2.2   Growth model for decagonal quasicrystals

The formation and growth of QCs belongs to the fundamental problems in quasi-crystal research, which are not fully solved yet. Crystal growth of compounds with simple structures is well understood and can be simply modeled. The situation is different for complex intermetallics, be they periodic or quasiperiodic. How do the atoms find their sites in the structure of $cF(23\,256 - x)$-Al$_{55.4}$Cu$_{5.4}$Ta$_{39.1}$ ($x = 122$, ACT-71) (see Subsection 7.4.4), for instance? One can assume that, when approaching the solidification temperature, already in the melt local ordering takes place, and the cluster preforms add their highly mobile atoms layer by layer to the growing crystals. Relative to the length scale of the constituting clusters (≈ 14 Å) and superclusters (≈ 40 Å), the lattice period of ≈ 70 Å is just a small multiple. On the scale of clusters, the complex structure formation is reduced to a rather simple cluster packing; a problem similar to close sphere packings, which all are periodic.

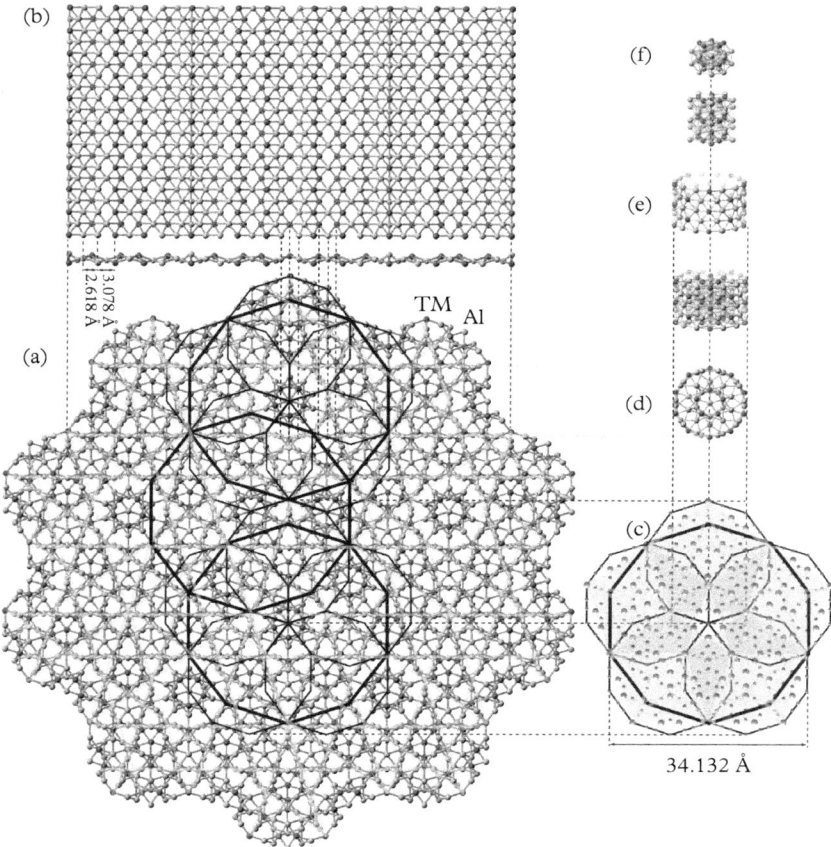

Fig. 9.4 *(a) One supercluster (Deloudi et al., 2011) with the traces of the puckered atomic layers marked by light-gray lines. (b) One of these puckered atomic layers is shown on top in two orthogonal projections. Note the pseudohexagonal arrangement of atoms. (c) One Hiraga-supercluster (Hiraga, 2001) (outlined in black), consisting of five Deloudi clusters, is depicted together with a ≈ 14 Å subcluster (d) in different projections and exploded view. The perfectly ordered hcp columnar cluster shell (e) is formed by pieces of the atomic layers shown in (a, b). So is the innermost cylindrical cluster shell. (f) Column of face-sharing pentagon-dodecahedra around vertex-connected, capped pentagonal bipyramids (Al. . . gray, TM. . . black). Reproduced from Steurer (2014a). Copyright ©2014 Elsevier Masson SAS. All rights reserved.*

The question is how quasiperiodicity is evolving during QC-crystal growth. Again, we can consider the QC structure resulting from the packing of clusters. The crucial point is that here the clusters can overlap only in a particular way, according to specific overlap rules. Overlap rules, however, are no growth rules. In the case of a covering, where all cluster overlaps obey the rules, one can be sure that it is quasiperiodic. However, a process adding cluster to cluster by obeying the overlap rules alone would run again and again into problems such as the formation of gaps in the packing or not-allowed overlaps. This problem can be solved by taking into account the flat atomic layers present in all QCs. Since each cluster defines such a set of atomic layers by its own structure, it provides global constraints to the growing structure beside the local overlap rules. It has been nicely demonstrated by cluster-based model calculations (Kuczera and Steurer, 2015) that the combination of local overlap rules with globally acting layer-continuation rules allows the growth of highly-perfect DQCs.

9.3   Icosahedral quasicrystals (IQCs)

Most of the QCs known so far are icosahedral quasicrystals (IQCs). They are quasiperiodic in all three dimensions, and their diffraction symmetry can be described by the icosahedral point group $m\bar{3}\bar{5}$ (Laue symmetry). Their structures, whose quasilattices are related to the 3D Ammann tiling, can best be described in 6D space, with space group symmetry $Pm\bar{3}\bar{5}$ or $Fm\bar{3}\bar{5}$, respectively.

The stability regions of most of the IQCs known so far are shown in Fig. 9.5. Similarly to DQCs, quite a few of the ternary IQCs show extended compositional stability regions, indicating the presence of some intrinsic substitutional disorder. Fortunately for the analysis of their structures, IQCs are often accompanied by rational approximants in their respective binary or ternary intermetallic systems.

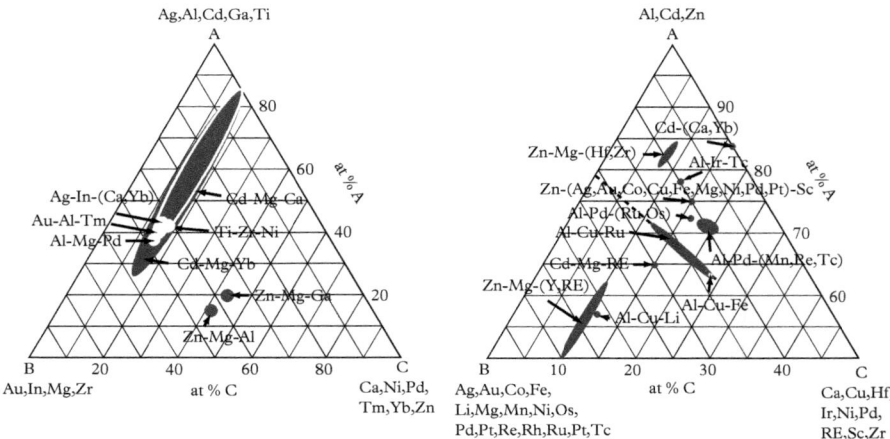

Fig. 9.5 *Approximate stability regions of some classes of icosahedral quasicrystals. RE denotes the rare earth metals Nd, Eu, Gd, Tb, Dy, Ho, Er, Tm, Yb, and Lu in the case of i-Cd–Mg–RE; La, Ce, Pr, Nd, Gd, Tb, Dy, Ho, Er, and Yb in the case of i-Zn–Mg–RE. Note that only the A-rich part (50 ≤ A ≤ 100 at.%) is shown in the right concentration diagram (from Steurer and Deloudi (2009)).*

IQCs are mostly classified according to the type of their constituting clusters: (i) Mackay-cluster based IQCs (type M), (ii) Bergmann-cluster based IQCs (type B), and (iii) Tsai-cluster based IQCs (type T).

9.3.1 Mackay-cluster based IQCs (Type M)

All stable IQCs of the M-type known so far are based on Al as main component, Cu or Pd as second element, and a TM element of group 7 or 8 as the third constituent. These IQCs are also sometimes called spd-IQCs due to the hybridization between the p states of Al and the d states of the TM atoms, which can lead to pseudo-gaps in the electron density of states. Their stability is related to their valence electron concentration, e/a, which should be in the range of $1.6 \leq e/a \leq 1.9$. The number of valence electrons is usually taken from Raynor (1949) (Al: 3, Fe: -2.66, Os: -2.66, Pd: 0, Ru: -2.66). The 3d electrons of Cu are considered core electrons since they do not significantly contribute to the Fermi energy. In Table 9.1 the stable M-type IQCs are listed together with their approximants.

The structure of a typical Mackay-cluster, as it appears in a 1/1-approximant, is illustrated in Fig. 9.6. It consists of three shells (Fig. 9.6, (a)–(c)) and sits in the body-center of α-$Al_{40}Mn_{10.1}Si_{7.4}$. The corners of the unit cell are occupied by so-called double-Mackay polyhedra (Fig. 9.6, (a)–(e)), which have two more cluster shells (see also Kuo (2002)). The innermost icosahedron shows Al/Si chemical disorder in both cases.

Even more substitutional disorder has been found in the IQC and its 1/1-approximant in the system Al–Cu–Fe (Fig. 9.5). The wide stability range leads to mixed Al/Cu and Al/Cu/Fe sites in the first shell of the Mackay-cluster in the body center and of the second cluster shell in the double-Mackay-cluster at the origin of the unit cell of the 1/1-approximant, and the related clusters of the IQC. In the case of 1/1-$Al_{57.3}Cu_{31.4}Ru_{11.3}$, the first disordered shells of both the clusters in the center and the origin of the unit cell are centered by Ru atoms.

9.3.2 Bergmann-cluster based IQCs (Type B)

Most of the type B ICQs only contain elements with s and p valence electrons, and are therefore called sp-QCs (Table 9.2). They are found for valence electron concentrations in the range of $2.1 \leq e/a \leq 2.4$ (Al: 3, Fe: -2.66, Os: -2.66, Pd: 0, Rh: -1.71, Ru: -2.66 (Raynor, 1949)). Another designation for these IQCs is Frank-Kasper IQCs, because their approximants belong to the class of Frank-Kasper (FK) phases. The structures of FK phases can be described as tetrahedrally close-packed (*tcp*). These structures only contain tetrahedral interstices and the coordination polyhedra are limited to essentially the four FK-polyhedra with coordination numbers (CN) 12, 14, 15, and 16 (Shoemaker and Shoemaker, 1986).

The constituting cluster of the B-type IQCs is the Bergman cluster. Its structure is illustrated in Fig. 9.7 on the example of the 1/1-approximant R-Al_5CuLi_3. The 160 atoms in the unit cell form a *bcc* packing of overlapping Bergman clusters.

Table 9.1 *Stable IQCs and approximants based on packings of Mackay-clusters. The structures are ordered with increasing (quasi)lattice parameter a_r (edge length of related 3D PT). The 6D and 3D space groups (SG), as well as the lattice parameters of the cubic approximants are given. Adapted from Steurer and Deloudi (2009).*

IQC	a_r [Å]	6D SG	Approximant	a [Å]	3D SG
i-Al$_{65}$Cu$_{20}$Ru$_{15}$	4.541	$Fm\bar{3}\bar{5}$	1/0-Al$_{71.5}$Cu$_{8.5}$Ru$_{20}$	7.745	$P2_13$
			γ-Al$_{55.1}$Cu$_{14.6}$Ru$_{20.2}$Si$_{10.1}$[a]	2×7.690	$Fm\bar{3}$
			1/1-Al$_{57.3}$Cu$_{31.4}$Ru$_{11.3}$	12.377	$Pm\bar{3}$
i-Al$_{65}$Cu$_{20}$Os$_{15}$	4.524	$Fm\bar{3}\bar{5}$			
i-Al$_{65}$Cu$_{20}$Fe$_{15}$	4.465	$Fm\bar{3}\bar{5}$	1/1-Al$_{55}$Cu$_{25.5}$Fe$_{12.5}$Si$_7$	12.329	$Pm\bar{3}$
i-Al$_{70.5}$Pd$_{21}$Mn$_{8.5}$	4.562	$Fm\bar{3}\bar{5}$	1/1-Al$_{67}$Pd$_{11}$Mn$_{14}$Si$_7$	12.281	$Pm\bar{3}$
			2/1-Al$_{70}$Pd$_{23}$Mn$_6$Si	20.211	$Pm\bar{3}$
i-Al$_{70}$Pd$_{20}$Re$_{10}$	4.617	$Fm\bar{3}\bar{5}$			
i-Al$_{70}$Pd$_{21}$Tc$_9$	4.606	$Fm\bar{3}\bar{5}$			
i-Al$_{71}$Pd$_{21}$Re$_8$	7.383	$Pm\bar{3}\bar{5}$			
i-Al$_{72}$Pd$_{17}$Ru$_{11}$		$Fm\bar{3}\bar{5}$	c-Al$_{68}$Pd$_{20}$Ru$_{12}$[a]	2×7.770	$P23$
i-Al$_{72}$Pd$_{17}$Os$_{11}$		$Fm\bar{3}\bar{5}$			
			c-Al$_{39}$Pd$_{21}$Fe$_2$[a]	2×7.758	$Fm\bar{3}$
			1/1-Al$_{40}$Mn$_{10.1}$Si$_{7.4}$	12.643	$Pm\bar{3}$
			2/1-Al$_{66.6}$Rh$_{26.1}$Si$_{7.3}$	19.935	$Pm\bar{3}$

[a] (2 × 2 × 2)-fold superstructure of a 1/0-approximant.

The first four shells form the 104-atom Samson cluster, which yields, together with the fifth shell, the 132-atom Pauling triacontahedron. In the *bcc* packing, the triacontahedra share one rhomb face in the [100] directions, and an oblate rhombohedron along [111] (for the packing of triacontahedra see also Fig. 4.5).

The packing can also be described as a vertex-decorated packing of oblate rhombohedra with edge length $a_r = a\sqrt{3}/2$. Compared to the oblate rhombohedron ($\alpha_r = 63.44°$), which is one of the two prototiles of the Ammann tiling, it appears slightly distorted ($\alpha_r = 70.53°$). Along the short diagonal (length $= a_{1/1}$) of the rhombic faces, the clusters share one of its rhombs. The neighboring clusters along the edges and the short body diagonal (length $a_r = a\sqrt{3}/2$) have an oblate rhombohedron as common volume.

In the case of the cubic 2/1-approximant, with lattice parameter $a_{2/1}$, essentially the same triacontahedral clusters occupy the points of the lattice complex generated by Wyckoff position $8c$ x, x, x in space group $Pa\bar{3}$. This point set can

α-Al$_{40}$Mn$_{10.1}$Si$_{7.4}$

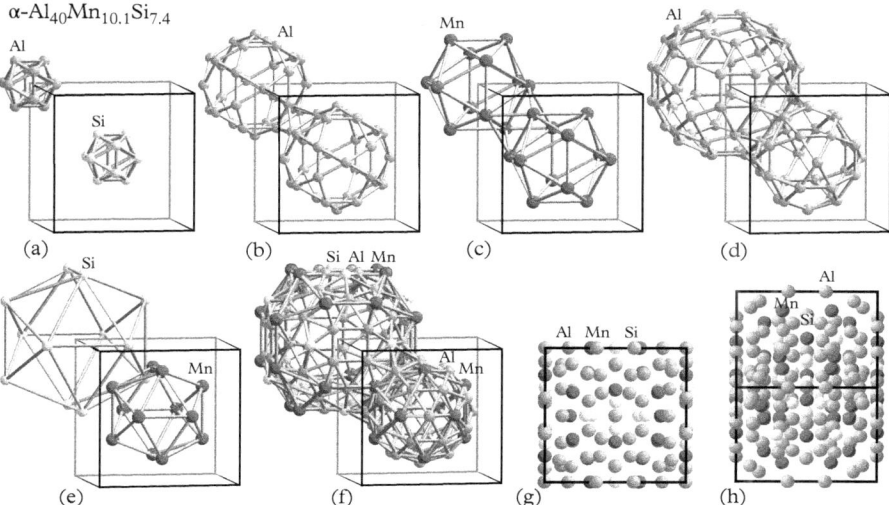

Fig. 9.6 *Shells of the Mackay-cluster (at the body center) and double-Mackay-cluster (at the origin) on the example of α-Al$_{40}$Mn$_{10.1}$Si$_{7.4}$ (a = 12.643 Å, Pm$\bar{3}$, a 1/1 = approximant of type M-IQCs. (a) (Al$_{0.65}$Si$_{0.35}$)$_{12}$ icosahedron (edge length a_r = 2.585 Å, diameter $\varnothing$=4.908 Å) connected via an octahedron to the icosahedron in the body center; (b) Al$_{30}$ origin-centered icosidodecahedron (a_r = 2.826-2.983 Å, $\varnothing$ = 9.301 Å); (c) Mn$_{12}$ icosahedron (a_r = 5.091 Å, $\varnothing$ = 9.648 Å) connected via an octahedron to the one in the body center; (d) Al$_{60}$ distorted rhombicosidodecahedron (a_r = 2.826-3.359 Å, $\varnothing$ = 13.361 Å) sharing a triangle face with the icosidodecahedron in the body-center; (e) (Al$_{0.01}$Si$_{0.99}$)$_{12}$ icosahedron (a_r = 7.324-7.775 Å, $\varnothing$ = 14.611 Å). (f) Combination of cluster shells (c)–(e); Si atoms cap all 12 pentagons, Mn all squares. The projections of one unit cell along [100] and [110] are shown in (g) and (h) (from Steurer and Deloudi (2009), Fig. 9.3. With kind permission from Springer Science+Business Media).*

also be described as the set of vertices of a 1:1 packing of oblate and prolate rhombohedra with edge length $a_r = a\sqrt{3}/(2\tau)$. The distortion of these rhombo-hedra, compared to the Ammann prototiles, is smaller than in the case of the 1/1-approximant as indicated by $\alpha_r = 69.83°$.

Based on the commonly used 6D lattice parameter $a_P = \sqrt{2}a_r$, the edge length of the Ammann rhombohedra in the approximants is larger by a factor τ^2. For exam-ple, the edge lengths of the rhombohedra in the structure of 2/1-Zn$_{47.3}$Mg$_{27}$Al$_{10.7}$ are 13.646 Å compared to $\tau^2 a_r = 13.535$ Å calculated from the icosahedral phase with $a_r = 5.17$ Å.

9.3.3   Tsai-cluster based IQCs (Type T)

The T-type IQCs are the largest class of IQCs with an even larger number of 1/1- and 2/1-approximants (Tables 9.3 and 9.4). They stand out against the M- and

Table 9.2 *Stable IQCs and approximants based on packings of Bergman clusters (Frank-Kasper type). a_r is the edge length of the 3D Penrose rhombohedra ($a_r = a_P/\sqrt{2} = A_F/(2\sqrt{2})$. The structures are ordered with increasing (quasi)lattice parameter a_r. Adapted from Steurer and Deloudi (2009).*

IQC	a_r [Å]	6D SG	Approximant	a [Å]	3D SG
i-$Zn_{76}Mg_{17}Hf_7$	5.011	$Fm\bar{3}\bar{5}$	$1/1$-$Zn_{77}Mg_{18}Hf_5$	13.674	$Pm\bar{3}$
i-$Zn_{84}Mg_7Zr_9$	5.031	$Pm\bar{3}\bar{5}$	$1/1$-$Zn_{77}Mg_{18}Zr_5$	13.709	$Pm\bar{3}$
i-Al_6CuLi_3	5.043	$Pm\bar{3}\bar{5}$	$1/1$-$Al_{88.6}Cu_{19.4}Li_{50.3}$	13.906	$Im\bar{3}$
i-$Mg_{43}Al_{42}Pd_{15}$	5.13	$Pm\bar{3}\bar{5}$			
i-$Zn_{40}Mg_{39.5}Ga_{25}$	5.133	$Pm\bar{3}\bar{5}$			
i-$Zn_{74}Mg_{15}Ho_{11}$	5.144	$Pm\bar{3}\bar{5}$			
i-$Zn_{41}Mg_{44}Al_{15}$ [a]	5.17	$Pm\bar{3}\bar{5}$	$1/1$-$Zn_{34.6}Mg_{40}Al_{25.4}$	14.217	$Im\bar{3}$
			$2/1$-$Zn_{37}Mg_{46}Al_{17}$	23.064	$Pm\bar{3}$
			$2/1$-$Zn_{47.3}Mg_{27}Al_{10.7}$	23.035	$Pa\bar{3}$
i-$Ti_{40}Zr_{40}Ni_{20}$	5.17	$Pm\bar{3}\bar{5}$	$1/1$-$Ti_{51}Zr_{33}Ni_{16}$	14.30	$Im\bar{3}$
i-$Zn_{56.8}Mg_{34.6}Tb_{8.7}$	5.173	$Fm\bar{3}\bar{5}$			
i-$Zn_{65}Mg_{26}Ho_9$	5.18	$Fm\bar{3}\bar{5}$			
i-$Zn_{64}Mg_{25}Y_{11}$	5.19	$Fm\bar{3}\bar{5}$			
i-$Zn_{55}Mg_{40}Nd_5$	5.25	$Pm\bar{3}\bar{5}$			
i-$Zn_{56.8}Mg_{34.6}Dy_{8.7}$		$Fm\bar{3}\bar{5}$			
			$1/1$-$Zn_{77}Mg_{17.5}Ti_{5.5}$	13.554	$Pm\bar{3}$
			$2/1$-$Zn_{61.4}Mg_{24.5}Er_{14.1}$	20.20	$F\bar{4}3m$
			$2/1$-$Zn_{73.6}Mg_{2.5}Sc_{11.2}$		$Pa\bar{3}$
			$3/2$-$2/1$-$2/1$-		
			o-$Zn_{40}Mg_{39.5}Ga_{16.4}Al_{4.1}$	$a = 36.840$	$Cmc2_1$
				$b = 22.782$	
				$c = 22.931$	

[a] possibly metastable.

B-type IQCs because of the existence of stable binary IQCs such as $Cd_{85}Ca_{15}$ and $Cd_{84}Yb_{16}$. Having only two constituents to consider makes the structure analysis simpler. Furthermore, substitutional disorder, which is quite common in ternary IQCs, is less likely due to the differences in electronegativity (Cd: 1.5, Ca: 1, Yb: 1.1) and atomic radii (Cd: 1.489 Å, Ca: 1.97.4 Å, Yb: 1.940 Å). This means that the entropy related to chemical disorder is not a necessary contribution to the

R-Al$_5$CuLi$_3$

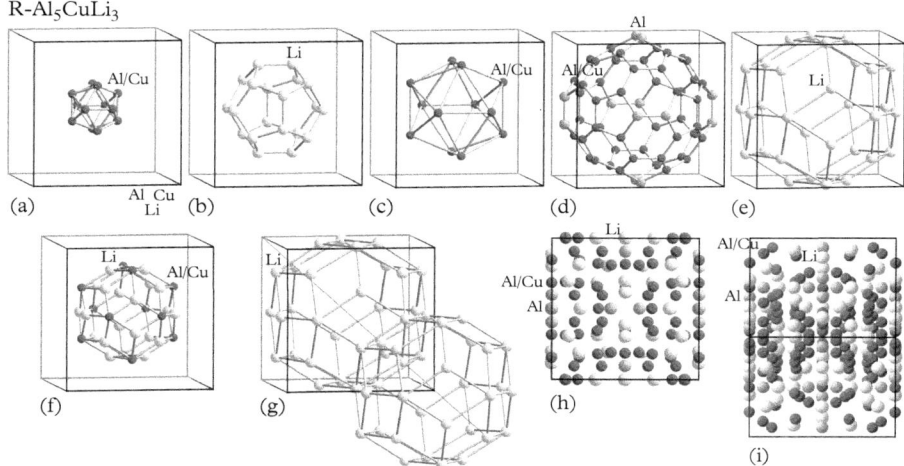

Fig. 9.7 *Shells of the Bergman cluster on the example of bcc* R-Al$_5$CuLi$_3$, *a 1/1-approximant of* i-Al$_6$CuLi$_3$. *(a)* (Al$_{0.89}$Cu$_{0.11}$)$_{12}$ *icosahedron (edge length* $a_r = 2.651$ Å, *diameter* $\varnothing = 5.033$ Å*); (b)* Li$_{20}$ *pentagonaldodecahedron* ($a_r = 3.226$ Å, $\varnothing = 9.078$ Å*); (c)* (Al$_{0.54}$Cu$_{0.46}$)$_{12}$ *icosahedron* ($a_r = 5.383$ Å, $\varnothing = 10.093$ Å*); (d)* (Al$_{0.89}$Cu$_{0.11}$)$_{48}$Al$_{12}$ *distorted truncated triacontahedron* ($a_r = 2.658$-2.861 Å, $\varnothing = 14.161$ Å*); (e)* Li$_{32}$ *triacontahedron* ($a_r = 5.020$-5.071 Å, $\varnothing = 16.238$ Å*). The shells in (b) and (c) can be combined to the* (Cu,Al)$_{12}$Li$_{20}$ *triacontahedron shown in (f). The ratio of the diameters of the large and the small triacontahedra depicted in (e) and (f) amounts to* τ. *The large triacontahedra are face-connected along the [100] directions and share an oblate rhombohedron along [111] (see also Fig. 3.16). The projections of one unit cell along [100] and [110] are shown in (h) and (i) (from Steurer and Deloudi (2009), Fig. 9.5. With kind permission from Springer Science+Business Media).*

stability of this kind of QC. The structure of the 1/1-approximant cI168-YbCd$_6$ shows a *bcc* packing of the constituting Tsai clusters (Fig. 9.8(a)–(g)), which are also the fundamental clusters of the T-type ICQs. The 158-atom triacontahedral clusters share one rhomb face along the [100] direction and an oblate rhombohedron along [111]. This is comparable to the packing of the also triacontahedral 160-atom Bergman clusters (see Subsection 9.3.2). The internal structure of the B- and T-type clusters, respectively, differ in the first and the third shell. The first shell is formed by a Cd tetrahedron, which is orientationally disordered according to the space available from the second shell, a Cd pentagonal dodecahedron. The probability distribution function of the Cd atoms at the tetrahedron corners has the shape of a truncated octahedron. The third shell corresponds to an Yb icosahedron. Its triangular faces form Yb octahedra with the corresponding faces along [111] (Fig. 9.8(c)). The fourth shell is a Cd icosidodecahedron, which is surrounded by a distorted, edge-centered Cd triacontahedron, the fifth cluster shell. Amazingly, the first two and the last two cluster shells consist entirely of Cd atoms, encapsulating the third shell, an Yb icosahedron.

Table 9.3 *Stable Tsai-type IQCs as a function of the Penrose rhombohedron edge length a_r. The structures are ordered with increasing quasilattice parameter. 6D space groups (SG) are given if available in literature. Adapted from Steurer and Deloudi (2009), Lin and Corbett (2010), Canfield et al. (2010), and Goldman et al. (2013).*

IQC	a_r [Å]	6D SG	IQC	a_r [Å]	6D SG
i-Cu$_{48}$Sc$_{15}$Ga$_{34}$Mg$_3$	4.906		i-Cd$_{65}$Mg$_{20}$Lu$_{15}$	5.571	
i-Cu$_{46}$Sc$_{16}$Al$_{38}$	4.921		i-Ag$_{42}$In$_{42}$Yb$_{16}$	5.590	$Pm\bar{3}\bar{5}$
i-Zn$_{84}$Ti$_8$Mg$_8$	4.966		i-Cd$_{87.9}$Tm$_{12.1}$	5.596	$Pm\bar{3}\bar{5}$
i-Zn$_{75}$Sc$_{15}$Ni$_{10}$	4.981		i-Cd$_{65}$Mg$_{20}$Tm$_{15}$	5.602	
i-Zn$_{72}$Sc$_{16}$Cu$_{12}$	4.996		i-Ag$_{42}$Ca$_{16}$In$_{42}$	5.606	$Pm\bar{3}\bar{5}$
i-Zn$_{75}$Sc$_{15}$Co$_{10}$	4.994		i-Au$_{44.2}$In$_{41.7}$Ca$_{14.1}$		$Pm\bar{3}\bar{5}$
i-Zn$_{75}$Sc$_{15}$Fe$_{10}$	5.008		i-Cd$_{65}$Mg$_{20}$Y$_{15}$	5.606	$Pm\bar{3}\bar{5}$
i-Zn$_{88}$Sc$_{12}$	5.017	$Pm\bar{3}\bar{5}$	i-Cd$_{88.4}$Ho$_{11.6}$	5.611	$Pm\bar{3}\bar{5}$
i-Zn$_{75}$Sc$_{15}$Mn$_{10}$	5.025		i-Cd$_{88.0}$Er$_{12.0}$	5.611	$Pm\bar{3}\bar{5}$
i-Zn$_{80}$Sc$_{15}$Mg$_5$	5.028	$Pm\bar{3}\bar{5}$	i-Cd$_{88.2}$Dy$_{11.8}$	5.621	$Pm\bar{3}\bar{5}$
i-Zn$_{75}$Sc$_{15}$Pt$_{10}$	5.029		i-Cd$_{65}$Mg$_{20}$Er$_{15}$	5.622	$Fm\bar{3}\bar{5}$
i-Zn$_{75}$Sc$_{15}$Pd$_{10}$	5.030		i-Cd$_{65}$Mg$_{20}$Ho$_{15}$	5.625	$Fm\bar{3}\bar{5}$
i-Zn$_{75}$Sc$_{15}$Au$_{10}$	5.057		i-Cd$_{88.2}$Y$_{11.8}$	5.625	$Pm\bar{3}\bar{5}$
i-Zn$_{75}$Sc$_{15}$Ag$_{10}$	5.054		i-Cd$_{88.5}$Tb$_{11.5}$	5.627	$Pm\bar{3}\bar{5}$
i-Zn$_{77}$Sc$_8$Ho$_8$Fe$_7$	5.066		i-Cd$_{65}$Mg$_{20}$Tb$_{15}$	5.628	$Fm\bar{3}\bar{5}$
i-Zn$_{77}$Sc$_7$Tm$_9$Fe$_7$	5.067		i-Cd$_{65}$Mg$_{20}$Dy$_{15}$	5.628	$Fm\bar{3}\bar{5}$
i-Zn$_{77}$Sc$_8$Er$_8$Fe$_7$	5.070		i-Cd$_{88.7}$Gd$_{11.3}$	5.637	$Pm\bar{3}\bar{5}$
i-Zn$_{56.8}$Er$_{8.7}$Mg$_{34.6}$	5.180	$Fm\bar{3}\bar{5}$	i-Cd$_{65}$Mg$_{20}$Gd$_{15}$	5.648	
i-Zn$_{76}$Yb$_{14}$Mg$_{10}$	5.211		i-Cd$_{84}$Yb$_{16}$	5.689	$Pm\bar{3}\bar{5}$
i-Au$_{46}$Al$_{38}$Tm$_{16}$	5.240	$Pm\bar{3}\bar{5}$	i-Cd$_{65}$Mg$_{20}$Yb$_{15}$	5.727	
i-Au$_{51}$Al$_{34}$Yb$_{15}$	5.267	$Pm\bar{3}\bar{5}$	i-Cd$_{85}$Ca$_{15}$	5.731	$Pm\bar{3}\bar{5}$
			i-Cd$_{65}$Mg$_{20}$Ca$_{15}$	5.731	$Pm\bar{3}\bar{5}$

The cluster structure of the 2/1-approximant of the T-type IQC is shown on the example of $cP712$-Ca$_{13}$Cd$_{76}$ (Fig. 9.8(h)–(l)). The triacontahedral clusters decorate the vertices of a rhombohedron ($a_r = 13.646$ Å, $\alpha = 69.83°$), which is close to the obtuse rhombohedron, one of the two prototiles of the Ammann tiling. Along the edges of the rhombohedra, the clusters overlap forming oblate rhombohedra. Along the short body diagonal they share a rhombohedron face. The rhombohedra are packed in a zigzag manner. The Cd atoms around the center

Table 9.4 *Stable 1/1- and 2/1-approximants based on the Tsai-cluster. The structures are ordered with increasing lattice parameter a. 3D space groups (SG) are given if available in literature. Adapted from Steurer and Deloudi (2009).*

Approximant	a [Å]	SG	Approximant	a [Å]	SG
1/1-Be$_{17}$Ru$_3$	11.337		1/1-Cd$_6$Gd	15.441	$Im\bar{3}$
1/1-Ga$_{3.85}$Ni$_{2.15}$Hf	13.319		1/1-Ag$_2$In$_4$Ca	15.454	$Im\bar{3}$
1/1-Ga$_{3.22}$Ni$_{2.78}$Zr	13.374		1/1-Ag$_{47.7}$In$_{38.7}$Ce$_{14.2}$	15.46	
1/1-Ga$_{3.64}$Ni$_{2.36}$Sc	13.440		1/1-Cd$_6$Dy	15.462	$Im\bar{3}$
1/1-Ga$_{2.3}$Cu$_{3.7}$Sc	13.472		1/1-Cd$_6$Y	15.482	
1/1-Ga$_{2.6}$Cu$_{3.4}$Lu	13.745		1/1-Cd$_6$Sm	15.589	$Im\bar{3}$
1/1-Zn$_{17}$Sc$_3$	13.843		1/1-Cd$_6$Nd	15.605	$Im\bar{3}$
1/1-Zn$_{17}$Yb$_3$	14.291		1/1-Cd$_6$Yb	15.661	$Im\bar{3}$
1/1-Au$_{48}$Al$_{38}$Tm$_{14}$	14.458	$Im\bar{3}$			
1/1-Au$_{51}$Al$_{35}$Yb$_{14}$	14.500	$Im\bar{3}$			
1/1-Ag$_{47}$Ga$_{38}$Yb$_{15}$	14.687	$Im\bar{3}$			
1/1-Ag$_{42.5}$Ga$_{42.5}$Yb$_{15}$	14.707		1/1-Ag$_{42.9}$In$_{43.6}$Eu$_{13.5}$	15.69	
1/1-Au$_{64}$Ge$_{22}$Yb$_{14}$	14.724	$Im\bar{3}$			
1/1-Au$_{50.5}$Ga$_{35.9}$Ca$_{13.6}$	14.731	$Im\bar{3}$	1/1-Cd$_6$Ca	15.702	$Im\bar{3}$
1/1-Au$_{61.2}$Sn$_{23.9}$Dy$_{15.2}$	14.90		1/1-Cd$_{25}$Eu$_4$ [a]	2×15.936	$Fd\bar{3}$
1/1-Au-Sn-Tb	14.91		1/1-Cd$_{19}$Pr$_3$	15.955	$Im\bar{3}$
1/1-Au$_{62.3}$Sn$_{23.1}$Gd$_{14.6}$	14.97		1/1-Cd$_6$Sr	16.044	
1/1-In$_{53}$Pd$_{33}$Ho$_{14}$	15.00	$Im\bar{3}$			
1/1-In$_{53}$Pd$_{33}$Dy$_{14}$	15.02	$Im\bar{3}$			
1/1-In$_{53}$Pd$_{33}$Y$_{14}$	15.04	$Im\bar{3}$			
1/1-In$_{53}$Pd$_{33}$Tb$_{14}$	15.06	$Im\bar{3}$			
1/1-Ag$_{42.2}$In$_{42.6}$Tm$_{15.2}$	15.05		2/1-Au$_{61.1}$Ga$_{25.0}$Ca$_{13.9}$	23.938	$Pa\bar{3}$
1/1-Au$_{47.2}$In$_{37.2}$Gd$_{15.6}$	15.07		2/1-Au$_{60.3}$Sn$_{24.6}$Yb$_{15.1}$	24.28	
1/1-Au$_{12.2}$In$_{6.3}$Ca$_3$	15.152	$Im\bar{3}$	2/1-Au$_{61.2}$Sn$_{24.3}$Ca$_{14.5}$	24.37	
1/1-Au$_{64.2}$Sn$_{21.3}$Pr$_{14.5}$	15.16		2/1-Au$_{42.9}$In$_{41.9}$Yb$_{15.2}$	24.63	$Pa\bar{3}$

Table 9.4 *continued*

Approximant	a [Å]	SG	Approximant	a [Å]	SG
1/1-Au$_{65}$Sn$_{20}$Ce$_{15}$	15.190	$Im\bar{3}$	2/1-Au$_{37}$In$_{39.6}$Ca$_{12.6}$	24.632	$Pa\bar{3}$
1/1-Ag$_{46.4}$In$_{39.7}$Gd$_{13.9}$	15.21		2/1-Ag$_{41.7}$In$_{43.2}$Yb$_{15.1}$	24.869	$Pa\bar{3}$
1/1-Au$_{49.7}$In$_{35.4}$Ce$_{14.9}$	15.28		2/1-Au$_{61.2}$Sn$_{24.5}$Eu$_{14.3}$	24.87	
1/1-Au$_{60.7}$Sn$_{25.2}$Eu$_{14.1}$	15.35		2/1-Ag$_{41}$In$_{44}$Yb$_{15}$	24.88	
1/1-Ag$_2$In$_4$Yb	15.362	$Im\bar{3}$	2/1-Ag$_{42}$In$_{45}$Ca$_{13}$	24.96	$Pa\bar{3}$
1/1-Ag$_{46.9}$In$_{38.7}$Pr$_{14.4}$	15.39		2/1-Cd$_{76}$Ca$_{13}$	25.339	$Pa\bar{3}$
1/1-Au$_{42}$In$_{42}$Yb$_{16}$	15.4	$Ia\bar{3}$	2/1-Ag$_{43.4}$In$_{42.8}$Eu$_{13.8}$	25.35	$Pa\bar{3}$

^{a} $(2 \times 2 \times 2)$ superstructure of a 1/1-approximant.

of such a rhombohedron leave a space in the form of a double Friauf polyhedron occupied by two Ca atoms along the long body diagonal.

The structure of i-YbCd$_{5.7}$ can be described as a 3D PT decorated by the Tsai clusters at the vertices, covering 93.8% of all atoms in this way. The remaining space is filled by different arrangements of acute and obtuse Penrose unit tiles. The acute rhombohedron corresponds to a double Friauf polyhedron with two Yb atoms along its long diagonal and Cd atoms on the vertices. The oblate rhombohedron is decorated by Cd on the vertices and edge centers. The structure shows τ^3 scaling symmetry. It is noteworthy that the structure cannot be described as a 3D Penrose tiling with uniquely decorated unit tiles while the decorating clusters have a unique structure.

9.3.4 Atomic layers in icosahedral quasicrystals

We have already seen in Subsection 9.2.1 that flat or slightly puckered atomic layers along the periodic direction are fundamental structural features in DQCs. This is also true for IQCs, where the layers, which have to obey the icosahedral point group symmetry, are crisscrossing each other. Only in IQCs, the icosahedral cluster symmetry and the quasiperiodic packing of these clusters are compatible with these atomic layers (Fig. 9.9). Each cluster already defines a subset of these "infinitely" extending flat layers and relates local with global order, guiding the growth of QCs.

9.4 Remarks on formation and stability of quasicrystals

When does a QC form in a given binary or ternary intermetallic system and when does only a rational approximant form, supposing that the formation of

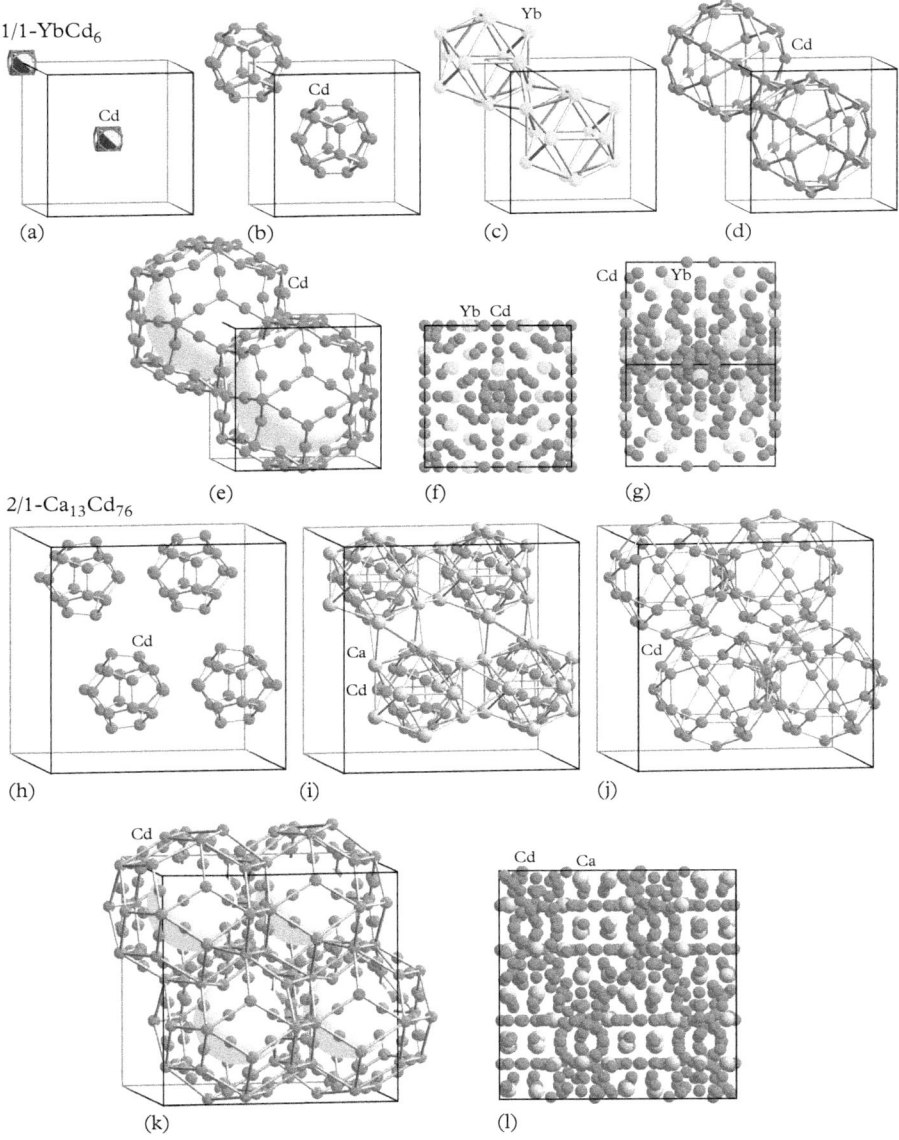

Fig. 9.8 *Shells of the Tsai-clusters on the example of bcc 1/1-YbCd₆ and 2/1-Ca₁₃Cd₇₆. (a) Orientationally disordered Cd₄ tetrahedron. The averaged electron density is smeared along the edges of a truncated octahedron; (b) Cd₂₀ pentagonaldodecahedron; (c) Yb₁₂ icosahedron; (d) Cd₃₀ icosidodecahedron; (e) Cd₉₂ distorted, edge-centered triacontahedron. The triacontahedra are face-connected along the [100] directions and share an oblate rhombohedron along [111]. The projections of one unit cell along [100] and [110] are shown in (f) and (g). (h) Cd₂₀ dodecahedron enclosing an orientationally disordered Cd₄ tetrahedron (not shown) similar to the 1/1-approximant; (i) Ca₁₂ icosahedron; (j) distorted Cd₃₀ icosidodecahedron connected via an octahedron to the neighboring cluster; (k) Cd₈₀ decorated triacontahedron sharing oblate rhombohedra with the overlapping other clusters; (l) Projection of one unit cell along [100] (from Steurer and Deloudi (2009), Figs. 9.8 and 9.9. With kind permission from Springer Science+Business Media).*

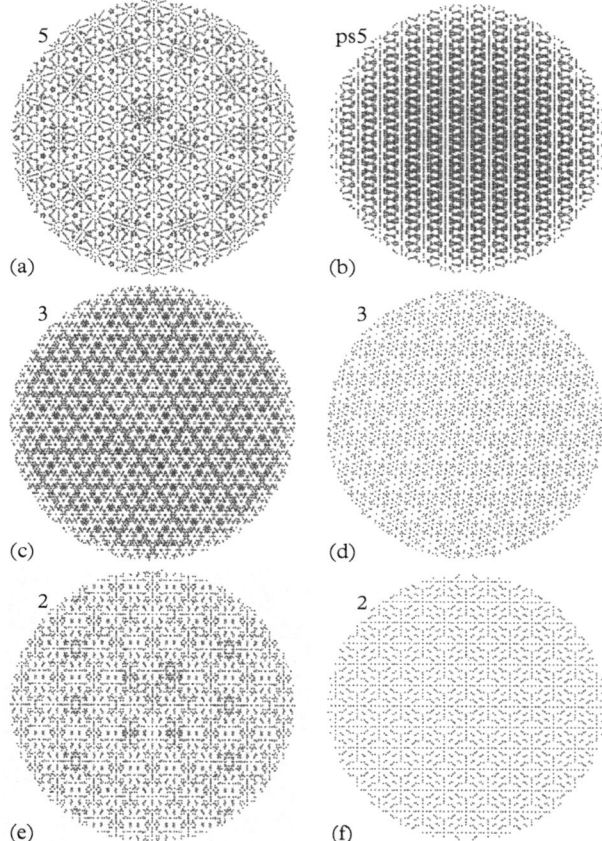

Fig. 9.9 *Projections of the structure of* i-Cd-Yb *(courtesy of H. Takakura) along (a) a five-fold axis, (c) a three-fold axis, and (e) a two-fold axis. In (b), (d), and (e), the corresponding projections of 1/1-Cd-Yb are depicted, i.e., along the pseudo-5-fold (ps5) and the 3- and 2-fold directions, respectively. The atomic layers form a network compatible with 5-fold symmetry only in (a) and not in (b). On top of it, almost all atoms are arranged in flat atomic layers (internal surfaces or interfaces), a kind of Ammann planes, which interpenetrate each other in a way, that is only possible in quasiperiodic structures. In all cases, projections of a spherical section (diameter 100 Å) of the structures are shown (from Steurer and Deloudi (2009), Fig. 10.2. With kind permission from Springer Science+Business Media).*

a particular kind of cluster is favorable (see also Steurer (2006*b*) and Steurer (2011*a*))? The crucial factor is the stoichiometry, which differs slightly for these two cases. Of course, there are also many systems where just an approximant exists and no QC or vice versa. In the case of IQCs there are more rational approximants than IQCs, in contrast to DQCs where the opposite is true.

The periodic and quasiperiodic structures of rational approximants and QCs, respectively, just differ by the ways their constituting clusters, which are the same in both cases, pack (Steurer and Deloudi, 2012). Consequently, one can assume that these clusters are energetically favorable subunits, which can be arranged in different ways. Their chemical composition generally differs from the overall composition of these compounds. By varying the kind of overlaps, only a few well-defined ones are possible; the overall chemical composition can be adjusted. For higher and higher approximants, the overall chemical composition should converge to that of the QC. Consequently, one could ask why we usually observe only rather low approximants, for instance 1/1, 2/1, and, at most, 3/2 for approximants to IQCs, instead of a devil's staircase of approximants. This is an open question. One hypothesis could be that the free-energy landscape shows a deep minimum just for the 1/1 and in some cases also for the 2/1 approximant. For higher approximants, the lattice energy may be very similar to that of the QC; however, the entropic contribution will be much higher for the QC mainly due to phasonic disorder on different length scales. Furthermore, since the cluster-overlapping rules have to be relaxed for the formation of approximants, it is not possible to create rational approximants of any order by just using the same type of constituting cluster without local changes and the need to fill gaps by "glue" atoms. This can turn low-order approximants into energetically quite favorable structures on their own, differing structurally, considerably, from QCs. In contrast, high-order approximants do not have a significant structure-based energetical advantage, if any, compared to QCs, which have higher entropy.

In this context one may ask why almost all structures of intermetallic phases in thermodynamic equilibrium are periodic with rather small unit cells. What is the advantage of periodicity compared to aperiodicity? In the case of small unit cells and well-defined atomic layers, the Bragg reflection density is low and all reflections relatively strong. This is favorable for the propagation of electrons and phonons yielding relatively simple band structures. In the case of IMSs and CSs the crucial factors hindering periodic structure formation are mutually incommensurate periodicities of substructures for otherwise favorable chemical compositions. The Bragg reflection density is infinite in at least one dimension. nD periodicity also leads to sharp Bragg reflections related to atomic layers that can diffract electron (Bloch) waves (see Hume-Rothery stabilization mechanism) in the case of intermetalllics.

In direct space, the crucial factor in the case of QCs is also the existence of two or more incommensurate length scales. In contrast to IMS and CS, respectively, they have specific fixed values. These values usually result from the structurally allowed distances between the constituting clusters, which have non-crystallographic symmetry in the case of all known quasiperiodic intermetallics.

Periodicity and quasiperiodicity, respectively, are the results of the energetically and entropically best possible packing of AETs and larger structural subunits with a narrow atomic (molecular) distances distribution and a minimum of different coordination polyhedra. In the case of small nanocrystals, for instance, where periodicity does not play a role anymore, a different type of packing (e.g., icosahedral atomic arrangements) can be energetically much more favorable than that taking place in the structures of larger crystals.

Crystal structures of intermetallics with large unit cells frequently have well-defined periodic average structures (PASs), related to large amplitudes of their Fourier modules (strong Bragg reflections), with a much smaller period (weak subperiod). The atomic positions in the actual (super)structure are slightly shifted away from the occupied sites in the PASs. While IMS also have PASs with a one-to-one relationship of atomic sites in the actual structure and in the PASs, this is not the case of PASs of QCs.

It should also be kept in mind that real crystals are never strictly periodic, not even in thermal equilibrium. They have finite size, with the atoms at the surface of the crystal having AETs differing from those in the bulk. There are atomic vibrations caused by phonons, thermal vacancies, and other point defects such as impurities (i.e., equilibrium defects). Furthermore, periodicity only exists on average in solid solutions (e.g., HEAs) and otherwise disordered crystal structures.

10

Structures and properties of functional intermetallics

The materials discussed in this chapter are not all intermetallic compounds according to our definition. Since their character, however, is metallic or at least half-metallic, we also include borides and silicides into our discussion in a few cases.

10.1 Ferromagnetic materials

A prerequisite for a strong permanent magnet is a high net magnetic moment of (some of) the constituting elements and the efficient coupling of the magnetic moments. Since the magnetic moment of an atom depends on the number of localized unpaired electrons, transition elements with up to five and rare earth elements with up to seven unpaired electrons in the d- and f-orbitals, respectively, are the most promising candidates. The atomic magnetic moment originates on one hand from the spin of the electrons and on the other hand from the orbital motion around the atomic cores. Both contributions add up vectorially. A strong spontaneous coupling of the parallel-aligned atomic magnetic moments by quantum mechanical exchange forces, which leads to the spontaneous polarization $\mathbf{J}$, is another prerequisite for a ferromagnetic material. The crystal structure governs this coupling and therewith the magnitude of the magnetocrystalline anisotropy, which is related to the energy needed to move $\mathbf{J}$ out of its self-chosen direction. At the Curie temperature, T_C, the thermal vibrations are strong enough to destroy the long-range order of the magnetic moments, and the ferromagnetic material becomes paramagnetic.

One has to distinguish between soft- and hard-magnetic materials (for a review see, e.g., Jiles (2003)). The former are characterized by a high permeability μ_r (a second rank tensor), low coercitivity, and low hysteresis loss; for the latter, a high coercitivity (which is connected with a low permeability) is important, i.e., a strong resistance to demagnetization. Soft-magnetic materials are employed for electrical transformers and due to their narrow hysteresis loop it does not

Intermetallics: Structures, Properties, and Statistics. First Edition. Walter Steurer and Julia Dshemuchadse.
© Walter Steurer and Julia Dshemuchadse 2016. Published in 2016 by Oxford University Press.

need much energy to invert the direction of the magnetic induction, $\mathbf{B} = \mu_r \mathbf{H}$. Hard-magnetic materials are used for permanent magnets, which are employed for providing strong constant magnetic fields. In both cases, the crystal structure plays a decisive role. Structures favoring a strong exchange correlation between the atomic magnetic moments cause a broader hysteresis loop. Amorphous and nanocrystalline alloys are the other extreme, the very weak coupling leading to soft-magnetic materials with a low Curie temperature.

It should be mentioned, however, that the constituting elements of a ferromagnet do not need to be *ferro*magnetic themselves. Ferromagnetic elements at ambient temperature are only: Co ($T_c = $ 1388 K), Fe ($T_c = $ 1043 K), Ni ($T_c = $ 627 K), and Gd ($T_c = $ 292 K). Examples are the Heusler phases, for instance, ferromagnetic $cF16$-Cu_2MnAl. Al and Mn are paramagnetic, Cu is diamagnetic. On the other hand, alloys consisting mainly of ferromagnetic elements exist that are just paramagnetic such as austenitic (*fcc*) stainless steels (*fcc* Fe–Cr–Ni alloys). Such steels can be used in strong magnetic fields, for instance, for magnetic resonance imaging (MRI). In contrast, martensitic and ferritic stainless steels are ferromagnetic.

Starting in the beginning of the twentieth century, the performance of hardmagnetic materials for the application as permanent magnets has been drastically improved, step by step. Magnets, which do not easily demagnetize due to their broad hysteresis loop, are called hard-magnetic. Their performance can be quantified by the maximum energy product $\mathbf{B} \cdot \mathbf{H}_{max}$, which is defined by the maximum product of B and H in the second quadrant of the hysteresis curve (demagnetization curve); it can be given in kJ/m^3 or Mega-Gauss-Oerstedt (MGOe).

The development of permanent magnets started with ferromagnetic steel (≈ 1 MGOe), an iron alloy containing C, Cr, W, Co, or other elements for impeding domain wall motion in order to get a low but significant coercitivity. The next magnet materials developed were ferrites ($\approx$3–4 MGOe), in particular barium-iron-, strontium-iron-, and cobalt-iron-oxides with the spinel structure type, which show a high magnetocrystalline anisotropy, and aluminum-nickeliron-cobalt alloys, Alnico ($\approx$5–10 MGOe). There, the high magnetocrystalline anisotropy results from the texture of the material, achieved by annealing the alloy in a strong external magnetic field. However, the coercitivity is rather small in both cases compared with the rare-earth-based magnetic materials discovered later.

The discovery of $hP6$-$SmCo_5$ and $hR19$-Sm_2Co_{17} ($\approx$ 20–30 MGOe) in the 1960s was a major breakthrough, topped only by $tP86$-$Nd_2Fe_{14}B$ ($\approx$ 56 MGOe), which was identified as an excellent hard-magnetic material in 1984 and optimized until the end of the 1990s. It is still the material underlying the by-far strongest permanent magnets. The high magnetocrystalline anisotropy is mainly caused by the rare-earth elements, while the transition metal atoms provide the main contribution to the high magnetization. Substitution of $\approx$ 4% Nd by Dy improves the linearity of the B-H curve as well as the intrinsic coercitivity and allows us to use this kind of permanent magnet for hybrid vehicle motors, for instance

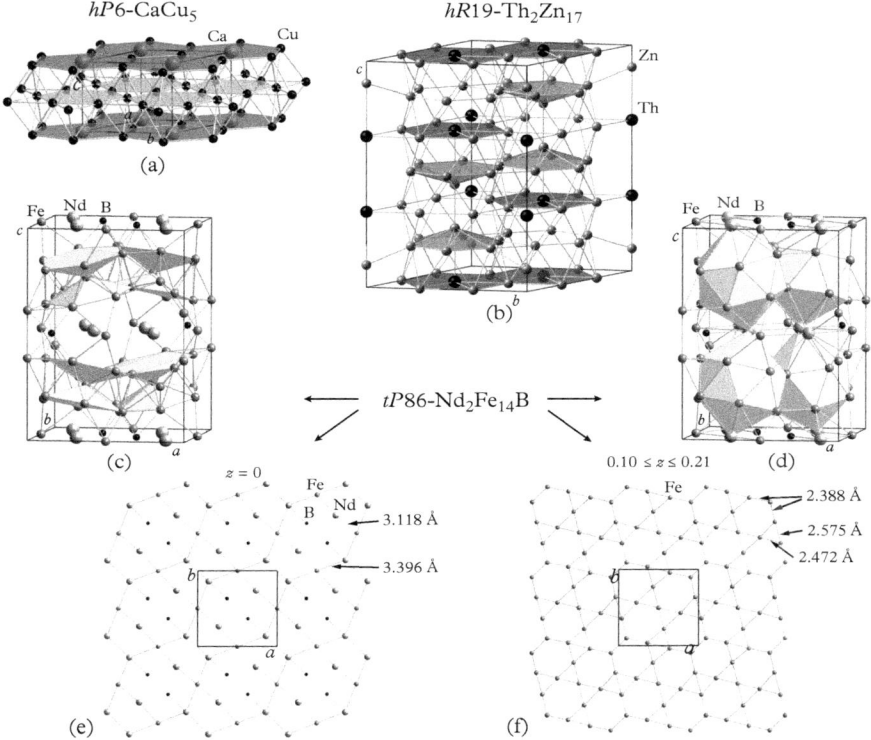

Fig. 10.1 *Structure types of the most important hard-magnetic materials: (a) the hP6-CaCu₅ type, (b) the hR19-Th₂Zn₁₇ type, and (c)-(d) the tP86-Nd₂Fe₁₄B type. In (a), the hexagons in the K-net of Cu atoms are shaded light gray, those containing Ca dark gray. In (b), the Th-containing hexagons are dark-gray, the Zn-capped Zn-hexagons are shown as pyramids. In (c) and (d), two different representations of the tP86-Nd₂Fe₁₄B structure are shown (Nd. . . light gray, Fe. . . gray, B. . . black). The polyhedra in (d) are CN14 FK-polyhedra. In (e) the flat Fe/Nd/B-layer (z = 0) is shown with some characteristic interatomic distances, in (f) the puckered Fe-layer (0.1 ≤ z ≤ 0.21). Subfigures (e, f) are on half the usual scale.*

(Gutfleisch *et al.*, 2011). It has been experimentally shown that Nd can be replaced in this structure type by all other RE elements except Pm and Eu; however, large energy products have been achieved for RE = Pr and Nd, only (Herbst and Croat, 1991; Goll and Kronmüller, 2000).

The structure of $hP6$-SmCo₅, which is of the $hP6$-CaCu₅ type, can be described as a stacking of h(exagon) layers and K(agomé) layers (Fig. 10.1(a)). Also in this binary system another good hard-magnetic material exists, $hR19$-Sm₂Co₁₇ ($hR19$-Th₂Zn₁₇ type) (Fig. 10.1(b)), which is the only kind of strong permanent magnet that can be used at elevated temperatures due to its high

Curie temperature of $T_C = 1189$ K. Its structure can be described in a similar way by h and K layers. The difference is that the h layers feature Sm-centered Co-hexagons, only, in the case of $hP6$-SmCo$_5$, and empty hexagons surrounded by Sm-centered ones in the case of $hR19$-Sm$_2$Co$_{17}$. The hexagons in the K layers are in both cases empty, however, partially capped by Co atoms in the case of $hR19$-Sm$_2$Co$_{17}$. It is amazing that the structures of these RE-based permanent magnets all have the honeycomb and K(agomé) nets in common, which are vertex-decorated with transition metal atoms and centered or capped, at least partially, with the larger RE elements. There are also similarities to the σ phase, $tP30$-Cr$_{46}$Fe$_{54}$ (see Fig. 7.13 in Subsection 7.4.2).

The structure of $tP86$-Nd$_2$Fe$_{14}$B can be described as a stacking of layers of face-connected tilted Fe$_{13}$Nd$_2$ CN14 FK-polyhedra (Fig. 10.1(d)) (Herbst *et al.*, 1984). The Fe-centered CN14 polyhedra in adjacent layers share their apical Nd atoms in $z = 0$ and 1/2, which themselves form flat layers together with the B atoms and Fe atoms outside the polyhedra (Fig. 10.1(e)). The B atoms center the trigonal Fe-prisms left in the open space in-between the CN14 polyhedra of two adjacent layers. Due to covalent bonding (Ching and Gu, 1987) the short Fe–B distances (≈ 2.1 Å) require a tilt of the CN14 polyhedra, which leads to a puckering of the Fe layers (Fig. 10.1(f)).

It should be mentioned that also in the Nd–Fe–B system a compound exists with the structurally related $hR19$-Th$_2$Zn$_{17}$ structure type. Compared to $tP86$-Nd$_2$Fe$_{14}$B, $hR19$-Nd$_2$Fe$_{17}$ has a lower Curie temperature, $T_C = 330$ K versus 585 K (Herbst *et al.*, 1985). Furthermore, the magnetic moments are oriented in the basal plane in the case of $hR19$-Nd$_2$Fe$_{17}$, while in $tP86$-Nd$_2$Fe$_{14}$, all magnetic moments are aligned parallel to the [001] direction. To the total magnetic moment of 41.61 μ_B per formula unit (f.u.) the local magnetic moments at the Fe sites contribute between 2.38 and 2.78 μ_B/f.u., and those at the Nd sites 3.39–3.41 μ_B/f.u. (Kitagawa and Asari, 2010).

According to Goll and Kronmüller (2000), $tP86$-RE$_2$Fe$_{14}$B, RE = Nd, Pr, permanent magnets have the highest maximum energy product, while $hP6$-CaCu$_5$ and $hR19$-Sm$_2$(Co,Cu,Fe,Zr)$_{17}$ supply the highest coercitive fields at ambient temperature as well above 1000 K, making them suitable for HT-applications.

The Heusler phases (see also Subsection 7.3.2) should be mentioned here as well, since they show interesting magnetic properties, which can be of interest, for instance, for spintronic applications. The ferromagnetic properties of $cF16$-Cu$_2$MnAl had been discovered in 1903 by Fritz Heusler. In the meantime, hundreds of Heusler and half-Heusler phases, with stoichiometries $cF16$-A$_2$BC and $cF12$-ABC, respectively, have been found with interesting properties (for a review see Graf *et al.* (2013)). A and B are mostly TM or RE elements, while C are main group elements. The Heusler phase can be seen as a $(2 \times 2 \times 2)$-fold superstructure of the $cI2$-W type. Removing four of the eight atoms centering the eighth-cubes leads to the half-Heusler phase. Magnetic half-Heusler compounds have one magnetic sublattice formed by B atoms, magnetic Heusler phases have two, one of A atoms and one of B atoms, which can couple ferromagnetically.

Heusler phases with compositions Co_2YZ and Mn_2YZ can be half-metallic ferro- or ferrimagnets, which behave like a metal for one spin direction and like an insulator for the other one.

Finally, a comment on the magnetic properties of quasicrystals (for a review see Stadnik (2013)). From symmetry analysis it was found that long-range ferromagnetic order is not possible in icosahedral QCs, while antiferromagnetic order is allowed. So far, experimental observations only show the existence of either paramagnetism, diamagnetism, or spin glass behavior for decagonal or icosahedral quasicrystals.

10.2 Magnetostrictive materials

Magnetostrictive materials reversibly change their dimensions under the influence of a magnetic field that changes their spontaneous magnetostrictive strain. The inverse magnetostrictive effect changes the magnetization of a material under mechanical stress. In contrast to these magnetoelastic materials, magnetoplastic materials do not simply recover after removal of the external magnetic field. Examples are the magnetic shape-memory materials, which will be discussed in Section 10.6.

The magnetostriction λ is defined as the relative length change under the influence of a magnetic field, $\lambda = \Delta l/l_0$. The changes are measured relative to the length l_0 of the material in the demagnetized state. Most magnetically ordered materials (ferro-, ferri-, and antiferromagnets) have $\lambda \approx 10^{-5} - 10^{-6}$, a few show a giant effect on the order of $\lambda \approx 10^{-3}$ (Jiles, 2003).

The technologically most important magnetostrictive material is $cF24$-$Dy_{1-x}Tb_xFe_2$ ($x \approx 0.3$), Terfenol D, named after its main constituents Tb and Fe, the place where it was developed in the seventies, the Naval Ordnance Laboratory (NOL), and its minor constituent Dy. Due to its large (positive) magnetostriction constant it is mainly used as an actuator, magnetomechanical sensor, and acoustic and ultrasonic transducer (e.g., for sonar systems). Its structure is that of the cubic Laves phase $cF24$-$MgCu_2$, and its Curie temperature is $T_C = 653$ K. The compound $cF24$-$SmFe_2$ crystallizes in the same structure type, and in contrast to Terfenol D it has a giant negative magnetostriction constant (Kawamura *et al.*, 2006).

The magnetostrictive effect also underlies the Invar effect, i.e., an anomalously low thermal expansion, which has been discovered by Guillaume (1897) in a fully disordered $cF4$-$Fe_{65}Ni_{35}$ alloy, and has later been found to exist as well in ordered compounds such as $cP4$-Fe_3Pt but also in elemental $hP2$-Gd (Ruban *et al.*, 2007). The negative magnetostrictive effect leads to a spontaneous expansion of the structure. With increasing temperature the magnetostrictive coupling gets more and more weakened by thermal motion of the atoms diminishing the thereby caused expansion and at the same time compensating thermal expansion therewith.

Magnetostriction has been observed in many other magnetic intermetallics such as $cF24$-YMn$_2$, $hP38$-Tm$_2$Fe$_{16}$Cr, and $hP38$-Er$_2$Fe$_{17}$, for instance, to name only a few (Chen *et al.*, 2015). It cannot only lead to almost zero thermal expansion (Invar effect) but also to negative thermal expansion as in $hP9$-GdAgMg, $hP38$-Y$_2$Fe$_{17}$, or $hR57$-Gd$_2$Fe$_{17}$, and related rare earth compounds (Chen *et al.*, 2015).

10.3 Magnetocaloric and magnetic barocaloric materials

Basically, all magnetic materials show the magnetocaloric effect (MCE); however, it may be arbitrarily small. In an adiabatic process, MC materials show a reversible change in temperature, ΔT_{ad}, upon application of a magnetic field. In the case of isothermal conditions, a reversible change in magnetic entropy, ΔS_M, takes place. ΔT_{ad} and ΔS_M are the two parameters fully characterizing an MC material, if mapped out as functions of the applied magnetic field and of temperature.

If the effect is large enough, as is the case for materials with giant magnetocaloric effect (GMCE), it can be practically used for cooling by adiabatic demagnetization, and in reverse for heating by adiabatic magnetization. Consequently, GMC materials with Curie temperatures close to the temperature of interest (in most cases at room temperature, $T_C \approx RT$), can be used for refrigeration and heat pumps.

A Carnot cycle for refrigeration, for instance, could work in the following way: first the magnetic contribution to entropy, S_M, is decreased by adiabatic magnetization of a superparamagnetic material or a ferromagnetic material above its Curie temperature, T_C (superparamagnetism can exist in small, single-domain ferromagnetic nanoparticles). In an adiabatic system the total entropy has to remain constant, a decrease in S_M has to be accompanied by a rise in the vibrational entropy, S_v, i.e., an increase of lattice vibrations (phonons) and therewith of the temperature by ΔT_{ad}. Then, after removal of the excess heat at ambient temperature, the reverse process, i.e., adiabatic demagnetization, restores the zero-field magnetic entropy again at the cost of the vibrational entropy. This leads to a decrease of the lattice vibrations and, therewith, of the temperature. A more detailed discussion of magnetic-refrigeration cycles can be found in the review by Romero Gómez *et al.* (2013).

According to the thermodynamic approach presented by Spichkin and Tishin (2005) (and references therein), the Gibbs free energy of a magnetic material, which can be transformed from one magnetic phase to another by a first order magnetic phase transition, can be written as:

$$G = G_{ex} + G_{me} + G_a - HM, \tag{10.1}$$

with G_{ex} the free energy of the exchange interaction, G_{me} that of the magnetoelastic interaction, G_a the anisotropy energy, and HM the magnetic energy

(H... magnetic field, M... magnetization). The entropy change at the magnetic transition follows from the equality of the potentials of the two phases, $G_1 = G_2$:

$$\Delta G_{ex} + \Delta G_{me} + \Delta G_a = H_{cr}\Delta M, G = G_{ex} + G_{me} + G_a - HM, \tag{10.2}$$

with H_{cr} the critical magnetic field of the transition. The isothermal magnetic entropy change at the transition results from equation 10.2 and the magnetic Clausius-Clapeyron equation:

$$\frac{dH_{cr}}{dT} = -\frac{\Delta S_M}{\Delta M} \tag{10.3}$$

yielding

$$\Delta S_M^{tr} = -\left(\frac{\partial \Delta G_{ex}}{\partial T} + \frac{\partial \Delta G_{me}}{\partial T} + \frac{\partial \Delta G_a}{\partial T} - H_{cr}\frac{\partial \Delta M}{\partial T}\right). \tag{10.4}$$

The adiabatic temperature change at the phase transition Δt^{tr} can be calculated using the equation

$$\Delta T = -\frac{T}{C_{p,H}}\Delta S_M, \tag{10.5}$$

with the heat capacity $C_{p,H}$, resulting in

$$\Delta T^{tr} = \frac{T}{C_{p,H}}\left(\frac{\partial \Delta G_{ex}}{\partial T} + \frac{\partial \Delta G_{me}}{\partial T} + \frac{\partial \Delta G_a}{\partial T} - H_{cr}\frac{\partial \Delta M}{\partial T}\right). \tag{10.6}$$

In Fig. 10.2(d), a T/S diagram is shown of an MC material undergoing a first-order phase transformation (Romero Gómez *et al.*, 2013). The transformation temperatures depend on the magnetic field, T_{t0} for $H_0 = 0$ and T_{t1} for $H_1 > H_0$. The isomagnetic curves show jumps in entropy, ΔS_M, at the respective transformation temperatures. The vertical dotted lines reflect the adiabatic temperature changes, ΔT, which have maximum values for the temperature range $T_m < T < T_{t0}$. In the case of a second-order phase transition, the isomagnetic curves would not show discontinuities, but the general schematic would be similar. However, the MCE would be much smaller due to the lack of a large change in the exchange energy due to a discontinuity in the interatomic distances.

The magnetic barocaloric effect (MBCE) leads to heating and cooling of magnetic materials upon variation of pressure. The main effect is based on the change in exchange interaction energy due to a significant change in interatomic distances at the phase transformation. According to Spichkin and Tishin (2005), the strong dependence of the exchange energy on interatomic distances should be reflected in the change of the Curie temperature with pressure, dT_C/dp (see below). The MBCE can also be used to enhance the performance of GMC materials (de Oliveira *et al.*, 2014).

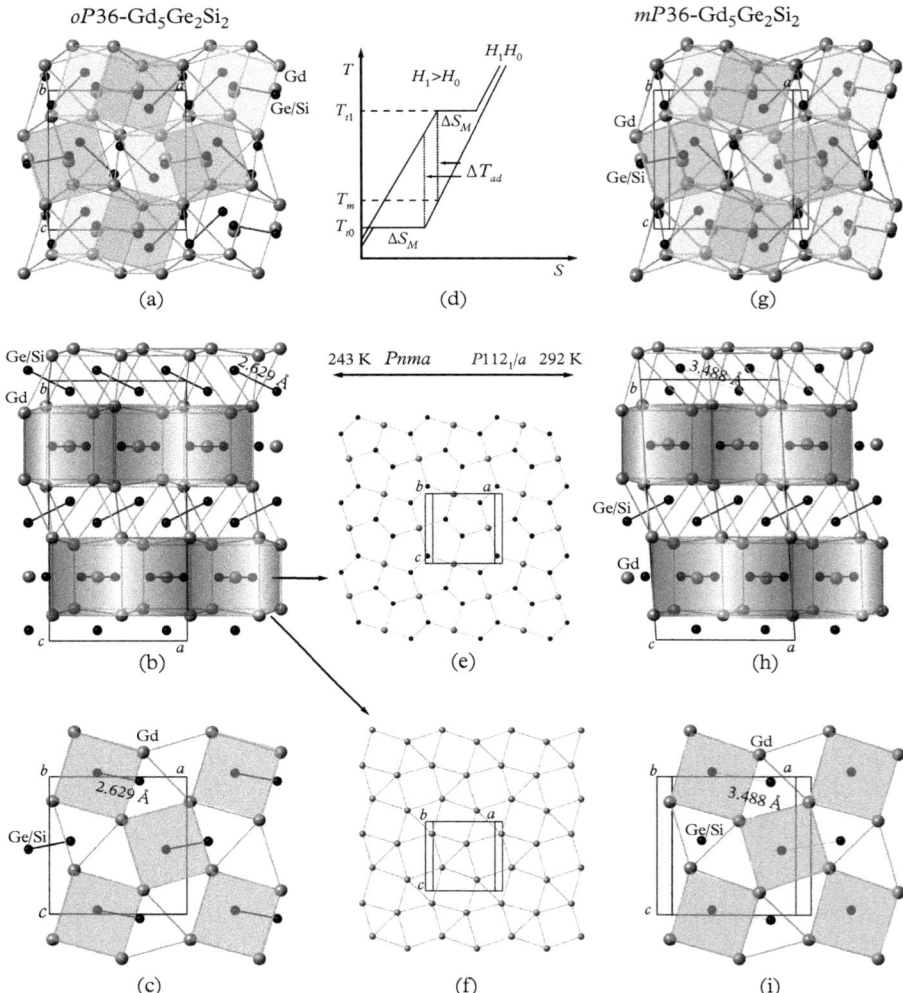

Fig. 10.2 *The structures of the (a)–(c), (e, f) LT-phase oP36-Gd₅Ge₂Si₂ (Pnma at 243 K) and the (g–i) HT-phase mP36-Gd₅Ge₂Si₂ (P112₁/a at 292 K) in different projections (Gd. . . gray, Si/Ge. . . black): (a, g) onto (101), with the slabs shaded gray; (b, h) along [001]; (c, i) bounded projections with −0.07 ≤ y ≤ 0.2; (e) bounded projection of the slightly puckered Gd/Si/Ge layer (0.24 ≤ y ≤ 0.26). The change in the interatomic distances in the dimers is indicated in (b, c) and (h, i). The nets shown in (c, f, i) correspond to the Archimedean snub square tiling 3².4.3.4, the net in (e) to its dual, the Catalan Cairo pentagon tiling V3².4.3.4. (d) Schematic T-S diagram of an MC material in two magnetic fields, H₀ and H₁, close to a first-order phase transformation (after Romero Gómez et al. (2013)).*

According to the reviews by Brück (2005) and Gschneidner Jr. *et al.* (2005), intermetallic GMC materials include:

- Gd ($\Delta S_M = -9$ J/kgK, $\Delta T_{ad} = 11.6$ K for $\Delta H = 5$ T) at the second-order magnetic transition temperature of 294 K (Fujieda *et al.*, 2003), and Gd-based solid solutions (mostly used so far in devices working close to room temperature); Tb, Dy, Er, $Tb_{0.5}Dy_{0.5}$; MBCE: $-0.5 \leq dT_C/d_p \leq 1.5$ K/kbar, depending on composition.

- $hP2$-$Fe_{0.49}Rh_{0.51}$ with $hP2$-CsCl structure type; magnetism based on itinerant electrons; highest magnetocaloric effect found so far: a magnetic field of $H = 2$ T at $T = 308.2$ K causes a $\Delta T = 12.9$ K (Annaorazov *et al.*, 1992); MBCE: $4.3 \leq dT_C/d_p \leq 5.75$ K/kbar, depending on composition.

- Lanthanoid cubic Laves phases, $cF24$-REM_2, with M either Al, Co, or Ni; magnetism based on localized electrons.

- $Gd_5(Si_{1-x}Ge_x)_4$ with $oP36$-Sm_5Ge_4, $mP36$-$Gd_5Ge_2Si_2$, or $oP36$-Gd_5Si_4 ($T_C = 336$ K) structure type; magnetism based on localized electrons; the magnetic properties are very sensitive to the Ge/Si ratio due to different atomic sizes of these isoelectronic atoms (chemical pressure); at the first-order structural transformation temperature of 278 K, $Gd_5Ge_2Si_2$ exhibits $\Delta S_M = -18$ J/kgK, $\Delta T_{ad} = 15.3$ K for $\Delta H = 5$ T (Fujieda *et al.*, 2003); MBCE: $dT_C/d_p = 3.79$ K/kbar; for a review on the magnetic properties of the whole class of the RE_5M_4 family (M... group 13–15 element) see Mudryk *et al.* (2011).

- $cF112$-$La(Fe_{13-x}Si_x)$ and $cF112$-$La(Fe_{13-x}Al_x)$ with $cF112$-$NaZn_{13}$ structure type; magnetism based on itinerant electrons; for $cF112$-$La(Fe_{0.9}Si_{0.1})_{13}$, $\Delta S_M = -30$ J/kgK, $\Delta T_{ad} = 12.1$ K for $\Delta H = 5$ T at the itinerant-electron metamagnetic transition temperature of 184 K (Fujieda *et al.*, 2003).

- $MnNi_2Ga$, with the LT-phase $tP4$-$Mn_{18.6}Ni_{55.2}Ga_{26.2}$ ($tP4$-AuCu structure type) and the HT-phase $cF16$-$Mn_{18.6}Ni_{55.2}Ga_{26.2}$ ($cF16$-$MnNi_2Ga$ structure type, Heusler phase); both the first-order (martensitic) structural transformation and the second-order metamagnetic (ferro- to paramagnetic) transformation have the same transition temperature for this composition, $T_{mart} = T_C = 315$ K; the entropy change amounts to $\Delta S_M = -20.4$ Jkg^{-1}K^{-1} at 317 K and $H = 5$ T (Zhou *et al.*, 2005).

- $hR19$-Nd_2Fe_{17} with $hR19$-Th_2Zn_{17} structure type; magnetism based on localized electrons.

A powerful magnetocaloric material for LT-refrigeration around 20 K is the type-VIII clathrate $cI54$-$Eu_8Ga_{16}Ge_{30}$ (Phan *et al.*, 2008). It undergoes a second-order ferro- to paramagnetic phase transition at ≈ 13 K, and exhibits a magnetic entropy change of $\Delta S_M = 11.4$ Jkg^{-1}K^{-1} at 3 T. This value is larger than that for Gd ($\Delta S_M = 10.2$ Jkg^{-1}K^{-1} at 5 T) and comparable to that of $Gd_5Ge_2Si_2$

$(\Delta S_M \approx 18 \text{ Jkg}^{-1}\text{K}^{-1}$ at 5 T). The cooling capacity amounts to ≈ 87 Jkg^{-1}. $Eu_8Ga_{16}Ge_{30}$ undergoes a phase transition from a type-VIII ($I\bar{4}3m$, $cI54$) into a type-I clathrate ($Pm\bar{3}n$, $cP54$) at 970 K. The crystal structure of the clathrate of type VIII is constituted by only one type of, partially defective, polyhedral cages, which can be derived from a pentagonaldodecahedron by adding three vertices on particular edges (Shevelkov and Kovnir, 2011).

As an example of the structure underlying an intermetallic GMC material, the LT- and HT-structures of $Gd_5Ge_2Si_2$ ($T_C = 276$ K), $oP36$-$Gd_5Ge_2Si_2$ (62 *Pnma* at 243 K), and $mP36$-$Gd_5Ge_2Si_2$ (14 $P2_1/c$ at 292 K), respectively, are illustrated in Fig. 10.2. Si/Ge atoms are dealt with as one type of atoms, since both occupy the same Wyckoff positions in the structure in a partially disordered way. According to Choe *et al.* (2000), the structures can be described as being composed of slab-like structural units stacked upon each other along the [010] direction, which themselves consist of vertex- and body-center-decorated Gd cubes. The first-order phase transformation between the paramagnetic monoclinic HT-phase (14 $P2_1/c$) and the ferromagnetic orthorhombic LT-phase (62 *Pnma*) can be induced by both temperature and magnetic field changes. Structurally, during the transformation the slabs are shifted against each other in the (101)-plane by half a cube's edge-length, breaking/forming one half of the covalent bonds in Ge/Si dimers capping the cubes, and the unit cell volume changes by 1%, the *a* lattice parameter even by 1.85%. This changes the electronic structure and the exchange interactions significantly. The atomic distances between the dimer atoms are expanded/shrunk by more than 30%. Each dimer atom sits in the center of a triangular Gd-prism filling the gaps between the slabs. The cube faces are part of a $3^2.4.3.4$ snub square tiling in the (101)-plane. The dimer atoms together with the Ge/Si atoms form a slightly puckered Catalan Cairo pentagon tiling, $V3^2.4.3.4$, dual to it.

10.4 Magnetooptic materials

Magnetooptic (gyromagnetic) materials can influence the polarization of transmitted (Faraday effect) or reflected (MO Kerr effect, MOKE) light by their quasistatic magnetic field. By spin-orbit coupling und the influence of the magnetic field, the permittivity tensor ε becomes anisotropic with complex off-diagonal components, leading to different velocities of left- and right-hand circularly polarized electromagnetic waves.

In most MO materials the effect is rather small. In some cases, however, a unique combination of physical properties provided by the specific crystal structure and chemical composition can lead to a giant MO effect. An important role play the magnitude of the 3d-magnetic moment, the spin-orbit coupling strength, the degree of hybridization in the chemical bonding, the density of states at the Fermi level, and the intraband plasma frequency, which is rather low in case of a half-metallic character of the material.

An example for a good MO material is the half-Heusler compound $cF12$-MnPtSb (Antonov *et al.*, 1997). It has an extremely large MOKE rotation of $-1.27°$ at ambient conditions. However, for the application as a recording material it has the drawback that it does not have the large magnetocrystalline anisotropy required for a magnetic orientation perpendicular to the material's surface.

The quadratic MOKE (QMOKE) has been found strong in the half-metallic Heusler alloys $cF16$-FeCo$_2$Si ($T_C = 1100$ K) and $cF16$-MnCo$_2$Ge ($T_C = 905$ K) (Hamrle *et al.*, 2007; Muduli *et al.*, 2009). The QMOKE is proportional to the mixed products $M_L M_T$ and $M_L^2 - M_T^2$, with the longitudinal and transversal magnetizations, M_L and M_T, respectively, relative to the plane of light incidence. The QMOKE from the surface of a $cF16$-FeCo$_2$Si-(100) thin film was found to be in the range up to $0.030°$ for 21 nm film thickness. This large signal has been attributed to an exceptionally large spin-orbit coupling of second or higher order (Hamrle *et al.*, 2007).

10.5 Thermoelectric materials

By applying a temperature gradient across a thermoelectric material, the thermal energy can be directly converted into electrical energy (Seebeck effect) and vice versa (Peltier effect). The thermoelectric performance is quantified by the dimensionless figure of merit (FOM),

$$ZT = \frac{\sigma S^2 T}{\kappa} \qquad (10.7)$$

with Z the actual figure of merit and T the temperature, the electrical conductivity σ, the thermal conductivity $\kappa = \kappa_{el} + \kappa_{phon}$, and S the Seebeck coefficient or thermopower (given in volts per Kelvin V/K),

$$S = -\frac{\Delta V}{\Delta T}. \qquad (10.8)$$

Consequently, a large FOM (> 1) can be achieved if the electrical conductivity is high, and the thermal conductivity low. The electrical conductivity of metals is high, however; this makes the electronic contribution to the thermal conductivity high as well. Furthermore, the electrical conductivity of metals decreases with increasing temperature. In contrast, it increases with increasing temperature in semiconductors. One way to improve the performance of materials is to decrease κ_{phon}, the phononic contribution to the thermal conductivity by inhibiting the propagation of phonons as much as possible (PGEG, "phonon glass, electron crystal" concept). This can be achieved by introducing structural disorder, and/or have "rattling atoms" in large voids as, for instance, in some clathrates and skutterudites. For reviews, e. g., see Sootsman *et al.* (2009) and Kleinke (2010).

Another important parameter for technological applications is the thermoelectric efficiency, which combines the Carnot efficiency, $\Delta T / T_{hot}$, and the FOM to

$$\eta = \frac{\Delta T}{T_{hot}} \frac{\sqrt{1 + ZT_{avg}} - 1}{\sqrt{1 + ZT_{avg}} + T_{cold}/T_{hot}}, \tag{10.9}$$

with T_{hot} and T_{cold} the temperatures of the hot and cold ends of the device and ΔT their difference; T_{avg} is its average temperature. For typical, currently available devices $ZT \approx 0.8$ and $\eta \approx 5\text{-}6\%$.

So far, the best room-temperature thermoelectric materials are semiconductors such as Bi_2Te_3 and Bi_2Se_3 with ZT between 0.8 and 1.0 in bulk materials and up to ≈ 2.4 for nanostructured materials. The highest FOM so far, ≈ 2.6 at 923 K, was found in SnSe single crystals along the [001] direction (Zhao *et al.*, 2014). Intermetallic materials are rare with the exception of half-Heusler phases, which can be used at higher temperatures. Intermetallics with very complex crystal structures such as $cF1140\text{-}Gd_{117}Co_{56}Sn_{112}$ (Schmitt *et al.*, 2012) or as quasicrystals, for instance, can also exhibit large Seebeck-coefficients due to the complex cluster structure, intrinsic disorder and/or their special electronic band structure and a spiky density of states (Macia, 2001).

In the following, we illustrate typical structures of intermetallic thermoelectric materials. First, as an example for cage compounds with "rattling" atoms, which can be regarded as PGEG compounds, we will discuss the structures of a narrow-bandgap semiconducting filled skutterudite and of a type-I clathrate, and then the structure of a half-Heusler phase.

Skutterudites, with the general chemical formula $cI32\text{-}AB_3$ (A... late TM, B... P, As, Sb) crystallize in the $cI32\text{-}CoAs_3$ structure type ($Im\bar{3}$), and can be classified as TM-Zintl phases (Fig. 10.3 (a)-(d)). The Co atoms center slightly distorted and vertex-connected As_6-octahedra, which leave large icosahedral voids around the corners and the center of the unit cell. The 12 As atoms, 3.118 Å apart from the center of such an icosahedron, together with the 8 Co atoms at a distance of 3.546 Å from the center, form a slightly distorted dodecahedron. If it is centered by a heavy ion M that is smaller than the void, such as a trivalent RE ion or Ba^{2+}, for instance, this leads to a filled skutterudite with the general formula $M_xA_4B_{12}$, $x \leq 1$. Such a rattling ion inhibits long-wavelength phonon propagation quite efficiently. An example of such a compound is $Yb_{0.19}Co_4Sb_{12}$ (Nolas *et al.*, 2000).

According to Shevelkov and Kovnir (2011), intermetallic clathrates crystallize in several different structure types, the most important of which for thermoelectric materials is the type-I clathrate, whose approximately 150 representatives can be considered as Zintl phases. Almost all of them have group 14 elements as the main constituents of the cage-forming tetrahedrally bonded (4-connected) framework, most of them are semiconducting, quite a few have metallic character. The ideal chemical formula is $cP54\text{-}A_2B_6C_{46}$, space group 221 $Pm\bar{3}n$, with A and B large cations as guest atoms in the cages, and C group-14 elements, constituting the

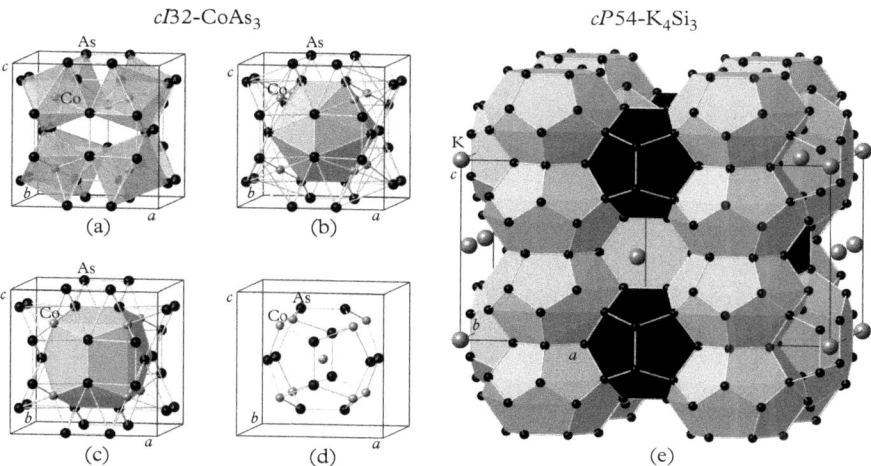

$cI32$-CoAs$_3$

$cP54$-K$_4$Si$_3$

(a) (b) (c) (d) (e)

Fig. 10.3 *Structures of the prototypic (a)-(d) skutterudite cI32-CoAs$_3$ and (e) the type-I clathrate cP54-K$_4$Si$_3$. In (a) the vertex-connected Co-centered As-octahedra are shown, and in (b) the distorted icosahedral void formed by them. Adding the cube of Co atoms (gray) to the As (black) icosahedron gives a distorted dodecahedron around another Co atom (c, d). (e) The packing of pentagon dodecahedra (black) and 24-vertex tetrakaidecahedra (gray), both decorated by Si (black) constituting the structure of the type-I clathrate cP54-K$_4$Si$_3$. These polyhedra are centered by K.*

framework; A . . . Na, K, Ba; B and C . . . Al, Ga, In, Si, Ge, Sn, for instance. The structure can be seen as an ordered $cP54$-K$_4$Si$_{23}$ structure type (Fig. 10.3 (e)).

The structure consists of a polyanionic tetrahedral framework of B and C atoms with dodecahedral cages, which can host large A cations that donate their electrons to the framework. This leads to a complete filling of the sp^3 orbitals of the framework atoms and a semiconducting character of the compound. An example is $cP54$-Ba$_8$In$_{16}$Ge$_{30}$, where the large Ba cations are located in off-center positions in the In/Ge cages and show large atomic displacement parameters in the time/space-averaged structure. Due to anharmonic potentials, the off-centering increases with temperature (Bentien *et al.*, 2005).

Since the ideal bond angle in such a network would be 109.45°, the abundance of the pentagon dodecahedron (5^{12}, 20 vertices) and related polyhedra such as the tetrakaidecahedron (5^{12}6^2, 24 vertices) is understandable, because in them the ideal bond angle amounts to 108°. The ratio between the frequency of pentagon dodecahedra and tetrakaidecahedra in type-I clathrate structures is 2:6. The hexagon-face-connected tetrakaidecahedra form a framework where the isolated pentagon dodecahedra are embedded.

In general, the electronic contribution to the thermal conductivity, κ_{el}, is below 10%, due to the specific interactions of the host and the "rattling" guest atoms increasing the scattering of phonons. The total thermal conductivity of some clathrates can be as low as 0.5 Wm^{-1}K^{-1} (Shevelkov and Kovnir, 2011). The power factor, σS^2, can be optimized by band structure tuning to some extent, leading to electrical conductivities of up to 10^5 Sm^{-1}. However, the maximization of the electrical conductivity and the minimization of the thermal conductivity can be reached by opposite measures only, therefore, only optimum values can be adjusted. Consequently, the best FOMs achieved so far were despite a mere 0.08 at room temperature, 0.4 at 400 K, 1.35 at 900 K, and 1.63 at 1100 K for Eu$_8$[Ga$_{16-x}$Ge$_{30+x}$], $15 < x < 0.16$ (Saramat *et al.*, 2006). Therefore, this material is better suited for converting heat into electrical energy instead for cooling applications around ambient temperature.

Heusler phases, which have already been discussed in great detail in Subsection 7.3.2, are another class of thermoelectric materials, in particular for applications at elevated temperatures up to 1500 K. Examples with a $ZT \approx 0.8$ at 1073 K are compounds of the type $cF12$-MNiSn (M . . . Ti, Zr, Hf), which are doped at the Sn site by Sb (Culp *et al.*, 2006). The empty sites in one half of the eighth-cubes of the unit cell give rise to narrow bands, resulting in d-orbital hybridization and a semiconducting character.

10.6 Thermo- and magnetomechanical materials: shape memory alloys

Thermomechanical materials (shape memory alloys, SMAs) are based on martensitic (displacive) phase transformations between a HT-austenitic phase and a LT-martensitic one. Cooling the austenite with a particular shape below the phase transformation temperature leads to a twinned martensite. It can then be deformed arbitrarily whereby it becomes partially detwinned. Heating the so-deformed martensite results in an austenitic phase, again in the original shape. Deformations of up to 10% can be achieved by mechanical stress in the martensitic state. Superelasticity (SE), which is a pseudoelasticity, is related to the shape memory effect (SME). A superelastic material in its martensitic state can recover high amounts of strain (up to $\approx 20\%$), the transformation into the austenitic phase can be triggered by mechanical stress.

Most commercially used SMAs, for instance for couplings, actuators, or smart materials, are either Ti–Ni-, Cu–Al-, or Cu–Zn-based. TiNi SMAs (Nitinol®) can be used at up to 373 K at most, while this limiting temperature is with 823 K much higher for Ti–Ni–Pd alloys (Otsuka and Ren, 1999). For some applications, the cheaper Cu-based SMAs are used. For applications at elevated temperatures (400 K), Cu–Zn–Al SMAs are the best choice. For a comprehensive review on materials for SMAs and their applications see, for instance, Ma *et al.* (2010).

In Fig. 10.4, the structural transformations of the SMA TiNi (one-way shape-memory effect, SME) as a function of temperature and stress are schematically shown. For the transition from the austenite $cP2$-TiNi (221 $Pm\bar{3}m$) to the

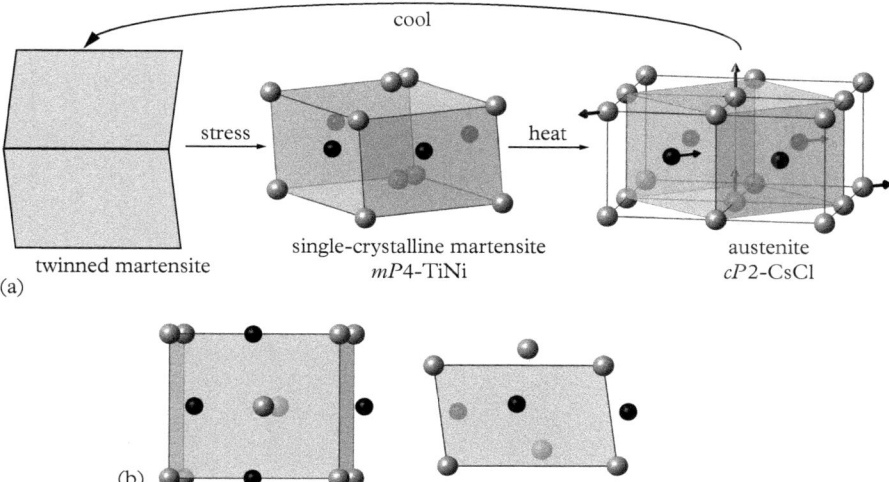

Fig. 10.4 *(a) Schematics of the structural transformations of the SMA TiNi (one-way SME). The parent HT-phase (austenite) has the cP2-CsCl type structure. Under cooling, it adopts the twinned mP4-TiNi type structure. By the application of stress, detwinning takes place. After heating, it transforms to single-crystalline austenite again. The arrows in the austenite structure indicate the shear directions for obtaining the monoclinic phase. In (b), projections along the unique b-axis and onto the (100) plane are shown. For clarity, twice the usual scale is applied.*

martensite $mP4$-TiNi (11 $P2_1/m$) by decreasing the temperature, two shears are necessary for the detwinned martensite: $\{110\}\langle1\bar{1}0\rangle$ and $\{001\}\langle1\bar{1}0\rangle$ (Otsuka and Ren, 1999). There are different twin laws known for this material, the most important one is $\langle011\rangle$ type-II twinning.

Apart non-magnetic materials, also magnetic ones show the shape-memory effect. Shape deformations of up to 5–10% can be reached on application of a magnetic field (Jiles, 2003). Magnetic SMAs can actuate at higher frequencies (up to ≈ 1 Hz) because the activation energy is provided by the magnetic field and not by the relatively slow heat transfer (Jani *et al.*, 2014). One example is the ferromagnetic Heusler phase $cF16$-Ni$_2$MnGa, $T_C = 376$ K, which undergoes the thermomechanical transformation to the martensitic state at $T_m = 202$ K. For a specific Mn$_{1-x}$/Ni$_{2+x}$ ratio of $x \approx 0.18$–0.20 these temperatures coincide, $T_m = T_C$. Such a material shows at the same time a structural transformation from the HT cubic to the LT tetragonal phase, and from the HT paramagnetic to the LT ferromagnetic state. Since in the tetragonal phase the ratio is $c/a = 0.94$, it is accompanied by a 6% deformation. The martensitic state consists of domains with three differently oriented tetragonal twin variants, each of them with a strong uniaxial magnetic anisotropy, with the easy axis along [001]. Application of a properly oriented magnetic field leads to the growth of the domains whose easy axis is aligned with the field, which is connected with large strains (Tickle and James, 1999).

10.7 Superconducting materials

Superconductivity of intermetallic phases strongly depends on the electronic band structure, electronic density of states, valence electron concentration, dimensionality, symmetry, and particular atomic interactions mediated by the crystal structure. Twenty nine (non-magnetic) metallic elements are superconducting at ambient pressures and temperatures up to 9.25 K for Nb, for instance. Most of them are transition metals from groups 3–9, the others mainly from groups 12–14, almost all of them with simple structures: *hcp*...13, *ccp*...6, and *bcc*...5 representatives. In case of the transition elements, the critical temperatures as a function of valence electron concentration peak at $N_e = 5$ and 7 (Poole Jr. *et al.*, 1995). Under pressure, 24 more elements (including non-metallic ones) become superconducting, and if they are already superconducting at ambient pressure, their critical temperature, T_c, can be increased after an initial decay by lattice stiffening (Fig. 10.5).

For Nb, the element with the highest critical temperature, T_c can be increased from 9.25 K to 9.9 K at 10 GPa, only. However, that of non-superconducting Ca jumps to remarkable 29 K at 216 GPa. The critical temperatures of Sc and Y are also close to 20 K for pressures beyond 100 GPa. A prerequisite for simple free-electron metals such as the alkali and alkaline earth elements to become superconducting are the structural changes with increasing pressure, which are connected with an s-d electron transfer. The high density of electronic states at the Fermi edge associated with the d electrons can stabilize superconductivity (Hamlin, 2015). This effect also applies to transition elements

Legend: Mendeleev number | 12 Li (element) ; maximum T_c at HP | 14(<1) K ; () T_c at ambient pressure ; required pressure | 30 GPa

1	2	3	4	5	6	7	8	9	10	11	12	13	14	15	16
12 Li 14(<1)K 30 GPa	77 Be (<1) K														
11 Na	73 Mg											80 Al (1.1) K			
10 K	16 Ca 29(*) K 216 GPa	19 Sc 19.6(*)K 106 GPa	51 Ti 3.4(<1) K 56 GPa	54 V 16.5(5.4)K 120 GPa	57 Cr	60 Mn	61 Fe 2.1 (*) K	64 Co	67 Ni	72 Cu	76 Zn (<1) K	81 Ga 7(1.1) K	84 Ge 5.4(*) K 11.5 GPa		
9 Rb	15 Sr 7(*) K 50 GPa	25 Y 19.5(*) K 115 GPa	49 Zr 11(<1) K 30 GPa	53 Nb 9.9(9.2) K 10 GPa	56 Mo (<1) K	59 Tc (7.8) K	62 Ru (<1) K	65 Rh (<1) K	69 Pd	71 Ag	75 Cd (<1) K	79 In (3.4) K	83 Sn 5.3(3.7) K 11.3 GPa	88 Sb 3.9(*) K 25 GPa	
8 Cs 1.3(*) K 12 GPa	14 Ba 5(*) K 18 GPa	33 La 13(6) K 15 GPa	50 Hf 8.6(<1) K 62 GPa	52 Ta 4.5(4.5) K 43 GPa	55 W (<1) K	58 Re (1.4) K	63 Os (<1) K	66 Ir (<1) K	68 Pt	70 Au	74 Hg (4.2) K	78 Tl (2.4) K	82 Pb (7.2) K	87 Bi 8.5(*) K 9.1 GPa	91 Po
7 Fr	13 Ra	48 Ac													

* Lanthanoids	32 Ce 1.7(*) K 5 GPa	31 Pr	30 Nd	29 Pm	28 Sm	18 Eu 2.8(*) K 142 GPa	27 Gd	26 Tb	25 Dy	24 Ho	23 Er	22 Tm	17 Yb	20 Lu 12.4(*)K 174 GPa
+ Actinoids	47 Th (1.4) K	46 Pa (1.4) K	45 U 2.4(*) K 1.2 GPa	44 Np	43 Pu	42 Am 2.2(<1) K 6 GPa	41 Cm	40 Bk	39 Cf	38 Es	37 Fm	36 Md	35 No	34 Lr

Fig. 10.5 *Critical temperatures, T_c, of elements, which are either already superconducting or become superconducting under high pressure, HP (based on Hamlin (2015)). (*) means that the element does not show ambient pressure superconductivity.*

and lanthanoids. In the case of further phase transitions with increasing pressure, the superconducting state can disappear again. For instance, in Li the onset of superconductivity is at about 25 GPa for $cF4$-Li, and it disappears with the transition to semiconducting $oC40$-Li at around 80 GPa.

In alloys, i.e., solid solutions of two or more elements, the critical temperatures can be lower (as in Nb–V) or, despite the disordered structure, higher (as in Nb–Zr, Ti–V, or Ti–Zr) than in their constituents. For instance, for Ti one has $T_c = 0.40$ K, for V $T_c = 5.13$ K and for a solid solution with composition TiV$_{0.89}$ $T_c = 6.86$ K (Hulm and Blaugher, 1961). Superconductivity has also been discovered in the HEA $cI2$-Ta$_{0.34}$Nb$_{0.33}$Hf$_8$Zr$_{0.14}$Ti$_{0.11}$, a type-II superconductor with $T_c \approx 7.3$ K (Koželj *et al.*, 2014).

The valence electron concentration plays an important role in this case. This is also true for amorphous alloys of transition metals, where T_c peaks at $6 < N_e < 7$. In the case of superconducting intermetallic compounds, the highest transition temperatures were found for representatives of the $cP8$-Cr$_3$Si (A15) structure type (221 $Pm\bar{3}n$, $2a\,6d$) (Fig. 10.6), for instance, $T_c = 23.2$ K in the case of sputtered films of $cP8$-Nb$_3$Ge. Although many non-intermetallic materials have been discovered with higher transition temperatures, in particular doped $cP8$-Nb$_3$Sn with $T_c = 17.9$ K is still the material of choice for superconducting magnets, i.e., everywhere where high critical currents are requested. For a recent review see Stewart (2015).

Another class of superconducting intermetallics are Laves phases with T_c up to ≈ 10 K, such as $cF24$-Zr$_{0.5}$Hf$_{0.5}$V$_2$ with $T_c = 10.1$ K. This compound shows a high critical field and is rather resistant to neutron irradiation and mechanical strain (Brown *et al.*, 1977). These properties make it a potentially interesting material for fusion reactor magnets.

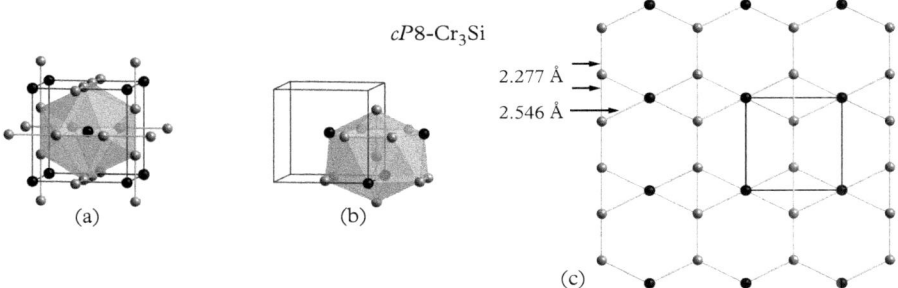

cP8-Cr$_3$Si

2.277 Å
2.546 Å

(a) (b) (c)

Fig. 10.6 *The structure of cP8-Cr$_3$Si in different representations: (a) one unit cell with the icosahedral AET of Cr atoms, Si@Cr$_{12}$, around the central Si atom (black) highlighted. The icosahedral Cr AETs are sharing 6 out of their 30 edges. The Cr atoms (gray) form 1D chains running along the main directions, and are responsible for superconductivity via electron–phonon coupling. In (b) the AET, a CN14 Frank-Kasper polyhedron around a Cr atom, Cr@Cr$_{10}$Si$_4$, is shown. A section of the structure at z = 0, corresponding to a two-uniform hexagon triangle tiling, $3^2.6^2$ is depicted in (c). Typical Cr–Cr and Cr–Si distances are marked.*

Conventional intermetallic superconductors all show inversion symmetry. It is well known that superconductivity is not restricted to compounds with centrosymmetric structures, where the superconducting state can be described based on even-parity spin-singlet Cooper pairs. In case of lacking inversion symmetry, the electrons are exposed to an electrical field gradient leading to antisymmetric spin-orbit coupling. Examples of such non-centrosymmetric superconducting intermetallics are: the heavy-Fermion compounds $tP5$-CeTM$_3$Si (TM = Rh, Ir, Pt) ($tP5$-CePt$_3$B structure type) (Bauer *et al.*, 2004), $mP16$-UIr (Akazawa *et al.*, 2004), and $cI22$-Ca$_3$Ir$_4$Ge$_4$ (von Rohr *et al.*, 2014), as well as the "normal" electronic intermetallic $tI10$-BaPtSi$_3$ ($tI10$-BaNiSn$_3$ type) (Bauer *et al.*, 2009). For a comprehensive discussion of all aspects of these compounds see Bauer and Sigrist (2012).

Another class of superconducting materials, most of which are of rather academic interest due to their radioactivity, are the heavy-fermion f-electron superconductors (for reviews see Pfleiderer (2009) and Griveau and Colineau (2014), for instance). One remarkable property is that, in contrast to conventional superconductors, in some of them the superconducting state co-exists with antiferromagnetism ($tP7$-CeCoIn$_5$, $hP6$-UPd$_2$Al$_3$, etc.), and in some with ferromagnetism ($oS12$-UGe$_2$, $oP12$-UCoGe, etc.). The by far highest f-electron superconducting transition temperature, $T_c = 18.5$ K, is achieved with the compound $tP7$-PuCoGa$_5$, followed by $tP7$-PuRhGa$_5$ ($T_c = 8.7$ K) and $tI16$-NpPd$_5$Al$_2$ ($T_c = 4.9$ K) (Fig. 10.7). Both $tP7$-PuCoGa$_5$ and $tP7$-PuRhGa$_5$ are of the rather frequent $tP7$-HoCoGa$_5$ structure type (123 $P4/mmm$,

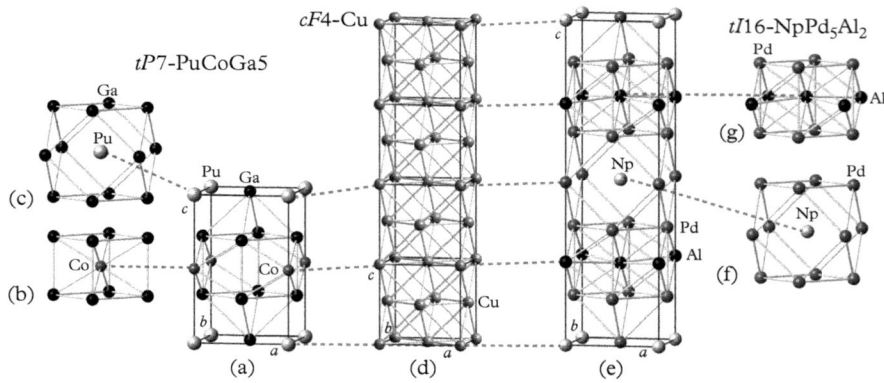

Fig. 10.7 *The structures of the heavy Fermion superconductors (a) $tP7$-PuCoGa$_5$ and (e) $tI16$-NpPd$_5$Al$_2$ in comparison with four unit cells of (d) cF4-Cu. (c) $tP7$-PuCoGa$_5$, and (g) $tI16$-NpPd$_5$Al$_2$ are two- and fourfold superstructures of cF4-Cu. The coordination polyhedra of (c) Pu and (f) Np are both cuboctahedra, Pu@Ga$_{12}$ and Np@Pd$_{12}$, respectively, as well as that of Al, Al@Al$_4$Pd$_8$, which is strongly compressed along [001]. The AET of (b) Co is a tetragonal prism, Co@Ga$_8$.*

1*abc* 4*i*, 82 representatives). *tI*16-NpPd$_5$Al$_2$ (139 *I*4/*mmm*, 2*ab* 4*e* 8*g*) has the *tI*16-ZrNi$_2$Al$_5$ structure type.

In *tI*16-NpPd$_5$Al$_2$, the 4d electrons of Pd are hybridized with the 5f electrons of Np leading to strongly correlated conduction electrons in the Np–Pd layers. The Fermi surface has a quasi-two-dimensional shape similar as it is the case for *tP*7-PuCoGa$_5$, together with the heavy Fermion state, enhancing the heavy Fermion superconductivity (Ōnuki *et al.*, 2014).

10.8 Highly-correlated electron systems

Intermetallic phases featuring highly-correlated electron systems can have technologically interesting magnetic and transport properties for applications in spintronics. The strongest correlations can be found in heavy-fermion materials. At low temperatures, they show enhanced interactions between f electrons and conduction electrons leading to an effective mass increase of the electrons by up to two orders of magnitude. This results, for instance, in a very large electronic specific heat as indicated by the Sommerfeld coefficient $\gamma \approx 8000$ mJmol^{-1}K^{-2} for the half-Heusler compound *cF*12-YbBiPt (*cF*12-MgAgAs structure type, also known as *cF*12-LiAlSi or *cF*12-MgCuSb structure type, 216 *F*$\bar{4}$3*m*), for instance. A large number of heavy-electron systems are listed with their structure type and physical properties in a review by Thomas *et al.* (2006*b*).

Other common structure types of these materials are *tI*10-ThCr$_2$Si$_2$ (CeCu$_2$Si$_2$, YbCu$_2$Si$_2$, URu$_2$Si$_2$,...), *tP*10-CaBe$_2$Ge$_2$ (CeCu$_2$Sb$_2$, CeIr$_2$Sn$_2$, UIr$_2$Si$_2$,...), *hP*8-Ni$_3$Sn (CeAl$_3$, UPt$_3$, ...), *hP*9-ZrNiAl (YbNiAl, YbPtIn, ...) (Fig. 10.8 (a)–(e)), *oP*12-TiNiSi (YbPtAl, YbNiSn, ...) (Fig. 10.8 (f)–(i)), *cP*4-AuCu$_3$ (CeIn$_3$, USn$_3$, ...), *cF*24-AuBe$_5$ (YbAgCu$_4$, UPdCu$_4$, ...), *tP*10-U$_3$Si$_2$ (U$_2$Ni$_2$In, U$_2$Pd$_2$Sn, ...), *cF*112-NaZn$_{13}$ (UBe$_{13}$, NpBe$_{13}$, ...) (Fig. 10.8 (m)–(o)), and *cF*184-CeCr$_2$Al$_{20}$ (YbFe$_2$Zn$_{20}$, ...), to name just a few.

The structure of *tP*10-U$_3$Si$_2$ (127 *P*4/*mbm*; U in 2*a* 0, 0, 0 and 4*h* 0.181, 0.681, 1/2; Si in 4*g* 0.389, 0.889, 0) is compared with the related structure *tI*12-CuAl$_2$ (140 *I*4/*mcm*; Cu in 4*a* 0, 0, 1/4; Al in 8*h* 0.1541,0.6541,0) in Fig. 7.26(a)–(d). Both structures are based on stackings of atomic layers corresponding to decorated 3^2.4.3.4 nets.

*oP*12-TiNiSi (62 *Pnma*; Ti in 4*c* 0.021, 0.180, 1/4; Ni in 4*c* 0.142, 0.561, 1/4; Si in 4*c* 0.765, 0.623,1/4) and its aristotype *oP*12-CeCu$_2$ (74 *Imma*; Ce in 4*e* 0, 1/4, 0.538; Cu in 8*h* 0, 0.051, 0.165) can be described based on 3D 4-connected polyanionic Ni–Si nets with the Ti cations in the large channels (Landrum *et al.*, 1998) (Fig. 10.8(f)–(i)).

The structure type *tP*10-ThCr$_2$Si$_2$ can be seen as and ordered variant of the *tP*10-BaAl$_4$ type (139 *I*4/*mmm*; Ba in 2*a* 0, 0, 0; Al in 4*d* 0, 1/2, 1/4 and 4*e* 0, 0, 0.38) (Fig. 10.8(k)–(l)).

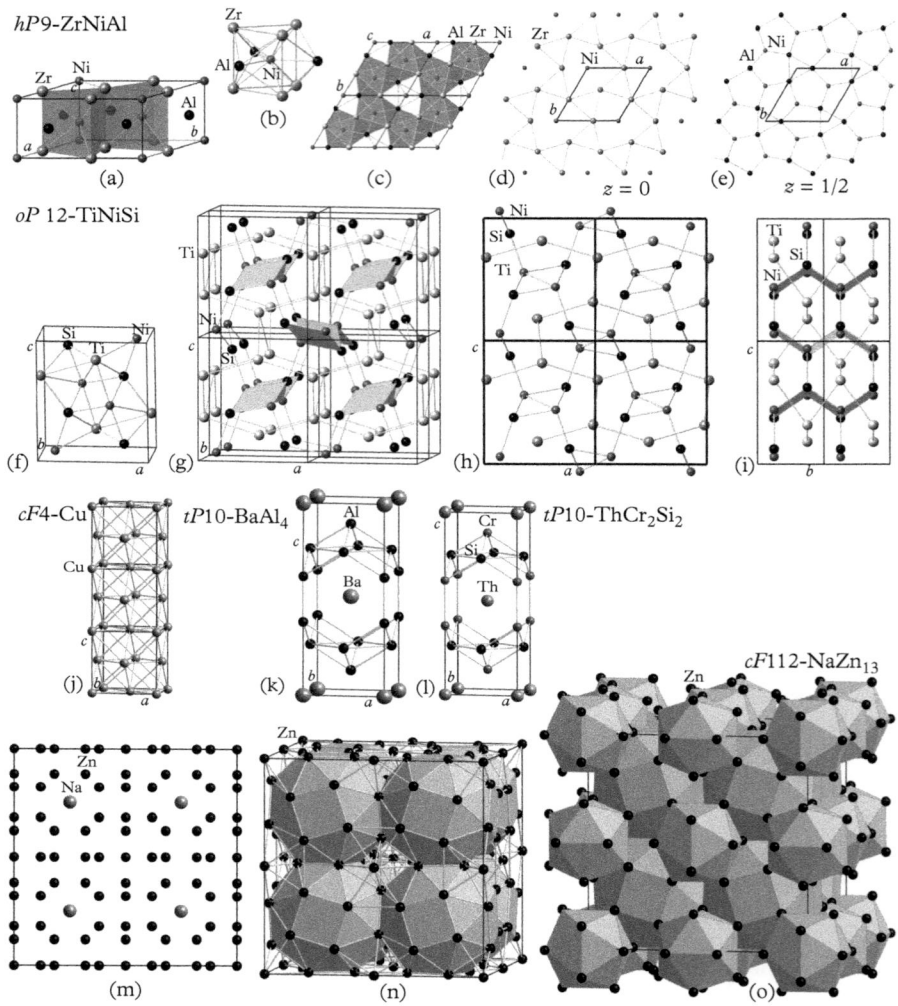

Fig. 10.8 *(a)–(e) Different views of the structure of hP9-ZrNiAl. One of the trigonal prismatic AETs around one kind of Ni sites shaded gray in (a) is shown in (b) together with the capping Al atoms. These structural units are highlighted in (c) as well. In (d) and (e), sections at z = 0 and z = 1/2 are depicted. (f)–(i) The structures of oP12-TiNiSi: (f) one unit cell, (g)-(i) 2 × 2 × 2 unit cells in different projections (Ti . . . light gray, Ni . . . gray, Si . . . black). In (j)–(m) the relationships between the structures of (j) cF4-Cu (three unit cells along [001]), (k) tP10-BaAl₄, and (l) tP10-ThCr₂Si₂ are shown. (m)–(o) Different representations of the cF112-NaZn₁₃ structure type: (m) projection along [100], (n) perspective view with the clusters around the Na atoms (gray), both enantiomorphs of the snub cube 3⁴.4, shaded. In (o) in addition to the snub cubes, the icosahdral AETs around the Zn atoms at the Wyckoff position 8b 0, 0, 0 are shown. All these polyhedra are slightly distorted.*

The structure of $hP9$-ZrNiAl (Fig. 10.8 (a)-(e)) can be seen as an ordered derivative of the $hP9$-Fe$_2$ type (P $\rightarrow$ Ni; Fe at $3f$ $x, 0, 0 \rightarrow$ Zr; Fe at $3g$ $x, 0, 1/2 \rightarrow$ Al). The atomic layer in $z = 0$ corresponds to a Zr-decorated shield/triangle tiling, with the shields centered by Ni atoms, while the one in $z = 1/2$ represents a pentagon/triangle tiling, with the 3-connected triangle vertices decorated by Zr, and the 3-connected pentagon vertices by Ni atoms. The shields and triangles constitute, together with squares, dodecagonal tilings, the pentagon-bands are structure motifs also abundant in decagonal tilings.

The structures of $cF4$-Cu, $tP10$-BaAl$_4$, and $tP10$-ThCr$_2$Si$_2$ are compared in Fig. 10.8 (j)-(l). Substituting the corner atoms of the tripled unit cell of $cF4$-Cu in (j) by Ba atoms (light gray) and the central octahedron by one further Ba atom, and letting the structure relax, then the $tP10$-BaAl$_4$ structure type results (k). The ternary derivative compound $tP10$-ThCr$_2$Si$_2$ (l) can be obtained by replacing Ba by Th (light gray), Al by Cr (gray), and Si (black) in an ordered way.

The $cF112$-NaZn$_{13}$ structure type (226 $Fm\bar{3}c$, Na in $8a$ 1/4, 1/, 4, 1/4; Zn in $8b$ 0, 0, 0, and $96i$ 0, 0.811, 0.119) (Fig. 10.8 (m)–(o)) can be described as a non-space-filling packing of square-face-sharing Na-centered NaZn$_{24}$ snub cubes and Zn-centered Zn$_{12}$ icosahedra, sharing triangle faces with the snub cubes, in between. This structure type is frequently found in hard-sphere self-assembled colloidal systems with size ratios around 0.49–0.63. For a size ratio of 0.58, a very high packing density of 0.748 results, and for a ternary system A$_{12}$BC, with a smaller diameter for the icosahedrally coordinated C, an even higher density of 0.771 can be obtained (Hudson, 2010).

Abbreviations and glossary

An Actinoids

AEM Alkaline earth metals

AET Atomic environment type

AM Alkali metals

AS Atomic surface

ASM American Society for Metals

AT Ammann tiling

ccp Cubic close-packed

CIM Complex intermetallic

CN Coordination number

CNn Coordination number n

CS Composite or host/guest structure

CSD Cambridge Structural Database

CSP Crystal structure prediction

DFT Density functional theory

DFT-CP Density-functional-theory chemical-pressure analysis

dhcp Double hexagonal closest packed

DOS Density of states

DQC Decagonal quasicrystal

e/a electron concentration

eDOS Electronic density of states

ELI Electron localizability indicator

ELF Electron localization function

F_v^f Fullerene with f faces and v vertices

FK Frank-Kasper

FK_f^v Frank-Kasper polyhedron with f faces and v vertices

FS Fibonacci sequence

GMCE Giant magnetocaloric effect

hbp Hexagonal bipyramid

hcp Hexagonal close-packed

HEA High-entropy alloy

HP High pressure

HT High temperature

IM Intermetallic

IMS Incommensurately modulated structure

IQC Icosahedral quasicrystal

LDA Local-density approximation

Ln Lanthanoids

LP Low pressure

LT Low temperature

M Mendeleev number

M Main group element

MBCE Magnetic barocaloric effect

MCE Magnetocaloric effect

M/M Mendeleev number vs. Mendeleev number

MO Molecular orbital

MOKE Magnetooptic Kerr effect

n**D** n-dimensional

OD Occupation domain

PAS Periodic average structure

PCD Pearson's Crystal Data

PLI Penrose local isomorphism

PPT Pentagon-Penrose tiling

PT Penrose tiling

QC Quasicrystal

QTAIM Quantum theory of atoms in molecules

RE Rare earth element

REME Equiatomic phase with RE a rare earth metal (in most cases), an actinoid or a group 1–4 element, M is a late transition metal from groups 8–12, and E is an element from groups 13–15

RPT Rhomb-Penrose tiling

RT Room temperature

SMA Shape-memory alloy

ST Structure type

TB-LMTO-ASA Tight-binding linear muffin-tin orbital atomic sphere approximation *tcp* Tetrahedrally / topologically close-packed

TM Transition metal element

VB Valence bond

VEC Valence electron concentration

Allotropes Different modifications of the chemical elements existing in a particular temperature and pressure range.

Alloy A phase-separated mixture of two or more (inter)metallic phases. Single-phase mixtures of the elements are denoted → *solid solutions*. See also → *high-entropy alloys (HEA)*.

Aperiodic crystal The signature of an ideal crystal is its pure-point Fourier spectrum. This means that its diffraction pattern shows sharp Bragg reflections only. We distinguish between periodic crystals and aperiodic crystals, where the structures of the former show translational periodicity in three dimensional (3D) space, and those of the latter only in the nD description, with $n > 3$. We distinguish between → *incommensurately modulated structure (IMS)*, → *host/guest or composite structure (CS)*, and → *quasiperiodic structure (QS)*.

Aristotype High-symmetry basic structure that can be seen as the idealized version of one or more different lower-symmetry derivative structures, the → *hettotypes*. For instance, the *cP*2-CsCl structure can be seen as a hettotype and that of *cI*2-W as its aristotype.

Atomic environment type (AET) The kind of coordination of an atom by its neighboring atoms (first coordination shell or polyhedron). The coordination number, → *CN*, corresponds to the number of atoms in the first coordination shell.

Atomic surface Volume in the $(N - d)$D perpendicular space defining an → *aperiodic crystal* structure, which results from a cut with the dD physical space.

Cluster We mean by cluster a polyhedral arrangement of atoms in several coordination (cluster) shells. In the case of complex intermetallic compounds and, in particular → *quasicrystals*, typical recurrent structure motifs (structural subunits) are frequently called clusters. It can be, but it is not necessarily the case that the atoms constituting these clusters differ from the surrounding structure in their chemical composition, bonding, or physical properties.

Composite or host/guest structure → *Aperiodic crystal* consisting of at least two substructures, which are incommensurate to each other. Its signature is a pure-point Fourier spectrum with at least two subsets of main reflections, sometimes accompanied by satellite reflections.

Cundy and Rollet symbol The vertex configuration of a vertex-transitive tiling can be described by n^m, meaning that m n-gons meet at a vertex. For instance, 6^3 means that 3 hexagons meet at a vertex. Its face configuration can be writen in the form $V.m_1.m_2. . . m_n$, meaning that each n-gonal unit tile has n vertices where $m_1, m_2, . . . , m_n$ n-gons meet at a vertex.

Derivative structure New structure obtained by specific modification of a basic (parent) structure (→ *aristotype*).

Frank-Kasper (FK) polyhedra Triangulated coordination polyhedra with coordination numbers 12, 14, 15, and 16 (CN12, CN14, CN15, and CN16) respectively, which allow → *topological close packing*. They show only 5- and 6-connected vertices. Larger FK polyhedra can occur, obeying the same rules with respect to their polyhedral faces and vertices.

Hettotype → *aristotype*

High-entropy alloys (HEAs) HEAs are → *solid solutions* of four or more multiprincipal elements crystallizing in simple crystal structures. The high mixing entropy of such solid solutions prevents the formation of intermetallic compounds at high-enough temperatures.

Homeotypic Two structures are homeotypic if one or more of the following conditions required for → *isotypism* are relaxed. For instance, $cI2$-W and $cP2$-NiAl are homeotypic according to (i) and (iii).

   (i)   Identical or enatiomorphic space-group types, allowing for group/subgroup or group/ supergroup relationships.

   (ii) Limitations imposed on the similarity of geometric properties, i.e., axial ratios, interaxial angles, values of adjustable positional parameters, and the coordination of corresponding atoms.

   (iii) Site occupancy limits, allowing given sites to be occupied by different atomic species.

Incommensurately modulated phase → *Aperiodic crystal* with incommensurately modulated structure (IMS). The displacive and/or substitutional modulation wave is incommensurate to the period of the underlying basic structure. The signature of an IMS is a pure-point Fourier spectrum with clearly distinguishable sets of main and satellite reflections.

Intermetallics Intermetallic compounds are chemical compounds between metallic elements with well-defined stoichiometry ("line compounds"), while intermetallic phases are the generic term for phases consisting of two or more metallic elements, which may have a narrow or extended compositional stability range. The term "intermetallics" may be used as short form for either of them.

Isoconfigurational Two structures are configurationally → *isotypic* if they are → *isopointal* and both the crystallographic point configurations (crystallographic orbits) and their geometrical interrelationships are similar; all geometrical properties, such as axial ratios, angles between crystallographic axes, values of corresponding adjustable positional parameters (x, y, z), and coordinations of corresponding atoms (AETs) are similar. Isoconfigurational structures belong to the same structure type. For instance, the isoconfigurational metallic phase

AlNi and the ionic compound CsCl are both representatives of the *cP*2-CsCl structure type, although they strongly differ in their atomic interactions.

Isopointal Two structures are isopointal if they have the same space-group type, and the atomic (Wyckoff) positions, occupied either fully or partially at random, are the same in both structures; as there are no limitations on the values of the adjustable parameters of the Wyckoff positions or on the cell parameters, isopointal structures may have locally different geometric arrangements and atomic coordinations (AETs) and may belong to different structure types.

Isotypic Two structures are crystal-chemically isotypic if they are → *isoconfigurational* and the corresponding atoms and bonds (interactions) have similar physical/chemical characteristics. For instance, *cP*2-CoAl and *cP*2-NiAl are isotypic.

Lattice complex A lattice complex is the set of all point configurations that may be generated within one type of Wyckoff set of a space group type. All Wyckoff positions, Wyckoff sets, and types of Wyckoff sets that generate the same set of point configurations are assigned to the same lattice complex (for details see Fischer and Koch (2002)).

Lattice planes and directions Miller indices in parentheses, (*hkl*), are used to denote lattice planes, and in braces, {*hkl*}, to denote sets of symmetrically equivalent lattice planes. Vector components in brackets, [*uvw*], indicate lattice directions.

Mendeleev numbers These numbers are related to a semiempirical chemical scale χ assigned to the chemical elements based on elemental properties such as electronegativity, etc. The Mendeleev numbers start with the least electronegative element, He, and end with one of the most electronegative ones, H. These numbers have been proven useful for structure maps, where they give an excellent separation of binary structures crystallizing in different structure types, for instance.

Metallic elements Those chemical elements that exhibit a metallic character at ambient temperature and pressure. Some elements at the boundary metallic/semimetallic/semiconducting can be included as well because some of their compounds show metallic character (see Fig. 6.1 in Chapter 6).

Occupation domain → *atomic surface.*

Pearson notation Shorthand characterization of a periodic compound or phase by a lower-case italic letter denoting the crystal family, an uppercase italic letter indicating the Bravais lattice type, and a number giving the number of atoms per unit cell. In the case of rhombohedral structures, the number *n* of atoms in the hexagonal setting ($a = b \neq c, \alpha = \beta = 90°$, and $\gamma = 120°$) is three times that in the rhombohedral unit cell ($a = b = c, \alpha = \beta = \gamma$). For instance, *hR*9-Sm means that there are three atoms in the rhombohedral unit cell and nine atoms in the hexagonal one. Unfortunately, in the literature some confusion persists, and the number n of atoms in the hexagonal setting is replaced by that in the rhombohedral one. See also Table 1.1.

Polymorphic An element or compound is polymorphic if it occurs in several structural modifications as a function of ambient conditions (temperature, pressure, external electric, magnetic, elastic fields, etc.). The different modifications are frequently called polymorphs and in the case of the chemical elements → *allotropes*.

Polytypic An element or compound is polytypic if it occurs in several structural modifications, each of which can be regarded as built up by stacking layers of (nearly) identical structure and composition, and if the modifications differ only in their stacking sequence. Polytypism is a special variant of → *polymorphism*. For instance, the structure types $hP2$-Mg, $cF4$-Cu, $hP4$-La, and $hR9$-Sm, are all different stacking variants of close-packed layers; their structures are polytypic variants of dense sphere packings in general.

Prototype Crystal structure of an element or compound used for the definition of a structure type, which represents the class of all materials with → *isotypic* structures.

Quasicrystal (QC) → *Aperiodic crystal* with quasiperiodic structure (QS), such as decagonal and icosahedral phases. Its signature is a pure-point Fourier spectrum with, in most cases, non-crystallographic point symmetry such as 5-, 8-, 10-, 12-fold, or icosahedral.

Repetivity Any bounded patch of a PT (a Gummelt decagon, for instance) can be found again in the PT within a distance of less than two diameters of that patch.

Schläfli symbol Notation in the form $\{p, q, r, \ldots\}$ for the characterization of regular polygons, vertex-transitive polyhedra, and polytopes, by describing the number of edges of each polygon meeting at the vertex of a tiling or solid. A Platonic solid is denoted by $\{p, q\}$, where p gives the number of edges each face has, and $\{q\}$ the number of faces meeting at each vertex. Its dual is designated by the reversed symbol $\{q, p\}$. The symbol $\{p\}$ characterizes a regular polygon with p edges for integer p, and a star polygon for rational p. For instance, a convex regular polygon such as a pentagon is given by $\{5\}$, and a nonconvex star polygon such as a pentagram is given by $\{5/2\}$. The rational value p/m describes a 2D polygon with p vertices where every m-th vertex is connected giving a p-gram. m is also the number of different polygons in a p-gram. In a generalized form, this notation can also be used for denoting face-transitive polyhedra by sequentially counting the number of faces meeting at each vertex around a face. For instance, $V(3.5)^2$ denotes a triacontahedron, where at the four vertices around each rhomb-face 3, 5, 3, 5 edges, respectively, meet.

Self-similarity A symmetry operation maps the vertices of a PT onto the vertices of a copy of the PT scaled by powers of τ^{-1}.

Solid solution Statistical mixture of two or more elements on a common underlying lattice. → *HEAs* are a special kind of solid solution, i.e., usually of five or more elements in more or less equal amounts.

Sphere packings Infinite sets of hard spheres with the property that any pair of spheres is connected by a chain of spheres with mutual contact. It is called homogenous if all spheres are symmetrically equivalent, otherwise heterogenous. Close(st) or dense packings have the maximum possible packing density of equal spheres. It amounts to $\pi/\sqrt{18} = 0.74048$ in the case of the cubic or hexagonal close packings (and their polytypes).

Structure type Prototypic crystal structure representing a whole class of → *iso-configurational* structures, i.e., structures with the same space group symmetry, occupied Wyckoff positions, and AETs.

Superspace group Space group describing the symmetry of an aperiodic crystal structure in nD superspace ($n > 3$).

Tiling A tesselation of the plane/space where the countable set of unit tiles (copies of prototiles) fills the plane/space without gaps or overlaps.

Topologically close packing (tcp) While *ccp* and *hcp* structures have both tetrahedral and octahedral interstices, topologically close packed (*tcp*) structures have tetrahedral voids only. These face-sharing tetrahedra are distorted (regular tetrahedra are not space filling). Typical examples of *tcp* structures are the Frank-Kasper (FK) phases.

Vertex configuration → *Cundy and Rollet symbol.*

Voronoi polyhedron Domain (Wigner-Seitz cell equivalent) containing all points closer to its central atom than to any other. It corresponds to the convex polyhedron resulting in the intersections of the planes bisecting the lines joining the central atom with its neighboring ones. The set of neighbors defining the Voronoi cell forms the coordination polyhedron (shell) or → *atomic environment type (AET)*.

Wyckoff position A Wyckoff position of a space group G consists of all points X for which the site-symmetry groups are conjugate subgroups of G. Each Wyckoff positon of a space group is labelled by a letter (Wyckoff letter).

$\mathbb{Z}$-module A vector module of rank n is an infinite set of vectors resulting from all linear combinations of its n basis vectors. It is called $\mathbb{Z}$-module in order to emphasize that all coefficients are elements of the set of integers $\mathbb{Z}$.

References

Abraham, N. L. and Probert, M. I. J. (2006). A periodic genetic algorithm with real-space representation for crystal structure and polymorph prediction. *Phys. Rev. B*, **73**, 224104.

Abraham, N. L. and Probert, M. I. J. (2008). Improved real-space genetic algorithm for crystal structure and polymorph prediction. *Phys. Rev. B*, **77**, 134117.

Ahuja, A. K., Dubrovinsky, L., Dubrovinskaia, N., Osorio Guillen, J. M., Mattesini, M., Johansson, B., and Le Bihan, T. (2004). Titanium metal at high pressure: Synchrotron experiments and ab initio calculations. *Phys. Rev. B*, **69**, 184102.

Akahama, Y., Fujihisa, H., and Kawamura, H. (2005*a*). New helical chain structure for scandium at 240 GPa. *Phys. Rev. Lett.*, **94**, 195503.

Akahama, Y., Fujihisa, H., and Kawamura, H. (2005*b*). New (distorted-bcc) titanium to 220 GPa. *Phys. Rev. Lett.*, **87**, 275503.

Akahama, Y., Nishimura, M., Kinoshita, K., Kawamura, H., and Ohisi, Y. (2006). Evidence of a fcc–hcp transition in aluminum at multimegabar pressure. *Phys. Rev. Lett.*, **96**, 045505.

Akazawa, T., Hidaka, H., Fujiwara, T., Kobayashi, T. C., Yamamoto, E., Haga, Y., Settai, R., and Onuki, Y. (2004). Pressure-induced superconductivity in ferromagnetic UIr without inversion symmetry. *J. Phys.: Condens. Matter*, **16**, L29–L32.

Akella, J., Smith, G. S., and Jephcoat, A. P. (1988). High-pressure phase-transformation studies in gadolinium to 106 GPa. *J. Phys. Chem. Sol.*, **49**, 573–576.

Akella, J., Weir, S. T., Vohra, Y. K., Prokop, H., Catledge, S. A., and Chesnut, G. N. (1999). High pressure phase transformations in neodymium studied in a diamond anvil cell using diamond-coated rhenium gaskets. *J.Phys.: Cond. Matt.*, **11**, 6515–6519.

Allan, D. R., Nelmes, R. J., McMahon, M. I., Belmonte, S. A., and Bovornratanaraks, T. (1998). Structures and transitions in strontium. *Rev. High Press. Sci. Technol.*, **7**, 236–238.

Allred, A. L. and Rochow, E. G. (1958). A scale of electronegativity based on electrostatic force. *J. Inorg. Nucl. Chem.*, **5**, 264–268.

Alvarez, S. (2006). Nesting of fullerenes and Frank-Kasper polyhedra. *Dalton Trans.*, **17**, 2045–2051.

Amsler, M. and Goedecker, S. (2010). Crystal structure prediction using the minima hopping method. *J. Chem. Phys.*, **133**, 224104.

Andersson, S. (1980). An alternative description of the structure of Cu_4Cd_3. *Acta Crystallogr. B*, **36**, 2513–2516.

Andrusyak, R. I. and Kotur, B. Ya. (1991). Phase equilibriums in the Sc–Mn–Ge and Sc–Fe–Ge systems at 870 K. *Izv. AN SSSR Met.*, **4**, 198–202.

Annaorazov, M. P., Asatryan, K. A., Myalikgulyev, G., Nikitin, S. A., Tishin, A. M., and Tyurin, A. L. (1992). Alloys of the Fe–Rh system as a new class of working material for magnetic refrigerators. *Cryogenics*, **32**, 867–872.

Antonov, V. N., Oppeneer, P. M., Yaresko, A. N., Perlov, A. Ya., and Kraft, T. (1997). Computationally based explanation of the peculiar magneto-optical properties of PtMnSb and related ternary compounds. *Phys. Rev. B*, **56**, 13012–13025.

Armbrüster, M. and Lidin, S. (2000). Reassessing the compound $CeCd_6$: the structure of Ce_6Cd_{37}. *J. Alloys Compd.*, **307**, 141–148.

Armbrüster, M., Schnelle, W., Cardoso-Gil, R., and Grin, Y. (2010). Chemical bonding in compounds of the $CuAl_2$ family: $MnSn_2$, $FeSn_2$ and $CoSn_2$. *Chemistry*, **16**, 10357–10365.

Armbrüster, M., Schnelle, W., Schwarz, U., and Grin, Y. (2007). Chemical bonding in $TiSb_2$ and VSb_2: A quantum chemical and experimental study. *Inorg. Chem.*, **46**, 6319–6328.

Arnberg, L., Jonsson, A., and Westman, S. (1976). The structure of the δ-phase in the Cu–Sn system. A phase of γ-brass type with an 18 Å superstructure. *Acta Chem. Scand. A*, **30**, 187–192.

Asahi, R., Sato, H., Takeuchi, T., and Mizutani, U. (2005). Verification of Hume-Rothery electron concentration rule in Cu_5Zn_8 and Cu_9Al_4 gamma-brasses by ab initio FLAPW band calculations. *Phys. Rev. B*, **71**, 1–8.

Atahan-Evrenk, S. and Aspuru-Guzik, A. (ed.) (2014). *Prediction and Calculation of Crystal Structures: Methods and Applications*. Volume 345, Topics in Current Chemistry. Springer International Publishing. ISBN: 978-3-319-05773-6.

Bader, R. F. W. (1994). *Atoms in Molecules: A Quantum Theory*. Oxford University Press.

Bahmann, S. and Kortus, J. (2013). EVO–evolutionary algorithm for crystal structure prediction. *Comp. Phys. Comm.*, **184**, 1618–1625.

Balakrishnarajan, M. M. and Jemmis, E. D. (2000). Electronic requirements of polycondensed polyhedral boranes. *J. Am. Chem. Soc.*, **122**, 4516–4517.

Baranov, A. I. and Kohout, M. (2008). Electron localizability for hexagonal element structures. *J. Comp. Chem.*, **29**, 2161–2171.

Baranov, A. I. and Kohout, M. (2011). Electron localization and delocalization indices for solids. *J. Comp. Chem.*, **32**, 2064–2076.

Bardwell, D. A., Adjiman, C. S., Arnautova, Y. A., Bartashevich, E., Boerrigter, S. X. M., Braun, D. E., Cruz-Cabeza, A. J., Day, G. M., Della Valle, R. G., Desiraju, G. R., van Eijck, B. P., Facelli, J. C., Ferraro, M. B., Grillo, D., Habgood, M., Hofmann, D. W. M., Hofmann, F., Jose, K. V. J., Karamertzanis, P. G., Kazantsev, A. V., Kendrick, J., Kuleshova, L. N., Leusen, F. J. J., Maleev, A. V., Misquitta, A. J., Mohamed, S., Needs, R. J., Neumann, M. A., Nikylov, D., Orendt, A. M., Pal, R., Pantelides, C. C., Pickard, C. J., Price, L. S., Price, S. L., Scheraga, H. A., van de Streek, J., Thakur, T. S., Tiwari, S., Venuti, E., and Zhitkov, I. K. (2011). Towards crystal structure prediction of complex organic compounds – a report on the fifth blind test. *Acta Crystallogr. B*, **67**, 535–551.

Barkan, K., Diamant, H., and Lifshitz, R. (2011). Stability of quasicrystals composed of soft isotropic particles. *Phys. Rev. B*, **83**, 172201.

Bauer, E., Hilscher, G., Michor, H., Paul, C., Scheidt, E. W., Gribanov, A., Seropegin, Y., Noel, H., Sigrist, M., and Rogl, P. (2004). Heavy fermion superconductivity and magnetic order in noncentrosymmetric $CePt_3Si$. *Phys. Rev. Lett.*, **92**, 027003.

Bauer, E., Khan, R. T., Michor, H., Royanian, E., Grytsiv, A., Melnychenko-Koblyuk, N., Rogl, P., Reith, D., and Podloucky, R. (2009). $BaPtSi_3$: A noncentrosymmetric BCS-like superconductor. *Phys. Rev. B*, **80**, 064504.

Bauer, E. and Sigrist, M. (ed.) (2012). *Non-centrosymmetric superconductors*. Volume 847, Lecture Notes in Physics. Springer, Berlin Heidelberg.

Becke, A. D. and Edgecombe, K. E. (1990). A simple measure of electron localization in atomic and molecular systems. *J. Chem. Phys.*, **92**, 5397–5403.

Bei, H. (2013). Multi-component solid solution alloys having high mixing entropy. Patent US2013/0108502A1, Ut-Batelle, Llc.

Beister, H.-J., Syassen, K., Deiseroth, H.-J., and Toelstede, D. (1993). Pressure-induced phase transition of the alkali metal amalgam KHg_2. *Z. Naturforsch.*, **48b**, 11–14.

Belin, C. (1983). Intermetallic phases of gallium and alkali metals: Synthesis and structure of the nonstoichiometric phase $K_3Li_9Ga_{28.83}$. *J. Solid State Chem.*, **50**, 225–234.

Belin, C. and Charbonnel, M. (1986). A new intermetallic phase $K_4Na_{13}Ga_{49.57}$: Synthesis and X-ray crystal structure. *J. Solid State Chem.*, **64**, 57–66.

Belin, C. H. E. and Belin, R. C. H (2000). Synthesis and crystal structure determinations in the Γ and δ phase domains of the iron-zinc system: Electronic and bonding analysis of $Fe_{13}Zn_{39}$ and $FeZn_{10}$, a subtle deviation from the Hume-Rothery standard? *J. Solid State Chem.*, **151**, 85–95.

Belin, R., Tillard, M., and Monconduit, L. (2000). Redetermination of the iron-zinc phase $FeZn_{13}$. *Acta Crystallogr. C*, **56**, 267–268.

Bentien, A., Nishibori, E., Paschen, S., and Iversen, B. B. (2005). Crystal structures, atomic vibration, and disorder of the type-I thermoelectric clathrates $Ba_8Ga_{16}Si_{30}$, $Ba_8Ga_{16}Ge_{30}$, $Ba_8In_{16}Ge_{30}$, and $Sr_8Ga_{16}Ge_{30}$. *Phys. Rev. B*, **71**, 144107.

Berger, R. F., Lee, S., Johnson, J., Nebgen, B., Sha, F., and Xu, J. Q. (2008). The mystery of perpendicular fivefold axes and the fourth dimension in intermetallic structures. *Chem. Eur. J.*, **14**, 3908–3930.

Berger, R. F., Walters, P. L., Lee, S., and Hoffmann, R. (2011). Connecting the chemical and physical viewpoints of what determines structure: From 1-d chains to γ-brasses. *Chem. Rev.*, **111**, 4522–4545.

Berliner, R. and Werner, S. A. (1986). The structure of the low-temperature phase of lithium metal. *Physica B*, **136**, 481–484.

Berns, V. M, Engelkemier, J., Guo, Y., Kilduff, B. J., and Fredrickson, D. C. (2014). Progress in visualizing atomic size effects with DFT-chemical pressure analysis: From isolated atoms to trends in AB_5 intermetallics. *J. Chem. Theory Comp.*, **10**, 3380–3392.

Berns, V. M. and Fredrickson, D. C. (2013). Problem solving with pentagons: Tsai-type quasicrystal as a structural response to chemical pressure. *Inorg. Chem.*, **52**, 12875–12877.

Berns, V. M. and Fredrickson, D. C (2014). Structural plasticity: How intermetallics deform themselves in response to chemical pressure, and the complex structures that result. *Inorg. Chem.*, **53**, 10762–10771.

Biehl, E. and Deiseroth, H. J. (1996). Crystal structure of potassium amalgam, KHg. *Z. Kristallogr. NCS*, **211**, 630.

Biehl, E. and Deiseroth, H. J. (1999). Rb_5Hg_{19}: Eine neue, geordnete Defektvariante des $BaAl_4$-Strukturtyps. *Z. Anorg. Allg. Chem.*, **625**, 389–394.

Blank, H. (1998). Fractional packing densities and fast diffusion in uranium and other light actinoids. *J. Alloys Compd.*, **268**, 180–187.

Blatov, V. A. (2006). A method for hierarchical comparative analysis of crystal structures. *Acta Crystallogr. A*, **62**, 356–364.

Blatov, V. A. (2012). Nanocluster analysis of intermetallic structures with the program package TOPOS. *Struct. Chem.*, **23**, 955–963.

Blatov, V. A. (2014). http://topos.ssu.samara.ru, http://www.topos.samsu.ru.

Blatov, V. A. and Ilyushin, G. D. (2010). New method for computer analysis of complex intermetallic compounds and nanocluster model of the samson phase Cd_3Cu_4. *Crystallogr. Rep.*, **55**, 1100–1105.

Blatov, V. A. and Ilyushin, G. D. (2011). Geometric and topological analysis of icosahedral structures of Samson Mg_2Zn_{11} ($cP39$) phases, $K_6Na_{15}Tl_{18}H$ ($cP40$), and $Tm_3In_7Co_9$ ($cP46$): Nanocluster precursors, self-assembly mechanism, and superstructure ordering. *Russ. J. Inorg. Chem.*, **56**, 729–737.

Blatov, V. A. and Ilyushin, G. D. (2012). New icosahedral nanoclusters in crystal structures of intermetallic compounds: Topological types of 50-atom deltahedra $D50$ in samson phases β-Mg_2Al_3 and ε-$Mg_{23}Al_{30}$. *Crystallogr. Rep.*, **57**, 885–891.

Blatov, V. A., Ilyushin, G. D., and Proserpio, D. M. (2010). Nanocluster model of intermetallic compounds with giant unit cells: β, β'-Mg_2Al_3 polymorphs. *Inorg. Chem.*, **49**, 1811–1818.

Blatov, V. A., Ilyushin, G. D., and Proserpio, D. M. (2011). New types of multishell nanoclusters with a Frank-Kasper polyhedral core in intermetallics. *Inorg. Chem.*, **50**, 5714–5724.

Blatov, V. A. and Shevchenko, A. P. (2015). http://topospro.com.

Blatov, V. A., Shevchenko, A. P., and Proserpio, D. M. (2014). Applied topological analysis of crystal structures with the program package ToposPro. *Cryst. Growth Des.*, **14**, 3576–3586.

Blatov, V. A., Shevchenko, A. P., and Serezhkin, V. N. (2000). TOPOS 3.2: a new version of the program package for multipurpose crystal-chemical analysis. *J. Appl. Cryst.*, **33**, 1193–1193.

Block, G. and Jeitschko, W. (1986). Ternary carbides $Ln_2Mn_{17}Cs_{3-x}$ (Ln = La, Ce, Pr, Nd, Sm) with filled Th_2Zn_{17} type structure. *Inorg. Chem.*, **25**, 279–282.

Bodak, O., Demchenko, P., Seropegin, Y., and Fedorchuk, A. (2006). Cubic structure types of rare-earth intermetallics and related compounds. *Z. Kristallogr.*, **221**, 482–492.

Bojin, M. D. and Hoffmann, R. (2003a). The REME phases – I. An overview of their structural variety. *Helv. Chim. Acta*, **86**, 1653–1682.

Bojin, M. D. and Hoffmann, R. (2003b). The REME phases – II. What's possible? *Helv. Chim. Acta*, **86**, 1683–1708.

Booth, M. H., Brandon, J. K., Brizard, R. Y., Chieh, C., and Pearson, W. B. (1977). γ-Brasses with F cells. *Acta Crystallogr. B*, **33**, 30–36.

Boucherle, J. X., Givord, F., and Lejay, P. (1988). Structures of Ce_2Sn_5 and Ce_3Sn_3, two superstructures of $CeSn_3$. *Acta Crystallogr. B*, **44**, 377–380.

Bovornratanaraks, T., Allan, D. R., Belmonte, S. A., McMahon, M. I., and Nelmes, R. J. (2006). Complex monoclinic superstructure in Sr-IV. *Phys. Rev. B*, **73**, 144112.

Broer, F. R. de and Pettifor, D. G. (1988). *Cohesion in Metals*. Volume I. North-Holland, Amsterdam.

Brown, B. S., Hafstrom, J. W., and Klippert, T. E. (1977). Changes in the superconducting critical temperature after fast-neutron irradiation. *J. Appl. Phys.*, **48**, 1759–1761.

Brück, E. (2005). Developments in magnetocaloric refrigeration. *J. Phys. D: Appl. Phys.*, **38**, R381–91.

Brühne, S., Uhrig, E., Gross, C., Assmus, W., Masadeh, A. S., and Billinge, S. J. L. (2005). The local atomic quasicrystal structure of the icosahedral $Mg_{25}Y_{11}Zn_{64}$ alloy. *J. Phys.: Condens. Matter*, **17**, 1561–1572.

Brunner, G. O. and Schwarzenbach, D. (1971). Zur Abgrenzung der Koordinationssphäre und Ermittlung der Koordinationszahl in Kristallstrukturen. *Z. Kristallogr.*, **133**, 127–133.

Brunton, G. (1969). Refinement of the crystal structure of β_1-K_2UF_6. *Acta Crystallogr. B*, **25**, 2163–2164.

Brylak, M. and Jeitschko, W. (1995). Ternary antimonides $LnTSb_3$ with $Ln = La–Nd$, Sm and $T = V$, Cr. *Z. Naturforsch.*, **50b**, 899–904.

Burdett, J. K. and Lee, S. (1985). Moments and the energies of solids. *J. Am. Chem. Soc.*, **107**, 3050–3063.

Burkhardt, A., Wedig, U., von Schnering, H. G., and Savin, A. (1993). Die Elektronen-Lokalisierungs-Funktion in closo-Bor-Clustern. *Z. Anorg. Allg. Chem.*, **619**, 437–441.

Bush, T. S., Catlow, C. R. A., and Battle, P. D. (1995). Evolutionary programming techniques for predicting inorganic crystal structures. *J. Mater. Chem.*, 5, 1269–1272.

Cahn, R. W. and Haasen, P. (ed.) (1996). *Physical Metallurgy* (Fourth edn). Volume 1. Elsevier.

Calvo, F., Pahl, E., Wormit, M., and Schwerdtfeger, P. (2013). Erklärung des niedrigen Schmelzpunkts von Quecksilber mit relativistischen Effekten. *Angew. Chem.*, **125**, 7731–7734.

Cambridge Crystallographic Data Centre, CCDC (2015). Cambridge structural database.

Canfield, P. C., Caudle, M. L., Ho, C. S., Kreyssig, A., Nandi, S., Kim, M. G., Lin, X., Kracher, A., Dennis, K. W., McCallum, K. W., and Goldman, A. I. (2010). Solution growth of a binary icosahedral quasicrystal of $Sc_{12}Zn_{88}$. *Phys. Rev. B*, **81**, 020201.

Cantor, B. (2014). Multicomponent and high entropy alloys. *Entropy*, **16**, 4749–4768.

Cantor, B., Chang, I. T. H., Knight, P., and Vincent, A. J. B. (2004). Microstructural development in equiatomic multicomponent alloys. *Mater. Sci. Eng. A*, **375–377**, 213–218.

Carrillo-Cabrera, W., Caroca-Canales, N., Peters, K., and von Schnering, H. G. (1993). $K_3Na_{26}In_{48}$: An intermetallic phase with large pseudo-icosahedral clusters and a Na_{46} clathrate-I network enveloping a covalent? $^3_\infty[In_{12}]$ cluster framework. *Z. Anorg. Allg. Chem.*, **619**, 1556–1563.

Carrillo-Cabrera, W., Caroca-Canales, N., and von Schnering, H. G. (1994). $K_{(21-\delta)}Na_{(2+\delta)}In_{39}$ ($\delta = 2.8$): A cluster-replacement clathrate-II structure with an alkali metal M_{136}-network. *Z. Anorg. Allg. Chem.*, **620**, 247–257.

Casper, F., Graf, T., Chadov, S., Balke, B., and Felser, C. (2012). Half-Heusler compounds: novel materials for energy and spintronic applications. *Semicond. Sci. Technol.*, **27**, 063001.

CCDC (2015, November). Blind Tests of Organic Crystal Structure Prediction Methods. https://www.ccdc.cam.ac.uk/Community/Initiatives/Pages/CSPBlindTests.aspx.

Cenzual, K., Chabot, B., and Parthé, E. (1985). Cubic $Sc_{57}Rh_{13}$ and orthorhombic $Hf_{54}Os_{17}$, two geometrically related crystal structures with rhodium- and osmium-centred icosahedra. *Acta Crystallogr. C*, **41**, 313–319.

Cenzual, K. and Parthé, E. (1986). $Zr_{21}Re_{25}$, a new rhombohedral structure type containing 12 Å-thick infinite $MgZn_2$ (Laves)-type columns. *Acta Crystallogr. C*, **42**, 261–266.

Cerenius, Y. and Dubrovinsky, L. (2000). Compressibility measurements on iridium. *J. Alloys Compd.*, **306**, 26–29.

Cerny, R., Renaudin, G., Fave-Nicolin, V., Hlukhyy, V., and Pöttgen, R. (2004). $Mg_{1+x}Ir_{1-x}$ ($x = 0$, 0.037 and 0.054), a binary intermetallic compound with a new orthorhombic structure type determined from powder and single-crystal X-ray diffraction. *Acta Crystallogr. B*, **60**, 272–282.

Cervellino, A. (2002). *Higher-dimensional modelling of decagonal quasicrystal structures.* Thesis no. 14023, ETH Zurich.

Cervellino, A., Haibach, T., and Steurer, W. (2002). Structure solution of the basic decagonal Al-Co-Ni phase by the atomic surfaces modelling method. *Acta Crystallogr. B*, **58**, 8–33.

Chahine, A., Tillard-Charbonnel, M., and Belin, C. (1994). Crystal structure of sodium copper cadmium gallium, $Na_{34}Cu_7Cd_6Ga_{92}$. *Z. Kristallogr.*, **209**, 542–543.

Charbonnel, M. and Belin, C. (1984). Synthesis and crystal structure of the new nonstoichiometric phase $Li_3Na_5Ga_{19.56}$. *Nouv. J. Chimie*, **8**, 595–599.

Charbonnel, M. and Belin, C. (1987). Synthesis and X-ray crystal structure of the new nonstoichiometric phase $Rb_{0.60}Na_{6.25}Ga_{20.02}$. *J. Solid State Chem.*, **67**, 210–218.

Che, G. C., Ellner, M., and Schubert, K. (1991). The *h*P1-type phases in alloys of cadmium, mercury, and indium with tin. *J. Mater. Sci.*, **26**, 2417–2420.

Chen, J., Hu, L., Deng, J. X., and Xing, X. R. (2015). Negative thermal expansion in functional materials: controllable thermal expansion by chemical modifications. *Chem. Soc. Rev.*, **44**, 3522–3567.

Chen, X. A., Jeitschko, W., Danebrock, M. E., Evers, C. B. H., and Wagner, K. (1995). Preparation, properties and crystal structures of Ti_3Zn_{22} and $TiZn_{16}$. *J. Solid State Chem.*, **118**, 219–226.

Cheng, Y., Chen, H. H., Xue, F. X., Ji, G. F., and Gong, M. (2013). Phase transition, elastic and thermodynamic properties of beryllium via first principles. *Intern. J. Modern. Phys. B*, **27**, 1350130.

Chesnut, G. N. and Vohra, Y. K. (1998). Phase transformation in lutetium metal at 88 GPa. *Phys. Rev. B*, **57**, 10221–10223.

Chesnut, G. N. and Vohra, Y. K. (1999). Structural and electronic transitions in ytterbium metal to 202 GPa. *Phys. Rev. Lett.*, **82**, 1712–1715.

Chesnut, G. N. and Vohra, Y. K. (2000). α-uranium phase in compressed neodymium metal. *Phys. Rev. B*, **61**, R3768–R3771.

Chieh, C. (1979). Archimedean truncated octahedron, and packing of geometric units in cubic-crystal structures. *Acta Crystallogr. A*, **35**, 946–952.

Chieh, C. (1980). The Archimedean truncated octahedron. 2. Crystal-structures with geometric units of symmetry $\bar{4}3m$. *Acta Crystallogr. A*, **36**, 819–826.

Chieh, C. (1982). The Archimedean truncated octahedron. 3. Crystal-structures with geometric units of symmetry *m*3*m*. *Acta Crystallogr. A*, **38**, 346–349.

Chieh, C., Burzlaff, H., and Zimmermann, H. (1982). The Archimedean truncated octahedron, and packing of geometric units in cubic-crystal structures – on the choice of origins in the description of space-groups – comments. *Acta Crystallogr. A*, **38**, 746–747.

Ching, W. Y. and Gu, Z. Q. (1987). Electronic structure of $Nd_2Fe_{14}B$. *J. Appl. Phys.*, **61**, 3718–3720.

Choe, W., Pecharsky, V. K., Pecharsky, A. O., Gschneidner Jr., K. A., Young Jr., V. G., and Miller, G. J. (2000). Making and breaking covalent bonds across the magnetic transition in the giant magnetocaloric material $Gd_5(Si_2Ge_2)$. *Phys. Rev. Lett.*, **84**, 4617–4620.

Chumak, I., Richter, K. W., and Ehrenberg, H. (2010). Redetermination of iron dialuminide, $FeAl_2$. *Acta Crystallogr. C*, **66**, i87–i88.

Cohen, M. L. (1989). Novel materials from theory. *Nature*, **338**, 291–292.

Conelly, N. G., Damhus, T., Hartshorn, R. M., and Hutton, A. T. (2005). *Nomenclature of Inorganic Chemistry. IUPAC Recommendations 2005*. RSC Publishing.

Conrad, M., Harbrecht, B., Weber, T., Jung, D. Y., and Steurer, W. (2009). Large, larger, largest – a family of cluster-based tantalum copper aluminides with giant unit cells. II. The cluster structure. *Acta Crystallogr. B*, **65**, 318–325.

Corbett, J. D. (1996). Diverse solid-state clusters with strong metal-metal bonding. In praise of synthesis. *Dalton Trans.*, **1996**, 575–585.

Corbett, J. D. (1997). Diverse naked clusters of the heavy main-group elements. Electronic regularities and analogies. In *Structural and Electronic Paradigms in Cluster Chemistry*, Volume 87, Structure and Bonding, pp. 157–193.

Corbett, J. D. (2000a). Exploratory synthesis in the solid state. Endless wonders. *Inorg. Chem.*, **39**, 5178–5191.

Corbett, J. D. (2000b). Polyanionic clusters and networks of the early p-element metals in the solid state: Beyond the Zintl boundary. *Angew. Chem. Intern. Ed.*, **39**, 670–690.

Cordier, G. and Müller, V. (1992). Crystal structure of potassium indium $(22 - x/39 + x)$ $(x = 0.67)$, $K_{21.33}In_{39.67}$. *Z. Kristallog.*, **198**, 302–303.

Cordier, G. and Müller, V. (1993). Crystal structure of sodium indium – $Na_{4.95}In_{9.22}$. *Z. Kristallogr.*, **205**, 306–308.

Cordier, G. and Müller, V. (1994). NaK_9Tl_{13}: Eine neue Verbindung an der Zintl-Grenze mit isolierten, von Thallium besetzten Tl_{17}-Ikosaedern. *Z. Naturforsch.*, **49b**, 935–938.

Cordier, G. and Müller, V. (1995). $K_{34}[Zn_{20}In_{85}]$: Eine ternäre Auffüllungsvariante des beta-rhomboedrischen Bors. *Z. Naturforsch.*, **50b**, 23–30.

Cordier, G., Müller, V., and Fröhlich, R. (1993). Crystal structure of potassium thallide (49/108), $K_{49}Tl_{108}$. *Z. Kristallogr.*, **203**, 148–149.

Corsan, J. M. and Cook, A. J. (1970). Specific heat and superconductivity of binary alloys containing V, Nb, and Ta. *Phys. Status Solidi*, **40**, 657–665.

Cottenier, S., Probert, M. I. J., Van Hoolst, T., Van Speybroeck, V., and Waroquier, M. (2011). Crystal structure prediction for iron as inner core material in heavy terrestrial planets. *Earth Planet. Sci. Lett.*, **312**, 237–242.

Coxeter, H. S. M (1973). *Regular Polytopes*. Dover Publications Inc, New York, USA.

Critchley, J. K. and Jeffery, J. W. (1965). A determination of the crystal structure of CdNi. *Acta Crystallogr.*, **19**, 674–676.

Culp, S. R., Poon, S. J., Hickman, N., Tritt, T. M., and Blumm, J. (2006). Effect of substitutions on the thermoelectric figure of merit of half-Heusler phases at 800 degrees C. *Appl. Phys. Lett.*, **88**, 042106.

Cundy, H. M. and Rollet, A. P. (1952). *Mathematical Models*. Clarendon Press, Oxford.

Cunningham, N. C., Qiu, W., Hope, K. M., Liermann, H. P., and Vohra, Y. K. (2007). Symmetry lowering under high pressure: Structural evidence for f-shell delocalization in heavy rare earth metal terbium. *Phys. Rev. B*, **76**, 212101.

Daams, J. L. C. and Villars, P. (1997). Atomic-environment classification of the tetragonal "intermetallic" structure types. *J. Alloys Compd.*, **252**, 110–142.

Daams, J. L. C. and Villars, P. (2000). Atomic environments in relation to compound prediction. *Eng. Appl. Artif. Intell.*, **13**, 507–511.

Dabos-Seignon, S., Dancausse, J. P., Gering, E., Heathman, S., and Benedict, U. (1993). Pressure-induced phase-transition in α-Pu. *J. Alloys Compd.*, **190**, 237–242.

d'Avezac, M. and Zunger, A. (2008). Identifying the minimum-energy atomic configuration on a lattice: Lamarckian twist on darwinian evolution. *Phys. Rev. B*, **78**, 064102.

Day, G. M., Cooper, T. G., Cruz-Cabeza, A. J., Hejczyk, K. E., Ammon, H. L., Boerrigter, S. X. M., Tan, J. S., Della Valle, R. G., Venuti, E., Jose, J., Gadre, S. R., Desiraju, G. R., Thakur, T. S., van Eijck, B. P., Facelli, J. C., Bazterra, V. E., Ferraro, M. B., Hofmann, D. W. M., Neumann, M. A., Leusen, F. J. J., Kendrick, J., Price, S. L., Misquitta, A. J., Karamertzanis, P. G., Welch, G. W. A., Scheraga, H. A., Arnautova, Y. A., Schmidt,

M. U., van de Streek, J., Wolf, A. K., and Schweizer, B. (2009). Significant progress in predicting the crystal structures of small organic molecules – a report on the fourth blind test. *Acta Crystallogr. B*, **65**, 107–125.

Day, G. M., Motherwell, W. D. S., Ammon, H. L., Boerrigter, S. X. M., Della Valle, R. G., Venuti, E., Dzyabchenko, A., Dunitz, J. D., Schweizer, B., van Eijck, B. P., Erk, P., Facelli, J. C., Bazterra, V. E., Ferraro, M. B., Hofmann, D. W. M., Leusen, F. J. J., Liang, C., Pantelides, C. C., Karamertzanis, P. G., Price, S. L., Lewis, T. C., Nowell, H., Torrisi, A., Scheraga, H. A., Arnautova, Y. A., Schmidt, M. U., and Verwer, P. (2005). A third blind test of crystal structure prediction. *Acta Crystallogr.*, **61**, 511–527.

de Oliveira, N. A., von Ranke, P. J., and Troper, A. (2014). Magnetocaloric and barocaloric effects: Theoretical description and trends. *Int. J. Refrig.*, **37**, 237–248.

Deemyad, S. and Schilling, J. S. (2003). Superconducting phase diagram of Li metal in nearly hydrostatic pressures up to 67 GPa. *Phys. Rev. Lett.*, **91**, 167001.

Degtyareva, O., McMahon, M. I., Allan, D. R., and Nelmes, R. J. (2004*a*). Structural complexity in gallium under high pressure: relation to alkali elements. *Phys. Rev. Lett.*, **93**, 205502.

Degtyareva, O., McMahon, M. I., and Nelmes, R. J. (2004*b*). High-pressure structural studies of group-15 elements. *High Press. Res.*, **24**, 319–356.

Degtyareva, V. F. and Degtyareva, O. (2009). Structure stability in the simple element sodium. *New J. Phys.*, **11**, 063037.

Degtyareva, V. F., Porsch, F., Khasanov, S. S., Shekhtman, V. S., and Holzapfel, W. B. (1997). Effect of pressure on structural properties of intermetallic LnM lanthanide compounds. *J. Alloys Compd.*, **246**, 248–255.

Deiseroth, H.-J. and Biehl, E. (1999). NaK$_{29}$Hg$_{48}$: A contradiction to or an extension of theoretical concepts to rationalize the structures of complex intermetallics? *J. Solid State Chem.*, **147**, 177–184.

Deiseroth, H.-J. and Rochnia, M. (1993). β-Na$_3$Hg: ein Feststoff mit geschmolzener Natriumteilstruktur im Temperaturbereich 36–60 °C. *Z. Anorg. Allg. Chem.*, **105**, 1556–1558.

Deiseroth, H.-J. and Rochnia, M. (1994). Einkristallstudien zur Temperaturabhängigkeit der Kristallstruktur von α-Na,Hg. *Z. Anorg. Allg. Chem.*, **620**, 1736–1740.

Deiseroth, H.-J. and Strunck, A. (1987). Square Hg$_4$ clusters in the compound CsHg. *Angew. Chem. Int. Ed. Engl.*, **26**, 687–688.

Deloudi, S., Fleischer, F., and Steurer, W. (2011). Unifying cluster-based structure models of decagonal Al–Co–Ni, Al–Co–Cu and Al–Fe–Ni. *Acta Crystallogr. B*, **67**, 1–17.

Demchyna, R., Leoni, S., Rosner, H., and Schwarz, U. (2006). High-pressure crystal chemistry of binary intermetallic compounds. *Z. Kristallogr.*, **221**, 420–434.

Diao, H. Y., Santodonato, L. J., Tang, Z., Egami, T., and Liaw, P. K. (2015). Local structures of high-entropy alloys (HEAs) on atomic scales: An overview. *JOM*, **67**, 2321–2325.

DiGennaro, M., Saha, S. K., and Verstraete, M. J. (2013). Role of dynamical instability in the ab initio phase diagram of calcium. *Phys. Rev. Lett.*, **111**, 025503.

Ding, Y., Ahuja, R., Shu, J., Chow, P., Luo, W., and Mao, H. K. (2007). Structural phase transition of vanadium at 69 GPa. *Phys. Rev. Lett.*, **98**, 085502.

Dogan, A. and Pöttgen, R. (2005). The ordered Laves phases CeNi$_4$Cd and RECu$_4$Cd (RE = Ho, Er, Tm, Yb). *Z. Naturforsch.*, **60b**, 495–498.

Dong, C., Wang, Q., Chen, W., Zhang, Q., Qiang, J. B., and Wang, Y. M. (2007). Cluster-based composition rules for ternary alloy systems. *J. Univ. Sci. Tech. Beijing*, **14**, 1–3.

Dong, Z. C. and Corbett, J. D. (1995). Unusual icosahedral cluster compounds: Open-shell $Na_4A_6Tl_{13}$ (A = K, Rb, Cs) and the metallic Zintl phase $Na_3K_8Tl_{13}$ (How does chemistry work in solids?). *J. Am. Chem. Soc.*, **117**, 6447–6455.

Dong, Z. C. and Corbett, J. D. (1996). CsTl: a new example of tetragonally compressed Tl_6^{6-} octahedra. Electronic effects and packing requirements in the diverse structures of ATl (A) Li, Na, K, Cs). *Inorg. Chem.*, **35**, 2301–2306.

Dotera, T. (2011). Quasicrystals in soft matter. *Isr. J. Chem.*, **51**, 1197–1205.

Dotera, T. (2012). Toward the discovery of new soft quasicrystals: From a numerical study viewpoint. *J. Polymer Sci. B*, **50**, 155–167.

Doye, J. P. K. and Wales, D. J. (1998). Thermodynamics of global optimization. *Phys. Rev. Lett.*, **80**, 1357–1360.

Dshemuchadse, J. (2013). *Structural building principles of complex intermetallics*. Ph.D. thesis, ETH Zurich.

Dshemuchadse, J., Bigler, S., Simonov, A., Weber, T., and Steurer, W. (2013). A new complex intermetallic phase in the system Al–Cu–Ta with familiar clusters and packing principles. *Acta Crystallogr. B*, **69**, 238–248.

Dshemuchadse, J., Jung, D. Y., and Steurer, W. (2011). Structural building principles of complex face-centered cubic intermetallics. *Acta Crystallogr. B*, **67**, 269–292.

Dshemuchadse, J. and Steurer, W. (2014). More of the 'fullercages'. *Z. Anorg. Allg. Chem.*, **640**, 693–700.

Dshemuchadse, J. and Steurer, W. (2015). More statistics on intermetallic compounds – ternary phases. *Acta Crystallogr. A*, **71**, 335–345.

Dubenskyy, V. P., Zaremba, V. I., Kalychaka, Y. M., Gulay, L. D., Kaczorowski, D., Stępień-Damm, J., and Wocłyrz, M. (2000). Crystal structure, magnetic and electrical properties of $M'M_2In$ (M' = Ti, Hf; M = Ni, Cu). *J. Alloys Compd.*, **306**, 21–25.

Dubrovinsky, L., Dubrovinskaia, N., Bykova, E., Bykov, M., Prakapenka, V., Prescher, C., Glazyrin, K., Liermann, H.-P., Hanfland, M., Ekholm, M., Feng, Q., Pourovskii, L. V., Katsnelson, M. I., Wills, J. M., and Abrikosov, I. A. (2015). The most incompressible metal osmium at static pressures above 750 gigapascals. *Nature*, **525**, 226–229.

Duwell, E. J. and Baenziger, N. C. (1960). The crystal structure of K_5Hg_7. *Acta Crystallogr.*, **13**, 476–479.

Ek, J. van, Sterne, P. A., and Gonis, A. (1993). Phase stability of plutonium. *Phys. Rev. B*, **48**, 16280–16289.

Engelkemier, J., Berns, V. M., and Fredrickson, D. C. (2013). First-principles elucidation of atomic size effects using DFT-chemical pressure analysis: Origins of $Ca_{36}Sn_{23}$'s long-period superstructure. *J. Chem. Theory Comp.*, **9**, 3170–3180.

Ernst, G., Artner, C., Blaschke, O., and Krexner, G. (1986). Low-temperature martensitic phase transition of bcc lithium. *Phys. Rev. B*, **33**, 6465–6469.

Errandonea, D., Boehler, R., Schwager, B., and Mezouar, M. (2007). Structural studies of gadolinium at high pressure and temperature. *Phys. Rev. B*, **75**, 014103.

Errea, I., Martinez-Canales, M., Oganov, A. R., and Bergara, A. (2008). Fermi surface nesting and phonon instabilities in simple cubic calcium. *High Press. Res.*, **28**, 443–448.

Evans, S. R., Loa, I., Lundegaard, L. F., and McMahon, M. I. (2009). Phase transitions in praseodymium up to 23 GPa: An X-ray powder diffraction study. *Phys. Rev. B*, **80**, 134105.

Evans, W. J., Lipp, M. J., Cynn, H., Yoo, C. S., Somayazulu, M., Häussermann, D., Shen, G., and Prakapenka, V. (2005). X-ray diffraction and Raman studies of beryllium: Static and elastic properties at high pressures. *Phys. Rev. B*, **72**, 094113.

Evers, J. and Oehlinger, G. (2000). After more than 60 years, a new NaTl type Zintl phase: KTl at high pressure. *Inorg. Chem.*, **39**, 628–629.

Fabbris, G., Matsuoka, T., Lim, J., Mardegan, J. R. L., Shimizu, K., Haskel, D., and Schilling, J. S. (2013). Different routes to pressure-induced volume collapse transitions in gadolinium and terbium metals. *Phys. Rev. B*, **88**, 245103.

Fadda, A. and Fadda, G. (2010). An evolutionary algorithm for the prediction of crystal structures. *Phys. Rev. B*, **82**, 104105.

Farr, J. D., Giorgi, A. L., Bowman, M. G., and Noney, R. K. (1961). The crystal structure of actinium metal and actinium hydride. *J. Inorg. Nucl. Chem.*, **18**, 42–47.

Fässler, T. E. and Hoffmann, S. (1999). Valence compounds at the border to intermetallics: alkali and alkaline earth metal stannides and plumbides. *Z. Kristallogr.*, **214**, 722–734.

Fässler, T. F., Hoffmann, S., and Kronseder, C. (2001). Novel tin structure motives in superconducting $BaSn_5$ – the role of lone pairs in intermetallic compounds. *Z. Anorg. Allg. Chem.*, **627**, 2486–2492.

Feng, J., Hoffmann, R., and Ashcroft, N. W. (2010). Double-diamond NaAl via pressure: Understanding structure through Jones zone activation. *J. Chem. Phys.*, **132**, 114106.

Ferro, R. and Saccone, A. (2008). *Intermetallic Chemistry*. Volume 13, Pergamon Materials Series. Elsevier, Amsterdam.

Feuerbacher, M., Heidelmann, M., and Thomas, C. (2015). Hexagonal high-entropy alloys. *Mater. Res. Lett.*, **3**, 1–6.

Feuerbacher, M., Thomas, C., Makongo, J. P. A., Hoffmann, S., Carrillo-Cabrera, W., Cardoso, R., Grin, Y., Kreiner, G., Joubert, J. M., Schenk, T., Gastaldi, J., Nguyen-Thi, H., Mangelinck-Noel, N., Billia, B., Donnadieu, P., Czyrska-Filemonowicz, A., Zielinska-Lipiec, A., Dubie, B., Weber, T., Schaub, G., Krauss, G., Gramlich, V., Christensen, J., Lidin, S., Fredrickson, D., Mihalkovic, M., Sikora, W., Malinowski, J., Brühne, S., Proffen, T., Assmus, W., de Boissieu, M., Bley, F., Chemin, J. L., Schreuer, J., and Steurer, W. (2007). The Samson phase, β-Mg_2Al_3, revisited. *Z. Kristallogr.*, **222**, 259–288.

Fischer, C. C., Tibbetts, K. J., Morgan, D., and Ceder, G. (2006). Predicting crystal structure by merging data mining with quantum mechanics. *Nat. Mater.*, **5**, 641–646.

Fischer, W. and Koch, E. (2002). Lattice complexes. In *International Tables for Crystallography* (Fifth edn) (ed. T. Hahn), Volume A. Kluwer Academic Publishers.

Fleming, M. A. (2014). ASM Alloy Phase Diagram Database. Private communication.

Florio, J. V., Rundle, R. E., and Snow, A. I. (1952). Compounds of thorium with transition metals. I. The thorium-manganese system. *Acta Crystallogr.*, **5**, 449–457.

Flot, D., Vincent, L., Tillard-Charbonnel, M., and Belin, C. (1997a). Crystal structure of sodium potassium cadmium gallium, $Na_{21}K_{14}Cd_{17}Ga_{84}$. *Z. Kristallogr. NCS*, **212**, 509–510.

Flot, D., Vincent, L., Tillard-Charbonnel, M., and Belin, C. (1997b). Crystal structure of sodium potassium cadmium gallium, $Na_{21}K_{14}Cd_{17}Ga_{84}$. *Z. Kristallogr. NCS*, **212**, 509–510.

Flot, D., Vincent, L., Tillard-Charbonnel, M., and Belin, C. (1998). $Na_{13}K_4Ga_{47.45}$: a New Sodium Potassium Gallide Phase Containing Trimeric Icosahedral Gallium Clusters. *Acta Crystallogr. C*, **54**, 174–175.

Fornasini, M. L. (1975). The crystal structure of Ba_4Al_5. *Acta Crystallogr. B*, **31**, 2551–2552.

Fornasini, M. L. (1987). Ca_8In_3, a structure related to the BiF_3 type. *Acta Crystallogr. C*, **43**, 613–616.

Fornasini, M. L., Chabot, B., and Parthé, E. (1978). The crystal structure of $Sm_{11}Cd_{45}$ with γ-brass and α-Mn clusters. *Acta Crystallogr. B*, **34**, 2093–2099.

Fornasini, M. L., Manfrinetti, P., Palenzona, A., and Dhar, S. D. (2003). R_2Sn_3 (R = La – Nd, Sm): A family of intermetallic compounds with their own triclinic structure. *Z. Naturforsch.*, **58b**, 521–527.

Fowler, P. W., Manolopoulos, D. E., Redmond, D. B., and Ryan, R. P. (1993). Possible symmetries of fullerene structures. *Chem. Phys. Lett.*, **202**, 371–378.

Franceschi, E. and Olcese, G. L. (1969). A new allotropic form of cerium due to its transition under pressure to the tetravalent state. *Phys. Rev. Lett.*, **22**, 1299–1300.

Frank, F. C. and Kasper, J. S. (1958). Complex alloy structures regarded as sphere packings. 1. definitions and basic principles. *Acta Crystallogr. A*, **11**, 184–190.

Frank, F. C. and Kasper, J. S. (1959). Complex alloy structures regarded as sphere packings. 2. Analysis and classification of representative structures. *Acta Crystallogr. A*, **12**, 483–499.

Frank-Cordier, U., Cordier, G., and Schäfer, H. (1982*a*). Die Struktur des Na_7Ga_{13}-I und ein Konzept zur bindungsmäßigen Deutung. *Z. Naturforsch.* **37b**, 119–126.

Frank-Cordier, U., Cordier, G., and Schäfer, H. (1982*b*). Neue Ga-Cluster-Verbände im Na_7Ga_{13}-II. *Z. Naturforsch.*, **37b**, 127–135.

Fredrickson, D.C., Lee, S., and Hoffmann, R. (2007). Interpenetrating polar and nonpolar sublattices in intermetallics: The $NaCd_2$ structure. *Angew. Chem. Intern. Ed.*, **46**, 1958–1976.

Fredrickson, D. C. (2011). Electronic packing frustration in complex intermetallic structures: The role of chemical pressure in Ca_2Ag_7. *J. Am. Chem. Soc.*, **133**, 10070–10073.

Fredrickson, D. C. (2012). DFT-chemical pressure analysis: Visualizing the role of atomic size in shaping the structures of inorganic materials. *J. Am. Chem. Soc.*, **134**, 5991–5999.

Fredrickson, D. C., Lee, S., Hoffmann, R., and Lin, J. (2004*a*). The Nowotny chimney ladder phases: Following the c_{pseudo} clue toward an explanation of the 14 electron rule. *Inorg. Chem.*, **43**, 6151–6158.

Fredrickson, D. C., Lee, S., Hoffmann, R., and Lin, J. (2004*b*). The Nowotny chimney ladder phases: Whence the 14 electron rule. *Inorg. Chem.*, **43**, 6159–6167.

Fredrickson, R. T., Kilduff, B. J., and Fredrickson, D. C. (2015). Homoatomic clustering in T_4Ga_5 (T = Ta, Nb, Ta/Mo): A story of reluctant intermetallics crystallizing in a new binary structure type. *Inorg. Chem.*, **54**, 821–831.

Frenking, G. and Shaik, S. (ed.) (2014). *The Chemical Bond: Fundamental Aspects of Chemical Bonding*. Wiley-VCH Verlag.

Frey, F. and Boysen, H. (1981). Disorder in cobalt single crystals. *Acta Crystallogr. A*, **37**, 819–26.

Fujieda, S., Fujita, A., and Fukamichi, K. (2003). Large magnetocaloric effects in $NaZn_{13}$-type $La(Fe_xSi_{1-x})_{13}$ compounds and their hydrides composed of icosahedral clusters. *Sci. Tech. Adv. Mater.*, **4**, 339–346.

Fujihisa, H., Nakamoto, Y., Sakata, M., Shimizu, K., Matsuoka, T., Ohisi, Y., Yamawaki, H., Takeya, S., and Gotoh, Y. (2013). Ca-VII: A chain ordered host-guest structure of calcium above 210 GPa. *Phys. Rev. Lett.*, **110**, 235501.

Fujihisa, H., Nakamoto, Y., Shimizu, K., Yabuuchi, T., and Gotoh, Y. (2008). Crystal structures of calcium IV and V under high pressure. *Phys. Rev. Lett.*, **101**, 095503.

Fujihisa, H. and Takemura, K. (1995). Stability and the equation of state of α-manganese under ultrahigh pressure. *Phys. Rev. B*, **52**, 13257–13260.

Gao, G. Y., Niu, Y. L., Cui, T., Zhang, L. J., Li, Y., Xie, Y., He, Z., Ma, Y. M., and Zhou, G. T. (2007). Superconductivity and lattice instability in face-centered cubic lanthanum under high pressure. *J. Phys.: Condens. Matter*, **19**, 425234.

Gardner, M. (1977). Mathematical games. *Sci. Amer.*, **236**, 110–121.

Gaspard, J. P. and Cyrot-Lackmann, F. (1973). Density of states from moments. Application to the impurity band. *J. Phys. C*, **6**, 3077–3096.

Gatti, C. (2005). Chemical bonding in crystals: New directions. *Z. Kristallogr.*, **220**, 399–457.

Ghandehari, K. and Vohra, Y. K. (1992). Onset of the structural transformation in thorium metal at high pressures. *Scr. Met. Mater.*, **27**, 195–199.

Gibney, E. (2015). Software predicts crystal structures. *Nature*, **527**, 20–21.

Giessen, B. C. and Grant, N. J. (1965). New intermediate phases in transition metal systems. II. *Acta Crystallogr.*, **18**, 1080–1081.

Gillespie, R. J. and Popelier, P. L. A. (2001). *Chemical Bonding and Molecular Geometry: From Lewis to Electron Densities*. Oxford University Press.

Girgis, K., Petter, W., and Pupp, G. (1975). The crystal structure of V_8Ga_{41}. *Acta Crystallogr. B*, **31**, 113–116.

Glass, C. W., Oganov, A. R., and N., Hansen. (2006). USPEX—Evolutionary crystal structure prediction. *Comp. Phys. Comm.*, **175**, 713–720.

Godwal, B. K., Vijayakumar, V., Sikka, S. K., and Chidambaram, R. (1986). Pressure-induced AlB_2 to $MgCu_2$-type structural transition in $ThAl_2$. *J. Phys. F: Met. Phys.*, **16**, 1415–1418.

Goedecker, S. (2004). Minima hopping: An efficient search method for the global minimum of the potential energy surface of complex molecular systems. *J. Chem. Phys.*, **120**, 9911–9917.

Goldberg, D. E. (1989). *Genetic Algorithms in Search, Optimization and Machine Learning* (1st edn). Addison-Wesley Longman Publishing Co., Inc., Boston, MA, USA. ISBN: 978-0-201-15767-3.

Goldman, A. I., Kong, T., Kreyssig, A., Jesche, A., Ramazanoglu, M., Dennis, K. W., Bud'ko, S. L., and Canfield, P. C. (2013). A family of binary magnetic icosahedral quasicrystals based on rare earths and cadmium. *Nat. Mater.*, **12**, 714–718.

Goll, D. and Kronmüller, H. (2000). High-performance permanent magnets. *Naturwissenschaften*, **87**, 423–438.

Gómez, C. P. and Lidin, S. (2004). Superstructure of Eu_4Cd_{25}: A quasicrystal approximant. *Chem. Eur. J.*, **10**, 3279–3285.

Gómez, C. P., Morita, Y., Yamamoto, A., and Tsai, A. P. (2009). Structure analysis of complex metallic alloys in the Eu-Ag-In system by X-ray diffraction. *J. Phys.: Conf. Ser.*, **165**, 012045.

Gómez, C. P., Ohhashi, S., Yammaoto, A., and Tsai, A. P. (2008). Disordered structures of the TM–Mg–Zn 1/1 quasicrystal approximants (TM = Hf, Zr, or Ti) and chemical intergrowth. *Inorg. Chem.*, **47**, 8458–8266.

Gourdon, O., Gout, D., Williams, D. J., Proffen, T., Hobbs, S., and Miller, G. J. (2007). Atomic distributions in the γ-brass structure of the Cu-Zn system: A structural and theoretical study. *Inorg. Chem.*, **46**, 251–260.

Graf, T., Casper, F., Winterlik, J., Balke, B., Fecher, G. H., and Felser, C. (2009). Crystal structure of new Heusler compounds. *Z. Anorg. Allg. Chem.*, **635**, 976–981.

Graf, T., Felser, C., and Parkin, S. S. P. (2011). Simple rules for the understanding of Heusler compounds. *Progr. Solid State Chem.*, **39**, 1–50.

Graf, T., Winterlik, J., Müchler, L., Fecher, G. H., Felser, C., and Parkin, S. S. P. (2013). Magnetic Heusler compounds. *Handbook Magn. Mater.*, **21**, 1–75.

Greer, A. L. (1993). Confusion by design. *Nature*, **366**, 303–304.

Gregoryanz, E., Lundegaard, L. F., McMahon, M. I., Guillaume, C. L., Nelmes, R. J., and Mezouar, M. (2008). Structural diversity of sodium. *Science*, **320**, 1054–1057.

Grin, Y., Savin, A., and Silvi, B. (2014). The ELF perspective of chemical bonding. In *The Chemical Bond: Fundamental Aspects of Chemical Bonding* (ed. G. Frenking and S. Sason), Chapter 10., pp. 345–382. WILEY-VCH Verlag.

Grin, Y. and Schuster, J. C. (2007). Crystal structure of rhenium aluminum (1:4.01), Re_8Al_{33-x} (x = 0.93), *the low-temperature phase* of $ReAl_4$. *Z. Kristallogr. NCS*, **222**, 85–86.

Grin, Y., Wagner, F. R., Armbrüster, M., Kohout, M., Leithe-Jasper, A., Schwarz, U., Wedig, U., and von Schnering, H.G. (2006). $CuAl_2$ revisited: Composition, crystal structure, chemical bonding, compressibility and Raman spectroscopy. *J. Solid State Chem.*, **179**, 1707–1719.

Griveau, J. C. and Colineau, E. (2014). Superconductivity in transuranium elements and compounds. *C.R. Physique*, **15**, 599–615.

Grünbaum, B. and Shephard, G. C. (1986). *Tilings and Patterns*. Freeman, W. H., New York, USA.

Gschneidner Jr., K. A., Pecharsky, V. K., and Tsokol, A. O. (2005). Recent developments in magnetocaloric materials. *Rep. Prog. Phys.*, **68**, 1479–1539.

Gu, Q. F., Krauss, G., Grin, Y., and Steurer, W. (2009). Experimental confirmation of the stability and chemical bonding analysis of the high-pressure phases Ca-I, II, and III at pressures up to 52 GPa. *Phys. Rev. B*, **79**, 134121.

Guillaume, C. L. (1897). Recherches sur les aciers au nickel. Dilatations aux temperatures elevees; resistance electrique. *CR Acad. Sci.*, **125**, 235–238.

Guillaume, C. L., Gregoryanz, E., Degtyareva, O., McMahon, M. I., Hanfland, M., Evans, S., Guthrie, M., Sinogeikin, S. V., and Mao, H. K. (2011). Cold melting and solid structures of dense lithium. *Nat. Phys.*, **7**, 211–214.

Gummelt, P. (1996). Penrose tilings as coverings of congruent decagons. *Geom. Dedic.*, **62**, 1–17.

Guo, S., Ng, C., Lu, J., and Liu, C. T. (2011). Effect of valence electron concentration on stability of fcc or bcc phase in high entropy alloys. *J. Appl. Phys.*, **109**, 103505.

Guo, Y., Stacey, T. E., and Fredrickson, D. C. (2014). Acid-base chemistry in the formation of Mackay-type icosahedral clusters: μ_3-acidity analysis of Sc-rich phases of the Sc–Ir system. *Inorg. Chem.*, **53**, 5280–5293.

Gupta, S. and Suresh, K. G. (2015). Review on magnetic and related properties of RTX compounds. *J. Alloys Compd.*, **618**, 562–606.

Gutfleisch, O., Willard, M. A., Brück, E., Chen, C. H., Sankar, S. G., and Liu, J. P. (2011). Magnetic materials and devices for the 21st century: Stronger, lighter and more energy efficient. *Adv. Mater.*, **23**, 821–842.

Haire, R. G. and Baybarz, R. D. (1979). Studies of einsteinium metal. *J. Phys. (France)*, **40**, C4-101–C4-102.

Haire, R. G., Heathman, S., Idiri, M., Le Bihan, T., Lindbaum, A., and Rebizant, J. (2003). Pressure-induced changes in protactinium metal: importance to actinoid-metal bonding concepts. *Phys. Rev. B*, **67**, 13401.

Haji-Akbari, A., Engel, M., Keys, A. S., Zheng, X. Y., Petschek, R. G., Palffy-Muhoray, P., and Glotzer, S. C. (2009). Disordered, quasicrystalline and crystalline phases of densely packed tetrahedra. *Nature*, 462, 773–777.

Hamlin, J. J. (2015). Superconductivity in the metallic elements at high pressures. *Physica C*, 514, 59–76.

Hamlin, J. J. and Shilling, J. S. (2007). Pressure-induced superconductivity in Sc to 74 GPa. *Phys. Rev. B*, 76, 012505.

Hamrle, J., Blomeier, S., Gaier, O., Hillebrands, B., Schneider, H., Jakob, G., Postava, K., and Felser, C. (2007). Huge quadratic magneto-optical Kerr effect and magnetization reversal in the Co_2FeSi Heusler compound. *J. Phys. D: Appl. Phys.*, 40, 1563–1569.

Hanfland, M., Syassen, K., Christensen, N. E., and Novikov, D. L. (2000). New high-pressure phases of lithium. *Nature*, 408, 174–178.

Harris, N. A., Hadler, A. B., and Fredrickson, D. C. (2011). In search of chemical frustration in the Ca–Cu–Cd system: Chemical pressure relief in the crystal structures of Ca_5Cu_2Cd and $Ca_2Cu_2Cd_9$. *Z. Anorg. Allg. Chem.*, 637, 1961–1974.

Häussermann, U., Wengert, S., Hofmann, P., Savin, A., Jepsen, O., and Nesper, R. (1994). Localization of electrons in intermetallic phases containing aluminum. *Angew. Chem. Intern. Ed.*, 33, 2069–2073.

Hawthorne, F. C. (1990). Crystals from first principles. *Nature*, 345, 297.

Hayashi, J., Shirotani, I., Adachi, T., Shimomura, O., and Kikegawa, T. (2004). Phase transitions of YbX (X = P, As and Sb) with a NaCl-type structure at high pressures. *Philos. Mag.*, 84, 3663–3670.

Hayashi, J., Shirotani, I., Tanaka, Y., Adachi, T., Shimomura, O., and Kikegawa, T. (2000). Phase transitions of *Ln*Sb (*Ln* = lanthanide) with NaCl-type structure at high pressures. *Solid State Comm.*, 114, 561–565.

He, J. Y., Liu, W. H., Wang, H., Wu, Y., Liu, X. J., Nieh, T. G., and Lu, Z. P. (2014). Effects of Al addition on structural evolution and tensile properties of the FeCoNiCrMn high-entropy alloy system. *Acta Mater.*, 62, 105–113.

Heathman, S., Haire, R. G., Le Bihan, T., Lindbaum, A., Idiri, M., Normile, P., Li, S., Ahuja, R., Johansson, B., and Lander, G. H. (2005). A high-pressure structure in curium linked to magnetism. *Science*, 309, 110–113.

Heathman, S., Haire, R. G., Le Bihan, T., Lindbaum, A., Litfin, K., Méresse, Y., and Libotte, H. (2000). Pressure induces major changes in the nature of americium's 5f electrons. *Phys. Rev. Lett.*, 85, 2961–2964.

Hecker, S. S. (2000). Plutonium and its alloys—from atoms to microstructure. In *Challenges in Plutonium Science* (ed. N. G. Cooper), Volume 26, Los Alamos Science, pp. 290–335. Los Alamos National Laboratory.

Hellner, E. and Pearson, W. B. (1987). An application of a symbolism for different description of the Cu_4Cd_3 structure type. *Z. Kristallogr.*, 179, 175–186.

Herbst, J. F. and Croat, J. J. (1991). Neodymium-iron-boron perment magnets. *J. Magn. Magn. Mater.*, 100, 57–78.

Herbst, J. F., Croat, J. J., and Pinkerton, F. E. (1984). Relationships between crystal structure and magnetic properties in $Nd_2Fe_{14}B$. *Phys. Rev. B*, 29, 4176–4178.

Herbst, J. F., Croat, J. J., and Yelon, W. B. (1985). Structural and magnetic properties of $Nd_2Fe_{14}B$. *J. Appl. Phys.*, 57, 4086–4090.

Hillebrecht, H., Kuntze, V., and Gebhardt, K. (1997). Synthese und Kristallstruktur von $Mo_7Sn_{12}Zn_{40}$ – einer kubischen Verbindung mit Ikosaedern aus Ikosaedern. *Z. Kristallogr.*, 212, 840–847.

Hiraga, K. (2001). A large columnar cluster of atoms in an Al–Cu–Rh decagonal quasi-crystal studied by atomic-scale electron microscopy observations. *Philos. Mag. Lett.*, **81**, 117–122.

Hiraga, K. and Yasuhara, A. (2013*a*). Arrangements of transition-metal atoms in three types of Al–Co–Ni decagonal quasicrystal studied by Cs-corrected HAADF-STEM. *Mater. Trans.*, **54**, 493–497.

Hiraga, K. and Yasuhara, A. (2013*b*). The structure of an AlCoNi decagonal quasicrystal in an $Al_{72}Co_8Ni_{20}$ alloy studied by Cs-corrected scanning transmission electron microscopy. *Mater. Trans.*, **54**, 720–724.

Hoch, C. and Simon, A. (2008). Cs_2Hg_{27}, das quecksilberreichste Amalgam -ein naher Verwandter der Bergman-Phasen. *Z. Anorg. Allg. Chem.*, **634**, 853–856.

Hodeau, J. L., Marezio, M., and Remeika, J. P. (1984). The structure of $[Er(1)_{1x},Sn(1)_x]Er(2)_4Rh_6Sn(2)_4Sn(3)_{12}Sn(4)_2$, a ternary reentrant superconductor. *Acta Crystallogr. B*, **40**, 26–38.

Hoffmann, R. (1963). An Extended Hückel Theory. I. Hydrocarbons. *J. Chem. Phys.*, **39**, 1397–1412.

Hoffmann, R. D. and Pöttgen, R. (2001). AlB_2-related intermetallic compounds – a comprehensive view based on group-subgroup relations. *Z. Kristallogr.*, **216**, 127–145.

Hong, T., Watson-Yang, T. J., and Freeman, A. J. (1990). Crystal structure, phase stability, and electronic structure of Ti-Al intermetallics: $TiAl_3$. *Phys. Rev. B*, **41**(18), 12462–12467.

Hu, C. E. Zeng, Z. Y., Zhang, L., Chen, X. R., and Cai, L. C. (2010). Phase transition and thermodynamics of thorium from first-principles calculations. *Solid State Comm.*, **150**, 393–398.

Huang, B. and Corbett, J. D. (1998). Two new binary calcium–aluminium compounds: $Ca_{13}Al_{14}$ with a novel two-dimensional aluminium network, and Ca_8Al_3, and Fe_3Al-type analogue. *Inorg. Chem.*, **37**, 5827–5833.

Hückel, E. (1931). Quantentheoretische Beiträge zum Benzolproblem. II. Quantentheorie der induzierten Polaritäten. *Z. Phys.*, **72**, 310–337.

Hudson, T. S. (2010). Dense sphere packing in the $NaZn_{13}$ structure type. *J. Phys. Chem. C*, **114**, 14013–14017.

Hulm, J. K. and Blaugher, R. D. (1961). Superconducting solid solution alloys of the transition elements. *Phys. Rev. B*, **123**, 1569–1581.

Hume-Rothery, W. and Powell, H. M. (1935). The theory of superlattice structure in alloys. *Z. Kristallogr.*, **91**, 23–47.

Hume-Rothery, W., Reynolds, P. W., and Raynor, G. V. (1940). Factors affecting the formation of 3/2 electron compounds in alloys of copper, silver, and gold. *J. Inst. Met.*, **66**, 191–207.

Husband, R. J., Loa, I., Stinton, G. W., Evans, S. R., Ackland, G. J., and McMahon, M. I. (2012). Europium-IV: An incommensurately modulated crystal structure in the lanthanides. *Phys. Rev. Lett.*, **109**, 095503.

Ilyushin, G. D. and Blatov, V. A. (2009). Structures of the $ZrZn_{22}$ family: Suprapolyhedral nanoclusters, methods of self-assembly and superstructural ordering. *Acta Crystallogr. B*, **65**, 300–307.

Ishikawa, T., Ichikawa, A., Nagara, H., Geshi, M., Kusakabe, K., and Suzuki, N. (2008). Theoretical study of the structure of calcium in phases IV and V via *ab initio* metadynamics simulation. *Phys. Rev. B*, **77**, 020101.

Ishikawa, T., Nagara, H., Suzuki, N., Tsuchiya, J., and Tsuchiya, T. (2010). Review of high pressure phases of calcium by first-principles calculations. *J. Phys.: Conf. Ser.*, **215**, 012105.

IUCr (2002). *International Tables for Crystallography.* Volume A. Kluwer Academic Publishers, Dordrecht/Boston/London.

IUPAC (1990). *Nomenclature of Inorganic Chemistry. IUPAC Recommendations 1990.* Blackwell Scientific Publications, Oxford, UK.

Ivanovic, N., Rodic, D., Koteski, V., Radisavljevic, I., Novakovic, N., Marjanovic, D., Manasijevic, M., and Koicki, S. (2006). Cluster approach to the Ti_2Ni structure type. *Acta Crystallogr. B*, **62**, 1–8.

Iwami, S. and Ishimasa, T. (2015). Dodecagonal quasicrystal in Mn-based quaternary alloys containing Cr, Ni and Si. *Phil. Mag. Lett.*, **95**, 229–236.

Iwasaki, H., Fujimura, T., Ichikawa, M., Endo, S., and Wakatsuki, M. (1985). Pressure-induced phase transformation in AgZn. *J. Phys. Chem. Sol.*, **46**, 463–468.

Jagodzinski, H. (1954). Polytypism in SiC crystals. *Acta Crystallogr.*, **7**, 300.

Jani, J. M., Leary, M., Subic, A., and Gibson, M. A. (2014). A review of shape memory alloy research, applications and opportunities. *Mater. Design*, **56**, 1078–1113.

Janssen, T., Chapuis, G., and de Boissieu, M. (2007). *Aperiodic crystals. From modulated phases to quasicrystals.* Number 20 in IUCr Monographs on Crystallography. Oxford University Press, Oxford, UK.

Jeitschko, W. (1969). The crystal structure of Fe_2AlB_2. *Acta Crystallogr. B*, **25**, 163–165.

Jemmis, E. D. and Balakrishnarajan, M. M. (2001). Polyhedral boranes and elemental boron: Direct structural relations and diverse electronic requirements. *J. Am. Chem. Soc.*, **123**, 4324–4330.

Jemmis, E. D., Balakrishnarajan, M. M., and Pancharatna, P. D. (2001). A unifying electron-counting rule for macropolyhedral boranes, metallaboranes, and metallocenes. *J. Am. Chem. Soc.*, **123**, 4313–4323.

Jemmis, E. D. and Jayasree, E. G. (2003). Analogies between Boron and Carbon. *Acc. Chem. Res.*, **36**, 816–824.

Jepsen, O., Burkhardt, A., and Andersen, O. K. (2000). *The Program TB-LMTO-ASA* (4.7 edn). Max-Planck-Institut für Festkörperforschung, Stuttgart, Germany.

Jiles, D. C. (2003). Recent advances and future directions in magnetic materials. *Acta Mater.*, **51**, 5907–5939.

Johansson, A. and Westman, S. (1970). Determination of the structure of cubic gamma-Pt,Zn; a phase of gamma brass type with an 18 Å superstructure. *Acta Chem. Scand.*, **24**, 3471–3479.

Johansson, B., Luo, W., Li, S., and Ahuja, R. (2014). Cerium; crystal structure and position in the periodic table. *Sci. Rep.*, **4**, 6398.

Kadir, K., Noréus, D., and Yamashita, I. (2002). Structural determination of $AlMgNi_4$ (where A = Ca, La, Ce, Pr, Nd and Y) in the $AuBe_5$ type structure. *J. Alloys Compd.*, **345**, 140–143.

Kammler, D. R., Rodriguez, M. A., Tissot, R. G., Brown, D. W., Clausen, B., and Sisneros, T. A. (2008). In-situ time-of-flight neutron diffraction study of high-temperature α-to-β phase transition in elemental scandium. *Met. Mater. Trans. A*, **39**, 2815–2819.

Kao, Y. F., Chen, T. J., Chen, S. K., and Yeh, J. W. (2009). Microstructure and mechanical property of as-cast, -homogenized, and -deformed $Al_xCoCrFeNi$ ($0 \leq x \leq 2$) high-entropy alloys. *J. Alloys Compd.*, **488**, 57–64.

Katrych, S., Weber, T., Kobas, A., Massüger, L., Palatinus, L., Chapuis, G., and Steurer, W. (2007). New stable decagonal quasicrystal in the system Al–Ir–Os. *J. Alloys. Comp.*, **428**, 164–172.

Katzke, H. and Toledano, P. (2005). Structural mechanisms and order-parameter symmetries for the high-pressure phase transitions in alkali metals. *Phys. Rev. B*, **71**, 184101.

Kauzlarich, S. M. (ed.) (1996). *Chemistry, structure and bonding of Zintl phases and ions.* VCH Publishers, New York, USA.

Kawamura, N., Taniguchi, T., Mizusaki, S., Nagata, Y., Ozawa, T. C., and Samata, H. (2006). Functional intermetallic compounds in the samarium-iron system. *Sci. Tech. Adv. Mater.*, **7**, 46–51.

Kazantsev, A. V., Karamertzanis, P. G., Adjiman, C. S., Pantelides, C. C., Price, S. L., Galek, P. T. A., Day, G. M., and Cruz-Cabeza, A. J. (2011). Successful prediction of a model pharmaceutical in the fifth blind test of crystal structure prediction. *Int. J. Pharm.*, **418**, 168–178. A priori performance predictions.

Kepler, J. (1619). *Harmonices mundi libri V.* Volume II. Forni.

Kim, S. J. and Fässler, T. F. (2009). Networks of icosahedra in the sodium–zinc–stannides $Na_{16}Zn_{13.54}Sn_{13.46(5)}$, $Na_{22}Zn_{20}Sn_{19(1)}$, and $Na_{34}Zn_{66}Sn_{38(1)}$. *J. Solid State Chem.*, **182**, 778–789.

Kirkpatrick, S., Gelatt, C. D., and Vecchi, M. P. (1983). Optimization by simulated annealing. *Science*, **220**, 671–680.

Kitagawa, I. and Asari, Y. (2010). Magnetic anisotropy of $R_2Fe_{14}B$ ($R = Nd$, Gd, Y): Density functional calculation by using linear combination of pseudo-potential-orbital method. *Phys. Rev. B*, **81**, 214408.

Kitaigorodsky, A. I. (1973). *Molecular crystals and molecules.* Volume 29, Physical Chemistry. Academic Press, New York and London.

Kleinke, H. (2010). New bulk materials for thermoelectric power generation: clathrates and complex antimonides. *Chem. Mater.*, **22**, 604–611.

Klemm, W. (1950*a*). Aus der Chemie der Übergangselemente. *Die Naturwissenschaften*, **37**, 172–177.

Klemm, W. (1950*b*). Aus der Chemie der Übergangselemente. *Die Naturwissenschaften*, **37**, 150–156.

Klemm, W. (1950*c*). Einige Probleme aus der Physik und der Chemie der Halbmetalle und der Metametalle. *Angew. Chem.*, **62**, 133–142.

Klemm, W. (1958). Metalloids and their compounds with the alkali metals. *Proc. Chem. Soc.*, **1958**, 329–341.

Knight, P. (1995). *Multicomponent alloys. BSc Part II Thesis.* Ph.D. thesis, University of Oxford, Oxford, UK.

Koch, E. and Fischer, W. (1992). Sphere packings and packings of ellipsoids. In *International Tables for Crystallography* (ed. A. Wilson), Volume C, Chapter 9.1, pp. 746–51. Kluwer Academic Publishers, Dordrecht/Boston/London.

Koch, W. and Holthausen, M. C. (2001). *A Chemist's Guide to Density Functional Theory* (second edn). Wiley-VCH Verlag GmbH.

Koželj, P., Vrtnik, S., Jelen, A., Jazbek, S., Jagličić, Z., Maiti, S., Feuerbacher, M., Steurer, W., and Dolinšek, J. (2014). Discovery of a superconducting high-entropy alloy. *Phys. Rev. Lett.*, **113**, 107001.

Kohout, M. (2004). A measure of electron localizability. *Int. J. Quant. Chem.*, **97**, 651–658.

Kohout, M. (2007). Bonding indicators from electron pair density functionals. *Faraday Discussions*, **135**, 43–54.

Kohout, M., Pernal, K., Wagner, F. R., and Grin, Y. (2004). Electron localizability indicator for correlated wavefunctions. I. Parallel-spin pairs. *Theor. Chem. Acc.*, **112**, 453–459.

Kohout, M., Pernal, K., Wagner, F. R., and Grin, Y. (2005). Electron localizability indicator for correlated wavefunctions. II. Antiparallel-spin pairs. *Theor. Chem. Acc.*, **113**, 287–293.

Kohout, M., Wagner, F. R., and Grin, Y. (2002). Electron localization function for transition-metal compounds. *Theor. Chem. Acc.*, **108**, 150–156.

Kohout, M., Wagner, F. R., and Grin, Y. (2006). Atomic shells from the electron localizability in momentum space. *Int. J. Quant. Chem.*, **106**, 1499–1507.

Kohout, M., Wagner, F. R., and Grin, Y. (2007). Electron localizability indicator for correlated wavefunctions. III: singlet and triplet pairs. *Theor. Chem. Acc.*, **119**, 413–420.

Komura, Y. and Kitano, Y. (1977). Long-period stacking variants and their electron-concentration dependence in the Mg-base Friauf-Laves phases. *Acta Crystallogr. B*, **33**, 2496–2501.

Komura, Y., Sly, W. G., and Shoemaker, D. P. (1960). The crystal structure of the R phase, Mo–Co–Cr. *Acta Crystallogr. B*, **13**, 575–585.

Kontio, A., Stevens, E. D., Coppens, P., Brown, R. D., Dwight, A. E., and Williams, J. M. (1980). New investigation of the structure of Mn_4Al_{11}. *Acta Crystallogr. B*, **36**, 435–436.

Kozak, R., Sologubenko, A., and Steurer, W. (2015). Single-phase high-entropy alloys – an overview. *Z. Kristallogr.*, **230**, 55–68.

Kreiner, G. and Schäpers, M. (1997). A new description of Samson's Cd_3Cu_4 and a model of icosahedral i-CdCu. *J. Alloys Compd.*, **259**, 83–114.

Kreiner, G. and Spiekermann, S. (1997). Investigations in the Ag–Mg and Ag–Al–Mg systems I. Models for cubic approximants of icosahedral quasicrystals in the Ag–Al–Mg system. *J. Alloys Compd.*, **261**, 62–82.

Krieger-Beck, P., A., Brodbeck, and Strähle, J. (1989). Synthese und Struktur von K_2Au_3, einer neuen Phase im System Kalium-Gold. *Z. Naturforsch.*, **44b**, 237–239.

Kuczera, P. and Steurer, W. (2015). Cluster-based growth algorithm for decagonal quasicrystals. *Phys. Rev. Lett.*, **115**, 085502.

Kuczera, P., Wolny, J., and Steurer, W. (2012). Comparative structural study of decagonal quasicrystals in the systems Al–Cu–Me (Me = Co, Rh, Ir). *Acta Crystallogr. B*, **68**, 578–589.

Kuczera, P., Wolny, J., and Steurer, W. (2014). High-temperature structural study of decagonal Al–Cu–Rh. *Acta Crystallogr. B*, **70**, 306–314.

Kuo, K. H. (2002). Mackay, anti-Mackay, double-Mackay, pseudo-Mackay, and related icosahedral shell clusters. *Struct. Chem.*, **13**, 221–230.

Laio, A. and Parrinello, M. (2002). Escaping free-energy minima. *Proc. Natl. Acad. Sci. USA*, **99**, 12562–12566.

Landrum, G. A., Hoffmann, R., Evers, J., and Boysen, H. (1998). The TiNiSi family of compounds: Structure and bonding. *Inorg. Chem.*, **37**, 5754–5763.

Lawson, A. C., Goldstone, J. A., Cort, B., Martinez, R. J., Vigil, F. A., Zocco, T. G., Richardson, J. W. Jr., and Mueller, M. H. (1996). Structure of ζ-Phase Plutonium-Uranium. *Acta Crystallogr. B*, **52**, 32–37.

Le Bail, A. (2010). Databases of virtual inorganic crystal structures and their applications. *Phys. Chem. Chem. Phys.*, **12**, 8521–8530.

Le Bail, A., Leblanc, M., and Audier, M. (1991). Crystalline phases related to the ico-sahedral Al–Li–Cu phase: a single-crystal X-ray diffraction study of the hexagonal Z-Al$_{59}$Cu$_5$Li$_{26}$Mg$_{10}$ phase. *Acta Crystallogr. B*, **47**, 451–457.

Leblanc, M., Le Bail, A., and Audier, M. (1991). Crystalline phases related to the icosahedral Al-Li-Cu phase. *Physica B:*, **173**, 329–355.

Lee, S. (1991). Elemental structures of the heavy main group atoms and the second moment scaling hypothesis. *J. Am. Chem. Soc.*, **113**, 8611–8614.

Lee, S., Hoffmann, R., and Fredrickson, D. C. (2007). Sich durchdringende polare und unpolare Untergitter in intermetallischen Phasen: die Struktur von NaCd$_2$. *Angew. Chem.*, **119**, 2004–2023.

Leger, J. M., Oki, K., Rossar-Mignot, J., and Vogt, O. (1985). Structural transition and volume compression of CeBi up to 20 GPa. *J. Phys. France*, **46**, 889–894.

Leger, J. M., Ravot, D., and Rossar-Mignot, J. (1984). Volume behaviour of CeSb and LaSb up to 25 GPa. *J. Phys. C*, **17**, 4935–4943.

Legut, D., Friák, M., and Šob, M. (2007). Why is polonium simple cubic and so highly anisotropic? *Phys. Rev. Lett.*, **99**, 016402.

Levine, D. J. and Steinhardt, P. J. (1986). Quasicrystals. I. Definition and structure. *Phys. Rev. B*, **34**, 596–616.

Li, B. and Corbett, J. D. (2003). Synthesis and characterization of the new cluster phase K$_{39}$In$_{80}$. Three K–In compounds with remarkably specific and transferable cation dispositions. *Inorg. Chem.*, **42**, 8768–8772.

Li, B. and Corbett, J. D. (2006). Electronic stabilization effects: Three new K–In–T (T = Mg, Au, Zn) network compounds. *Inorg. Chem.*, **45**, 8958–8964.

Li, J., Dong, X., Jin, Y., and Fan, C. Z. (2014). Mechanical properties, electronic properties and phase stability of Mg under pressure: A first-principles study. *Intern. J. Modern. Phys. B*, **28**, 1450200.

Li, M. R., Hovmöller, S., Sun, J. L., Zou, X. D., and Kuo, K. H. (2008). Crystal structure of the 2/1 cubic approximant Ag$_{42}$In$_{42}$Yb$_{16}$. *J. Alloys Compd.*, **465**, 132–138.

Lidin, S., Jacob, M., and Larsson, A.-K. (1994). (Fe, Ni)Zn$_{6.5}$, a superstructure of γ-brass. *Acta Crystallogr. C*, **50**, 340–342.

Lifshitz, R. (2014). Explaining complex metals with polymers. *Proc. Natl. Acad. Sci. USA*, **111**, 17698–17699.

Lin, Q. S. and Corbett, J. D. (2010). Development of an icosahedral quasicrystal and two approximants in the Ca-Au-Sn system: Syntheses and structural analyses. *Inorg. Chem.*, **49**, 10436–10444.

Lin, Q. S. and Corbett, J. D. (2006). The 1/1 and 2/1 approximants in the Sc–Mg–Zn quasicrystal system: Triacontahedral clusters as fundamental building blocks. *J. Am. Chem. Soc.*, **128**, 13628–13273.

Lin, Q. S. and Corbett, J. D. (2008). Li$_{14.7}$Mg$_{36.8}$Cu$_{21.5}$Ga$_{66}$: An intermetallic representative of a type IV clathrate. *Inorg. Chem.*, **47**, 10825–10831.

Lindbaum, A., Heathman, S., Bihan, T. Le, Haire, R.G., Idiri, M., and Lander, G.H. (2003). High-pressure crystal structures of actinide elements. *J. Phys.: Condens. Matter*, **15**, S2297–S2303.

Lindbaum, A., Heathman, S., Kresse, G., Rotter, M., Gratz, E., Schneidewind, A., Behr, G., Litfin, K., LeBihan, T., and Svoboda, P. (2000a). Structural stability of LaCu$_2$ and YCu$_2$ studied by high-pressure X-ray diffraction and ab initio total energy calculations. *J. Phys.: Condens. Matter*, **12**, 3219–3228.

Lindbaum, A., Heathman, S., Le Bihan, T., and Rogl, P. (2000b). Pressure-induced orthorhombic distortion of UMn$_2$. *J. Alloys Compd.*, **298**, 177–180.

Ling, R. G. and Belin, C. (1982). Structure of the intermetallic compound $Na_{22}Ga_{39}$ ($\approx$ 36.07% Na). *Acta Crystallogr. B*, **38**, 1101–1104.

Liu, Y., Su, X. P., Yin, F. C., Li, Z., and Liu, Y. H. (2008). Experimental determination and atomistic simulation on the structure of $FeZn_{13}$. *J. Phase Equilib. Diff.*, **29**, 488–492.

Liu, Z. L. (2014). Muse: Multi-algorithm collaborative crystal structure prediction. *Comp. Phys. Comm.*, **185**, 1893–1900.

Liu, Z. L., Cai, L. C., Zhang, X. L., and Xi, F. (2013). Predicted alternative structure for tantalum metal under high pressure and high temperature. *J. Appl. Phys.*, **114**, 073520.

Loa, I., Nelmes, R. J., Lundegaard, L. F., and McMahon, M. I. (2012). Extraordinarily complex crystal structure with mesoscopic patterning in barium at high pressure. *Nat. Mater.*, **11**, 627–632.

Lommerse, J. P. M., Motherwell, W. D. S., Ammon, H. L., Dunitz, J. D., Gavezzotti, A., Hofmann, D. W. M., Leusen, F. J. J., Mooij, W. T. M., Price, S. L., Schweizer, B., Schmidt, M. U., van Eijck, B. P., Verwer, P., and Williams, D. E. (2000). A test of crystal structure prediction of small organic molecules. *Acta Crystallogr. B*, **56**, 697–714.

Lucas, M. S., Mauger, L., Muñoz, J. A., Xiao, Y., Sheets, A. O., Semiatin, S. L., Horwath, J., and Turgut, Z. (2011). Magnetic and vibrational properties of high-entropy alloys. *J. Appl. Phys.*, **109**, 07E307.

Lucas, M. S., Wilks, G. B., Mauger, L., Munoz, J. A., Senkov, O. N., Michel, E., Horwath, J., Semiatin, S. L., Stone, M. B., Abernathy, D. L., and Karapetrova, E. (2012). Absence of long-range chemical ordering in equimolar FeCoCrNi. *Appl. Phys. Lett.*, **100**, 251907.

Luck, J. M., Godréche, C., Janner, A., and Janssen, T. (1997). The nature of the atomic surfaces of quasiperiodic self-similar structures. *J. Phys. A: Math. Gen.*, **26**, 1951–1999.

Lundegaard, L. F. Stinton, G. W., Zelazny, M., Guillaume, C. L., Proctor, J. E., Loa, I., Gregoryanz, E., Nelmes, R. J., and McMahon, M. I. (2013). Observation of a reentrant phase transition in incommensurate potassium. *Phys. Rev. B*, **88**, 054106.

Lundegaard, L. F., Gregoryanz, E., McMahon, M. I., Guillaume, C. L., Loa, I., and Nelmes, R. J. (2009a). Single-crystal studies of incommensurate Na to 1.5 Mbar. *Phys. Rev. B*, **79**, 064105.

Lundegaard, L. F., Marqués, M., Stinton, G., Ackland, G. J., Nelmes, R. J., and McMahon, M. I. (2009b). Observation of the *oP*8 crystal structure in potassium at high pressure. *Phys. Rev. B*, **80**, 020101.

Lux, R., Kuntze, V., and Hillebrecht, H. (2012). Synthesis and crystal structure of cubic $V_{11}Cu_9Ga_{46}$ – A 512-fold super structure of a simple *bcc* packing. *Solid State Sci.*, **14**, 1445–1453.

Lv, J., Wang, Y. C., Zhu, L., and Ma, Y. M. (2011). Predicted novel high-pressure phases of lithium. *Phys. Rev. Lett.*, **106**, 015503.

Lyakhov, A. O., Oganov, A. R., Stokes, H. T., and Zhu, Q. (2013). New developments in evolutionary structure prediction algorithm USPEX. *Comp. Phys. Comm.*, **184**(4), 1172–1182.

Lyakhov, A. O., Oganov, A. R., and Valle, M. (2010). How to predict very large and complex crystal structures. *Comp. Phys. Comm.*, **181**, 1623–1632.

Ma, J., Karaman, I., and Noebe, R. D. (2010). High temperature shape memory alloys. *Intern. Mater. Rev.*, **55**, 257–315.

Ma, S. G., Zhang, S. F., Gao, M. C., Liaw, P. K., and Zhang, Y. (2013). A successful synthesis of the $CoCrFeNiAl_{0.3}$ single-crystal, high-entropy alloy by Bridgman solidification. *JOM*, **65**(12), 1751–1758.

Ma, Y. M., Eremets, M., Oganov, A. R., Xie, Y., Trojan, I., Medvedev, S., Lyakhov, A. O., Valle, M., and Prakapenka, V. (2009). Transparent dense sodium. *Nature*, **458**, 182–185.

Ma, Y. M., Oganov, A. R., and Xie, Y. (2008). High-pressure structures of lithium, potassium, and rubidium predicted by an ab initio evolutionary algorithm. *Phys. Rev. B*, **78**, 014102.

Macaluso, R. T., Wells, D. M., Sykora, R. E., Albrecht-Schmitt, T. E., Mar, A., Nakatsuji, S., Lee, H., Fisk, Z., and Chan, J. Y. (2004). Structure and electrical resistivity of CeNiSb$_3$. *J. Solid State Chem.*, **177**, 293–298.

Macia, E. (2001). Theoretical prospective of quasicrystals as thermoelectric materials. *Phys. Rev. B*, **64**, 094206.

Maddox, J. (1988). Crystals from first principles. *Nature*, **335**, 201.

Maiti, S. (2015). *Local Structure and Related Properties of Refractory High-Entropy Alloys. PHD Thesis ETH Zurich.* Ph.D. thesis, ETH Zurich.

Mao, H. K., Wu, Y., Shu, J. F., Hu, J. Z., Hemley, R. J., and Cox, D. E. (1990). High-pressure phase-transition and equation of state of lead to 238 GPa. *Solid State Commun.*, **74**, 1027–1029.

Mao, W. L., Wang, L., Ding, Y., Yang, W., Liu, W., Kim, D. Y., Luo, W., Ahuja, R., Meng, Y., Sinogeikin, S., Shu, J., and Mao, H. K. (2010). Distortions and stabilization of simple-cubic calcium at high pressure and low temperature. *Proc. Natl. Acad. Sci. USA*, **107**, 9965–9968; corrections on p. 12734.

Marples, J. A. C. (1965). On the thermal expansion of protactinium metal. *Acta Crystallogr.*, **18**, 815–817.

Marqués, M., McMahon, M. I., Gregoryanz, E., Hanfland, M., Guillaume, C. L., Pickard, C. J., Ackland, G. J., and Nelmes, R. J. (2011). Crystal structures of dense lithium: A metal-semiconductor-metal transition. *Phys. Rev. Lett.*, **106**, 095502.

Marsh, R. E. and Slagle, K. M. (1988). On the structure of Ca$_8$In$_3$. *Acta Crystallogr. C*, **44**, 395–396.

Martoňák, R., Laio, A., Bernasconi, M., Ceriani, C., Raiteri, P., Zipoli, F., and Parrinello, M. (2005). Simulation of structural phase transitions by metadynamics. *Z. Kristallogr.*, **220**, 489–498.

Martoňák, R., Laio, A., and Parrinello, M. (2003). Predicting crystal structures: The Parrinello-Rahman method revisited. *Phys. Rev. Lett.*, **90**, 075503.

Marx, D. and Savin, A. (1997). Topological bifurcation analysis: Electronic structure of CH$_5^+$. *Angew. Chem. Intern. Ed.*, **36**, 2077–2080.

Masakova, Z., Patera, J., and Zich, J. (2005). Classification of Voronoi and Delone tiles of quasicrystals: III. Decagonal acceptance window of any size. *J. Phys. A Math. Gen.*, **38**, 1947–1960.

Massalski, T. B. (1990). *Binary Alloy Phase Diagrams*. Volume 1–3. ASM International, USA.

Massalski, T. B. and Mizutani, U. (1978). Electronic structure of Hume-Rothery phases. *Progr. Mater. Sci.*, **22**, 151–262.

McHargue, C. J. and Yakel, H. L. (1960). Phase transformations in cerium. *Acta Met.*, **8**, 637–646.

McMahan, A. K. (1985). Alkali-metal structures above the s-d transition. *Phys. Rev. B*, **29**, 5982–5985.

McMahon, M. I., Bovornratanaraks, T., Allan, D. R., Belmonte, S. A., and Nelmes, R. J. (2000). Observation of the incommensurate barium-IV structure in strontium phase V. *Phys. Rev. B*, **61**, 3135–3138.

McMahon, M. I., Degtyareva, O., and Nelmes, R. J. (2007). Incommensurate modulations of Bi-III and Sb-II. *Phys. Rev. B*, **75**, 184114.

McMahon, M. I., Lundegaard, L. F., Hejny, C., Falconiand, S., and Nelmes, R. J. (2006*a*). Different incommensurate composite crystal structure for Sc-II. *Phys. Rev. B*, **73**, 134102.

McMahon, M. I. and Nelmes, R. J. (1997). Different results for the equilibrium phases of cerium above 5 GPa. *Phys. Rev. Lett.*, **78**, 3884–3887.

McMahon, M. I. and Nelmes, R. J. (2006). High-pressure structures and phase transformations in elemental metals. *Chem. Soc. Rev.*, **35**, 943–963.

McMahon, M. I., Nelmes, R. J., and Rekhi, S. (2001). Complex crystal structure of cesium-III. *Phys. Rev. Lett.*, **87**, 255502.

McMahon, M. I., Nelmes, R. J., Schwarz, U., and Syassen, K. (2006*b*). Composite incommensurate K-III and a commensurate form: Study of a high-pressure phase of potassium. *Phys. Rev. B*, **74**, 140102.

McMahon, M. I., Rekhi, S., and Nelmes, R. J. (2006*c*). Pressure dependent incommensuration in Rb-IV. *Phys. Rev. Lett.*, **87**, 055501.

Mei, Q. S. and Lu, K. (2007). Melting and superheating of crystalline solids: From bulk to nanocrystals. *Progr. Mater. Sci.*, **52**, 1175–1262.

Meredig, B., Agrawal, A., Kirklin, S., Saal, J. E., Doak, J. W., Thompson, A., Zhang, K., Choudhary, A., and Wolverton, C. (2014). Combinatorial screening for new materials in unconstrained composition space with machine learning. *Phys. Rev. B*, **89**, 094104.

Méresse, Y., Heathman, S., Rijkeboer, C., and Rebizant, J. (1999). High pressure behaviour of PuBi studied by X-ray diffraction. *J. Alloys Compd.*, **284**, 65–69.

Miao, M. S. and Hoffmann, R. (2014). High pressure electrides: A predictive chemical and physical theory. *Acc. Chem. Res.*, **47**, 1311–1317.

Michel, K. J. and Wolverton, C. (2014). Symmetry building Monte Carlo-based crystal structure prediction. *Comp. Phys. Comm.*, **185**, 1389–1393.

Miedema, A. R. (1976). On the heat of formation of solid alloys. II. *J. Less-Common Met.*, **46**, 67–83.

Mihalkovic, M. and Widom, M. (2012). Structure and stability of Al_2Fe and Al_5Fe_2: First-principles total energy and phonon calculations. *Phys. Rev. B*, **85**, 014113.

Miller, G. J. (1996). Structure and bonding at the Zintl border. In *Chemistry, Structure and Bonding of Zintl Phases and Ions* (ed. S. M. Kauzlarich), pp. 1–59. VCH Publishers, New York, USA.

Miller, G. J., Schmidt, M. W., Wang, F., and You, T. S. (2011). Quantitative advances in the Zintl-Klemm formalism. *Structure and Bonding*, **139**, 1–55.

Mingos, D. M. P. (1972). A general theory for cluster and ring compounds of the main group and transition elements. *Nature*, **236**, 99–102.

Mingos, D. M. P. (1984). Polyhedral skeletal electron pair approach. *Acc. Chem. Res.*, **17**, 311–319.

Mizutani, U., Noritake, T., Ohsuna, T., and Takeuchi, T. (2010). Hume-Rothery electron concentration rule across a whole solid solution range in a series of gamma-brasses in Cu–Zn, Cu–Cd, Cu–Al, Cu–Ga, Ni–Zn and Co–Zn alloy systems. *Philos. Mag.*, **90**, 1985–2008.

Mizutani, U., Sato, H., Inukai, M., and Zijlstra, E. S. (2014). Theoretical foundation for the Hume-Rothery electron concentration rule for structurally complex alloys. *Acta Phys. Pol. A*, **126**, 531–534.

Montgomery, J. M., Smudrala, G. K., Tsoi, G. M., and Vohra, Y. K. (2011). High-pressure phase transitions in rare earth metal thulium to 195 GPa. *J. Phys.: Condens. Matter*, **23**, 155701.

Moore, K. T. and van der Laan, G. (2009). Nature of the 5f states in actinide metals. *Rev. Mod. Phys.*, **81**, 235–298.

Moreau, J. M., Paccard, D., and Gignoux, D. (1974). The crystal structure of Er_3Ni_2. *Acta Crystallogr. B*, **30**, 2122–2126.

Moreau, J. M., Paccard, D., and Parté, E. (1976). The tetragonal crystal structure of R_3Rh_2 compounds with R = Gd, Tb, Dy, Ho, Er, Y. *Acta Crystallogr. B*, **32**, 1767–1771.

Motherwell, W. D. S., Ammon, H. L., Dunitz, J. D., Dzyabchenko, A., Erk, P., Gavezzotti, A., Hofmann, D. W. M., Leusen, F. J. J., Lommerse, J. P. M., Mooij, W. T. M., Price, S. L., Scheraga, H., Schweizer, B., Schmidt, M. U., van Eijck, B. P., Verwer, P., and Williams, D. E. E. (2002). Crystal structure prediction of small organic molecules: A second blind test. *Acta Crystallogr. B*, **58**, 647–661.

Mudryk, J., Pecharsky, V. K., and Gschneidner Jr., K. A. (2011). Extraordinary responsive intermetallic compounds: The R_5T_4 family (R = rare earth, T = group 13-15 element. *Z. Anorg. Allg. Chem.*, **637**, 1948–1956.

Muduli, P. K., Rice, W. C., He, L., Collins, B. A., Chu, Y. S., and Tsui, F. (2009). Study of magnetic anisotropy and magnetization reversal using the quadratic magnetooptical effect in epitaxial $Co_xMn_yGe_z(111)$ films. *J. Phys.: Condens. Matter*, **21**, 296005.

Mukherjee, D., Sahoo, B.D., Joshi, K.D., and Gupta, S. C. (2015). High pressure phase transition in Zr–Ni binary system: A first principle study. *J. Alloys Compd.*, **648**, 951–957.

Mulliken, R. S. (1934). A new electroaffinity scale; together with data on valence states and on valence ionization potentials and electron affinities. *J. Chem. Phys.*, **2**, 782–793.

Nakamoto, Y., Sakata, M., and Shimizu, K. (2010). Ca-VI: A high-pressure phase of calcium above 158 GPa. *Phys. Rev. B*, **81**, 140106.

Nasch, T. and Jeitschko, W. (1999). Niobium and molybdenum compounds with high zinc content: $NbZn_3$, $NbZn_{16}$, and $MoZn_{20.44}$. *J. Solid State Chem.*, **143**, 95–103.

Nelmes, R. J., Allan, D. R., McMahon, M. I., and Belmonte, S. A. (1999). Self-hosting incommensurate structure of barium IV. *Phys. Rev. Lett.*, **83**, 4081–4084.

Nelmes, R. J., Liu, S. A., Belmonte, S. A., Loveday, J. S., Allan, D. R., and McMahon, M. I. (1996). *Imma* phase of germanium at 80 GPa. *Phys. Rev. B*, **53**, R2907–2909.

Nelmes, R. J., McMahon, M. I., Loveday, J. S., and Rekhi, S. (2002). Structure of Rb-III: Novel modulated stacking structures in alkali metals. *Phys. Rev. Lett.*, **88**, 155503.

Nesper, R. (2014). The Zintl-Klemm Concept – A Historical Survey. *Z. Anorg. Allg. Chem.*, **640**, 2639–2648.

Nolas, G. S., Kaeser, M., Littleton IV, R. T., and Tritt, T. M. (2000). High figure of merit in partially filled ytterbium skutterudite materials. *Appl. Phys. Lett.*, **77**, 1855–1857.

Nyman, H., Carroll, C. E., and Hyde, B. G. (1991). Rectilinear rods of face-sharing tetrahedra and the structure of β-Mn. *Z. Kristallogr.*, **196**, 39–46.

Oers, T., Takakura, H., Abe, E., and Steurer, W. (2014). The quasiperiodic average structure of highly disordered decagonal Zn–Mg–Dy and its temperature dependence. *Acta Crystallogr. B*, **70**, 315–330.

Oesterreicher, H. (1973). X-ray and neutron diffraction study of ordering on crystallographic sites in rare-earth-base alloys containing alumninium and transition metals. *J. Less-Comm. Met.*, **33**, 25–41.

Oganov, A.R., Ma, Y., Glass, C.W., and Valle, M. (2007). Evolutionary crystal structure prediction: Overview of the USPEX method and some of its applications. *Psi-k Newsletter*, **84**, 142–171.

Oganov, A.R., Ma, Y. M., Lyakhov, A. O., Valle, M., and Gatti, C. (2010*a*). Evolutionary crystal structure prediction and novel high-pressure phases. In *High-Pressure Crystallography: From Fundamental Phenomena to Technological Applications* (ed. E. Boldyreva and P. Dera), NATO Science for Peace and Security Series B: Physics and Biophysics, pp. 293–323. Springer Netherlands.

Oganov, A. R. (ed.) (2011). *Modern Methods of Crystal Structure Prediction.* WILEY-VCH Verlag & Co. KGaA. ISBN: 978-3-527-40939-6.

Oganov, A. R. and Glass, C. W. (2006). Crystal structure prediction using *ab initio* evolutionary techniques: Principles and applications. *J. Chem. Phys.*, **124**, 244704.

Oganov, A. R., Lyakhov, A. O., and Valle, M. (2011). How evolutionary crystal structure prediction works – and why. *Acc. Chem. Res.*, **44**, 227–37.

Oganov, A. R., Ma, Y. M., Lyakhov, A. O., Valle, M., and Gatti, C. (2010*b*). Evolutionary crystal structure prediction as a method for the discovery of minerals and materials. *Rev. Min. Geochem.*, **71**, 271–298.

Oganov, A. R., Ma, Y. M., Xu, Y., Errea, I., Bergara, A., and Lyakhov, A. O. (2010*c*). Exotic behavior and crystal structures of calcium under pressure. *Proc. Natl. Acad. Sci. USA*, **107**, 7646–7651.

Oganov, A. R. and Valle, M. (2009). How to quantify energy landscapes of solids. *J. Chem. Phys.*, **130**, 104504.

Okamoto, N. L., Tanaka, K., Yashuara, A., and Inui, H. (2014). Structure refinement of the δ_{1p} phase in the Fe–Zn system by single-crystal X-ray diffraction combined with scanning transmission electron microscopy. *Acta Crystallogr. B*, **70**, 275–282.

O'Keeffe, M. and Hyde, B. G. (1980). Plane nets in crystal chemistry. *Phil. Trans. R. Soc. A*, **295**, 553–618.

O'Keeffe, M. and Hyde, B. G. (1996). *Crystal Structures. I. Patterns and Symmetry.* Mineralogical Society of America, Washington, USA.

Oliynyk, A. O., Stoyko, S. S., and Mar, A. (2015). Many metals make the cut: Quaternary rare-earth germanides $RE_4M_2InGe_4$ (M = Fe, Co, Ni, Ru, Rh, Ir) and $RE_4RhInGe_4$ derived from excision of slabs in RE_2InGe_2. *Inorg. Chem.*, **54**, 2780–2792.

Ōnuki, Y., Settai, R., Haga, Y., Machida, Y., Izawa, K., Honda, F., and Aoki, D. (2014). Fermi surface, magnetic, and superconducting properties in actinide compounds. *C.R. Physique*, **15**, 616–629.

Ormeci, A. and Grin, Y. (2011). Chemical bonding in Al_5Co_2: The electron localizability-electron density approach. *Isr. J. Chem.*, **51**, 1349–1354.

Ormeci, A., Simon, A., and Grin, Y. (2010). Structural topology and chemical bonding in Laves phases. *Angew. Chem. Int. Ed.*, **49**, 8997–9001.

Otsuka, K. and Ren, X. B. (1999). Recent developments in the research of shape memory alloys. *Intermetallics*, **7**, 511–528.

Otto, F., Yang, Y., Bei, H., and George, E. P. (2013). Relative effects of enthalpy and entropy on the phase stability of equiatomic high-entropy alloys. *Acta Mater.*, **61**, 2628–2638.

Pankova, A. A., Blatov, V. A., Ilyushin, G. D., and Proserpio, D. M. (2013). γ-Brass polyhedral core in intermetallics: The nanocluster model. *Inorg. Chem.*, **52**, 13094–13107.

Pankova, A. A., Ilyushin, G. D., and Blatov, V. A. (2012). Nanoclusters based on pentagon-dodecahedra with shells in the form of D32, D42, and D50 deltahedra in crystal structures of intermetallic compounds. *Crystallogr. Rep.*, **57**, 1–9.

Pannetier, J., Bassas-Alsina, J., Rodriguez-Carvajal, J., and Caignaert, V. (1990). Prediction of crystal structures from crystal chemistry rules by simulated annealing. *Nature*, **346**, 343–345.

Pauling, L. (1932). The nature of the chemical bond. IV. The energy of single bonds and the relative electronegativity of atoms. *J. Am. Chem. Soc.*, **54**, 3570–3582.

Pauling, L. (1960). *The Nature of the Chemical Bond*. Cornell University Press, New York.

Pauling, L. (1985). Apparent icosahedral symmetry is due to directed multiple twinning of cubic crystals. *Nature*, **317**, 512–514.

Pauling, L. (1989). Icosahedral quasicrystals of intermetallic compounds are icosahedral twins of cubic crystals of three kinds, consisting of large (about 5000 atoms) icosahedral complexes in either a cubic body-centered or a cubic face-centered arrangement or smaller (about 1350 atoms) icosahedral complexes in the β-tungsten arrangement. *Proc. Natl. Acad. Sci. USA*, **86**, 8595–8599.

Pavlovich, A. and Klèman (1987). Generalised 2D Penrose tilings: Structural properties. *J. Phys. A: Math. Gen.*, **20**, 687–702.

Pearson, R. G. (1985). Absolute electronegativity and absolute hardness of Lewis acids and bases. *J. Am. Chem. Soc.*, **107**, 6801–6806.

Pearson, R. G. (1988). Absolute electronegativity and hardness: Application to inorganic chemistry. *Inorg. Chem.*, **27**, 734–740.

Pearson, W. B. (1972). *The crystal chemistry and physics of metals and alloys*. Wiley-Interscience, New York, USA.

Pecharskii, V. K., Bodak, O. I., Bel'skii, V. K., Starodub, P. K., Mokra, I. R., and Glady-shevskii, E. I. (1987). Crystal structure of $Tb_{117}Fe_{52}Ge_{112}$. *Sov. Phys. Crystallogr.*, **32**, 194–196.

Pendás, A. M., Kohout, M., Blanco, M. A., and Francisco, E. (2012). Beyond standard charge density topological analyses. In *Modern Charge-Density Analysis* (ed. C. Gatti and P. Macchi), Chapter 9, pp. 303–358. Springer Netherlands, Dordrecht.

Peng, B. Y. and Fu, X. J. (2015). Configurations of the Penrose tiling beyond nearest neighbors. *Chin. Phys. Lett.*, **32**, 056101.

Penrose, R. (1974). The rôle of aesthetics in pure and applied mathematical research. *Bull. Inst. Math. Appl.*, **10**, 266–271.

Perez-Mato, J. M., Elcoro, L., Aroyo, M. I., Katzke, H., Tolédano, P., and Izaola, Z. (2006). Apparently complex high-pressure phase of gallium as a simple modulated structure. *Phys. Rev. Lett.*, **97**, 115501.

Pettifor, D. G. (1984). A chemical scale for crystal-structure maps. *Solid State Commun.*, **51**, 31–34.

Pettifor, D. G. (1986). The structure of binary compounds: I. Phenomenological structure maps. *J. Phys. C*, **19**, 285–313.

Pettifor, D. G. (1988). Structure maps for pseudobinary and ternary phases. *Mater. Sci. Techn.*, **4**, 675–691.

Pettifor, D. G. (1995). *Bonding and Structure of Molecules and Solids*. Oxford University Press, New York.

Pettifor, D. G. and Podloucky, R. (1984). Microscopic theory of the structural stability of pd-bonded *AB* compounds. *Phys. Rev. Lett.*, **53**, 1080–1083.

Pettifor, D. G. and Podloucky, R. (1985). Pettifor and Podloucky respond. *Phys. Rev. Lett.*, **55**, 261.

Pettifor, D. G. and Podloucky, R. (1986). The structure of binary compounds: II. Theory of the pd-bonded AB compounds. *J. Phys. C*, **19**, 315–330.

Pfleiderer, C. (2009). Superconducting phases of f-electron compounds. *Rev. Mod. Phys.*, **81**, 1551–1624.

Phan, M. H., Woods, G. T., Chaturvedi, A., Stefanoski, S., Nolas, G. S., and Srikanth, H. (2008). Long-range ferromagnetism and giant magnetocaloric effect in type VIII $Eu_8Ga_{16}Ge_{30}$ clathrates. *Appl. Phys. Lett.*, **93**, 252505.

Pickard, C. J. and Needs, R. J. (2010). Aluminium at terapascal pressures. *Nat. Mater.*, **9**, 624–627.

Pickard, C. J. and Needs, R. J. (2011). Ab initio random structure searching. *J. Phys.: Condens. Matter*, **23**, 053201.

Ponou, S., Fässler, T. F., and Kienle, L. (2008). Structural complexity in intermetallic alloys: Long-periodic order beyond 10 nm in the system $BaSn_3/BaBi_3$. *Angew. Chem. Int. Ed.*, **47**, 3999–4004.

Poole Jr., C. P., Farach, H. A., and Creswick, R. J. (1995). *Superconductivity*. Academic Press, San Diego, California, USA.

Porsch, F. and Holzapfel, W. B. (1993). Novel reentrant high pressure phase transtion in lanthanum. *Phys. Rev. Lett.*, **70**, 4087–4089.

Pöttgen, R., Gravereau, P., Darriet, B., Chevalier, B., Hickey, E., and Etourneau, J. (1994). Crystal structure of the ternary silicide U_2RuSi_3: A new ordered version of the hexagonal AlB_2-type structure. *J. Mater. Chem.*, **4**, 463–467.

Pöttgen, R. and Johrendt, D. (2000). Equiatomic intermetallic europium compounds: Syntheses, crystal chemistry, chemical bonding, and physical properties. *Chem. Mater.*, **12**, 875–897.

Rabson, D. A., Mermin, N. D., Rokhsar, D. S., and Wright, D. C. (1991). The space-groups of axial crystals and quasi-crystals. *Rev. Mod. Phys.*, **63**, 699–733.

Ranganathan, S. (2003). Multimaterial cocktails. *Curr. Sci.*, **85**, 1404–1406.

Ranganathan, S. and Inoue, A. (2006). An application of Pettifor structure maps for the identification of pseudo-binary quasicrystalline intermetallics. *Acta Mater.*, **54**, 3647–3656.

Range, K. J., Grosch, G. H., Rau, F., and Klement, U. (1994). Hochdrucksynthese und kristallstruktur von Rb_3Au_7 [1]. *Z. Naturforsch.*, **49b**, 27–30.

Rawn, C. J., Schneibel, J. H., Hoffmann, C. M., and Hubbard, C. R. (2001). The crystal structure and thermal expansion of Mo_5SiB_2. *Intermetallics*, **9**, 209–216.

Raynor, G. V. (1949). Progress in the theory of metals. *Prog. Met. Phys.*, **1**, 1–76.

Reed, S. K. and Ackland, G. J. (2000). Theoretical and computational study of high-pressure structures in barium. *Phys. Rev. Lett.*, **84**, 5580–5583.

Rohrer, F. E., Lind, H., Eriksson, L., Larsson, A. K., and Lidin, S. (2000). On the question of commensurability – the Nowotny chimney-ladder structures revisited. *Z. Kristallogr.*, **215**, 650–660.

Rohrer, F. E., Lind, H., Eriksson, L., Larsson, A. K., and Lidin, S. (2001). Incommensurately modulated Nowotny chimney-ladder phases $Cr_{1-x}Mo_xGe_{1.75}$ with $x = 0.65$ and 0.84. *Z. Kristallogr.*, **215**, 650–660.

Romero Gómez, J., Ferreiro Garcia, R., De Miguel Catoira, A., and Romero Gomez, M. (2013). Magnetocaloric effect: A review of the thermodynamic cycles. *Ren. Sust. Energy Rev.*, **17**, 74–82.

Rousseau, B., Xie, Y., Ma, Y., and Bergara, A. (2011). Exotic high pressure behavior of light alkali metals, lithium and sodium. *Eur. Phys. J. B*, **81**(1), 1–14.

Rousseau, R., Uehara, K., Klug, D. D., and Tse, J. S. (2005). Phase stability and broken-symmetry transition of elemental lithium up to 140 GPa. *Chem. Phys. Chem.*, **6**, 1703–1706.

Ruban, A.V., Khmelevskyi, S., Mohn, P., and Johansson, B. (2007). Magnetic state, magnetovolume effects, and atomic order in $Fe_{65}Ni_{35}$ invar alloy: A first principles study. *Phys. Rev. B*, **76**, 014420.

Ruck, M. (1996). Kristallstruktur und Zwillingsbildung der intermetallischen Phase β-Bi_2Rh. *Acta Crystallogr. B*, **52**, 605–609.

Sahu, P. C., Shekar, N. V. C., Subramanian, N., Yousuf, M., and Rajan, K. G. (1995). Crystal Structure of UAl_2 above 10 GPa at 300 K. *J. Alloys Compd.*, **223**, 49–52.

Salamakha, P., Sologub, O., Bocelli, G., Otani, S., and Takabatake, T. (2001). $Dy_{117}Co_{57}Sn_{112}$, a new structure type of ternary intermetallic stannides with a giant unit cell. *J. Alloys Compd.*, **314**, 177–180.

Salamat, A., Briggs, R., Bouvier, P., Petitgirard, S., Dewaele, A., Cutler, M. E., Cora, F., Daisenberger, D., Gabarino, G., and McMillan, P. F. (2013). High-pressure structural transformations of Sn up to 138 GPa: Angle-dispersive synchrotron X-ray diffraction study. *Phys. Rev. B*, **88**, 104104.

Salamat, A., Gabarino, G., Dewaele, A., Bouvier, P., Petitgirard, S., Pickard, C. J., McMillan, P. F., and Mezouar, M. (2011). Dense close-packed phase of tin above 157 GPa observed experimentally via angle-dispersive X-ray diffraction. *Phys. Rev. B*, **84**, 140104.

Salamon, P., Sibani, P., and Frost, R. (2002). *Facts, Conjectures, and Improvements for Simulated Annealing*. Society for Industrial and Applied Mathematics.

Samson, S. (1962). Crystal structure of $NaCd_2$. *Nature*, **195**, 259–262.

Samson, S. (1964). A method for the determination of complex cubic metal structures and its application to the solution of the structure of $NaCd_2$. *Acta Crystallogr.*, **17**, 491–495.

Samson, S. (1965). The crystal structure of the phase β-Mg_2Al_3. *Acta Crystallogr.*, **19**, 401–413.

Samson, S. (1967). The crystal structure of the intermetallic compound Cu_4Cd_3. *Acta Crystallogr.*, **23**, 586–600.

Samudrala, G. K., Thomas, S. A., Montgomery, J. M., and Vohra, Y. K. (2011). High-pressure phase transitions in rare earth metal erbium to 151 GPa. *J. Phys.: Condens. Matter*, **23**, 315701.

Samudrala, G. K., Tsoi, G. M., and Vohra, Y. K. (2012). Structural phase transitions in yttrium under ultrahigh pressure. *J. Phys.: Condens. Matter*, **24**, 362201.

Samudrala, G. K. and Vohra, Y. K. (2012). Crystallographic phases in heavy rare earth metals under megabar pressures. *J. Phys.: Conf. Ser.*, **377**, 012111.

Sanati, M., Saxena, A., and Lookman, T. (2001). Domain wall modeling of bcc to hcp reconstructive phase transformation in early transition metals. *Phys. Rev. B*, **64**, 092101.

Sanderson, M. J. and Baenziger, N. C. (1953). The crystal structure of $BaCd_{11}$. *Acta Crystallogr.*, **6**, 627–631.

Saramat, A., Svensson, G., Palmqvist, A. E. C., Stiewe, C., Mueller, E., Platzek, D., Bryan, J. D., and Stucky, G. D. (2006). Large thermoelectric figure of merit at high temperature in Czochralski-grown clathrate $Ba_8Ga_{16}Ge_{30}$. *J. Appl. Phys.*, **99**, 023708.

Savin, A. (2005). On the significance of ELF basins. *J. Chem. Sci.*, **117**, 473–475.

Savin, A., Jepsen, O., Flad, J., Andersen, O. K., Preuss, H., and von Schnering, H. G. (1992). Electron localization in solid-state structures of the elements: The diamond structure. *Angew. Chem. Int. Ed.*, **31**, 187–188.

Savin, A., Nesper, R., Wengert, S., and Fässler, T. E. (1997). ELF: The Electron Localization Function. *Angew. Chem. Int. Ed. Engl.*, **36**, 1808–1832.

Savin, A., Silvi, B., and Coionna, F. (1996). Topological analysis of the electron localization function applied to delocalized bonds. *Canad. J. Chem.*, **74**, 1088–1096.

Schmitt, D. C., Haldolaarachchige, N., Xiong, Y. M., Young, D. P., Jin, R. Y., and Chan, J. Y. (2012). Probing the lower limit of lattice thermal conductivity in an ordered extended solid: $Gd_{117}Co_{56}Sn_{112}$, a phonon glass electron crystal system. *J. Am. Chem. Soc.*, **134**, 5965–5973.

Schön, J. C. and Jansen, M. (1996). First step towards planning of syntheses in solid-state chemistry: Determination of promising structure candidates by global optimization. *Angew. Chem. Int. Ed. Engl.*, **35**, 1286–1304.

Schulte, O. and Holzapfel, W. B. (1993). Phase diagram for mercury up to 67 GPa and 500 K. *Phys. Rev. B*, **48**, 14009–14012.

Schulte, O. and Holzapfel, W. B. (1996). Effect of pressure on the atomic volume of Zn, Cd, and Hg up to 75 GPa. *Phys. Rev. B*, **53**, 569–580.

Schuster, J. C. and Parthé, E. (1987). Triclinic $ReAl_4$, a periodic domain structure variant of the monoclinic WAl_4 type. *Acta Crystallogr. C*, **43**, 620–623.

Schwarz, U., Bräuninger, S., Burkhardt, U., Syassen, K., and Hanfland, M. (2001a). Structural phase transition of $GdGa_2$ at high pressure. *Z. Kristallogr.*, **216**, 331–336.

Schwarz, U., Bräuninger, S., Grin, Y., and Syassen, K. (1996). Structural phase transition of $TmGa_2$ at high pressure. *J. Alloys Compd.*, **245**, 23–29.

Schwarz, U., Bräuninger, S., Grin, Y., Syassen, K., and Hanfland, M. (1998a). Structural phase transition of $HoGa_2$ at high pressure. *J. Alloys Compd.*, **268**, 161–165.

Schwarz, U., Bräuninger, S., Syassen, K., and Kniep, R. (1998b). Pressure-induced phase transformation of LiIn and LiCd: From NaTl-type phases to β-brass-type alloys. *J. Solid State Chem.*, **137**, 104–111.

Schwarz, U., Giedigkeit, R., Niewa, R., Schmidt, M., Schnelle, W., Cardoso, R., Hanfland, M., Hu, Z., Klementiev, K., and Grin, Y. (2001b). Pressure-induced oxidation state change of ytterbium in $YbGa_2$. *Z. Anorg. Allg. Chem.*, **627**, 2249–2256.

Schwarz, U., Grzechnik, A., Syassen, K., Loa, I., and Hanfland, M. (1999a). Rubidium-IV: A high pressure phase with complex crystal structure. *Phys. Rev. Lett.*, **83**, 4085–4088.

Schwarz, U., Syassen, K., Grzechnik, A., and Hanfland, M. (1999b). The crystal structure of rubidium-VI near 50 GPa. *Solid State Commun.*, **112**, 319–322.

Schwarz, U., Takemura, K., Hanfland, M., and Syassen, K. (1998c). Crystal structure of cesium-V. *Phys. Rev. Lett.*, **81**, 2711–2714.

Seipel, M., Porsch, F., and Holzapfel, W. B. (1997). Characterization of the fcc-distorted fcc-structural transition in lanthanum in an extended pressure and temperature range. *High Press. Res.*, **15**, 321–330.

Senkov, O. N., Scott, J. M., Senkova, S. V., Meisenkothen, F., Miracle, D. B., and Woodward, C. F. (2012). Microstructure and elevated temperature properties of a refractory TaNbHfZrTi alloy. *J. Mater. Sci.*, **47**, 4062–4074.

Senkov, O. N., Scott, J. M., Senkova, S. V., Miracle, D. B., and Woodward, C. F. (2011). Microstructure and room temperature properties of a high-entropy TaNbHfZrTi alloy. *J. Alloys Compd.*, **509**, 6043–6048.

Senkov, O. N., Wilks, G. B., Miracle, D. B., Chuang, C. P., and Liaw, P. K. (2010). Refractory high-entropy alloys. *Intermetallics*, **18**, 1758–1765.

Sevov, S. C. (2002). Zintl Phases. In *Intermetallic Compounds Principles and Practice* (ed. J. H. Westbrook and R. L. Fleischer), Volume 3, Chapter 6, pp. 113–132.

Sevov, S. C. and Corbett, J. D. (1992). Synthesis, characterization, and bonding of indium clusters: $Na_7In_{11.8}$, a novel network structure containing closo-In_{16} and nido-In_{11} clusters. *Inorganic Chemistry*, **31**, 1895–1901.

Sevov, S. C. and Corbett, J. D. (1993*a*). Carbon-free fullerenes: Condensed and stuffed anionic examples in indium systems. *Science*, **262**, 880–883.

Sevov, S. C. and Corbett, J. D. (1993*b*). Synthesis, characterization, and bonding of indium cluster phases: $Na_{15}In_{27.4}$, a network of In_{16} and In_{11} clusters; Na_2In with isolated indium tetrahedra. *J. Solid State Chem.*, **103**, 114–130.

Sevov, S. C. and Corbett, J. D. (1993*c*). Synthesis, characterization, and bonding of indium clusters: $A_3Na_{26}In_{48}$ (A = K, Rb, Cs) with a novel cubic network of arachno- and closo-In_{12} clusters. *Inorg. Chem.*, **32**, 1612–1615.

Sevov, S. C. and Corbett, J. D. (1996). A new indium phase with three stuffed and condensed fullerane-like cages: $Na_{172}In_{197}Z_2$ (Z = Ni, Pd, Pt). *J. Solid State Chem.*, **123**, 344–370.

Shechtman, D., Blech, I., Gratias, D., and Cahn, J. W. (1984). Metallic phase with long-range orientational order and no translational symmetry. *Phys. Rev. Lett.*, **53**, 1951–1953.

Shekar, N. V. C., Subramanian, N., Kumar, N. R. S., and Sahu, P. C. (2004). Phase transformation in $CeGa_2$ under high pressure. *Phys. Status Solidi (b)*, **241**, 2893–2897.

Sheldon, R. I. and Peterson, D. E. (1985*a*). The Np–Pu (neptunium-plutonium) system. *Bull. Alloy Phase Diagrams*, **6**, 215–217.

Sheldon, R. I. and Peterson, D. E. (1985*b*). The Np–U (neptunium-uranium) system. *Bull. Alloy Phase Diagrams*, **6**, 217–219.

Shen, Y. R., Kumar, R. S., Cornelius, A. L., and Nicol, M. F. (2007). High-pressure structural studies of dysprosium using angle-dispersive X-ray diffraction. *Phys. Rev. B*, **75**, 064109.

Shevchenko, V. Y., Blatov, V. A., and Ilyushin, G. D. (2009). Intermetallic compounds of the $NaCd_2$ family perceived as assemblies of nanoclusters. *Struct. Chem.*, **20**, 975–982.

Shevchenko, V. Y., Blatov, V. A., and Ilyushin, G. D. (2013). New types of two-layer nanoclusters with an icosahedral core. *Glass Phys. Chem.*, **39**, 229–234.

Shevelkov, A. V. and Kovnir, K. A. (2011). Zintl clathrates. *Struct. Bond.*, **139**, 97–144.

Shirotani, I., Hayashi, J., Yamanashi, K., Ishimatsu, N., Shimomura, O., and Kikegawa, T. (2001). Pressure-induced phase transitions in lanthanide monoantimonides with a NaCl-type structure. *Phys. Rev. B*, **64**, 132101.

Shoemaker, C. B. and Shoemaker, D. P. (1969). Structural properties of some σ-phase related phases. In *Developments in the Structural Chemistry of Alloy Phases* (ed. B. C. Giessen), pp. 107–139. Plenum Press, New York - London.

Shoemaker, D. P. and Shoemaker, C. B. (1986). Concerning the relative numbers of atomic coordination types in tetrahedrally close packed metal structures. *Acta Crystallogr. B*, **42**, 3–11.

Sichevych, O., Kohout, M., Schnelle, W., Borrmann, H., Cardoso-Gil, R., Schmidt, M., Burkhardt, U., and Grin, Y. (2009). $EuTM_2Ga_8$ (TM = Co, Rh, Ir) – a contribution to the chemistry of the $CeFe_2Al_8$-type compounds. *Inorg. Chem.*, **48**, 6261–6270.

Sikka, S. K. (2005). A high pressure distorted α-uranium (*Pnma*) structure in plutonium. *Solid State Commun.*, **133**, 169–172.

Sikora, W., Malinowski, J., Kuna, A., and Pytlik, L. (2008). Symmetry analysis in the investigation of clusters in complex metallic alloys. *J. Phys.: Conf. Ser.*, **104**, 012023.

Silverberg, J. L., Na, J. H., Evans, A. A., B., Liu., Hull, T. C., Santangelo, C. D., Lang, R. J., Hayward, R. C., and Cohen, I. (2015). Origami structures with a critical transition to bistability arising from hidden degrees of freedom. *Nat. Mater.*, **14**, 389–393.

Silvi, B. (2002). The synaptic order: A key concept to understand multicenter bonding. *J. Mol. Struc.*, **614**, 3–10.

Silvi, B. and Savin, A. (1994). Classification of chemical bonds based on topological analysis of electron localization functions. *Nature*, **371**, 683–686.

Simak, S. I., Häussermann, U., Ahuja, R., Lidin, S., and Johansson, B. (2000). Gallium and indium under high pressure. *Phys. Rev. B*, **85**, 142–145.

Skriver, H. L. (1989). Crystal structure from one-electron theory. *Phys. Rev. B*, **31**, 1909–1923.

Smetana, V., Babizhetskyy, V., Hoch, C., and Simon, A. (2007a). Icosahedral Li clusters in the structures of $Li_{33.3}Ba_{13.1}Ca_3$ and $Li_{18.9}Na_{8.3}Ba_{15.3}$. *J. Solid State Chem.*, **180**, 3302–3309.

Smetana, V., Babizhetskyy, V., Vajenine, G. V., Hoch, C., and Simon, A. (2007b). Double-icosahedral Li clusters in a new binary compound $Ba_{19}Li_{44}$: A reinvestigation of the Ba–Li phase diagram. *Inorg. Chem.*, **46**, 5425–5428.

Söderlind, P., Ahuja, R., Eriksson, O., Wills, J. M., and Johansson, B. (1994). Crystal structure and elastic-constant anomalies in the magnetic 3d transition metals. *Phys. Rev. B*, **50**, 5918–5927.

Söderlind, P., Eriksson, O., Johansson, B., Wills, J. M., and Boring, A. M. (1995a). A unified picture of the crystal structures of metals. *Nature*, **374**, 524–525.

Söderlind, P., Johansson, B., and Eriksson, O. (1995b). Theoretical zero-temperature phase diagram for neptunium metal. *Phys. Rev. B*, **52**, 1631–1639.

Solokha, P., De Negri, S., Pavlyuk, V., and Saccone, A. (2009). Crystal chemical peculiarities of rare earth (R) rich magnesium intermetallic compounds in R–T–Mg (T = transition element) systems. *Chem. Met. Alloys*, **2**, 39–48.

Sootsman, J. R., Chung, D. Y., and Kanatzidis, M. G. (2009). New and old concepts in thermoelectric materials. *Angew. Chem. Intern. Ed.*, **48**, 8616–8639.

Spichkin, Y. I. and Tishin, A. M. (2005). Magnetocaloric effect at the first-order magnetic phase transitions. *J. Alloys Compd.*, **403**, 38–44.

Stacey, T. E. and Fredrickson, D. C. (2012a). http://www.chem.wisc.edu/ danny/software/ehtuner/.

Stacey, T. E. and Fredrickson, D. C. (2012b). Perceiving molecular themes in the structures and bonding of intermetallic phases: The role of Hückel theory in an ab initio era. *Dalton Trans.*, **41**, 7801–7813.

Stacey, T. E. and Fredrickson, D. C. (2012c). The μ_3 model of acids and bases: Extending the Lewis theory to intermetallics. *Inorg. Chem.*, **51**, 4250–4264.

Stacey, T. E. and Fredrickson, D. C. (2013). Structural acid-base chemistry in the metallic state: How μ_3-neutralization drives interfaces and helices in $Ti_{21}Mn_{25}$. *Inorg. Chem.*, **52**, 8349–8359.

Stadnik, Z. M. (2013). Magnetic properties of quasicrystals and their approximants. *Handbook Magn. Mater.*, **22**, 77–130.

Staun-Olsen, J. and Gerward, L. (1994). A high-pressure study of thallium. *J. Appl. Cryst.*, **27**, 1002–1005.

Stein, F., Palm, M., and Sauthoff, G. (2004). Structure and stability of Laves phases. Part I. Critical assessment of factors controlling Laves phase stability. *Intermetallics*, **12**, 713–720.

Stein, F., Palm, M., and Sauthoff, G. (2005). Structure and stability of Laves phases part II – structure type variations in binary and ternary systems. *Intermetallics*, **13**, 1056–1074.

Steinhardt, P. J., Jeong, H. C., Saitoh, K., Tanaka, M., Abe, E., and Tsai, A. P. (1998). Experimental verification of the quasi-unit-cell model of quasicrystal structure. *Nature*, **396**, 55–57.

Steurer, W. (2006*a*). Reflections on symmetry and formation of axial quasicrystals. *Z. Kristallogr.*, **221**, 402–411.

Steurer, W. (2006*b*). Stable clusters in quasicrystals: Fact or fiction? *Philos. Mag*, **86**, 1105–1113.

Steurer, W. (2011*a*). On a realistic growth mechanism for quasicrystals. *Z. Anorg. Allg. Chem.*, **637**, 1943–1947.

Steurer, W. (2011*b*). Quasicrystals: Sections of hyperspace. *Angew. Chem. Int. Ed.*, **50**, 2–6.

Steurer, W. (2012). Why are quasicrystals quasiperiodic? *Chem. Soc. Rev.*, **41**, 6719–6729.

Steurer, W. (2014*a*). Decagonal quasicrystals – what has been achieved? *CR Physique*, **15**, 40–47.

Steurer, W. (2014*b*). *Physical Metallurgy* (Fifth edn), Volume 1, Chapter Crystal structure of metallic elements and compounds, pp. 1–103. Elsevier, Amsterdam.

Steurer, W. and Deloudi, S. (2008). Fascinating quasicrystals. *Acta Crystallogr. A*, **64**, 1–11.

Steurer, W. and Deloudi, S. (2009). *Crystallography of Quasicrystals. Concepts, Methods and Structures*. Number 126 in Springer Series in Materials Science. Spinger, Heidelberg, Dordrecht, London, New York.

Steurer, W. and Deloudi, S. (2012). Cluster packing from a higher-dimensional perspective. *Struct. Chem.*, **23**, 1115–1120.

Stewart, G. R. (2015). Superconductivity in the A15 structure. *Physica C*, **514**, 28–35.

Strutz, A., Yamamoto, A., and Steurer, W. (2010). Basic Co-rich decagonal Al–Co–Ni: Superstructure. *Phys. Rev. B*, **82**, 064107.

Sugiyama, K., Kaji, N., Hiraga, K., and Ishimasa, T. (1998). Crystal structure of a cubic $Al_{70}Pd_{23}Mn_6Si$; a 2/1 rational approximant of an icosahedral phase. *Z. Kristallogr.*, **213**, 90–95.

Sun, J. Lee, S. and Lin, J. (2007). Four-dimensional space groups for pedestrians: Composite structures. *Chem. Asian. J.*, **2**, 1204–1229.

Takemura, K. (1994). High-pressure structural study of barium to 90 GPa. *Phys. Rev. B*, **50**, 16238–16246.

Takemura, K. (1997). Structural study of Zn and Cd to ultrahigh pressures. *Phys. Rev. B*, **56**, 5170–5179.

Takemura, K., Christensen, N. E., Novikov, D. L., Syassen, K., Schwarz, U., and Hanfland, M. (2000). Phase stability of highly compressed cesium. *Phys. Rev. B*, **61**, 14399–14403.

Takemura, K., Fujihisa, H., Nakamoto, Y., Nakano, S., and Ohisi, Y. (2007). Crystal structure of the high-pressure γ phase of mercury: A novel monoclinic distortion of the close-packed structure. *J. Phys. Soc. Jpn.*, **76**, 023601.

Takemura, K., Schwarz, U., Syassen, K., Christensen, N. E., Hanfland, M., Novikov, D. L., and Loa, I. (2001). High-pressure structures of Ge above 10 GPa. *Phys. Status Solidi (b)*, **223**, 385–390.

Takemura, K., Shimomura, O., and Fujihisa, H. (1991). Cs(VI): A new high-pressure polymorph of cesium above 72 GPa. *Phys. Rev. Lett.*, **66**, 2014–2017.

Takemura, K. Kobayashi, K. and Masao, A. (1998). High-pressure bct-fcc phase transition in Ga. *Phys. Rev. B*, **58**, 2482–2486.

Tappe, F., Schwickert, C., and Pöttgen, R. (2012). Ternary ordered Laves phases $RENi_4Cd$. *Intermetallics*, **24**, 33–37.

Tebbe, K. F., von Schnering, H. G., Rüter, B., and Rabeneck, G. (2007). Li$_3$Al$_2$, eine neue Phase im System Li/Al. *Z. Naturforsch.*, **28b**, 600–605.

Thiede, V. M. T., Fehrmann, B., and Jeitschko, W. (1999). Ternary rare earth metal palladium and platinum aluminides R$_4$Pd$_9$Al$_{24}$ and R$_4$Pt$_9$Al$_{24}$. *Z. Anorg. Allg. Chem.*, **625**, 1417–1425.

Thiede, V. M. T., Jeitschko, W., Niemann, S., and Ebel, T. (1998). EuTa$_2$Al$_{20}$, Ca$_6$W$_4$Al$_{43}$ and other compounds with CeCr$_2$Al$_{20}$ and Ho$_6$Mo$_4$Al$_{43}$ type structures and some magnetic properties of these compounds. *J. Alloys Compd.*, **267**, 23–31.

Thimmaiah, S., Richter, K. W., Lee, S., and Harbrecht, B. (2003). γ_1-Pt$_5$Zn$_{21}$ – a reappraisal of a γ-brass type complex alloy phase. *Solid State Sci.*, **5**, 1309–1317.

Thomas, E. L., Gautreux, D. P., and Chan, J. Y. (2006a). The layered intermetallic compound LaPdSb$_3$. *Acta Crystallogr. E*, **62**, i96–i98.

Thomas, E. L., Millican, J. N., Okudzeto, E. K., and Chan, J. Y. (2006b). Crystal growth and the search for highly correlated intermetallics. *Comm. Inorg. Chem.*, **27**, 1–39.

Tickle, R. and James, R. D. (1999). Magnetic and magnetomechanical properties of Ni$_2$MnGa. *J. Magn. Magn. Mater.*, **195**, 627–638.

Tillard-Charbonnel, M., Belin, C., and Chouaibi, N. (1993). Crystal structure of sodium gold gallium, Na$_{128}$Au$_{81}$Ga$_{275}$. *Z. Kristallogr.*, **206**, 310–312.

Tillard-Charbonnel, M., Chahine, A., and Belin, C. (1994). Crystal structure of lithium cadmium gallium (58/16/128), Li$_{58}$Cd$_{16}$Ga$_{128}$. *Z. Kristallogr.*, **209**, 280.

Tillard-Charbonnel, M., Chouaibi, N., Belin, C, and Lapasset, J. (1992a). Synthesis and crystal structure determination of the new ternary intermetallic phase Na$_{17}$Cu$_6$Ga$_{46.5}$ (Na$_{102}$Cu$_{36}$Ga$_{279}$). *J. Solid State Chem.*, **100**, 220–228.

Tillard-Charbonnel, M., Chouaibi, N. E., and Belin, C. (1992b). Synthèse et étude structurale de la nouvelle phase ternaire stœchiométrique Na$_{17}$Zn$_{12}$Ga$_{40.5}$. *C. R. Acad. Sci. Paris*, **315**, 661–665.

Todorov, E. and Sevov, S. C. (1997). Synthesis, characterization, electronic structure, and bonding of heteroatomic deltahedral clusters: Na$_{49}$Cd$_{58.5}$Sn$_{37.5}$, a network structure containing the first empty icosahedron without a group 13 element and the largest closo-deltahedron. *J. Am. Chem. Soc.*, **119**, 2869–2876.

Todorov, E. and Sevov, S. C. (2000). Synthesis and structure of the alkali-metal amalgams A$_3$Hg$_{20}$ (A = Rb, Cs), K$_3$Hg$_{11}$, Cs$_5$Hg$_{19}$, and A$_7$Hg$_{31}$ (A = K, Rb). *J. Solid State Chem.*, **149**, 419–427.

Tonkov, E. Y. (1992-1996). *High Pressure Phase Transformations: a Handbook*. Volume 1–3. Gordon and Breach Science Publishers, Philadelphia, USA.

Tonkov, E. Y. and Ponyatovsky, E. G. (2005). *Phase Transformations of Elements Under High Pressure*. CRC Press, Boca Raton, USA.

Trimarchi, G. and Zunger, A. (2007, Mar). Global space-group optimization problem: Finding the stablest crystal structure without constraints. *Phys. Rev. B*, **75**, 104113.

Troparevsky, M. C., Morris, J. R., Daene, M., Wang, Y., Lupini, A. R., and Stocks, G. M. (2015). Beyond atomic sizes and hume-rothery rules: Understanding and predicting high-entropy alloys. *JOM*, **67**, 2350–2363.

Trudel, S., Gaier, O., Hamrle, J., and Hillebrands, B. (2010). Magnetic anisotropy, exchange and damping in cobalt-based full-Heusler compounds: An experimental review. *J.Phys. D: Appl. Phys.*, **43**, 193001.

Tsai, A. P. (2003). "Back to the future" – An account discovery of stable quasicrystals. *Acc. Chem. Res.*, **36**, 31–38.

Tsai, A. P. (2013). Discovery of stable icosahedral quasicrystals: progress in understanding structure and properties. *Chem. Soc. Rev.*, **42**, 5352–5365.

Tsai, K. Y., Tsai, M. H., and Yeh, J. W. (2013). Sluggish diffusion in Co–Cr–Fe–Mn–Ni high-entropy alloys. *Acta Mater.*, **61**, 4887–4897.

Tsai, M. H. and Yeh, J. W. (2014). High-entropy alloys: A critical review. *Mater. Res. Lett.*, **2**, 107–123.

Tse, J. S., Desgreniers, S., Ohisi, Y., and Matsuoka, T. (2012). Large amplitude fluxional behaviour of elemental calcium under high pressure. *Sci. Rep.*, **2**, 372.

Tursina, A., Noel, H., Murashova, E., Morozova, Y., and Seropegin, Y. (2014). Crystal structures of the new intermetallics Ce_3Pt_5Al and $Ce_3Pt_5Al_2$. *Chem. Met. Alloys*, **7**, 15–19.

van Smaalen, S. (2007). *Incommensurate Crystallography*. Number 21 in IUCr Monographs on Crystallography. Oxford University Press, Oxford, UK.

van Smaalen, S. and George, T. F. (1987). Determination of the incommensurately modulated structure of α-uranium below 37 K. *Phys. Rev. B*, **35**, 7939–7951.

Velisavljevic, N. and Vohra, Y. K. (2004). Distortion of α-uranium structure in praseodymium metal to 311 GPa. *High Press. Res.*, **24**, 295–302.

Venturini, G., Ijjaali, I., and Malaman, B. (1999). RGe_{2x} compounds (R=Y, Gd–Ho) with new ordered $ThSi_2$-defect structures. *J. Alloys Compd.*, **285**, 194–203.

Verma, A. K. and Modak, P. (2008). Structural phase transitions in vanadium under high pressure. *Europhys. Lett.*, **81**, 37003.

Verma, A. K., Modak, P. and. Rao, R. S., Godwal, B. K., and Jeanloz, R. (2007). High-pressure phases of titanium: First-principles calculations. *Phys. Rev. B*, **75**, 014109.

Verstrate, M. J. (2010). Phases of polonium via density functional theory. *Phys. Rev. Lett.*, **104**, 035501.

Villars, P. and Calvert, L. D. (1991). *Pearson's Handbook of Crystallographic Data for Intermetallic Phases*. Volume 1–4. ASM, Materials Park, Ohio, USA.

Villars, P. and Cenzual, K. (2011a). *Pearson's Crystal Data: Crystal Structure Database for Inorganic Compounds – User Manual, Release 2011/12*. ASM International ®, Materials Park, Ohio, USA.

Villars, P. and Cenzual, K. (2011b). *Pearson's Handbook of Crystallographic Data for Intermetallic Phases*. ASM International, Materials Park, Ohio, USA.

Vincent, A. J. B. (1981). *A study of three multicomponent alloys. BSc Part II Thesis*. Ph.D. thesis, University of Oxford, Oxford, UK.

Vohra, Y. K. and Beaver, S. L. (1999). Ultrapressure equation of state of cerium metal to 208 GPa. *J. Appl. Phys.*, **85**, 2451–2453.

von Rohr, F., Ni Ni, H. L., Wörle, M., and Cava, R. J. (2014). Superconductivity and correlated fermi liquid behavior in noncentrosymmetric $Ca_3Ir_4Ge_4$. *Phys. Rev. B*, **89**, 224504.

Wade, K. (1971). The structural significance of the number of skeletal bonding electron-pairs in carboranes, the higher boranes and borane anions, and various transition-metal carbonyl cluster compounds. *J. Chem. Soc. D: Chem. Comm.*, **1971**, 792–793.

Wade, K. (1976). Structural and bonding patterns in cluster chemistry. *Adv. Inorg. Chem. Radiochem.*, **18**, 1–66.

Wagner, F. R., Baranov, A. I., Grin, Y., and Kohout, M. (2013). A position-space view on chemical bonding in metal diborides with AlB_2 type of crystal structure. *Z. Anorg. Allg. Chem.*, **639**, 2025–2035.

Wagner, F. R., Bezugly, V., Kohout, M., and Grin, Y. (2007). Charge decomposition analysis of the electron localizability indicator: A bridge between the orbital and direct space representation of the chemical bond. *Chemistry*, **13**, 5724–5741.

Wagner, F. R., Kohout, M., and Grin, Y. (2008). Direct space decomposition of ELI-D: interplay of charge density and pair-volume function for different bonding situations. *J. Phys. Chem. A*, **112**, 9814–9828.

Wales, D. J. (2004). *Energy Landscapes: Applications to Clusters, Biomolecules and Glasses.* Cambridge Molecular Science. Cambridge University Press. ISBN: 978-0-521-81415-7.

Wales, D. J. and Doye, J. P. K. (1997). Global optimization by basin-hopping and the lowest energy structures of Lennard-Jones clusters containing up to 110 atoms. *J. Phys. Chem. A*, **101**, 5111–5116.

Wales, D. J. and Scheraga, H. A. (1999). Global optimization of clusters, crystals, and biomolecules. *Science*, **285**, 1368–1372.

Wang, F. and Miller, G. J. (2011). Revisiting the Zintl-Klemm concept: Alkali metal trielides. *Inorg. Chem.*, **50**, 7625–7636.

Wang, W. R., Wang, W. L., Wang, S. C., Tsai, Y. C., Lai, C. H., and Yeh, J. W. (2012*a*). Effects of Al addition on the microstructure and mechanical property of $Al_xCoCrFeNi$ high-entropy alloys. *Intermetallics*, **26**, 44–51.

Wang, Y., Gabe, E. J., Calvert, L. D., and Taylor, J. B. (1976). The crystal structure of Y_5Bi_3 and its relation to the Mn_5Si_3 and the Yb_5Sb_3 type structures. *Acta Crystallogr. B*, **32**, 1440–1445.

Wang, Y. C., Lv, J., Zhu, L., and Ma, Y. M. (2010). Crystal structure prediction via particle-swarm optimization. *Phys. Rev. B*, **82**, 094116.

Wang, Y. C., Lv, J., Zhu, L., and Ma, Y. M. (2012*b*). CALYPSO: A method for crystal structure prediction. *Comp. Phys. Comm.*, **183**, 2063–2070.

Wang, Y. C. and Ma, Y. M. (2014). Perspective: Crystal structure prediction at high pressures. *J. Chem. Phys.*, **140**, 040901.

Weber, S. and Yamamoto, A. (1998). Noncentrosymmetric structure of decagonal $Al_{70}Mn_{17}Pd_{13}$ quasicrystal. *Acta Crystallogr. B*, **54**, 997–1005.

Weber, T., Dshemuchadse, J., Kobas, M., Conrad, M., Harbrecht, B., and Steurer, W. (2009). Large, larger, largest – a family of cluster-based tantalum copper aluminides with giant unit cells. I. Structure solution and refinement. *Acta Crystallogr. B*, **65**, 308–317.

Wendorff, M. and Röhr, C. (2006). Polar binary Zn/Cd-rich intermetallics: Synthesis, crystal and electronic structure of $A(Zn/Cd)_{13}$ (A = alkali/alkaline earth) and $Cs_{1.34}Zn_{16}$. *J. Alloys Compd.*, **421**, 24–34.

Wenk, H. R., Kaercher, P., Kanitpanyacharoen, W., Zepeda-Alarcon, E., and Wang, Y. (2013). Orientation relations during the $\alpha - \omega$ phase transition of zirconium: *in situ* texture observations at high pressure and temperature. *Phys. Rev. Lett.*, **111**, 195701.

Wilson, A. J. C. and Prince, E. (ed.) (1999). *International Tables for Crystallography.* Volume C. Kluwer Academic Publishers, Dordrecht/Boston/London.

Wolff, M. W., Niemann, S., Ebel, T., and Jeitschko, W. (2001). Magnetic properties of rare-earth transition metal aluminides $R_6T_4Al_{43}$ with $Ho_6Mo_4Al_{43}$-type structure. *J. Magn. Magn. Mater.*, **223**, 1–15.

Wolny, J., Kozakowski, B., Duda, M., and Kusz, J. (2008). Stacking of hexagonal layers in the structure of β-Mg_2Al_3. *Philos. Mag. Lett.*, **88**, 501–507.

Woodley, M. S., Battle, D. P., Gale, D. J., and Catlow, R. A. C. (1999). The prediction of inorganic crystal structures using a genetic algorithm and energy minimisation. *Phys. Chem. Chem. Phys.*, **1**, 2535–2542.

Woodley, S. M. (2004). Prediction of crystal structures using evolutionary algorithms and related techniques. In *Applications of Evolutionary Computation in Chemistry* (ed. R. L. Johnston), Volume 110, Structure and Bonding, pp. 95–132. Springer-Verlag Berlin Heidelberg.

Woodley, S. M. and Catlow, R. (2008). Crystal structure prediction from first principles. *Nat. Mater.*, **7**, 937–946.

Xie, Y., Ma, Y. M., Cui, T., Li, Y., Qiu, J., and Zou, G.T. (2008). Origin of bcc to fcc phase transition under pressure in alkali metals. *New J. Phys.*, **10**, 063022.

Xie, Y., Oganov, A. R., and Ma, Y. M. (2010). Novel high pressure structures and superconductivity of $CaLi_2$. *Phys. Rev. Lett.*, **104**, 177005.

Xiong, D. B., Zhao, Y. F., Schnelle, W., Okamoto, N. L., and Inui, H. (2010). Complex alloys containing double-Mackay clusters and $(Sb_{1-\delta}Zn_\delta)_{24}$ snub cubes filled with highly disordered zinc aggregates: Synthesis, structures, and physical properties of ruthenium zinc antimonides. *Inorg. Chem.*, **49**, 10788–10797.

Yamamoto, A., Takakura, H., Ozeki, T., Tsai, A.-P., and Ohashi, Y. (2004). Structure refinement of i-Al–Pd–Re quasicrystals by synchrotron radiation data. *J. Non-Cryst. Solids*, **334/335**, 151–155.

Yao, Y., Martonak, R., Patchkovskii, S., and Klug, D. D. (2010). Stability of simple cubic calcium at high pressure: A first-principles study. *Phys. Rev. B*, **82**, 094107.

Yao, Y. S., Klug, D. D., Sun, J., and Martoňák, R. (2009). Structural prediction and phase transformation mechanisms in calcium at high pressure. *Phys. Rev. Lett.*, **103**, 055503.

Yatsenko, S. P., Hladyschewsky, R. E., Sitschewitsch, O. M., Belsky, V. K., Semyannikov, A. A., Hryn, Y. N., and Yarmoluk, Y. P. (1986). Kristallstruktur von Gd_3Ga_2 und isotypen Verbindungen. *J. Less-Comm. Met.*, **115**, 17–22.

Yatsenko, S. P., Hladyschewsky, R. E., Tschuntonow, K. A., Yarmoluk, Y. P., and Hryn, Y. N. (1983). Kristallstruktur von Tm_3Ga_5 und analoger Verbindungen. *J. Less-Comm. Met.*, **91**, 21–32.

Yeh, J. W. (2002). High entropy alloys, for direct production of tool, contains specific metallic elements, each of which have specific mole number. Patent JP2002173732-A; JP4190720-B2, Univ Qinghua.

Yeh, J. W. (2015). Physical metallurgy of high-entropy alloys. *JOM*, **67**, 2254–2261.

Yeh, J. W., Chen, S. K., Gan, J. Y., Lin, S. J., Chin, T. S., Shun, T. T., Tsau, C. H., and Chang, S. Y. (2004*a*). Formation of simple crystal structures in Cu–Co–Ni–Cr–Al–Fe–Ti–V alloys with multiprincipal metallic elements. *Met. Mater. Trans. A*, **35**, 2533–2536.

Yeh, J. W., Chen, S. K., Lin, S. J., Gan, J. Y., Chin, T. S., Shun, T. T., Tsau, C. H., and Chang, S. Y. (2004*b*). Nanostructured high-entropy alloys with multiple principal elements: novel alloy design concepts and outcomes. *Adv. Eng. Mater.*, **6**, 299–303.

Yoo, C. S., Cynn, H., Söderlind, P., and Iota, V. (2000). New β(fcc)-cobalt to 210 GPa. *Phys. Rev. Lett.*, **84**, 4132–4135.

Young, D. A. (1991). *Phase Diagrams of the Elements*. University of California Press, Berkeley.

Yurko, G. A., Barton, J. W., and Parr, J. G. (1959). The crystal structure of Ti_2Ni. *Acta Crystallogr.*, **12**, 909–911.

Yurko, G. A., Barton, J. W., and Parr, J. G. (1962). The crystal structure of Ti_2Ni. (A correction). *Acta Crystallogr.*, **15**, 1309.

Zaddach, A. J., Niu, C., Koch, C. C., and Irving, D. L. (2013). Mechanical properties and stacking fault energies of NiFeCrCoMn high-entropy alloy. *JOM*, **65**, 1780–1789.

Zaremba, R., Rodewald, U. C., Hoffmann, R. D., and Pöttgen, R. (2007). The rare earth metal-rich indides RE$_4$RhIn (RE=Gd–Tm, Lu). *Monatsh Chem.*, **138**, 523–528.

Zhang, F., Zhang, C., Chen, S. L., Zhu, J., Cao, W. S., and Kattner, U. R. (2014a). An understanding of high entropy alloys from phase diagram calculations. *Calphad*, **45**, 1–10.

Zhang, H., Baitinger, M., Fang, L., Schnelle, W., Borrmann, H., Burkhardt, U., Ormeci, A., Zhao, J. T., and Grin, Y. (2013). Synthesis and properties of type-I clathrate phases Rb$_{8-x-t}$K$_{xt}$Au$_y$Ge$_{46-y}$. *Inorg. Chem.*, **52**, 9720–9726.

Zhang, J., Zhao, Y., Hixson, R. S., and Gray, G. T. (2008). Thermal equation of state for titanium obtained by high pressure-temperature diffraction studies. *Phys. Rev. B*, **78**, 054119.

Zhang, Y., Yang, X., and Liaw, P. K. (2012). Alloy design and properties optimization of high-entropy alloys. *JOM*, **64**, 830–838.

Zhang, Y. and Zhou, Y. (2007). Solid solution formation criteria for high entropy alloys. *Mater. Sci. Forum*, **561–565**, 1337–1339.

Zhang, Y., Zuo, T. T., Tang, Z., Gao, M. C., Dahmen, K. A., Liaw, P. K., and Lu, Z. P. (2014b). Microstructures and properties of high-entropy alloys. *Prog. Mater. Sci.*, **61**, 1–93.

Zhao, L. D., Lo, S. H., Zhang, Y. S., Sun, H., Tan, G. J., Uher, C., Wolverton, C., Dravid, V. P., and Kanatzidis, M. G. (2014). Ultralow thermal conductivity and high thermoelectric figure of merit in SnSe crystals. *Nature*, **508**, 373–377.

Zhao, Y. C., Porsch, F., and Holzapfel, W. B. (1994). Irregularities of ytterbium under high pressure. *Phys. Rev. B*, **49**, 815–817.

Zhou, H. (1991). Room temperature section of the phase diagram of the Cu–Fe–C ternary system. *J. Less-Comm. Met.*, **171**, 113–118.

Zhou, X. Z., Li, W., Kunkel, H. P., and Williams, G. (2005). Relationship between the magnetocaloric effect and sequential magnetic phase transitions in Ni–Mn–Ga alloys. *J. Appl. Phys.*, **97**, 10M515.

Zhu, Q., Oganov, A. R., Lyakhov, A. O., and Yu, X. X. (2015). Generalized evolutionary metadynamics for sampling the energy landscapes and its applications. *Phys. Rev. B*, **92**, 024106.

Zhu, Q., Oganov, A. R., and Zhou, X. F. (2014). Crystal structure prediction and its application in earth and materials sciences. In *Prediction and Calculation of Crystal Structures* (ed. S. Atahan-Evrenk and A. Aspuru-Guzik), Volume 345, Topics in Current Chemistry, pp. 223–256. Springer International Publishing.

Zintl, E. (1939). Intermetallische Verbindungen. *Angew. Chem.*, **52**, 1–6.

Zintl, E. and Brauer, G. (1933). Über die Valenzelektronenregel und die Atomradien unedler Metalle in Legierungen. *Z. Phys. Chem. B*, **20**, 245–271.

Zintl, E. and Dullenkopf, W. (1932). Über den Gitterbau von NaTl und seine Beziehung zu den Strukturen vom Typus des β-Messings. *Z. Phys. Chem. B*, **16**, 195–205.

Zou, Y., Maiti, S., Steurer, W., and Spolenak, R. (2014). Size-dependent plasticity in a Nb$_{25}$Mo$_{25}$Ta$_{25}$W$_{25}$ refractory high-entropy alloy. *Acta Mater.*, **65**, 85–97.

Zurek, E. and Grochala, W. (2015). Predicting crystal structures and properties of matter under extreme conditions via quantum mechanics: the pressure is on. *Phys. Chem. Chem. Phys.*, **17**, 2917–2934.

Zurek, E. and Yao, Y. S. (2015). Theoretical predictions of novel superconducting phases of BaGe$_3$ stable at atmospheric and high pressures. *Inorg. Chem.*, **54**, 2875–2884.

Index